INDUSTRIAL GUMS

Polysaccharides and Their Derivatives

Second Edition

CONTRIBUTORS

Mark F. Adams
Emil N. Alter
Jack B. Batdorf
J. N. BeMiller
E. Bjerre-Petersen
W. A. P. Black
J. Christensen
Otto Christensen
A. J. Desmarais
E. T. Dewar
Robert J. Dobbins
Landis W. Doner
Bruce V. Ettling
Thomas Gerard
Martin Glicksman
Arthur M. Goldstein
G. K. Greminger, Jr.
P. Hemmingsen
Erling T. Hjermstad
Donald J. Iwinski
L. Jullander
Kenneth S. Kang
S. Lindenfors
William H. McNeely
George Meer
William A. Meer
P. T. Murphy
David J. Pettitt
Eugene L. Powell
P. S. Rao
Nelson E. Rodgers
Franco Rol
James M. Rossman
Ralph E. Sand
Robert W. Satterthwaite
A. B. Savage
James K. Seaman
H. H. Selby
H. C. Srivastava
Gordon A. Towle
Roy L. Whistler
W. H. Wynne

INDUSTRIAL GUMS

Polysaccharides and Their Derivatives

SECOND EDITION

Editor

ROY L. WHISTLER
Department of Biochemistry
Purdue University
Lafayette, Indiana

Assistant Editor

JAMES N. BEMILLER
Department of Chemistry and Biochemistry
Southern Illinois
University at Carbondale
Carbondale, Illinois

ACADEMIC PRESS
A Subsidiary of Harcourt Brace Jovanovich, Publishers
New York London Toronto Sydney San Francisco

ACADEMIC PRESS, INC.
111 Fifth Avenue, New York, New York 10003

United Kingdom Edition published by
ACADEMIC PRESS, INC. (LONDON) LTD.
24/28 Oval Road, London NW1

Library of Congress Cataloging in Publication Data

Whistler, Roy Lester.
Industrial gums.

Includes bibliographical references.
1. Polysaccharides. 2. Gums and resins.
I. BeMiller, James N., joint author. II. Title.
[DNLM: 1. Polysaccharides. TP979.5.P60 W576i]
TP979.5.P6W5 1973 668'.37 72-82652
ISBN 0-12-746252-x

PRINTED IN THE UNITED STATES OF AMERICA

83 84 85 9 8 7 6 5 4 3

CONTRIBUTORS

Numbers in parentheses indicate the pages on which the authors' contributions begin.

MARK F. ADAMS, *College of Engineering Research Division, Washington State University, Pullman, Washington* (415)

EMIL N. ALTER, *Stein, Hall & Co., Inc., Long Island City, New York* (273, 303)

JACK B. BATDORF, *Hercules Incorporated, Wilmington, Delaware* (695)

J. N. BEMILLER, *Department of Chemistry and Biochemistry, Southern Illinois, University at Carbondale, Carbondale, Illinois* (339, 545)

E. BJERRE-PETERSEN, *Litex Industri, Denmark* (123)

W. A. P. BLACK, *Inveresk Research International, Inveresk, Musselburgh, Midlothian, Scotland* (137)

J. CHRISTENSEN, *Litex Industri, Denmark* (123)

OTTO CHRISTENSEN, *The Copenhagen Pectin Factory Ltd., Lille Skensved, Denmark* (429)

A. J. DESMARAIS, *Hercules Incorporated, Wilmington, Delaware* (649)

E. T. DEWAR, *Inveresk Research International, Inveresk, Musselburgh, Midlothian, Scotland* (137)

ROBERT J. DOBBINS, *American Cyanamid Co., Stanford, Connecticut* (19)

LANDIS W. DONER, *Department of Biochemistry, Purdue University, Lafayette, Indiana* (115)

BRUCE V. ETTLING, *College of Engineering Research Division, Washington State University, Pullman, Washington* (415)

THOMAS GERARD, *Meer Corporation, North Bergen, New Jersey* (265, 289)

MARTIN GLICKSMAN, *General Foods Corp., Tarrytown, New York* (147, 197)

ARTHUR M. GOLDSTEIN, *Stein, Hall & Co., Inc., Long Island City, New York* (273, 303)

G. K. GREMINGER, JR., *The Dow Chemical Company, Midland, Michigan* (619)

P. HEMMINGSEN, *Litex Industri, Denmark* (123)

ERLING T. HJERMSTAD, *Penick and Ford Ltd., Cedar Rapids, Iowa* (601)

DONALD J. IWINSKI, *Technical Service Department, Industrial Division of CPC International Inc., Argo, Illinois* (577)

I. JULLANDER, *Swedish Forest Products Laboratory, Stockholm, Sweden,* (673)

KENNETH S. KANG, *Kelco Company, San Diego, California* (473)

S. LINDENFORS, *Research and Development Department, MoDoKemi AB, Ornskoldsvik, Sweden* (673)

WILLIAM H. MCNEELY, *Kelco Company, San Diego, California* (49, 473)

GEORGE MEER, *Meer Corporation, North Bergen, New Jersey* (265, 289)

WILLIAM A. MEER, *Meer Corporation, North Bergen, New Jersey* (265, 289)

P. T. MURPHY, *Department of Biochemistry, Purdue University, Lafayette, Indiana* (513)

DAVID J. PETTITT, *Kelco Company, San Diego, California* (49)

EUGENE L. POWELL, *American Maize-Products Company, Hammond, Indiana* (567)

P. S. RAO, *Forest Research Laboratory, Bangalore, India* (369)

NELSON E. RODGERS, *The Pillsbury Company, Minneapolis, Minnesota* (499)

FRANCO ROL, *Cesalpinia S. P. A., Bergamo, Italy (Subsidiary of Hercules, Inc., Wilmington, Delaware)* (323)

JAMES M. ROSSMAN, *Hercules Incorporated, Wilmington, Delaware* (695)

RALPH E. SAND, *Anderson Clayton Foods, Richardson, Texas* (147, 197)

ROBERT W. SATTERTHWAITE, *Technical Service Department, Industrial Division of CPC International Inc., Argo, Illinois* (577)

A. B. SAVAGE, *The Dow Chemical Company, Midland, Michigan* (619)

JAMES K. SEAMAN, *Stein, Hall and Co., Inc., Long Island City, New York* (303)

H. H. SELBY, *American Agar and Chemical Company, San Diego, California* (29)

H. C. SRIVASTAVA, *Ahmedabad Textile Industry's Research Association, Ahmedabad, India* (369)

GORDON A. TOWLE, *The Copenhagen Pectin Factory Lts., Lille Skensved, Denmark (Subsidiary of Hercules, Inc., Wilmington, Delaware)* (83, 429)

ROY L. WHISTLER, *Department of Biochemistry, Purdue University, Lafayette, Indiana* (2, 115, 465, 513)

W. H. WYNNE, *American Agar and Chemical Company, San Diego, California* (29)

PREFACE

Growth of the gum industry has continued rapidly over the years since publication of the first edition of "Industrial Gums." The first printing sold out within a few years and a second printing filled the need for a time. However, with new uses, new gums, and new technology constantly emerging, it became logical that a completely new book appear. As the first, this book is written by experts in the field, and gives comprehensive coverage of gum sources, prices, production, properties, and uses. It will be seen in this edition that several new gums are finding industrial acceptance and that new biochemical approaches to the modification and production of gums are rapidly evolving. In some instances, little new information is available. A few gums described in the first edition have failed to find a practical place in industry, and these have been deleted or are treated only briefly. The space provided is well filled by descriptions of several new gums. A few gums are described which are not yet commercially produced, but which are included because of their expected commercial usefulness. It is hoped that the chapters have been so organized that potential gum users can make a rational selection of the gum best suited for their particular product or application.

Some gums are primarily food ingredients, others go primarily into nonfood uses, and some into both areas of application. An attempt has been made to give proper emphasis to each area.

Analytical methods and specific chemical modifications have not been included, but many useful manipulations can be found in "Methods in Carbohydrate Chemistry," particularly in Volumes III–VI (Volumes I–VI, published by Academic Press). This treatise can be a useful companion to "Industrial Gums."

Although documentation of material is quite complete, descriptions should not be construed as indicating that the use of the procedures or processes described are free from patent restrictions.

I am grateful to the authors for their ready response to the request to contribute and for their kindness and understanding in accepting my editorial efforts.

Professor James N. BeMiller, a close friend and long-time associate, is particularly thanked for the detailed, conscientious, and skillful manner in which he handled his portion of the editorial work.

ROY LESTER WHISTLER

CONTENTS

PART A: NATURAL GUMS

GENERAL

SEAWEED EXTRACTS

Part A: Natural Gums

General (Chapters I, II)

Seaweed Extracts (Chapters III–IX)

Plant Exudates (Chapters X–XIII)

Seed Gums (Chapters XIV–XVII)

Plant Extracts (Chapters XVIII, XIX)

Animal Extract (Chapter XX)

GENERAL

CHAPTER I

FACTORS INFLUENCING GUM COSTS AND APPLICATIONS

ROY L. WHISTLER

Department of Biochemistry, Purdue University, Lafayette, Indiana

I. INTRODUCTION

Gums, in the form of natural, biosynthetic, or modified polysaccharides, are consumed in tremendous quantities by industry. Over a billion pounds are consumed annually within the United States, where the growth of gum usage exceeds 8–10% per year. Gums seldom constitute an entire finished product but are mainly used as additives to improve or control the properties of a commodity. Their magnitude of use results from the low cost of many gums and from the important properties that they contribute to products, even when they are present in low concentrations. Some gums exert their effect at very low concentrations indeed. Okra gum at 25 parts per million in water reduces friction in fluid flow by 80%.

Commercial gums are water-soluble or water-dispersible hydrocolloids. Their aqueous dispersions usually possess suspending, dispersion, and stabilizing properties; or the gums may act as emulsifiers, have gelling characteristics, and be either adhesive or mucilaginous. They may, on occasion, act as coagulants, binders, lubricants, or film formers.

An industrialist giving thought to the manufacture of a gum or to the use of a gum in his products will give serious consideration to the numerous factors that affect his taking a sound position. Among these factors are (1) physical and chemical properties required; (2) gum cost; (3) constancy of cost, particularly if world labor rates continue to rise; (4) constancy of supply and composition; and (5) possibility of eventual replacement of the selected gum by another. If the gum is intended for use in the food industry, consideration must be given to acceptability by government agencies. Some of these factors will be discussed in this chapter as other general information on gums is presented.

II. Definition of Gum

In practical terms, gums are either hydrophobic or hydrophilic high-molecular-weight molecules, usually with colloidal properties, that in an appropriate solvent or swelling agent produce gels, highly viscous suspensions or solutions at low dry substance content. Thus, the term gum is applied to a wide variety of substances with "gummy" characteristics and cannot be precisely defined. Hydrophobic substances often called gums are high-molecular-weight hydrocarbons and other petroleum products, rubbers, certain synthetic polymers, chicle for chewing gum, and the resinous saps that often exude from evergreens and that are sometimes commercially tapped yielding, for example, gum balsam and gum resin. Incense gums, such as myrrh, huataco, and frankincense, are fragrant plant exudates that are mixtures of resins and carbohydrates and are gummy because of their content of hydrophobic resin.

Most commonly, however, the term gum as technically employed in industry refers to plant or microbial polysaccharides or their derivatives that are dispersible in either cold or hot water to produce viscous mixtures or solutions. Thus, modern usage includes the water-soluble or water-swellable derivatives of cellulose and the derivatives and modifications of other polysaccharides that are insoluble in the natural form. Usage would classify as gums all polysaccharides or their derivatives that when dispersed in water at low dry substance content swell to produce gels or highly viscous dispersions or solutions. This definition does not require that gums have the property of tackiness, and consequently, such a definition includes as gums those polysaccharides and derivatives which are slimy or mucilaginous. Some authors have tried to classify separately these slimy substances from plants into a category called mucilages. However, it is more logical to consider tackiness and sliminess as the expression of two different physical properties of gums. Hence, there are tacky gums and slimy or mucilaginous gums. Tackiness and sliminess are manifestations of two somewhat controllable physical properties. It is possible to modify a gum so that tacky properties are withdrawn and mucilaginous properties introduced, yet the gum remains hydrophilic and capable of giving high viscosity to its dispersions even at low concentrations. Consequently, the mucilaginous property is distinctive and useful but a category of mucilages has no chemical significance and should be abandoned.

III. Factors Affecting Raw Gum Costs

Because about three-quarters of the dry weight of plants consists of polysaccharides, it is not unexpected that polysaccharides are abundant and obtainable at relatively low costs. Plant tissues contain mixtures of polysaccharides from which individual components can be separated, but usually the separation process is chemically and mechanically intricate. Costs are thereby so elevated that use is made of only those plants or plant parts which are rich in a particular polysaccharide and which can be simply processed.

Many microorganisms produce polysaccharides, and some produce extracellular polysaccharides in such quantities that it is economical to culture the organisms and harvest the polymers produced.

1. Exudate Gums

Gums of the ancient world were largely plant exudates. Most plant families include species that exude gums in greater or lesser degree. Gums can be exuded only in very small quantity and not be readily discernible, or they can be produced very copiously, forming large, conspicuous incrustations. Those produced in large amounts constituted the gums of the ancient world and even today constitute a significant segment (10–25%) of the natural gums of commerce. Plants that produce commercial gums are usually shrubs or low-growing trees from which the gums exude as vermiform or tear shapes and may build up in thickened layers. Harvesting is by hand picking, usually by native workers, in countries where labor costs are very low. After collection in sufficient quantities, exudate gums are taken to central collection areas where they are sorted by hand into grades, packaged, and shipped.

It is evident that a large proportion of the initial cost of exudate gums is in the labor of collection. As long as labor costs remain low in the warm countries where they are produced, raw gum prices will remain low. However, if local wage rates rise in these countries, collection costs for exudate gums will undergo a significant increase. Although wage rates for tropical laborers have always been low and may remain low for some years, it must be recognized that all tropical countries are entering a phase of economic and cultural growth that will eventually raise standards of living and per capita income requirements. As this occurs, the cost of labor for manual picking and collection of exudate gums may increase their raw prices so as to reduce their present competitive position.

2. *Seaweed Gums*

Gum extraction from seaweeds, originally practiced in oriental countries, has spread to many different parts of the world where shallow waters and seaweeds are abundant. Because sound engineering principles are used in the comparatively simple extraction processes, which remove a large portion of the dry weight of the weeds as rather pure material, ultimate costs of seaweed extracts are distributed between processing and harvesting. When harvesting is done by hand, as with agar weed in the Orient, or when the seaweeds are picked from beaches where they have been deposited by the tide, labor costs become important in the determination of final product costs. Presently, labor costs in the Orient and in those areas where beach weed is gathered are low, but sharp changes in labor prices could severely affect industries dependent upon these methods of collection. Consequently, much developmental work is in progress to devise suitable mechanical pickers for certain types of seaweeds. In Scandinavian waters, harvesting is done by dragging a trawl net on the ocean floor. However, the best engineered harvesting of seaweeds is that applicable to the giant kelps, *Macrocystis pyrifera,* which are found in large beds from Point Conception, California, southward to distant Mexican waters. These beds are periodically trimmed by a barge with an alfalfa cutter attached to the prow about 3–6 ft. (1–2 meters) under water. Cut kelp is lifted by a hay-loading rake to the barge top where it is stacked with a claw on a drag line. By such simple mechanical means, many tons of trimmings can be harvested rapidly and brought to the processing plant on the coast. Rapid growth of the kelp beds permits retrimming within a few months. Such standard mechanical practice lowers harvesting cost and tends to stabilize weed cost at the extraction point.

3. *Seed Gums*

Seeds were also an ancient source of gums. Most seeds contain starch as the principal food stored for use by the embryonic plant in its initial growth. Almost all the food plants used by man produce seeds with starch as the carbohydrate reserve. All these starches, which differ from each other in physical appearance and in physical behavior, have been used in gum applications when pasted in water. Today, only the lowest-priced starches find wide application because of vigorous economic competition.

Many seeds contain polysaccharide food reserves that are not starch, and some of these seeds are harvested to produce the seed gums of industry. The more ancient seed gums were extracted from quince, psyllium, flax, and locust seeds, and some of these are still quite important today. The yearly import into the United States of locust bean gum is over 14 million lb and of psyllium seed, over 3 million lb. However, here also, changes in labor costs and methods of agronomic production have been felt and will continue to influence market availability of the gums. Those gum seeds which are amenable to normal agricultural production should be lowest in price. These are the seeds from annual plants with a normal growing season, which can be grown on agricultural land by normal methods, and which can be planted and harvested by standard agricultural machinery. An example is guar. This plant was not used for its seed gum until World War II. The plant was grown as a cattle feed in Asia, particularly in India and Pakistan. With the discovery that its gum was similar to that from locust beans, its cultivation was increased in western Asia and initiated in the United States. The plant is an annual legume that resembles the soybean plant in appearance with seed pods along a vertical stem. Planting and harvesting can be done with ordinary farm machinery. It is apparent, therefore, that guar seeds can be produced by conventional agricultural practices and can thus be obtained at low prices. This accounts for the observation that guar seeds have taken over most of the expanding market for galactomannan gum. In the United States in 1970, about 22 million lb of guar gum was consumed, an amount more than one and a half times the imports of locust bean gum. As agonomists and geneticists improve the yield per acre of guar seeds, the price of guar gum will continue to decrease even below its present reasonable levels.

Psyllium is also amenable to agricultural production, and its use may further develop. Although seeds from trees, such as locust bean, are still holding up well against competition from annual plants, some tree seeds cannot be obtained at sufficiently low cost to be marketable. Perhaps an extreme example of a tree crop that cannot be produced economically is ivory nuts from the tagua palm, which yield a mannan with many industrially useful properties.

4. *Starch and Cellulose Derivatives*

Starch and cellulose are two readily available polysaccharides produced in great volume and are excellent starting materials for the production of gums. Both polysaccharides undergo chemical modification rather easily by heat, oxidation, or derivatization. Through proper control of the modification, a great variety of products is possible, and it is conceivable that, as more is learned about the relation of structure to physical properties of a polymer, gum properties can be custom tailored into starch and cellulose molecules so they will more closely match the properties desired in special gum applications. Starch at $0.06–0.09 a pound and high alpha-cellulose pulps at $0.09–0.14 a pound look inviting as starting materials for new product development. However, it must be remembered that, whereas developed chemical procedures may properly modify a polysaccharide to the desired product, the modification process may be so involved or expensive in reagent requirements that the products cannot compete in price with natural gums. Development of low-cost modification practices offers a stimulating challenge.

5. *Microbial Gums*

Extracellular polysaccharides produced by microorganisms offer a variety of useful and potentially low-cost industrial gums. Such gums fashioned by selected, and perhaps carefully mutated, organisms growing on low-cost energy sources, such as grain or molasses, can be low priced. A host of gums with widely different properties is available and more will be found. It is expected, therefore, that fermentation gums will eventually develop to fulfill many industrial needs and the cost will decrease, expectedly, as volume of production rises.

Incidental to the development of fermentation gums will appear techniques for producing gums by the use of microbial enzyme systems. Initially, homoglycans consisting of but one type of sugar unit will be made, as exemplified in the production of dextran. However, enzyme-catalyzed synthesis of polysaccharides from simple sugars and enzyme modification of existing polysaccharides will eventuate and become commonplace and low cost as manipulative techniques and knowledge advance.

IV. Industrially Valuable Properties of Gums

The properties of gums that give them useful performance in the food and nonfood areas are caused by physical effects, although sometimes part of the value added to food products is a consequence of organoliptic improvements. Physical effects derive from the interaction of the polysaccharide molecules with themselves and with the molecules of their environment. Environmental molecules with which polysaccharides may interact vary widely, ranging from various ingredients in processed foods and pharmaceuticals to inorganic particles in clay slips and oil-well muds. In most industrial applications, however, polysaccharides are used in an environment rich in water molecules. Thus, although polysaccharides must interact appropriately and beneficially with proteins, lipids, and other environmental molecules commonly met in industrial usage, they must foremost and continuously interact with water molecules.

To visualize, perhaps more clearly, the behavior of polysaccharide gums in water, it is helpful to briefly review how structural characteristics of gum molecules contribute to gum properties.

Polysaccharides are either branched or linear. Some polysaccharides are naturally anionic, possessing acidic functions, such as carboxyl groups, sulfate groups, or phosphate groups. Other natural polysaccharides possess amino groups, which are usually monoacetylated or, in one instance, sulfated. Such polysaccharides may have some free amino groups and show cationic properties.

1. *Linear Neutral Polysaccharides*

In the solid state, all polysaccharides have regions where molecules or chain segments are in a jumbled or disorganized arrangement with intermolecular forces and intermolecular hydrogen bonding only partially satisfied because of the random spatial arrangement. These amorphous regions, consequently, have numerous unsatisfied hydrogen bonding positions that can avidly hydrate. Perfectly dry polysaccharides have a small but very strong affinity for water. At normal humidities, polysaccharides contain 8 to 10% water molecules as water of hydration, the water occupying hydrogen bonding positions not otherwise involved in intra- and intermolecular bonding of the polysaccharide molecules. When a soluble polysaccharide is placed in water, the abundant water molecules quickly penetrate amorphous regions and bind to available polymer sites, competing for and eventually reducing to negligible numbers still other interpolysaccharide bonds. Segments of a polysaccharide chain become fully solvated and by kinetic action move away, tearing apart more interpolysaccharide bonds, which are immediately solvated. Soon, many sections of the polysaccharide chains are fully solvated and are solubilized, while a lessening number of segments are still attached to other polysaccharide chains, themselves not yet completely solvated. This intermediate stage in the dissolution of a polymer molecule represents a transient gel state and must portray a universal stage in the dissolution of all polysaccharides. Soluble polysaccharides continue to hydrate until molecules become completely surrounded by an atmosphere of partially immobilized water molecules and the polymer is monodispersed. Many polysaccharides do not hydrate beyond the intermediate gel stage and remain as incompletely dispersed molecules, forming gels whose physical properties depend upon the extent of hydration or, as said in another way, the extent of remaining interpolysaccharide bonding.

Perfectly linear polysaccharides are either insoluble or can be solubilized only under special conditions. Such molecules are homoglycans, which are polysaccharides composed of a single type of sugar, wherein all sugar units are linked uniformly with neighboring units. These molecules, although they have the same solvation capacity as other polysaccharides, also have the capability of fitting snugly together, with most of their intramolecular forces embracing each other. Large regions may be in such perfect order as to be crystalline. Such homoglycans remain insoluble and do not dissolve in water, even when they are thermally energized at elevated temperatures. Some of these polymers may be dissolved by use of high temperatures or by overcoming hydrogen bonds through the use of such agents as strong alkalis. However, should these polysaccharides be dissolved, it has been observed that lowering the temperature of their solutions or neutralizing the alkalinity of their solutions cause them to precipitate. Precipitation results from the collision of molecular segments wherein the solvating water molecules are sheared away, allowing chain segments to cling to each other. Intramolecular organization can be increased as thermal motion brings into alignment and register still additional chain segments, thereby further increasing molecular order. In such manner, particles form and grow through colloidal dimensions to precipitable size, whereupon the entire polymer is observed to come out of solution. Illustrative of such polymers are starch amylose and certain xylans. Perfectly homogeneous glycans of hexose sugars tend to develop intramolecular bonds from one sugar ring to the next along the chain. This helps to control the orientational relation of one sugar ring to the next along the chain and results in some restriction to the rotation of sugar rings relative to each other. A more ribbonlike structure results. The increased regularity further facilitates the fitting together of chains, resulting in less solubility of the polysaccharide. Cellulose is an extreme example of the insolubilization resulting from extensive regions where molecules fit together perfectly and, hence, because of mutual bonding remain insoluble in water.

Ionization of soluble neutral polysaccharides remains roughly constant over a wide range of pH values and is greatly altered only in strongly alkaline solutions. Therefore, these polysaccharides maintain their shape and absorbed water atmospheres essentially intact at pH's near neutrality; and, except in strongly alkaline conditions, their solution viscosities vary only slightly with changes in pH.

Salts at low concentrations do not greatly alter the viscosity of neutral gum molecules provided that they do not form a complex with them. At high concentrations, salts, because of the affinity of their ions for water, tend to remove the absorbed solvent molecules from around the neutral polysaccharide molecules and thereby reduce solubility, sometimes to the extent that precipitation occurs.

Because of their strong powers for adsorption, linear neutral polysaccharides are good coating agents for textiles and paper. Like other linear molecules, they usually can be cast into selfsupporting sheets and may have use as films or as packaging material. Only rarely are the molecules sufficiently long that the polysaccharides can be spun into fibers.

In general, linear, neutral polysaccharides are not tacky.

2. Branched Neutral Polysaccharides

It is easy to envision that neutral, linear, and branched polysaccharides will differ greatly in their ease of dissolution and in their ability to affect viscosity. Linear molecules will dissolve slowly, if at all. Branched molecules will dissolve more readily but will form solutions that, at equal concentrations with linear molecules of the same molecular weight, have much lower viscosity. Both will be equally hydrated in solution. But as linear molecules gyrate in solution they sweep out greater volumes than branched molecules, as seen in Fig. 1. Consequently, linear molecules interfere with each other and solution components more effectively than branched molecules, which sweep out smaller volumes of space. Linear molecules are, therefore, particularly sought after when high viscosities are to be produced by low concentrations of additive.

When limbs of branched molecules collide in solution, they may entangle, if the concentration is high enough, so that a gel results. Extensive association cannot form because it is not possible for the bushlike molecules to fit closely upon each other. Hence, solutions of branched molecules are stable and do not undergo spontaneous precipitation as do solutions of linear molecules.

Partially because of entangling with one another in solution, branched polysaccharides can produce tacky pastes if concentrations are sufficiently high.

When solutions of branched polysaccharides are dried, the molecules still do not form strong intermolecular associations; hence, if water is again added to the dry coating, the polysaccharide easily redissolves or reconstitutes, softening quickly to a tacky layer. This property makes many of the branched polysaccharides useful as remoistening gums or adhesives. Highly branched polysaccharides, like other highly branched polymers, form only brittle films and consequently have no application as selfsupporting sheets, although in low concentration with linear molecules they may have limited application.

Some polysaccharides are long chains with numerous very short branches. These molecules have many of the properties of both linear and highly branched molecules. Good examples are guar gum and locust bean gum. The former,

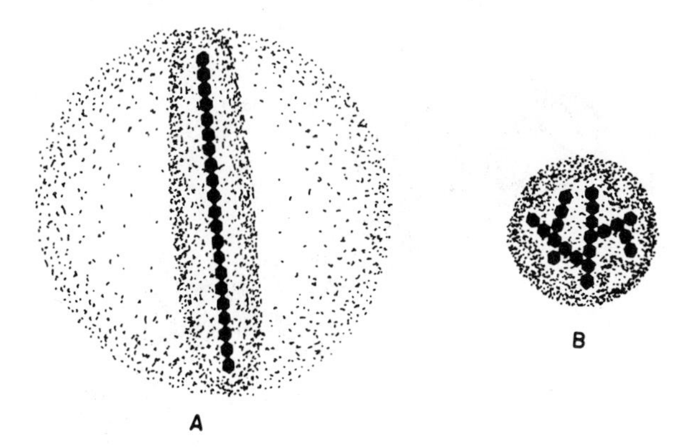

FIG. 1.—Artist's conception of the space occupied by the gyration of an extended linear polysaccharide, A, and a branched polysaccharide, B, of equal molecular weight.

guaran, is a long chain of D-mannose units with single D-galactose units joined to alternate units of the main chain (Fig. 2). It may be looked upon as a linear mannan derivatized uniformly with D-galactopyranosyl units to a degree of substitution (DS) of 0.5. Because it is a long polysaccharide, its solutions are highly viscous. However, the D-galactose side groups fend off other colliding molecules and prevent the associative lining up of molecules as would occur in solutions of purely linear molecules. Therefore, guar gum solutions are stable. Likewise, when its solutions are dried or when the polysaccharide is precipitated from solution, extensive intermolecular association is prevented and redissolution occurs rather readily. The short side chains do not entirely prevent molecular association in concentrated solutions but allow sufficient intermolecular orientation to insure that cast films are quite strong and flexible. Because the molecules adsorb readily and strongly on cellulose, guaran is an excellent paper and textile size.

It is obvious from the above considerations that the solubility of a linear polysaccharide would be improved by changes in the molecule that would reduce the fit of one molecule to another or by introducing charges, which by coulombic repulsion would facilitate dissolution of the molecules and, in solution, prevent their coming together so extensively as to produce a precipitate.

As we shall see, reduced intermolecular fit and hence solubility can be given to molecules such as cellulose by partial methylation or hydroxyethylation to make methylcellulose or hydroxyethylcellulose or by the production of other but similar derivatives. Coulombic forces, or charges, may be introduced into neutral molecules by carboxymethylation with chloroacetic acid, by sulfation or phosphorylation, or by other means.

3. *Polysaccharides with Carboxyl Groups*

Numerous carboxyl groups in a polysaccharide convey special characteristics, such as are found in algin and sodium pectate. Alkali metal salts of these linear

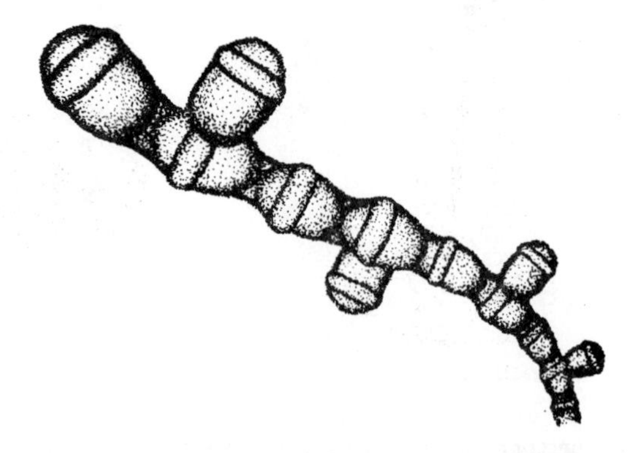

FIG. 2.—Artist's conception of guaran, where rings are space filling.

molecules are highly ionized in solution, and the distribution of ionic charge along the molecule tends to keep it in an extended form by reason of coulombic repulsion. Furthermore, this same coulombic repulsion between the carboxylate anions repels one polysaccharide molecule from another to promote a high degree of solution stability. The extended independent polysaccharide chains produce highly viscous solutions. Extensive hydration, not only of the polysaccharide as a whole but of the carboxylate anion in particular, leads to both improved solution stability and increased solution viscosity. In the presence of multivalent cations, such as calcium, salt bridges are established intra- and intermolecularly. Cross-linking between polysaccharide molecules leads to gel formation and, if extensive, to precipitation. Although the salts formed from monovalent cations are stable over the center range of the pH scale, molecular instability occurs at very high pH values due to the normal process of alkali degradation by the β-elimination mechanism. As the pH is lowered to about 3, carboxylic acid groups form; and, as the hydrogen ion concentration of the solution increases, their ionization is repressed. Under conditions of repressed ionization, the uncharged carboxylic acid groups lose a great amount of associated water; the polysaccharide molecules no longer repel each other and, being linear, readily associate to form a gel or precipitate.

4. Polysaccharides with Strong Acid Groups

Polysaccharides with numerous strong acid groups attached are highly hydrated, bear ionic charges, and are hence both highly viscous and stable in solution. Examples are certain polysaccharides derived from seaweeds, such as furcellaran and carrageenan, which are esterified with sulfuric acid in the half-ester form. Ionization of such strongly acidic groups is not repressed extensively at low pH levels, so the polysaccharide molecule maintains its ionic charges with their high degree of hydration and is stable in solutions of low pH.

5. Polysaccharides with Basic Groups

No naturally occurring polysaccharide containing basic groups is sold commercially, but work on chitin shows signs of incipient application. A linear basic molecule, such as deacetylated chitin, is soluble in strong acids and shows particularly tenacious adsorbance on negative surfaces. Such a molecule in solution remains extended. Because of its molecular linearity, the polysaccharide can form strong, semiflexible sheets or surface coatings. It will also combine in solution with anionic polysaccharides or other negative polymer molecules to produce a precipitate by mutual neutralization and aggregation. The polysaccharide might in some instances act as a flocculation agent for negative colloids. Introduction of cationic groups into otherwise neutral polysaccharides has produced useful gums and more will be developed in the future.

V. Modified Gums

Chemical modification of polysaccharides is useful for altering chemical and physical properties to give the polysaccharide new applications. Sometimes modification of a low-cost polysaccharide introduces qualities that make it a suitable replacement for a more expensive gum. In general, the properties of neutral polysaccharides are altered to a remarkable degree by the introduction of very small amounts of substituent groups of either neutral or ionic type. Frequently, derivatization to a DS of 0.01–0.04 is sufficient to produce entirely new properties and open new applications. However, polysaccharides bearing formal charges usually have their properties but little affected by low levels of derivatization. Perhaps this is not so unexpected if the view is taken that charged polysaccharides are derivatized neutral polysaccharides in which small additional charges do not greatly change the overall structural and charge effect.

1. Introduction of Neutral Groups

Placement of neutral groups on linear polysaccharides leads to increased solution viscosity and stability. In effect, addition of methyl, ethyl, hydroxyethyl, and similar groups tends in a broad sense to give linear polysaccharides the behavior patterns of guar or locust bean gums. It is apparent that random introduction of substituent groups to a DS of 0.5–1.5 will produce sufficient irregularity in a linear polysaccharide that molecules will not easily fit over long segments of their length. Consequently, they will dissolve easily to form stable solutions whose viscosity will be high because the molecules will tend to remain extended. Solubility characteristics are enhanced more by use of hydroxylated substituent groups, such as hydroxyethyl, than by completely hydrophobic groups, such as ethyl or propyl. Water solubility diminishes and solubility in organic solvents increases as hydrophobic groups are introduced beyond a DS of about 1.0.

2. Introduction of Acidic Groups

Carboxyl groups may be introduced into a neutral polysaccharide by oxidation, for example with dinitrogen tetraoxide, or by appropriate derivatization, for example with chloroacetic acid. Sulfation and phosphorylation are commercially useful means for the introduction of strongly ionized groups. Polysaccharide molecules with such groups readily dissolve to form viscous solutions that may be quite mucilaginous. Sometimes the moistened derivatives are almost salvelike in consistency.

3. Introduction of Basic Groups

Basic groups are introduced directly into the sugar rings of a polysaccharide with great difficulty because it is hard to find a convenient reaction by which hydroxyl groups are replaced with amino groups. However, it is quite easy to derivatize a polysaccharide with a group that contains a free or substituted amino group. It is also possible to react an ester of a carboxylated polysaccharide with ammonia to form the amide.

Flocculation of negative colloids with basic polysaccharides has been examined and promises many applications. Coating of negative surfaces also may become practical.

4. *Graft Polymers*

A fertile field of research with strong potentialities for industrial application is the grafting of linear or branched polymers to polysaccharides. This almost totally unexplored field offers interesting problems in the many avenues by which it may be approached.

From information in starch dextrin chemistry dating back many years, it is evident that, when a polysaccharide is heated in the presence of a small amount of an acidic catalyst, glycosidic bonds break and reform. It is not necessary that bonds be completely opened but only that a transfer reaction be catalyzed. As the process is random, the reformation of bonds most likely occurs at a new location. If an internal bonds cleaves and reforms, the result is that of chain transfer. A linear molecule tends to become branched and a branched molecule tends to become still more highly branched. However, with proper control, preferred chain transfers may be possible. If two different kinds of polysaccharides are reacted, it is conceivable that they may become joined into a single polymer. It is also likely that, with proper control, a polysaccharide could be derivatized with single sugar units. Much more work is needed in this promising field.

Still a quite different approach to grafting is the attachment to polysaccharides of synthetic polymers of different lengths and with different reactive groups. Perhaps such grafts can be formed most easily by introduction into a polysaccharide of the desired number of monomer units on which polymer growth can be controlled to the extent wanted. Present interesting information on the modification of properties brought about by production of such graft polymers gives assurance that they will be the subject of much further work.

5. *Other Chemical Modifications of Natural Polysaccharides*

For many years, polysaccharides have been modified by thermal dextrinization, partial hydrolysis, and mild oxidation. Control of properties through proper choice of kind and amount of substituent groups is quite well known. Although almost limitless possibilities exist for the chemical modification of polysaccharides, much work is needed to correlate physical properties with structure. Except for a few instances, information is not available that would allow a chemist to predict accurately the physical properties that may be imparted to a polysaccharide as the result of a chemical modification. The ultimate goal of the chemist is to custom-modify low-cost polysaccharides to fit new needs and applications. To this end, more fundamental investigations are needed on the relation between the structure of polysaccharides and their physical properties.

VI. Acceptability of Gums for food Usage

Carbohydrates are innocuous edible components of foods. Many supply significant portions of the caloric requirement of man. Of all the polysaccharides, starch is the principal metabolizable energy source. The vast number of other polysaccharides consumed normally as natural components of edible vegetable, fruit, and other plant materials are not metabolized in man but pass through the gastrointestinal tract with little or no change, for no enzymes exist in the human gastrointestinal tract for hydrolyzing most polysaccharides. Furthermore, the acidity of the stomach is neither strong enough nor is the time of residence of polysaccharides in the stomach sufficiently long to cause significant chemical cleavage.

However, as the undigested and hence unmetabolized polysaccharides reach the large intestines, they come into contact with the normal intestinal flora of microorganisms, some of which may produce enzymes that may catalyze hydrolysis of certain polysaccharides or certain parts of polysaccharide molecules. The consequence of this is that some polysaccharides which are not cleaved in the upper intestinal tract may undergo some degree of cleavage within the large intestines. Usually such cleavage is nonexistent or occurs to a minor and insignificant extent. Where it does occur, the result is a lowering of the molecular weight of the polysaccharides. If sugar units are split from the polysaccharide chain, they are mostly metabolized by the intestinal organisms, although some small fraction of the sugars produced may be transported through the intestinal wall and be metabolized in the human system.

Although many experimental investigations have been conducted in which the fate of polysaccharides has been followed in the gastrointestinal tract of man and animals, there has never been observed an instance of acute toxicity from the oral intake of reasonable levels of natural polysaccharides. On the contrary, most polysaccharides are beneficial in providing bulk to intestinal content. Excessive amounts of certain very hydrophilic polysaccharides may produce diarrhea as a consequence of their water binding capacity. In lower concentration, they may be beneficially demulcent. The innocuous character of natural polysaccharides has given them wide usefulness as food components and as food additives. There is reason to believe that they will continue as useful additives to improve food characteristics.

Present evidence indicates that chemical and biochemical modifications of high-molecular-weight polysaccharides are likewise innocuous and mainly nonmetabolizable dietary components when used in reasonable quantities in human food.

CHAPTER II

SOLUTE-SOLVENT INTERACTIONS IN POLYSACCHARIDE SYSTEMS

ROBERT J. DOBBINS

*American Cyanamid Co., Stamford, Connecticut**

I. INTRODUCTION

Since the time of the first edition of this book, a major research effort in the field of physical chemistry has involved solute–solvent interactions, and the role of these relatively weak (1–5 Kcal/mole) forces in determining the stability of macromolecular systems in aqueous solution.

Water can no longer be regarded simply as an omnipresent, homogeneous solvent system of no greater significance to the reaction mechanism than the reaction vessel itself. A variety of spectroscopic techniques have shown that there is a dynamic equilibrium between clusters of associated molecules $(H_2O)_n$, and discrete, highly reactive monomeric H_2O units. Many solutes, including polysaccharide systems, appear to be very sensitive to this equilibrium, which is influenced in turn by the addition of any third component. For this reason, when attempts are made to explain the reactivity or lack of reactivity of a solute upon a polymer system, it is often necessary to consider the effect of the species upon the bulk water properties, and of this effect in turn upon the polymer interface.

II. STRUCTURE OF WATER

A number of excellent reviews have discussed the experimental data and cur-

*Present address: St. Regis Technical Center, West Nyack, New York.

rent theories on the structure of water.[1,2] Nearly all current theories consider the solvent as a mixture of nonbonded, highly reactive monomers, similar to gaseous water vapor, and rigid, structured clusters similar to those found in ice.

In common ice I, each water molecule participates in four hydrogen bonds. The pair of hydrogen atoms and the two electron pairs are arranged tetrahedrally about the oxygen atoms. Each hydrogen points toward the electron pair of an adjacent water oxygen (OH→:O), and each electron pair points toward a neighboring hydrogen. Because there are equal numbers of hydrogen bond donors (the hydrogens bound to the oxygen) and acceptors (the lone pairs of electrons), a highly structured, nonreactive system is the result.

Many of the physical properties of liquid water have been attributed to its partial retention of the ice crystalline structure. X-ray diffraction studies[3] of liquid water show a high probability of finding adjacent water molecules at distances of 2.82A, 4.5A, and 7.0A from a central water molecule, and these are precisely the distances that correspond to the tetrahedral structure of ice I. Equally significant, at distances greater than 8A, the probability values are nearly equal to the bulk density values, which is good evidence for an extensive short-range tetrahedral orientation of water molecules, with very little of the longer-range ordering that is found in the solid phase.

Frank and Wen[4] have proposed a "flickering cluster" model of water structure, consisting of short-lived (10^{-10} to 10^{-11} sec), icelike clusters of varying sizes, in dynamic equilibrium with the nonhydrogen bonded water that makes up the balance of the solvent. Over time intervals longer than 10^{-11} sec, the clusters dissolve into chaotic groups of nonbonded molecules, and these rapidly coalesce to form a new cluster with a size and shape different from that of the preceding cluster. At any given moment, 70 to 90% of the water molecules are believed to be incorporated into clusters.

The energies associated with these hydrogen bonds are low (1.3–2.5 Kcal/mole as compared to 25–100 Kcal/mole for many covalent bonds), so that the formation of a cluster can be initiated merely by a local energy fluctuation that creates "cold" or more icelike regions. Electron shifts that increase the acidity of the hydrogen make the oxygen correspondingly more basic, so that the formation of clusters becomes a cooperative phenomenon, the formation of one bond often leading to a cluster, and the breaking of one bond within a given cluster tending to break all the bonds that are associated with that same unit.

III. Effects of Solutes on Water Structure

The introduction of a solute into an aqueous system inevitably changes the number and the size of the icelike clusters. The effect is largely determined by the specific attractive force between the solute and the neighboring water molecules. In the case of simple electrolytes, three distinct classifications can be formulated.

Small or highly charged ions attract water through strong ion–dipole forces, and the resulting solute–water bonds are stronger than the normal water–water hydrogen bonds discussed above. Cations such as hydrogen, lithium, or aluminum ions act as nuclei for cluster formation by immobilizing the surrounding water molecules. This attractive force extends beyond the four to six nearest neighbors, and the resulting effect on the bulk water is to produce regions of high local structure centered about the solute ion. Ionic mobility data show, for example, that lithium and sodium ions retain their solvent sheaths while undergoing transport phenomena in solution.

A second distinct class of electrolytes are those with relatively weak attractive forces between solute and the surrounding water. The electronic fields around ions such as potassium, ammonium, or rubidium are weak and few water molecules are immobilized in the area of these ions. They do exert a torque on the electronic fields of the water molecules in solution however and, in so doing, interfere with the electronic shifts necessary for the initiation and propagation of clusters by the cooperative hydrogen bonding network in the water. Because cluster making forces are buffered, with little or no solute-centered clustering, there is a decrease in the number of water clusters and a proportional gain in the concentration of free unbound water.

The effect of these strong and weak field electrolytes upon the solvent properties is reflected in a comparison of solution viscosities. At 25°, a molar concentration of rubidium iodide has a viscosity 0.921 times that of pure water, whereas a molar concentration of lithium chloride has a viscosity 1.142 times that of pure water.[5]

The third class involves large ions having considerable organic, hydrophobic character. These ions are characterized by repulsion rather than attraction to the nearest water molecules. The adjacent solvent molecules are oriented and hydrogen bonded by the water beyond them in the bulk solution, with no counterattraction toward the solute ion. This results in a tightening of the water structure around the solute on the internal surface of the solvent cavity. The larger and more hydrophobic the molecule, the greater is this effect. Tetraalkylammonium ions and many alcohols are solvated in this manner.

A phenomenon that is common and specific to all such hydrophobic electrolytes is described by Diamond[6] as a "water structure enforced association." The repulsive solvent–solute interactions can be minimized by forcing the solute molecules to associate in a common solvent cavity, rather than by having each remain fully dissociated and thereby maximizing the solute–solvent repulsive forces. The associative process is largely one of entropy; two separate cavities require more ordering of the surrounding solvent than does one slightly large cavity holding the associated molecules.

IV. Solvation of Polysaccharides

The high-molecular-weight, ethylenic backbone, and lack of polar substituents places alpha-cellulose in the hydrophobic class of compounds,[7] whereas the addition of amino groups, as in chitin, significantly alters the pattern of reactivity in aqueous solution.

Giles and Arshid[8] used a refractive index technique to determine the extent of inter- and intramolecular hydrogen bonding and, thus, to measure the extent of adsorption of a series of chemically related solutes onto cellulose and chitin.[9] Methanol and phenol are both adsorbed in the absence of water, but only chitin will adsorb them from aqueous solution. A series of four compounds of similar molecular size containing different hydrogen bonding sites, 2-hydroxyanthraquinone, anthraquinono-2-sulfonic acid, sulfonated alizarin, and the corresponding azo dye, show varying degrees of adsorptivity onto chitin, according to the hydrogen bonding affinity of their substituents. The sulfonic acids of benzene, anthracene, and naphthalene are not adsorbed by chitin to a measurable extent.

Cellulose shows a significantly different reactivity. The failure to adsorb phenol is surprising, because it is known to hydrogen bond with each hydroxyl group in a straight chain polyhydric alcohol, such as mannitol. Benzenesulfonic acid is not adsorbed, but the larger anthracene- and naphthalenesulfonic acids are readily adsorbed. The four quinone derivatives, differing in substituents but possessing a common size and hydrophobic character, are adsorbed to a similar extent.

It is apparent that substitution of the amino group for a secondary hydroxyl substantially alters the mechanism of adsorption by the otherwise structurally similar polymers. Solutes bind to chitin by hydrogen bonds, the extent of the adsorption varying with the polarity of the adsorbant and inversely with molecular size. In contrast, adsorption by cellulose is similar to the "water structure enforced association" that is characteristic of hydrophobic compounds. Cellulose itself tends to produce a hydrophobic pattern of hydration in aqueous solution, and another hydrophobic solute can be readily adsorbed in order to reduce the net solute–solvent repulsive forces.

By a similar process, the adsorption of cetyltrimethylammonium chloride to algin, agar, pectin, or dextrans can be considered a hydrophobic type of interaction, a bonding produced to reduce the total solvent–solute repulsive forces by incorporating the polymer and electrolyte into a common solvent cavity. In support of this hypothesis, Scott[10] has found that substitution of the polar sulfate groups into the polymers, or alternatively, addition of a strong field electrolyte, such as sodium chloride, reduces the extent of this polymer–electrolyte bonding.

V. Effects of Ions upon Polysaccharide Solvation

It has long been recognized that some electrolytes are able to stabilize certain marcromolecular conformations in aqueous solution. Various workers in unrelated fields have found the same relative order of cation and anion effectiveness, the well-known lyotropic series of Hofmeister.[11]

The swelling, viscosity, and flocculation of amylose is one such example. Lithium iodide has a solubilizing effect, whereas magnesium sulfate has a flocculating effect and is employed in the commercial fractionation of starch. Similarly, some salts (potassium thiocyanate and calcium chloride) are potent destabilizers of the ordered structures found in proteins, whereas others (potassium hydrogen phosphate and ammonium sulfate) are excellent stabilizers of these ordered conformations. These effects and the same specific ranking of ionic effectiveness appear to apply to all systems in which there is a net transfer of polymer segments from an unsolvated to a fully solvated environment.

Recent work by von Hippel and Schleich[12] has clarified the mechanism whereby these salts influence the stability of macromolecules in water. Many such molecules consist of a hydrophobic portion that forms an "inside," consisting of segments not in contact with the solvent, and an "outside," consisting of the more hydrophilic groups in contact with the water. The free energies of these states are the sums of the individual group–solvent and group–group interactions, and a change of a kilocalorie or less per mole in one of the stabilizing interactions involved in maintaining the delicate balance is enough to trigger a cooperative transition to a different conformation.

Working with polyacrylamide as a typical example of a polymer combining a hydrophobic backbone with pendent amide dipoles, they were able to demonstrate that those ions known to be good destabilizers (iodide, perchlorate, thiocyanate, and calcium) were able to bind to the gel more firmly than the solvent molecules and, thus, to stabilize the extended, unfolded conformations. Inert ions (sodium, potassium, and chloride) were found to bind to the gel with a force approximately as strong as the gel–solvent forces, and ions known to be excellent structure stabilizers (sulfate and fluoride) were bound less firmly to the polymer than the solvent and, thus, tended to stabilize the rigid macromolecular conformations.

This data supports the hypothesis that the Hofmeister series is related to the ability of a particular ion to bind to a relatively hydrophobic site, changing the hydrophobic segment–solvent repulsive forces and the delicate balance of intermolecular interactions that determine macromolecular stability in solution.

VI. Effect of Salts on Polysaccharide Stability

It is well known that metal ions, even when present at the concentrations commonly found in natural water systems, have specific and often unpredictable effects on the stability of polysaccharides in solution. Electrolytes increase the viscosity of guar and algin but decrease the viscosity of hypnean, carrageenan, and gum arabic. Clearly, these systems cannot be adequately described by the simple coulombic attraction of cations to the polybasic polymers but are more likely to involve a change in the hydrophilic–hydrophobic balance of the polysaccharide by the added electrolyte.

In the case of amylose, with a macromolecular structure and charge density similar to that of cellulose, it is apparent that ions that preferentially replace water in the hydration sphere of the polysaccharide (lithium and iodide) make the aggregate more soluble than native amylose. Ions that are bound less strongly than the waters of hydration (magnesium and sulfate) not only fail to increase polymer stability, but also decrease the concentration of free water by their strong hydration demands and thus promote flocculation.

Algin is known to be a long, relatively rigid, linear chain. The small changes in the polymer–solvent energies produced by a preferential adsorption of electrolytes have little effect on such a rigid polymer conformation, and as a result, solution viscosities are relatively independent of salt concentration. Calcium seems capable of influencing the viscosity and gelation through a specific bonding to the carboxyl and/or cis hydroxyl groups.

In contrast, guar gum has regularly spaced single D-galactopyranosyl branches along its uncharged mannan backbone. The presence of these side chains means that interchain hydrophobic bonding is possible where adjacent polymeric chains have branched groups hydrated in common solvent cavities. Thus, the viscosity is linearly related to guaran concentration only at very low polymer levels, but non-Newtonian behavior is found at concentrations above 0.5%.

Gum arabic, with its highly branched stiff coil configuration, reflects this same interchain hydrophobic bond behavior. In dilute polysaccharide solutions, with little interchain behavior, added electrolytes are effective in reducing the viscosity by binding to the polymer and reducing the repulsive segment–solvent forces. At higher concentrations of gum arabic (35%), however, where there is extensive hydrophobic bonding of side chain segments of adjacent polymers, added salts have no effect on solution viscosities, the hydrophobic bonding being stronger than the gum–electrolyte forces that occur at dilute concentrations.

Hypnean is characterized by a different order of cation effectiveness. Both the gel strength and the temperature of gelation are sensitive to cations, the relative order being cesium > rubidium > potassium > ammonium, and for potassium salts, acetate > chloride > sulfate > bromide.[13] Sugar and alcohols, which are known to have correspondingly small coulombic fields, promote gel formation. Pending a detailed study of the configuration of hypnean, the exact mechanism of polymer–solute interaction for this gum remains uncertain.

Carageenan is also sensitive to the presence of potassium, ammonium, rubidium, and cesium salts. It has been found that the specific gelling temperature can be related to the potassium ion concentration and is independent of the carageenan concentration. This suggests that there is no change in conformational stability produced by ionic adsorption because this would be dependent upon the polysaccharide concentration. Gelation in this case may be sensitive to the cluster ⇄monomer equilibium in the solvent water because this is dependent upon potassium ion concentration but is independent of the relative concentration of polymer to cation. Because potassium increases the concentration of monomeric, "free water" in solution, its salts promote gelation of the carrageenan.

Obviously, a great deal of work remains to be done in the field of polysaccharide–solvent–solute interactions. Studies on the structure and conformation of polymers, on the complex electrolyte–solvent interactions, and the delicate balance of inter- and intramolecular attractions and repulsions that determine polysaccharide stability in aqueous solution have given some reasonable explanation for some of the well-established reactions in these systems. It is hoped that further work in the fields of polymer and biochemistry will shed additional light upon these complex, yet significant, phenomena.

VII. References

(1) D. Eisenberg and W. Kauzmann, "The Structure and Properties of Water," Oxford University Press, New York, 1969.

(2) E. Wicke, *Angew. Chem., Int. Ed. Engl.,* **5,** 106 (1966).

(3) A. H. Narten, M. D. Danford, and H. A. Levy, *Discuss Faraday Soc.,* **43,** 97 (1967).

(4) H. S. Frank and W. Y. Wen, *Discuss. Faraday Soc.,* **24,** 133 (1957).

(5) "International Critical Tables of Numerical Data, Physics, Chemistry, and Technology," McGraw Hill, New York, Vol. 5, 1929.

(6) R. M. Diamond, *J. Chem. Phys.,* **67,** 2513 (1963).

(7) R. J. Dobbins, *Tappi,* **53,** 2284 (1970).

(8) F. M. Arshid, C. H. Giles, E. C. McLure, A. Ogilvie, and T. J. Rose, *J. Chem. Soc.,* 67 (1955).

(9) F. M. Arshid, C. H. Giles, and S. K. Jain, *J. Chem. Soc.,* 559 (1956).

(10) J. E. Scott, *Chem. Ind. (London),* 168 (1955).

(11) M. Hofmeister, *Arch. Exp. Pathol. Pharmakol.,* **24,** 247 (1888).

(12) P. H. von Hippel and T. Schleich, *Accounts Chem. Res.,* **2,** 257 (1969).

(13) G. E. Osborne and C. O. Lee, *Drug Stand,* **19,** 13 (1951).

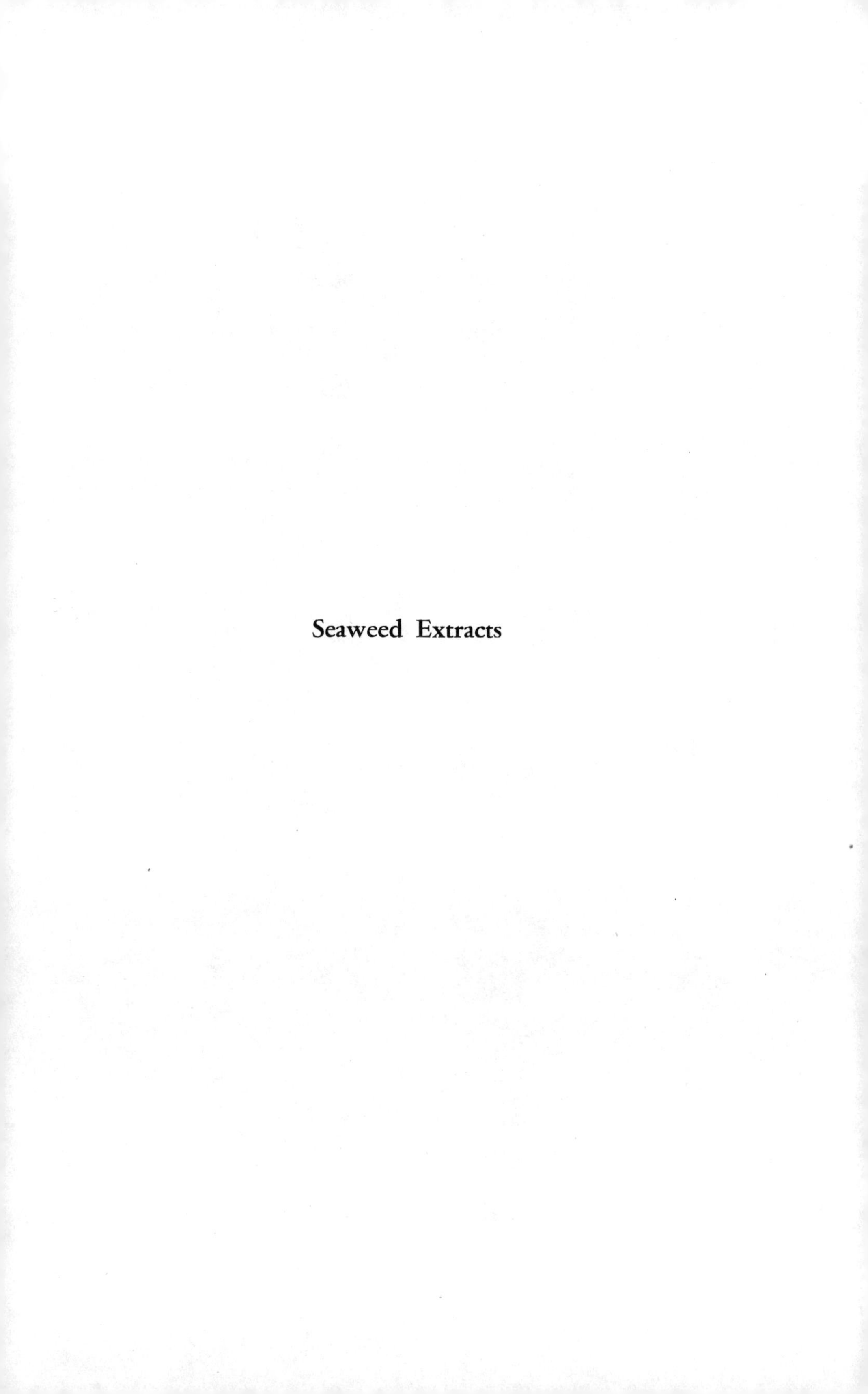

Seaweed Extracts

CHAPTER III

AGAR

H. H. SELBY AND W. H. WYNNE

American Agar and Chemical Company, San Diego, California

I. INTRODUCTION

According to the U.S. Pharmacopeia,[1] agar is a hydrophilic colloid extracted from certain marine algae of the class Rhodophyceae. Agar is insoluble in cold water but soluble in boiling water. A 1.5% solution is clear, and when cooled to 32°–39°, it forms a firm, resilient gel that does not melt below 85°. In the interest of explicitness, the term agar will be reserved here for polysaccharides so defined. Other gums resembling agar, but not meeting all the specifications of this definition, will be referred to as *agaroids*.

II. Source

1. Raw Material

Agar is obtained from various genera and species of the red-purple seaweeds, class Rhodophyceae, where it occurs as a structural carbohydrate in the cell walls and probably also performs a function in ion-exchange and dialysis processes.

Species, habitat, and commercial importance.—Agarophytes of major commercial value are listed in Table I. These seaweeds are found from the intertidal zone to depths of more than 40 m. Growth is most prolific in areas of surface turbulence and of marked top to bottom exchange. The plants grow from holdfasts attached to rocky substrates and attain lengths of from 0.1 to 2 m. Gelidiales

TABLE I

Algae Yielding Agar

Name	Location	Remarks [a]
Acanthopeltis japonica	Japan	S
Gelidiella acerosa	Japan, India	P
Gelidium amansii	Japan	P
Gelidium arborescens	S. Calif., U.S.A.	T
Gelidium cartilagineum	U.S.A., Mexico, S. Africa	P
Gelidium caulacanthum	New Zealand	S
Gelidium corneum	S. Africa, Portugal, Spain, Morocco	P
Gelidium coulteri	Mexico	T
Gelidium crinale	Japan	S
Gelidium devaricatum	Japan	S
Gelidium japonicum	Japan	S
Gelidium liatulum	Japan	P
Gelidium lingulatum	Chile	P
Gelidium nudifrons	California, U.S.A.	T
Gelidium pacificum	Japan	P
Gelidium pristoides	S. Africa	P
Gelidium pusillum	Japan	S
Gelidium sesquipedale	Portugal, Morocco	P
Gelidium spinulosum	Morocco	S
Gelidium subfastigiatum	Japan	S
Gelidium vagum	Japan	S
Gracilaria confervoides	S. Africa	P
Pterocladia capillacea	Egypt, Japan, New Zealand	P
Pterocladia densa	Japan	S
Pterocladia lucida	New Zealand	P
Pterocladia nana	Japan	S
Pterocladia tenuis	Japan	S

[a] Key to remarks. P, Primary commercial value; S, secondary commercial importance; T, tertiary commercial importance.

usually show an alternation of isomorphic generations, and propagation may be either by spores or stolons.

The botanical names given in Table I are matters of some controversy in taxonomic circles and are subject to change. The nomenclature of the Rhodophyceae according to the Schmitz—Kylin system has enjoyed long and wide acceptance, but Dixon[2, 3] and others suggest extensive revision.

Harvesting (collection).—Harvesting from Mexican and some Japanese waters is done by divers in full pressure suits who tear the individual plants from their holdfasts and place them in rope bags in which they are raised to the surface. A diver is usually served by two helpers in a boat that is provided with an air compressor, emergency gear, weed storage space, and motive power. The diver must often work under the formidable handicaps of jagged topography, strong currents, and low illumination. Despite these conditions, he must maintain proper buoyancy and select only suitable plants as he moves with or against a bottom current or wave surge. In this sort of operation, an expert diving team can gather from 200 to 1000 kg of wet weed per 6-hr diving day. When dried, this output will range between 40 and 320 kg.

In other parts of the world, a preponderence of weed is harvested by waders at low tide, raked from rowboats, or picked by skin divers, who may wear either unventilated face masks or be supplied with compressed air from the surface in the manner of the full pressure suit divers.

After the seaweed is gathered, it is freed of unwanted plants, stones, and other detritus; washed with fresh water when available; and spread in 1–5 cm layers on sand, grass, or racks to dry. In average weather, with daily inversion of the layers, the moisture content will have dropped to less than 20% in 4 days and the weed will have been partially bleached by solar radiation. At this point, most weed is baled, although some may be washed and redried until thoroughly bleached before baling. The baled weed, if the moisture content is maintained below 20%, will have a storage life of approximately 5 years.

Price.—The major commercial weeds, which may contain 30 to 45% agar, are frequently in short supply, which causes their market prices to undergo substantial fluctuations. The raw material used in the United States is purchased directly from foreign and domestic agents on the basis of analysis of representative samples taken at time of delivery. The prices paid are based on the agar content, gel strength, and the percentage of contaminants and may vary from $500 to $1250 per 1000 kg, delivered. The minor varieties, usually containing 15 to 30% agar, will follow these variations and will command from $350 to $700 per 1000 kg.

Variability.—The agar content of seaweeds varies intraspecifically and interspecifically. Carbon dioxide concentration, oxygen tension, water temperature, and intensity of solar radiation might be significant in a given species. The presence of bryozoan epiphytes markedly affects the yield of commercially collected seaweed.

Gelidium cartilagineum harvested in Mexico contains 15–30% agar on a dry basis. The better beds are found in the cooler, more turbulent locations. Seasonal effects are noted in *G. cartilagineum,* the agar content being more than 30% higher in July than in January, with the fluctuations correlating well with the curve of solar ultraviolet intensity. In Spain, the agar content of *G. spinulosum* and *G. sesquipedale* peaks in July–August and again in November–December.[4] In Japan, the May and June collections of agar weeds are known to give greater yields than later crops.

Gelation temperatures and gel strength of the extracted agar from the Mexican *G. cartilagineum* show no such seasonal change, although yield, gelation temperature, and gel strength vary substantially in *Gracilaria confervoides* from North Carolina.

Resource conservation.—Perpetuation of the weed beds is assured by certain practices conducted by the diver during the picking operation. The plants are severed approximately 10 cm above the holdfasts, and in Mexico, the picked plants are shaken vigorously to release whatever viable spores may be available in the plant. In this manner, the plant will propagate and be ready for another harvesting in the next year. In localities that enjoy particularly good growing conditions, two crops may be taken each year.

Large scale cultivation of agarophytes is not practiced, but some Japanese beds are maintained and enlarged by dumping broken stone and concrete in areas of sparse growth. When the new stone has become encrusted with natural growths and deposits, it is seeded with spores and stolons.

2. *Processing*

Agar is insoluble in cold water but is colloidally dispersible in water above 90°. When agar gels are frozen, the agar skeleton contracts toward the center of mass as a membrane, leaving the ice as a separate phase. The ice will, in general, contain ten times more soluble salts, sugars, and simpler gums than is retained by the agar membrane. Hence, most commercial agar is manufactured by hot water extraction followed by freezing for purification. Other methods are possible, such as extraction with glycerol, anhydrous ammonia, or other solvents, and the use of alcohols and other flocculants to avoid the freezing operation.

In the traditional process, great care is exercised in blending the various algae chosen for a batch in an attempt to obtain a product of the desired flexibility, luster, density, and surface smoothness and of the usual physical properties of gel solidity and resilience. Usually six or seven types of weed are used.

Operation begins soon after freezing nights are assured. The weeds are washed in batches of 8 kg and pounded for 20 min to remove any sand and epiphytes.

For each 200 kg of seaweed, about 2200 liters of water are boiled in an open iron caldron over a pine log fire. The tougher types of algae are introduced into the caldron first and the softest last. When the intermediate weeds have been added, the mixture is treated with 1 g of sulfuric acid or 0.3 g of polyphosphate per kg of seaweed to adjust the pH to 5–6 and extraction continues at 80° for 8–9 hr; at that time, weak liquor from the previous day is added. At approximately the twelfth hour, calcium hypochlorite or sodium bisulfite is introduced for bleaching at a rate of about 2 g per kg of weed, and the cooking is continued to the fifteenth hour.

The entire cook is strained through cloth of 3 mm mesh and the cake is pressed and retained for recooking. The liquor containing 1% of agar is cleared somewhat by sedimentation, after which it is allowed to solidify in 170 × 30 × 1-cm wooden trays.

The gel is cut into strips, laid on straw mats outdoors and allowed to freeze, the strips being covered if the temperature is too low. Each day some of the night-formed ice melts, taking with it some salts, nitrogenous material, and residual color. Sprinkling is used when needed to prevent excessively rapid drying. After 5–6 days, the racks carrying the mats are orientated with respect to the sun in such a manner that final drying is complete in another 15–30 days.

In order to prepare an acceptable product, the manufacturer must be constantly alert to counter the effects of climatic changes. Rapid freezing causes poor luster and feel. The surface ice must be broken by tapping to prevent the formation of sharp, fragile corners. Poor color results if either freezing or drying is too rapid.

The traditional method is essentially a cottage industry, most installations being one or two cooker ventures and many being one-family enterprises.

Since 1945, several firms throughout the world have begun a more scientific agar manufacture. Each manufacurer generally uses various locally developed modifications of the basic extraction–freezing–thawing–drying method. In the main, the newer methods employ countercurrent and cascade multiple extraction, centrifugation, plate-and-frame press filtration, artificial freezing, chemical bleaching, drying with hot air by drum and spray methods, and grinding.

In the United States, the following sequence of operations is employed: (a) cleaning raw material, (b) chemical pretreatment, (c) pressure extraction, (d) chemical posttreatment, (e) filtration, (f) gelation, (g) freezing, (h) posttreatment, (i) washing, (j) drying, (k) sterilization, (l) bleaching, (m) washing, (n) drying.

Trade intelligence suggests that enzymic predigestion is used in Denmark and dialytic desalting in Spain and Japan. Some firms are said to have employed electrodialysis, synthetic resin decolorization, and alcohol precipitation and predrying of gel by pressing.

Several promising areas of technology have received attention in the recent past; for example, caustic pretreatment of *Gracilaria* has enabled agaroid processors to increase substantially the gel strength of their products.[5]

Irradiation of dried species of *Gelidiella, Gelidium, Gracilaria,* and *Hypnea* by 1000 Ci of cobalt-60 in the dosage range of 0.9–6.4 $\times$ 10^4 r/g has been found to improve yield, gel strength, and stability of the dry extracts.[6] Similar treatment of agarophytes and agar elsewhere has, however, been negative.

Pretreatment of agarophytes with cellulolytic enzymes has in some cases acclerated extraction rates, improved yields, and increased gel strength.[7,8]

III. Finished Product

1. *Specification of Type, Grade, and Purity*

Type and grade specifications.—Apparently only the one North American manufacturer and the Japanese industry employ grading systems and publish specifications. Three types of American agar are produced. All types are offered as three dimensional flakes, approximately 3 mm on a diagonal. A large percentage of the output is ground.

In Japan, local inspection bureaus certify individual lots of agar on the basis of visual and tactile examination as No. 1, No. 2, or No. 3 with subgrades A and B indicating relative admixture of lower quality material. Agar for export is further graded by the Price Assessment Committee of the Japan Agar-Agar Association under a government ordinance promulgated in 1949.[9]

The older geographic names attached to grade numbers, such as Kobe, Yokohama, and Sakhalin, have lost their significance. Agar shipped from the seaport of Kobe was once considered superior, and practically all Japanese agar produced by the traditional process is now called Kobe No. 1, No. 2, or No. 3 if exported.

Traditional process agar is exported as string or 3 $\times$ 320-mm noodle-shaped pieces and as 15 $\times$ 320-mm square bars. Modern process agars are available as chopped shreds, granules, grounds, sheets, and powder.

Official Specifications.—Agar monographs in the U.S. Pharmacopeia[1] and the Food Chemicals Codex[10] give the requirements (paraphrased) shown in Table II.

The U.S. Pharmacopeia[1] requires freedom from *Salmonella* species.

The Society of American Bacteriologists (now the American Society for Microbiology) have adopted specifications for agar for microbiological use. The latest (1958) requirements[11] are shown in Table III.

Specifications of purity.—Interest in deleterious or undesirable substances present in agars began to increase in about 1937, when Araki resolved agar into low-sulfate, pyruvic acid-free agarose (agaran) and high-sulfur, high-ash agaropectin.[12] By the late 1950's, organic sulfate and trace metal content had been

TABLE II

	Maximum	Minimum
Gelation temperature, 1.5%	39°	32°
Gel melting temperature, 1.5%	—	85°
Moisture	20%	—
Ash	6.5%	—
Ash, acid-insoluble	0.5%	—
Foreign organic matter	1.0%	—
Foreign insoluble matter	1.0%	—
Foreign starch	0	—
Gelatin	0	—
Water absorption	—	5 times its weight
Arsenic	3 ppm	—
Lead	10 ppm	—
Other heavy metals	40 ppm	—

TABLE III

	Maximum	Minimum
Total solids	—	78%
Solubility, cold	2.0%	—
Solubility, hot	—	99.8%
Gelation temperature, 1.5%	39°	33°
Gel melting temperature, 1.5%	—	70°
Rate of dissolution, 1.5%	15 min	—
Sol turbidity, 1.5%	10 ppm	—
Threshold gel concentration	0.25%	—
Protein nitrogen	0.32%	—
Reducing substances as galactose	10%	—
Chlorides as sodium chloride	1.5%	—
Viable spores	3/g	—
Debris count	30/g	—

shown to have adverse effects in several fields of use. Interest in the use of agarose (agaran) for bacteriology,[13] virology, and clinical electrophoresis is particularly high.

Agar is predominantly the calcium salt of strongly ionized, acidic polysaccharides. The agars of commerce frequently have sodium or magnesium ions in place of some calcium ions. Hence, alkali metals and alkaline earth metals are not considered impurities except for the cultivation of some magnesium-intolerant microorganisms.

The principal impurities in commercial agars are nonagar gums, nitrogenous compounds, soluble and insoluble salts, free sugars, heavy metals, and occasionally floridean starch[14] (Table II). Mannitol has been detected in the methanol extract of an agar.[15] Diatoms, pericarpial rice cells (from bran used as an antifoaming

TABLE IV

Summary of Results of Analyses of Agar, Agaroids, and Agaroses[a]

	Agars			
	Domestic	Foreign	Agaroids	Agaroses
Number of sources	1	32	14	4
Aluminum, ppm	0–50	50–1200	200–8000	0–230
Arsenic, ppm	0	0–2	0–9	—
Barium, ppm	0–3	3–50	3–70	0–13
Bismuth, ppm	0	0–5	0	—
Boron, ppm	20–110	10–300	20–240	—
Cadmium, ppm	0–0.8	0–2.1	0–1.5	0.2–1.3
Calcium, %	0.15–0.3	0.02–0.9	0.7–3.0	0.2
Chromium, ppm	0–2	0–10	0–5	0–2
Cobalt, ppm	0–2	0–10	0–15	0–2
Copper, ppm	0–2	2–66	0–140	0–10
Iron, ppm	5–25	24–800	40–9000	25–140
Lead, ppm	0–0.8	0.3–100	0–60	2–8
Magnesium, %	0.04	0.02–0.62	0.1–2.1	0–0.02
Manganese, ppm	0.1–1	2–150	3–200	1–2
Nickel, ppm	0.1–4	0–10	0–1	0–1
Rubidium, ppm	0	0	0–40	—
Sodium, %	1.5	0.01–4.2	0.1–0.8	—
Strontium, ppm	8–25	0–400	10–200	—
Tin, ppm	0	0–50	0	—
Titanium, ppm	0–10	0–200	20–200	—
Zinc, ppm	2–15	4–2000	20–2000	4–20
Ash, %	3.5–4.4	3.0–7.4	7–15	0.6–2.8
Protein nitrogen, %	0.1–0.2	0–2.0	0.5–3	0–0.2
Reducing substances, %	0–1	0.4–21	0.5–32	0–8
Sulfate, organic, %	0.6–1.3	0.9–4.2	2.5–5	0.1–0.4
Sulfate, inorganic, %	1–3	0.6–2.7	1.5–8	0.1–0.3
Uronic acids, %	0–0.3	0–4	0–10	0–0.2
Moisture, as purchased, %	20	11–24	10–22	9–16
Clarity, 1.6% sol, 45°, mm	900	50–850	20–300	200–600
Gelation temp, 1.6% sol, °C	35–37	36–39	38–63	31–35
Melting temp, 1.6% gel, °C	86–88	60–88	60–76	64–90
Insolubles, %	0–0.1	0.2–2.8	0.3–1.5	0–0.4
Thermoduric spores, No./g	0–1	0–1000	0–2000	0–2

[a] 118 samples; 51 sources; about 5000 determinations; 1950–1971; dry basis.

agent), starch granules, pollens, insoluble salts, and dead microorganisms often contribute to the turbidity of agar dispersions.

Viable spores of thermoduric bacteria are present in many agars, a few agarose (agaran) preparations, and some agaroids. This type of contamination, most undesirable in microbiology, pharmaceuticals manufacture, impression materials, and the production of moist products, was once almost universal,[16] but it is less widespread today.

Resins, waxes, sterols, vitamins, and ureides are trace impurities of doubtful importance and are found only occasionally.

Because the agarose (agaran) fraction of agar is lower in sulfate and is less ionized than the agaropectin portion, it is more satisfactory than agar in gels used for electrophoretic separations, particularly of globulins. Also, in work involving the development of viral plaques and in gel filtration and gel chromatography, the nonionic nature of pure agarose is often an advantage. It is, therefore, possible that high organic sulfate content may become a valid index of impurity for some uses.

In some work, which may be adversely affected by agaropectin, agar use can be made more economical by the incorporation of an agaropectin inactivator, such as aminoalkyldextrans,[17] protamine, or a cationic detergent, provided that it has no adverse effect on the system under study.

2. *Volume of Production and Consumption*

Estimates of the amounts of agar made and consumed in a decade are presented in Table III. Foreign production and U.S. consumption changed little from 1968 into 1972. Although Japan is still the largest producer, Spanish interests are operating with great energy and Spain's exports now equal Japan's.

Agarose (agaran) production is still reckoned in kilograms rather than in tons. It is estimated that world production will surpass 200 kg (440 lb) in 1970.[18] The largest producer is probably Marine Colloids, Inc., Rockland, Maine.

3. *Price*

Large-quantity market prices of agar have changed little since 1966 with average annual prices ranging from ~\$3.00 to ~\$3.25/lb. Agarose (agaran) was sold for \$20 to \$100 per 100 g in 1972.[18]

IV. Chronology

1. *Discovery*

Legend has it that in about 1660, Minoya Tarozaemon,[19] a Japanese innkeeper, threw some surplus seaweed jelly into the winter night, expecting it to thaw in the morning sun and to disappear into the soil. He found, however, after several

TABLE V

Foreign Production and American Consumption of Agar and Agaroids

	Foreign Output, Metric Tons	
	1958	1968
Japan	1500	1500
Spain	300	1000
Nationalist China	200	200
Korea	300	500
Morocco	300	300
Chile	50	200
Portugal	100	200
South Africa	20	50
France	0	50
New Zealand	30	50
Australia	50	50
Indonesia	20	50
Others	100	200
	Uses, U.S.A., lb.	
Microbiology	200,000	400,000
Baked goods	200,000	200,000
Confectionery	90,000	100,000
Meat and poultry	70,000	100,000
Desserts and beverages	50,000	80,000
Laxative and health foods	50,000	50,000
Pet foods	—	50,000
Impression materials	30,000	30,000
Pharmaceuticals	20,000	20,000
Miscellaneous	30,000	30,000
Total, approximate	700,000	1,000,000

days of alternate freezing and thawing, a porous mass that could be reboiled in water and cooled to yield a gel equal to the original. He had discovered agar.

2. *Manufacture*

At Shimizu-mura, Japan, a monument[19] commemorates the first commercial manufacture of agar by a relative of Tarozaemon, Miyata Hanbei of Aza Shiroyama. The Japanese industry expanded slowly until, in 1940, there were 400 processors. Manufacture of agar started in China and on the Malay Peninsula in 1850. Irregular production began in India, Ceylon and Australia in about 1880–1900, and in about 1915, Indonesian production was established.

Chokichi Matsuoka attempted in 1919 to make agar in Tropico (Glendale),

California, using mechanical refrigeration and sun drying.[20] In 1933, John Becker[21] established the first of a series of agar companies in San Diego, California, where production continues.

The advent of World War II found the United States with but one small California producer and the other Allied Nations with token plants or none. In the United States, eight plants attempted agar manufacture during the war; and Mexico, South America, South Africa, Australia, New Zealand, France, Spain, Portugal, and Russia produced some agar or agaroids. Since 1953, there has been but one producer in the United States.

3. *Use*

Agar became popular in areas close to production as a food, food ingredient, and medicine. Popularity spread steadily and rapidly, and in about 1866, European use of agar for food began.[22]

Frau Fanny Hesse suggested in 1881 the use of agar for growing colonies of bacteria to her husband, Dr. Walther Hesse.[23] Hesse told Koch, who discovered *Mycobacterium tuberculosis* with its aid. Poller[24] discovered in 1924 that agar gels had the requisite qualities for a moulage material capable of reproducing fine detail with great accuracy, thus opening the field of impression material use to criminologists, museum preparators, dentists, plastic surgeons, and artists. From 1935 onward, dental prosthesis made wide use of agar compositions in precise inlay, crown, and bridge work, as well as in ordinary denture molds.[25] In the latter, alginate gels have largely supplanted agar in recent years.

Agar, and particularly agarose, are now used in differentiating proteins, lipoproteins, enzymes, and other high-molecular-weight compounds by electrophoresis, gel-bead filtration, and gel chromatography.

4. *Derivatives*

Fairbrother and Mastin[26] found that the calcium of native agar could be replaced by other cations. Hoffman and Gortner[27] showed agar to behave like a salt of a highly ionized acid capable of salt formation by neutralization with bases. Other exchanges[28] and derivatives, such as the acetate[29, 30] and methyl ether[30] were soon investigated. Partial precipitation of agar by quaternary ammonium compounds was found to be a useful separation method.[31]

A method of obtaining agarose (agaran) by precipitation of agaropectin by a quaternary ammonium salt and its separation from agarose by centrifugation became generally useful.[32, 33] Interest in this fraction is increasing, and other agarose separation methods based on poly(ethylene glycol),[34] enzymes and sequestrants,[35] ammonium sulfate,[36] aluminum chloride,[37] methyl sulfoxide,[38] EDTA and aluminum hydroxide,[39] and buffer extraction or electrophoretic purification followed by anion-exchange[40] have been devised.

For making sodium agarose, sodium iodide has been found to have certain advantages.[41]

V. Present Applications

Microbiology.—Agar is most valuable in microbiology, although outside the United States larger quantities are used in other applications.

The ideal agar is low in metabolizable or inhibitory substances, debris, and thermoduric spores; has a gelation temperature of 35°–40° and a gel-melting temperature of 75°–85°; is readily soluble, and has good gel firmness, resilience, clarity, and stability. Agar concentrations of 1–2% are commonly used for this purpose.

In low concentrations, agar prevents the entry of oxygen into liquid media, making the cultivation of anaerobes feasible in air-exposed broths. The usual range of agar concentrations used in liquid media is 0.007–0.08%.

Few organisms metabolize agar or elaborate agarolytic enzymes. *Vibrio pupureus,*[42] *V. agarliquefaciens,*[43] a flavobacterium,[44] a pseudomonad,[45] *Cytophaga* species,[45] and certain diatoms[46] are exceptions.

Impression materials.—In prosthetic dentistry, criminology, tool making, and other fields, it is necessary to make accurate casts of intricate undercut objects. Poller[24] and others developed moulage compositions utilizing the low gelation temperature, high gel strength, and superior gel resilience of agar. Restriction of the use of agar to microbiology, including antibiotic manufacture, by the U.S. government during World War II caused substitutes to be sought. Some were found and have supplanted agar in the less critical areas. Alginate gels and natural and synthetic elastomers, such as silicone rubbers, are examples. Agar concentration used for impression materials ranges from 6 to 14%.

Foods.—Although agar is practically indigestible,[47] it is used in many food products where its emulsifying, stabilizing, and gelling properties and the heat resistance of its gels are useful.

In sherbets and ices, agar–tragacanth stabilizers have been used for years. Agar–locust bean gum–gelatin is superior with respect to texture, syneresis, incrustation, and flavor stability. The optimum stabilizer concentrations have 0.12% of agar, 0.07% of locust bean gum, and 0.20% of gelatin.

Agar is employed at levels of 0.05–0.85% in Neufchatel-type processes and cream cheeses and in fermented milk products, such as yoghurt. Improved texture and stability are conferred.

In confectionery, agar jelly candies are staples, with fig–agar and marshmallows as leaders. In South America, a sweet potato–sugar–agar confection has wide popularity. Agar percentage varies with the texture desired; 0.3–1.8% is common.

Manufacturers of baked goods use agar in cookies, cream shells, piping gels, pie fillings, icings, and meringues as an antitackiness ingredient. It has been used with success as an antistaling agent in breads and cakes. The range of concentration is 0.1–1%.

The use of agar is well established in vegetarian and health food products, including prepared cereals, meat substitutes, and desserts.

Poultry and meat canners employ agar as a gelling agent at 0.5–2.0% of the broth weight, thereby eliminating transit damage of fragile tissues.

Medicine and pharmaceuticals.—Agar has been widely used as a laxative for several decades. When well hydrated, agar furnishes the smooth nonirritating bulk that appears to be necessary for normal peristalsis; its use appears to be nonhabitforming and is said to be not contraindicated even in cases of uncomplicated peptic ulcer.[48] Medicinal-type agar especially prepared in the form of thin flakes designed to prevent the formation of obstructive masses[49] and to absorb 12–15 times its weight of fluid is well received professionally.

Agar is used as a suspending agent for barium sulfate in radiology, as an ingredient of slow-release capsules, in suppositories, in surgical lubricants, in emulsions of many types, and as a carrier of topical medicaments. It is used as a disintegrating agent and an excipient in tablets. Its use as an anticoagulant was patented,[50] and sulfated agar has been shown to have antilipemic activity equal to that of heparin. The ingestion of agar appears to increase the excretion of fats and might promote the intestinal synthesis of niacin. It has been reported to inhibit the aerobic oxidation of vitamin C.

Laboratory.—The accuracy of particle-size determinations can be increased with agar as can turbidimetric determinations involving suspensions of dense solids. (High-clarity agar is preferred.) Certain dyes, such as methylene blue, toluidine blue, thionine, and pinacyanol, can be reversibly polymerized by agar. In the analytical coagulation of calcium sulfate, arsenic sulfide, ferric hydroxide, and barium sulfate, agar can be employed to advantage.

Microtomy of plant tissues is facilitated when a 5% agar gel is used as the imbedding medium. Agar gels containing 20–25% of potassium chloride and 2–2.5% of agar have been used for many years as antidiffusion, conductive bridges in connection with calomel electrodes. The electrophoretic migration of proteins through agar gels has been used to resolve ferritin, ovalbumin, hemoglobin, and pepsin.

Sodium and ammonium agar, agarose (agaran), and sodium agarose have proved valuable in globulin electrophoresis, immunodiffusion diagnostic techniques, gel filtration, and gel chromatography. The work in these fields is too voluminous to be referred to in detail.

Agar has been found to stabilize cholesterol solutions.[51]

Dried agar gel slices are used for obtaining infrared absorption spectra of amino acids.[52]

Agar and agarose (agaran) derivatives incorporating epichlorhydrin-linked carboxymethyl and dimethylaminoethyl units[53] have been used in the electrophoresis and chromatography of proteins.

Agar and agarose (agaran) gel techniques are being used to separate bacterial toxins, to estimate the chain length of nucleic acids, for fractionation of antibiotics and antimycoplasma substances, for classification of viral particle sizes, and for separation and purification of enzymes.

Miscellaneous and potential.—Agar has been found suitable for use in photographic stripping films and papers when esterified with succinic or phthalic anhydride and after enzymic hydrolysis. The moisture content and dimensions of agar films change promptly and reproducibly with ambient humidity. Such films are useful as sensing elements in mechanical and electrical hygrometers. Tough transparent films can be made from agar acetylated in pyridine.

The use of agar in solidified alcohol fuel, dyed coatings for paper, textiles and metals, and in pressure-sensitive tape adhesives in conjunction with a phthaloyl resin has been disclosed. As a flash inhibitor in sulfur mining explosives, agar–ammonium chloride gels have been used. The foaming properties of dodecyl and hexadecyl detergents are enhanced by agar, and its detergent power alone, as well as in synergistic mixture with other detergents, has been noted.

Better density, brightness, and adhesion of copper plated on aluminum when agar is employed are claimed. The use of agar in dry cell separators has been suggested. Agar is an ingredient of some cosmetic creams and lotions. As a corrosion inhibitor for aluminum, agar has been used in caustic solutions with some success. The activity of nicotine plant sprays is increased and action is prolonged when agar is incorporated. Agar–shellac–wax mixtures are popular in shoe and leather polishes and dressings. The use of agar media in orchid culture is standard. Advantage can also be taken of the cationic exchange ability of agar.

For further information on matters treated in foregoing sections, see references in the first edition of "Industrial Gums."[54]

Some newer uses for agar and its derivatives include agar as a setting inhibitor for deep-well cements;[55] agar as an adhesive in gloss finishing of paper products;[56] agar as an ingredient in Fe–Ni–Cr–Ti electrolytes in the alloy-plating of steel;[57] agar as an iron corrosion inhibitor in citric and malic acid solutions;[58] agar as a temporary neutron absorber in reactors;[59] and agar as an inhibitor of corrosion of iron and lead by distilled water.

VI. Structure and Function

Whistler and Smart[60] stated in 1953 ". . . the complete structure of agar is not established." This is still true and promises to remain until unanimity is reached in defining what agar is. In this chapter, the definition adopted for agar has been a pragmatic one based on origin and physical properties. Elsewhere, it is common to call almost any phycocolloid agar if it exhibits sol–gel–sol hysteresis. The specific galactanolytic enzyme approach[61] to definition and identification may one day help.

Agar, as the term is here used, and agarose (agaran) are probably[62] predominantly repeating units of alternating β-D-galactopyranosyl and 3,6-anhydro-α-L-galactopyranosyl units coupled $1\rightarrow3$. Uronic acids are absent or are present in amounts less than 1%.

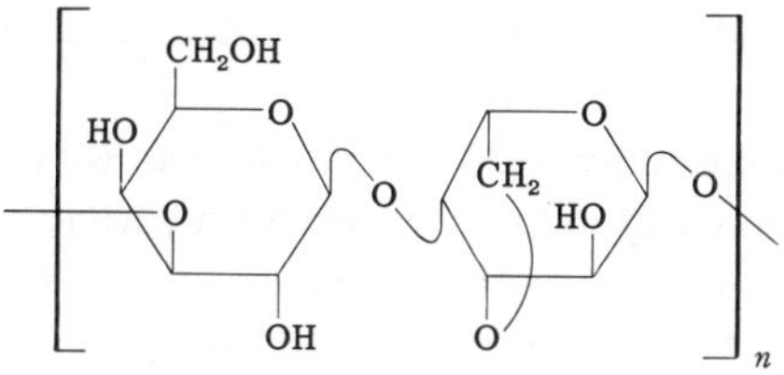

IGB A-3, sf 1

In contrast, purified "agar" produced in Archangel from *Ahnfeltia plicata* is said to contain 12 D-galactopyranosyluronic acid and 3 D-galactopyranosyl 6-sulfate units per 100 chain units.[63]

The heterogeneous nature of agar has been emphasized by Izumi, who using anion-exchange chromatography,[64] separated agar into four fractions. Similar fractions have been obtained by Duckworth and coworkers[65, 66] who later found *Gracilaria* agaroids to contain from one to five components.

An avenue of particular promise for structural analysis seems to be investigation of circular dichroism in regions of the Cotton effect[67, 68] in the infrared part of the spectrum, using pleated selenium film polarizers.[69]

Specificity of identification may be improved through future work using cellulose acetate electrophoresis of the dyed polysaccharides.[70]

In nature, agar may exist as an protoagar rather than as agar as extracted. *Gelidium cartilagineum,* reduced to a 5-μ powder, yields no agar when extracted for 30 min at 100°. Extraction for 2 hr at 110° gives a normal yield however. This suggests the precursor hypothesis because, with cell walls disrupted, normal agar should be relatively free to disperse.

VII. Properties

1. Solid

Solubility.—At 25°, agar of high purity is practically insoluble in water. It is very slightly soluble in ethanolamine and soluble in formamide.

If agar, agarose, or *Gracilaria* agaroid is flocculated from a warm dispersion by five to ten volumes of ethanol and drained but not allowed to dry, it is soluble in water at 25° and will form a gel without the need for heating. Agar so flocculated is also soluble at 25° in other solvents, but gelation will not occur until water is added. Dry agar is colloidally dispersible in water and in other liquids at 97°–100°, and moist agar flocculated by ethanol, 2-propanol or acetone, or salted out by high concentrations of electrolytes, is soluble in a number of liquids.

Dispersions of 0–5% are conveniently made by heating on a steam or boiling water bath with occasional agitation, on a hot plate or over a low flame with constant stirring, by stagnant autoclaving at 100°–120° or, best, by the use of spherical containers with segmental agitators heated by electric mantles or steam jackets.

For the preparation of heavier dispersions, such as 8–14% moulage mixtures, a vertical, jacketed, metal cylinder, provided with packing glands at both ends and a perforated piston agitator mounted in the center of a piston rod passing through both glands, is convenient. The solid ingredients are introduced; the cylinder is closed, evacuated to 1 mm or less, completely filled with the desired liquid by suction, and sealed. By translational movement of the piston and the use of steam at atmospheric pressure in the jacket, rapid dispersion without foaming can be effected.

2. *Sols*

Viscosity.—The viscosity of agar and agaroid dispersions is markedly influenced by the type of raw material and the processing conditions employed. The relative viscosities of agar and *Gracilaria* agaroid at 1% and 1.5% concentration have been reported.[54, 71] The viscosity of an agar dispersion at 45° is relatively constant from pH 4.5 to pH 9 and is not greatly affected by age or ionic strength within the limits of pH 6.0–8.0. Once gelation begins, however, viscosity at constant temperature increases with time.

Gelation temperature.—Agar is unique among polysaccharides in that gelation occurs at a temperature relatively far below the gel-melting temperature. Many uses of agar depend upon this high hysteresis. Agars and agaroids from different species have markedly different gelation temperatures, each of which is practically constant, with the exception of the agaroid from American *Gracilaria confervoides,* which exhibits seasonal variations of more than 20°. A gum from *Hypnea musciformis,* in which gelation temperature can be increased 60° by the addition of potassium chloride, is known.[72]

The gelation temperature of agarose (agaran) sols is correlated with the methoxyl content of the gum.[73]

Coagulants.—Agar is quantitatively flocculated in the presence of electrolytes by ten volumes of ethanol, 2-propanol, or acetone. It is salted out by near saturation with sodium sulfate, magnesium sulfate, or ammonium sulfate. Prior to drying, such flocculated agar exists in a metastable state, in which it is dispersible in cold water and in other solvents. In general, the higher the temperature of flocculation and the higher the concentration of electrolytes, the less soluble is the floc.

Many quaternary ammonium compounds cause turbidity and agaropectin precipitation, as does silicotungstic acid. The most sensitive precipitants for agar appear to be tannic, phosphotungstic, and phosphomolybdic acids when used at pH 1.5–2.5.

Miscellaneous.—Viscosity, diffraction, and gel strength studies show that the chain length of agar is reduced by ultrasonic vibrations[74] and strong gamma radiation, as well as by intensive agitation and high temperature. The infrared absorption spectra of agar, agarose (agaran), and other gums have been published.[75, 76]

3. *Gels*

Agar and agarose (agaran) are among the most potent gel-forming agents known, for gelation is perceptible at concentrations as low as 0.04%. Threshold gels are valuable for their protective action, diffusion prevention, and texture enhancement effects. Stronger gels are of value because of their strength, resilience, elasticity, relative transparency, relative permanence, and reversibility. Agarose gels are firmer but less elastic than gels from the parent agar.

Melting temperature.—The melting temperature of an agar gel is a function of concentration, and molecular weight. Agar and agaroid gels with 1.5% solids melt from 60° to 97°.

Strength.—The threshold gel concentration (TGC) is determined by the concentration of agar solids necessary for the formation of a particular gel under standard conditions. The results correlate well with emulsifying, stabilizing, and protective ability and with gel strength in the 0.2–2.0% range.

Rupture at Constant Stress (CS) Method: In Japan, the Nikkankyo method has been used in official grading. Ten plungers loaded in an ascending series of mass are simultaneously lowered on ten replicate samples of a 1.5% gel (air-dry basis) that have been aged 15 hr at 20°. The maximum stress in g/cm^2 withstood for 20 sec without rupture is reported as the solidity of the sample.

Stress–Strain Curve (SSC) method: This development from Stoloff's method[77] yields information on gel strength, tenacity, and resilience. Screw-topped, 65-ml, ointment jars are filled to the 50 ml level with a 1.6% solids sol at 45°, sealed and held 1 hr in a stirred water bath maintained at 19.5°–20.5°. A 1-cm^2, cylindrical plunger is pressed into the gel without lateral movement at 2 mm/sec. Load and depth of gel depression are simultaneously recorded to rupture. Rupture stress is a measure of strength; depression depth at failure varies as tenacity, and curve slope is a function of elasticity.

In the United States, three gel samples with 1.5% of solids aged 15 hr at 4° are subjected to different stresses estimated from previous SSC determinations and the rupture times are graphed. The stress in g/cm^2 required to effect rupture in 20 sec is taken from the curve.

Compatibility.—Near neutrality, agar is compatible with most other polysaccharide gums and with proteins in the sense that flocculation or marked degradation does not occur when their dispersions are mixed. An exception is gum kino.

Near pH 3, flocculation occurs when warm agar and gelatin dispersions are mixed. Such mixtures are used to excellent advantage, however, in sherbets and ices by delaying the additions of acids until freezing has begun.

Sodium alginate and starch decrease the strength of agar gels, whereas dextrin and sucrose cause increases. Locust bean gum has a marked synergistic effect on the strength of agar gels. The incorporation of 0.15% of locust bean gum can increase the rupture strain of an agar gel 50–200%, depending on conditions. Iceland moss extractive (lichenan) and carboxyethylcellulose show similar action to a lower degree. Gelatin, Russian isinglass, and gum karaya tend to weaken agar gels, but only slightly. When added to warm agar dispersions, most salts, glycerol, sorbitol, the alkanolamines, and 1,2,6-hexanetriol have little effect on the strength of the gels.

Miscellaneous.—Pure agar–water gels are rather stable. Those made with high-strength agar appear to be as stable as dry agar itself if sterile and hermetically stored. Low-strength agars, however, deteriorate more rapidly in the gel than in the solid form. The fact that few microorganisms metabolize agar or elaborate enzymes that degrade it might explain the generally greater stability of agar gels in comparison with gels of other natural colloids.

Agar gels age slightly. After 1 hr at 25°, the strength increases about 1%/hr for about 8 hr, then increases at a decreasing rate and becomes stable in 15 hr. Agar gels have micelle structure, and their elasticity is energy elasticity rather than entropy elasticity.

VIII. References

(1) "The Pharmacopeia of the United States of America," Mack Publ. Co., Easton, Pa., 18th Ed., 1970, p. 17.

(2) P. Dixon, *Ann. Rev. Oceanogr. Marine Biol.*, **1,** 177 (1963).

(3) E. Y. Dawson, "Marine Botany," Holt, Rinehart and Winston, New York, 1966, p. 175.

(4) R. Establier Torregrosa, *Invest. Pesquera,* **26,** 165 (1964); *Chem. Abstr.,* **62,** 3111 (1965).

(5) T. Yanagawa, *Bull. Japan Soc. Sci. Fisheries,* **6,** 274 (1938).

(6) Y. A. Doshi and P. S. Rao, *Nature,* **216,** 931 (1967); Y. A. Doshi, S. T. Talreja, and P. S. Rao, *Indian J. Technol.,* **6,** 275 (1968).

(7) M. Hachiga and K. Hayashi, *Hakko Kogaku Zasshi,* **42,** 207 (1964); *Chem. Abstr.,* **64,** 20592 (1966).

(8) K. C. Hong and W. Yaphe, *Proc. Int. Seaweed Symp., 6th, Santiago de Campostela, Spain, 1968* (1969).

(9) "Guide to Japan's Exports," No. 20, Foreign Trade Press, Tokyo, 1949, pp. 33,34.

(10) "Food Chemicals Codex," (Publ. 1406), Natl. Acad. Sci.—Natl. Res. Council, Washington, D.C., 1st Ed., 1966, pp. 17, 18.

(11) "Manual of Microbiological Methods," McGraw-Hill, New York, 1957, pp. 64-67.

(12) C. Araki, *Nippon Kagaku Zasshi,* **58,** 1338 (1937).

(13) C. Burns, "Metals and Microorganisms" (Bibliographies), American Agar and Chemical Co., San Diego, Calif. 1966, 1967, 1968, 1969, 1970, 1971.

(14) P. O'Colla, *Proc. Roy Irish Acad.,* **55B,** 321 (1953).

(15) F. Scheffer, E. Schlimme, and R. Kickuth, *Naturwissenschaften,* **52,** 518 (1965); *Chem. Abstr.,* **63,** 18537 (1965).

(16) C. T. Townsend and T. L. Zuch, *J. Bacteriol.,* **46,** 269 (1943).

(17) N. M. Tauraso, *J. Bacteriol.,* **93,** 1559 (1967).

(18) K. B. Guiseley, Marine Colloids, Inc., personal communication, 1972.

(19) Ref. 9, p. 1.

(20) C. K. Tseng, *Food Ind.,* **17,** 11 (1945).

(21) H. D. Mackinnon, *Food Ind.,* **2,** 123 (1930).

(22) C. K. Tseng, *in* "Colloid Chemistry," J. Alexander, ed., Rheinhold Publ. Corp., New York, Vol. 6, 1946, p. 630.

(23) A. P. Hitchens and M. C. Leikind, *J. Bacteriol,* **37,** 491 (1939).

(24) A. Poller, U.S. Patent 1,672,776 (1929); *Chem. Abstr.,* **22,** 2644 (1928).

(25) N. H. Larson, *Dental Survey,* **26,** 674 (1950).

(26) F. Fairbrother and H. Mastin, *J. Chem. Soc.,* **123,** 1412 (1923).

(27) W. F. Hoffman and R. A. Gortner, *J. Biol. Chem.,* **65,** 376 (1925).

(28) H. DeWaele, *Ann. Physiol. Physiochim. Biol.,* **5,** 877 (1929).

(29) E. G. V. Percival and W. S. Sim, *Nature,* **137,** 997 (1936).

(30) C. Araki, *Nippon Kagaku Zasshi,* **58,** 1351 (1937).

(31) M. Yenson, *Rev. Fac. Sci. Univ. Instanbul,* **13A,** 97 (1948); *Chem. Abstr.,* **42,** 6560 (1948).

(32) S. Hjerten, *Biochim. Biophys. Acta,* **62,** 445 (1962).

(33) J. Blethen, U.S. Patent 3,281,409 (1966); *Chem. Abstr.,* **66,** 12116 (1967).

(34) B. Russell, T. H. Mead, and A. Polson, *Biochim. Biophys. Acta,* **86,** 169 (1964).

(35) Hyland Laboratories, Brit. Patent 1,070,770 (1967); *Chem Abstr.,* **67,** 65704k (1967).

(36) G. Y. Azhitskii and B. V. Kobozev, *Lab. Delo,* 143 (1967); *Chem. Abstr.,* **66,** 106120s (1967).

(37) E. Bjerre-Beterson, Litex, Inc., personal communications, 1968.

(38) S. Hirase and C. Araki, *Chem. Soc. Japan.* (*Nippon Kagaku*), *Abstr. Papers,* **15,** 176 (1962).

(39) S. J. Barteling, *Clin. Chem.,* **15,** 1002 (1969).

(40) S. Hjerten, *J. Chromatog.,* **61,** 73 (1971).

(41) K. Katsuura, T. Fuse, and K. Kano, *Kogyo Kagaku Zasshi,* **68,** 205 (1965).

(42) M. Yoshikawa and K. Watanabe, *Hyoko Noka Daigaku Kenkyu Hokoku,* **3,** 53 (1957); *Chem. Abstr.,* **52,** 19198 (1958).

(43) K. Ishimatsu, K. Minami, and I. Fujita, *Kagaku To Kogyo* (*Osaka*), **35,** 429 (1961); *Chem. Abstr.,* **57,** 14276 (1962).

(44) T. Hidaka and M. Shameshima, *Kogoshima Daigaku Suisan Gakubakiyo,* **3,** 158 (1953).

(45) H. Veldkamp, *J. Gen. Microbiol.,* **26,** 331 (1961).

(46) R. A. Lewin, *Proc. Int. Seaweed Symp., 6th, Santiago de Campostela, Spain, 1968* (1969).

(47) H. W. Nilson and J. W. Schaller, *Food Res.,* **6,** 461 (1941).

(48) J. Gerendasy, *J. Med. Soc. New Jersey,* **43,** 84 (1946).

(49) A. I. Friedman and A. A. Alessi, *J. Amer. Med. Ass.,* **154,** 1273 (1954).

(50) H. Elsner, Germ. Patent 667,279 (1938); *Chem. Abstr.,* **33,** 22852 (1939).

(51) L. S. Tarasova, *Biokhimiya,* **26,** 736 (1961); *Chem. Abstr.,* **56,** 1741 (1962).

(52) M. L. Tarver and L. M. Marshall, *Anal. Chem.,* **36,** 1401 (1964).

(53) V. Ghetie, D. Motet-Grigoras, and H. Schell, Romanian Patent 48,707 (1967); *Chem. Abstr.,* **69,** 19482e (1968).

(54) "Industrial Gums," R. L. Whistler, ed., Academic Press, New York, 1st Ed., 1959, pp. 28–31.

(55) S. L. Lantsevitskaya and A. V. Vimberg, *Trudy Azerbaidzhan, Nauch.-Issledovatel. Inst. Dobyshe Nefti,* No. **10,** 328 (1960); *Chem. Abstr.,* **56,** 4374 (1962).

(56) J. C. Rice, U.S. Patent 3,028,258 (1962); *Chem. Abstr.,* **57,** 1132 (1962).

(57) Z. Sugahara, Japan. Patent 13,401 (1961); *Chem. Abstr.,* **57,** 14888 (1962).

(58) L. E. Chernenko, E. P. Gilinskaya, I. N. Patilova, and I. N. Smirnova, *Izv. Vyssh. Ucheb. Zaved., Pishch. Tekhnol.,* 49 (1968); *Chem. Abstr.,* **69,** 53824c (1968).

(59) Etablissements Lemer & Cie., Fr. Patent 1,385,276 (1965); *Chem. Abstr.,* **62,** 14150 (1965).

(60) R. L. Whistler and C. L. Smart, "Polysaccharide Chemistry," Academic Press, New York, 1953, p. 211.

(61) W. Yaphe, *Proc. Int. Seaweed Symp., 5th, Halifax, Nova Scotia, 1965,* 333 (1966).

(62) C. Araki, *Proc. Int. Seaweed Symp., 5th, Halifax, Nova Scotia, 1965,* 3 (1966).

(63) V. A. Evtushenko, M. A. Antonova, and G. V. Makhnova, *Zhur. Priklad. Khim.,* **40,** 1767 (1967).

(64) K. Izumi, *Carbohyd. Res.,* **17,** 227 (1971).

(65) M. Duckworth and W. Yaphe, *Carbohyd. Res.,* **16,** 189 (1971).

(66) M. Duckworth, K. C. Hong, and W. Yaphe, *Carbohyd. Res.,* **18,** 1 (1971).

(67) A. Cotton, *C.R.H. Acad. Sci.,* **120,** 989 (1895).

(68) G. L. Henderson, *J. Chem. Ed.,* **45,** 515 (1968).

(69) R. G. Greener, K. W. Adolph, and G. H. Emmons, *Appl. Optics,* **5,** 1468 (1966).

(70) W. F. Duncan and C. T. Bishop, *Can. J. Chem.,* **46,** 3079 (1968).

(71) L. S. Stoloff, *Fishery Market News,* **5,** No. 11, 4 (1943).

(72) W. S. DeLoach, O. C. Wilton, H. J. Humm, and F. A. Wolf, *Duke Univ., Marine Station Bull.,* No. **3,** 31 (1946).

(73) K. B. Guiseley, *Carbohyd. Res.,* **13,** 247 (1970).

(74) W. F. Head, Jr., and W. M. Lauter, *J. Amer. Pharm. Ass.,* **46,** 617 (1957).

(75) S. H. Newburger, J. H. Jones, and G. R. Clark, *Toilet Goods Ass., Proc. Sci. Sect.,* **19,** 25 (1953).

(76) Y. Tsuchiya and K. C. Hong, *Proc. Int. Seaweed Symp., 5th, Halifax, Nova Scotia, 1965,* 317 (1966).

(77) L. S. Stoloff, *U.S. Fish Wildlife Serv., Fishery Leaflet,* No. **306** (1948).

CHAPTER IV

ALGIN

WILLIAM H. MCNEELY AND DAVID J. PETTITT

Kelco Company, San Diego, California

I. INTRODUCTION

As a polysaccharide that is nearly a pure polymer of uronic acids arranged in a long, linear chain, algin is a unique water-soluble gum. Algin is found in all species of brown algae, in which it is an important constituent of the cell walls, but only a few species are commercial sources.

The term *algin* is the generic designation for the salts of alginic acid. The most common algin is sodium alginate.

Algin's thickening, suspending, emulsifying, stabilizing, gel-forming, and film-forming properties have resulted in its use in a very large number of applications. The chemistry and properties of alginates are described in several reviews.[1–7]

II. Production

1. Source

Plant source.—Algin is found in all species of Phaeophyceae, brown algae. However, the major part of commercial algin is produced from *Macrocystis pyrifera, Laminaria hyperborea, Laminaria digitata,* and *Ascophyllum nodosum. Ecklonia maxima, Eisenia bicyclis, Ecklonia cava,* and *Laminaria japonica* are also used for algin manufacture.[4] Minor amounts of algin have been produced from *Nereocystis leutkeana* and *Cystoseira barbata,* the littoral or rock weeds *Fucus vesiculosus* and *Fucus serratus,* and the sublittoral seaweed *Laminaria saccharina.* The distribution and composition of the various species of brown algae have been investigated.[8–11]

Many species are not economical sources of algin owing to thin stands, harvesting difficulties, or processing problems.

Macrocystis pyrifera is the principal source of the world's supply of algin. This giant kelp is found in the temperate zones of the Pacific Ocean. The commercially harvested beds, however, are found off Southern California and Australia where it grows in stands from 50 ft (15 m) to 1 mile (1.6 km) in width and several miles in length.[12–14] The plant grows on rocky bottoms on exposed coasts where there is a continual swell. Most kelp grows in water ranging from 25 to 80 ft (8 to 25 m) in depth. Each plant has a holdfast, up to 4 ft (1.2 m) in diameter, attached to a rock or rocky ledge on the ocean floor. A group of stipes, ranging up to 100 for large plants, grows from the holdfast. After reaching the surface of the water, stipes spread out for a considerable distance. Individual mature stipes vary from 0.25 to 0.75 in (0.6 to 2.0 cm) in diameter and range from 50 to 200 ft (15 to 60 m) in length. Each stipe bears, at intervals along its entire length, a series of fronds or leaves, each having at its base a small hollow float bulb about 2 in (5 cm) in length and 1 in (2.5 cm) in diameter. The fronds have a serrated edge and are about 4 in (10 cm) wide and 12 in (30 cm) long. *Macrocystis pyrifera* is a perennial plant that lives from 8 to 10 years. The holdfast continually sends up new stipes, which grow up to 12 in (30 cm) per day. Mechanical harvesting, done between two and four times a year, is beneficial to the plants.[13, 14]

Laminaria hyperborea is the most important of the smaller kelps that are used as a source of algin. It is a sublittoral seaweed that grows on rocky or hard bottoms extending from the lower edge of the *Laminaria digitata* zone [approximately 4 ft (1.2 m) below low spring tide level] to the 60 ft (18 m) zone.[3, 15] *Laminaria hyperborea* has a sturdy stipe that tapers from the base upward and is 5–10 ft (1.5–3 m) long. The surface of the stipe is finely ridged. At the base, a single stipe is attached to a rock or a hard surface by haptera or crampons to form a small holdfast. The stipe is relatively rigid except at the top, so the plant stands upright in the water. The flexible top of the stipe terminates in a broad, flat frond or lamina, which is split lengthwise into many finger-like segments that are somewhat shorter than the stipe. After fructification has been completed, new tissue develops below the narrowest part of the old frond and carries the old frond forward. Splits that appear in the broadening new frond divide it into primary segments that are still held together by the base of the old frond. In late spring or early summer, the old fronds break loose from the stipes and are cast on the beaches, resulting in what is called the Maycast. In some areas, these fronds are gathered and used for the production of algin and kelp meal. In the late fall or early winter, storms break loose numbers of entire plants and cast them on the beaches. These large, tough tangles, which consist primarily of stipes, are collected for algin and kelp meal production. For the most part, fall and early winter casts are used for this purpose. Because a practical mechanical harvesting method has not been developed for *Laminaria hyperborea,* only small amounts of the growing seaweed are harvested. Although the plant is perennial and lives 10–15 years, harvesting damages it. Thus, harvesting must be a cyclic process that can only be repeated every fourth year.

Laminaria digitata is a small, perennial, sublittoral kelp[3, 15, 16] that is closely related to *Laminaria hyperborea.* However, the stipe of *Laminaria digitata* is smooth and flexible and does not taper. Its broad, flat frond is split into numerous finger-like segments. In older plants, the rate at which the tips of the fronds wear away approximately equals the growth rate so that the frond does not increase in size. The plants are 8–14 ft (2.4–4.2 m) long and grow in a narrow zone, especially on steep coasts, on either side of the low water mark. In some regions, plants extend into the littoral zone of *Fucus serratus. Laminaria digitata* is harvested at low tide by hand from small boats. Because harvesting damages the plants, they also must be cut on a 4 year cycle.

Ascophyllum nodosum is the most plentiful littoral or rock weed. This brown alga competes for a wide area in the middle of the littoral zone, which is exposed by tidal changes.[3, 15, 16] It is attached by a disk on rocky beaches. Normally, the plant reaches a length of only 3–5 ft (0.9–1.5 m) because of breaking of the branches in the surf, but it might range from 1 to 10 ft (0.3–3 m) in length. *Ascophyllum nodosum* has an almost cylindrical frond that forks at the major shoots; it also has lateral branches. The stem contains a linear series of hollow enlargements that serve as float bulbs. In older plants, laterals develop near the base. The basal and side laterals often give the older plants a bushy appearance. This perennial plant can live for many years if not broken loose. It is hand-harvested from small boats or from the shore at low tide. If cut at a point 6–8 in (15–20 cm) above the disk, the plant will usually regenerate and can be harvested on a 4–5 year cycle. It grows 6–8 in (15–20 cm) per year.

Areas.—The largest proportion of the world's supply of algin is produced in the United States.[17] Nearly all this algin originates in mechanically harvested beds of *Macrocystis pyrifera* along the Southern California coastline.[6]

Algin production in Great Britain is second to that in the United States.[2, 17] The British industry depends mainly on *Laminaria hyperborea* and *Ascophyllum nodosum* as raw materials.[2, 8] *Ascophyllum nodosum* is hand-harvested from small boats or from the shore. *Laminaria hyperborea* is collected after storms from beaches on which tangles of the plants have been deposited.

Norway has large reserves of *Laminaria hyperborea* and *Ascophyllum nodosum* and lesser amounts of *Laminaria digitata,* because the latter usually grows in a very narrow zone. Seaweed meal and dried kelp are exported from Norway for the production of algin and for use in animal feeds.[18] The domestic Norwegian algin industry relies primarily on *Laminaria digitata* as a raw material for algin manufacture.[2]

Algin production in France is the second largest in Europe[2] and is based mainly on *Laminaria digitata.*

Japanese production of algin is divided among a number of small producers.[4] Although large quantities of Kombu, a mixture of various species of *Laminaria,* are collected by hand, most of it is used for the production of food and iodine. *Ecklonia cava* and *Eisenia bicyclis* are the major species utilized for algin production.[4]

In Canada, *Ascophyllum nodosum* is utilized as a commercial source of algin.

A manufacturing facility has been established near Australia.[19] *Macrocystis pyrifera* is the raw material.

Small scale algin manufacture is done in the U.S.S.R., Chile, and Morocco.[2, 3, 8]

2. *Seasonal Variations*

The alginic acid content of the stipes of commercially important *Laminaria* species is relatively constant throughout the year in mature plants. *Laminaria* fronds have a seasonal variation in alginic acid content if calculated on a dry solids basis,[20–22] but the alginic acid level remains nearly constant in fresh algae.[2]

Fluctuations in the alginic acid content of dried *Laminaria* fronds apparently are caused by changes in the levels of other constituents, such as laminaran and mannitol.[1, 2] Dried *Ascophyllum nodosum* shows only minor and irregular variations that are independent of season.

Alginic acid contents[2] of some Norwegian algae on a dry solids basis are *Ascophyllum nodosum,* 22–30%; *Laminaria digitata* fronds, 25–44%; *Laminaria digitata* stipes, 35–47%; *Laminaria hyperborea* fronds, 17–33%; and *Laminaria hyperborea* stipes, 25–38%.

3. *Harvesting*

Because *Macrocystis pyrifera* is perennial and grows very rapidly, the beds can be harvested three to four times per year. Removing kelp from the surface prevents it from being eventually broken loose by storms and washed ashore. Harvesting prevents development of excessive temperatures in matted surface kelp that causes rotting of the plants during summer. Beds that are periodically harvested have been found to have more luxuriant growth beneath the surface of the ocean, as light can then penetrate to the depths.[13, 14] Harvesting is done by large motor barges equipped with underwater cutting blades that cut the kelp approximately 3 ft (0.9 m) below the surface of the water. The cut kelp is automatically conveyed into the barge by a moving belt.

Because *Laminaria hyperborea* is a sublittoral seaweed that grows on rocky bottoms at depths of 4–60 ft (1.2–18 m) below the low tide level, it is not readily adaptable to mechanical harvesting. Much effort has been expended to develop a workable mechanical harvester for this species,[23] but none is now in commercial use. Only a small amount of this seaweed can be gathered by the use of grappling hooks or knives. However, it is cast ashore during storms, and most used commercially is gathered from the beaches. Most of the algin production originates from stipes cast ashore in the fall; a minor amount is produced from fronds washed onto the beaches in the spring. This collected kelp is partially air-dried locally to prevent spoilage and then shipped to central stations for final drying.[24]

Laminaria digitata grows in shallow water and, therefore, can be harvested from small boats at low tide. In some places, the entire year's supply must be collected in about 30 days. Usually, *Laminaria digitata* is also partially air-dried and then sold to collection stations where drying is completed. The seaweed is harvested on a 4 year cycle in order to allow time for the growth of new plants.

Ascophyllum nodosum is harvested by hand either from small boats at high tide or from shore at low tide. It is cut at least 6–8 in (15–20 cm) above the base so that the plant will not die. The cut plants are air-dried and sold to central stations, at which they are milled and dried further. Much of this sea plant is used to produce seaweed meal rather than algin. Because the plant grows very slowly, it is harvested on a 4–5 year cycle.

4. *Processing*

Algin.—Many reports and patents concerning the production of algin have been published.[1, 4, 7, 25–27] However, details of the successful commercial processes have not been made known. *Macrocystis pyrifera* is processed while wet because it is mechanically harvested, whereas all other species usually go through an intermediate drying process. The kelp is washed and then digested with an alkali, generally sodium carbonate. After removal of a small amount of cellulose, algin is recovered from solution by precipitation as the calcium salt or as alginic acid. If the algin is recovered as the calcium salt, the precipitate is washed with an acid to convert it to alginic acid. Alginic acid is then treated with a base to afford the desired salt. Alternatively, electrolysis can be used for recovering the alginate from solution.[28] It can be bleached at any one of several steps in the process.[29] After recovery, it is dried, milled, tested, and blended. In a continuous process, algin can be produced in several different viscosity grades within relatively narrow limits.

Algin Derivatives.—Propylene glycol alginate is the only organic alginate derivative manufactured in large quantities. The ester is prepared by reaction of propylene oxide under moderate pressure with a partially neutralized alginic acid which is in a fibrous condition because of the presence of a controlled amount of water.[30, 31] Certain monovalent and divalent ions can catalyze the reaction.[32, 33]

Other esters of both the carboxyl and hydroxyl groups, for example, algin acetate[34] and algin sulfate,[35] have been described in the literature, but none has as yet found commercial application.

Alginamides are formed by the action of ammonia or amines on alginic acid esters.[36, 37] Treatment of sodium alginate with chloroacetic acid and alkali yields carboxymethylalginate.[38]

By simply neutralizing alginic acid with an amine, a wide variety of amine salts of alginic acid have been prepared.[39, 40] Similarly, quaternary ammonium derivatives, such as benzyltrimethylammonium alginate, are formed from sodium alginate or alginic acid and the appropriate quaternary ammonium compound.[41, 42]

5. *Grades*

Sodium alginate is the principal commercial, water-soluble salt of alginic acid. Other water-soluble salts that are commercially available are potassium alginate and ammonium alginate. Water-insoluble compounds that are marketed are calcium alginate and alginic acid. The only commercially important organic derivative is propylene glycol alginate.

Food grades of algin are free of cellulose and are bleached so that their color is light tan to white. Pharmaceutical grades are also free of cellulose and bleached to off-white to white. Industrial grades are usually cellulose-free, although some might contain cellulose; they vary in color from brown to white. Other physical properties also vary with the method of manufacture: pH, 3.5–10.0; viscosity of a 1% by weight water solution, 10–5,000 cps; moisture content, 5–20%; particle size, 10–200 U.S. Standard mesh.

6. *Industrial Importance*

Sustained production of algin first began in the United States in 1929. Since that time, algin has developed into an important natural water-soluble gum, produced in the United States, Great Britain, France, Norway, and Japan. Data on total algin production has not been released by the algin industry, but it has been estimated at greater than 10,000 tons (9×10^6 kg) annually.[8, 17, 43]

Because much of the algin production is based on continuous processing in modern plants with mechanized harvesting of a rapidly growing raw material, prices of algin do not fluctuate like those of some other natural water-soluble gums. Prices in the U.S. range from about $1.13 to $2.05 per pound[44] depending on the grade and composition. Considerable variation in world prices may result from different tariff levels.

7. *Potential Amount*

A considerable amount of effort has been expended on obtaining estimates of seaweed resources,[4, 8] primarily by or with the assistance of governmental agencies. These sources have estimated total wet seaweed supply in the areas surveyed at approximately 80 million tons.[45, 46] It is difficult to estimate seaweed supply, although improvements in the methods of estimation have been made.[47] It is the authors' experience that estimates of potential seaweed production are unreliable unless based on actual harvesting records over a period of years. The published estimates of harvestable seaweed along the Pacific Coast of North America, based on governmental surveys before harvesting was commenced, are greatly exaggerated. In general, seaweed processors take a more conservative approach to available seaweed resources than is indicated from the published estimates.[4, 8, 23] However, resources of harvestable brown algae appear to be adequate to support a reasonable expansion of the industry.

III. Applications

1. *Introduction*

Algin was discovered by Stanford[48, 49] in 1880 while he was searching for useful products from kelp. Stanford produced some algin and attempted to develop its applications, but the company that he formed failed. Sustained commercial production of algin was first begun in 1929 by the Kelco Company in

California. Initial production was for can-sealing and boiler compounds. Milk-soluble algin,[50, 51] which had been developed by 1934, became the most important commercial ice cream stabilizer. Propylene glycol alginate was originated in 1944, and soon afterward commercial production processes were developed.[30] Sustained manufacture of algin began in Great Britain on a limited scale in 1934, and commercial production began in 1939.[6] Norway's industry originated during World War II.[6] Additional algin-producing firms have been established in Great Britain,[2, 52] France, Norway,[2] the United States,[45] Canada,[2] Japan,[2, 5] and Australia.[19] Some algin is also produced in Morocco, the U.S.S.R., and Chile.[2]

Whenever thickening, suspending, emulsifying, stabilizing and gel-forming applications are involved, the use of algin is indicated. Applications in additional specialized fields will be suggsted from the properties of algin discussed in the following section. In the sections on food, pharmaceutical, cosmetic, and industrial applications, only representative uses are discussed. Reviews of algin applications have been published.[6, 7, 53, 54]

2. *Mode of Action*

Algin forms water-soluble salts with monovalent cations and low-molecular-weight amines, as well as with magnesium ion. Because algin is a linear polymer[55, 56] with a high molecular weight,[57–61] its highly hydrated, water-soluble salts have high viscosities at low concentrations in water. Algin is, therefore, very effective as a thickener. In many cases, it acts as a suspending agent because of its ability to reduce sedimentation rates of suspensions by thickening the solutions.[62] In other suspending applications, the effectiveness of algin is caused by its negative charge and colloidal size, which enable it to form a protective coating over particles.

Viscosity of the aqueous phase, interfacial tension, and the presence of protective films influence the stability of oil in water emulsions. Propylene glycol alginate has both lipophilic groups and hydrophilic groups and, thus, combines in one molecule the properties of a true emulsifier with the thickening and protective-colloid properties of a water-soluble gum.[30] The effectiveness of sodium alginate as an emulsion stabilizer is due mainly to its thickening ability although, in some cases, its ability to form protective coatings is also important.

The term *stabilizer,* as applied to water-soluble gums, is used loosely. Sometimes it refers to an emulsion stabilizer; more often it refers to an ice cream or dairy product stabilizer. In ice cream, the action of a stabilizer is complex; by its water-binding properties and protective-colloid action, milk-soluble algin[50, 51, 63] prevents the formation of coarse ice crystals in ice cream which is repeatedly subjected to partial thawing and refreezing.[64] Its colloidal size also helps to give the proper body and smoothness to ice cream and other dairy product desserts.

Because calcium ions are divalent, they are able to combine with two carboxyl groups and join algin molecules. Secondary bonding may exist between calcium ions and hydroxyl groups on the polymer.[65] Thus, the addition of a soluble calcium salt to a solution of a water-soluble alginate results in the formation of a cross-linked aggregate that is insoluble in water. Depending on the method of addition and formulation, gels or water-insoluble fibers or films can be produced. Other divalent and trivalent metal ions also form gels, films, and fibers with algin.

Uncharged, free carboxylic acid groups of alginic acid are not hydrated as effectively as the negatively charged carboxylate ions. Therefore, the addition of excess acids tends to insolubilize algin. Long chain quaternary ammonium salts or amines form insoluble adducts with algin because of a lipophilic surface around the resulting polymer molecule.[66]

Either acids or water-soluble divalent or trivalent metal salts, or a combination of these, can be employed to form algin gels that set at room temperature. Gel textures ranging from soft and tender to hard and tough can be produced by adjusting the ratios of metal salts or acids.

Formation of a uniform gel requires that all the ingredients be mixed before the onset of gelation. An acid-forming reagent accompanied by a sparingly soluble calcium salt that becomes more soluble in acid may be included in the formulations. For example, D-glucono-1,5-lactone, which slowly forms acid in aqueous media, may be combined with calcium phosphate and sodium alginate.[67] Other useful gel-forming compositions consist of adipic acid, fumaric acid, or citric acid in combination with a soluble alginate salt and calcium carbonate, calcium phosphate, or calcium tartrate.[68–70] Also, a slightly soluble calcium salt, such as calcium sulfate, which slowly releases calcium ions, can be mixed with powdered algin to form powdered compositions that dissolve in water at room temperature and gel upon standing.[71]

A polyphosphate sequestering agent or a soluble anion capable of forming insoluble components with polyvalent metal ions can be added to control the setting time of formulations containing slightly soluble metal salts.[68, 70] The sequestering or precipitating agent reacts with the metal ions that are released initially and delays gelation until all the agent is used.

Because of its high molecular weight and linear structure, algin forms strong films. The presence of hydrophilic carboxylate and hydroxyl groups causes its films to resist penetration by oil and grease. When algin solutions are applied to the surface of paper or paperboard, a thin oil-and grease-proof film forms. This film produces a smoother surface and bridges the openings between surface fibers.[72,73]

3. *Foods*

Applications of alginates in the food industry have been reviewed.[74–76]

DAIRY PRODUCTS

Ice cream.—Algin-based stabilizers are widely used in the production of ice cream. They produce a natural-bodied ice cream that maintains a smooth texture and a creamy consistency. Algin prevents the formation of large ice crystals in ice cream which is repeatedly subjected to partial thawing and refreezing. Ice cream that has been stabilized with algin has good flavor release without masking and a clean attractive meltdown.[77] Milk-soluble algin[50] and propylene glycol alginate[78] are used for this purpose. Stabilizer concentrations in ice cream range from 0.1 to 0.5%. A variety of algin-based stabilizers and stabilizer–emulsifier combinations are available to meet the requirements of the various markets.

Algin is listed as an optional ingredient in the Standards of Identity for Ice Cream under Title 21 of the United States Code of Federal Regulations, Section 20.1.

Ice milk.—Frozen desserts that contain 2–7% of butterfat are classified as ice milk. They are sold either as hard-frozen or as soft-serve ice milk; in the latter case, the product is dispensed directly from the freezer. Algin-based stabilizers developed for this purpose give good dryness and stiffness and a slow meltdown to soft-serve ice milk. Delayed churning and proper viscosity control of the soft-serve mix is also accomplished. The uses and advantages of algin in hard-frozen ice milk are similar to those in ice cream. Stabilizer concentrations range from 0.2 to 0.5% of the weight of the ice milk.

Algin is listed as an optional ingredient in the Standards of Identity for Ice Milk under Title 21 of the United States Code of Federal Regulations, Section 20.3.

Milk-shake mixes.—Hard-frozen ice milk that is to be used in milk-shake mixes requires stabilization. Milk-soluble, algin-based stabilizers provide good secondary overrun and creamy, thick milk shakes. Emulsifiers are detrimental to secondary overrun and are, therefore, not included in the mix. Stabilizer concentrations of 0.25–0.5% are used.

Sherbets and water ices.—Sherbets are frozen desserts that contain 1–2% of butterfat and less than 5% of total milk solids. Sherbets stabilized with propylene glycol alginate[78] have clean flavor, smooth texture, and good body without crumbliness or sugar syrup separation. Uniform overrun is obtained, and the sherbet is protected against growth of large ice crystals during repeated partial thawing and freezing cycles. Stabilizer concentrations that are used for sherbets and water ices are 0.3–0.5% of the weight of the mix.

Sodium alginate and propylene glycol alginate are listed as optional ingredients in the Standards of Identity for Fruit Sherbets (Title 21, Section 20.4) and

Water Ices (Title 21, Sections 20.5, 20.6, 20.7) in the United States Code of Federal Regulations.

Chocolate milk.—Milk-soluble, algin–carrageenan compositions are used to suspend cocoa fibers and to give a smooth, uniform-viscosity chocolate milk product. Stabilizer concentrations are 0.25% or less.

Miscellaneous dairy products.—In yogurt, sour cream, and imitation dairy products, algin is used as a bodying agent for viscosity control.

BAKERY PRODUCTS

Icings.—Bakery icings that have a soft gel consistency and a light body and smooth texture are formulated with small amounts of algin.[79, 80] These icings have reduced tack so that they do not stick to Cellophane or wrapping paper. The water-holding properties of algin reduce the tendency of the icings to dry under conditions of low humidity or to soak into the bakery product at high humidities. Algin concentrations range from 0.1 to 0.5%.

Cake fillings and toppings.—Incorporation of small amounts of algin into fillings and toppings produces products having a tender body and smooth texture. The products are well stabilized against syrup separation. Upon aging, the fillings and toppings retain their texture and do not become tough or rubbery. Algin is used in concentrations of 0.3–0.5%.

Bakery jellies.—A freeze-thaw stable, bakery jelly that will also withstand baking temperatures contains 0.25–0.75% algin[81, 82] along with, for example, a calcium phosphate, a sodium phosphate, and adipic acid.

Meringues.—Liquid egg white meringues and dry meringue powders that contain propylene glycol alginate whip rapidly to desired stiffness.[75] A fine-textured product is obtained in which bleeding is reduced. The algin derivative is used at a concentration of about 0.2%.

Glazes.—Algin–sugar combinations produce a glaze for raised doughnuts and other bakery goods that resists sweating and does not become brittle. Algin is used at concentrations of 0.3–0.5%.

Pie fillings.—Small amounts of algin that are added to cooked starch pie fillings prevent separation and cracking. The filling has a soft, smooth gel body with improved clarity and gloss.[83] Recommended algin concentrations range from 0.3 to 0.5% of the total filling weight.

In instant neutral or acid-type chiffon pie fillings, sodium alginate is used at 0.7–1.5% in combination with a whipping agent.[83–85] A lipid-based, aerated, gelled filling is prepared using 1.25–6% algin.[86]

OTHER FOOD PRODUCTS

Dietetic foods.—Algin has a caloric value of about 1.4 cal/g. As most applications require less than 1% of algin, the number of calories contributed by algin to dietetic foods is very low. Through its bodying or gelling action, algin improves the palatability of low-sugar products. Many different formulations are available for use in low-calorie salad dressings,[87] imitation French dressing, and dietetic jellies, jams, syrups, puddings, sauces, icings, and candies.[75]

French dressing.—The unique combination of emulsifying and thickening properties in propylene glycol alginate has resulted in its widespread utilization in French dressings.[75, 88] Dressings made with this algin derivative have a longer shelf-life and resist breakdown, whether exposed to high room temperatures or kept in refrigerated storage. Propylene glycol alginate is readily soluble in water without heating, and it has excellent compatibility with acidic solutions. The uniformity of propylene glycol alginate results in uniform emulsion stability, body, and flow properties of French dressing. The Federal Definition and Standard of Identity for French dressings specifies propylene glycol alginate as one of the optional emulsifying ingredients (Title 21, Section 25.2). Ordinarily, 0.5% or less of the algin derivative is required in these products.

Salad dressings.—Soft, smooth-textured salad dressings can be made by using algin derivative as a corn starch modifier.[75] As little as 0.1–0.2% (of the salad dressing weight) of propylene glycol alginate produces a desirable gel body that resists cracking and oil separation even after standing for long periods. Propylene glycol esters of alginic acid are listed as optional ingredients in the Standards of Identity for Salad Dressing under Title 21 of the United States Code of Federal Regulations, Section 25.3.

Syrups and toppings.—Fountain syrups, variegated syrups for ice cream, flavor syrups, and chocolate and fruit toppings that are prepared with algin maintain excellent body, smooth texture, and a natural appearance and color. Algin protects these products from syrup drainage, fruit separation, and loss of fruit structure. Superior flavor release is obtained in algin-based products. Different algin compositions are available for both fruit and nonfruit types of fountain syrups and toppings.

Dessert puddings.—Algin formulations make excellent puddings with milk or milk solids.[76, 89] These puddings are quickly prepared without heating and set to a soft, smooth dessert. Algin-based puddings give full flavor release, are not pasty, and have no starchy taste. The surface of the puddings remains soft and does not harden into a skin. Dry, packaged formulations containing skim milk powder can be produced so that only water needs to be added at the time of use. The algin content of the finished pudding is less than 0.5%.

Dessert gels.—Attractive dessert gels can be quickly prepared with algin gelling compositions.[67, 82, 90] The gels are clear and firm, and can be easily molded. Their strength can be varied by adjusting the formulation. These gels set rapidly either in a refrigerator or at room temperature. They are nonmelting at room temperature. Algin concentrations range from 0.4 to 1.0% in the finished dessert gel.

Candy gels.—Combinations of algin with calcium salts or food acids are used to make candy gels ranging from soft tender types to chewy bodied gels.[67, 91] These candies are clear and have a long shelf life. Algin-based candy gels have excellent moisture retention. Algin is used in concentrations of 0.1–0.7%.

Beer foam stabilization.—When added to beer at concentrations of 40–80 ppm (1–2 lb/100 bbl, 1 mg/liter), propylene glycol alginate produces a more stable, longer-lived, creamier foam.[92, 93] An attractive, white, nonsticky, natural-appearing lace is formed. Beer foam containing this small amount of algin derivative is particularly resistant to breakdown in the presence of traces of soap, grease, or bactericidal washing compounds often encountered in spite of precautions.

Noncarbonated fruit-flavored drinks.—Propylene glycol alginate has excellent suspending action in drinks that contain fruit pulp. The slight extra body imparted to these drinks by the algin derivative results in a smoother-tasting product with better flavor release. In dietetic beverages, it furnishes the body that is characteristic of a high sugar content. This results in a smooth, full-bodied dietetic drink that is comparable in flavor to one that contains sugar. Concentrations used are 0.10–0.25%.

Canning.—In canning foods containing sauce or gravy, a reduction in heat-process time is made possible by replacing most of the starch normally used by 0.3–0.8% of algin.[94, 95] Delayed release of calcium keeps the viscosity low enough to permit convection heating. As the temperature drops after heat processing, the calcium ion reacts with the algin, causing the viscosity to increase so as to achieve the desired final value.

Fabricated foods.—Artificial cherries and glacé fruits can be made from a mixture of sugar, flavoring, and other ingredients encased in a firm calcium alginate gel. These are used in cakes, pastry, and, to a lesser degree, in ice cream.

4. *Pharmaceuticals and Cosmetics*

Suspensions.—Algin is used as a suspending agent and as a protective colloid in many pharmaceuticals. In some systems, such as those containing penicillin, sulfa drugs,[96, 97] and colloidal sulfur, it provides exceptional stability because of its action as a protective colloid. Algin concentrations are 0.25–2.0%. The bioavailability of certain pharmaceuticals apparently is enhanced by sodium alginate.[98]

Jellies.—Smooth-flowing, nontacky surgical lubricants and medicated jellies are produced from algin. These have good lubricating and emollient properties. Calcium salts can be added to give the desired type of flow properties ranging from long flowing through short flowing to a soft gel.[99]

Ointments.—Algin imparts body and emulsion stability to pharmaceutical ointments. Ointments that contain algin have excellent spreading properties.[100, 101] Required concentrations of algin range from 0.5 to 3.5%.

Emulsions.—Propylene glycol alginate in a concentration of 0.5–1% provides excellent stability in pharmaceutical emulsions.[30] These algin derivative-based emulsions are easily prepared.

Tablet disintegrating agent.—Alginic acid or sodium–calcium alginate added in relatively small quantities to tablets, either before or after granulation, greatly accelerates the rate of disintegration.[102] As the alginic acid particles absorb water and swell, the tablet is broken apart. Alginic acid is used in amounts that range from 0.5 to 5% of the weight of the tablet.[103]

Tablet binder.—Sodium alginate is an effective tablet-binding agent and can be added to the medicament either in the dry state or as a solution. Uniform tablets are formed with reduced incidence of capping. When added as a powder, 1–2% of the tablet weight is used. When added as a solution, 1–5% concentrations are employed. A mixture of a water-insoluble algin, such as calcium alginate, and a soluble algin is useful as a combination tablet binder and disintegrating agent.[104] Improved shelf life and excellent spreading characteristics are obtained. These creams and lotions are easily prepared because algin is soluble in either hot or cold water. Algin concentrations of 0.5–2% are used.

Liquid shampoos.—Algin is an effective thickener for liquid shampoos[104] when used at a concentration of 0.5–1.5%. Improved pouring and handling properties are obtained. Algin is compatible with most anionic and nonionic synthetic detergents.

Dental impression materials.—Algin-based dental impression materials are easy to use, have a controlled setting time, and set at room or mouth temperatures without prior heating. The resulting gels are tough and elastic and have excellent dimensional reproducibility. These impression materials contain algin, a filler, a slightly soluble divalent or trivalent metal salt as a gelling agent, and a phosphate or polyphosphate buffer as a timing agent.[105–108]

Molding compounds.—Compositions similar to those used in dental impression materials are employed in industrial, art, and surgical molding or modeling compositions. The tough, resilient, algin-based gels provide excellent reproduction of fine details. The algin content of these compositions range from 10% to 15% of the dry weight of the material.

5. *Industrial Applications*

Paper Products

Surface sizing.—The unique filming and penetration controlling properties of algin, when used as a surface size, make it possible to produce paper and paperboard with improved surface smoothness and uniformity and controlled surface density.[72, 73, 109] Algin lays down the surface fibers and forms a thin, dense film at the surface of the sheet that resists penetration by oil, ink, wax, and grease.[110] As a result, algin gives exceptional results at low concentrations in gloss ink, wax, and pigmented sizings. Surface coverage ranges from 20,000 to 60,000 ft^2/lb (850 to 2500 m^2/kg) of algin.

Coatings.—Addition of less than 0.5% of algin to a water-based coating reduces and controls the penetration of the coating into the web of the paperboard or paper.[111–115] In pigmented coatings, the binder is held at the surface with the pigment, which results in a more uniform and scuff-resistant coating and a saving in both binder and pigment. Algin increases the slip of pigmented coatings, especially when clay is used as the pigment, and thus allows the employment of higher-viscosity, but more fluid, coatings. The film-forming property of algin is very useful in the application of materials such as casein, starch, and resin emulsions to porous surfaces, such as paperboard.

Adhesives.—Algin controls the penetration and stabilizes the viscosity of starch- and latex-type adhesives that are used in the manufacture of corrugated boards. The presence of 0.1–0.2% of algin in the adhesive improves bonding, increases machine speed, reduces warping, and cuts scrap losses.[72, 115]

Textile Products

Printing.—Print pastes that are thickened with algin are used for both machine and silk-screen printing.[116–118] Algin-based print pastes are easily prepared and give sharp lines without bleeding. Color yields are good. Because of the complete solubility of algin in water, no screen plugging is encountered. The flow properties of these print pastes can be varied from short to long by the addition of polyphosphates. After printing, the algin is easily removed with cold water and leaves a fabric that has a soft hand. Sodium alginate, unlike other natural gums, is inert toward many reactive dyes.[116, 117] Algin, therefore, finds application as a thickener for reactive dye printing pastes. Algin concentrations range from 1.5 to 3% in the final pastes.

Finishing.—As an additive to starch-based warp sizes, algin controls penetration and provides a uniform, easily removed film on the yarn.[116]

Rubber Products

Latex creaming.—Ammonium alginate is an efficient creaming agent for rubber latex[119, 120] and other polymer latices such as poly(vinyl halides)[121] and perfluorohaloolefin polymers.[122] Less than 0.1% of ammonium alginate causes natural rubber latex to separate into a high-rubber-solids cream layer and a low-rubber-solids serum layer.

Latex thickening.—Whereas small amounts of algin cause natural rubber latex to cream, additional amounts stabilize synthetic latices or creamed natural latex and increase their viscosities. The algin content can be varied to maintain production lots of latex within a narrow viscosity range. Algin-thickened latices have improved spreading and film-forming characteristics. These compositions are used in forming dipped, molded,[123] and foamed rubber products. Algin concentrations for thickening purposes vary from 0.05 to 0.2% of the rubber latex.

OTHER INDUSTRIAL USES

Paints.—In water-based paints, algin is used to suspend the pigments and control the viscosity.[124] Its water-holding ability helps to maintain a wet edge and to improve the brushability of the paint. The film-forming properties of algin help to hold the pigments on the surface and, thus, to improve the coverage of the paint. If pigments settle in the container during lengthy storage, algin makes resuspension of the pigments easier. Concentrations of algin in the paints are 0.05–0.15%.

Ceramic refractories.—Algin improves the wet strength and plasticity of ceramic bodies.[125] In ceramic glazes and slips, algin is used as a suspending, viscosity-controlling, and drying agent.[126] It reduces penetration of the slips into porous ceramics and, as a filming agent, prevents pinholes and aids release of the casting.[127]

Insecticides and herbicides.—In insecticidal and herbicidal sprays, algin serves as a suspending and filming agent.[128] The plasticity and wet strength of seed coatings are improved by the incorporation of algin. After the coating dries, algin acts as a binding agent.

Flocculants.—Algin is an efficient flocculating agent in some systems. As an example, in water treatment it is effective as a flocculant and improves the settling characteristics of the particles.[129, 130]

Welding-rod coatings.—The inclusion of algin in welding-rod coatings promotes greater wet strength and improved lubrication.[131] This results in smoother, more uniform coatings.

Boiler feedwater compounds.—As a protective colloid, algin facilitates the formation of a soft sludge rather than a hard scale in boilers. This sludge can be removed more easily than can scale by blowdown or cleaning. Algin also helps to reduce foaming. It is easily incorporated into the feedwater compound and is economical because of the small amount required for this purpose.

Sodium alginate and ammonium alginate have been approved for use in boiler water for the preparation of steam that will contact food.[132]

Auto polishes and cleaners.—Liquid polishes and cleaners or combination cleaner-polishes of excellent stability are produced with algin as the stabilizing and suspending agent. The pouring properties and the viscosity can be adjusted by controlling the algin concentration between 0.4% and 0.75%.

Photography.—Propylene glycol alginate, when combined with gelatin and then treated at high pH, provides water-insoluble, but water-permeable, films that are useful as photographic layers.[133] Sodium alginate finds application in silver recovery from waste photographic solutions.

IV. Structure

Impure alginic acid was first prepared and analyzed by Stanford, who established it as a weak organic acid that readily forms salts with bases.[134, 135] Quantitative decarboxylation analyses of alginic acid determined that uronic acids were its principal components.[136] In 1929, Nelson and Cretcher[137] hydrolyzed alginic acid from *Macrocystis pyrifera* and isolated the salt of D-mannuronic acid.

The x-ray and chemical evidence obtained prior to 1955 suggested that alginic acid was a high-molecular-weight polymer made up of D-mannopyranosyluronic acid units connected by β-D-(1→4) linkages.[1] Then Fisher and Dörfel, using paper chromatographic techniques, identified L-guluronic acid as well as D-mannuronic acid in the hydrolyzate of alginic acid from European algae.[138]

Subsequent reports verified the presence of L-guluronic acid as a component of alginic acid.[139–141] More recent work has provided strong evidence that the polymer molecule is entirely linear and consists exclusively of β-D-(1→4)-linked mannuronic acid units and α-L-(1→4)-linked guluronic acid units.[55, 56, 142, 143]

Models constructed from x-ray diffraction data indicate that the D-mannuronic acid units are in the C1 chair conformation, whereas the L-guluronic acid units have the 1C conformation.[142]

Oligomers containing residues of both uronic acids have been isolated.[144–146] Thus, algin is a copolymer and not simply a mixture of a mannuronan and a guluronan.

The uronic acids were shown to be stable with respect to epimerization at C-5 in alkaline media. This eliminates the possibility that L-guluronic acid units could have been formed from D-mannuronic acid units during isolation of the algin.[147]

Precise determination of the uronic acid composition of algin was made possible by refinements in the methods for hydrolysis, separation, and analysis.[2, 148] A rapid semiquantitative estimation of the composition of alginates can be made by infrared spectroscopy.[149]

Table I lists the ratio of D-mannuronic acid to L-guluronic acid in the more important commercial algae. Nearly all commercial algins (column *a*) are similar in composition and are rich in D-mannuronic acid. The exception is algin obtained from *Laminaria hyperborea* stipes, which has a high L-guluronic acid content.

The range in composition of mature plants (column *b*) is not as extreme as that found between young and old plants. The ratio of the two uronic acids varies in different parts of the same plant as well.[2]

Controlled hydrolysis of alginic acid leads to the formation of three detectable

TABLE I

Ratio of D-*Mannuronic Acid (M) to* L-*Guluronic Acid (G) in Commercial Algae*

Algae Species	M/G Ratio	
	a	*b*
Macrocystis pyrifera	1.56	—
Ascophyllum nodosum	1.85 (1.1)	1.40–1.95
Laminaria digitata	1.45	1.40–1.60
Laminaria hyperborea stipes	0.45	0.40–1.00
Ecklonia cava and *Eisenia bicyclis*	1.60	—

[a] Data of Haug[2] and Haug and Larsen[148] for commercial algin samples. Of the two ratios shown for *Ascophyllum nodosum,* the algin sample manufactured in Canada had the higher M/G value; the lower ratio was given by a European sample.

[b] Data of Haug[2] showing the range in composition for mature algae collected in different locations and at different times.

kinds of polymer fragments.[150–152] Of the two hydrolysis-resistant fragments, one is greatly enriched in D-mannuronic acid units and the second contains a similarly high L-guluronic acid content. Enrichment in one kind of uronic acid increases with the extent of hydrolysis until, at a degree of polymerization (DP) of 15–20, fractions of nearly pure D-mannuronan and L-guluronan can be obtained. The third type of fragment is more rapidly hydrolyzed, and as the hydrolysis proceeds, the ratio of monomer residues approaches unity in the hydrolyzate. The solubility–pH profile of the fragments compared to whole alginate indicates that all three types of fragments are present in the same polymer molecule.[152]

Thus, the alginate molecule may be regarded as a copolymer containing blocks of D-mannopyranosyluronc acid units and blocks of L-gulopyranosyluronic acid units. These are linked together by segments that have a predominantly alternating copolymer composition (Fig. 1). Block segments are more crystalline and are, therefore, more slowly hydrolyzed than the alternating copolymer portions of the polymer.

The relative amounts of the three different building elements vary with the algae species and the type and age of the plant tissue.[152, 153] *Ascophylum nodosum* receptacles, for example, contain, as an intercellular substance, essentially pure polymannuronic acid. However, the whole mature algae consists of a large fraction (greater than 50%) of alternating copolymer segments and a lesser number of block copolymer segments. In alginate from *Laminaria hyperborea* stipes, L-guluronic acid block segments predominate and only a relatively small proportion of the other two building elements are present. Alginate from *Lamin-*

(1) →4)-β-D-ManpA(1→4)-β-D-ManpA(1→4)β-D-ManpA(1→4)-β-D-ManpA(1—
(2) →4)-α-L-GulpA(1→4)-α-L-GulpA(1→4)-α-L-GulpA(1→4)-α-L-GulpA(1—
(3) →4)-β-D-ManpA(1→4)-α-L-GulpA(1→4)-β-D-ManpA(1→4)α-L-GulpA(1—

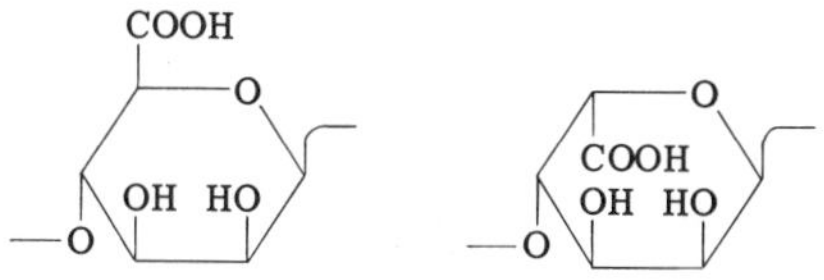

COOH, O, OH HO, —O — β-D-Mannopyranosyluronic acid unit (β-D-ManpA)

O, COOH, OH HO, —O — α-L-Gulopyranosyluronic acid unit (α-L-GulpA)

FIG. 1.—The alginate molecule is a linear copolymer of β-D-(1→4)-linked mannopyranosyluronic acid units (ManpA) and α-L-(1→4)-linked gulopyranosyluronic acid units (GulpA). Monomers are arranged along the polymer chain as block segments of ManpA (1), block segments of GulpA (2) and alternating copolymer segments (3). All three types of building elements are present in the same polymer molecule.

aria digitata has an intermediate composition, i.e., about the same proportion of alternating and block copolymer segments.

A statistical description of the structure of algin is best obtained by assuming that the copolymer is formed according to the penultimate-unit theory of addition copolymerization.[154, 155] Accordingly, in the biosynthesis of the polymer molecule,[156, 157] in which chain growth appears to take place by stepwise addition of uronic acid residues to an acceptor molecule, the unit penultimate to the end of the acceptor chain must influence or determine the addition of new uronic acid units or segments to the growing chain.[154]

V. Properties

1. *Dissolution*

Water-soluble salts of alginic acid include those of the alkali metals, ammonia, and low-molecular-weight amines and quaternary ammonium compounds.

Alginic acid and most salts with polyvalent metals are essentially insoluble. An exception is magnesium alginate, which is soluble because of the absence of strong intermolecular linkages.[158] Some mixed monovalent–divalent salts are soluble and have important applications.[69]

Solutions can be made readily in either hot or cold water. Alginates dissolve most easily when sifted into water that is vigorously agitated. High-speed stirring and gradual addition of the solid increases the rate of dissolution and decreases the tendency toward lumping of the finer-mesh products. When the formulation includes sucrose, dextrin, salts, or fillers, lumping can be prevented by dry-blending one of these materials with the fine-mesh algin before adding the mixture to water. Rapid solution rates can be obtained by first wetting the algin with ethanol, glycerol, or a similar water-miscible liquid before it is added to water or by preparing a slurry of the algin in the liquid and then rapidly adding the slurry to water with vigorous stirring. With good dispersion and agitation, dissolution should be complete in a few minutes.

2. *Solution Properties*

General.—The properties of algin solutions can be changed readily to provide either long, short, or intermediate types of flow.[159] Algin solutions that contain no divalent or trivalent metal ions and those that contain polyphosphates to sequester the polyvalent metal ions have long flow properties. Increasing the amounts of polyvalent metal ions, among which calcium is particularly important, raises the viscosity and shortens the flow properties.

Sodium alginate solutions decrease in apparent viscosity with increasing shear rate but low-molecular-weight (low DP) alginate approaches Newtonian behavior. Alginates show increasingly non-Newtonian (pseudoplastic) flow characteristics with increased degree of polymerization, concentration, and substitution of calcium for sodium[159] (Fig. 2). At low levels of calcium, the effect of calcium in causing increased viscosity is more pronounced in the case of alginates with

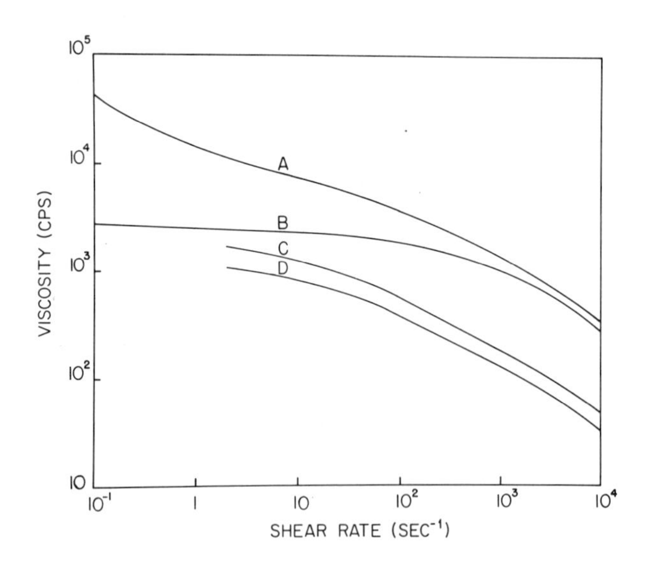

FIG. 2.—Flow curves for a high-viscosity grade and a low-viscosity grade of sodium alginate containing low levels of calcium ion: A, 10% solution of low-viscosity sodium alginate before addition of a sequestering agent; B, solution A after addition of a polyphosphate sequestering agent; C, 1.5% solution of high-viscosity sodium alginate before addition of a sequestering agent; D, solution C after addition of a polyphosphate sequestering agent.

higher D-mannuronic acid contents.[159, 160] Sodium alginate solutions are not thixotropic, but thixotropy is increasingly evident as sodium is replaced by calcium. The properties of sodium–calcium alginate solutions become essentially identical to those of calcium-free sodium alginate upon addition of a sequestering agent, such as polyphosphate or EDTA, as well as phosphate and carbonate.

Addition of calcium salts to propylene glycol alginate solutions causes a viscosity increase that is dependent on the degree of esterification. At high esterification levels, the effect of calcium diminishes.

At sodium alginate concentrations below about 3%, the presence of added inorganic sodium or potassium salts depresses the viscosity. This counterion effect is typical of polyelectrolytes, such as alginates.[161] Above 3% concentration, however, an additional increase in ionic strength caused by the presence of monovalent inorganic salts has little or no effect.[162]

Freezing and thawing sodium alginate solutions containing an appreciable amount of calcium can result in a marked increase in viscosity.[163] Polyols and calcium sequestering agents, such as citrate and polyphosphate, inhibit this viscosity change.

Concentration.—Algin is available in a number of viscosity ranges with the viscosity of each product controlled within a relatively narrow range. A high-viscosity algin has a viscosity of at least 2000 cps at a concentration of 1% in water, whereas a very low-viscosity product will have a viscosity of less than 10 cps at the same concentration (Fig. 3).

Temperature.—Algin solutions behave like other fluids in their dependence of viscosity on temperature. Over a limited range, the viscosity of algin solutions decreases approximately 2.5% for each degree centigrade increase in temperature. This decrease is reversible at short heating times. If high solution temperatures are maintained for extended periods, however, there will be a progressive decrease in viscosity caused by a partial deploymerization of the molecule. Algin solutions do not coagulate upon heating nor do they gel upon cooling, but they maintain their smooth flow properties over a wide temperature range.

pH.—The viscosities of solutions of water-soluble algin salts change only slightly with changes in pH in the range of 4 to 10. There is a slightly higher viscosity near neutrality (pH 6–8) because of the repulsive effects of the negatively charged carboxyl groups that extend the chain and increase its water-binding capacity. The viscosity increases below pH 4.5, reaches a maximum (gelation) at pH 3–3.5, and decreases as the pH is lowered further. This viscosity increase apparently results from a combination of hydrogen bonding and electrostatic effects. It is not observed at low concentration (0.25% or lower) and is inhibited by high levels of urea and salt.[164]

In highly alkaline solutions, a gel forms at a pH of 11.5–12 if the gum concentration is above 1–2%, depending on the molecular weight of the algin. Algin

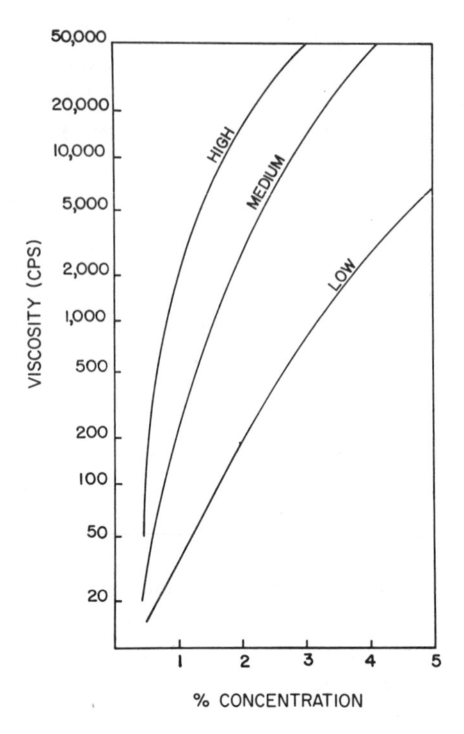

FIG. 3.—Effect of concentration on the viscosity of dispersions of representative high-, medium-, and low-viscosity sodium alginates at 20°.

solutions that contain calcium ions decrease in viscosity as the pH is raised above 9 because of the ability of hydroxides and carbonates to sequester calcium ions. However, these solutions also gel at a pH of 11.5–12.

Propylene glycol alginate is used in acidic solutions. This algin remains soluble and can be used down to about pH 2.

Stability.—Solutions of salts of alginic acid are stable at pH 5 to 10 for long periods at room temperature. Degradation becomes significant above or below this pH region.[2, 165, 166] In the pH range 1–4, alginic acid, in common with other glycuronans, is hydrolyzed more rapidly than neutral polysaccharides, whereas under more acidic conditions the trend is reversed.[167, 168]

The viscosity stability of sodium alginate is inversely related to the degree of polymerization. Small amounts of calcium greatly increase the stability of sodium alginate solutions.[169] Sodium chloride has a less marked beneficial effect.[170]

Propylene glycol alginate shows good stability at pH 3–4 and room temperature and may be used up to about pH 6. Below pH 2, propylene glycol alginate solutions will lose considerable viscosity if kept for several days at room temperature; the rate is dependent on both pH and temperature. In alkaline media, alginate esters are depolymerized by a pH dependent elimination reaction.[166] Under carefully controlled high pH conditions, however, the viscosity of a propylene glycol alginate solution increases dramatically.[171, 172]

Alginates, like other polysaccharides and certain synthetic polymers, are deploymerized in the presence of a number of autoxidizable substances. These include phenolic compounds, ascorbic acid, thiols, and certain metal ions.[2, 173–175] The degradation results from attack of free-radical species which are formed by autoxidation of the reducing substance.

Although algin is relatively resistant to attack by microorganisms, its solutions are subject to bacterial action during prolonged storage. In some industrial applications, it is advisable to use 0.1–0.2% (based on the weight of the solution) of a suitable preservative, such as formaldehyde, an ester of *p*-hydroxybenzoic acid, or a phenol derivative. In some foods, a 0.1% concentration of sodium benzoate (based on the weight of the solution) can be used as a preservative.

3. Gels

Algin formulations gel at room temperature, and their setting time can be controlled. Water gels have excellent clarity. By altering the formulations, gels can be varied in texture from those which are soft and tender to those which are tough and elastic. As a result, algin gels are widely used for food, pharmaceutical, and industrial purposes. Algin gels are of three major types; calcium or other di- or trivalent metal salt gels, acid gels, and combinations of calcium salt and acid gels. In general, gels are formed by the gradual and uniform release of either calcium or hydrogen ions, or a combination of the two, throughout an algin solution. Setting time can be controlled by the addition of a limited amount of a compound, such as a phosphate or polyphosphate, that is capable of combining with a di- or trivalent metal ion.

Gel strength increases with increasing concentration and degree of polymerization of the alginate. The algae species from which the alginate originated also affects gel properties. High L-guluronic acid content causes abrupt gelation after addition of a relatively large amount of a calcium salt to the alginate solution. Transition from solution to gel is more gradual when a calcium salt is added to a solution of alginate having a high D-mannuronic acid content.[53]

Formulations are available in which all the ingredients are combined in a single powder. As a large number of formulations are available for various purposes, the algin manufacturer should be contacted for specific recommendations.

Partial acetylation inhibits gelation of algin by calcium. The acetyl groups may sterically hinder association of the polymer chains[1] or prevent coordination of calcium with the hydroxyl groups.[65]

Propylene glycol alginate forms soft gels in the presence of calcium salts. Alginamide does not require divalent ions to cause gel formation.[36, 176] These gels are thermoreversible.

4. *Films*

A wide range of water-insoluble and -soluble films can be prepared from algin. Films prepared from sodium alginate are water soluble, but calcium alginate films are insoluble, although they do swell in water. Many of these films are clear, tough, and flexible. They resist penetration by greases, oils, fats, waxes, and organic solvents, but transmit water vapor. Algin films are compatible with plasticizers, such as glycerol and sorbitol. Films are prepared by four basic methods: (*a*) evaporation of water from a cast film of a soluble alginate; (*b*) extrusion of a solution of a soluble alginate into a precipitating bath, which yields an insoluble alginate; (*c*) treatment of a soluble film with a di- or trivalent metal salt solution or an acidic solution to form an insoluble film; or (*d*) drying a cast film to drive off the ammonia from a solution of a di- or trivalent metal alginate that is soluble in excess ammonium hydroxide.

Films can be made water resistant by including in the formulation urea–formaldehyde resins or a polymerizable monomer and catalyst[177] which insolubilizes the film when heated. They can also be made water resistant by treatment with solutions of zinc chloride, zirconium oxychloride, or aluminum salts.

Water-insoluble films that are at the same time highly permeable result from combining propylene glycol alginate with gelatin or a polyfunctional amine.[133, 172]

5. *Compatibilities*

Algin in solution is compatible with a large number of compounds; a partial list is given in Table II. Propylene glycol alginate is compatible with acids and, in many cases, with alkaline-earth metal salts. Deionized solutions of alginic acid, at concentrations below 0.5%, are soluble in acidic solutions above pH 1.5.

6. *Algin in Foods*

Algin has been eaten for hundreds of years as a constituent of kelp in specialty foods,[4, 8] and algin itself has been used in foodstuffs for 40 years. Extensive animal feeding tests on sodium alginate and propylene glycol alginate have shown that they are wholesome, edible products and not allergenic.[178–181]

Algin appears on the GRAS (Generally Recognized as Safe) list of the Food and Drug Administration of the United States under Title 21 of the Code of Federal Regulations, Section 121.101(d)(7).

Propylene glycol alginate (algin derivative) was approved by the Food and Drug Administration of the United States on June 29, 1960, for general use as a food additive in foods for which standards of identity do not preclude such use.

TABLE II

Compatibilities of Sodium Alginate in Solution

	Compatible Compounds
Carbohydrates	Starch, dextrin, sucrose, D-glucose, invert sugar, neutral and anionic soluble cellulose derivatives
Water-soluble gums	Acacia, carrageenan, guar, karaya, locust bean, pectin, tragacanth
Polyhydric alcohols	Glycerol, ethylene glycol, propylene glycol, sorbitol, mannitol
Resins	Phenolic, urea–formaldehyde, melamine–formaldehyde, melamine–formaldehyde resin emulsions
Proteins	Casein (acid-precipitated), gelatin, soy bean, egg albumin
Pigments and fillers	China clay, chrome oxides, cobalt blue, iron oxides, lithopone, titanium dioxide
Dyes	Vat dyes, rapidogens, pharmasols, indigosols, acid azos, directs, acetates
	Partially Compatible Compounds
Water-miscible solvents	Ethanol (about 25% by volume), 2-propanol, acetone
Salts	Sodium sulfate (10–12% of solution weight), sodium chloride (about 4%), ammonium chloride, most alkali salts
	Incompatible Compounds
Salts	Ionized salts of alkaline earths (except magnesium), heavy metal salts
Acids	Those strong enough to bring the pH below 3.5
Amines	Long-chain amines, quarternary ammonium salts having eight carbon atoms or more in a straight chain, salts of polyamines

The specific food additive regulation authorizing this use appears under Title 21 of the Code of Federal Regulations, Section 121.1015.

Alginates have been specified as optional ingredients in Federal Standards of Identity for several foods. These include ice cream, sherbets, water ices, creamed cottage cheese, cream cheese, Neufchatel cheese, cheese spreads, French dressing, salad dressing, and artificially sweetened jelly, jams, and preserves.

7. *Other Properties*

Dissociation constant.—The pK_a, or negative logarithm, of the dissociation constant of alginic acid is equal approximately to the pH of a half neutralized alginic acid solution; pK_a values for alginic acid may range from 3.4 to 4.4. Factors that affect the pK_a are alginic acid concentration, the amount of dissolved salts, and the ratio of D-mannuronic acid units (M) and L-guluronic acid units (G) in the polymer.[2]

In distilled water, the pK_a of *L. digitata* alginic acid (M/G 1.40) is 3.93 and that of *L. hyperborea* alginic acid (M/G 0.45) is 4.41. These same alginic acids, when dissolved in 0.1*N* NaCl, have pK_a values of 3.42 and 3.74, respectively.[2]

Equivalent weight.—The equivalent weight of commercial alginic acid (as an acid) ranges from 194 to 215. The theoretical equivalent weight for a hexuronoglycan is 176. However, the equivalent weight does not approach the theoretical value even if the solids have been determined by drying at 105° for 3 hr. A major part of this difference might be due to water which is tenaciously held by the alginic acid. Attempts to remove this water with rigorous drying conditions degrades the alginic acid and, in some cases, introduces additional acidic groups. Introduced acidic groups in a degraded alginic acid might account for some of the reports of equivalent weights that approximate the theoretical value.

Molecular weight.—In kelp, algin has a very high molecular weight. By varying the processing conditions, algin can be produced that covers a wide range of viscosities and molecular weights. However, several different methods[57–61] indicate that commercial sodium alginates have molecular weights between 32,000 and 200,000 and a degree of polymerization of 180–930. Laboratory samples with a molecular weight in excess of one million have been prepared.[57, 182]

Optical rotation.—Reports of the $[\alpha]^{20}_D$ of sodium alginate solutions vary from −113° to −148°.[137, 183] Specific optical rotation has little relation to quality and no relation to viscosity.

Ion-exchange properties.—Insoluble metal alginates behave as typical ion-exchange resins. When an alginate is exposed to a mixture of cations, certain of the ions will be more strongly bound by the alginate.

The affinities of divalent metal ions are dependent on the relative amounts of D-mannuronic and L-guluronic acid units in the alginate. For example, alginate prepared from *Laminaria digitata,* which is rich in D-mannuronic acid, has a slightly different affinity series from that for alginate prepared from *Laminaria hyperborea* stipes, which is rich in L-guluronic acid.[184] The affinity for divalent ions decreases in the following order:

L. digitata alginate: Pb > Cu > Cd > Ba > Sr > Ca > Co, Ni, Zn, Mn > Mg
L. hyperborea alginate: Pb > Cu > Ba > Sr > Cd > Ca > Co, Ni, Zn, Mn > Mg

The concentration of divalent cations required to bring about gel formation and precipitation is the same for the sodium alginate prepared from *Laminaria digitata* and *Laminaria hyperborea* and increases in the order:[185]

Ba < Pb < Cu < Sr < Cd < Ca < Zn < Ni < Co < Mn, Fe < Mg

Thus, the precipitation effect of a metal ion is determined not only by the affinity of the metal for the alginate, but also by the gel-forming ability of the metal.

Additional studies of the interaction of polyvalent metal ions with alginic acid indicate that the hydroxyl groups on the polymer may play some role in ion binding.[65, 186]

The selectivity coefficient describing the calcium–potassium ion-exchange equilibrium indicates that the affinity of alginate for calcium increases in relation to the L-guluronic acid content of the polymer. The monomeric uronates do not bind calcium. Algin's calcium binding properties are apparently determined by the steric arrangement of the active groups in the polymer chain.[187] Calcium selectivity is not influenced by molecular weight in the range 350,000–700,000.

The relatively high selectivity for strontium appears to have practical importance in preventing intestinal absorption of radiostrontium in case of exposure.[188, 189]

Ion-exchange properties of alginic acid have been used for the separation and determination of metal ions by column chromatography and thin-layer chromatography.[190, 191] The retention capacity is apparently influenced by interactions of the metal ions and the two vicinal hydroxyl groups on each monomeric uronic acid.

8. *Analytical Methods*

Choice of analytical method will depend primarily on the level of alginate present, the sensitivity required, and the type and quantity of interfering components. Separation of algin from a mixture by precipitation as the insoluble calcium salt frequently can be accomplished by adding a calcium chloride solution to the dissolved sample. Once the alginate is separated as calcium alginate, decarboxylation can be employed as a quantitative method.[192,193] A minimum of about 100 mg of algin is required. Alternatively, the calcium alginate can be washed with dilute acid, and the resulting alginic acid is then determined by titration.[2]

Colorimetric methods are more sensitive. Once the algin is freed from interfering components, carbazole,[193] naphthoresorcinol,[194] or phenol–sulfuric[195] acid may be used. The results vary somewhat with the plant source of the algin because color development can be influenced by the uronic acid composition of the alginate.[2]

Recently, a modification of the ferric hydroxide–sulfuric acid (Schroeder–Racicot)[196] method was proposed as a quantitative analysis for algin in foods.[197] After isolation of calcium alginate from a papain digest of the sample, the algin (50–250 mg) is heated with the color-forming reagent under carefully controlled conditions.

A number of qualitative tests have been suggested including Night Blue dye[198] and ferric hydroxide–sulfuric acid.[196] The latter has been recommended for detecting algin in a variety of foods.[199] Naphthoresorcinol also has been recommended.[200] A number of techniques for eliminating interfering components have been reported.[198, 201]

VI. References

(1) E. Percival and R. H. McDowell, "Chemistry and Enzymology of Marine Algal Polysaccharides," Academic Press, New York, 1967.

(2) A. Haug, "Composition and Properties of Alginates," Norwegian Inst. Seaweed Res., Trondheim, Norway, Rept. No. **30,** 1964.

(3) T. Levring, H. A. Hoppe, and O. J. Schmid, "Marine Algae," Cram, De Gruyter & Co., Hamburg, 1969.

(4) A. Okazaki, "Seaweeds," a report concerning Japanese seaweed and its utilization, personal communication, 1964 (revised 1970).

(5) F. Smith and R. Montgomery, "The Chemistry of Plant Gums and Mucilages," Rheinhold, New York, 1959.

(6) A. B. Steiner and W. H. McNeely, *Advan. Chem. Ser.,* **11,** 68 (1954).

(7) H. Maass, "Alginsäure und Alginate," Strassenbau Chemie & Technik Verlagsgesellschaft, Heidelberg, 1959.

(8) V. J. Chapman, "Seaweeds and Their Uses," Methuen and Co., Ltd., London, 2nd Ed., 1970.

(9) A. Haug and A. Jensen, "Seasonal Variations in the Chemical Composition of *Alaria esculenta, Laminaria saccharina, Laminaria hyperborea* and *Laminaria digitata* from Northern Norway, Norwegian Inst. Seaweed Res., Oslo, Norway, Rept. No. **4,** 1954.

(10) I. Munda, *Botan. Marina,* **7,** No. 1–4, 76 (1964).

(11) B. Larsen and A. Haug, *Proc. Int. Seaweed Symp., 4th, Biarritz, Fr., 1961,* 335 (1964).

(12) A. B. Steiner, *in* "Encyclopedia Americana," Americana Corp., New York, Vol. 1, 1952, p. 397.

(13) "The Seaweed Story," Dept. Fish Game, State of Calif., Sacramento, Calif., 1954.

(14) W. J. North and C. L. Hubbs, "Utilization of Kelp-bed Resources in Southern California," *Fish Bull.* No. **139,** Dept. Fish Game, State of Calif., 1968.

(15) F. E. Fritsch, "The Structure and Reproduction of The Algae," Cambridge Univ. Press, London, Vol. II, 1952, p. 19.

(16) J. M. Kingsbury, "Seaweeds of Cape Cod and the Islands," Chatham Press, Chatham, Mass., 1969.

(17) F. N. Woodward, *Proc. Int. Seaweed Symp., 5th, Halifax, N.S., 1965,* 55 (1966).

(18) *Chem. Trade J.,* **141,** 1430 (1957).

(19) W. A. Scholes, *Ocean Ind.,* **4,** No. 3, 69 (1969).

(20) W. A. P. Black, *J. Soc. Chem. Ind., London,* **67,** 165, 355 (1948); **69,** 6, 161 (1950).

(21) W. A. P. Black and E. T. Dewar, *J. Mar. Biol. Ass. U. K.,* **28,** 673 (1949).

(22) W. A. P. Black, *J. Mar. Biol. Ass. U. K.,* **29,** 45 (1950).

(23) "Seaweed Prospects," Inst. Seaweed Res., Inveresk, Midlothian, Scotland, 1956.

(24) H. Printz, *Norwegian Trade Post,* **5,** 45 (1950).

(25) V. G. Heydt, *Stärke,* **11,** 38 (1959).

(26) L. A. Bashford, R. S. Thomas, and F. N. Woodward, *J. Soc. Chem. Ind., London,* **69,** 337 (1950).

(27) W. A. P. Black and F. N. Woodward, *Advan. Chem. Ser.,* **11,** 83 (1954).

(28) Taiho Industrial Co., Brit. Patent 856,416 (1962); *Chem. Abstr.,* **56,** 11366 (1962).

(29) Henkel & Cie, G.m.b.H., Fr. Patent 1,360,504 (1964); *Chem. Abstr.,* **62,** 4654 (1965).

(30) A. B. Steiner and W. H. McNeely, *Ind. Eng. Chem.,* **43,** 2073 (1951).

(31) Kelco Co., Brit. Patent 676,618 (1952).

(32) A. B. Steiner and W. H. McNeely, U.S. Patent 2,463,824 (1949); *Chem. Abstr.,* **43,** 4291 (1949).

(33) E. Nishide, *Kogyo Kagaku Zasshi,* **66,** 458 (1963); *Chem. Abstr.,* **60,** 3090 (1964).

(34) R. G. Schweiger, *J. Org. Chem.,* **27,** 1786 (1962).

(35) R. G. Schweiger, U.S. Patent 3,349,078 (1967); *Chem. Abstr.,* **68,** 22907g (1968).

(36) R. Köhler and W. Dierichs, U.S. Patent 2,881,161 (1959); *Chem. Abstr.,* **53,** 14387 (1959).

(37) K. W. Kirby, Ph.D. Dissertation, Purdue Univ., Lafayette, Ind., 1958; *Diss. Abstr.,* **19,** 945 (1958).

(38) W. H. McNeely and J. J. O'Connell, U.S. Patent 2,902,479 (1959); *Chem. Abstr.,* **54,** 2201 (1960).

(39) W. H. McNeely, U.S. Patent 2,688,598 (1954).

(40) J. L. Boyle, Brit. Patent 835,009 (1960); *Chem. Abstr.,* **54,** 22397 (1960).

(41) W. H. McNeely, U.S. Patent 2,979,499 (1961).

(42) National Lead Co., Germ. Patent 1,103,336 (1961); *Chem. Abstr.,* **57,** 4522 (1962).

(43) R. H. McDowell, *CIBA Rev.,* No. **1,** 3 (1969).

(44) *Oil, Paint Drug Rep.,* **198,** No. 25, 40 (1970).

(45) B. Idson, *Chem. Week,* **79,** No. 29, 57 (1956).

(46) W. A. P. Black, *Chem. Prod.,* **16,** 139 (1953).

(47) B. Grenager and E. Baardseth, *Proc. Int. Seaweed Symp., 5th, Halifax, N.S., 1965,* 129 (1966).

(48) E. C. C. Stanford, Brit. Patent 142 (1881).

(49) T. Dillon, *Abstr. Int. Seaweed Symp., 3rd, Galway, Ireland, 1958,* 41 (1958).

(50) H. J. Lucas, U.S. Patent 2,097,228 (1937); U. S. Patent 2,097,231 (1937).

(51) A. B. Steiner, U.S. Patent 2,485,934 (1949); *Chem. Abstr.,* **44,** 1622 (1950).

(52) E. Booth, *Chem. Trade J.,* **142,** 19 (1958).

(53) R. H. McDowell, *Rev. Pure Appl. Chem.,* **10,** 1 (1960).

(54) C. T. Blood, *CIBA Rev.,* No. **1,** 36 (1969).

(55) E. L. Hirst and D. A. Rees, *J. Chem. Soc.,* 1182 (1965).

(56) D. A. Rees and J. W. B. Samuel, *J. Chem. Soc., C,* 2295 (1967).

(57) O. Smidsrød and A. Haug, *Acta Chem. Scand.,* **22,** 797 (1968).

(58) B. Buchner, R. E. Cooper, and A. Wassermann, *J. Chem. Soc.,* 3974 (1961).

(59) L. Langmaack, *Kolloid-Z.,* **156,** 32 (1958).

(60) W. Diemair and H. H. Weichel, *Deut. Lebensm.-Rundsch.,* **54,** 51, 76 (1958).

(61) F. G. Donnan and R. C. Rose, *Can. J. Res.,* **28B,** 105 (1950).

(62) S. J. Werbin, *Advan. Chem. Ser.,* **25,** 5 (1960).

(63) A. B. Steiner and G. D. Sperry, U.S. Patent 2,485,935 (1949); *Chem. Abstr.,* **44,** 1622 (1950).

(64) N. H. Nash, *Advan. Chem. Ser.,* **25,** 45 (1960).

(65) R. G. Schweiger, *J. Org. Chem.,* **27,** 1789 (1962).

(66) J. E. Scott, *Biochem. J.,* **84,** 270 (1962).

(67) A. B. Steiner, U.S. Patent 2,441,729 (1948); *Chem. Abstr.,* **42,** 6962 (1948).

(68) A. Miller and J. K. Rocks, U.S. Patent 3,455,701 (1969); *Chem. Abstr.,* **71,** 100668h (1969).

(69) J. C. Freedman, U.S. Patent 3,349,079 (1967); *Chem. Abstr.,* **68,** 4191q (1968).

(70) K. F. Gibsen, U.S. Patent 2,918,375 (1959); *Chem. Abstr.,* **54,** 7926 (1960).

(71) S. W. Wilding, Brit. Patent 518,596 (1940); *Chem. Abstr.,* **35,** 7591 (1941).

(72) V. V. Vallandigham, A. L. Magnuson, and A. Miller, *Paper Ind.,* **33,** 788 (1951); **33,** 1176 (1952).

(73) K. Hilton, *World Paper Trade Rev.,* **161,** 1702 (1964).

(74) M. Glicksman, "Gum Technology in the Food Industry," Academic Press, New York, 1969, p. 239.

(75) F. X. McDermott, *Food Eng.,* **34,** No. 5, 66 (1962).

(76) T. R. Andrew and W. C. MacLeod, *Food Prod. Develop.,* **4,** No. 5, 99 (1970).

(77) S. Stistrup and J. Andreasen, *Proc. Int. Dairy Congr., 16th, Copenhagen, 1962,* **3,** 29 (1962); *Chem Abstr.,* **65,** 6188 (1966).

(78) G. D. Sperry, *Ice Cream Rev.,* **37,** No. 3, 74 (1953).

(79) R. L. Edlin, *Baker's Dig.,* **41,** No. 6, 49 (1967).

(80) P. E. Ellis, U.S. Patent 3,332,784 (1967); *Chem. Abstr.,* **67,** 74711n (1967).

(81) B. T. Messina, U.S. Patent 3,352,688 (1967); *Chem. Abstr.,* **68,** 11824 (1968).

(82) D. Muncaster and B. Messina, *Food Process.,* **22,** No. 11, 62 (1961).

(83) F. X. McDermott, *Baker's Dig.,* **37,** No. 3, 66 (1963).

(84) A. R. Hunter, U. S. Patent 3,016,302 (1962).

(85) A. R. Hunter, U.S. Patent 2,987,400 (1961).

(86) A. R. Hunter, U.S. Patent 3,365,305 (1968); *Chem. Abstr.,* **68,** 67930a (1968).

(87) S. Szczesniak and E. Engel, U.S. Patent 3,300,318 (1967); *Chem. Abstr.,* **66,** 64528y (1967).

(88) A. B. Steiner, U.S. Patent 2,455,820 (1948); *Chem. Abstr.,* **43,** 1577 (1949).

(89) A. R. Hunter and J. K. Rocks, U.S. Patent 2,949,366 (1960); *Chem. Abstr.,* **55,** 1964 (1961).

(90) A. Miller and J. K. Rocks, U.S. Patent 3,266,906 (1966); *Chem. Abstr.,* **65,** 12786d (1966).

(91) H. F. Angermeier, U.S. Patent 2,536,708 (1951); *Chem. Abstr.,* **45,** 6772 (1951).

(92) H. E. Nordman and W. H. Mohr, *Amer. Brewer,* **96,** 22 (1963).

(93) A. B. Steiner, Can. Patent 546,646 (1957).

(94) F. X. McDermott, U.S. Patent 3,257,214 (1966).

(95) B. T. Messina and D. Pape, *Food Eng.,* **38,** No. 4, 48 (1966).

(96) W. L. Kerr and L. C. Zopf, *J. Amer. Pharm. Ass., Pract. Pharm. Ed.,* **14,** 426 (1953).

(97) J. Lazarus and J. Cooper, U.S. Patent 3,142,621 (1964); *Chem. Abstr.,* **61,** 8141 (1964).

(98) G. Levy and B. K. Rao, *J. Pharm. Sci.,* **61,** 279 (1972).

(99) A. B. Steiner and L. B. Rothe, U.S. Patent 2,635,067 (1953); *Chem. Abstr.,* **47,** 7168 (1953).

(100) H. O. Doell, U. S. Patent 3,258,398 (1966); *Chem Abstr.,* **65,** 8685 (1966).

(101) F. H. Schwarzkopf, Germ. Patent 1,158,212 (1963); *Chem. Abstr.,* **60,** 9098 (1964).

(102) C. T. Blood, *Chem. Prod.,* **22,** 304 (1959).
(103) T. G. Gerding and H. G. Dekay, *Drug Stand.,* **23,** 132 (1955).
(104) E. Thomsen, *Amer. Perfum. Cosmet.,* **78,** No. 10, 45 (1963).
(105) S. W. Wilding, U.S. Patent 2,249,694 (1941); *Chem. Abstr.,* **35,** 6704 (1941).
(106) S. Buchan and R. W. Peggie, *J. Dent. Res.,* **45,** 1120 (1966).
(107) S. Higashi and K. Taguchi, U.S. Patent 3,246,998 (1966); *Chem. Abstr.,* **65,** 3680 (1966).
(108) J. C. Morrell, U.S. Patent 3,282,710 (1966); *Chem. Abstr.,* **66,** 14049x (1967).
(109) C. T. Blood, *Chem. Ind. (London),* 690 (1968).
(110) A. D. Bonnano, *Paper Mill News,* **85,** No. 8, 21 (1962).
(111) W. P. Fairchild and J. E. Robison, U.S. Patent 3,298,862 (1967); *Chem. Abstr.,* **66,** 56928b (1967).
(112) W. P. Fairchild, U.S. Patent 3,255,028 (1966); *Chem. Abstr.,* **65,** 4086h (1966).
(113) A. Miller and A. L. Magnuson, U.S. Patent 2,913,364 (1959); *Chem. Abstr.,* **54,** 3954 (1960).
(114) H. W. Bartlett and A. L. Magnuson, *Tappi,* **39,** 214 (1956).
(115) M. H. O'Grady, U.S. Patent 3,620,801 (1971).
(116) K. A. Hilton, *CIBA Rev.,* No. **1,** 19 (1969).
(117) A. J. Hall, *Text. World,* **114,** No. 8, 82 (1964).
(118) E. L. Kelley, U.S. Patent 3,282,717 (1966); *Chem. Abstr.,* **66,** 3759v (1967).
(119) I. H. Duckworth, *Plant. Bull. Rubber Res. Inst. Malaya,* No. **74,** 111 (1964).
(120) C. A. Uraneck and D. F. Dodgen, U.S. Patent 2,883,351 (1959); *Chem. Abstr.,* **53,** 13645 (1959).
(121) Monsanto Co., Fr. Patent 1,449,285 (1966); *Chem. Abstr.,* **66,** 66107j (1967).
(122) M. R. Buffington, U.S. Patent 2,878,196 (1959); *Chem. Abstr.,* **53,** 11861 (1959).
(123) R. G. Madeline and J. F. Pathus-Labour, Fr. Patent 78,308 (1962); *Chem. Abstr.,* **57,** 15358 (1962).
(124) E. P. Budewitz, U.S. Patent 2,780,555 (1957); *Chem. Abstr.,* **51,** 11732 (1957).
(125) T. A. Smith, *Trans. Brit. Ceram. Soc.,* **61,** 523 (1962).
(126) E. L. Thellmann, U.S. Patent 3,216,841 (1965); *Chem. Abstr.,* **64,** 4739 (1966).
(127) P. E. Rempes, Jr., U.S. Patent 2,990,292 (1961); *Chem. Abstr.,* **55,** 26310 (1961).
(128) Produits Chimiques Pechiney Saint-Gobain, Fr. Patent 1,345,690 (1963); *Chem. Abstr.,* **61,** 1191 (1964).
(129) G. J. Holland, *Proc. Soc. Water Treat. Exam.,* **15,** No. 1, 11 (1966); *Chem. Abstr.,* **65,** 13406 (1966).
(130) Imperial Chemical Industries, Brit. Patent 1,045,599 (1966); *Chem. Abstr.,* **66,** 13979p (1967).
(131) C. A. H. M. Smit and A. C. M. Smit, Brit. Patent 887,502 (1962); *Chem. Abstr.,* **57,** 10874 (1962).
(132) *Fed. Regist.,* **27,** 6232 (June 30, 1962); *Chem. Abstr.,* **57,** 10293 (1962).
(133) H. V. Rintelen, U.S. Patent 3,378,373 (1968).
(134) E. C. C. Stanford, *J. Soc. Chem. Ind.,* **5,** 218 (1886).
(135) T. Dillon, *Abstr. Int. Seaweed Symp., 3rd, Galway, Ireland, 1958,* 41 (1958).
(136) L. H. Cretcher and W. L. Nelson, *Science,* **67,** 537 (1928).
(137) W. L. Nelson and L. H. Cretcher, *J. Amer. Chem. Soc.,* **51,** 1914 (1929).
(138) F. G. Fisher and H. Dorfel, *Hoppe-Seyler's Z. Physiol. Chem.,* **302,** 186 (1955).
(139) R. L. Whistler and K. Kirby, *Z. Physiol. Chem.,* **314,** 46 (1959).
(140) E. L. Hirst, *Proc. Chem. Soc.,* 177 (1958).

(141) D. W. Drummond, E. L. Hirst, and E. Percival, *Chem. Ind. (London)*, 1088 (1958).
(142) E. D. T. Atkins, W. Mackie, and E. E. Smolko, *Nature,* **225,** 626 (1970).
(143) E. D. T. Atkins, W. Mackie, K. D. Parker, and E. E. Smolko, *Polymer Lett.,* **9,** 311 (1971).
(144) D. I. Vincent, *Chem. Ind. (London)*, 1109 (1960).
(145) E. L. Hirst, E. Percival, and J. K. Wold, *Chem. Ind. (London)*, 257 (1963).
(146) E. L. Hirst, E. Percival, and J. K. Wold, *J. Chem. Soc.,* 1493 (1964).
(147) D. W. Drummond, E. L. Hirst, and E. Percival, *J. Chem. Soc.,* 1208 (1962).
(148) A. Haug and B. Larsen, *Acta Chem. Scand.,* **16,** 1908 (1962).
(149) W. Mackie, *Carbohyd. Res.,* **20,** 413 (1971).
(150) A. Haug, B. Larsen, and O. Smidsrød, *Acta Chem. Scand.,* **20,** 183 (1966).
(151) A. Haug, B. Larsen, and O. Smidsrød, *Acta Chem. Scand.,* **21,** 691 (1967).
(152) A. Haug, S. Myklestad, B. Larsen, and O. Smidsrød, *Acta Chem. Scand.,* **21,** 768 (1967).
(153) A. Haug, B. Larsen, and E. Baardseth, *Proc. Int. Seaweed Symp., 6th, Santiago de Compostela, Spain, 1968,* 443 (1969).
(154) B. Larsen, T. Painter, A. Haug, and O. Smidsrød, *Acta Chem. Scand.,* **23,** 355 (1969).
(155) T. Painter, O. Smidsrød, B. Larsen, and A. Haug, *Acta Chem. Scand.,* **22,** 1637 (1968).
(156) T. Y. Lin and W. Z. Hassid, *J. Biol. Chem.,* **241,** 3284, 5284 (1966).
(157) W. Z. Hassid, *Science,* **165,** 137 (1969).
(158) P. Buchner, R. E. Cooper, and A. Wasserman, *J. Chem. Soc.,* 3974 (1961).
(159) R. H. McDowell, *J. Soc. Chem. Ind., London,* Monogr. No. **24,** 19 (1966).
(160) O. Smidsrød and A. Haug, *Acta Chem. Scand.,* **19,** 329 (1965).
(161) H. Morawetz, "Macromolecules in Solution," Interscience, New York, 1965, Chap. 7.
(162) A. Haug and O. Smidsrød, *Acta Chem. Scand.,* **16,** 1569 (1962).
(163) G. Levy and T. W. Schwartz, *J. Amer. Pharm. Ass.,* **47,** 455 (1958).
(164) J. S. Yudelson and R. E. Mack, *J. Polymer Sci., Part A,* **2,** 4683 (1964).
(165) A. Haug, B. Larsen, and O. Smidsrød, *Acta Chem. Scand.,* **17,** 1466 (1963).
(166) A. Haug, B. Larsen, and O. Smidsrød, *Acta Chem. Scand.,* **21,** 2859 (1967).
(167) O. Smidsrød, A. Haug, and B. Larsen, *Acta Chem. Scand.,* **20,** 1026 (1966).
(168) O. Smidsrød, B. Larsen, T. Painter, and A. Haug, *Acta Chem. Scand.,* **23,** 1573 (1969).
(169) G. Levy, *J. Pharm. Sci.,* **50,** 429 (1961).
(170) K. Bolewski, *Makromol. Chem.,* **66,** 1 (1963).
(171) R. H. McDowell, J. L. Boyle, W. Copes, and W. Bryden, Brit. Patent 1,135,856 (1968); *Chem. Abstr.,* **70,** 56480u (1969).
(172) R. H. McDowell, *J. Soc. Cosmet. Chem.,* **21,** 441 (1970).
(173) O. Smidsrød, A. Haug, and B. Larsen, *Acta Chem. Scand.,* **17,** 1473, 2628 (1963).
(174) O. Smidsrød, A. Haug, and B. Larsen, *Acta Chem. Scand.,* **23,** 2955 (1969).
(175) O. Smidsrød, A. Haug, and B. Larsen, *Carbohyd. Res.,* **5,** 482 (1967).
(176) R. Kohler and W. Dierichs, U.S. Patent 2,919,198 (1959); *Chem. Abstr.,* **54,** 7926 (1960).
(177) H. Thiele, Brit. Patent 918,626 (1963); *Chem. Abstr.,* **58,** 14253 (1963).
(178) H. W. Nilson and J. A. Wagner, *Proc. Soc. Exptl. Biol. Med.,* **76,** 630 (1951).
(179) R. A. Ouer, *Ann. Allergy,* **7,** 681 (1949).

(180) G. Woodward, unpublished data (1959).

(181) Woodward Research Corporation, "Safety Evaluation of Kelcoloid and Related Algin Products by Repeated Oral Administration to Laboratory Animals: A Review," unpublished report (1964).

(182) O. Smidsrød, *Carbohyd. Res.*, **13,** 359 (1970).

(183) W. A. P. Black, W. J. Cornhill, and E. T. Dewar, *J. Sci. Food Agr.*, **3,** 542 (1952).

(184) A. Haug, *Acta Chem. Scand.*, **15,** 1794 (1961).

(185) A. Haug and O. Smidsrød, *Acta Chem. Scand.*, **19,** 341 (1965).

(186) R. G. Schweiger, *Kolloid-Z.*, **196,** 47 (1964).

(187) A. Haug and O. Smidsrød, *Acta Chem. Scand.*, **22,** 3098 (1969).

(188) D. Waldron-Edward, *Can. Med. Ass. J.*, **99,** 986 (1968).

(189) A. Hodgkinson, B. E. C. Nordin, J. Hambleton, and C. B. Oxby, *Can. Med. Ass. J.*, **97,** 1139 (1967).

(190) T. Takahashi and S. Emura, *Bunseki Kagaku,* **7,** 568 (1958); *Chem. Abstr.*, **54,** 16253 (1960).

(191) D. Cozzi, P. G. Desideri, and T. Lepri, *J. Chromatog.*, **40,** 130 (1969).

(192) "Food Chemicals Codex," Publ. 1406, Natl. Acad. Sci.–Natl. Res. Council, Washington, D.C., 1st Ed., 1966, p. 718.

(193) B. L. Browning, "Methods of Wood Chemistry," Interscience, New York, Vol. II, 1967, p. 632.

(194) W. Wagner, *Anal. Chim. Acta,* **29,** No. 21, 182 (1963).

(195) H. D. Graham, *J. Dairy Sci.*, **52,** 443 (1969).

(196) C. W. Schroeder and P. A. Racicot, *Ind. Eng. Chem., Anal. Ed.*, **13,** 165 (1941).

(197) H. D. Graham, *J. Food Sci.*, **35,** 494 (1970).

(198) R. H. McDowell, *Proc. Int. Seaweed Symp., 2nd, Trondheim, Norway, 1955,* 131 (1956).

(199) "Official Methods of Analysis A.O.A.C.," W. Horwitz, ed., Association of Official Analytical Chemists, Washington, D.C., 11th Ed., 1970, pp. 196, 209, 281, 519.

(200) Ref. 192, p. 20.

(201) H. Schulzen, *Text.-Prax.*, **19,** 67 (1964).

CHAPTER V

CARRAGEENAN

GORDON A. TOWLE

The Copenhagen Pectin Factory Ltd., Lille Skensved, Denmark
(Subsidiary of Hercules, Inc. Wilmington, Delaware)

I. Introduction

1. *General Relations*

The vegetation of the sea, in contrast to that of the land, consists almost entirely of various forms of algae. These range from the unicellular organisms, which form the important first stage of the food cycle, up to the giant kelps, which may attain sizes of the order of their terrestrial counterparts, the trees. The unicellular algae are abundant in the upper layers of sea water, whereas larger species are found in continental shelf areas where a firm attachment to the sea bottom is possible up to depths at which light is still available.

Classification of the algae has been aided by differences in photosynthetic pigmentation, which generally parallel morphological differences. Four classes encompass most species of algae. The blue-green algae of the classes Chlorophyceae and Cyanophyceae include both fresh- and salt-water species. These occupy the lower niches on the evolutionary scale and are generally unicellular. The algae of the class Rhodophyceae and Phaeophyceae comprise the larger life forms found almost entirely in a salt water environment that are collectively referred to as seaweeds. The Rhodophyceae (red seaweeds) derive their red coloration from bileproteins, whereas members of the Phaeophyceae (brown seaweeds) derive their brown coloration from the xanthophyll, fucoxanthin.

The algae represent a more primitive life form than their terrestrial counterparts, the land plants, and in addition thrive under quite different environmental conditions. Thus, it is not surprising to find considerable biochemical diversity between algae and terrestrial plants. A main difference is found in the polysaccharides employed as ground or support substances. Whereas land plants use cellulose as a support media, higher forms of algae rely upon heteropolysaccharides composed of D- and L-galactose and 3,6-anhydro-D- and L-galactose or of D-mannuronic and L-guluronic acid. Members of the class Rhodophyceae contain large amounts of the former polysaccharide usually containing a relatively large amount of sulfate ester groups, whereas members of the Phaeophyceae contain the latter heteroglycuronan (Chapt. IV).

2. *Classification of Polysaccharides of the Rhodophyceae*

Structural polysaccharides of the Rhodophyceae may generally be grouped into several categories on the basis of structural characteristics (Table I). All are members of a family of galactans composed of varying proportions of D- and L-galactopyranosyl and 3,6-anhydro-D- and L-galactopyranosyl units linked alternately $(1 \rightarrow 3)$ and $(1 \rightarrow 4)$ and often substituted with sulfate half-ester groups.

Agar and agarlike polysaccharides are derived chiefly from the agarophytes, *Gelidium* and *Gracilaria* (Chapt. III). A second type of extractive of the red seaweeds includes the carrageenans and related polysaccharides. The name *carrage-*

TABLE I

Monomer Units of Red Seaweed Galactans

Galactan	Predominant Monosaccharide Units
Agarose (agaran) and agaropectin	D-galactose 3,6-anhydro-L-galactose
Porphyran	D-galactose 6-O-methyl-D-galactose D-galactose 6-sulfate 3,6-anhydro-L-galactose
Iota-carrageenan	D-galactose 4-sulfate 3,6-anhydro-D-galactose 2-sulfate
Kappa-carrageenan	D-galactose 4-sulfate 3,6-anhydro-D-galactose
Lambda-carrageenan	D-galactose 2-sulfate D-galactose 2,6-disulfate
Mu-carrageenan	D-galactose 4-sulfate D-galactose 6-sulfate 3,6-anhydro-D-galactose
Nu-carrageenan	D-galactose 4-sulfate D-galactose 2,6-disulfate 3,6-anhydro-D-galactose
Furcellaran	D-galactose D-galactose 2-sulfate D-galactose 4-sulfate D-galactose 6-sulfate 3,6-anhydro-D-galactose
Hypnean	D-galactose D-galactose sulfate 3,6-anhydro-D-galactose

enan has been derived from the older term carrageenin originally used to designate extractives of *Chondrus* and *Gigartina* species. Galactans of similar structure have since been found in other related species, including the genera *Furcellaria, Eucheuma, Hypnea, Iridea,* and *Polyides* (Chapt. V).

Carrageenan can be induced to form thermally reversible gels in the presence of specific cations, such as potassium and ammonium. Another property of carrageenan that has contributed greatly to its industrial importance is its ability at low concentrations to cause agglomeration of protein solutions, in particular the casein particles in cow's milk. The carrageenan–protein reaction is quite specific and is influenced by the amount and location of *O*-sulfate groups on the molecule and molecular shape.

Milk reactivity is low or absent in a third class, which has been given the tentative name of gelans. Members of this group, which includes furcellaran (Danish agar) and hypnean, although showing physical properties similar to those of the carrageenans, differ structurally in having fewer sulfate ester groups per monomer unit.

Galactans, termed porphyrans, have been found in *Porphyra* and *Laurentia* species and in *Bangia fuscopurpurea.* The porphyrans, like agar, contain 3,6-anhydro-L-galactose and 6-*O*-methyl-D-galactose and like carrageenan contain galactose 6-sulfate but as the L rather than the D isomer. The composition of porphyrans varies widely depending upon seasonal and environmental effects, with the amounts of sulfate ester present varying between 6 and 11%.

A number of other sulfated galactans, although not strictly amenable to the foregoing classification, have been divided into these groups as a matter of convenience. These include the galactan sulfate from *Gloiopeltis furcata*[1, 2] having similarities to both carrageenan and agar, highly branched galactan sulfates from *Dilsea edulis*[3] and *Dumontia incrassata,*[4] and a highly complex mucilage from *Polysephonia fastigiata.*[5] It has been suggested[6] that such polysaccharides be regarded as belonging to a family separate from those of agar, carrageenan, gelan, and porphyran types. A list of those sea plants of the Rhodophyceae reported as sources of the various types of galactan sulfates is presented in Table II.

3. *Historical Aspects*

Seaweeds and their extractives have found use as foods, medicinal agents, sources of potash and iodine, and fertilizers for some 5000 years. In retrospect, the "sea-farming" of Irish moss and the industry that has evolved around this seaweed and its close relatives is a recent one.

The Irish are given credit for the discovery of the weed and its extract. Some 600 years ago, shore residents of County Carragheen on the south Irish coast first used the plant, or Irish moss as it became known, in foods, in medicines, and as fertilizer and ultimately noted its milk reactivity. Irish settlers coming to America in the 1700's brought with them a taste for Irish moss, and it was soon recognized as a component of the natural flora off the coast of Massachusetts.

Isolation of the polysaccharide extract of Irish moss was reported[7] in 1837, and a patent for its purification appeared as early[8] as 1871. Names, such as carrageen and carrageenin, used in the early literature to describe the polysaccharide extract have now been dropped on recommendation of the Polysaccharide Nomenclature Committee, a subcommittee of the Committee on Carbohydrate Nomenclature of the American Chemical Society, and the name *carrageenan* adopted.

TABLE II

Sea Plants Recorded as Sources of Agar, Carrageenan, Gelan, Porphyran and Miscellaneous Galactans

Agars[a]	
Acanthopeltis japonica	*Gelidium pristoides*
Ahnfeltia plicata	*Gelidium pulchellum*
Campylaephora hypnaeoides	*Gelidium pulvinatum*
(Ceramium hypnaeoides)	*Gelidium pusillum*
Ceramium boydenii	*Gelidium sesquipedale*
Corallopsis salicornia	*Gelidium subcostatum*
Digenea simplex	*Gracilaria armata*
Endocladia muricata	*Gracilaria blodgettii*
Gelidiella acerosa	*Gracilaria caudata*
(Gelidium rigidum)	*Gracilaria cornea*
Gelidium amansii	*Gracilaria henriquesiana*
Gelidium arborescens	*Gracilaria lichenoides*
Gelidium arbuscula	*Gracilaria multipartita*
Gelidium attenuatum	*Gracilaria taenioides*
Gelidium cartilagineum	*Gracilaria verrucosa*
Gelidium corneum	*(G. confervoides)*
Gelidium crinale	*Phyllophora nervosa*
Gelidium japonicum	*Phyllophora rubens*
Gelidium latifolium	*Pterocladia lucida*
Gelidium nudifrons	*Pterocladia pinnata*
Gelidium pacificum	*(P. capillacea)*

Carrageenans	
Aeodes orbitosa	*Gigartina decipiens*
Agardhiella tenera	*Gigartina deptohyncos*
Chondrus canaliculatus	*Gigartina pistillata*
Chondrus crispus	*Gigartina radula*
Chondrus ocellatus	*Gigartina skottsbergii*
Eucheuma cottonii	*Gigartina stellata*
Eucheuma edule	*(G. mamillosa)*
Eucheuma isiforme	*Gloiopeltis coliformis*
Eucheuma muricatum	*Gloiopeltis furcata*
(E. spinosum, E. denticulatum)	*Gloiopeltis tenax*
Eucheuma uncinatum	*Gymnogongrus furcellatus*
Gigartina acicularis	*Gymnogongrus norvegicus*
Gigartina asperifolia	*Gymnogongrus patens*
Gigartina atropurpurea	*Iridaea laminarioides*
Gigartina canaliculata	*Iridaea capensis (Iridophycus capensis)*
Gigartina chauvinii	*Iridaea flaccida (Iridophycus flaccidum)*

[a] See also Chapter III.

TABLE II, *Continued*

Gelans[b]	Porphyrans[b]
Furcellaria fastigiata	*Bangia fuscopurpurea*
Hypnea cervicornis	*Laurencia pinnatifida*
Hypnea musciformis	*Porphyra capensis*
Hypnea spicifera	*Porphyra perforata*
Suhria vittata	*Porphyra umbilicalis*

Unclassified[b]
Ahnfeltia duruillaei
Corallina officinalis
Dilsea edulis
Dumontia incrassata
Polysephonia fastigiata

[b] See also Chapter IX.

Although trade in Irish moss flourished in the 19th century, it was predominantly on sales of bleached whole or pulverized weed, and it was not until well into the 1900's that commercial production of the extract was attempted. Several fledgling firms on the Eastern coast of the United States began producing the extract in the 1930's.

Success of the first firms was limited, and not until the early 1940's did the industry show growth when shortages and demands of the war years resulted in sharply increased production. With peacetime, newly created demands for carrageenan persisted, making possible continual growth of the new industry. Industrial growth has been augmented by the entry of several European firms into carrageenan production, and at present, the carrageenan industry is divided nearly equally between United States' and European–Asian interests.

II. Production

1. *Geographical Distribution*

The algal flora indigenous to a particular area is influenced by a number of factors, chief of which are temperature, tidal amplitude, and the position and flow of major ocean currents.[9] Those areas having favorable combination of factors and supporting the greatest number of algal types are the Japanese, south Australian, South African, northeast Pacific Coastal, and Mediterranean regions.

Chondrus crispus and *Gigartina stellata* continue to serve as chief sources of carrageenan. Large stands of weed are found in the North Atlantic coastal region from Norway to the coast of North Africa. Respective American and European processors rely chiefly on harvests of raw material indigenous to their particular areas, although there is a tendency to transport weed greater distances because of increased demands and competition for raw materials and to obtain certain desired types. Red seaweeds other than *Chondrus* and *Gigartina* are harvested in several additional areas including South America, Malaya, Japan, East Africa, Indonesia, and the Philippines. The latter three regions are noted for their stands of *Eucheuma* types.

2. *Growth*

Sea plants, by virtue of their being bathed in a relatively constant nutrient media that is subject to relatively little temperature variation, are influenced by fewer growth adversities than are terrestrial plants. Growth rate is determined largely by water temperature and the availability of nutrients and sunlight. Favorable growth conditions are usually at a maximum in spring and early summer months, at which time growth is most rapid. At this time, plants may attain growth increases of over 5% per day, the stands of weed doubling in size in one to two months.

To meet increasing demands for raw material, work is progressing on cultivation of seaweeds. This involves the starting of new plants, fertilization, or a combination of both. The value of such operations has yet to be proven, although such methods are now used in Japan for cultivation of *Laminaria, Porphyra, Undaria,* and *Gelidium* species.

3. *Harvesting of Seaweeds*

Few advances have been made in harvesting methods since the inception of the industry. Collection is still done using whatever methods and implements local harvesters have found most applicable to the task. Much of the raw material is harvested by workers in boats using special rakes to loosen and gather the weed.

Considerable amounts of weed are obtained from the shore following storms, which may loosen and deposite plants upon the beach. Attempts at more sophisticated gathering procedures, such as the use of power driven harvesting devices or use of scuba divers, have met with some success, although local legislation and conflict with fishing interests have hampered development of such mechanized methods.

The collected seaweed is dried before shipment to processing plants to improve storage qualities and reduce transport weight. Formerly drying was accomplished by simply spreading the weed in the open. Drying is now accomplished under controlled conditions in plants situated near points of collection. The latter drying operation avoids the bleaching action of sunlight as well as contamination by foreign material, resulting in a more uniform raw material. Other operations that remove various foreign materials may be conducted along with the drying operation, resulting in a considerable upgrading of the quality.[10]

The necessity for separating mixtures of seaweed prior to extraction depends upon the proportion of types present, the properties of their extracts, and the feasibility of such separation. For example, *Chondrus crispus* and *Gigartina stellata,* two morphologically similar species often found growing together, fortuitously yield extracts of similar properties and are processed without a prior separation. *Eucheuma spinosum* and *Eucheuma cottonii,* however, yield extracts of different properties and are usually separated prior to processing. This separation is made before drying by native workers skilled in recognizing the slight morphological differences of the two species.

4. Extraction and Recovery

Technology in the industry with respect to extraction and recovery techniques has advanced considerably in recent years. Coupled with this is a far better understanding of raw materials and types of carrageenans with the end result that, by careful blending and extraction procedures, products of nearly any desired property or combination of properties can be produced.

Methods used in extraction and recovery processes as well as in selection and blending of raw materials vary with producer and are for the most part still closely guarded trade secrets. Some information is available from the patent literature, however. Most operations, following a blending of seaweeds, if necessary, begin with a washing operation to remove foreign material, such as salts and stone particles. The weed is then extracted with water, usually under slightly alkaline conditions at temperatures approaching 100° for as long as is necessary for exhaustive extraction. This time ranges from one to several hours. The alkalinity has two functions. It first aids in a more complete extraction of polysaccharide from the plant material, and secondly, it catalyzes an elimination of 6-sulfate groups from appropriate monomer units with formation of 3,6-anhydro-D-galactose residues, which generally serves to increase the water gel strength and protein reactivity of the extract.[11]

The amount of extractant liquid used per given weight of dry seaweed must be sufficient to avoid excessive viscosity in the resulting extract. A weight of extractant equal to 30–40 times the weight of dry weed is usually employed and yields an extract containing about 1% carrageenan. The extract is filtered in the presence of a filter aid, such as cellulose or diatomaceous earth, and concentrated in evaporators as far as is practicable, which is usually not higher than 3% solids. The hydrocolloid is finally recovered in a drying or precipitation operation involving either the use of drum-type driers or alcohol precipitation and drying. Drum drying yields an amorphous product of flaky consistency containing those soluble ballast materials not removed by filtration. Alcohol precipitation offers the advantage of removing the majority of such components, although in practice a certain amount becomes occluded and adsorbed into the precipitate. Methanol and 2-propanol are the most widely used precipitants. A volume of alcohol of from 1.5 to 4 times that of the extract is required. The amount will vary with the water content of the alcohol used and the carrageenan extract. Extracts containing mainly *kappa*-carrageenan require less alcohol for precipitation than do those containing a high proportion of *lambda*-carrageenan. The precipitated polysaccharide is pressed to remove excess solvent, dried in drying ovens under reduced pressure, and ground to a specific particle size. The average particle size affects rates of hydration of the finished product and is established by requirements of the carrageenan user.

5. *Standardization*

Standardization procedures have been developed to correlate with particular usage and are not uniform throughout the industry. Product specifications agreed upon by producer and purchaser are useful only in a specific sense and are of little value in comparing or establishing uniform grades among producers.

Carrageenans fall roughly into four major categories: (*a*) viscosity builders for water systems, (*b*) gelling agents for water systems, (*c*) stabilizers for milk systems, and (*d*) gelling agents for milk systems. It is important that methods used in the standardization of carrageenans agree closely with the use intended.

As an example, a carrageenan to be employed as a stabilizer for chocolate milk is usually standardized with respect to the amount required per unit volume of milk to prevent settling of cocoa particles at a specific combination of temperature and time. Through such standardization procedures, shelf life conditions are approximated with the result that effective but economical levels of stabilizers may be used.

Carrageenans designed for both water and milk systems are standardized to some arbitrary value by blending procedures either before or after extraction and/or by addition of an inert diluent, such as sucrose or D-glucose (dextrose). Potassium chloride and potassium citrate are also used as standardizing agents in arriving at products of desired gel strength. Such standardized products may then be used with confidence in a prescribed manner to produce a desired effect.

6. *Market Volume and Price*

World production of carrageenan in 1971 was approximately 4500 metric tons, representing nearly a twofold increase in annual production over the period from 1965 to 1971. Of this total, it is estimated that 2300 metric tons were produced in the United States with the bulk of the remainder attributed to Danish and French firms. Small amounts of carrageenan are also produced in Spain, Japan, and England.

Prices for carrageenan have risen only moderately from the previously reported average figure of $1.50/lb.[12] Various grades now bring a price of from $1.50 to $2.50/lb ($3.30 to $5.60/kg) with an average price of near $1.75/lb ($3.90/kg).

Expansion of production facilities over the past decade has been adequate to meet increased demands and improved technology in the field and has resulted in significant improvements in production and introduction of new carrageenan types. There is some concern at present regarding the availability of raw materials, caused partly by increasing competition among producers and partly by natural shortages. There appears to be some danger of damage to existing seaweed beds through overharvesting practices,[13] leading to the view that conservation practices are necessary in future years to maintain adequate seaweed supply.

III. Applications

1. *Introduction*

Carrageenan is used in a wide variety of food as well as nonfood applications, with new uses arising regularly. Approximately 80% of the present production of carrageenan goes to food and food-related industries with a large portion of the remainder being used in pharmaceutical and cosmetics applications.

Carrageenan finds applicability for its attributes as gelling agent, viscosity builder, and stabilizing and emulsifying agent. Often these properties are realized at very low concentrations, making its use feasible despite a somewhat higher cost per unit weight than many competing gums. Many requirements include the need of an additive having a combination of properties for which a specific carrageenan can usually be prepared through a proper combination of raw materials selection and chemical treatment during extraction and recovery. In many applications, carrageenan is used to balance and improve the properties of other gums such as carboxymethylcellulose and locust bean gum.

Major uses for carrageenan have developed as a result of its ability to complex with other hydrocolloids and specifically with ampholytes such as proteins. Most such uses evolve around milk systems in which carrageenan may be used as a stabilizer to prevent separation and settling, as a viscosity builder, or as a gelling agent.

The choice of a particular industrial gum in filling a specific application is often a difficult one. Frequently, not one but several gums vie for a particular use. A careful consideration of cost, concentration required, and end result is then necessary in choosing one over several others. Because of the large number of types and blends of carrageenan available, it is advisable when considering use of this gum to consult with service laboratory personnel in the industry for help in choosing a correct type for a particular application. In the following section are listed some representative uses for carrageenan in both food and nonfood applications.

2. *Foods*

Dairy Products

Chocolate milk, pasteurization process.—*Kappa*-carrageenan used in a concentration of 0.02–0.03% per weight of milk[14, 15] holds cocoa particles in suspension and prevents cream separation. The carrageenan my be incorporated in a flavoring syrup[16] or powder. Chocolate beverages prepared from nonfat dry milk can be stabilized by carrageenan in combination with dioctyl sodium sulfosuccinate.[17]

Chocolate milk, cold process.—*Lambda*-carrageenan at a concentration of 0.05% per weight of milk produces thickening and stabilization.[18]

Ice cream and sherbet.—*Kappa*-carrageenan as a secondary stabilizer at a concentration of 0.01–0.05% in combination with a primary stabilizer such as a galactomannan or carboxymethylcellulose adds creaminess and prevents syneresis and crystal formation under freeze–thaw conditions.[19–23] Carrageenan also counters the undesirable tendency of the primary stabilizer to cause whey separation in the unfrozen ice cream mix.

Milk shakes and instant breakfast powders.—*Lambda*-carrageenan at a concentration of 0.03–0.05% produces thickening and stabilization and imparts a richer mouth feel.

Cooked milk puddings.—Light bodied custard or *blanc mange* desserts may be prepared by incorporation of 0.1–0.3% *kappa*-carrageenan, sometimes in combination with furcellaran, locust bean gum, or phosphate salts in a suitable milk base formulation.[24–26] By incorporation of starch and smaller amounts of carrageenan (0.05–0.1%), heavier-type puddings are produced. *Iota*-carrageenan has been used for the preparation of freezable milk puddings.[27]

Instant milk puddings.—*Lambda*-carrageenan at a concentration of 0.2-1.0% produces "instant" gelling in cold milk systems.[28] Gels formed show good flavor release and freedom from syneresis.

Whipped dessert mix.—*Kappa*-carrageenan in a concentration of 0.15–0.5% gives whipped mixes of improved stability.[29–31]

Frozen whipped topping.—Incorporation of mixed *kappa/lambda*-carrageenan at a concentration of 0.03–0.05% results in a product having improved body and showing reduced syneresis under freeze–thaw conditions.[32, 33]

Cottage and creamed cheese products.—*Kappa*-carrageenan at a concentration of 0.01–0.05%, usually in combination with locust bean gum, induces curd formation, imparts shape retention, and prevents syneresis.[34]

Concentrated milk.—*Kappa*-carrageenan at a concentration of 0.01% prevents fat separation and imparts smoothness.[35] Carrageenan is useful also in the preparation of calcium enriched milk for infant feeding. Small amounts of carrageenan added to calcium enriched milk prevent the insoluble calcium from settling out on storage.[36]

Liquid coffee whitners.—Addition of mixed *lambda/kappa*-carrageenan at a concentration of 0.3% in combination with carboxymethylcellulose, sodium alginate, or refined galactomannan improves colloidal solubility of protein resulting in a better shelf life.[37]

BAKERY PRODUCTS

Bread.—Sodium *lambda*- or sodium *kappa*-carrageenan at a concentration of 0.01–0.035% in combination with hydroxylated lecithin or hydroxylated phosphatide gives improved dough characteristics and allows incorporation of higher levels of nonfat milk solids.[38–41]

Sweet dough products.—Incorporation of 0.1% carrageenan in flour results in improved texture and appearance of cakes. Fat pickup is reduced in fried items.

WATER GELS

Dessert gels.—*Kappa*- and *iota*-carrageenan at a concentration of 0.1–0.5% in combination with a refined galactomannan and potassium salts, dissolved by heating, forms a clear, resilient gel that is stable at room temperature.[42–46] Such gels are ideally suited for use as dessert gels and in the preparation of gelled fruit baby foods. Gels prepared with *iota*-carrageenan are stable under freeze–thaw conditions.[27]

Cold-water-soluble dessert gels.—Sodium *iota*-carrageenan at a concentration of 0.6–1.2% in combination with an acid and calcium or potassium salt dissolves in cold water to yield a gel. Powdered formulations that can be added to cold water to give instant dessert gels are possible.[47]

MEAT PRODUCTS

Canned meat products.—*Kappa*- and *iota*-carrageenans at a concentration of 0.2–0.5% with potassium salts and sometimes in combination with locust bean gum are effective as a gel binder in ground meat packs, such as pet foods, or as a gel coating for whole meat packs.[48, 49]

Spun protein products.—Incorporation of 1% carrageenan in protein solutions prior to extrusion and spinning gives fibers of improved quality.[50] Carrageenan can also be used as a component of a binder liquor used in the binding of such fibers.[51]

MISCELLANEOUS

Low-sugar jams and jellies.—*Kappa*- and *iota*-carrageenan at a concentration of 0.5–1.2% in combination with potassium salts and sodium citrate give gels that are not sugar or calcium dependent. Locust bean gum or low methoxyl pectin may be added to modify gel properties.

Icing base.—Incorporation of 3–4% *kappa*-carrageenan in icing base mixes gives an icing that sets more quickly and shows a greater resistance to drying out.[52]

Macaroni.—Addition of 0.1–0.2% *lambda*-carrageenan allows incorporation of up to 25% nonfat milk solids for added nutritional value and enhanced flavor.[53, 54]

Molded, edible product.—Incorporation of 2.5% carrageenan in preparations for use in the making of novelty edible, molded products stabilizes the premold liquid and increases the strength of the product.[55, 56]

Beer clarification.—Carrageenan added to beer as an auxiliary fining agent accelerates and improves clarification.[57, 58]

Other uses.—Carrageenan has been suggested for a number of additional uses, including incorporation in various fountain products, dressings and relishes, as a coating for frozen products, and as a batter mix ingredient. Most such uses represent novel applications for the stabilizer and do not constitute important industrial outlets.

3. Nonfoods

Pharmaceutical Products

Toothpastes and powders.—Combinations of *lambda-*, *kappa-*, and *iota*-carrageenan used at concentrations of 0.8–1.2% impart short texture and good rinseability characteristics to pastes. Shape retention is also enhanced. Addition of carrageenan to powders provides body and better foam characteristics.[59]

Mineral oil and insoluble drug preparations.—Use of *iota*-carrageenan at concentrations of from 0.1–0.5% gives stable emulsions or suspensions.[60–62]

Other Uses

Industrial suspensions.—*Iota*-carrageenan at concentrations of 0.25–0.8% in polishing and cutting abrasives, ceramic glaze compositions, water base paints, etc., gives stable emulsions or suspensions, allowing more uniform application and ease of handling.

Water-insoluble herbicides.—Incorporation of *iota*-carrageenan at concentrations of 0.25–0.8% gives a stable emulsion that can be applied more uniformly to the leaf surface.

Air treating gel.—A mixed *iota/kappa*-carrageenan at a concentration of 1.5% in combination with locust bean gum, carboxymethylcellulose, and potassium chloride is used in making an air treating gel.[63] Volatile air treatment components incorporated in the gel are released uniformly from the gel surface.

IV. Structure

Carrageenan is a complex mixture of several polysaccharides, the relative amounts and compositions of which vary considerably with seaweed source. Fractional precipitation of a dilute aqueous solution of carrageenan with potassium ion results in separation of two major components, an insoluble fraction, termed *kappa*-carrageenan, and a soluble fraction, termed *lambda*-carrageenan.[64] This classical fractionation method does not always give sharply defined fractions, however, and better separation can usually be achieved by an extraction or leaching procedure in which the *lambda* component is leached from the dried, powdered extract with a solution of potassium chloride.[65, 66] The relative proportion of the two fractions is not constant and varies considerably with the seaweed species and geographical origin. The effect of seasonal variation on the *kappa/lambda* ratio is of less significance than originally thought,[67] although there is evidence that a more weakly gelling *kappa*-subfraction exists in the spring than in the autumn.[66] The *kappa*- and *lambda*-carrageenan fractions are not homogeneous as was earlier believed, but may be further separated.[68] A minor fraction termed *mu*-carrageenan exists as part of the *lambda*-fraction,[69, 70] whereas *kappa*-carrageenan may be subfractionated into a series of polysaccharides of different chemical composition.[68, 71]

1. Kappa-*Carrageenan*

The *kappa* fraction of carrageenan is made up of D-galactose, 3,6-anhydro-D-galactose, and ester sulfate groups in the approximate ratio[72] of 6:5:7. Analysis by fragmentation has shown the backbone structure to be composed of a nearly perfect alternating sequence of O-4 substituted 3,6-anhydro-α-D-galactopyranosyl and O-3 substituted β-D-galactopyranosyl units.[73, 74] Nearly all the O-3 substituted β-D-galactopyranosyl units are sulfated at O-4,[69, 74] whereas the O-4 substituted 3,6-anhydro-α-D-galactopyranosyl units may be sulfated at O-2 or replaced by α-D-galactopyranosyl 6-sulfate or α-D-galactopyranosyl 2,6-disulfate units[69, 75–77] to give a "masked repeating" structure (Fig. 1).

Structural variations of *kappa*-carrageenans from different species of seaweed appear to be caused largely by differences in the relative amounts of (1→4)-linked units that occur as unsulfated and 2-sulfated 3,6-anhydro-D-galactopyranosyl units and 6-sulfated and 2,6-disulfated D-galactopyranosyl units. *Kappa*

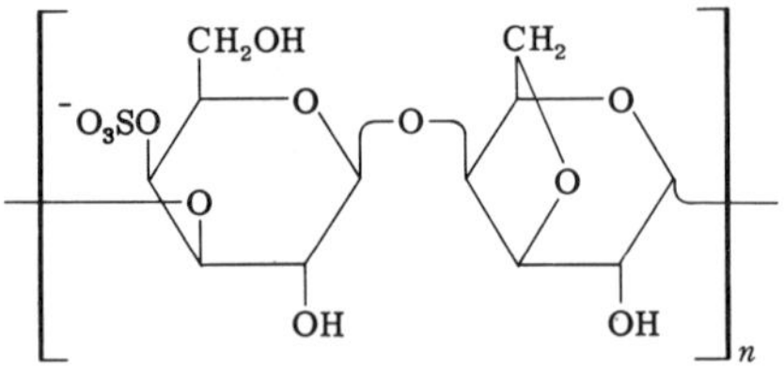

FIG. 1.—*Kappa*-carrageenan, repeating unit. Repeating structure is not perfect as shown, but consists of a masked repeating structure in which some 3,6-anhydro-α-D-galactopyranosyl units are replaced by 3,6-anhydro-α-D-galactopyranosyl 2-sulfate, α-D-galactopyranosyl 6-sulfate, or α-D-galactopyranosyl 2,6-disulfate units.

carrageenans of *Chondrus* species have a lower content of both types of 2-sulfated units than those from *Gigartina* species.[78]

2. Lambda-*Carrageenan*

The fraction not precipitated in the presence of potassium ion, termed *lambda*-carrageenan, is a mixture of highly sulfated galactans in admixture with small amounts of other contaminating polysaccharides, including floridean starch and xylan. The latter polymers may be removed by fractional precipitation with alcohol to yield a series of pure galactans.[79] Small, variable amounts of 3,6-anhydro-D-galactose are also present in the *lambda*-fraction and are considered to arise largely from *mu*-carrageenan.

A similar "masked repeating" structure based on an alternating sequence of β-D-(1→4) and α-D-(1→3)-linked galactopyranosyl units has been shown for *lambda*-carrageenan, but with several evident differences. α-D-Galactopyranosyl units substituted at O-4 are present mainly as the 2,6-disulfate in contrast to the 3,6-anhydro form that predominates in *kappa*-carrageenan, whereas the O-3 substituted units are mainly 2-sulfated.[80–83] On the basis of such differences, it has been suggested that the term *lambda*-carrageenan be limited to molecules lacking both 4-sulfate and 3,6-anhydro-D-galactopyranosyl units.[82] The repeating structure of *lambda*-carrageenan is shown in Figure 2.

3. Mu-*Carrageenan*

Treatment of crude *lambda*-carrageenan with alkaline borohydride, followed by addition of potassium chloride, results in the precipitation of a third component, termed *mu*-carrageenan.[69, 70] This polysaccharide comprises approximately 10% of the *lambda*-fraction and contains all the D-galactose 4-sulfate and 3,6-anhydro-D-galactose present in crude *lambda*-carrageenan. *Mu*-carrageenan has a repeating structure (Fig. 3) closely analogous to that of *kappa*-carrageenan and can be converted to a polymer having the properties of *kappa*-carrageenan by alkali treatment in which 6-sulfate groups are eliminated via an intramolecular S_N2 displacement reaction to yield 3,6-anhydro-D-galactose residues.

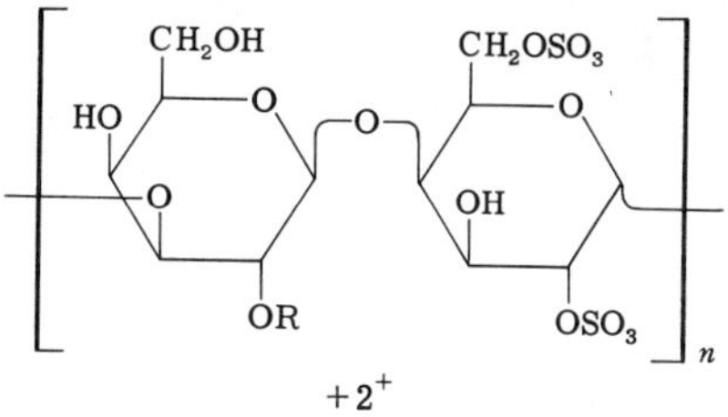

FIG. 2.—*Lambda*-carrageenan, repeating unit. R = H (~30%) and SO_3^- (~70%). Some α-D-galactopyranosyl 2,6-disulfate units are replaced by α-D-galactopyranosyl 6-sulfate and D-galactopyranosyl 2-sulfate units to give a masked repeating structure.

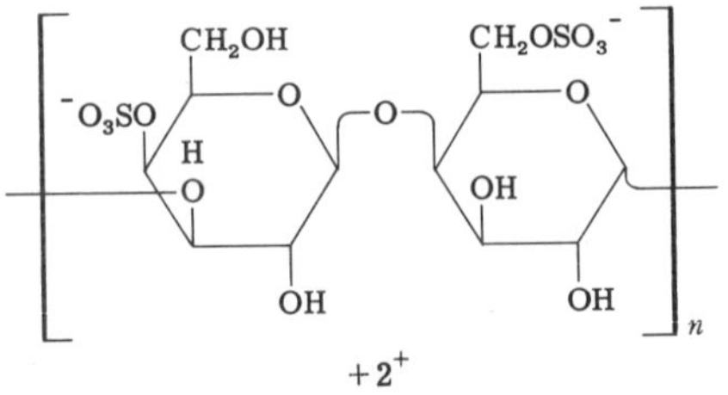

$$+2^{+}$$

FIG. 3.—*Mu*-carrageenan, repeating unit. Repeating structure is not perfect as shown, but consists of a masked repeating structure in which some α-D-galactopyranosyl 6-sulfate units are replaced by D-galactopyranosyl 2-sulfate, α-D-galactopyranosyl 2,6-disulfate, 3,6-anhydro-α-D-galactopyranosyl or 3,6-anhydro-α-D-galactopyranosyl 2-sulfate units.

The close structural similarity of *mu*-carrageenan to that of the *kappa*-form has led to the suggestion that the *mu*-form may be a biological precursor of *kappa*-carrageenan in which conversion is achieved through enzyme-catalyzed elimination of 6-sulfate groups.[75] It follows that *kappa*- and *lambda*-, and perhaps also *iota*-carrageenan, would be regarded as divergent end products of biosynthesis.

4. Iota- *and* Nu-*Carrageenan*

Carrageenan fractions have been isolated from *Aghardhiella tenera* and *Eucheuma spinosum* that are made up of β-D-galactoypranosyl 4-sulfate and 3,6-anhydro-α-D-galactopyranosyl 2-sulfate[84] units with the latter units replaced by about 10% α-D-galactopyranosyl 2-sulfate units.

Named *iota*-carrageenan (Fig. 4), this fraction differs from *kappa*-carrageenan with respect to the much larger proportion of 3,6-anhydro-α-D-galactopyranosyl 2-sulfate units. The presence of the 2-sulfate group results in a marked difference in physical properties between *kappa*- and *iota*-carrageenan attributable mainly to the greater hydrophilicity of the *iota*-molecule.

This structural difference is also seen through comparison of the infrared spectra of *kappa*- and *iota*-carrageenan. Spectra of *iota*-carrageenan show an

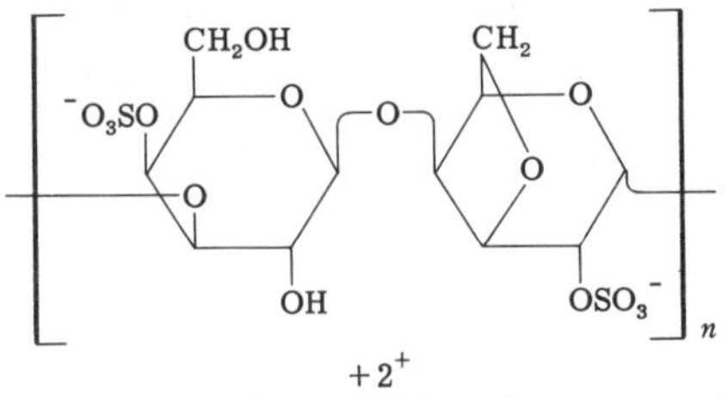

$$+2^{+}$$

FIG. 4.—*Iota*-carrageenan, repeating unit. Repeating structure is not perfect as shown, but consists of a masked repeating structure in which some 3,6-anhydro-α-D-galactopyranosyl units are replaced by α-D-galactopyranosyl 6-sulfate and α-D-galactopyranosyl 2,6-disulfate units.

absorption at 805–810 cm^{-1} attributable to the 2-sulfate group. Absorption in this band may thus be used to estimate the proportion of 3,6-anhydro-D-galactopyranosyl 2-sulfate units in a particular carrageenan and serves as a useful aid in comparing the structures of various carrageenan extracts.

A biological precursor of *iota*-carrageenan, similar to that for *kappa*-carrageegan, is suggested by work on a carrageenan extract from *Eucheuma uncinatum*.[66] A fraction of this extract, soluble in 3M potassium chloride solution and somewhat analogous to *lambda*-carrageenan, can be further fractionated after alkali treatment into two components on the basis of solubility in 0.3*M* potassium chloride solution. The insoluble component has gel properties and infrared spectra similar to that of alkali-treated *iota*-carrageenan. It is suggested that this fraction, termed *nu*-carrageenan, differs from *mu*-carrangeenan in that the O-4 substituted α-D-galactopyranosyl units are sulfated at O-2 as well as at O-6 (Fig. 5).

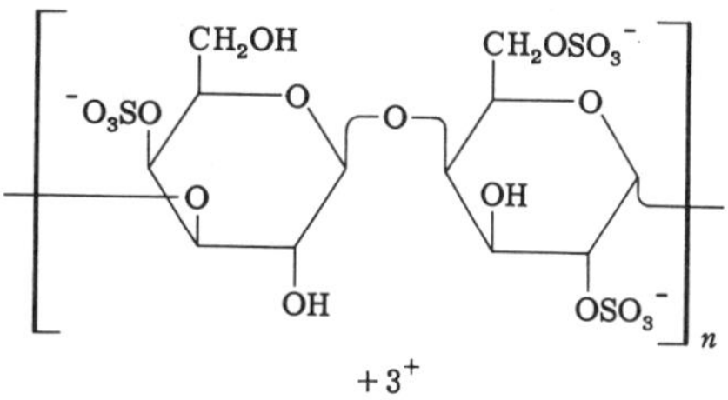

FIG. 5.—*Nu*-carrageenan, repeating unit. Structure may be of masked repeating type on which α-D-galactopyranosyl 2,6-disulfate units are replaced by α-D-galactopyranosyl 6-sulfate, α-D-galactopyranosyl 2-sulfate, 3,6-anhydro-α-D-galactopyranosyl and 3,6-anhydro-α-D-galactopyranosyl 2-sulfate units.

5. *Conformation and Molecular Weight*

Carrageenans of the *kappa/iota* and *lambda* types, although both composed of an alternating α-D-(1→3) β-D-(1→4) system of monomer units, show important conformational differences largely as a result of the presence of the 3,6-anhydro bridge in the former types. *Kappa-* and *iota*-carrageenans, in which the glycosidic oxygen atom is always equatorial to both sugar rings, exist in orientated fibers as right-handed, threefold helices that have been shown to reversibly form double helices in the solid state[85] and in aqueous solution.[84, 86] This property, long noted in the case of polypeptides and DNA, involves intramolecular cohesion between sugar residues resulting in a stabilization effect. x-Ray studies agree with this interpretation and show repeat distances in a complete turn of each single chain of 24.6A for *kappa*-carrageenan and 26.0A for *iota*-carrageenan.[85]

Lambda-carrageenan, on the other hand, resembles a ribbon that zigzags from side to side, the 1e,4a linkage imparting a ribbon quality and the 1a,3e linkage introducing a zigzag character into the molecule.[87] Such differences in chain conformation may be important in influencing specific properties involved in biological functions, such as hydration and interactions with cations.

Molecular weight values of $2.6–3.2 \times 10^5$ and $3.3–7.9 \times 10^5$ have been reported for *kappa*- and *lambda*-carrageenans based on ultracentrifugation and viscosimetric methods.[65, 88–90] Molecular weights are dependent upon the extraction conditions used. The highest molecular weight samples are obtained by short extraction periods at 100°.

6. *Biosynthesis*

The biosynthetic route to carrageenan in red seaweeds is still largely a matter of speculation. A route suggested,[75] analogous to that reported for porphyran biosynthesis,[91] is believed to involve a transferase mediated addition of galactopyranosyl units from uridine diphosphate D-galactose or guanosine diphosphate D-galactose to a growing galactan chain. Sulfate ester groups would be introduced by transfer of sulfate from adenosine 3′-phosphate 5′-sulphatophosphate, a commonly observed donor of sulfate groups[92] with the aid of appropriate enzymes into specific positions of the molecule. An enzyme system similar to that observed in *Porphyra*[93] may then catalyze conversion of D-galactopyranosyl 6-sulfate units into 3,6-anhydro-D-galactopyranosyl units at the polysaccharide level.

V. Properties

1. *Solubility*

Water is the principal solvent for carrageenan, although anhydrous hydrazine also is reported as being a good solvent. Carrageenan is sparingly soluble in formamide and methyl sulfoxide and swells, but does not dissolve, in *N,N*-dimethylformamide.

The solubility of carrageenan in water is influenced by several factors, including temperature, the presence and type of counter ion associated with the polymer, the presence of other water-soluble organic compounds and salts, and the type of carrageenan. Most important in controlling water solubility is the hydrophilicity of the molecule, that is sulfate half-ester groups, and the galactopyranosyl unit, as opposed to the presence of the more hydrophobic 3,6-anhydrogalactopyranosyl unit.[76] Thus, *lambda*-carrageenan, being void of 3,6-anhydrogalactopyranosyl units and having a high percentage of sulfate ester groups, is cold-water soluble in all salt forms. *Kappa*-carrageenan, having less hydrophilic sulfate ester groups and containing a larger proportion of hydrophobic 3,6-anhydrogalactopyranosyl units, is cold-water soluble only in the sodium salt form. In the presence of hydrophobic cations, such as potassium ion, the delicate balance of solubility versus insolubility is tipped and gel formation ensues. Thus, a spectrum of solubility exists, depending upon the balance of hydrophilicity versus hydrophobicity inherent in the primary structure of various carrageenans. That the primary structure alone is responsible for the solubility characteristics of various carrageenans is, however, an oversimplification. Other factors, such as the location of groups on the polymer chain and the molecular conformation, are also deciding factors in determining solubility characteristics.

Less polar, water-miscible organic compounds, such as the lower mono- and polyhydric alcohols, acetone, sugars, and polysaccharides, retard hydration of carrageenan to various degrees. Salts also exert a retarding effect on hydration, with potassium chloride and sodium chloride limiting hydration at levels of approximately 0.25*M* and 0.75*M*, respectively.

Hydration of carrageenan is more rapid at lower pH levels, with the rate of hydration being considerably slower at pH levels of 6 and above.[94] Acid-catalyzed hydrolysis of glycosidic linkages becomes significant at lower pH levels, especially at pH 3.0 and below. This depolymerization process is much faster at higher temperatures, and for this reason there are practical limitations beyond which the increased rate of hydration offered by lower pH and higher temperature is offset by a significant drop in solution viscosity.

2. *Viscosity*

Dilute aqueous dispersions of carrageenan are viscous, the viscosity being dependent on concentration, temperature, the presence of other solute molecules, the type of carrageenan, and the molecular weight. As the concentration of a carrageenan solution is increased, the viscosity increases nearly logarithmically (Fig. 6). This is a general property of linear polymers containing charged groups and is a result of increasingly greater interchain interaction. This relationship can be expressed empirically by the relation $mc = \log \eta - 0.89$, where η is the intrinsic viscosity, c is the concentration, and m is a constant as determined for a particular polymer type.

Viscosity measurements, to be valid, must be made at concentrations below and temperatures above those at which aggregation or gel formation occurs. Thus, viscosities are usually reported for 0.5%, 1%, or 1.5% solutions of carrageenan at temperatures of 25°, 40°, or 75° in the absence of ions, such as potassium, that promote gel formation. The most widely used instruments for viscosity measurements are rotational viscometers of the Brookfield type.

The viscosity of carrageenan solutions decreases progressively with increases in temperature. This change is generally reversible provided that heating is done at or near the stability optimum at pH 9 and that conditions are such that no thermal degradation occurs.

Salts of monovalent cations have little effect on the viscosity of carrageenan solutions. Divalent cations, however, tend to reduce the viscosity significantly at higher concentration, but may produce a viscosity increase at low levels. This may

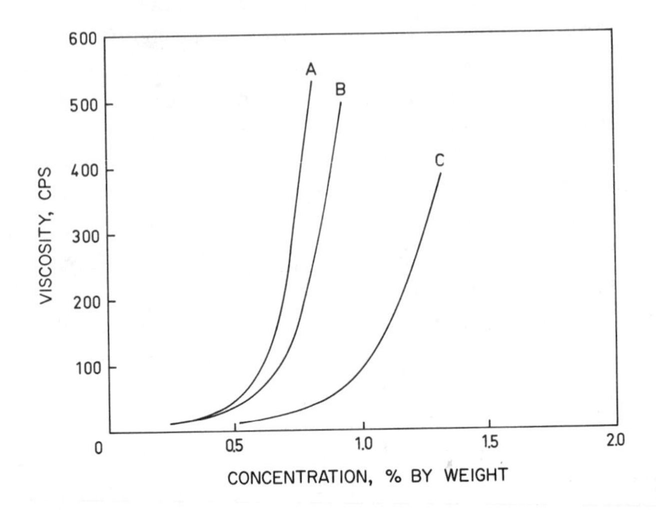

FIG. 6.—Concentration viscosity relations for extract of *Eucheuma spinosum* (A), *Eucheuma spinosum* and *Chondrus crispus,* 2:1 (B), and *Chondrus crispus* (C). Viscosities were measured at 40° with a 48-cm disk rotating at 20 rpm in an infinite container.

be caused by a contraction of the molecule as a result of the ionic binding of pairs of suitably positioned sulfate groups by the divalent ion.

Solutions of carrageenan display non-Newtonian, thixotropic flow characteristics, typlified mainly by a decrease in viscosity with increasing shear or agitation[95] and a return to normal viscosity with ceasation of agitation. In addition, carrageenan solutions display plastic flow characteristics attributable to reversible gel–fluid–gel transition. Such a system possesses a nonzero yield value and thus shows good long term suspending properties. The degree of thixotropy can be altered through selection and blending of carrageenan types as well as through addition of certain other noncarrageenan polymers. An increase in thixotropy generally parallels a tendency toward gel formation, and it is within this region that the rather intangible properties, termed "mouthfeel" and texture, become important.

A high viscosity at relatively low concentration is not always a desirable attribute of a carrageenan solution. On the contrary, certain gelling types should display a low solution viscosity up to the point of gelation to facilitate handling. Fortunately, gel formation and viscosity prior to gelation are not related, and products containing a high content of *kappa*-carrageenan having high gelling capacity often show acceptably low solution viscosity.

3. *Gelation*

Kappa-carrageenan at concentrations as low as 0.3% forms thermally reversible water gels in the presence of ammonium, potassium, rubidium, or cesium salts. Members of this group of monovalent cations all have approximately the same ionic diameter and influence gel formation in a similar manner. Sodium and lithium salts do not behave in this manner, perhaps as a result of the greater size of these hydrated cations.

Reversible gel formation is a general phenomena shown by many natural and synthetic polymers and is important to the functioning of most biological systems. The manner in which certain polymer molecules in solution interact to form gels is not yet entirely clear, although some underlying principles have emerged.[96, 97]

In general, a transition from the sol to gel state is envisioned as involving an association of polymer chain segments at junction regions to give a three-dimensional framework. Entrained in this network are solvent molecules, usually comprising in excess of 99% of the gel.

Various types of junction regions are possible as determined by the nature of the polymer system. There is evidence[85, 86] to suggest that gels of *kappa*- and *iota*-carrageenan, and possibly gels of the agar type, form via double helix formation and alignment of helices into junction zones.

This mechanism requires a rather high degree of structural regularity, and the fine structure of the polymer is therefore important in establishing gelling tendencies. Any structural modification of *kappa*- or *iota*-carrageenan that tends to increase the degree of regularity, such as conversion of 6-sulfated units into 3,6-anhydro units, increases gel strength. A kink in the helix occurs at those positions where a 6-sulfated galactose unit replaces the 3,6-anhydro unit. For this reason, gel strength is more sensitive to the presence of 6-sulfate groups than to 2-sulfate or 4-sulfate groups, as the former exert a much greater effect on the conformational regularity of the polysaccharide.

The consistency of carrageenan gels may be varied from hard and brittle to soft and elastic types. This texture depends upon several variables, including the nature of the carrageenan, its concentration, the type of counter ion present, and the presence of other solutes and nongelling hydrocolloids. Gels of *kappa*-carrageenan in the presence of potassium ion tend to be more brittle than those of *iota*-carrageenan, the increased elasticity of the *iota*-type gel being due to increased amounts of 2-sulfated residues in the polymer.[97] Various carrageenan yielding seaweeds differ widely in the ratio of types of carrageenans[67, 98] and in the finer structural detail of these types. Thus, by selection and blending, carrageenan preparations yielding a variety of gel types are possible. Carrageenans having water gelling properties that are generally enhanced by alkali treatment are derived mainly from *Chondrus, Furcellaria,* and *Hypnea* species. Carrageenan from *Eucheuma cottonii,* although not water gelling as extracted, can be converted through alkali treatment to a form having very high water gel strength.

The strength of a carrageenan gel is most often expressed in terms of the force required to depress a plunger of given diameter, set on the gel surface, a distance of 4 mm. This force is measured by means of an instrument known as the Bloom gelometer. The method is rapid and convenient and affords a reasonably accurate and reproducible indication of gel strength. Several types of Bloom gelometers, including sophisticated types that register force electronically on a digital readout scale, are now commercially available.

A second method of measurement, and one that serves as a better indicator of elasticity, is a break force measurement in which a plunger of given diameter is driven at a constant rate of speed into the gel surface causing a rupture of the gel. The force required to the point of rupture is registered on a scale. The method, known as the Kobe method, has been in use in the agar industry for a number of years. Break force measurements give a somewhat more satisfactory measurement of gel strength because a gel that is strong but very elastic would show a low Bloom reading but would register a considerably higher break force reading.

Gel strength may also be expressed by the extent to which an unsupported gel sags in a given time period. This method, used originally in the measurement pectin gels,[99–102] has been modified and used in this laboratory for measurement of certain carrageenan gel types. A carrageenan gel prepared under set conditions in a standard Ridgelimeter glass is allowed to stand in a cooling bath at 5° for 2.5 hr, after which the gel is inverted from the glass and allowed to stand 2 min. The amount of sag in this time is measured by means of a Ridgelimeter, a device consisting essentially of a vertically mounted micrometer head. This value, expressed in millimeters, serves as a useful measure of cohesiveness of a carrageenan gel.

Lambda-carrageenan, although nongelling alone, will in admixture with *kappa*-carrageenan reduce brittleness and syneresis of the latter gel. Locust bean gum exerts a synergistic effect on carrageenan gels, resulting in an increase in gel strength and cohesiveness. Guar gum and carboxymethylcellulose reduce the strength of *kappa*-carrageenan gels, but increase cohesiveness. Corn and wheat starches produce a moderate increase in gel strength, whereas starches of potato and tapioca are without effect.

Potassium chloride is most used of the various nontoxic potassium salts in formulation of gelling type carrageenans. Addition of potassium chloride results in a nearly linear rise in gel strength (Fig. 7). Certain limits on the amount of potassium chloride that may be used, however, are imposed by taste considerations. The upper limit in delicately flavored foods is from 0.1–0.2% potassium

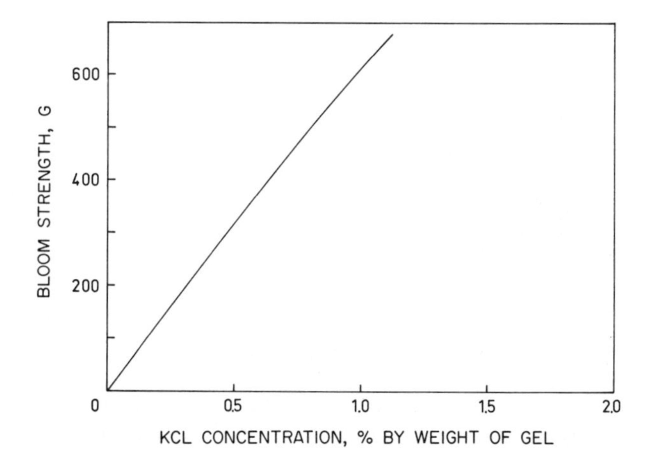

FIG. 7.—Effect of potassium chloride concentration on strength of carrageenan gel. Gel consists of 1.26% potassium carrageenan of the *kappa* type.

chloride whereas in certain salty foods, up to 0.5% potassium chloride may replace sodium chloride without detection. Other potassium salts, such as potassium phosphate, may be used in place of potassium chloride in applications requiring higher levels of potassium ion.

Salts other than those of potassium have various effects on gel strength. Sodium salts impart a short, brittle consistency to potassium-carrageenan gels and, at higher concentration, reduce the gel strength. Calcium salts strengthen potassium-carrageenan gels and increase brittleness. Ammonium salts are less effective than potassium salts in promoting gel strength and, in replacing potassium salts, reduce gel strength.

Carrageenan gels are thermally reversible under conditions that do not cause depolymerization and exhibit a sharp setting temperature as determined largely by the polymer composition, the type and amount of salts in solution, and the presence of other solutes. The effect of various concentrations of potassium chloride (Fig. 8) and sucrose (Fig. 9) serve as illustrations. In both cases, the melting and gelling temperatures parallel one another, differing by some 10°–15°, and increase with increasing salt or sugar concentration.

4. *Protein Reactivity*

Carrageenan retains a strong negative charge over the normal pH range and reacts with other polyelectrolytes, particularly those with a positive charge. Proteins, being ampholytes, will precipitate in the presence of carrageenan when the pH of the solution is below the isoelectric point of the protein. This pH depend-

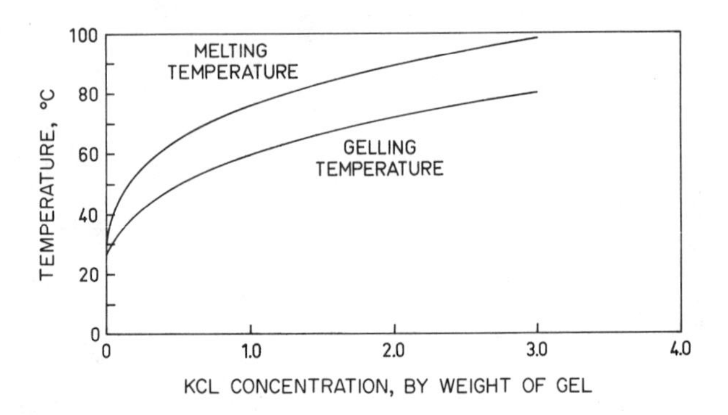

FIG. 8.—Effect of potassium chloride concentration on melting and gelling temperatures of a carrageenan gel. Gel consists of 1.26% potassium carrageenan of the *kappa* type.

ency has been used as a means of separating similar proteins of different isoelectric point.[103]

The reaction of carrageenan with proteins may be direct or may be mediated by the presence of polyvalent cations, such as calcium, which act as a bridge or alter the charge characteristics of the participating species by removal of charge from carboxyl groups.[104] In addition to ionic bonding, hydrogen bonding and van der Waal's forces may also be important in the formation of carrageenan–protein complexes.

The protein reactivity of carrageenan might be used to advantage in several industrial processes. Yields of protein from soybean extraction are enhanced by addition of small amounts of *kappa*-carrageenan.[105, 106] Carrageenan will precipitate the transglucosidase component in a mixture of transglucosidase and

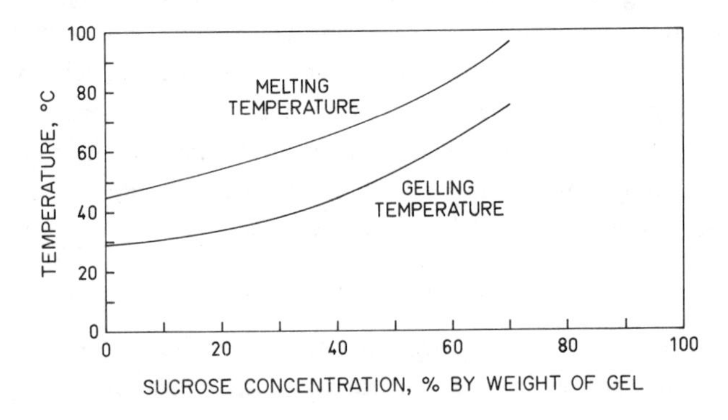

FIG. 9.—Effect of sucrose concentration on melting and gelling temperatures of a carrageenan gel. Gel consists of 0.65% *kappa*-carrageenan containing 0.13% potassium chloride.

glucoamylase, thereby resulting in a purer preparation of the latter enzyme for use in the production of glucose from starch.[107]

A special type of protein reactivity is exemplified by the agglomeration of milk proteins by carrageenan. Both *lambda-* and *kappa*-carrageenan possess milk reactivity, but to different degrees.[108] *Lambda*-carrageenan produces thickening and stabilization in cold milk, whereas the *kappa* type, to be effective, must be dissolved by heating to about 50°. Carrageenan made up predominantly of the *lambda*-type molecule is used in products where a thickening and stabilization of cold milk is required, whereas *kappa* types find use in pasteurized products such as chocolate milk.

The reaction of carrageenan with milk proteins[109, 110] is rather complex and as yet is not thoroughly understood. Carrageenan apparently acts much in the same manner as *kappa*-casein in that it reacts with α_s- and β-caseins to form complexes that are stable against precipitation by calcium ions and show improved heat stability.[111, 112] *Kappa*-carrageenan is roughly twice as effective as the *lambda*-type in this respect. Carrageenan also reacts with *kappa*-casein, but in the absence of calcium ion.[113]

An ionic bonding of contrasting unspecific nature is shown by the reaction of carrageenan, as well as other anionic polysaccharides, with β-lactoglobulin of milk.[114] This bonding is maximal at pH 4.0 and is influenced by the ionic strength as well as the polysaccharide: protein ratio.

The structural requirements for stabilization of calcium sensitive caseins appear to be the presence of sulfate groups, their location, and the size of the carrageenan molecule. The presence of sulfate groups on C-2 and C-4 but not on C-6 positions appear to enhance casein reactivity. The presence of C-6 sulfate groups may act in an antagonistic manner. 3,6-Anhydro-D-galactosyl units, although important to the water gelling characteristics of carrageenan, do not appear to be important for stabilization of caseins.[112, 115]

The binding of carrageenan to proteins presents a special problem when its extraction must be effected for analytical purposes. This binding is sufficiently strong at neutral pH to prevent complete separation of the carrageenan and protein. In milk systems, this problem can be circumvented by digesting the milk proteins with a proteolytic enzyme preparation, after which isolation of the carrageenan is possible.[116–118] In meat samples containing carrageenan, the stabilizer may be removed by extraction with hot aqueous salt solution at pH 10–11.[49]

5. *Chemical and Enzymic Modification*

Carrageenan undergoes depolymerization in acid solution with hydrolysis of 3,6-anhydro-D-galactosidic linkages.[67] The rate of depolymerization is dependent upon pH and temperature.[94, 119, 120] Carrageenan solutions are most stable at pH 9 and, at this pH, undergo strong heating for short periods without significant loss of viscosity. Lowering the pH reduces the stability to heating (Fig. 10), and in general, carrageenan should not be processed below pH 3.5. Sulfate ester groups are stable to acid treatment and so are not removed under the slightly acidic conditions sometimes encountered in processing.

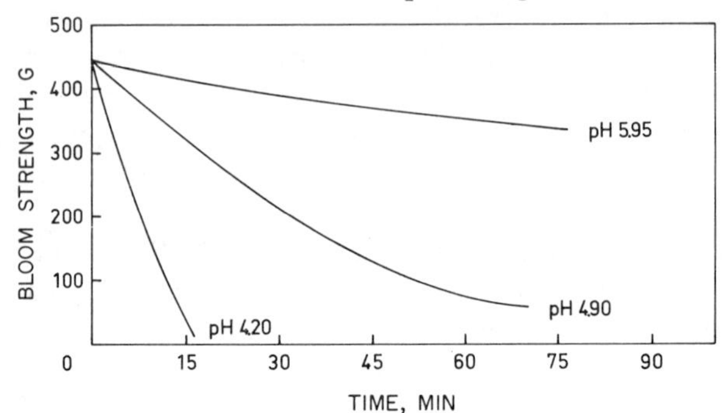

FIG. 10.—Effect of acid on the gel strength of carrageenan at 58°. Gel consists of 1.2% *kappa*-carrageenan containing 8.0% sucrose and 0.2% potassium citrate and sufficient citric acid to give pH values shown.

Alkali treatment results in an elimination of 6-sulfate groups from (1→4)-linked residues with formation of 3,6-anhydro residues.[81] There is evidence of a slight loss of other sulfates with alkali treatment, although this has not been established with certainty. Alkali treatment generally converts a fraction originally insoluble at intermediate potassium chloride concentrations, which may correspond to the *mu*-component, to one having solubility properties more like those of the *kappa*-fraction.[68] Carrageenans from different seaweed sources vary markedly in amounts of this intermediate component. Those from *Chondrus crispus* and *Gigartina tenella* contain only small amounts of this intermediate fraction, whereas extracts of *Gigartina stellata* and *Gigartina skottsbergii* contain substantial amounts.

Enzymes have been found that either remove sulfate ester groups from the molecule or hydrolyze glycosidic linkages. A sulfatase preparation from the marine gastropod *Charonia lampas* was found to split sulfate ester groups from carrageenan, but at a rate slower than that observed for a number of other synthetic and natural substrates.[121]

An enzyme has been isolated from cultures of *Pseudomonas carrageenovora* that hydrolyzes β-D-(1→4) linkages in *kappa*-carrageenan.[73] In addition to this *kappa*-carrageenase, a *lambda*-carrageenase has also been reported.[74]

6. *Physiological Action*

Carrageenan has been used as a natural food additive for over 600 years and is today recognized as a harmless food additive without nutritional value.[122]

Numerous tests on experimental animals, particularly the rat, have failed to show any harmful effects from the feeding of high dose levels of native carrageenan other than an expected decrease in weight gain.[123, 124] Rats fed carrageenan in the diet at levels of 2–20% excrete the polysaccharide quantitatively in the feces.[125]

Carrageenan has been reported to be of benefit in the treatment of ulcerative colitis and peptic ulceration.[126–131] Several pharmaceutical preparations based on acid-degraded carrageenan have been used for some time in the treatment of such conditions. Its effectiveness may be caused by an antipeptic action of the sulfated polysaccharide,[132, 133] although at levels normally used in foods there is no interference with protein digestion.[134]

It has recently been demonstrated that ingestion of large amounts of degraded carrageenan through drinking water by guinea pigs[135] and rabbits[136] results in a condition in the colon described as ulcerative colitis. This condition, which appears to be unlike ulcerative colitis in man, is not observed in rats, mice,[137] pigs,[138] hamsters, ferrets, and squirrel monkeys[139, 140] under similar dosage conditions.

Subcutaneous injection of carrageenan results in a reversible stimulation of connective tissue growth, with the *lambda*-fraction being the more active in this respect.[141] This response may be of an immunological nature, as both *kappa*- and *lambda*-carrageenan have been shown to elicit the formation of precipitating antibodies.[142] It is interesting to note that antibodies against *lambda*- and *kappa*-types show a high degree of specificity for their homologous antigen.

Carrageenan is acutely toxic to experimental animals if administered intravenously.[143, 144] Rabbits (2.5–3.0 kg body weight) are killed within 48 hr by a single 50 mg intravenous injection. A dose of 20 mg/kg is fatal to guinea pigs. The primary effect, which leads to the death of the animal, appears to be the production of renal lesions, the severity of which increases with the dose level.

Carrageenan shows blood anticoagulant activity, with the more highly sulfated types showing the greatest activity.[145–147] Undegraded carrageenan forms insoluble complexes with fibrinogen, whereas degraded carrageenan, like heparin, yields a soluble complex with fibrinogen and is significantly less toxic than the undegraded polymer.

VI. References

(1) S. Hirase, C. Araki, and T. Ito, *Bull. Chem. Soc. Jap.*, **29,** 985 (1956).
(2) S. Hirase, C. Araki, and T. Ito, *Bull. Chem. Soc. Jap.*, **31,** 428 (1958).
(3) V. C. Barry and J. E. McCormick, *J. Chem. Soc.*, 2777 (1957).
(4) T. Dillon and J. McKenna, *Nature,* **165,** 318 (1950).
(5) S. Peat and J. R. Turvey, *Fortschr. Chem. Org. Naturst.*, **23,** 1, (1965).

(6) E. Percival and R. H. McDowell, "Chemistry and Enzymology of Marine Algal Polysaccharides," Academic Press, New York, 1967, p. 145.

(7) J. Pereira, "Elements of Materia Medica and Therapeutics," Sea and Blanchard, Philadelphia, 1843, Vol. 1, p. 35.

(8) G. Bourgade, U.S. Patent 112,535 (1871).

(9) E. Y. Dawson, "Marine Botany," Holt, Rinehart and Winston, New York, 1966, p. 284.

(10) H. D. Johnston, *Proc. Int. Seaweed Symp., 6th, Santiago de Compostela, Spain, 1968,* 691 (1969).

(11) N. F. Stanley, U.S. Patent 3,094,517 (1963); *Chem. Abstr.,* **59,** 6603 (1963).

(12) L. Stoloff, *in* "Industrial Gums," R. L. Whistler, ed., Academic Press, New York, 1st Ed., 1959, p. 90.

(13) S. Lund and J. Christensen, *Proc. Int. Seaweed Symp., 6th, Santiago de Compostela, Spain, 1968,* 699 (1969).

(14) R. C. Rose and W. H. Cook, *Can. J. Res.,* **27,** 324 (1949).

(15) G. Andersen, *Milchwissenschaft,* **17,** 75 (1962).

(16) L. H. Berndt and R. A. Klein, U.S. Patent 2,830,903 (1958); *Chem. Abstr.,* **52,** 14024 (1958).

(17) E. B. Hotelling, U.S. Patent 3,486,905 (1969).

(18) M. P. Hess and A. E. Sichrs, U.S. Patent 2,462,398 (1949); *Chem. Abstr.,* **43,** 3542 (1949).

(19) S. J. Werbin, U.S. Patent 2,502,397 (1950); *Chem. Abstr.,* **44,** 5496 (1950).

(20) *J. R. Moss, Ice Cream Trade J.,* **51,** 22 (1955).

(21) J. L. Boyle, *Food Technol. Aust.,* **11,** 543 (1959).

(22) W. S. Steinitz, U.S. Patent 2,823,129 (1958); *Chem. Abstr.,* **52,** 7669 (1958).

(23) T. A. Nickerson, *J. Dairy Sci.,* **45,** 354 (1962).

(24) L. Stoloff, *Food Eng.,* **24,** 114 (1952).

(25) E. Polya and J. Green, U.S. Patent 3,409,443 (1968); *Chem. Abstr.,* **70,** 27783q (1969).

(26) A. L. Moirano, U.S. Patent 3,443,968 (1968); *Chem. Abstr.,* **71,** 37614v (1969).

(27) M. Glicksman, U.S. Patent 3,250,621 (1966); *Chem. Abstr.,* **65,** 2916 (1966).

(28) A. L. Moirano, U.S. Patent 3,499,768 (1970); *Chem. Abstr.,* **72,** 110017e (1970).

(29) L. H. Berndt, U.S. Patent 3,010,830 (1962); *Chem. Abstr.,* **64,** 7762 (1962).

(30) E. B. Rodgers, U.S. Patent 3,350,209 (1967); *Chem. Abstr.,* **68,** 11834e (1968).

(31) Lenderink en Co. N. V., Neth. Patent 69.07,938 (1969); *Chem. Abstr.,* **72,** 99390y (1970).

(32) D. Downs, B. Wilkinson, and T. I. Hedric, *Mich. State Univ., Agr. Expt. Sta., Quart. Bull.,* **42,** 871 (1960).

(33) G. J. Lorant, Brit. Patent 1,173,827 (1969); *Chem. Abstr.,* **72,** 53905s (1970).

(34) V. H. Nielsen, *Amer. Milk Rev.,* **26,** 54 (1964).

(35) D. F. Wilcox, U.S. Patent 2,845,350 (1958); *Chem. Abstr.,* **52,** 18957 (1958).

(36) C. D. Bauer, R. M. Marks, and E. A. Bernardoni, U.S. Patent 2,871,123 (1959); *Chem. Abstr.,* **53,** 9514 (1959).

(37) W. H. Knightly, *Food Technol.,* **23,** 1 (1969).

(38) E. F. Glabe, P. W. Anderson, and E. C. Jertson, *Cereal Sci. Today,* **9,** 300 (1964).

(39) J. J. R. Andt, U.S. Patent 3,271,164 (1966); *Chem. Abstr.,* **65,** 15993 (1966).

(40) E. C. Jertson and E. F. Glabe, U.S. Patent 3,234,027 (1966); *Chem. Abstr.*, **65,** 13296 (1966).

(41) E. F. Glabe and E. C. Jertson, U.S. Patent 3,411,919 (1968); *Chem. Abstr.*, **70,** 56486a (1969).

(42) A. Frieden and S. J. Werbin, U.S. Patent 2,427,594 (1947); *Chem. Abstr.*, **41,** 7580 (1947).

(43) G. L. Baker, U.S. Patent 2,466,146 (1949); *Chem. Abstr.*, **43,** 5132 (1949).

(44) G. L. Baker, J. W. Carrow, and C. W. Woodmansee, *Food Ind.*, **21,** 617 (1949).

(45) Standard Brands Inc., Brit. Patent 841,973 (1960); *Chem. Abstr.*, **55,** 844 (1961).

(46) S. E. Foster and A. L. Moirano, U.S. Patent 3,342,612 (1967); *Chem. Abstr.*, **68,** 2140y (1968).

(47) M. Glicksman, E. Farkas, and R. E. Klose, U.S. Patent 3,502,483 (1970); *Chem. Abstr.*, **72,** 2347d (1970).

(48) F. Wirth, *Fleischwirtschaft*, **6,** 848 (1970).

(49) G. A. Towle, *Fleischwirtschaft*, **6,** 1669 (1970).

(50) C. Giddey, U.S. Patent 2,947,644 (1960); *Chem. Abstr.*, **55,** 2128 (1961).

(51) C. Giddey, U.S. Patent 2,952,542 (1960); *Chem. Abstr.*, **55,** 1965 (1961).

(52) American Molasses Co., Brit. Patents 930,461, 930,462 (1963).

(53) *Fed. Regist.*, **31,** 5618 (1966); *Chem. Abstr.*, **64,** 18304 (1966).

(54) E. F. Glabe, P. W. Anderson, and P. F. Goldman, *Cereal Sci. Today*, **12,** 510 (1967).

(55) J. W. Ryan, E. O. Stastney, and E. Burns, U.S. Patent 3,493,382 (1970); *Chem. Abstr.*, **72,** 99383y (1970).

(56) J. W. Ryan, A. L. Stone, and E. T. Martin, U.S. Patent 3,493,383 (1970); *Chem. Abstr.*, **72,** 89058g (1970).

(57) J. V. MacDonough, U.S. Patent 2,658,829 (1953); *Chem. Abstr.*, **48,** 2982 (1954).

(58) W. Griffiths, Brit. Patent 912,492 (1962).

(59) E. Thomsen, *Amer. Perfum. Cosmet.*, **78,** 45 (1963).

(60) B. W. Fitzgerald and D. M. Skauen, *J. Amer. Pharm. Ass.*, **44,** 358 (1955).

(61) L. A. Wasson, *J. Amer. Pharm. Ass.*, **44,** 284 (1955).

(62) T. W. Schwarz, *Amer. Perfum. Cosmet.*, **77,** 85 (1962).

(63) M. Lanzet, U.S. Patent 2,927,055 (1960); *Chem. Abstr.*, **54,** 14504 (1960).

(64) D. B. Smith and W. H. Cook, *Arch. Biochem. Biophys.*, **45,** 232 (1953).

(65) D. A. I. Goring and E. G. Young, *Can. J. Chem.*, **33,** 480 (1955).

(66) D. J. Stancioff and N. F. Stanley, *Proc. Int. Seaweed Symp., 6th, Santiago de Compostela, Spain, 1968,* 597 (1969).

(67) W. A. P. Black, R. R. Blakemore, J. A. Colquhoun, and E. T. Dewar, *J. Sci. Food Agr.*, **16,** 573 (1965).

(68) O. Smidsrød, B. Larsen, A. J. Pernas, and A. Haug, *Acta Chem. Scand.*, **21,** 2585 (1967).

(69) T. C. S. Dolan, Ph.D. Thesis, Edinburgh, Scotland, 1965.

(70) N. S. Anderson and D. A. Rees, *Proc. Int. Seaweed Symp., 5th, Halifax, Nova Scotia, 1965,* 243 (1966).

(71) A. S. Cerezo, *J. Chem. Soc., C,* 992 (1967).

(72) A. N. O'Neill, *J. Amer. Chem. Soc.*, **77,** 6324 (1955).

(73) J. Weigl and W. Yaphe, *Can. J. Microbiol.*, **12,** 939 (1966).

(74) J. Weigl, J. R. Turvey, and W. Yaphe, *Proc. Int. Seaweed Symp., 5th, Halifax, Nova Scotia, 1965,* 329 (1966).

(75) D. A. Rees, *Ann. Rep. Prog. Chem., 1965,* **62,** 469 (1966).

(76) T. J. Painter, *Proc. Int. Seaweed Symp., 5th, Halifax, Nova Scotia, 1965,* 305 (1966).

(77) N. S. Anderson, T. C. S. Dolan, and D. A. Rees, *J. Chem. Soc., C,* 596 (1968).

(78) N. S. Anderson, T. C. S. Dolan, A. Penman, D. A. Rees, G. P. Mueller, D. J. Stancioff, and N. F. Stanley, *J. Chem. Soc., C,* 602 (1968).

(79) D. B. Smith, W. H. Cook, and J. L. Neal, *Arch. Biochem. Biophys.,* **53,** 192 (1954).

(80) K. Morgan and A. N. O'Neill, *Can. J. Chem.,* **37,** 1201 (1959).

(81) D. A. Rees, *J. Chem. Soc.,* 1821 (1963).

(82) T. C. S. Dolan and D. A. Rees, *J. Chem. Soc.,* 3534 (1965).

(83) C. J. Lawson and D. A. Rees, *J. Chem. Soc., C,* 1301 (1968).

(84) A. A. McKinnon, D. A. Rees, and F. B. Williamson, *Chem. Commun.,* 701 (1969).

(85) N. S. Anderson, J. W. Campbell, M. M. Harding, D. A. Rees, and J. W. B. Samuel, *J. Mol. Biol.,* **45,** 85 (1969).

(86) D. A. Rees, I. W. Steele, and F. B. Williamson, *J. Polymer Sci., Part C,* **28,** 261 (1969).

(87) D. A. Rees, *J. Chem. Soc., B,* 217 (1969).

(88) W. H. Cook, R. C. Rose, and J. R. Colvin, *Biochim. Biophys. Acta,* **8,** 595 (1952).

(89) C. R. Masson and G. W. Caines, *Can. J. Chem.,* **32,** 51 (1954).

(90) D. B. Smith, A. N. O'Neill, and A. S. Perlin, *Can. J. Chem.,* **33,** 1352 (1955).

(91) J. C. Su and W. Z. Hassid, *Biochemistry,* **1,** 468 (1962).

(92) P. W. Robbins, *in* "The Enzymes," P. D. Boyer, H. Lardy, and K. Myrbäck, ed., Academic Press, New York, Vol. 6, 1962, p. 363.

(93) D. A. Rees, *Biochem. J.,* **81,** 347 (1961).

(94) L. Stoloff, *Food Technol.,* **4,** 138 (1950).

(95) C. R. Masson and D. A. I. Goring, *Can. J. Chem.,* **33,** 895 (1955).

(96) D. A. Rees, "The Shapes of Molecules," Oliver and Boyd, Edinburgh, 1967, p. 119.

(97) D. A. Rees, *Advan. Carbohyd. Chem.,* **24,** 267 (1969).

(98) W. A. P. Black, *in* "The Chemistry and Rheology of Water Soluble Gums and Colloids, S.C.I. Monogram No. 24," Society of Chemical Industry, London, Gordon and Breach, Science Publishers, London, 1966, p. 33.

(99) H. C. Lockwood and R. S. Hayes, *J. Soc. Chem. Ind.,* **50,** 145T (1931).

(100) R. F. Cox and R. H. Higby, *Food Ind.,* **16,** 441 (1944).

(101) G. H. Joseph and W. E. Baier, *Food Technol.,* **3,** 18 (1949).

(102) P. E. Christensen, *Food Res.,* **19,** 163 (1954).

(103) K. J. Turner and B. J. Magnusson, *Nature,* **194,** 451 (1962).

(104) A. MacMullan and F. R. Eirich, *J. Colloid Sci.,* **18,** 526 (1963).

(105) A. C. Eldridge, A. M. Nash, and A. K. Smith, U.S. Patent 3,069,327 (1962); *Chem. Abstr.,* **70,** 6129 (1963).

(106) A. K. Smith, A. M. Nash, A. C. Eldridge, and W. J. Wolf, *J. Agr. Food Chem.,* **10,** 302 (1962).

(107) J. Corman, U.S. Patent 3,345,268 (1967); *Chem. Abstr.,* **67,** 118389v (1967).

(108) D. B. Smith, *Can. J. Technol.*, **31,** 209 (1953).

(109) G. A. Muck and J. Tobias, *J. Dairy Sci.*, **45,** 481 (1962).

(110) A. L. Gordon, J. J. Jones, and M. N. Overholt, *Proc. Int. Seaweed Symp., 5th, Halifax, Nova Scotia, 1965,* 377 (1966).

(111) P. M. T. Hansen, *J. Dairy Sci.*, **51,** 192 (1968).

(112) C. F. Lin and P. M. T. Hansen, "Stabilization of Calcium Caseinate by Carrageenan," paper presented at the 63rd annual meeting of the American Dairy Science Association, 1968.

(113) J. Grindrφd and T. A. Nickerson, *J. Dairy Sci.*, **51,** 834 (1968).

(114) J. Hidalgo and P. M. T. Hansen, *J. Agr. Food Chem.*, **17,** 1089 (1969).

(115) C. F. Lin and and P. M. T. Hansen, *Macromolecules,* **3,** 269 (1970).

(116) H. D. Graham, *J. Dairy Sci.*, **49,** 1102 (1966).

(117) H. D. Graham, *J. Food Sci.*, **33,** 390 (1968).

(118) H. D. Graham, "Colorimetric Determination of Sulfated Polysaccharides," paper presented at the 65th annual meeting of the American Dairy Science Association, 1970.

(119) D. A. Goring, *J. Colloid Sci.*, **9,** 141 (1954).

(120) C. R. Masson, *Can. J. Chem.*, **33,** 597 (1955).

(121) N. Takahashi and F. Egami, *Biochem. J.*, **80,** 384 (1961).

(122) "Toxicological Evaluation of Some Food Colours, Emulsifiers, Stabilizers, Anticaking Agents and Certain Other Substances," FAO Nutritional Meetings Report Series No. 46A, 1969, p. 93.

(123) H. W. Nilson and J. A. Wagner, *Food Res.*, **24,** 235 (1959).

(124) P. L. Carey, Ph.D. Thesis, Purdue University, Lafayette, Indiana (1958).

(125) W. W. Hawkins and W. Yaphe, *Can. J. Biochem.*, **43,** 479 (1965).

(126) B. P. Babkin and S. A. Komarov, *Can. Med. Ass. J.*, **27,** 463 (1932).

(127) S. Levey and S. Sheinfield, *Gastroenterology,* **27,** 625 (1954).

(128) W. Anderson and J. Watt, *J. Pharm. Pharmacol.*, **11,** 173T (1959).

(129) J. C. Houck, J. Bhayana, and T. Lee, *Gastroenterology,* **39,** 196 (1960).

(130) W. Anderson and J. Watt, *J. Physiol.*, **147,** 52 (1969).

(131) W. van W. Robertson and B. J. Schwartz, *J. Biol. Chem.*, **201,** 689 (1953).

(132) W. Anderson and J. Watt, *J. Pharm. Pharmacol.*, **11,** 318 (1959).

(133) D. W. Piper and B. Fenton, *Gastroenterology,* **40,** 638 (1961).

(134) O. W. Vaughan, L. J. Filer, Jr., and H. Churella, *J. Agr. Food Chem.*, **10,** 517 (1962).

(135) R. Marcus and J. Watt, *Lancet,* No. 7618, 489 (Aug. 30, 1969).

(136) J. Watt and R. Marcus, *J. Pharm. Pharmacol.*, **22,** 130 (1970).

(137) M. Maillet, S. Bonfils, and R. E. Lister, *Lancet,* No. 7669, 414 (Aug. 22, 1970).

(138) J. Tournut and C. Labie, National Veterinary School, Toulouse, France, personal communication.

(139) M. Sharratt, paper presented at the meeting of the Pathological Society of Great Britain, London, Jan. 9, 1971.

(140) The British Industrial Biological Research Association, Carlston, England, unpublished report, 1971.

(141) E. L. McCandless and J. Lehoczky-Mona, *Growth,* **28,** 143 (1964).

(142) K. H. Johnston and E. L. McCandless, *J. Immunol.*, **101,** 556 (1968).

(143) J. C. Houck, R. D. Morris, and E. J. Lazaro, *Proc. Soc. Exp. Biol. Med.*, **96,** 528 (1957).

(144) J. C. Morard, A. Fray, A. Abadic, and L. Robert, *Nature,* **202,** 401 (1964).

(145) W. W. Hawkins and V. G. Leonard, *J. Lab. Clin. Med.*, **60,** 641 (1962).

(146) W. W. Hawkins and V. G. Leonard, *Can. J. Biochem. Physiol.*, **41,** 1235 (1963).

(147) W. Anderson and J. G. C. Duncan, *J. Pharm. Pharmacol.*, **17,** 647 (1965).

CHAPTER VI

FUCOIDAN

LANDIS W. DONER AND ROY L. WHISTLER

Department of Biochemistry, Purdue University, Lafayette, Indiana

I. INTRODUCTION

Fucoidan is a water-soluble polysaccharide that occurs in all brown algae (Phaeophyceae) in varying amounts.[1] It was first isolated by Kylin[2, 3] from *Laminaria digitata* and *Fucus vesiculosus* and was named fucoidin when L-fucose (6-deoxy-L-galactose) was identified in a hydrolyzate. The original name was changed to fucoidan.[4]

Fucoidan is present in the intercellular tissues[5] and most strikingly in the droplets that exude from the surface of fronds of *Laminaria digitata,*[6] *Ascophyllum nodosum,*[7] and *Macrocystis pyrifera;*[8] algin has also been found in the exudate of the latter plant.

Fucoidan is very hygroscopic and may serve to retain moisture and prevent dessication of the plant on exposure to air at low tide.[9] This idea is supported by observations that littoral Phaeophyceae, for example *Ascophyllum nodosum,* or floating algae, such as *Macrocystis pyrifera,* contain more fucoidan than sublittoral algae;[10] those growing highest on the shore, such as *Pelvetia canaliculata,*[11] have the highest fucoidan content. Seasonal fluctuations in the fucoidan content of brown algae have been reported,[12, 13] and *Laminaria, Fucaceae,* and *Chordana* species, which range in fucoidan content from 5 to 20%, have maximum amounts in the autumn and early winter months. In British Columbia, *Laminaria* has been found to have its highest fucoidan content in March.[12]

II. Isolation and Purification

Methods thus far employed to obtain fucoidan have either involved water or acid extraction of the dried kelp or alcohol precipitation of the exudate from the fresh weed. The latter method was used to obtain fucoidan from *Laminaria*. The product had mucilagenous properties[6] at very low concentrations but contained small amounts of algin and proteins. Methods suitable for the removal of these impurities destroyed the colloidal properties. Even drying the kelp markedly impairs the colloidal properties of fucoidan and results in a chemically stable polymer that has practically no mucilagenous properties, even at high concentrations in water.[14] Fucoidan samples obtained for structural analyses have been quite pure, but in all cases, its native mucilagenous properties were lost. Procedures given in the following paragraphs must be modified if mucilagenous properties are desired.

A procedure for the isolation of fucoidan with the aim of freeing it from contaminating protein and algin has been developed.[15, 16] The dried milled seaweed is extracted with hot water, and the dissolved fucoidan is separated from soluble protein and algin by precipitation with lead acetate and barium hydroxide. The resulting lead hydroxide–fucoidan complex is treated with dilute sulfuric acid and dialyzed to produce a purified fucoidan.

Fucoidan has been extracted[17] from dried milled kelp with a hydrochloric acid solution at pH 2–2.5. Neutralization and concentration of the extract produced crude fucoidan as a brown residue. This residue, on dissolution in water, was treated with ethanol to 30% concentration to precipitate impurities. The filtrate was brought to 60% ethanol content, and partially purified fucoidan was obtained in a 50–60% yield. Treatment of an aqueous solution of this partially purified material with formaldehyde yielded a glassy solid from which fucoidan could be extracted with water, because most of the impurities remained insoluble. Fucoidan can be isolated from aqueous solution by precipitation with 70% ethanol and can be dried with ethanol and ether.

Extensive purification of fucoidan from *Macrocystis pyrifera* was achieved[18] by slowly pouring the exudate from the fresh kelp through a sieve into 2-propanol. The precipitate obtained was dried, ground, and extracted in a Soxhlet extractor with 90% methanol. This precipitation with isopropanol and drying with methanol was repeated several times to obtain a fucoidan of suitable purity for structural work.

III. Structure

The fucoidan content of brown algae is obtained[19] by multiplying the fucose content by 1.75. Boiling the seaweed with hydrochloric acid liberates L-fucose from fucoidan and converts it to 5-methyl-2-furfuraldehyde, which is distilled into a solution of phloroglucinol (1,3,5-benzenetriol). The resulting precipitate is weighed and the L-fucose content calculated.

A second method involves hydrolysis of fucoidan with 2.5% sulfuric acid at 100°. The L-fucose liberated is determined either gravimetrically as the methylphenylhydrazone[20] or by periodate oxidation, whereby the terminal —CHOH—CH_3 group is converted into acetaldehyde. The acetaldehyde is estimated either by a modification of the procedure of Nicollet and Shinn[21] or colorimetrically by the method of Fromageot and Heitz.[22, 23]

An examination[24] of the sugar content of fucoidan preparations from various species of brown algae show that L-fucose is always present in the highest amount (31–72%), but galactose (5–31%), mannose (4–20%), xylose (3–29%), and arabinose (0–25%) are usually also present. There is likely not just one type of fucoidan, but a series of related fucoidans.

Fucoidan preparations have been reported to contain from 30–38.3% sulfate.[15, 16] Calcium is generally found associated with fucoidan,[16, 25] but in one instance,[6] sodium was reported as the principal inorganic ion.

The finding that only 10% of the sulfate groups on fucoidan from *Fucus vesiculosus* are alkali labile[26] indicates that only this proportion of the sulfate groups are linked to C-2 or C-3. Those linked to C-4 are alkali stable, as a vicinal trans relationship of hydroxyl groups is necessary for alkali liability.[27] Thus, it can be concluded that the glycosidic linkages of the L-fucopyranosyl units cannot be through the hydroxyl on C-4, as this would necessitate sulfate groups on C-2 or C-3. Methylation and hydrolysis of fucoidan gave 3-*O*-methyl-L-fucose as 60% of the total sugar in the hydrolyzate. Thus, fucoidan is predominately (1→2) linked and carries a sulfate ester on C-4. Twenty percent of the sugar isolated after methylation and hydrolysis was free L-fucose, which may arise from a disulfated residue or from a branch point. 2,3-Di-*O*-methyl-L-fucose also accounted for 20% and may have arisen from nonreducing end-units.

Direct proof for the existence of the sulfate ester at C-4 has recently been obtained.[28, 29] Fucoidan from *Pelvetia wrightii* was subjected to mild acid hydrolysis, and from the hydrolyzate, two sugar sulfates were isolated. These were characterized as L-fucose 4-sulfate and 2-*O*-α-L-fucopyranosyl-L-fucose with an ester sulfate at C-4 of the nonreducing moiety.

Strongly negative rotations[16, 25, 30] ranging from −75° to −140° provide evidence that the linkages in fucoidan are alpha.

From the proportions of the various derivatives produced by methylation and hydrolysis it can be calculated that there is an average of one branch in every five

L-fucose units. The following formula[31] conforms with the requirements established thus far.

```
                                           4SO3⁻
                                             |
                                   →2) α-L-Fucp
                                             1
                                             ↓
                                             3
→2)-α-L-Fucp(1→2)-α-L-Fucp(1→2)-α-L-Fucp(1→2)-α-L-Fucp(1→4)-α-L-Fucp(1—
         |               |               |                           |
       4SO3⁻           4SO3⁻           4SO3⁻                   2 or 3SO3⁻
```

This structure is supported by the findings of O'Neill[32] and Cote[33] in their examination of the products formed by the mild acetolysis of fucoidan from *Fucus vesiculosus.* O'Neill deacetylated the mixture, reduced the sugars to alcohols, and then reacetylated. He isolated and characterized 2-O-α-L-fucopyranosyl-L-fucitol. Cote separated not only the (1→2)-linked disaccharide but also the (1→4)- and a small amount of (1→3)-linked disaccharide. The latter could derive from the branch points in the proposed structure.

Schweiger[18] found that an extensively purified fucoidan from *Macrocystis pyrifera* had a ratio of fucose to galactose of about 18 to 1 and that this ratio remained constant through a variety of purification treatments. He found the xylose to vary from 0.5 to 2.5% depending on the degree and method of purification. He concluded that the major part of the fucoidan from *M. pyrifera* consists of a heteropolymer with L-fucose and D-galactose residues in a ratio of 18:1. The presence of a small amount of xylogalactofucan is also possible.

Additional evidence for the existence of more than one molecular species of fucoidan has been obtained. That from *Fucus vesiculosus* has been shown to consist of two species by free boundary electrophoresis,[31] chromatography,[34] and fractional precipitation.[35] By these methods, fucoidan from *Ascophyllum nodosum* has been shown to consist of at least three polysaccharides.[36, 37] This cannot be explained simply on the basis of different degrees of sulfation, as different proportions of sugars are found in the three fractions.[19, 36]

Molecular weight data on fucoidan preparations from *Fucus vesiculosus* are inconsistent. Osmometric determinations[32] gave a value of 133,000 ± 20,000 but ultracentrifuge measurements[35] gave a value of 78,000. Bernardi and Springer[35] suggest that fucoidan can best be depicted as a highly charged random coil in salt solution, with branching contributing to rigidity. That the L-fucopyranose units exist in the $^4C_1$(L) conformation is indicated by infrared measurements showing that axial sulfate substitutents (C-4) predominate.

IV. Properties

When freshly prepared by alcohol precipitation, fucoidan is a tan to off-white powder. It is soluble in water and insoluble in organic solvents. Fucoidan has been prepared[6] that is readily soluble in water and gives aqueous solutions which are viscous and exhibit dilatent properties with pronounced mucilagenous or stringy flow.[4] Within a few days or weeks, the viscosity drops and some precipitation occurs. Treatments such as lowering the pH to below 4 or raising it above 9, dialysis, repeated precipitation from water, and excessive dehydration impair or destroy the viscosity imparting ability of fucoidan samples.

Solutions of fucoidan in 65–70% sucrose syrup show pronounced mucilagenous properties, even in low concentrations, and it is possible to pull out threads several feet long from these solutions. Such solutions are stable and retain their properties for years. This marked stability of fucoidan solutions containing polyhydroxy compounds indicates that the loss of viscosity in water is caused by the formation of fucoidan aggregates through hydrogen bonding, which have reduced solubility in water. Excessive drying may carry the aggregation further, leading to an insoluble product.

Treatment of fucoidan with dilute acids or bases quickly results in irreversible changes in its properties. Salts of the products of acid treatment are soluble in water at high concentrations, but the solutions have practically no mucilagenous properties and possess low viscosities.[14] This indicates that considerable degradation or aggregation of the polymer has occurred.

Some marine bacteria are capable of utilizing fucoidan as their sole carbon source.[38] *Pseudomonas atlanticum* can utilize 31.5% of the fucoidan in the medium, and *P. carrageenova* 29.9%. Cell-free extracts of these organisms converted fucoidan to L-fucose in yields of 24.2% and 15.9% respectively. An α-L-fucosidase with similar properties has been found in snail midgut.[39] Enzyme extracts from other marine organisms were completely devoid of activity toward fucoidan.[40] More recently, α-L-fucosidases have been isolated from various species of abalone. The enzyme from *Haliotus rufescens* and *H. corrugata* hydrolyzed fucoidan[41] without desulfation to form oligosaccharides from two to ten units and L-fucose. Ester groups are cleaved only later in the hydrolysis period. The enzyme from the abalone *H. gigantea* completely hydrolyzes fucoidan[42] and does not form oligosaccharides. This enzyme may be an exofucoidanase, whereas that from the other two species of abalone mentioned above are endofucoidanases.

V. Uses

Methyl L-fucoside has been produced[8] in high yields by methanolysis of fucoidan. The sulfate groups are quantitatively removed during the methanolysis.

A fucoidan fraction from *Fucus vesiculosus* having 40–70% of the blood anticoagulant activity of heparin has been prepared.[43] Other fucoidan fractions are less active as anticoagulants but are potent lipemia clearers.[44]

If there should be success in producing fucoidan in a stable form, it could be of economic significance because of its excellent physical properties, including its very high viscosity.

VI. References

(1) W. A. P. Black, *Chem. Prod.,* **16,** 139 (1953).

(2) H. Kylin, *Hoppe-Seyler's Z. Physiol. Chem.,* **83,** 171 (1913).

(3) H. Kylin, *Hoppe-Seyler's Z. Physiol. Chem.,* **94,** 357 (1915).

(4) W. H. McNeely, *in* "Industrial Gums," R. L. Whistler, ed., Academic Press, New York, 1959, p. 117.

(5) M. E. McCulley, *Can. J. Bot.,* **43,** 1001 (1965).

(6) G. Lunde, E. Heen, and E. Oy, *Hoppe-Seyler's Z. Physiol. Chem.,* **247,** 189 (1937).

(7) T. Dillon, K. Kristensen, and C. O'hEocha, *Proc. Roy. Irish Acad., Sect. B,* **55,** 189 (1953).

(8) R. G. Schweiger, *J. Org. Chem.,* **27,** 4267 (1962).

(9) W. A. P. Black, *J. Sci. Food Agr.,* **5,** 445 (1954).

(10) T. Levering, H. A. Hoppe, and O. J. Schmid, "Marine Algae," Cram, Degruyter and Co., Hamburg, 1969, p. 330.

(11) W. A. P. Black, *Ann. Repts. Prog. Chem. 1953,* **50,** 322 (1954).

(12) E. G. Young, *in* "Handbuch der Pflanzenanalyse," W. Ruhland, ed., Springer Verlag, New York, 1958.

(13) P. S. O'Colla and R. A. Lewin, "Physiology and Biochemistry of Algae," Academic Press, New York, 1962.

(14) "Seaweed Prospects," Institute of Seaweed Research, Inveresk, Midlothian, Scotland, 1956, p. 24.

(15) G. N. Bird and P. Haas, *Biochem. J.,* **25,** 403 (1931).

(16) E. G. V. Percival and A. G. Ross, *J. Chem. Soc.,* 717 (1950).

(17) W. A. P. Black, E. T. Dewar, and F. N. Woodward, *J. Sci. Food Agr.,* **3,** 122 (1952).

(18) R. G. Schweiger, *J. Org. Chem.,* **27,** 4270 (1962).

(19) E. Percival and R. H. McDowell, "Chemistry and Enzymology of Marine Algal Polysaccharides," Academic Press, New York, 1967, p. 156.

(20) E. Percival, *Methods Carbohyd. Chem.,* **1,** 197 (1962).

(21) M. C. Cameron, A. G. Ross, and E. G. V. Percival, *J. Soc. Chem. Ind., London,* **67,** 161 (1948).

(22) C. Fromageot and P. Heitz, *Mikrochim. Acta,* **3,** 52 (1938).

(23) W. A. P. Black, W. J. Cornhill, E .T. Dewar, E. G. V. Percival, and A. G. Ross, *J. Soc. Chem. Ind., London,* **69,** 317 (1950).

(24) M. Quillet, *Colloq. Int. Cent. Nat. Rech. Sci.,* **103,** 145 (1961).

(25) D. R. Hoagland and L. L. Lieb, *J. Biol. Chem.,* **23,** 287 (1915).

(26) J. Conchie and E. G. V. Percival, *J. Chem. Soc.,* 827 (1950).

(27) E. G. V. Percival, *Quart. Rev. Chem. Soc.,* **3,** 369 (1949).

(28) K. Anno, N. Seno, and M. Ota, *Proc. Int. Seaweed Symp., 6th, Santiago de Compostela, Spain 1968,* 421 (1969).

(29) K. Anno, N. Seno, and M. Ota, *Carbohyd. Res.,* **13,** 167 (1970).

(30) W. L. Nelson and L. H. Cretcher, *J. Biol. Chem.,* **94,** 147 (1931).

(31) E. Percival, *Oceanogr. Mar. Biol.,* **6,** 137 (1968).

(32) A. N. O'Neill, *J. Amer. Chem. Soc.,* **76,** 5074 (1954).

(33) R. H. Côté, *J. Chem. Soc.,* 2248 (1959).

(34) K. O. Lloyd, Ph.D. Thesis, University of Wales (1960).

(35) G. Bernardi and G. F. Springer, *J. Biol. Chem.,* **237,** 75 (1962).

(36) B. Larsen and A. Haug, *Acta Chem. Scand.,* **17,** 1646 (1963).

(37) B. Larsen, A. Haug, and T. L. Painter, *Acta Chem. Scand.,* **20,** 219 (1966).

(38) W. Yaphe and K. Morgan, *Nature,* **183,** 761 (1959).

(39) D. R. Galli and A. C. Giese, *J. Exp. Zool.,* **140,** 415 (1954).

(40) W. A. M. Duncan, D. J. Manners, and A. G. Ross, *Biochem. J.,* **63,** 44 (1956).

(41) N. M. Thanassi and H. I. Nakada, *Arch. Biochem. Biophys.,* **118,** 172 (1967).

(42) K. Tanaka and S. Sorai, *FEBS Lett.,* **9,** 45 (1970).

(43) G. F. Springer, H. A. Wurzel, G. M. McNeal Jr., N. J. Ansell, and M. F. Doughty, *Proc. Soc. Exp. Biol. Med.,* **94,** 404 (1957)

(44) W. Schuler and G. F. Springer, *Naturwissenschaften,* **44,** 26 (1957).

CHAPTER VII

FURCELLARAN

E. BJERRE-PETERSEN, J. CHRISTENSEN, AND P. HEMMINGSEN

Litex Industri, Denmark

I. INTRODUCTION

Furcellaran (Danish agar) is an extract of the red alga *Furcellaria fastigiata* found along many coasts of the northern part of the Atlantic Ocean.

The presence of large, pure beds of *Furcellaria* were the basis of the commercial production of furcellaran, which started in Denmark in 1943. World War II had cut off Europe from its normal supplies of agar from East Asia, and the search for a substitute gelling agent led to the commercialization of furcellaran, at that time called *Danish agar*.

The new seaweed extract maintained its market after the war and has continued to grow in importance. From its beginning as a substitute for agar, it has found its own applications as a gelling and thickening agent, mainly in the food industry, and is now a recognized member of the seaweed extracts group. Three companies in Denmark are presently involved in furcellaran production. Most of the gum is used in Europe, although small amounts are imported and used in the U.S.

II. Production

1. *Source*

Furcellaria fastigiata (Huds.) Lamour. is found along many coasts of the North Atlantic and its adjacent seas, for example the coasts of Norway, Sweden, and Denmark, the Baltic Sea, around the British Isles, along parts of the French and Spanish coast, and even in the Mediterranean and the Black Seas. In addition, it occurs in the Arctic Sea and along the coasts of the Atlantic provinces of Canada. However, beds that are sufficiently extensive for commercial utilization are limited.

The most obvious occurrence of *Furcellaria* is in inner Danish waters where it is one of the most common algae.[1] In some localities it grows in unattached form (*Forma aegagropila* Reinke),[2] in which it is sterile and reproduces only vegetatively. A large body of unattached *Furcellaria* in the Kattegat, near the coast of Jutland, was for many years the sole source of raw material for furcellaran production.[3, 4] During the last 5 years, several other growing areas of *Furcellaria* have been found and utilized.

In the Atlantic provinces of Canada, especially in the Northumberland Strait[5] and in the Gulf of St. Lawrence, rich beds of *Furcellaria* have been found. After storms and hard on-shore winds, large amounts may be gathered on the coasts of Prince Edward Island and Nova Scotia. In some locations, *Furcellaria* grows alone, but in most places it grows in mixed beds together with *Chondrus crispus*. The unattached form of *Furcellaria* has not been found in Canada.

2. *Harvesting or Collecting*

Denmark and Canada are the only countries in which *Furcellaria* is collected or harvested in any major quantity.

In Denmark, *Furcellaria* may be found within a depth of 2–30 m. For many years, harvesting has been from boats using special trawls in water depths from 4 to 16 m. Large deposits of *Furcellaria* are often found in local sea areas. These beds, formed by favorable stream or bottom conditions, consist of unattached *Furcellaria* or of originally attached *Furcellaria* that has been detached by the action of waves. They can be harvested year round, except in severe winters when ice stops the harvesting. About 320,000 metric tons of wet *Furcellaria* were harvested in the period 1946 to 1971, out of which 235,000 tons were taken from one single locality in the Kattegat.[6]

Besides harvesting at sea (Fig. 1), collection on the shore has gained considerable importance. In certain places in Denmark, large amounts of *Furcellaria* are thrown up on the beach by winds and may be gathered. In Canada, *Furcellaria*, or a mixture of *Furcellaria* and *Chondrus*, has been collected in this way since about 1964. The weed is dried and is exported in bales or bags.

FIG. 1.—*Furcellaria* is harvested by trawling.

3. *Processing*

In the processing plant, fresh *Furcellaria* is washed with water to remove sand and mud. The seaweed is then treated in concrete tanks with an alkaline solution for one or more weeks. This alkaline treatment removes coloring matter and some proteins and makes the gum more easily extractable. After this pretreatment, the seaweed is washed and extracted or is dried for storage. In recent years, an increasing part of the seaweed is dried for storage.

The seaweed is extracted in boiling water either in open vessels or in pressure cookers. The extract is separated from the residue by centrifugation and filtration.

The viscous, clear yellow extract is normally concentrated by vacuum evaporation and is then precipitated as gelled threads by spraying the extract into a cold 1–1.5% solution of potassium chloride.

The threads are drained to a higher concentration (2–4%) and are frozen in a brine freezer for 20–30 hr. During this process, which is similar to that used for agar, furcellaran concentrates in the middle of each thread, leaving frozen liquid around it. After a subsequent thawing in a potassium chloride solution, the gel precipitate is pressed or centrifuged to a fibrous mass containing about 15% dry substance. This material is dried, ground, and sifted, yielding a light, free-flowing furcellaran powder.

Based on laboratory analyses, standardized products are made by blending different batches and additives, such as sucrose, carrageenan, and locust bean gum.

All commercial furcellaran products are potassium furcellaran. They are soluble in hot water, and their solutions gel on cooling. Furcellaran is available in different types, and it is common to produce milk gelling and water gelling types.

Typical laboratory analyses used to characterize the product are determination of the contents of moisture, ash, potassium chloride, sulfate, and 3,6-anydro-D-galactose. Evaluations of viscosity of furcellaran preparations is normally determined in a 1.5% water solution at 90°, using Brookfield, Emila, Epprecht, or similar rotational viscometers.

To determine gel strength, a FIRA-tester, Bloom gelometer, Penetrometer, Ridgelimeter, or other instrument may be used, but normally a modified Kobe apparatus is used. This gives the water-gel strength as the weight in grams required to force a cylindrical plunger with a cross sectional area of 1 cm^2 through the surface of a 1.5% furcellaran water gel at 20° (See also Chapt. V).

The milk gelling properties of furcellaran are normally determined in 0.4% furcellaran milk gels at 10°, using a Bloom gelometer (milk reactivity value), Ridgelimeter (% sag), or a special milk pudding tester expressing the weight required to break a 0.4% unmolded standard pudding (See also Chapt. V).

4. *Industrial Importance*

Furcellaran has gained importance in a number of specific fields, because of its unique properties. In some instances, it has replaced other gums or proteins, such as starch, gelatin, or egg yolks. In others, especially in the food industry, it has generated new products.

The annual world consumption is about 1200 metric tons, and the price varies between $3.20 and $5.00 per kg. This production volume places furcellaran third in importance among red algal extracts, following agar and carrageenan.

III. Structure

The polysaccharides present in *Furcellaria fastigiata* probably exhibit heterogeneity as is found with the polysaccharides in carrageenan, for example. However, because the manufacture of furcellaran includes precipitation in potassium chloride, some fractionation of the natural polysaccharide occurs, and the term *furcellaran* usually refers to the commercial KCl-precipitated portion of the extract.

Following the first work on the chemical composition of *Furcellaria* extract,[7] several investigators have worked on the structure of furcellaran.[8, 9] Furcellaran consists mainly of D-galactose (46–53%), 3,6-anhydro-D-galactose (30–33%), the half-ester sulfate (16–20% as SO_4—) of these sugars, and the corresponding cations. The corresponding molar ratio is approximately 1.5 : 1.0 : 0.7.

If analysis is based on commercial furcellaran, lower percentages may be found for sulfate and 3,6-anhydro-D-galactose because of the presence of free potassium chloride.

It has been shown that "carrabiose" (4-*O*-β-D-galactopyranosyl-3,6-anhydro-D-galactose) is present as a structural unit in furcellaran,[9] although the molar ratio between D-galactose and 3,6-anhydro-D-galactose indicates that "carrabiose" can only partially account for the structure.

A common feature of the red algal polysaccharides is the alternating sequence of (1→3)- and (1→4)-linked galactose or galactose derivatives,[10] and it seems likely that furcellaran follows this scheme with alternating 3-*O*-substituted-β-D-galactopyranosyl and 4-*O*-substituted-3,6-anhydro-D-galactopyranosyl units. Some 4-*O*-substituted-D-galactose can also be expected.

In its main structure, furcellaran is very much like *kappa*-carrageenan (Chapt. V), the main difference being the amount of half-ester sulfate present. Furcellaran contains one sulfate group per three to four monomer units, whereas *kappa*-carrageenan contains about one sulfate group per two monomer units. D-Galactose 2-sulfate, D-galactose 4-sulfate, D-galactose 6-sulfate, and 3,6-anhydro-D-galactose 2-sulfate have been identified as components of furcellaran.[11] The distribution of sulfate along the molecular chain is not known.

Furcellaran, like carrageenan,[12] may be modified by treating it with hot alkali to produce some increase in gel strength. This effect is probably caused by a desulfation of some D-galactose 6-sulfate groups, accompanied by formation of 3,6-anhydro-D-galactose units.

Although the structural difference between *kappa*-carrageenan and furcellaran seems slight, there is a distinct difference in some of the chemical properties. For example, precipitation of furcellaran with potassium chloride occurs at much lower concentrations than does precipitation of carrageenan.[13]

A branched structure has been considered, and methylation analysis[8] has indicated that some of the D-galactopyranosyl units are substituted at both C-3 and C-6.

IV. Properties

1. *General*

All furcellaran on the world market comes from two Danish plants using the same principal production method and the same raw material, *Furcellaria fastigiata.* Because of the production method, that is, potassium chloride precipitation, freezing, and thawing, all commercial furcellaran is in the potassium form and has a free potassium chloride content of 8–15%.

The innate potassium chloride content is of functional importance for the gelling properties of furcellaran.

2. *Solubility*

Furcellaran is easily dispersed in cold water to a homogeneous suspension without lumps. The furcellaran particles hydrate, swell, and become more or less invisible, but they do not dissolve. Dissolution requires heating of the suspension to 75°–80° or at least heating to above the gel melting temperature. Hydration and dissolution are retarded by salts, particularly potassium salts. A furcellaran in the sodium salt form is cold-water-soluble. Light boiling gives ready solubility in milk.

3. *Viscosity*

The viscosity of hot water solutions of furcellaran are related to its gelling properties, especially gel texture; hence, viscosity is an important factor in the evaluation and standardization of furcellaran. The viscosity of a 1.5% solution in water 90° ranges from 5–10 cps for low-viscosity furcellaran up to 40–50 cps for high-viscosity furcellaran at a shear rate of 2000 sec^{-1}. Hot water solutions of furcellaran are non-Newtonian and exhibit pseudoplastic flow characteristics. The apparent viscosity depends on the shear rate. Measurements at low shear rates may yield results several times higher than measurements at high shear rates (Fig. 2).

The relation between viscosity and concentration of furcellaran may be expressed with reasonable accuracy by the equation:

$$\log \frac{\eta_1 - \eta_w}{\eta_2 - \eta_w} = \alpha \log \frac{c_1}{c_2}.$$

in which η_w is the viscosity of water, and η_1 and η_2 are the viscosities at the concentrations c_1 and c_2. The value of α is about 1.6–1.7.

The viscosity depends strongly on the temperature (Fig. 3). At temperatures far above the setting point, the dependency may be expressed by the formula:

$$\log \frac{\eta_1}{\eta_2} = k \left(\frac{1}{T_1} - \frac{1}{T_2}\right)$$

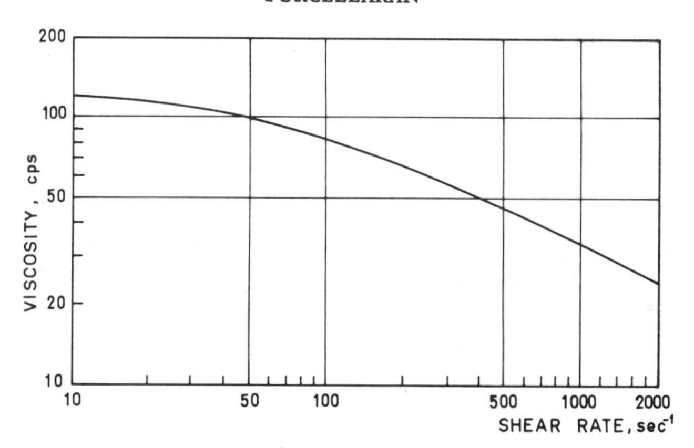

FIG. 2.—Apparent viscosity of a 1.5% furcellaran solution at 90° measured at different rates of shear.

in which η_1 and η_2 are the viscosities at two different absolute temperatures, T_1 and T_2. The value of k will depend on the shear rate in the measurement. For a shear rate of 2000 sec^{-1}, k will have a value of about 750.

4. Gelation

Upon cooling, a water solution of furcellaran will set to a gel showing some opalescence, especially at high concentrations. Gel formation occurs over a range of a few degrees at a temperature depending on the furcellaran concentration. Increased concentration is followed by a marked increase in setting temperature (Fig. 4). Furcellaran gels are thermoreversible and remelt when heated to a temperature 15°-20° above their setting temperature.

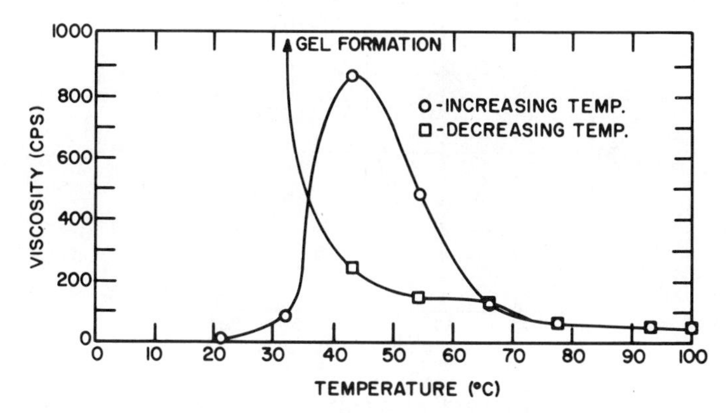

FIG. 3.—Viscosity of a 1.5% furcellaran dispersion when it is heated to 100° and then cooled until a gel is formed. (Courtesy of Duché Uni-Gum Corp. and Ralph E. Sand and Martin Glicksman.)

The gel strength of a furcellaran-water gel will rise more than linearly with the furcellaran concentration (Fig. 4).

The gelation properties of furcellaran are linked, in part, to its content of 3,6-anhydro-D-galactose, its degree of polymerization, and its salt content. Potassium ions (Fig. 5) as well as ammonium ions, rubidium ions, and cesium ions are effective in producing high gel strength. Calcium ions have a lesser effect and sodium ions do not produce gelation. The different effect of these cations can be explained by their different ion radii. A furcellaran with high degree of polymerization, i.e. high viscosity, forms an elastic gel, whereas a low-viscosity furcellaran forms a more brittle gel.

In practical use, furcellaran gels will incorporate a great variety of ingredients, such as salts, acids, sugar and proteins, which all influence the gelling properties.

The effect of different cations has been mentioned. It should be further noted that the normal potassium content of furcellaran is too low to produce maximum gel strength. Addition of 0.1% of potassium chloride to a 1% solution of furcellaran almost doubles the gel strength. Gels of lower furcellaran concen-

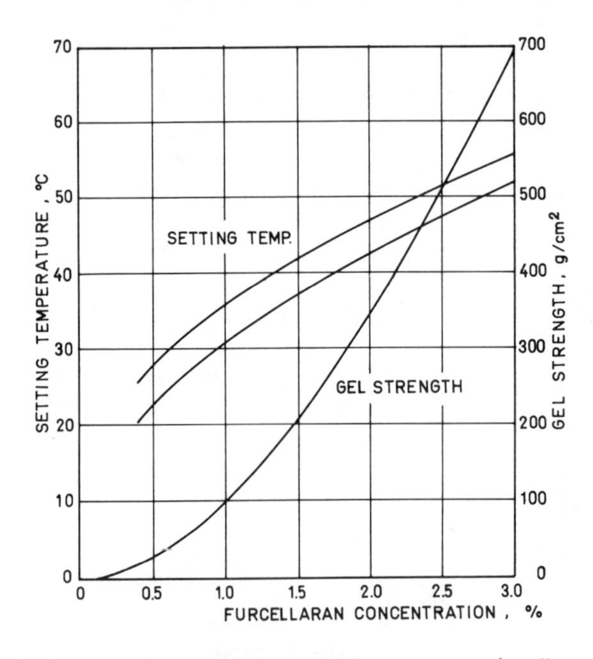

FIG. 4.—Setting temperature and gel strength for an average furcellaran–water gel at different concentrations. The gel strength in g/cm² is expressed as the weight required to force a plunger with a cross-sectional area of 1 cm² through the surface of a 1.5% furcellaran gel cooled to 20°.

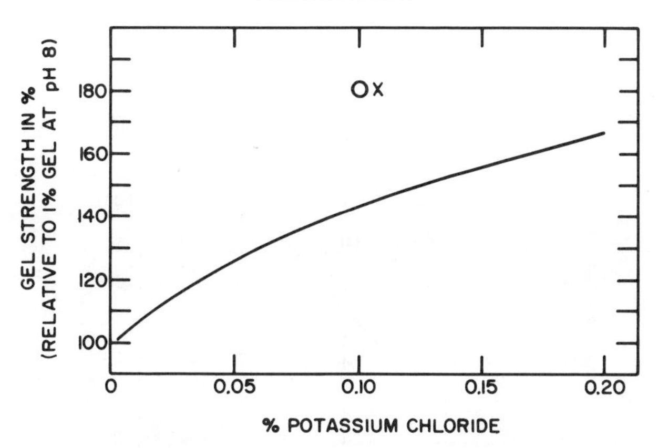

FIG. 5.—Effect of replacing some furcellaran with potassium chloride on the gel strength of a 1% furcellaran dispersion (relative to a standard 1% gel at pH 8). The point "X" indicates the relative gel strength of a gel containing, 1% of furcellaran and 0.1% of potassium chloride. (Courtesy of Duché Uni-Gum Corp. and Ralph E. Sand and Martin Glicksman.)

trations benefit especially from added potassium salt, although gel texture and taste qualities are lowered and increased syneresis occurs.

Sugar has some practical effects on a furcellaran gel. Whereas a furcellaran–water gel shows opalescence, gels containing sugar can be made water clear. Higher concentrations of sugar also increase the gel strength and the setting and melting temperatures of furcellaran gels (Fig. 6). The elevated setting temperatures of furcellaran sugar gels and their unretarded setting are the bases for their use in quick-setting jellies.

Glycerol, like sugar, increases the clarity and strength of furcellaran gels.

Furcellaran is sensitive to acids, especially at higher temperatures (Fig 7). Acid-catalyzed hydrolysis causes a loss of viscosity and gel strength. The effect depends on the pH, the heating time, and the temperature. Decrease in viscosity is not accompanied by an equal decrease in gel strength. Furcellaran may be hydrolyzed to a rather low viscosity and still maintain a high gel strength. In this respect, furcellaran is more stable to acids than is carrageenan but is less stable than agar.

If furcellaran is brought into solution with protein, and the pH of the solution is lower than the isoelectric point of the protein, the two materials will form a coprecipitate. In a pH range near the isoelectric point, only cloudiness forms in the solution. At pH values above the isoelectric point, furcellaran–protein gels can be formed without precipitation or loss in clarity and without the decrease in gel strength that follows the precipitation.

Furcellaran is easily soluble in milk at 80°–100°, and the solution sets by cool-

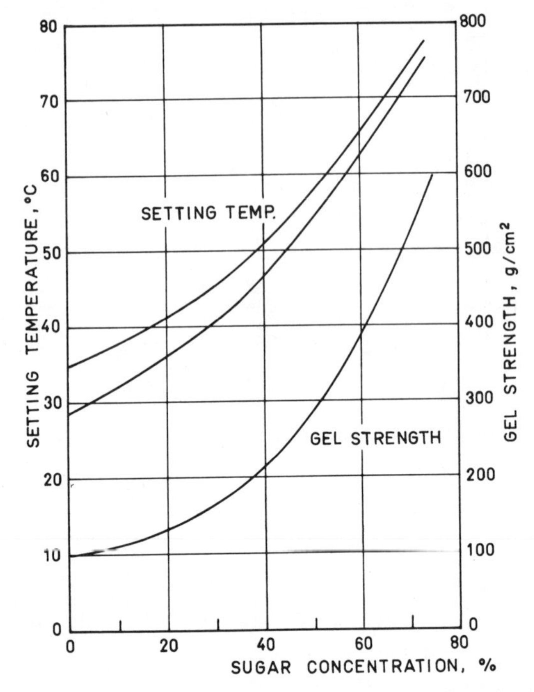

FIG. 6.—Setting temperature and gel strength for a 1% furcellaran gel at different sugar concentrations. The gel strength in g/cm² is expressed as the weight required to force a plunger with a cross-sectional area of 1 cm² through the surface of a 1.5% furcellaran gel cooled to 20°.

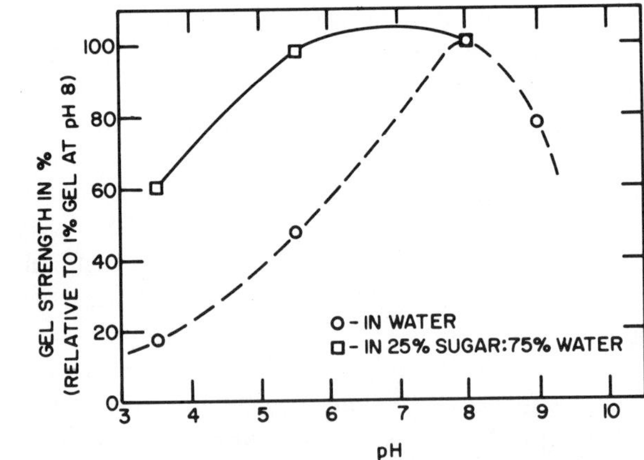

FIG. 7.—Effect of pH on the gel strength (relative to a standard 1% gel at pH 8) of 1% furcellaran gels; the broken line is for a water dispersion, and the solid line, for a 25% sucrose dispersion. (Courtesy of Duché Uni-Gum Corp. and Ralph E. Sand and Martin Glicksman.)

ing to a milk gel having a higher gel strength than a water gel with a similar concentration of furcellaran. The higher milk-gel strength is mainly produced by the calcium and potassium salts in the milk, particularly the potassium salts present in a concentration of $\sim$0.04 mole/liter of milk. Milk casein and serum proteins contribute to the texture and body of the milk gel, but it is questionable whether they have a direct effect on the milk gel strength expressed by the milk reactivity value. In any case, reaction of the furcellaran with the milk proteins is far less pronounced than that of the carrageenans. In a furcellaran–milk gel, the milk taste is fully present, whereas in a carrageenan–milk gel it is almost eliminated.

A milk gel prepared from furcellaran and fresh milk normally has a pH of 6.65–6.70. The strength of a milk gel is greatly influenced by alterations of its pH value. An increase of the pH, achieved by the addition of sodium phosphate, reduces the gel strength and may also turn a normal clear syneresis into a milky one. If the pH is lowered by adding an acid or a soluble calcium salt (0.2–0.5 g $CaCl_2$/liter of milk), the gel strength is increased. The latter effect is normally accompanied by an increased separation or curdling of the milk, and it is possible that the increased gel strength is an effect of a casein precipitation.

An intensified search for gelling agents or stabilizers with specific properties has led to the combination of furcellaran with other gums or thickening agents. Quite favorable effects are obtained with furcellaran and locust bean gum (carob). A pure furcellaran gel has a rather short texture and shows some syneresis. The incorporation of locust bean gum produces a much more elastic, gummy gel with increased gel strength and markedly reduced syneresis. This gel type reproduces the texture of gelatin gels very well. Furthermore, the furcellaran–locust bean gum gel has improved stability to acids. A disadvantage is the taste-covering effect of the locust bean gum. Guar gum increases viscosity and reduces syneresis in the same way but does not enhance the elasticity and strength of furcellaran gels.

Cellulose derivatives, particularly carboxymethylcellulose, have found limited use in blendings with furcellaran where a weakly gelling stabilizer is desired.

V. Applications

The fact that furcellaran originally started as a wartime substitute for agar and acquired the name Danish agar may still lead to some confusion and misuse of the product. The properties of furcellaran place it in a middle position between agar and carrageenan, with most resemblance to carrageenans of the gel forming *kappa* type.

Furcellaran is produced in food grades only, complying with existing food and drug standards. More than 90% of furcellaran production is used in food products, with the remainder employed mainly by the pharmaceutical industry.

The fields of application are widespread and comprise the following products and product groups:

- Food products
  - Flan jelly, *Tortenguss,* and other cake-covering jellies
  - Piping jelly and decorating jelly
  - Jams and other fruit preserves
  - Fruit juices (stabilizing effect)
  - Certain confectionery products
  - Meat products (jellied veal, minced meat, pie fillings)
  - Milk puddings and custards
  - Chocolate milk
  - Beer (clarification)
- Pharmaceutical and cosmetic products
  - Suspensions, emulsions, and foams
  - Tablets (disintegration)

The general properties of furcellaran and its reactions with some essential product ingredients have been described in the previous section.

Flan or pudding powders and ready-to-eat puddings are prepared with furcellaran types standardized for milk gelling. The standard formula for a pudding powder to be cooked with 0.5 liter (1 pint) of milk is as follows:

Furcellaran	2–2.5 g
Sugar	40–50 g
Starch	0–10 g
Flavor and color	as needed

The pudding powder is dissolved in milk by moderate heating, but it is normal and good practice to prescribe boiling for 1–3 min to ensure complete dissolution. Upon cooling, the solution sets quickly to form a smooth and full-bodied milk gel that is quite homogeneous without tendency of the cream to separate. A tendency to syneresis can be eliminated by the addition of a few grams of starch to the pudding powder.

A furcellaran pudding has its best texture and appearance, as well as the most negligible milk separation, at a pH of 6.60 to 6.70 as provided by furcellaran and normal fresh milk. However, some pudding powder ingredients, such as cocoa, soluble coffee and starch, may effect the pudding pH and require corrections.

The exceptional eating quality of furcellaran milk puddings has made furcellaran a preferred gelling agent in this field.

Flan jelly, *Tortenguss, apricoture,* and *nappage* are all quick-setting cake-covering jellies containing furcellaran in concentrations of 0.4–0.8%.

Tortenguss is sold to European housewives in the form of small powders for boiling with 250 ml (0.5 pint) of water or fruit juice. The basic formula for a *Tortenguss* powder is as follows:

Furcellaran	1.5–2.0 g
Sugar	40–50 g
Starch	0–10 g
Acidifying agent	as needed

The quick-setting property of furcellaran in sugar solutions together with its relatively good resistance to acids has made furcellaran a preferred gelling agent for this group of jelly products.

Jams, marmalades, and other fruit preserves are easily thickened with furcellaran. Pectin, the traditional gelling agent in jams and jellies, requires a minimum sugar content of 50–60% and a specific acidity. Furcellaran gels independently of sugar and acid and can be used for jams and jellies with lower sugar contents or even without added sugar. The concentration of furcellaran needed is 0.2–0.5% varying with sugar content and stiffness required.

To prevent unnecessary hydrolysis of the furcellaran, it should be added to the hot, ready-cooked fruit–sugar mixture as a 2–3% solution in water, just before tins or glasses are filled.

Fruit pulps and tomato purees are stabilized and thickened in a similar way.

Fruit juices are stabilized with 0.05–0.1% of furcellaran to maintain an even suspension of the fine fruit pulp throughout the juice.

In meat products, furcellaran is used as a stabilizing and gelling agent in such products as minced meat, meat paste, and pie fillings. Furcellaran should not, however, be used in meat products unless its compatibility with the proteins of the product has been examined in advance.

Furcellaran for stabilization of suspensions is subject to some limitations or special measures: (*a*) furcellaran is used in concentrations equal to or lower than, the minimum concentration needed for gelation (as in chocolate milk); (*b*) the gelling effect is reduced by acid-catalyzed hydrolysis (as in piping jelly); (*c*) the furcellaran gel is transformed to a flowable system by mechanical treatment; (*d*) furcellaran is used in a swollen form, not dissolved by heating, and the swollen granules are disintegrated mechanically. The latter method has found extended use in some pharmaceutical suspensions; it produces highly thixotropic, viscous liquids.

In the brewing and fermentation industry, furcellaran has found application as a fining added during hop boiling to assist coagulation and precipitation of proteins from the wort and as an auxiliary fining added at the end of fermentation to promote flocculation. The final filtration of the beer is facilitated, and the beer can be stored with less risk of cloudiness.

VI. References

(1) L. K. Rosenvinge, "The Marine Algae of Denmark," Pt. II, D. Kgl. Danske Vidensk. Selsk., Copenhagen, 1917.

(2) A. P. Austin, *Hydrobiologia,* **14,** (3–4), 255 (1960).

(3) S. Lund and E. Bjerre-Petersen, *Proc. Int. Seaweed Symp., 1st, Edinburgh, 1952,* 85 (1953).

(4) S. Lund and E. Bjerre-Petersen, *Proc. Int. Seaweed Symp., 4th, Biarritz, 1961,* 410 (1964).

(5) C. I. MacFarlane, *Abstr. Int. Seaweed Symp., 3rd, Galway, 1958,* 28 (1958).

(6) S. Lund and J. Christensen, *Proc. Int. Seaweed Symp., 6th, Santiago de Compostela, 1968,* 699 (1969).

(7) H. Kylin, *Kgl. Fysiogr. Saellsk. Lund, Foerh.,* **13,** 51 (1943).

(8) M. J. Clancy, K. Walsh, T. Dillon, and P. S. O'Colla, *Sci. Proc. Roy. Dublin Soc., Ser. A,* **1,** No. 5, 197 (1960).

(9) T. J. Painter, *Can. J. Chem.,* **38,** 112 (1960).

(10) N. S. Anderson, T. C. S. Dolan, and D. A. Rees, *Nature,* **205,** 1060 (1965).

(11) T. J. Painter, *Proc. Seaweed Symp., 5th, Halifax, 1965,* 305 (1966).

(12) D. A. Rees, *J. Chem. Soc.,* 1821 (1963).

(13) O. Smidsrød, B. Larsen, A. J. Pernas, and A. Haug, *Acta Chem. Scan.,* **21,** 2585 (1967).

CHAPTER VIII

LAMINARAN

W. A. P. BLACK AND E. T. DEWAR

Inveresk Research International, Inveresk, Musselburgh, Midlothian, Scotland

I. SOURCE

Although β-D-$(1\rightarrow 3)$-linked glucans are known to be widely distributed in nature, for example, in fungi, yeasts, flagellates, and diatoms,[1–3] laminaran of potential economic importance occurs only in brown seaweeds (Phaeophyceae).

Brown algae, unlike the red and green algae, do not synthesize starch-type polysaccharides; instead they store carbohydrate as laminaran. The polymer is found mainly in the frond (leaf) of the *Laminaria* genus, notably *Laminaria hyperborea, L. digitata,* and *L. saccharina,* and to a lesser extent in the Fucaceae, for example, *Fucus serratus.*[4] It is not present in the stipe (stem) of these marine plants. The soluble laminarans isolated from *Eisenia bicyclis*[5, 6] and *Ishige okamurai,*[7] which are also brown seaweeds, appear to differ both chemically and physically from the laminarans from the Laminariales and will not be discussed here.

Macrocystis pyrifera, the giant brown seaweed of the Pacific and the source of most of the U.S. production of alginate, contains little or no laminaran. The *Laminaria species* that store laminaran occur in temperate and cool coastal waters off Norway, Scotland, Ireland, France, Spain, and the islands of eastern Canada. Only a few quantitative surveys of areas in which brown seaweeds are prolific have been made, and the accuracy of some of these is open to question. The *Laminaria* resources of Norway are believed to be about 20 million tons (fresh weight), and those of Scotland about half this amount.

The *Laminaria* seaweeds of Scotland[8] and Norway[9] are subject to marked seasonal variation in chemical compostion. Although the time and extent of the maximum laminaran content in the fronds are susceptible to environmental and other factors, in general the maximum is reached in the autumn and early winter. Laminaran is absent from *Laminaria* stipes throughout the year and from the fronds during late winter and early spring.

Maximum laminaran contents of the dried fronds of the commonest *Laminaria* species, *L. hyperborea, L. digitata,* and *L. saccharina,* reported from different locations in Scotland[8] and Norway[9] are between 20 and 35% (on a dry weight basis). The freshly harvested algae must be processed immediately or dried under carefully controlled conditions to prevent decomposition of the laminaran by microorganisms. Maximum laminaran contents are obtained from plants growing in unpolluted waters around islands and lochs, where fast currents are absent and nitrogen and phosphorus are frequently reduced to almost zero levels, for the presence of nutrients in the water tends to prevent the buildup of laminaran.

Because laminaran has not yet been produced in commercial quantities, the economics of harvesting *Laminaria* frond have been difficult to assess accurately. The main sources of alginate, *M. pyrifera* (mechanically harvested), *Ascophyllum nodosum* (hand cut), and *L. hyperborea* stipe (cast or driftweed), contain almost no laminaran. Even in the case of *L. digitata,* which is frequently collected by hand-operated grapnels from small boats, the seaweed is partially or completely air-dried on the shore for fairly long periods with consequent loss of laminaran content. In order to overcome this difficulty, the British Institute of Seaweed Research developed a mechanical harvester capable of collecting *L. hyperborea* and similar brown algae directly from the seabed.[10] Trials of their prototype harvester off the Orkney Islands in the North of Scotland have demonstrated that a harvesting vessel, with twin harvesters costing about £13,000 complete, could be expected to collect over 5000 tons of this type of seaweed in an average year and that the cost of the freshly harvested weed should be about £1 per wet ton.

II. Production

Laminaran occurs in two forms distinguished by their solubility in cold water. They are therefore referred to as insoluble and soluble laminaran, although both forms dissolve readily in hot water. Insoluble laminaran is present in the fronds of *L. hyperborea* to the extent of 55–65% of the total laminaran present and to a much lesser extent in *L. saccharina* fronds. *Laminaria digitata* fronds and *F. serratus* contain only the soluble form.[4]

The only attempt yet made to isolate laminaran on a large scale is that of Le Gloahec and Herter,[11] who treated the seaweed with an alkaline earth metal salt solution, such as calcium or barium chloride, separated the solution, and precipitated the laminaran by means of lead subacetate $[Pb(OAc)_2 \cdot 2Pb(OH)_2]$ and sodium hydroxide. The following methods, however, are simpler and fit in well with the recovery of other valuable components of brown seaweeds such as mannitol, fucoidan, and alginate.

1. *Insoluble Laminaran*[4, 12]

Minced or finely shredded, freshly harvested *L. hyperborea* fronds (250 g) are stirred with 0.125N hydrochloric acid (500 ml) and 40% formaldehyde (1 ml) at 70° for 60 min, and the weed residue is removed by filtration through cloth and washed with warm water (2 × 100 ml). The filtrate and washings are stirred for 3 hr and allowed to stand at 20° for 2 days. The laminaran that precipitates is collected by filtration, washed with ethanol and ether, and dried under diminished pressure to a moisture content of less than 10%. In general, this procedure removes 80–90% of the total laminaran from the weed and about 60% of this is recovered in the insoluble form (ash, about 1%).

Instead of freshly harvested weed, dried weed (milled to pass a 60-mesh screen) can be used equally well, but the pH of the mixture must never be allowed to fall below 2.4 during the extraction. If desired, insoluble laminaran can be recrystallized from a 10% aqueous solution, giving an 80–90% recovery of a pure white powder; $[\alpha]_D$ −12° to −14° (water).

2. *Soluble Laminaran*[4, 12]

Dried *L. digitata* frond (25 g, milled to pass a 60-mesh screen) is stirred with 0.09N hydrochloric acid (250 ml) at 20° for 2 hr. The mixture (pH 2.5) is centrifuged; the weed residue is washed with 0.05*N* hydrochloric acid (2 × 50 ml), and the supernatant and washings are made up to 85% with ethanol. The crude precipitate, containing 70–75% of laminaran, is readily purified by solution in water and passage through a strongly acidic cation-exchange column, such as one of Zeo-Karb 225(H^+). The effluent and washings from the column are treated with ethanol to 85% concentration, and the precipitate is coagulated by the addition of a small amount of sodium chloride solution. The soluble laminaran is centrifuged, washed with ethanol and ether, and dried under diminished pressure to a white powder; $[\alpha]_D$ −11° to −13° (water), ash content ~ 1%, yield 50–60% of the total laminaran in the original seaweed.

3. *Analysis of Laminaran Content*[13]

Determination of the laminaran content of seaweeds is based[13] on the reducing power of hydrolyzates. Because other polysaccharides, such as alginate and fucoidan, also produce reducing sugars on hydrolysis with acids, it is necessary to measure the reducing power before and after removal of the D-glucose. This may be done by the copper micromethod on as little as 1 g of ground seaweed, the glucose being removed with an excess of washed yeast cells.

III. Structure

1. *Historical and Early Studies*

Laminaran was first described by Schmiedeberg[14] in 1885 and it has since been studied by Krefting and Torup,[15] Kylin,[16] Gruzewska,[17] Colin and Ricard,[18] Lunde,[19] Nisizawa,[20] Le Gloahec and Herter,[11] and Barry.[21]

Barry showed that laminaran was composed entirely of D-glucose residues, and his methylation studies, together with the low negative rotation, indicated that the polymer is an essentially linear β-D-(1$\rightarrow$3)-linked glucan. This general structure has been confirmed by many later workers.

2. *Fine Structure*

During the past 30 years, a great deal of work has been done on the fine structure of laminaran and this has been amply reviewed.[2, 22, 23] Only a few major comments can be made in a book of this kind, and these will be directed mainly to (*a*) the nature of the end-groups, (*b*) the presence of (1$\rightarrow$6)-linkages, (*c*) the degree of branching, and (*d*) attempts to explain the difference between insoluble and soluble laminarans.

An important advance in our understanding of the structure of laminaran was the discovery[24] of 2–3% of D-mannitol as a constitutent of both insoluble and soluble laminaran as an end-group. Thus there are two types of laminaran molecules, namely, those that are terminated by a mannitol residue (M-chains) linked through a primary alcohol group, and those that are terminated by reducing glucose residues (G-chains) linked through C-3 (Figs. 1 and 2).

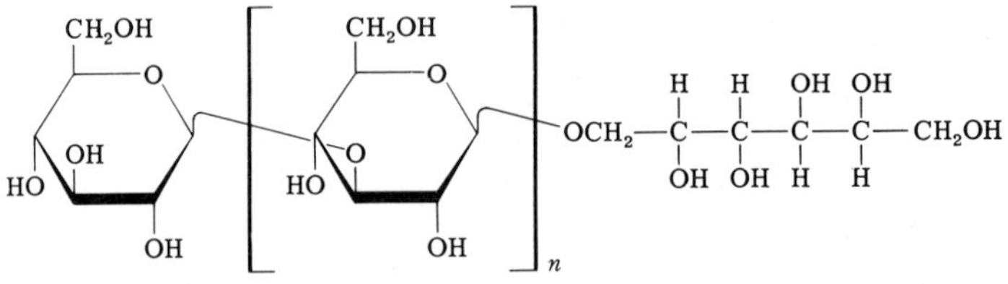

Fig. 1.—Laminaran M-chains.

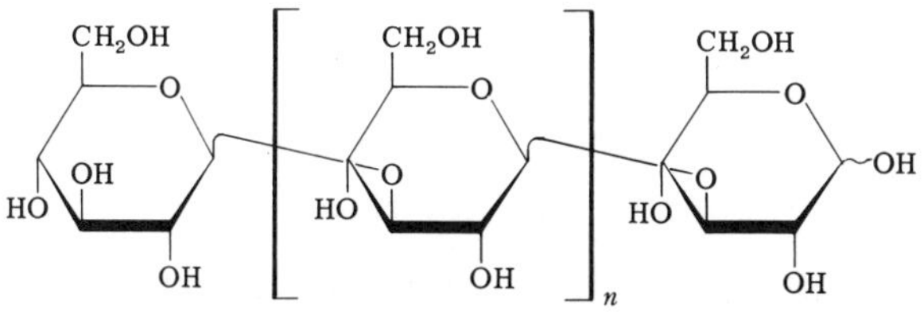

FIG. 2.—Laminaran G-chains.

Treatment of laminaran with borohydride converts the reducing D-glucose units (G-chains) into D-glucitol (sorbitol)[25, 26] and the product is known as laminaritol, which thus comprises a mixture of mannitol-terminated chains (M-chains) and 3-*O*-substituted sorbitol-terminated chains (S-chains). Similarly, oxidation of laminaran with bromine converts the reducing glucose units into D-gluconic acid units without affecting the M-chains.[27]

In addition to β-D-(1→3) linkages, both laminarans contain small but definite amounts of (1→6) linkages, and much controversy has arisen as to whether the few (1→6) linkages occur as interresidue linkages in an essentially linear chain or whether they are present as branch points.[2, 22, 23] The most recent evidence favors the presence of (1→6) branch points.

Attempts to explain the difference in solubility between laminarans from *L. hyperborea* and *L. digitata* on a structural basis have also proved difficult. Manners[2] believes that the essential difference between the two forms of laminaran is in the degree of branching. Six samples of insoluble laminaran had average chain-lengths (CL) of 15–19 D-glucose residues and degrees of polymerization (DP) of 16–21; therefore, the molecules were almost linear.[26] In contrast, four soluble laminarans had CL values of 7–10 and DP's of 26–31, indicating the presence of two to three branch points per molecule. The essentially linear molecules can pack more closely to form insoluble aggregates, whereas this is not possible with the branched molecules. It is, therefore, probable that, in different algae, a range of polymers that differ in degree of branching, and hence in solubility, is synthesized.

A stereospecific chemical synthesis of a linear G-chain of laminaran has recently been reported;[28] the polymer had a DP of 30 and a specific rotation of –9° (water) and is thus closely related to native laminaran.

IV. HYDROLYSIS

1. *Acid-catalyzed Hydrolysis*

Laminaran is hydrolyzed by acids only slightly less rapidly than potato starch.[29, 30] The method of production of crystalline D-glucose from laminaran is closely related to the commercial production of dextrose by the acid-catalyzed hydrolysis of starch, but obviously laminaran cannot compete with starch on economic grounds.

2. *Enzyme-catalyzed Hydrolysis*[1, 23]

The enzymes that hydrolyze laminaran are termed laminarases (laminaranases) and Bull and Chesters[1] have recently prepared an excellent review on the biochemistry of laminaran and the nature of laminarase in which they suggest that this term should be used to describe the whole enzyme complex that degrades laminaran and includes exo- and endo- β-D-(→3)-glucanases and β-D-glucosidases.

Laminarase occurs in bacteria, fungi, algae, higher plants, and molluscs. These enzymes are involved in the intracellular mobilization of food reserves [not only in algae, but also in higher plants and fungi, all of which synthesize β-D-(1→3)-glucans] and are also encountered in the extracellular breakdown of plant debris and in the digestive metabolism of invertebrates. Enzyme preparations have been reported from cereals, hyacinth bulbs, and potato tubers;[31] from brown, red, and green seaweeds;[32] from *Euglena*;[33] from marine bacteria; from many fungi;[34–36] and from the digestive juices of molluscs.[37, 38]

The isolation, method of assay, specificity, and activity of laminarases are discussed elsewhere in excellent reviews.[1, 2, 23] Both soluble and insoluble laminarans give the same hydrolysis pattern. Endwise attack (exoenzyme) removes D-glucose units from the nonreducing ends of the chains, while endo-β-glucanases attack the molecule randomly and release a series of laminaridextrins.[39] Mannitol containing oligosaccharides are very resistant to further hydrolysis.[35]

V. Properties, Uses and Derivatives

Laminaran is a white, odorless, tasteless powder with the characteristic properties of a low-molecular-weight polysaccharide. Although the insoluble form is practically insoluble in cold water, it is readily soluble in hot water (22 g in 100 g of water at 90°). The specific optical rotation of both forms in water is normally $-12° \pm 2°$.

Average molecular weights, calculated from sedimentation, diffusion, and viscosity data,[40] are ~5300 (DP ~ 33) and ~3500 (DP ~ 22) for soluble and insoluble laminaran, respectively, in good agreement with the results from chemical end-group methods.[26] Other physical measurements of molecular weight have been done on certain derivatives of laminaran, but these are less reliable because of possible depolymerization during derivification. Both laminarans are heterogeneous and polymolecular. The intrinsic viscosity of both forms in water is close to 0.10 dl/g.[40]

Blaine[41] has shown that laminaran is a safe surgical dusting powder. On treatment with lime water, laminaran gives a 40–50% yield of D-glucometasaccharinic acids (3-deoxy-D-*ribo*- and 3-deoxy-D-*arabino*-hexonic acids) and this reaction constitutes a most convenient source of these acids,[42] which are readily converted into the biologically important 2-deoxy-D-*erythro*-pentose (2-deoxy-D-ribose) by oxidative degradation.

There has been recent interest, particularly in Japan, in the antitumor activity of β-D-(1→3)-glucans,[43–46] and laminaran has been found to have tumor inhibiting power against subcutaneously implanted sarcoma-180 in mice.[47]

Laminaran readily yields partially and fully substituted derivatives, such as the hydroxyethyl, hydroxypropyl, benzyl, and β-aminoethyl[48] ethers and acetate, benzoate, and carbanilate esters.[49]

Some years ago, there was considerable interest in the neutral sodium salts of sulfated laminaran and its aminoethyl ether as blood anticoagulants,[48, 50–53] and a whole series of half-sulfate esters with different degrees of substitution (DS) were synthesized and tested *in vivo* in animals. Results proved, however, that the esters are too toxic for therapeutic use in humans.[51, 54, 55] However, laminaran sulfate esters with a lower DS, which possess little or no anticoagulant activity, act like heparin in clearing alimentary lipemia[56] and are of obvious interest as antilipemic agents in ischemic heart disease.[57, 58]

VI. References

(1) A. T. Bull and C. G. C. Chesters, *Advan. Enzymol.*, **28,** 325 (1966).

(2) D. J. Manners, *Ann. Rep. Prog. Chem.*, **63,** 590 (1967).

(3) P. A. J. Gorin and J. F. T. Spencer, *Advan. Carbohyd. Chem.*, **23,** 367 (1968).

(4) W. A. P. Black, W. J. Cornhill, E. T. Dewar, and F. N. Woodward, *J. Appl. Chem.*, **1,** 505 (1951).

(5) N. Handa and K. Nisizawa, *Nature,* **192,** 1078 (1961).

(6) M. Maeda and K. Nisizawa, *J. Biochem. (Tokyo),* **63,** 199 (1968).

(7) M. Maeda and K. Nisizawa, *Carbohyd. Res.*, **7,** 97 (1968).

(8) W. A. P. Black, *J. Soc. Chem. Ind., London,* **67,** 165, 169, 172, 355 (1948); **68,** 183 (1949); *J. Marine Biol. Ass. U.K.*, **29,** 45 (1950).

(9) *Norwegian Inst. Seaweed Res.* (Trondheim, Norway), *Repts.* No. **2** (1953); No. **4** (1954); No. **14** (1956).

(10) P. Jackson, *Engineer,* **203,** 400, 439 (1957).

(11) V. C. E. Le Gloahec and J. R. Herter, U.S. Patent 2,188,092 (1940); *Chem. Abstr.*, **34,** 3849 (1940).

(12) W. A. P. Black, *Methods Carbohyd. Chem.*, **5,** 159 (1965).

(13) M. C. Cameron, A. G. Ross, and E. G. V. Percival, *J. Soc. Chem. Ind., London,* **67,** 161 (1948).

(14) J. E. O. Schmiedeberg, *Gesellschaft Deutsch. Naturforscher Ärzte, Strassburg, Tageblatt Versammlung,* No. **58,** 427 (1885).

(15) A. Krefting and S. Torup, *Pharmacia (Estonia),* **6,** 151, 153 (1909).

(16) H. Kylin, *Hoppe-Seyler's Z. Physiol. Chem.*, **83,** 171 (1913); **94,** 337 (1915); **101,** 236 (1918).

(17) Z. Gruzewska, *Bull. Soc. Chim. Biol.*, **5,** 216 (1923).

(18) H. Colin and P. Ricard, *C. R. Acad. Sci.*, **188,** 1449 (1929); *Bull. Soc. Chim. Biol.*, **12,** 88 (1930).

(19) G. Lunde, *Z. Angew. Chem.*, **50,** 731 (1937).

(20) K. Nisizawa, *Sci. Rep. Tokyo Bunrika Daigaku, Sect. B,* **5,** 9 (1940).

(21) V. C. Barry, *Sci. Proc. Roy. Dublin Soc.*, **21,** 615 (1938); **22,** 59 (1939).

(22) S. Peat and J. R. Turvey, *Fortschr. Chem. Org. Naturst.*, **23,** 1 (1965).

(23) E. Percival and R. H. McDowell, "Chemistry and Enzymology of Marine Algal Polysaccharides," Academic Press, New York, 1967, p. 53.

(24) S. Peat, W. J. Whelan, and H. G. Lawley, *J. Chem. Soc.*, 724, 729 (1958).

(25) M. Abdel-Akher, J. K. Hamilton, and F. Smith, *J. Amer. Chem. Soc.*, **73,** 4691 (1951).

(26) M. Fleming and D. J. Manners, *Biochem. J.*, **94,** 17P (1965).

(27) I. J. Goldstein, F. Smith, and A. M. Unrau, *Chem. Ind. (London),* 124 (1959).

(28) N. K. Kochetkov and A. F. Bochkov, *Carbohyd. Res.*, **9,** 61 (1969).

(29) W. A. P. Black, E. T. Dewar, and F. N. Woodward, *J. Sci. Food Agr.*, **4,** 58 (1953).

(30) J. Szejtli, *Acta Chim. Acad. Sci. Hung.*, **45,** 141 (1965).

(31) T. Dillon and P. O'Colla, *Nature,* **166,** 67 (1950).

(32) W. A. M. Duncan, D. J. Manners, and A. G. Ross, *Biochem. J.*, **63,** 44 (1956).

(33) J. Fellig, *Science,* **131,** 832 (1960).

(34) C. G. C. Chesters and A. T. Bull, *Biochem. J.*, **86,** 28 (1963).

(35) C. G. C. Chesters and A. T. Bull, *Biochem. J.*, **86,** 31 (1963).

(36) C. G. C. Chesters and A. T. Bull, *Biochem. J.*, **86,** 38 (1963).

(37) V. C. Barry, *Sci. Proc. Roy. Dublin Soc.*, **22,** 423 (1941).

(38) M. Quillet, *C. R. Acad. Sci.*, **246,** 812 (1958).

(39) M. Fleming, D. J. Manners, and A. J. Masson, *Biochem. J.*, **104,** 32P (1967).

(40) M. H. G. Friedlaender, W. H. Cook, and W. G. Martin, *Biochim. Biophys. Acta,* **14,** 136 (1954).

(41) G. Blaine, *Med. Press,* **226,** 611 (1951).

(42) W. M. Corbett and J. Kenner, *J. Chem. Soc.*, 1431 (1955).

(43) H. Nakayoshi, *Nippon Saikingaku Zasshi,* **22,** 641 (1967); *Chem. Abstr.*, **69,** 10649 (1968).

(44) S. Shibata, Y. Nishikawa, T. Takeda, M. Tanaka, F. Fukuoka, and M. Nakanishi, *Chem. Pharm. Bull.*, **16,** 1639, 2362 (1968).

(45) G. Chihara, J. Hamuro, Y. Maeda, Y. Arai, and F. Fukuoka, *Nature,* **225,** 943 (1970).

(46) J. Hamuro, Y. Yamashita, Y. Ohsaka, Y. Y. Maeda, and G. Chihara, *Nature,* **233,** 486 (1971).

(47) S. Sakai, S. Takada, T. Kamasuka, Y. Momoki, and J. Sugayama, *Gann,* **59,** 507 (1968); *Chem. Abstr.*, **70,** 45863 (1969).

(48) A. N. O'Neill, *Can. J. Chem.*, **33,** 1097 (1955).

(49) W. A. P. Black and E. T. Dewar, *J. Sci. Food Agr.*, **5,** 176 (1954).

(50) E. T. Dewar, *Proc. Int. Seaweed Symp., 2nd, Trondheim, Norway, 1955,* 55 (Publ. 1956).

(51) C. C. Burt, *Abstr. Int. Seaweed Symp., 3rd, Galway, Ireland, 1958,* 43 (1958).
(52) W. W. Hawkins and A. N. O'Neill, *Can. J. Biochem. Physiol.,* **33,** 545 (1955).
(53) W. W. Hawkins and V. G. Leonard, *Can. J. Biochem. Physiol.,* **36,** 161 (1958).
(54) S. S. Adams and H. M. Thorpe, *J. Pharm. Pharmacol.,* **9,** 459 (1957).
(55) S. S. Adams, H. M. Thorpe, and L. E. Glynn, *Lancet,* No. 7047, 618 (Sept. 20, 1958).
(56) R. A. Gollin, R. Michaelis, D. Walker, and S. S. Adams (to Boots Pure Drug Co. Ltd.), Brit. Patent 911,484 (1962); *Chem. Abstr.,* **58,** 12662 (1963).
(57) E. M. M. Besterman and J. Evans, *Brit. Med. J.,* **1,** 310 (1957).
(58) H. Murata, *J. Atheroscler. Res.,* **10,** 371 (1969).

CHAPTER IX

SEAWEED EXTRACTS OF POTENTIAL ECONOMIC IMPORTANCE

RALPH E. SAND

Anderson Clayton Foods, Richardson, Texas

and

MARTIN GLICKSMAN

General Foods Corp., Tarrytown, N. Y.

I. Introduction

Three major types of seaweed extracts are produced—agar (Fig. 1), carrageenans (Fig. 2), and alginates. A fourth, furcellaran, is manufactured in Denmark and is finding expanding usage. (See Chapt. VII.) There are also several other lesser-known seaweed species whose potential value warrants their inclusion in a discussion of commercial seaweed gums. Some of these seaweeds are already being used to a small extent.

Production of seaweed hydrocolloids rose in the United States from a small specialized business grossing about $3,000,000 a year at the end of World War II to about $10,000,000 in 1956, then to $20,000,000 in 1970. It is expected to be a $30,000,000 a year business by 1980. Whereas present sales depend on the established uses of agar, alginates, and carrageenans, future growth of seaweed extracts must include not only new sources of established extracts but also the economic development and exploitation of new seaweeds and new seaweed extracts.

Names of seaweed extracts used in the past have often been of a confusing and contradictory nature. For a long time, seaweeds were known only by their local names or were classified by botanists who were mainly concerned with the physical appearance and morphology of the plant. It has only been in recent years, coinciding with the growth of the seaweed industry, that attempts have been made to define and clarify the terminology peculiar to it. Whistler,[1]

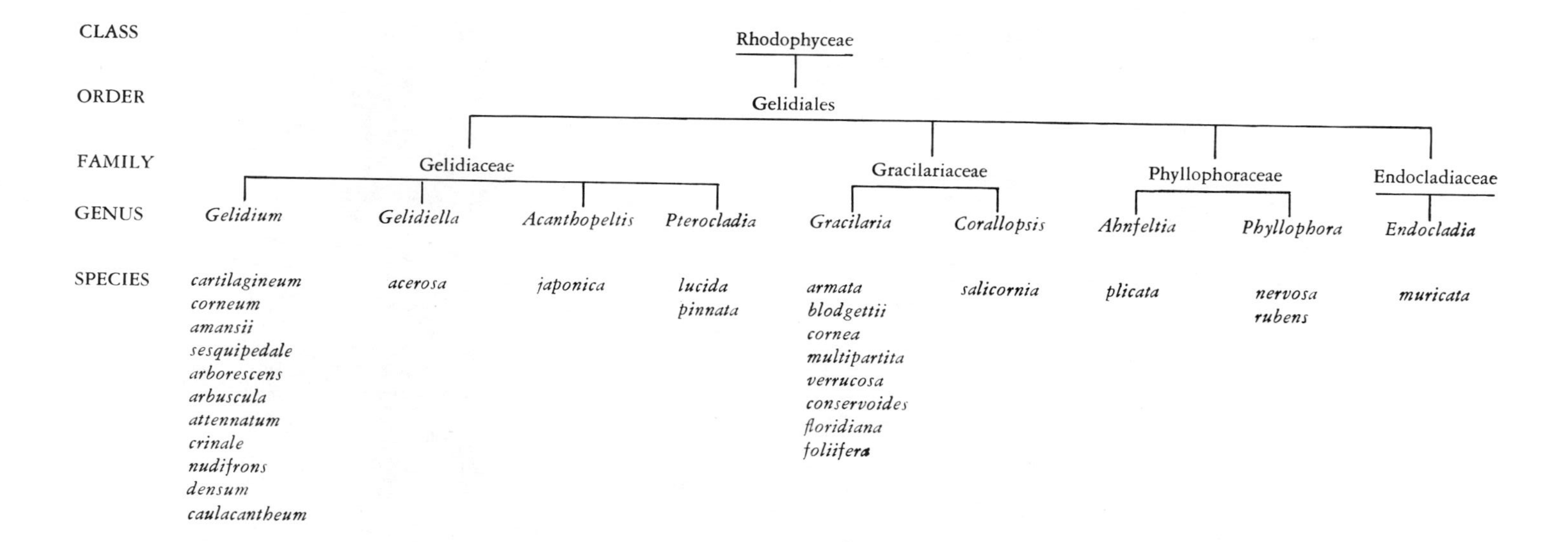

FIG. 1.—Agar-yielding seaweed species.

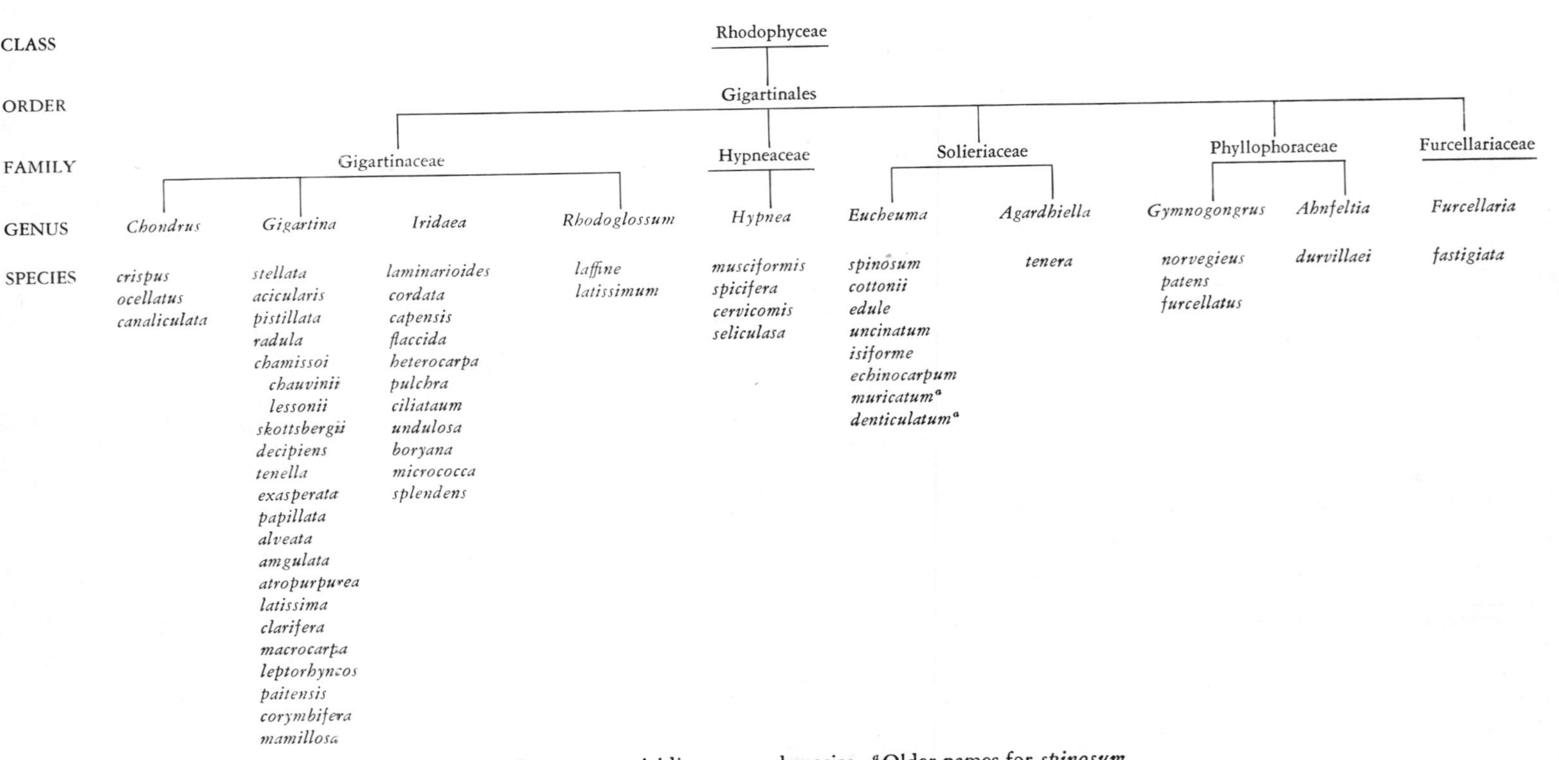

FIG. 2.—Carrageenan-yielding seaweed species. [a] Older names for *spinosum*.

Tseng,[2] Humm,[3] Stoloff,[4] Smith and Montgomery,[5] Glicksman,[6–8] and Schachat,[7] have led the attempt to systematize the nomenclature of marine algae. In this chapter, modern names suggested by these authors are used. In cases where the extracts have been nameless, the authors have chosen names in the same style as those given to similar types of extracts.

Because polysaccharides are carbohydrate polymers that are described by replacing the suffix *-ose* in the basic sugar unit by the suffix *-an,* Whistler[1] suggested that the *-an* ending be substituted for the undesirable *-in* ending in many of the former common names of polysaccharides, such as carrageenin and laminarin. This later became a general recommendation of the Polysaccharide Nomenclature Subcommittee of the Committee on Carbohydrate Nomenclature of the Division of Carbohydrate Chemistry of the American Chemical Society. The authors have extended this nomenclature of algae extracts in suggesting names for some seaweed extracts discussed in this chapter.

Some of the most important seaweed extracts are those of the *Eucheuma* species. Different species of *Eucheuma* yield various types of gelling and thickening extracts similar to agar and carrageenan. The original Malayan word *agar-agar* referred to certain East Indian seaweeds, particularly *Eucheuma muricatum.* The term eventually found its way to Japan and today is used only for extracts of the species of *Gelidium* and *Gracilaria* that meet the specifications for agar. The extracts of the various species of *Eucheuma* are actually carrageenan-type polymers, but in addition to the *kappa-* and *lambda-carrageenans,* a third, most unusual polymer, *iota*-carrageenan, has been isolated and is widely used.[6] Thus, today these extracts are known and sold under the more meaningful designation of *carrageenan* rather than the previously suggested name of *eucheuman.*[7]

The name *hypnean* was suggested for the agar-like extract of *Hypnea musciformis* and related species found in the United States and various other parts of the world.[7]

The extract of *Iridaea,* formerly called *Iridophycus,* has been established in the literature as *iridophycin* and was changed by Whistler[1] to *iridophycan* in order to conform to standardized nomenclature.

Gloiopeltis has long been utilized by the Japanese for the production of a hydrocolloid called *funori.* In keeping with the systemitized terminology, we suggested that the name *funoran* be used to retain the well-known identity of the material, while following present day nomenclature.[7]

The lesser-known genera of *Rhodymenia* and *Porphyra* in the United States are commonly known as *dulse* and *laver,* respectively. To avoid confusion between the genera of *Rhodymenia* and *Rhodomela* and to avoid the use of an unwieldy name, the authors recommended the name *dulsan* for the extract of *Rhodymenia* rather than the alternative *rhodymenian.*[7] The extract of *Porphyra* is less apt to be confusing and can readily be called *porphyran.*[7]

It is interesting that of all the vast families and genera of red algae, only a few are utilized to any extent for the extraction of hydrocolloid materials.

It has been difficult to select representative seaweeds that have not been commercially utilized. Sometimes only one species of a class shows promise. Thus, *Gigartina mamillosa* and *Gigartina stellata* are often used interchangeably with *Chondrus crispus* in the manufacture of carrageenan. However, *Gigartina acicularis* produces an extract that is quite different from the *Chondrus* extract and is important in its own right. This material, available from Portugal in commercial quantities, has been investigated by at least one U.S. seaweed extraction company, but very little information has been released.

The authors have chosen to discuss extracts from the following genera of algae: *Eucheuma, Gloiopeltis, Gracilaria, Hypnea, Iridaea, Porphyra,* and *Rhodymenia.* It is not implied that these are the only promising seaweeds for increased exploitation. They are representative only of that which is potentially available.

II. Hypnea and Hypnean

The extract of *Hypnea musciformis* and other *Hypnea* seaweed species is known as *hypnean.* Its chief value is its strong gelling ability. Furthermore, the physical properties of hypnean gels can be controlled to a considerable extent. Therefore, inasmuch as *Hypnea* species are rather plentiful throughout the world, hypnean could become an extract of economic importance.

1. Source and Producing Areas

Hypnea seaweeds are members of the class Rhodophyceae, order Gigartinales, and family Hypneaceae.[9] They are widely distributed, and the species found depends largely on the prevailing water temperature. Thus, 3–5 *Hypnea* species are found in the Caribbean Sea, but only one (*H. musciformis*) is found north of Florida.[10] Botanists have identified about 25 species of the *Hypnea* genus in the tropical seas of the world. One of the most important producing areas is along the Atlantic coast of the United States between Florida and Nantucket.

The weed is seasonal and is most abundant on rocky breakwaters during the summer and fall. Southern Florida (Tampa Bay), the Caribbean Sea, and, in general, the Gulf of Mexico are good sources of this alga. *Hypnea* weeds found in Australian and New Zealand waters show promise as substitutes for agar.[11] The Hawaiian Islands do not have *H. musciformis,* but they have a good supply of a closely related species. *H. nidifica,* referred to by the natives as *huna.*[12] Large quantities of *H. specifera,* known locally as *green tips,* are found along the coast of the Union of South Africa. The bright olive to green plant (3–16 in, 7.5–40 cm, in height) is of possible economic importance because, under optimum conditions, it forms very dense and almost pure communities. On some parts of the coast, it is the dominant plant; in other regions, it is found mixed with other weeds.[12, 13]

The ocean in the vicinity of Bali is the locale of *H. cervicornis.* It is collected and used by the natives, who have named it *boeloeng djadja.*[13] This weed is also found near the Philippines, where it is known as *culot.*[10] Commercial quantities of *H. musciformis* have been reported along the east coast of Brazil between the São Francisco River and Natal, inside the reefs. The densest weed population is north of Recife, near Cabadelo. Considerable amounts are also found south of Recife, between Recife and Aracaju.[3] *Hypnea cenomyce* is found near the Philippines, Indonesia, and Australia. *Hypnea divaricata* is found in the waters around the Philippine Islands, Indonesia, New Guinea, and in most warm seas. It is known as *arien* in Amboina.[9] Little is known about the available quantities of these weeds.

2. *Seasonal Effects*

Hypnea species are usually seasonal, although in some locations, the weed is found throughout the year. In the Beaufort, North Carolina area, the optimum season is May through July, whereas in Tampa Bay, Florida, the weed grows continually, but reaches a peak during the winter months from November to early February. Brazilian seaweed is not seasonal, but it can be collected on a large scale only during the dry season (October through March); the wet season is impractical for harvesting because the weed must be spread out for drying, and excessive rain would leach out considerable quantities of the extract, which is relatively soluble in fresh water.[3, 12]

Hypnea weeds grow on the ocean floor or attached to other plants or animals. They can be collected with long-handled rakes or grapnels in a manner similar to that used for collection of *Chondrus crispus.* Floating, detached plants sometimes accumulate in sufficient quantities to warrant collection.[10]

3. *Extraction and Purification*

Collected weeds are often stored as long as a week in fenced pounds, for if they are kept covered with fresh sea-water, they do not deteriorate appreciably. The processing starts by placing them on large wire-mesh racks, about 4×10 ft (1.2×3.0 m). The weeds are washed briefly with sprays of fresh water to remove salt and other impurities and then are completely dried, if they are to be stored.

Extraction of the cleaned weed is accomplished with water 50°–121°; the hotter the water, the more complete is the extraction. Potassium chloride is added to reduce the viscosity, and the extract is filtered. Hypnean is recovered from the solution by a freezing–thawing process which separates it from the potassium chloride and water-soluble carbohydrates.[10]

Extraction of small quantities of weed is easily done. Dry *Hypnea* weed is carefully but briefly washed in distilled water and then minced. It is heated at 100° for 30 min in distilled water in an open vessel. Addition of 0.2–0.5% of potassium chloride is helpful. The pH is adusted to 6, and insoluble matter is removed by centrifugation. Addition of several parts of ethanol precipitates hypnean. If it is undesirable to heat the extraction solution to boiling, the extraction can be accomplished at lower tempertures, using longer times and small amounts of acid. Thus, 1 part of weed in 100 parts of water is extracted in 1.5–2 hr at 45°–60° and yields 40–50% of hypnean on a dry-weight basis. A second extraction should be made using about one-half the original quantity of water. Purification procedures are the same as for high-temperature extraction. Addition of small amounts of acetic acid is helpful not only for low-temperature extraction, but also for weeds collected during the latter part of the season.[14] However, with extraction temperatures of 60°–95°, the pH should be between 5 and 6. With extractions at 95° or higher, the pH should be 6–7.

The yield of hypnean is low compared to that of agar. For example, *Gelidium* can be expected to yield about 6% (wet basis) of extract compared to about 2% for *Hypnea*.[3] On a dry basis, *Hypnea* weeds give between 30 and 40% of extract; a typical batch of *H. musciformis* from Brazil gave 35–40% of extract. A water to seaweed ratio of 100 : 1 produces a maximum yield of hypnean at 40°–60°. The optimum time at 50° is 45–60 min,[10] and the best extraction pH is 5.5–6.0.

4. *Commercial Aspects*

Hypnea was collected in North Carolina on a commercial scale in 1945 and 1946. Dry-weight production figures were 4000 and 27,000 lb (1800 and 12,000 kg), respectively; the price of raw weed was $0.10/lb. Although it has not been manufactured in the United States on a commercial scale since 1946, it remains an excellent raw material for potential utilization.[15] An early extractor of hypnean was Sperti Chemical Co., Beaufort, North Carolina.

5. *Structure*

Although the structure of hypnean is not fully known, it has been shown that the extract of *H. specifera* is similar to agarose (agaran) and *kappa*-carrageenan in that it contains a high percentage of 3,6-anhydrogalactose. Enzymic studies of the extract of *H. musciformis* have supported the presence of 85% of *kappa*-carrageenan-like material.[16] Paper chromatography and electrophoresis of hydrolyzates supported the belief that the polysaccharides of *H. musciformis* and

Chondrus crispus are identical.[17] The ester sulfate content[18] of hypnean is 12–16%.

6. *Properties*

Hypnean is quite similar to *kappa*-carrageenan or furcellaran in its properties, principally in that it forms a stiff brittle gel with potassium salts.[19–21] These gels are like agarose (agaran) in their short-break gel structure and synersis.[18]

Hypnean is also soluble in cold distilled water and resembles sodium carrageenate in this respect.[3] However, hypnean is quite insoluble in water containing an electrolyte. It is soluble in warm glycerol and ethylene glycol, and the solution forms a firm gel upon cooling. Aqueous hypnean dispersions have much greater viscosities than have equal concentrations of extracts of *Gelidium* or *Gracilaria* species. As in the case of other seaweed extracts, addition of electrolytes lowers the viscosity. However, the viscosity change of hypnean is much greater than that which occurs with other seaweed gums. The effect of salts on the viscosity of hypnean is shown in Figures 3 and 4.

Hypnean dispersions can form very strong gels. A 1% hypnean dispersion containing 0.5% of potassium chloride forms a stronger gel than does agar. The presence of an electrolyte is required for gelation, and hypnean dispersions made by extracting the weed with hot distilled water will not gel even upon cooling to 0°. However, addition of an electrolyte solution to such hypnean dispersions causes immediate gelation. Heating dissolves the hypnean gel; cooling produces a smooth uniform gel.[3]

7. *Effect of Nonelectrolytes on Hypnean Gels*

Nonelectrolytes can induce the gelation of hypnean dispersions, although they are not as effective as are electrolytes. Thus, hypnean dispersions can gel in the presence of sugars, glycerol, methanol, glycols, glycol monoalkyl ethers, and similar compounds.[10] Only small quantities of sugars are required to induce gel formation. Glycerol– and ethylene glycol–hypnean gels are notable for their toughness and resiliency. A typical gel of this type, made by dissolving 4% of hypnean in warm glycerol and cooling it, forms a tough gel that is quite permanent, regardless of ambient conditions. It is not subject to deterioration by evaporation or by bacterial action.

Humm[10, 15] examined the effect of mono- and polyhydric alcohols on gelation of hypnean. The weakest gel was produced with glycerol, a trihydric alcohol, and the strongest with monohydric alcohols. Whereas the gel strengths of gels made without added alcohols were too weak to measure, those made with 0.5 mole of a monohydric alcohol, 3.0 moles of water, and 1% of hypnean were as strong as, or much stronger than, 1% agar gels.[7]

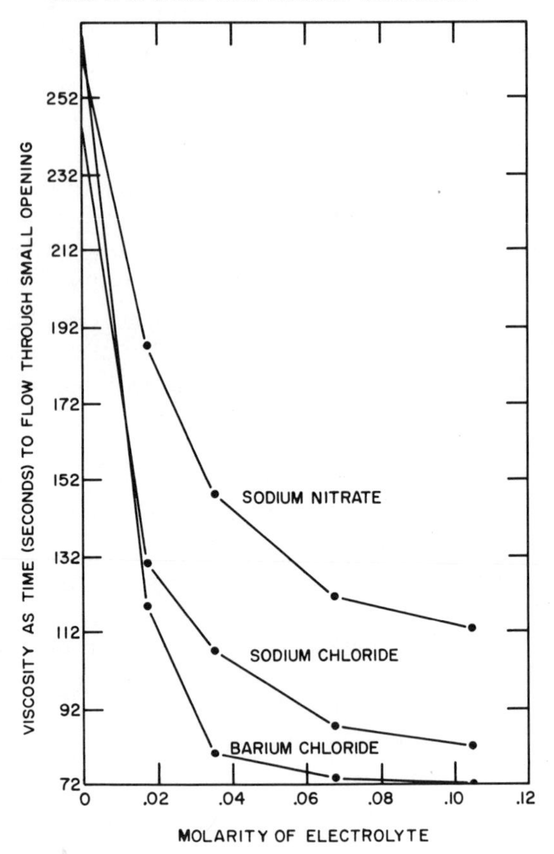

FIG. 3.—Effect of salts on viscosity of hypnean. (Courtesy of H. J. Humm)

8. *Effect of Electrolytes on Hypnean Gels*

In general, the effect of electrolytes on increasing the strength of hypnean gels is greater than that of the nonelectrolytes. Gel strength and temperature of gelation vary directly with added solute level. In general, the effect of electrolytes on hypnean gels is in the order of the lyotropic series. The effect on *Gelidium* gels is the reverse of that on hypnean gels.[3, 12] The cation is more important than the anion. Cation effectiveness, in decreasing order, is as follows (for chlorides): $Cs^+ > Rb^+ > K^+ > NH_4^+ >> Na^+$. Anions are effective in the following decreasing order (Potassium is the cation in each case.): $OAc^- > Cl^- > SO_4^{2-} > Br^- > HPO_4^{2-} > NO_3^-$.

Potassium chloride specifically influences gel strength, gelling temperature, and melting temperature of hypnean gels.[10] Addition of potassium chloride to

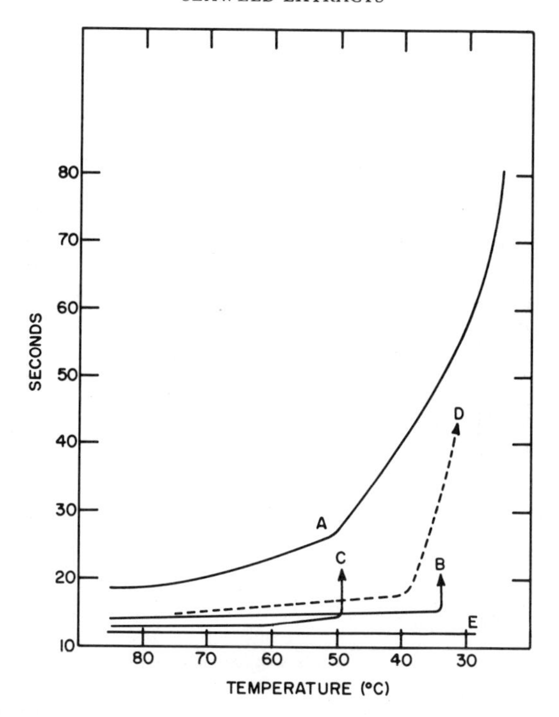

FIG. 4.—Relative viscosity of hypnean agar as a function of temperature (Courtesy of H. J. Humm) A, hypnean, 1%, no salts added; B, hypnean, 1%, with 0.2% KCl; C, same, with 6% KCl; D, *gelidium* agar, 1%; E, distilled water.

a 1% hypnean dispersion causes a lineal increase of gel strength up to 1.5% of salt. Additional potassium chloride has no further effect. The melting and gelling points of a hypnean dispersion are increased, at first rapidly and later more slowly, by addition of potassium chloride up to about 3%, and the two points remain a fairly constant 16° apart (Fig. 5). This temperature difference varies with the salt used. For cesium and rubidium chlorides it is 18°, and for monohydrogen potassium phosphate[10] it is 12°. Increasing the level of hypnean produces a strong gel and slightly elevates the temperature of gelation (Fig. 6).

Hypnean gels can be sterilized repeatedly by heating in an autoclave without appreciable loss in gel strength, provided the pH is adjusted to neutrality after each treatment. If such an adjustment is not made, the pH will progessively drop during heating. The pH of a 1% hypnean dispersion originally at about pH 7, dropped to 5.2 after 3 hr of heating in an autoclave. There was also a decided drop in ultimate gel strength.[10]

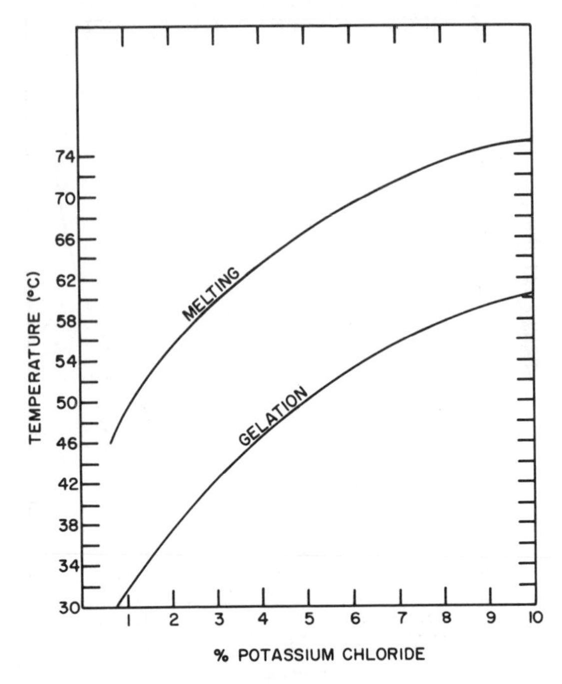

FIG. 5.—Effect of potassium chloride on melting and gelling temperatures of a 1% hypnean dispersion.[10]

Syneresis occurs in hypnean gels. Syneresis in 1% hypnean gels can be reduced with a potassium chloride content of 1.5% or more. With more extractive (1.5%), only 0.5% of potassium chloride is needed to keep syneresis low. Greater syneresis is produced by use of potassium chloride, potassium bromide, or potassium dihydrogen phosphate than with potassium nitrate, potassium sulfate, cesium sulfate, or ammonium chloride.[10]

9. *Uses*

Hypnean has excellent commercial possibilities because of the great range of its properties and their degree of control. Gel strength can be varied by changing the solute or the concentration of extractive; gelation temperature is controlled by varying the salt concentration; melting temperature can be regulated by the electrolyte used. Specific illustrations of uses for hypnean follow.

Hypnea musciformis is used as a vermifuge in Indonesia (where it is known as *worm roos*), Greece, and Turkey.[12] In Australia, it serves as an agar substitute for bacterial work. In the Philippines, the fresh weed is eaten raw or boiled and sieved to make jellied dishes.[9] A rather unusual use for *H. musci-*

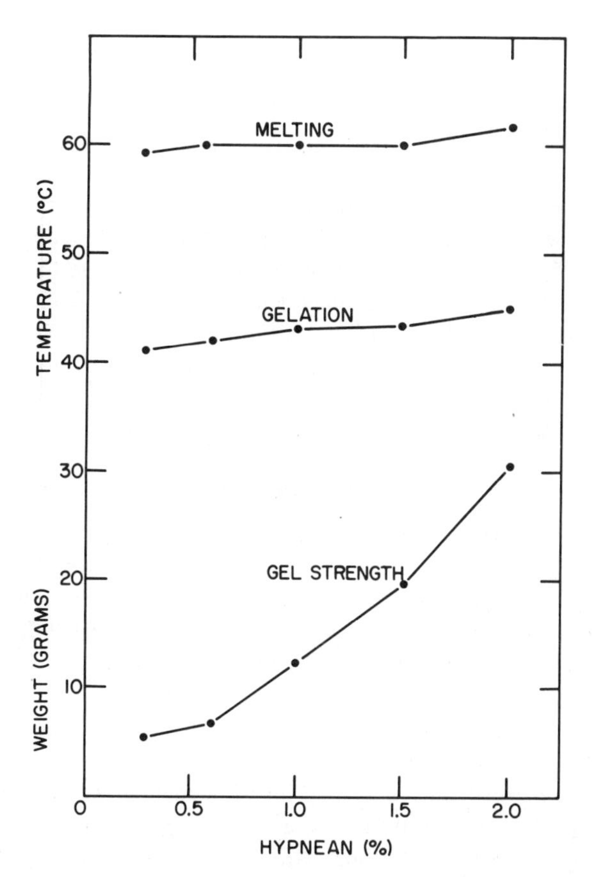

FIG. 6.—Effect of hypnean concentration on melting and gelling temperatures (in °C) and gel strength (in g) of its dispersions.[10]

formis is in the cultivation of blue-green algae.[22] Brazilian fishermen in the Natal region use considerable amounts of seaweed (much of it *H. musciformis*) as manure for young coconut palms in the arid, sandy soil near the shore. The palm roots are covered with seaweeds to keep the surrounding soil humid until the plant has developed an extensive and efficient root system.[7]

Hypnea cenomyce is eaten in Indonesia, and *H. cervicornis,* in Bali. The dried, bleached weed is boiled and filtered. After cooling, the jelly is eaten with palm sugar and coconut.[9] *Hypnea divaricata* is a food in Amboina. In Timor, it is used to prepare gels for bacterial studies. *Hypnea nidifica,* a common remedy for stomach disorders in the Hawaiian Islands, is also considered a desirable food. It is usually served as a relish on meats or poi. *Hypnea nidifica* is about

56% utilized by dogs and 10% by man.[12] *Hypnea specifera* from the South African coast is used to make both bacterial agar and a jelly for food use.

III. *Eucheuma* Extracts

The genus *Eucheuma* gained early prominence in the seaweed industry because the initial source of agar was *E. muricatum* (also known as *E. spinosum* or *E. denticulatum*).[23] *Eucheuma* extracts are considered to be carrageenan-type products rather than agar and are so classified legally, as well as technically.[6] Carrageenans in general are differentiated from agars in that they have a much higher ester sulfate content (20–50% calculated as SO_3Na).[23] In addition, the 3,6-anhydro-α-L-galactopyranosyl units of the agar structure are replaced by 3,6-anhydro-α-D-galactopyranosyl units in *Eucheuma* extracts.

1. *Domestic Sources*

Eucheuma is a red alga of the order Gigartinales and family Solieriaceae. At least two species of this genus are found in the United States, principally along the west coast of Florida. The common species found is *Eucheuma isiforme,* which is sometimes known as *E. spinosum* or *E. denticulatum*.[24, 25] In general, the taxonomic identification of the various *Eucheuma* species has been somewhat conflicting, and it is difficult to give a positive identification of these species because of the lack of adequate literature and herbarium reference specimens.

Experimental work to determine the economic potential of *E. isiforme* as a source of agar-like materials was done by Williams[26] in 1945. Pressure extraction of the weed gave only 0.43% of a nongelling extract based on the fresh weight of the plant; the indigenous *Gracilaria blodgetti* species yielded 2.9% of extract under the same conditions.

Another common *Eucheuma* species found in Florida is similar to *E. isiforme* but is sufficiently different that they can be distinguished from each other. This unidentified species is a red plant that grows to an average height of about 1 ft (30 cm) in water ranging in depth from 8 to 50 ft (2.4 to 15 m). It is very similar in appearance to *Gracilaria* and can be readily mistaken for *Gracilaria cornea.* Autoclaving a 3% slurry of the dried plant in water for 15 min at a pressure of 15 psig (1 atm gauge pressure) gave an average yield of 54% of extract based on the dried plant weight. An examination of extraction conditions showed that a gum having higher viscosity was extracted at lower temperatures or with shorter heating times. The extract is cold-water soluble and does not gel upon heating and cooling over a wide range of temperature.

Domestic species of *Eucheuma* are not used for commercial purposes at this time. Processors of seaweed rely instead on dried seaweed imported from various foreign countries.

2. *Commercial Extracts*

The properties of carrageenan from the genus *Eucheuma* differ from species to species, and from an applications standpoint, these various *Eucheuma* extracts offer unique properties. The two most commonly used species are *E. spinosum* (*E. muricatum* or *E. denticulatum*) and *E. cottonii,* which are found in substantial quantities in the warmer seas off the coasts of Southeast Asia, Southeast Africa, and the Philippine Islands. Several other species are used noncommercially in Asia.[19] One of the reasons that these species have not reached industrial significance is that the taxonomic identification of the various *Eucheuma* species is very difficult, in part because of the lack of a systematic study of the seaweeds of the areas where *Eucheuma* abounds and in part because of the vast number of forms inherent in the genus, some of which are only growth forms in various stages of development and others of which are truly different species.

In an effort to uplift the depressed economics of the nations of Southeast Asia, the United Nations through its Food and Agriculture Organization (FAO) has attempted to classify and organize information pertaining to marine algae that might have an economic potential. Several of these interesting *Eucheuma* species have been evaluated by Zaneveld[9] and Chapman[12] (Table I).

Small amounts of *Eucheuma* seaweeds have been used in various parts of the world for specific local applications. In Australia and New Zealand, *E. speciosum* was once used for the production of a gelling type of extract[11] but was replaced by *Gracilaria confervoides,* which gave a better product.[12] In Southeast Asia, *E. gelatinosa* has been used for the production of gelling materials and as a creaming agent in the production of natural rubber.[12]

Eucheuma papillosum (also termed *Kallymenia dentata*) is found in quantity along the Japanese coast where it is known to the natives as *tosaka-nori.* It grows on submerged reefs off the island of Kozu and along the shores of the provinces of Ize, Shima, and Higo. During the seasonal storms in August and September, large quantities of the weed are cast onto the beaches, where it is gathered and dried for use as a condiment or as an ingredient in soybean sauce. It is also collected by divers.[12]

Another species of *Eucheuma* that is common in Japanese waters is *E. spinosum.* This is of special interest because it is very similar to *Gloiopeltis* in both physical and chemical properties. Nakamura[27, 28] investigated the gum from *E. spinosum* and found that the ash content, nitrogen analysis, and other analyses indicated that it was identical with the extract of *Gloiopeltis.* Alkali treatment caused marked changes in the pentose, pentan, reducing sugar, total sulfate, and ash-sulfate contents. Nakamura believes that these substances are closely related to the gel-forming properties of the extract. A cold-water extract of the

TABLE I

Important Species of Eucheuma[9, 12]

Species	Common Names	Distribution	Uses	Remarks
E. cottonii		South Africa	Manufacture of carrageenan	Extract is *kappa-* (not *iota-*) carrageenan
E. edule	*Agar-agar besar* (Indonesia) *Agar-agar halus* (Celebes)	Indonesia, New Caledonia	For preparation of jellies	
E. gelatinae (*gelatinosa*)		India, Philippines, Indonesia, Japan, New Caledonia, Bourbon Isles	In Philippines for making agar. In Netherlands Indies to replace algin as creaming agent in manufacture of natural rubber	Made in the 1950's in Netherland Indies for use as agar
E. horridum		Malay Peninsula, Indonesia, Mauritius Isles, Bourbon Isles	For preparation of jellies	
E. isiforme		Antigua, Barbados	Food jellies	
E. Muricatum f. depauperata (*E. spinosum*) (*E. denticulatum*)	Rupruppuuc (Philippine Islands) *Agar-agar halus* (Indonesia) *Agar-agar kasar* *Agar-agar gesev*	India, Philippines, Indonesia, Mauritius Isles, Bourbon Isles, Japan, New Calendonia, Australia, Lord Howe Isles, Singapore	In general for food jellies. Natives of Malay, Indonesia, and Philippines use it for home preparation of agar alone or mixed	Not suitable for commercial agar. Considerable quantities exported to Japan as adulterant for agarophytes. Also exported to China.

	Agar-agar pulu *Agar-agar seru laut* *Karang laut* *Garanggang* Macassar agar (Malay) Java agar (Malay) *Algal-algal* (Malay) East Indian carragheen (Malay) *Chilints'ai* (unicorn vegetable) (China)		with agarophytes (e.g. *Gelidium* and *Gracilaria*) Eaten in Amoy. Commercially, used for preparation of *iota*-carrageenan	The extract has been identified as *iota*-carrageenan
E. populosa (Kallymenia dentata)	*Tosaki-nori*	Island of Kozu and along provinces of Ize, Shima, and Higo	Food (China), e.g. as a condiment or in soybean sauce	Grows on submerged reefs at depths of 8 ft (2.4 m) or more where it is collected by divers. Loose weed thrown on beaches by storms is also collected
E. serra	*Bulung djukut* *Lelipan* (Bali)	India, Indonesia, Mauritius Isles	Preparation of agar, as a vegetable, and as a medicine	
E. speciosum		Western Australia (Dongarra)	Preparation of agar	More recently partially replaced by *Gracilaria verrucosa* (manufactured in New South Wales)
E. uncinatum				*Nu*-carrageenan (precursor of *iota*-carrageenan) has been identified in the extract

air-dried weed had a pH of 6.85 and gave positive ketone and negative pentose and 6-deoxyhexose reactions.[27, 28] Acid-catalyzed hydrolysis of the cold-water extract released D-glucuronolactone, 3,6 anhydro-D-galactose, and D-galactose in a molar ratio 1 : 3 : 12.

Eucheuma muricatum (*E. spinosum*) from Indonesia was extracted with water at various temperatures (30°–100°) and also under dilute acid and alkaline conditions.[29] The various fractions, extracted successively at 30°, 50°, and 100°, differed significantly in their gelling and thickening properties and found distinctive applications as stabilizers in food and cosmetic products.

3. *Properties*

Both *Eucheuma spinosum* and *Gracilaria confervoides* were found to be sensitive to heat, mild acid, and gamma-ray irradiation, all of which caused depolymerization and, hence, lowered viscosities.[23, 30, 31] However, alkali treatment greatly increased the gel strength if the heating time was kept short.

Inasmuch as *iota*-carrageenan has a relatively low order of reactivity with potassium ions, seaweeds containing this extract, such as *Eucheuma spinosum,* cannot be fractionated using potassium ions. The solution may thicken when potassium salts are added, but even with relatively high concentrations of potassium (0.5 *M*), no precipitation occurs. This indicates that *iota*-carrageenan-containing weeds do not contain a *lambda*-carrageenan component comparable to that found in *Chondrus* and *Gigartina* seaweed species. An extract of *E. spinosum* may show the following characteristics and composition:[23] yield 55.5%; $[\alpha]_D$ +35°; D-galactose, 30.3%; SO_3Na, 37.8%; 3,6-anhydro-D-galactose, 19.0%.

Black and coworkers[23] describe some chemical modifications of *iota*-carrageenan such as partial acid-catalyzed hydrolysis and hypochlorite oxidation, where depolymerization occurs with good retention of sulfate, and mercaptolysis to produce pure 3,6-anyhdro-D-galactose diethyl dithioacetal.

4. *Structure*

Hydrolysis of *Eucheuma muricatum* extract yielded 3.27% glucuronic acid, about 16.5% galactose sulfate and 80% neutral sugars. The component sugars were xylose (1 part), anhydrogalactose (3.5 parts) and galactose (12.5 parts).[32]

In general, *Eucheuma* extract is structurally related to *kappa*-carrageenan but has a higher sulfate ester content and therefore cannot be fractionated with potassium salts as can *kappa*-carrageenan.[23, 33]

Extracts from various species of *Eucheuma* differ, although they are all basically carrageenan-type extracts. In the extract of *E. spinosum,* almost all the 3,6-anhydro-D-galactose groups are sulfated, whereas the extract of *E. cottonii* is identical with *kappa*-carrageenan, which has little or no sulfated anhydrogalactose units. The extract from *E. spinosum* has different properties from either *kappa-* or *lambda*-carrageenan. Although the original name suggested for *Eucheuma* extracts was *eucheuman,* the elucidation of their basic structures has now shown them to be similar to various carrageenan fractions, and hence, they can now be called *carrageenan* also.

The *E. spinosum* extract, which is different from most of the others, has been designated *iota*-carrageenan.[6, 18] It has now been suggested that various extracts be renamed on the basis of the chemical structure rather than on the basis of botanical derivation used previously.[18]

5. *Uses*

Eucheuma extracts are used in ice cream, sherbets, meats, and fish and also for cosmetic purposes.[29] The use frequently is dependant upon the method of extraction. Thus, crushed *Eucheuma* extracted with water at 40° gives a thin, nongelling extract useful for stabilizing ice cream. Extraction of the residue with water at 75° yields an extract suitable as a chocolate milk stabilizer. Extraction of the residue with boiling water gives an extract useful in frozen desserts or in tablet form as a laxative.

Water-gel desserts and starch–milk puddings based on *Eucheuma* extracts and shown to be capable of withstanding severe freeze–thaw conditions without textural breakdown were described by Glicksman.[34] The thawed gelled products retained a smooth unbroken texture, exhibited no syneresis, and were far superior to conventional desserts and puddings based on gelatin and starches, respectively.

An excellent dessert gel system resembling gelatin-type desserts based on *E. spinosum* extracts has been prepared by Foster and Moirano.[35] The product had good clarity, strength and textural qualities.

V. *Gracilaria* Extracts

The term *agar* has been applied for many years to the extracts of *Gelidium amansii* (Japan), *Gelidium cartilagineum* (California), and other *Gelidium* species. Nations poor in *Gelidium* have sought to produce agar from available native sea plants. Thus, Russia was able to produce high-grade agar from *Ahnfeltia plicata;* New Zealand used *Pterocladia lucida,* and Australia and South Africa, *Gracilaria confervoides.*[3] (See also Chapt. III.)

Of these seaweeds, the *Gracilaria* genera is of particular interest for several reasons. In the first place, large quantities of several species of *Gracilaria,* some of which have already been successfully utilized on a commercial basis, are available off the shores of the United States. Second, *Gracilaria* is interesting in that some species yield a good grade of agar, whereas others have agaroids that have different, but unique and useful, properties. With the growth of knowledge of the chemical structure and physical properties of these hydrocolloids, it might be that new uses will be found for these potentially valuable *Gracilaria* extracts. Common *Gracilaria* species are summarized in Table II.

1. Domestic Species

In the United States, the agar-yielding *Gracilaria confervoides* is present on both the west coast of California and the east coast of North Carolina and Florida.[38] The less desirable, weaker gel-yielding *G. multipartita* (also known as *G. lacinulata* or *G. foliifera*) and *G. blodgettii* are found only on the east coasts of North Carolina and Florida.[12, 38] Florida is also reported to be the source of *G. cornea* and *G. floridiana*.[38] The comparative properties of the extracts of these species were determined and compared to those of some similar foreign species.[38, 39] As can be seen in Table V, the different *Garcilaria* species yield extracts with various gelling temperatures and gel strengths, *G. confervoides* being the best of the domestic species (Table III). It is interesting to note that the domestic species *G. cornea* gives an extract that forms an extremely weak gel, whereas the Japanese species of the same plant is reported to give high-grade agar.

During World War II, the shortage of Japanese agar was instrumental in encouraging the production of agar-type extracts from available domestic seaweeds. Of all the domestic *Gracilaria* species, the most important was *G. confervoides*. The Beaufort Chemical Co., formerly the Van Sant Co., was established in Beaufort, North Carolina in 1943, and it began operations with the production of agar from *G. confervoides*. In 1945, another factory was established at Jensen, Florida for small-scale production of agar from *G. foliifera* from the Indian River. Later, this factory closed, and all available seaweed was processed at Beaufort. By 1949, the Beaufort factory had so expanded that it was capable of processing 1 ton (900 kg) of dry seaweed daily to yield 350–400 lb (160–180 kg) of agar. In subsequent years, because of unfavorable economic conditions, this plant was also forced to discontinue the production of agar.[3]

The commercial yield of agar from *Gracilaria confervoides* and *G. foliifera* was 15–20% on a dry basis (2% on a fresh basis) although laboratory yields of carefully prepared raw material ranged[3] as high as 40–45%. Lee and Stoloff also reported a 26–28% yield on a dry basis from *G. confervoides* from North Carolina.[39]

In Japan, recent production figures of *Gracilaria* have been estimated to be about 15,000 kg monthly. In comparison, production of *Gelidium* is about 40,000 kg monthly. *Gracilaria* beds are found off the coast of Venezuela and may be present in sufficient quantity to justify exportation to the United States for processing.[40]

2. Extraction

Dried, unbleached *Gracilaria confervoides* and *G. foliifera* were processed separately, but in the same manner. *Gracilaria foliifera* yielded a weak extract that could be used in certain applications in which a high gel strength was not wanted. The seaweed was covered with water in a wooden tank, and the mixture was heated to the boiling point with live steam for a short time to remove salts and pigments. The water was discarded, and the wet seaweed was cooked for 1 hr in a cooker-tank with fresh water in a 20 : 1 ratio. The solid residue was screened off, and the liquid extract was treated with a filter-aid and passed through a filter-press into 300 lb (135 kg) ice cans. The cans were cooled in ice and water and then frozen in brine. The seaweed residue was extracted a second time, and this extract was treated in the same way. The frozen agar gels were cut into blocks and then shaved into flakes, which were passed on trays through a hot-air drying tunnel. The final dry agar pieces were hammer-milled and packed.[3]

Although the extract of North Carolina *G. confervoides* was of excellent quality, it was found to be unsuitable as a substitute for agar in bacteriological culture media because of its high viscosity, excessive syneresis, and nonuniform gelation characteristics.[38] However, for other purposes, it could be readily used in place of agar. Successes in these other applications led to attempts at planting and cultivating *G. confervoides* in certain Florida and Carolina waters, but this venture was not successful because of the unfavorably high water temperature (30°–32°) in this area.[26]

The *G. confervoides* available on the California coast yields an agar of lower gel strength than that of the North Carolina species.[3] For this reason, it has not been collected in recent years. On the Pacific coast, another *Gracilaria* species, *G. sjostedtii,* grows in large quantities attached to rocks on sandy beaches. This species could also be utilized by industry if extensive beds are found.[40, 41] In the continuing search for agar-bearing seaweeds, two other indigenous Florida *Gracilaria* species were investigated, *G. blodgettii* and *G. lacinulata.* The extract of *G. lacinulata* has properties that approach those of commercial agar, whereas *G. blodgettii* is inferior to agar and the properties of its extract are markedly influenced by the processing treatment employed.[26] In more recent publications, it has been reported that *G. lacinulata* is actually the same species as both *G. foliifera* and *G. multipartita* and that it is essentially the same species that has been used for the production of agar in North Carolina.[42]

The conditions used in extracting either *Gracilaria* or *Gelidium* seaweed have an important bearing on the quality of the product. Usually, the clean dry

TABLE II

Common Gracilaria *Species*[12, 36, 37]

Species	Common Names	Distribution	Use	Remarks
G. blodgettii		Florida	Agar substitute (inferior product)	
G. caudata		Cuba, Puerto Rico	Potential agar source	1% sol is liquid. 2% sol gels at 31° and melts at 70° (Cuban variety)
G. cervicornis		Cuba, Puerto Rico	Potential agar source	1% sol is liquid. 2% sol gels at 35° and melts at 82° (Cuban variety)
G. compressa		England and Wales	Picked and eaten	
G. confervoides (*verrucosa*)	Gulaman Guraman Gulaman dagat (Philippines) Ceylon moss (British Commonwealth) *Raucau* *Xoa-xoa* *Ogo-nori* (Formosa) Chinese moss	Nearly cosmpolitan: Cuba, Puerto Rico, Atlantic, N. America, esp. Calif., S. America, India, Ceylon, Japan (esp. S. Kyushu), Australia (esin Botany Bay), S. Africa, China, Formosa, Philippines, Amoy, Pei-Tai-Ho, Wei-Hai-Wei,	It is used as food in 2 ways: (a) washed and eaten as a salad, raw or cooked; (b) sunbleached, dried, and marketed as a cheap substitute for gelatin. Eaten raw in Philippines. Used commercially as a source of agar although the agar is of lower quality than that obtained from gelidium weeds	1% sol is liquid. 2% sol gels at 37° and melts at 78° (Cuban). 1.5% sol gels at 28–30° and melts at 53°–54° (Puerto-Rican). In S. Africa, grows below low tide mark on Atlantic side of the Cape. No. Carolina weed yields 55–65% agar (dry basis).

G. cornea		Cuba, Puerto Rico, Florida, Brazil	Strong gel (Japanese species). (U.S. weeds give weak gel.) Used as gelling agent for food in W. Indies	1% sol is liquid. 2% sol gels at 33° and melts at 52° (Cuban). Gives 35% yield of agar (dry wt.)
G. coronopifolia	*Guraman* *Gulaman* *Limu manauea*	Hawaii	Food; composition given as 12.85% H_2O, 7.9% raw protein, 0.05% fat, 58.4% starch and sugar, 3.0% fiber, and 17.8% ash (30–33% utilized by dog and man)	
G. crassa	*Conji parsi* (India) *Susueldot-baybay* (Philippines) *Rau cau chung vit* (Viet-Nam)	Indian Ocean, India, Philippines, Indonesia	Eaten in Philippines	
G. crassissima		Cuba, Puerto Rico	Potential agar source	1% sol is liquid. 2% sol gels at 34° and melts at 78° (Cuban)
G. cylindrica		Cuba, Puerto Rico	Potential agar source	1% sol is liquid. 2% sol gels at 36° and melts at 79° (Cuban)
G. damaecornis		Cuba, Puerto Rico		1% sol is liquid. 2% sol gels at 37° and melts at 81° (Cuban).
G. debilis			Yields gel of exceptionally high strength	

TABLE II (Continued)

Species	Common Names	Distribution	Use	Remarks
G. eucheumoides	*Cauot-cauot* (Philippines) *Kauot-kauot* (Philippines)	Malay, Philippines, Indonesia, Japan, Indian Ocean, Pacific Ocean	In Philippines, eaten raw as a salad or cooked as a vegetable	Contains up to 59% agar
G. ferox		Cuba, Puerto Rico	Potential food gel	1% sol is a solid gel. 2% sol gels at 39° and melts at 73° (Cuban)
G. floridiana		Florida	Agar substitute	
G. foliifera		Cuba, Puerto Rico, Atlantic coast of U.S. (from Fla. to Mass.)	Agar substitute (poor gel strength)	1% sol is liquid. 2% sol gels at 27° and melts at 62°. 43–57% (dry wt.) of weed is polysaccharide
G. henriquesiana		Gold Coast of Africa, South Africa		2% sol sets at 46°–70° and has gel strength of 63 g/cm^2. Jelly melts at 87°
G. lacinulata		Atlantic, N. America	Extract produces agar of good properties	Reported to be same species as *G.foliifera*
G. lichenoides	*Conji parsi* (India) *Chan-chow parsi* (Ceylon) *Gargararao* (Philippines) *Agar-agar karang* *Djanggut dujung*	Southeast Asia, Indian Ocean, Japan, most Pacific Islands, Australia, Tasmania, Ceylon, Lingga (Indonesia), Hawaii	Used in various food preparations such as soups, puddings and jellies, and as medical poultice for certain ailments.	Gel strength variable (52–98 g) compared with 75 g for Japanese agar

	Bulung embulung *Bulung tombong puti* *Sajor karang* (Indonesia) *Agar-agar karang* (Indonesia) *Djanggut monjet* (Indonesia) *Doejoeng* (Indonesia) *Aysana* *Aytsana* *Lotu-lotu putih* *Rume yar waccar* *Callocane* *Dongi-dongi* *Gose* *Sango-sango* *Limu manauea* Bengal-isinglass Ceylon moss		Use of agar from this weed has been explored in India, proved acceptable, and found comparable with Japanese weed. In Hawaii, eaten both raw and cooked.	
G. mamillaris		Cuba, Puerto Rico	Food gel	1% sol is a solid gel. 2% sol gels at 39° and melts at 77° (Cuban)
G. multipartitia		Atlantic, N. America	Agar substitute	Reported to be same species as *G. foliifera*
G. taenioides	*Djanggut dujung* *Doejoeng* (Indonesia)	Malay peninsula, Indonesia, Ceylon, Korea, Japan, Australia, Bangka (Indonesia)	Used in Singapore to make an agar-type preparation	

TABLE III

Gelling Properties of Common Gracilaria *Species*[38]

	Gelling Temperature	Concentration, %	Breaking Load[a]
Commercial strip agar	37°	1	80
	37°	1.5	270
Japanese agar	38°	1	70
	38°	1.5	80
G. confervoides (North Carolina)	43°	0.5	113
	43°	1	407
	59°	1.5	650
G. confervoides (California)	47°	1	120
	47°	1.5	190
G. cornea (Florida)	46°	1	0
	46°	2	10
	46°	3	15
G. floridiana (Florida)	48°	1	20
	48°	2	110
	48°	3	210
G. confervoides (South Africa)	37°	1	187
	37°	1.5	337
G. confervoides (Australia)	46°	1	95
	46°	1.5	167

[a] Grams per cm of plunger circumference.

seaweed is wetted, ground to a paste, washed, extracted with boiling water at atmospheric pressure, treated with activated carbon, filtered, and allowed to set into a gel. Purification is by freezing, thawing, and drying. A clean white product in 30–40% yield (based on dry seaweed) is obtained.

Recent advances have been made in extraction methods in order to obtain products with high potential gel strength. Two papers describe a procedure in which *G. confervoides* is preheated in alkali. Typically, the weed was soaked in 60% alcohol containing 5% sodium hydroxide, boiled 4 hr, washed with water, and dried.[43, 44]

Gamma-ray irradiation of *Gracilaria* increases the gel strength of the extract. A small initial dose of 2.3×10^{18} eV per gram increases the gel strength, over a control, 1.3–2.5-fold.[45]

3. *Japan*

In Japan, the various species of *Gracilaria* have been used in combination with *Gelidium* for the production of agar for many years. *Gracilaria* may be added because it is cheaper or because the resulting gel is softer, more elastic, and more suitable for food uses.[3] By far the most important *Gracilara* species in Japan (and, for that matter, in the world) is *G. confervoides*. Results of work on a semiindustrial plant scale have given 57% yields on a dry weight basis.[46] The crude agar was of good quality and had a gel strength of over 400 g/cm^2.

A great deal of work has also been done on the improvement of extraction procedures.[47, 48] Studies have been made on the seasonal variation of the chemical composition of *G. confervoides*[49] as well as on the structure of the extract itself.[50, 51] Attempts have also been made to use this weed as a source of D-galactose.[52]

In addition to utilizing *G. confervoides* for agar production, the Japanese also eat it as a garnish called *ogi-nori*.[12]

4. *Indo-Pacific Area*

The various species of *Gracilaria* that are utilized by the natives of south and east Asia are listed in Table II.

5. *Australia*

The shortage of agar in World War II also caused Australia to look for it in her own natural resources. The large quantities of *Gracilaria confervoides* that are distributed for 850 miles (1400 km) along the southeastern Australian coastline served as a source of raw material. Washed-up weed was used initially, but bulk harvesting with grapple irons was much more practical and was adopted. After collection, the weed is laid on wire-netting racks to dry in the sun. Eight tons of wet weed produce 1 ton of bleached weed and 7 tons of wet weed produce 1 ton of dry dark weed; 3 tons of dried weed yield 1 ton of agar which is equivalent in grade to Japanese agar and has a high setting temperature of 47°–48°.[11, 12, 53]

6. *South Africa*

During World War II, South African seaweeds were examined as potential agar sources and *G. confervoides* was selected for use as the most abundant. Known in South Africa as *sea string,* the weed grows below the low-tide level on the Atlantic side of the Cape; much of it is cast ashore after winter storms.[12] Agar derived from the South African *Gracilaria* can readily be used for culture media because of its gel temperature (35°–37°), its high gel strength, and its relatively low viscosity when in solution.[3] *Gracilaria corneum* is also available in South Africa but has not been utilized there.

7. *India*

Gracilaria confervoides found off the eastern coast of South India can be extracted to yield an agar suitable for use in culture media, food products, and related industrial applications.[54]

The most important Indian species, however, is *G. lichenoides,* Ceylon moss, which has been used in Asia for centuries for the production of agar-type extracts. Large quantities of this weed are found both in Indian waters and along the coast of Ceylon.[55]

An improved process for the extraction of agar from this species has been reported.[56] The seaweed is first washed and soaked for 2 hr in fresh water and then sun dried with aeration. It is then recleaned three times with water and finally leached with water (10 : 1) for 24 hr. The mixture is agitated for 10 min and filtered through cloth, and the residual pulp is sun dried on the cloth. The pulp is extracted with 75 parts of mother liquor for 15–30 min at 90°–95°. The extract is allowed to cool slowly to its gelation point of 43°; the gel is cut into strips and sun dried.

8. *Other Useful Species*

Buzzonetti[57] has reported the use of various species of *Gracilaria* for the production of agar in Italy. *Gracilaria cornea* from Brazil is reported to give a 35% yield of agar, based on the dry weight of the weed. This extract has a lower gel strength than commercial agar but is comparable to it with respect to viscosity, gelling temperature, and melting temperature. In a 2% concentration, it can be used as a culture medium but is more suitable for other uses.

Chilean *G. lemaeniformis* is reported to yield large quantities of an agarlike product.[58] In the West Indies, dry *G. cornea* is sold as a gelling agent for foods.[3] *Gracilaria henriquesiana,*[59] found off the African Gold Coast, yields an agarlike gel that has a melting point of 86°–87° and a gelation point of 46°–47° for 1 and 2% solutions. A 0.5% solution at 50° has a relative viscosity of 1.9; a 2% solution at 30°–35° has a gel strength of 63 g/cm². Using 50 parts of water to 1 part of seaweed at pH 8, a 10–20% yield can be obtained, based on the dry weight of the plant.

9. *Structure and Composition*

Gracilaria is known to consist of at least two or more fractions. Separation can be achieved by acetylation with acetic anhydride and pyridine, and fractionation, with chloroform into soluble agarose (agaran) acetate and insoluble agaropectin acetate. Deacetylation of the fractions is achieved with alcoholic alkali.

Agar from *Gelidium amansii,* for example, is believed to consist of alternating 3-*O*-substituted β-D-galactopyranosyl and 4-*O*-substituted 3,6-anhydro-α-L-galactopyranosyl units.[41] The extract from *Gracilaria verrucosa* probably has the same structure. Some residues are 6-*O*-methylated. Thus, it appears that all agaroses (agarans) have much the same structure, namely

$$[\rightarrow 3)\text{-}\beta\text{-D-Gal}(1\rightarrow 4)\text{-}\alpha\text{-L-AGal}(1\text{—}]_x\ [\rightarrow 3)\text{-6-}O\text{-Me-}\beta\text{-D-Gal}(1\rightarrow 4)\text{-}\alpha\text{-L-AGal}(1\text{—}]_y$$

where AGal is 3,6-anhydrogalactopyranosyl and $x : y$ is about 50 : 1 for *G. amansii.*

Analysis of *Gracilaria verrucosa* shows 61% agarose (agaran), 36.3% D-galactose, 16.3% 6-*O*-methyl-D-galactose, 44.0% 3,6-anhydro-L-galactose, and 2.1% L-galactose.[12]

Less is known about the agaropectin component. However, *Gelidium* agaropectin (which may be similar to *Gracilaria* agaropectin) appears to be a polymer of agarose (agaran) with uronic acid units. It may consist of a mixture of 3-*O*-substituted β-D-galactopyranosyl units joined to 4-*O*-substituted 3,6-anhydro-L-galactopyranosyl units plus a small proportion of sulfate, D-glucuronic acid, and perhaps, some pyruvic acid.

The higher the ratio of agarose (agaran) to agaropectin, the lower the gel strength. In *Gelidium* agars, the ratio is about 1.5 : 1, whereas in *Gracilaria,* it runs as high[12] as 20 : 1.

V. *Porphyra* AND PORPHYRAN

1. Distribution

The *Porphyra* genus of seaweeds is abundant in the United States, but the plants are used for food rather than as a source of a polysaccharide extractive (Table IV). The plant commonly harvested is *Porphyra perforata.* Called *California laver, purple laver* or *red laver,* it grows on rocks in sheltered places between 3.5 and 2 ft (1.0 and 0.6 m) tide levels and has blades that are about 3 ft (1 m) in length. It is common from Alaska to Southern California. Locally, from Washington to California, two other *Porphyra* species are found in much smaller quantities. These species are *P. naiadum,* also known as *red fringe* because of its habit of hanging from the edges of eel grass or other marine plants like a delicate fringe, and *P. lanceolata* (*red jabot laver*).[60] A large amount of *P. perforata* is also available. Laver is usually pulled by hand from the rocks on which it grows and then is dried in the sun in rectangular sheets.

The most extensive use of *Porphyra* is made in Japan, where its cultivation, which began in about 1700 in Tokyo Bay, is an extremely important and highly developed industry. In 1901, 2000 acres (800 hectares) of ocean floor were cultivated, while by 1944 the acreage was more than 12,000 acres (4800 hectares) with the annual production valued at about $6,000,000. The most intensely cultivated species is *P. tenera.*

The *Porphyra* plant reproduces by microscopic single-celled spores, which are shed by the thousands. These spores can only germinate by attaching themselves to a rough surface in a suitable tidal zone. By increasing the amount of surface available for spore attachment, the number of germinating spores can

TABLE IV

Common Porphyra *Species*

Species	Common Names	Distribution	Use	Remarks
P. angustata	*Kosujinori*	Japan	Food	
P. capensis	Laver	S.W. Africa	Food	
P. columbina	*Karengo* (New Zealand)	New Zealand, Chile	Steamed and used as a delicacy by Maoris	
P. dentata	*Oni-amanori* (Japan)	Japan	Eaten in China and Japan	
P. kunieda	*Marubanori*	Japan (wild species)	Food	
P. kunthiana	*Luche apanodo fresco* (mixture with *Ulva latissima*)	Chile	Food	
P. laciniata	*Slack, Sloke, Slouk, Sloukaen*, or *Sloukaum* (Scotland)	Great Britain	Eaten in W. Europe, England, Wales and Alaska	Eaten as a salad, but more usually cooked as a breakfast dish
P. naiadum	Red fringe	U.S. (Pacific) and Orient	Food	Contains vitamin C
P. nereocystis			Food	Contains vitamins B_1, B_{12}, and C
P. okamurai	*Kuro-nori*	Japan	Food	
P. onoi	*Ono-nori*	Japan	Food	

P. perforata (*Wildemannia perforata*)	Purple laver, red laver, California laver	Pacific Coast of U.S.	Eaten in W. Europe and Western U.S.	Over 300,000 lbs (135,000 kg) dry weight has been collected in one year. Contains vitamins B_1, B_{12} and C
P. pseudolinearis	*Uruppui-nori*	Japan	Food	
P. suborbiculata		Amoy	Eaten in China	
P. tenera	*Nori, amanori, hoshinori, asakusa-nori*	Cultivated in Japan	Protein rich food. 75% of the weed digestable	Contains about 21.8% of water-soluble nitrogenous substances. Contains 18 parts per million iodine (dry basis). Contains vitamin B_2
P. umbilicalis	Laver (England and Wales) *Chishima-kuronori* (Japan)	Europe, S. America. Cultivated or collected in Hokkaido (Japan)	Eaten in England and Wales	Contains vitamin C
P. vulgaris		Japan	Food	A form of *P. umbilicalis*
P. yezoensis	*Susabinori*	Japan	Food	

be greatly increased. The Japanese do this by placing rows of bamboo or oak brushwood in the intertidal zone. These brushwood areas soon become covered with *Porphyra* spores, which germinate and grow into adult plants. The plants are gathered, chopped, and washed; later, with a wooden scoop, they are transferred to sectioned wooden frames standing on bamboo mats. After the water has drained away, the dark-brown, membranous sheets are dried and packed for sale. The final dried product, called *nori,* can be toasted and eaten hot as a delicacy, used to wrap rice balls, or used as fiavoring.[61]

Porphyra lacinata, or laverbread, is abundant on the Welsh coast and has been used as a food in Wales since the 15th century.[61]

Porphyra is also cultivated in South Korea to produce a quality product known as *doug-laver, phoo-laver,* and *poug-laver.*[12]

2. *Structure*

It has long been known that the carbohydrate cell wall of *Porphyra umbilicalis* consists of a galactan composed of D- and L-galactose, 6-*O*-methyl-D-galactose, 3,6-anhydro-L-galactose, and ester sulfates[62, 63] in the molar ratio of 6 : 3 : 2 : 4. The principal ester sulfate was identified as L-galactose 6-sulfate. The carbohydrate content[64] is variable depending upon the species and season but usually is D- and L-galactose, 24–45%; 6-*O*-methyl-D-galactose, 3–28%; 3,6-anhydro-L-galactose, 5–19%; and ester sulfates 6–11%.

Recent work on structure indicates that porphyran is a linear chain of 3-*O*-substituted β-D-galactopyranosyl units and 4-*O*-substituted α-L-galactopyranosyl units in alternating sequence. However, the regularity is interrupted by the occurrence of 6-*O*-methyl-D-galactopyranosyl units and 3,6-anhydro-L-galactopyranosyl and L-galactopyranosyl 6-sulfate units.[65, 66]

There is also evidence of an associated xylan and a β-D-$(1\rightarrow4)$ mannan.[67]

VI. *Gloiopeltis* AND FUNORAN

Funoran is the solubilized polymer of the seaweed genera *Gloiopeltis* (Table V). *Gloiopeltis* plants, termed *funori* in Japan and *hailo* in China, have many uses in the Orient. In Japan, more than 100 producers market about $1,000,000 worth of the material per year.[3]

Gloiopeltis weeds are marketed as the dried, and frequently bleached, whole seaweed. Extraction of the plant is not required, for merely mixing the fresh or dried plant with lukewarm or hot water causes the entire plant to dissolve, giving a clear, viscous, nongelling colloidal solution with virtually no extraneous suspended matter.[68, 69] The structure and properties of funoran are not known. It is of potential use to industry.[70]

TABLE V

Common Gloiopeltis *Species*

Species	Common Name	Distribution	Use	Remarks
G. coliformis		North China Malaysia Indonesia	Laundry starch. Production of agarlike compound	May be a form of *G. furcata*
G. complanata	*Hana-funori*	Japan	Adhesive	
G. furcata	*Funori* (Japan) *Fukoro-funori* (Japan) *Hailo* (China)	Japan, North China	Adhesive. Sizing. Source of iodine. Constituent of hair-dye formulations. Used for settling of suspended matter and as a paste former for dynamite	
G. intricata			Adhesive	May be a form of *G. furcata*
G. tenax	*Ma-funori* *Yanagi-funori*	North China, Japan (Hizen and Isushima areas)	Viscosity producer in many applications. Source of vitamin B complex	

1. *Available Species and Producing Areas*

Funoran is prepared from several species of *Gloiopeltis;* the most important are *G. tenax* and *G. furcata.*[5] Most funoran comes from *G. furcata* (*funori* or *fukoro-funori*); lesser amounts are collected from *G. tenax* (*ma-funori*) and *G. complanata* (*hana-funori*).

The habitat of the *Gloiopeltis* genus is the shallow, rocky waters off the coast of Japan, principally near Osaka. The most abundant growths are usually found in warmer waters.[3, 12, 71] Undetermined quantities of *Gloiopeltis* are present along the coast of California.[38] In fact, originally the United States' agar industry was based on *Gloiopeltis* weed and used *Gelidium* only casually.[12]

2. *Collection and Manufacture*

The plant is collected at any time of the year by means of long-handled rakes or hooks that are used to detach it from the rocks on which it grows. That growing in deeper areas beyond the reach of rakes is collected by divers.[12] After collection, the seaweed is dried, cleaned, and sprinkled with water or softened by steaming. Drying with occasional water sprays to prevent excessive curling produces a bleached material. The weed is packed in thin layers on large shallow trays or mats and dried in sheets. Several sheets are packed together for sale in the form of a roll ~3 ft (~1 m) long and ~6 in (~15 cm) in diameter. The user merely breaks a chosen amount from the roll and dissolves it in warm water.[3, 12]

Japan is the major producer of funoran and produced an annual average of 770 tons during 1940 to 1949.[72] The Peoples' Republic of China is the world's second largest producer of funoran but imports some from Japan. Typical Japanese production is about 5000 tons of wet weed a year, whereas Chinese production is 500–1000 tons a year, mostly from the Chechiang area.[68, 69]

Several grades of funoran are commercially available and range in price from $0.60 to $1.00/lb ($1.32 to $2.20/kg), depending upon quality.

3. *Structure*

The exact structure of funoran is unknown, but there are indications that it is mainly a galactan sulfate.[12, 69, 72, 73] Early work on its hydrolysis gave 61.3% of reducing sugars consisting of L-fucose (6-deoxy-L-galactose), D-galactose, and L-arabinose.[72] Later, it was found that funoran was a disulfuric half-ester of a polysaccharide consisting of an equimolar mixture of D-galactose and a hexuronic acid.[74, 75] More recent work states that funoran hydrolyzate contains 3,6-anhydro-L-galactose, D-galactose, and L-galactose in the molar ratio 8 : 12 : 1. Sulfate is also present in large quantities, about 18.2% on a dry basis in funoran from *G. tenax* and 15.4% in the extract from *G. furcata.*[76]

Gloiopeltis contains a mixture of polysaccharides. Two other polysaccharide sulfates have been isolated from the water extract of *G. furcata*. Both of these contain D-galactose, D-xylose, D-*arabino*-hexulosonic acid, and 3,6-anhydro-L-galactose.[77]

Funoran and the extract from *Euchema spinosum* contain the same principal sugar units.[22] Funoran contains about the same sulfate content as *kappa*-carrageenan but its properties are different. This may be because of the 3,6-anhydro-L-galactose in funoran, typical of agarose (agaran), rather than D-form, typical of carrageenan. Stancioff[78] considers funoran to be a sulfated analog of agarose (agaran). Chapman[12] also classifies funoran as an agaran, thus catagorizing it with extracts of *Gracilariia* and *Gelidium*.

Funoran contains 53 ppm of iodine.[12] It contains practically no lipids but is rich in protein and carbohydrate.

Mita[79] determined the inorganic components of the plant and the purified, ethanol-precipitated gum of *Gloiopeltis furcata*. The contents based on the ash of the seaweed and the gum, respectively, are sodium, 6.6%, 11.9%; potassium, 1.8%, 3.2%; silicon, 0.9%, 3.1%; phosphorus, 0.7%, 0.2%; aluminum, 0.4%, 0.1%; iron, 0.7%, 0.4%; calcium, 4.3%, 1.7%; magnesium, 8.2%, 4.2%; sulfur, 19.5%, 17.6%; total ash, 13.4%, 16.1%.

4. *Properties*

The viscosity-producing ability of funoran is dependent upon a number of factors, such as species, plant size, degree of branching, time of boiling, impurities present, and storage conditions. Funoran from *G. tenax* has a viscosity that is almost 50% higher than that from *G. furcata,* even though extraction procedures are the same. Boiling lowers the viscosity of funoran from both species (Fig. 7). Premium-grade funoran is obtained from large, slightly branched plants (Fig. 8). Generally, storage of funoran increases its viscosity. Each species of *Gloiopeltis* appears to have an optimum heating temperature for maximum viscosity. This temperature is about 38° for *G. furcata* and *G. tenax* but is about 80° for funoran (Fig. 9).

Acids, bases, and salts reduce the viscosity of funoran solutions. They all have approximately the same effect and this lowering of viscosity appears to be common to electrolytes of all types. However, addition of water-soluble ketones or alcohols to aqueous funoran solutions raises their viscosities. Maximum viscosity occurs with 30% methanol, 30% ethanol, 30% 2-propanol, 30% acetone, 40% formalin, and 100% ethylene glycol. Phenol has no effect on viscosity up to a concentration of 5%. Sodium peroxide, used as a bleach, does not affect viscosity up to a concentration of 0.1%; above this concentration, it causes a reduction in viscosity. Sodium bicarbonate lowers viscosity and adhesive power.[69]

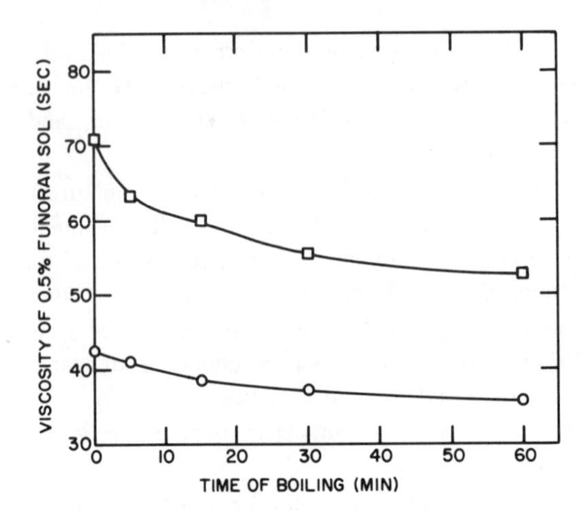

FIG. 7.—Boiling times versus viscosity of 0.5% solutions of *G. furcata* (○) and *G. tenax* (□) at 24°. Viscosity is given in seconds as determined by a Redwood viscometer.[80]

Any fermentation that occurs during processing destroys linkages in the molecule and results in a product with a lower viscosity.[83]

5. *Uses*

The many uses of funoran center around its excellent sizing and adhesive properties. Some representative uses are illustrated in the following examples.

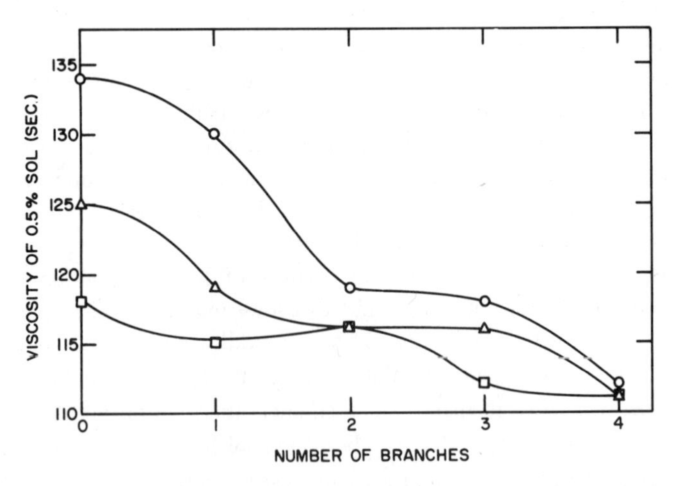

FIG. 8.—Effect of branching and size of the *Gloiopeltis* weed[81] on viscosity of the resulting funoran solution given in Redwood seconds at 24°. (○) Large-sized group; (□) small-sized group; (△) medium-sized group.

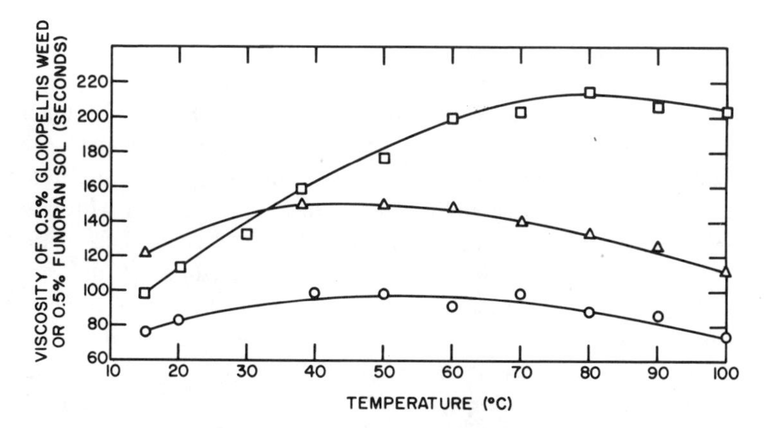

FIG. 9.—Effect of temperature at time of dissolution on the viscosities at 15°, in Redwood seconds, of 0.5% solutions of two species of *Gloiopeltis* and that of the prepared bleached product, funoran.[82] (○) *G. furcata;* (Δ) *G. tenex;* (□) funoran.

Hair waving and dyeing preparations—One of the most important uses of funoran in Japan is that of a constitutent in hair-waving preparations.[84] A typical formula is composed of a solution of 20 parts of sodium alkyl sulfonate, 0.7 parts of funoran, 4 parts of ethanol, 0.3 parts of perfume, and 75 parts of water.

Hair dyeing formulas also use funoran to impart sufficient viscosity so that the dye will stay on the hair longer.[85] A typical formulation consists of 1 g of 1,4-diamino-2-benzenesulfonic acid, 1.2 g of sodium perborate dihydrate and 3 g of funoran. The mixture is kept in a sealed container and dissolved in about 20 ml of water before use. Other amino compounds can be used in place of the 1,4-diamino-2-benzenesulfonic acid, for example, phenylenediamine-*N*-sulfonic acid or *p*-aminobenzenesulfonic acid.

Adhesives.—*Gloiopeltis furcata* yields a mucilaginous extract useful as a household adhesive.[3, 12] Some workers recommend autolyzing the weed at 10°–70° and then making the extract slightly alkaline with sodium bicarbonate.[86] About 50 g of *G. furcata* is immersed in 200 ml of 1% sulfuric acid, and the mixture is heated at 50° for 10 min. The solution is made alkaline with 100 ml of 17% sodium hydroxide. About 10 g of carbon disulfide is added, and the mixture is heated for 3 hr to produce a strong, somewhat elastic glue with lasting adhesive properties that is especially suitable for use with wood.[87]

In China, funoran glue is used in art work. Delicate Chinese paintings made on very thin paper are pasted onto thick, tough paper with these glues.[69] The glues are also used on cement walls and tiles.[12] In addition, funoran can be used in forming pellets of iron ore for processing.[88]

Sizing.—Textile and paper sizing is one of the most important applications of funoran.[3, 69] Its chief value as a size lies in its excellent solubility and short solution time. In the textile industry, it is used for glazing and stiffening fabrics (especially silk) and for stiffening threads;[12, 68] 5 lb (11 kg) of funoran will size a batch of cloth 1 yd (0.9 m) wide and 280 yd (250 m) long.[69] A typical cloth sizing is prepared by dissolving *G. furcata* in boiling water and bleaching the solution with sodium peroxide. The product is then mixed with a wax, a starch, and a wetting oil,[69] and the mixture is dried and powdered for use.[89]

Paints.—Funoran produces excellent water-based paints of the calcimine variety. For this, it is sometimes first combined with plaster.[3, 69] A related use is in shoe polish or shoe dye, in which funoran helps to retain the coloring matter that is otherwise worn off quickly.[76] Funoran paints are also used to decorate porcelain. The coloring agent is bonded to the porcelain by the funoran, which is later lost in the heating process without forming any off-colors or tones.[12]

General.—Funoran serves as a starch-like laundry agent in Japan and China.[3, 90] Funoran has the ability to function as a blood coagulation inhibitor, both *in vivo* and *in vitro*.[74, 91] Its effectiveness varies with the type of animal tested and decreases in the following order: man $>$ rabbit $>$ guinea pig $>$ dog $>$ cat. Heat does not destroy the inhibiting effect.[91] A resemblance to heparin was quite apparent.[91] One report suggests that calcium influences the coagulation inhibition to some extent.[92] In tests with rabbits, some degree of toxicity was found; but dilute solutions showed no toxicity. The lasting effect was better than that of heparin, because 24 hr were required for blood to regain its normal clotting time.[92]

Aqueous seaweed extracts serve as excellent culture media for growing yeasts.[93] For this use, some polysaccharide fractions are hydrolyzed with various enzymes, but whole funoran may be used with its full complement of nutrient minerals and protein.

Funoran is an important ingredient of a preparation use to coat clay and to cement casting molds.[94]

Gloiopeltis furcata extracts were found to be useful as binders for industrial explosives[95] and to be very effective in settling suspended matter in bleach liquors.[96]

VII. *Rhodymenia* AND DULSAN

1. *Distribution*

Rhodymenia palmata, a red alga, is commonly known as *dulse* or *red kale,* and because of its resemblance to a ribbon, it is also called *Neptune's girdle.* The botanical name indicates that it is shaped like the palm of a hand. The plant is composed of dull red, forked, irregularly divided, rubberlike blades that originate in a short, inconspicuous stipe. It grows to a height of 5–15 in (12.5–38 cm) and a width of 1–3 in (2.5–7.5 cm).

Rhodymenia palmata is widely distributed througout the world. In American waters it is found from California to Alaska in tide pools or on rocks near the low-water mark, where it often forms a distinct zone. It can also attach itself to the blades of the ribbon kelp, *Nereocystis luetkeana,* which is prevalent on the Pacific coasts of Canada and Alaska. On the Atlantic coast, it is found in New England, Canada, and Iceland.[60]

Another species, *R. pertusa,* known as *red eyelet silk,* is found on the Pacific coast in the northern, subarctic regions. This is a sublittoral species that is usually found near or below the low-tide mark.

In European waters, *R. palmata* is prevalent near Lapland, Norway, France, Ireland, and the British Isles, and it is also collected in the Mediterranean Sea. It is often found in abundance on the stems of *Laminaria cloustoni,* and considerable quantities of the latter frequently occur in the drift cast up on shore.[12] In south and east Asia as well as Indonesia, *R. palamata* is found on rocks and dead coral in the littoral region, together with *R. indica.* Two species of *Rhodymenia* are also found in deep water off the coasts of South America.

Rhodymenia palmata (dulse), unlike most seaweeds, is completely digestible by man and has long been used as a food rather than as a source of seaweed extract.[12] For many years, it was gathered on the New England shores and sold for food. As early as 1876, quantities of dulse were imported from Canada and sold in the United States. Today most of it comes from Canada, but the Indians and other local inhabitants of the Puget Sound area still gather dulse and use it as a relish. In numerous parts of the world it is still used as food, fodder, or medicine. As a food, it is commonly eaten raw, chewed like gum, eaten with fish and butter, or boiled with milk and rye flour. In the countries of the Mediterranean region, dulse is used in ragouts and other prepared dishes. It adds attractive red color and thickening to the food. In Kamchatka, Alaska, the natives ferment dulse and produce an alcoholic beverage.

In Ireland, dulse is an important food stuff. During the famine years, dulse and potatoes were the staple foods of the people inhabiting the coastal regions. Young plants of *R. palmata* were particularly preferred; these were known as *crannogh.* In Iceland, *Rhodymenia* species are known as *sol* and are eaten with dried fish, butter, and potatoes and are baked into bread.[26]

Rhodymenia is most widely used as fodder.[61] Icelanders also use dulse for stall feeding cattle; it does not affect the smell or taste of milk as do most other seaweeds.[12] In both Norway and Lapland, it is also commonly known as *sea-devil* or *horse seaweed.* It is used largely as cattle fodder in Normandy and

Brittany, and in those areas it is known as *cow seaweed* (*goemon a vache*) or *animal seaweed* (*goemon a bestiaux*).[12]

An analysis of the composition of *R. palmata* and other related seaweeds has been done.[97] The analytical results are as follows: 21.2% ash; 0.7% sulfate; 3.5% nitrogen; 2.4% fats, oils, and pigments soluble in carbon tetrachloride; 7.9% substances soluble in alcohol; 3.3% uronic anhydride; 2.4% cellulose; 36.2% pentose (as anhydro units); 3.5% hexose (as anhydro units).

In Nova Scotia and New Brunswick, production increased from ~71,000 lb (32,000 kg) dry weight in 1940 to 724,000 lb (325,000 kg) in 1966.[26]

2. *Dulsan*

Dulsan does not gel.[98] The principal sugar released by acid-catalyzed hydrolysis of *R. palmata* is D-xylose,[97] but dulsan differs from xylans in land plants in that it is soluble in water. Dillon[99] showed that it contained both (1→3) and (1→4) linkages and could readily be extracted from the plant by treatment with dilute acid.

3. *Structure*

The extract contains a number of fractions. An *alpha*-cellulose fraction (2–7% of dry wt) is composed of about equal parts of glucose and xylose residues.[100] Two other fractions containing xylose have been found; both are linear, containing β-D-(1→3) and β-D-(1→4) linkages in a ratio of about 1 : 2.[101]

In *R. palmata,* 44.21% of the organic matter is a mixture of hemicellulose, dextrin, starch, and cellulose types of polysaccharides. The monosaccharide fraction contained galactose. Water-soluble polysaccharide fractions contained galactose, xylose, and glucose.[102]

VIII. *Iridaea* AND IRIDOPHYCAN

The extract of species of the *Iridaea* or *Iridophycus* genera is known as *iridophycan.* The structure and properties of this phycocolloid are similar to those of carrageenan and funoran.[3, 69]

The nomenclature of this genus of seaweeds is confusing; for example, *Iridaea flaccidum, Iridaea laminarioides,* and *Iridophycus flaccidum* actually refer to the same seaweed species.[3, 12] Of these names, the one most frequently encountered by the authors is *Iridophycus flaccidum.* Recently, however, there has been a growing tendency to replace the genus name *Iridophycus* with the shorter *Iridaea.* Both of these names can be considered correct.

The *Iridophycus* genus is most prevalent in the ocean near central California, northward to Oregon. The plants are found mostly in the intertidal zone, often in very dense colonies.[24] Since 1945, a California firm prepared iridophycan from two closely related species of *Iridophycus, I. flaccidum* and *I. splendens.*[40] Companies in Oregon have also started commercial production of iridophycan, but no yearly production figures are available. *Iridophycus* weeds are relatively easy to collect. Plants are hand-picked and about 200–300 lb are collected during a 2–3 hr low tide. Commercial *Iridophycus* species are obtained near Crescent City, Pt. Arena, and Jenner, California and south of Coos Bay, Oregon. The harvesting season is June to December. Commercial gathering began in the summer of 1944 with about 30 tons (27 metric tons) of fresh seaweed (78% water content) being collected. The cost of gathering weed is about $100/ton (about $110/metric ton) (wet weight) or about $700/ton (about $770/metric ton) (dry weight, including drying costs).[40] Japanese production of *Iridophycus* weeds during the period 1940–1949 was 1200 metric tons/year (1300 tons/year), dry weight.

1. *Extraction*

The treatment of *Iridophycus* weeds is somewhat similar to that used with *Gloiopeltis* species. After gathering, the weed is washed with seawater, spread on racks, and then rewashed with fresh water. The weed can be dried if it is to be stored for any appreciable length of time. Otherwise, it is immediately extracted by soaking for 1 hr in boiling water. Some agitation is required, either by mechanical mixers or compressed air lines. A steam hose used to keep the water at the boiling point can supply sufficient agitation, but care must be taken that no foreign odors or colors are introduced if this is done. The extract is filtered to remove foreign matter and can be used directly; for example, it can be added in solution form to syrups to aid in stabilization.[103]

Mori and Tsuchiya[104] reported improved extraction of *I. flaccidum* by pressure cooking for 3 hr at 120°. The extract was concentrated, and barium chloride was added. About 25–40% dry-basis yields can be expected.[105,106] Iridophycan can be separated from salts by precipitation with methanol. Whether the increased yield obtained at higher temperatures is sufficient to offset the danger of thermal degradation and the increased cost of the required equipment is as yet undecided.

2. *Structure*

Weeds of the *Iridophycus* genus contain no cellulose. The principal cell wall constituent is iridophycan (~40% dry basis). Hassid[106, 107] reported that iridophycan was a galactan sulfate that gave only D-galactose upon hydrolysis; the ash content of the extract was more than 25% (chiefly sodium sulfate). Hydrolysis of the extract of *I. flaccidum* gave 60% of reducing sugars, namely D-galactose, L-arabinose, and another sugar termed floridose.[72]

Tadokoro and Saito[108] considered iridophycan to be a mixture or combination of fucan and galactan, contaminated with some arabinoglycan. Conclusive proof of this structure is lacking. Some additional carbohydrates in the weed were also determined by Tadokoro and coworkers.[109, 110] They extracted the polysaccharides from *Iridaea* with 35% ethanol, hot water, and 2% sodium hydroxide. Each of the products was hydrolyzed. In this manner, the 35% ethanol extract gave L-arabinose, another pentose, and a 6-deoxyhexose. The hot water extract gave the same products as the ethanol extract and, in addition, D-galactose, D-glucose, and L-fucose. The alkaline extract gave DL-erythrose.[111]

In a similar experiment in which the polysaccharide(s) of a hot water extract were precipitated with 70% ethanol and hydrolyzed with 1% hydrochloric acid, D-galactose was the principal product; small amounts of D-glucose, D-fructose, and L-arabinose were also detected.[111]

The presence of D-glucose in the plant has been established.[112] Galactitol can be isolated from both *Iridophycus flaccidum* and *Bostrychia scorpiodes*. It is not known to occur in other red algae.[106, 113, 114]

Two proteins have also been isolated from *Iridophycus* weeds.[115] One, a chromoprotein, is soluble in salt solutions and in 30–50% ethanol. The other protein is soluble in hot water and 2% sodium hydroxide and has properties similar to those of a prolamine. Extraction of *I. flaccidum* with 45% ethanol at 50°, followed by addition of acetone, gives a precipitate of the chromoprotein and polysaccharides.

Even though there is a large amount of other water-soluble substances present, it is relatively easy to prepare the pure galactan sulfate. The *Iridophycus* weed is first treated with methanol and ethanol to remove chlorophyll. After hot-water extraction, the solution is poured into 90% ethanol. The process is repeated with 95% ethanol as a precipitant and finally with absolute ethanol. The fibrous sodium galactan sulfate prepared in this way is reported to be relatively pure.[107]

Mori and Fumoto[116] found that iridophycan contains α-D-galactopyranosyl units. They also found that reaction of purified iridophycan with chlorotriphenylmethane gives a very low yield of trityl ether. Their conclusion was that there is no free primary alcohol group, and hence, that the sulfate ester is attached to C-6. Pure iridophycan, recovered as the barium salt, analyzed as follows:[105] 10.06% sulfur, 20.80% barium, 52.2% total aldoses, 50.0% total sugars (as galactose), and 41.8% D-galactose. The analysis shows that iridophycan is similar to carrageenan from *Chondrus ocellatus*.

Iridophycan is a galactan sulfate in which each D galactopyranosyl unit contains one sulfate half-ester. On methylation, it yields a dimethyl derivative, which still contains a sulfate group[107] and produces dimethyl-D-galactose upon hydrolysis. Acetylation forms a diacetyl derivative, which, when hydrolyzed, yields a sulfate-free galactan. Classical methylation analysis of this galactan gives 2,4,6,tri-*O*-methyl-α-D-galactopyranose.[105] Therefore, present data indicate that iridophycan is an α-D-(1→4)-linked galactan 6-sulfate (Fig. 10).

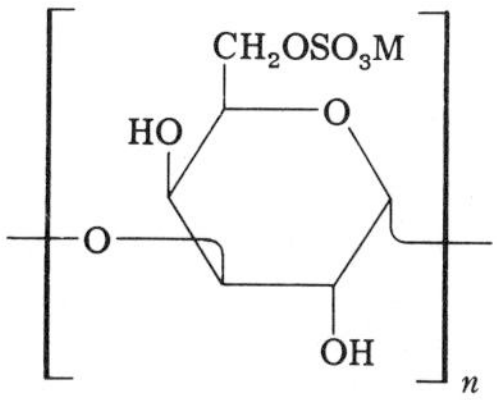

FIG. 10.—Probable repeating unit of iridophycan. M = metallic ion.

3. *Properties*

In general, the properties of iridophycan resemble those of carrageenan. However, detailed viscosity and gel data are not available. An iridophycan extract, isolated as the barium salt by precipitation with methanol, is remarkable in that a rapid decrease in viscosity occurs as the dispersion is stored, even at 10°.

An extract of the gut of the sea urchin, *Strongylocentrotus purpuratus,* is capable of digesting iridophycan.[116] Agar- and iridophycan-digesting bacteria are present in the second loop of its intestine in large numbers. These bacteria completely digest *I. flaccidum* within 1 week. However, it is observed that isolated bacteria grown on the algae do not liberate sugar into the culture medium.

Pure iridophycan (sodium salt) has a fibrous structure. Its aqueous solutions[104] have a pH of about 6.8 and a specific optical rotation of 69°. Dialysis results in loss of sodium but not of sulfur. The dialysis end-product is the free acid known as iridophycanic acid.[107]

4. *Uses*

The uses of iridophycan generally center around its ability to form a highly ionized anion of low equivalent weight while remaining in the colloidal state. Most uses are based on the resemblance of its properties to those of carrageenan and funoran; and it is often used as a substitute or, more generally, as a diluent for them. Specific illustrations of applications of iridophycan are given here to show the extreme versatility of this valuable extract.

Beverage stabilizer.—Iridophycan is popular as a cold-mix stabilizer in California.[3] It is not necessary to use purified gum for this; it is usually simpler to take dried or fresh *Iridophycus flaccidum* and, after a brief washing with fresh water, to boil the weed for ~1 hr in enough water to cover. After filtration, the clear extract can be added directly to chocolate or other beverage syrups. The syrup is heated to boiling with agitation and then dried or directly packaged as desired; cooling is not required. The product will now remain in suspension much longer than without the iridophycan.[97] Quantities required vary so much in individual cases that no general recommendation can be given. Some experimentation is needed to establish the optimum amount.

Food uses.—Besides being used as a beverage stabilizer, *Iridophycus* weeds can be eaten directly; for example, in Scotland, *I. edulis* is gathered for food. This species is called *dulse* because of its resemblance to *Rhodymenia,* but it is a thicker and brighter red plant than the latter. In Japan, where it is known as *ginnanso, Iridophycus* is also used as food. Principally, it is used together with *Gloiopeltis* weeds to prepare *funori.* However, inasmuch as it is a larger and coarser species, it must be steamed and chopped during processing.[3, 12]

Sizes and glues.—In Japan, iridophycan finds wide application as a paper and cloth size. This is a result of its adhesive nature and of the moderate viscosity of its aqueous solutions. The extract also serves as an excellent all-purpose glue. In addition, the Japanese use iridophycan as a cementing material for walls, to make calcimine paints more permanent and, mixed with plaster and coloring agents, to make wall paints.[69, 108]

Anticoagulant.—An extremely dilute water extract of *I. flaccidum* can prevent coagulation of blood. This property is shared by extracts of certain other red algae, for example, *Delesseria sanguinea.*[12, 117]

Preparation of ^{14}*C-labeled compounds.*—An unusual use of *I. flaccidum* is in the preparation of [^{14}C]D-galactose and [^{14}C]glycerol. The preparation is dependent upon the fact that the alga contains 1–4% of alcohol-soluble 2-glyceryl α-D-galactopyranoside. The compound, upon hydrolysis (for example, with yeast galactosidase) gives D-galactose and glycerol. Of the carbon dioxide activity initially used, approximately 25% is incorporated into the D-galactopyranosyl unit and 12% into the gylcerol.[118, 119]

5. *Iridophycanic Acid*

Electrodialysis of iridophycan results in the production of colloidal iridophycanic acid.[2, 69, 107] The product, a rather strong acid, can be dried to an amorphous granular powder that is soluble in water. The pH of a 1% solution is 3.6.

IX. References

(1) R. L. Whistler and C. L. Smart, "Polysaccharide Chemistry," Academic Press, New York, 1953.

(2) C. K. Tseng, *Science,* **101,** 597, 610 (1945).

(3) H. J. Humm, *in* "Marine Products of Commerce," D. K. Tressler and J. M. Lemon, eds., Reinhold Publ. Corp., New York, 1951, p. 50.

(4) L. Stoloff and P. Silva, *Econ. Bot.,* **11,** 327 (1957).

(5) F. Smith and R. Montgomery, "The Chemistry of Plant Gums and Mucilages," Reinhold Publ. Corp., New York, 1959.

(6) M. Glicksman, "Gum Technology in the Food Industry," Academic Press, New York, 1969.

(7) R. Schachat and M. Glicksman, *in* "Industrial Gums," R. L. Whistler, ed., Academic Press, New York, 1959, p. 135.

(8) M. Glicksman, *Advan. Food Res.*, **11,** 109 (1962).

(9) J. S. Zanefield, *Indo-Pacific Fish. Council (Bangkok, Thailand)*, Special Publ., No. **3,** 2 (1955).

(10) W. S. DeLoach, O. C. Wilton, H. J. Humm, and F. A. Wolf, *Duke Univ. Marine Station Bull.*, No. **3,** 31 (1946).

(11) H. L. Jensen, *Aust. J. Sci.*, **7,** 94 (1944).

(12) V. J. Chapman, "Seaweeds and Their Uses," Methuen and Co., London, 1950, pp. 68, 88–222, 116, 119, 129–130, 163–164, 174, 180–189, 250–252.

(13) W. M. Isaac and F. Hewitt, *Proc. Int. Seaweed Symp., 1st, Edinburgh, Scotland, 1952,* 14 (1955).

(14) H. J. Humm, U.S. Patent 2,446,091 (1948); *Chem. Abstr.*, **42,** 8999 (1948).

(15) H. J. Humm and L. G. Williams, *Amer. J. Bot.*, **35,** 287 (1948).

(16) W. Yaphe, *Can. J. Bot.*, **37,** 751 (1959).

(17) R. D. Hamilton and J. J. Carroll, *Nature,* **196,** 1200 (1962).

(18) S. D. Upham, "The Structure of Red Seaweed Polysaccharides," Tech. Bull. No. **3,** Marine Colloids, Inc., Rockland, Maine, 1967.

(19) K. G. Guiseley, "Seaweed Colloids," *in* "Kirk-Othmer Encyclopedia of Chemical Technology," Interscience, 2nd Ed., Vol. 17, 1968, p. 768.

(20) D. B. Smith and W. H. Cook, *Arch. Biochem. Biophys.*, **45,** 232 (1953).

(21) E. Percival and R. H. McDowell, "Chemistry and Enzymology of Marine Algal Polysaccharides," Academic Press, New York, 1967, pp. 137–138, 144.

(22) V. Krishnapillai, *Curr. Sci.*, **24,** 21 (1955).

(23) W. A. P. Black, W. R. Blakemore, J. A. Colquhoun, and E. T. Dewar, *J. Sci. Food Agr.*, **16,** 573 (1965).

(24) H. J. Humm, personal communication, 1956.

(25) E. Y. Dawson, "How to Know the Seaweeds," W. C. Brown Co., Dubuque, Iowa, 1956, pp. 40, 153.

(26) R. H. Williams, *Quart. J. Fla. Acad. Sci.*, **8,** 161 (1945).

(27) T. Nakamura, *Nippon Suisan Gakkaishi,* **20,** 501 (1954); *Chem. Abstr.*, **49,** 7877 (1955).

(28) T. Nakamura, *Nippon Suisan Gakkaishi,* **21,** 1053 (1956); *Chem. Abstr.*, **50,** 12516 (1956).

(29) S. L. Tjoa, U.S. Patent 2,811,451 (1957); *Chem. Abstr.*, **52,** 3200 (1958).

(30) K. Katsuura, S. Suzuki, and T. Okano, *Kogyo Kagaku Zasshi,* **65,** 1076 (1962).

(31) K. Katsuura, S. Suzuki, and S. Takahashi, *Kogyo Kagaku Zasshi,* **63,** 332 (1960).

(32) T. Nakamura, *Nippon Suisan Gakkaishi,* **24,** 285 (1958–59); *Chem. Abstr.*, **53,** 7329 (1959).

(33) W. A. O. Black, *SCI (Soc. Chem. Ind. London), Monogr.*, **24,** 33 (1966).

(34) M. Glicksman, U.S. Patent 3,250,621 (1966); *Chem. Abstr.*, **65,** 2916 (1966).

(35) S. E. Foster and A. L. Moirano, U.S. Patent 3,342,612 (1967); *Chem. Abstr.*, **68,** 2140y (1968).

(36) P. H. Ho and L. C. Kinh, *Khao-Cu u Nien-San Khao-Hoc Dai-Hoc Du ong,* 351 (1962); *Chem. Abstr.*, **62,** 2950 (1965).

(37) C. S. Kim and H. J. Humm, *Bull. Marine Sci.*, **15,** 1036 (1965); *Chem. Abstr.*, **64,** 14591 (1966).

(38) L. S. Stoloff and C. F. Lee, *U.S. Fish Wildl. Ser., Fish. Leaflet,* **173** (1946).

(39) C. F. Lee and L. S. Stoloff, *U.S. Fish Wildlife Ser., Spec. Sci. Rep., Fish. Ser.,* No. **37,** (1946).

(40) C. K. Tseng, *Econ. Bot.,* **1,** 69 (1947).

(41) "The Seaweed Story," Calif. Dept. Fish and Game.

(42) R. H. Williams and G. L. Voss, *Fla. State Board Educ., Educ. Ser.* No. **7** (revised), (1955).

(43) R. Kiyoda, Japan. Patent 7676 (1958); *Chem. Abstr.,* **54,** 3795 (1960).

(44) F. Ota and T. Tanaka, *Kagoshima Daigaku Suisan Gakubu Kiyo,* **13,** 38 (1964).

(45) Y. A. Doshi and P. S. Rao, *Indian J. Chem.,* **5,** 342 (1967); *Chem. Abstr.,* **68,** 674b (1968).

(46) Y. Kojima, J. Kusakabe, and K. Funaki, *Nippon Suisan Gakkaishi,* **18,** 245 (1952); *Chem. Abstr.,* **48,** 1604 (1954).

(47) S. Ushiyama and H. Koike, Japan. Patent 2286 (1953); *Chem. Abstr.,* **48,** 6619 (1954).

(48) S. Ushiyama and S. Sato, Japan. Patent 4037 (1953); *Chem. Abstr.,* **48,** 8989 (1954).

(49) Y. Kojima and J. Kusakabe, *Nippon Suisan Gakkaishi,* **17,** 345 (1952); *Chem. Abstr.,* **48,** 8181 (1954).

(50) S. Akiya and A. Tanimura, *Yakugaku Zasshi,* **74,** 125 (1954); *Chem. Abstr.,* **48,** 6076 (1954).

(51) S. Akiya, S. Tejima, and H. Harada, *Yakugaku Zasshi,* **71,** 730 (1951); *Chem. Abstr.,* **45,** 10509 (1951).

(52) J. Mizuguchi, S. Sizuki, and S. Misono, *Nippon Nogei Kagaku Kaishi,* **25,** 461 (1951); *Chem. Abstr.,* **47,** 6687 (1953).

(53) *Food Manf.,* **23,** 248 (1948).

(54) D. Chakraborty, *J. Proc. Inst. Chem., Calcutta,* **17,** 188 (1945).

(55) *Commer. Fish. Rev.,* **19** (1957).

(56) K. Pillai, *J. Sci. Ind. Res., Sect. B,* **14,** 473 (1955).

(57) G. C. Buzzonetti, *Chimica,* **12,** 157 (1956); *Chem. Abstr.,* **50,** 17253 (1956).

(58) G. B. Marina-Bettólo and J. Ibanez, *Ann. Chim. Appl.,* **38,** 390 (1958).

(59) G. W. Lawson, *Amer. J. Bot.,* **41,** 212 (1954).

(60) M. L. Guberlet, "Seaweeds At Ebb Tide," Univ. Wash. Press, Seattle, 1956, pp. 76–80, 122–124.

(61) L. Newton, "Seaweed Utilisation," Sampson Low, London, 1951, pp. 32, 99, 100.

(62) J. R. Turvey and D. A. Rees, *Nature,* **189,** 831 (1961).

(63) J. R. Turvey and D. A. Rees, *J. Chem. Soc.,* 1590 (1961).

(64) D. A. Rees and R. Conway, *Nature,* **195,** 398 (1962).

(65) J. R. Turvey and T. P. Williams, *Proc. Int. Seaweed Symp., 4th, Biarritz, France, 1961,* 370 (1964); *Chem. Abstr.,* **61,** 5952 (1964).

(66) N. S. Anderson and D. A. Rees, *J. Chem. Soc.,* 5880 (1965).

(67) C. C. Su, T. M. Chang, and C. C. Wang, *Chung Kuo Nung Yeh Hua Hsueh Hui Chih, Spec. Issue,* 55 (1965); *Chem. Abstr.,* **65,** 12555 (1966).

(68) C. K. Tseng, *Sci. Monthly,* **59,** 37 (1944).

(69) C. K. Tseng, *in* "Colloid Chemistry," J. Alexander, ed., Reinhold Publ. Corp., New York, Vol. 6, pp. 643, 660, 718 (1946).

(70) G. B. Rigg, *Pacific Northwest Ind.,* **10,** 208 (1951).

(71) *Sea Secrets,* Univ. Miami Marine Lab., Coral Gables, Fla., 1956.

(72) E. Takahashi, *J. Coll. Agr., Hokkaido Imp. Univ.,* **8,** 183 (1920).

(73) S. Suto, *Proc. Int. Seaweed Symp., 1st, Edinburgh, Scotland, 1952,* 98 (1955).

(74) M. Suzuki, *J. Biochem. (Tokyo),* **33,** 245 (1941); *Chem. Abstr.,* **36,** 1960 (1942).

(75) J. K. N. Jones and F. Smith, *Advan. Carbohyd. Chem.,* **4,** 275 (1949).

(76) S. Hirase, C. Araki, and T. Ito, *Bull. Chem. Soc. Japan,* **29,** 985 (1956); *Chem. Abstr.,* **52,** 8972 (1958).

(77) I. S. Takemori, *Hirosaki Igaku,* **8,** 749 (1957); *Chem. Abstr.,* **52,** 9402 (1958).

(78) D. J. Stancioff and N. F. Stanley, *Proc. Int. Seaweed Symp., 6th, Santiago de Compestela, Spain, 1968,* 595 (1969).

(79) K. Mita, *Nippon Suisan Gakkaishi,* **22,** 558 (1956–57); *Chem. Abstr.,* **52,** 7444 (1958).

(80) K. Aoki, *Nippon Suisan Gakkaishi,* **3,** 359 (1935).

(81) K. Aoki, *Nippon Suisan Gakkaishi,* **6,** 88 (1937).

(82) K. Aoki, *Nippon Suisan Gakkaishi,* **6,** 145 (1937).

(83) T. Yanagigawa, *Osaka Kogyo Shikenjo Hokoku,* **18,** No. **2,** 29 (1937).

(84) S. Asai, Japan. Patent 3250 (1952); *Chem. Abstr.,* **47,** 7170 (1953).

(85) M. Iwashiro, Japan. Patent 3100 (1952); *Chem. Abstr.,* **47,** 10816 (1953).

(86) J. Nakamura, Japan. Patent 6742 (1952); *Chem. Abstr.,* **47,** 4012 (1953).

(87) T. Kuji, Japan. Patent 1135 (1951); *Chem. Abstr.,* **46,** 11780 (1952).

(88) H. Kozuka, Japan. Patent 2206 (1953); *Chem. Abstr.,* **48,** 5064 (1954).

(89) K. Kosaka, Japan. Patent 133,489 (1939); *Chem. Abstr.,* **35,** 4224 (1941).

(90) K. Katsuura, *Kogyo Kagaku Zasshi,* **56,** 903 (1953); *Chem. Abstr.,* **48,** 14263 (1954).

(91) M. Kawamura, *Hokkaido Igaku Zasshi,* **18,** 418, 567 (1940); *Jap. J. Med. Sci. 2, Biochem.,* **4,** No. 4, 107 (1941).

(92) K. Mita, *Yakugaku Zasshi,* **73,** 198 (1953); *Chem. Abstr.,* **47,** 4551 (1953).

(93) E. O. Morris, *J. Sci. Food Agr.,* **6,** 611 (1955).

(94) T. Kitamura, Japan. Patent 175,318 (1947); *Chem. Abstr.,* **44,** 6598 (1950).

(95) N. Nakashiro and K. Takizawa, Japan. Patent 5849 (1958).

(96) Y. Yamaguchi, Y. Onoe, T. Okada, and Y. Fujii, Japan. Patent 8014 (1956); *Chem. Abstr.,* **52,** 14183 (1958).

(97) A. C. Ross, *J. Sci. Food Agr.,* **4,** 333 (1952).

(98) S. M. Marshall, L. Newton, and A. P. Orr, "A Study of Certain British Seaweeds and Their Utilization in the Preparation of Agar," Charles Birchall, Liverpool, England, 1949, p. 63.

(99) T. Dillon, *Proc. Int. Seaweed Symp., 1st., Edinburgh, Scotland, 1952,* 45 (1955).

(100) A. Myers and R. D. Presont, *Proc. Roy. Soc., Ser. B,* **150,** 447 (1959); *Chem. Abstr.,* **53,** 20245 (1959).

(101) H. Bjorndal, K. E. Eriksson, P. J. Garegg, B. Lindberg, and B. Swan, *Acta Chem. Scand.,* **19,** 2309 (1966).

(102) G. P. Serekkov and I. V. Zlochevakaya, *Nauchn. Dokl. Vyssh. Shk., Biol. Nauki,* **4,** 151 (1962); *Chem. Abstr.,* **58,** 8236 (1963).

(103) K. W. Karnopp, U.S. Patent 2,448,599 (1948); *Chem. Abstr.,* **43,** 332 (1949).

(104) T. Mori and Y. Tsuchiya, *Nippon Nogei Kagaku Kaishi,* **14,** 609 (1938).

(105) T. Mori, *Nippon Nogei Kagaku Kaishi,* **19,** 297 (1943).

(106) W. Z. Hassid, *Plant Physiol.,* **11,** 461 (1936).

(107) W. Z. Hassid, *J. Amer. Chem. Soc.,* **55,** 4163 (1933); **57,** 2046 (1935).

(108) T. Tadokoro and T. Saito, *Kogyo Kwagaku Zasshi,* **38,** Suppl. Binding, 270

(1935); *Chem. Abstr.*, **29**, 6388 (1935).
(109) T. Tadokoro and K. Yoshimura, *Nippon Kagaku Kaishi,* **55,** 617 (1934).
(110) T. Tadokoro, K. Yoshimura, and M. Yanase, *Nippon Kagaku Kaishi,* **56,** 188 (1935).
(111) T. Tadokoro and K. Yoshimura, *Nippon Kagaku Kaishi,* **56,** 655 (1935).
(112) R. C. Bean and W. Z. Hassid, *Science,* **124,** 171 (1956).
(113) P. Haas and T. G. Hill, *Biochem. Z.,* **25,** 1470 (1931); **26,** 987 (1932).
(114) W. Z. Hassid, *Plant Physiol.,* **8,** 480 (1933).
(115) T. Tadokoro and K. Yoshimura, *Nippon Kagaku Kaishi,* **55,** 525 (1934).
(116) R. Lasker and A. C. Giese, *Biol. Bull.,* **106,** 328 (1954).
(117) H. Elsner, U.S. Patent 2,284,566 (1942); *Chem. Abstr.,* **36,** 6755 (1942).
(118) R. C. Bean and E. W. Putnam, *J. Biol. Chem.,* **204,** 169 (1953).
(119) E. W. Putnam and W. Z. Hassid, *J. Amer. Chem. Soc.,* **76,** 2221 (1954).

Plant Exudates

CHAPTER X

GUM ARABIC

MARTIN GLICKSMAN
General Foods Corp., Tarrytown, N. Y.

AND

RALPH E. SAND
Anderson Clayton Foods, Richardson, Texas

I. INTRODUCTION

Gum arabic was used at least 4000 years ago when it was shipped as an article of commerce by Egyptian fleets. Ancient Egyptian inscriptions make frequent mention of gum arabic, called *kami,* which was used largely in painting as an adhesive for mineral pigments.[1] Eventually the gum found its way into Europe through various Arabian ports and acquired the name *gum arabic* after its place of origin or port of export. During the Middle Ages, the trade was carried on through ports controlled by the Turkish Empire, and the name *Turkey gum* was given to this material. At one time a reexport trade also developed in India, chiefly around Bombay. Here the gum acquired the name *East Indian gum* or *Indian gum,* but it was not of Indian origin as it was imported from Africa and Arabia and then reexported to Europe and America. Indian gum should not be confused with India gum which is an entirely different product, more commonly known as *gum ghatti.* Gum arabic is also known by various names pertaining to the local area where it is collected. In addition, other names relating to the color and grade have been given to it so that there is a great deal of confusion concerning the relationship between the name, the actual gum, and the species of *Acacia* from which it is derived.[2-7] In Table I, the various names, localities and species of gum arabic that are known to exist in the industry are tabulated.

TABLE I

Varieties of Gum Arabic

Name	Geographical Origin	Botanical Source
Abysssinia	Abyssinia	*Acacia abyssinica*
Acacia	Africa, Near East	*Acacia* species
Adad	Somalia	*Acacia senegal*
Aden	East Africa	*Acacia senegal* (*Acacia verek*)
Amrad	India	*Acacia arabica*
Amrawatti	India	*Acacia arabica*
Angra pequena	South Africa	*Acacia karroo* (*Acacia horrida*)
Australian Black Wattle	New South Wales	*Acacia pycanatha, decurrens, dealbata, sentis,* or *homalophylla*
Babool	India	*Acacia arabica*
Babul	India	*Acacia arabica*
Berbera	West Africa	*Acacia senegal*
Black wattle	Australia, New South Wales	*Acacia pycanatha, decurrens, dealbata, sentis,* or *homalophylla*
Blanche gomme	West Africa	*Acacia senegal*
Blonde gomme	West Africa	*Acacia senegal*
Brown barbary	Morocco	*Acacia gummifera, spirocarpa,* or *arabica*
Cape	South Africa	*Acacia karroo* (*Acacia horrida*)
Catechu	India, Burma, Ceylon	*Acacia catechu*
Cutch	India, Burma, Ceylon	*Acacia catcheu*
East India	India (domestic)	*Acacia* species
East Indian	India (from Africa)	*Acacia senegal*
Falli	Nigeria	*Acacia senegal*
Gedda	Africa	*Acacia* species
Geddaref	Sudan, Egypt	*Acacia senegal*
Gehzirah	Africa	*Acacia seyal*
Gomme blanche	West Africa	*Acacia senegal*
Gomme blonde	West Africa	*Acacia senegal*

TABLE I—*Continued*

Name	Geographical Origin	Botanical Source
Gomme de galam	West Africa	*Acacia senegal*
Gomme de podor	West Africa	*Acacia senegal*
Gomme de Tombouctou	West Africa	*Acacia senegal*
Gomme du bas du fleuve	West Africa	*Acacia senegal*
Gomme du haut du fleuve	West Africa	*Acacia senegal*
Gomme fabrique	West Africa	*Acacia senegal*
Gomme friable	West Africa	*Acacia senegal*
Hashab	Sudan	*Acacia senegal*
Hashab geneina	Sudan	*Acacia senegal*
Hashab wady	Sudan	*Acacia senegal*
Hobloho	East Africa	*Acacia* species
Jalefan	Somalia	*Acacia* species
Jeddah	Morocco	*Acacia senegal*
Khair	India	*Acacia catechu*
Khartoum	Sudan	*Acacia senegal*
Khor	India	*Acacia senegal*
Kolkol	Nigeria	*Acacia senegal*
Kordofan	Sudan	*Acacia senegal*
Kumta	India	*Acacia senegal*
Mamrah	India	*Acacia modesta* and *Acacia arabica* mixed
Marrons et bois	Africa	*Acacia* species
Marrua	Nigeria	*Acacia seyal* and *Acacia xanthophloea*
Mimosa	South Africa	*Acacia* species
Mogador	Morocco	*Acacia gummifera, spirocarpa,* or *arabica*
Morocco	Morocco	*Acacia gummifera, spirocarpa,* or *arabica*
Nujal	East Africa	*Acacia* species
Ondurman	Egypt	*Acacia* species
Sadra Beida	West Africa	*Acacia senegal*
Salabreda	West Africa	*Acacia senegal*
Savakin	East Africa near Red Sea	*Acacia verek, seyal, fistula, stenocarpa,* or *procera*
Senegal	West Africa	*Acacia senegal*
Sennaar	Central America	*Acacia* species
Somali	Somalia	*Acacia* species
Somaliland	Somalia	*Acacia* species
Suakim	East Africa, Red Sea Area	*Acacia verek, seyal, fistula, stenocarpa,* or *procera*
Sudan	Sudan	*Acacia senegal*

TABLE I—*Continued*

Name	Geographical Origin	Botanical Source
Sunt	Sudan	*Acacia arabica*
Tahl	Sudan	*Acacia seyal*
Talba	Sudan	*Acacia seyal*
Talca	Sudan	*Acacia seyal*
Talha	Sudan	*Acacia seyal*
Talki	Sudan	*Acacia seyal*
Tripoli	North Africa	*Acacia* species
Tunis	North Africa	*Acacia* species
Turic	Africa	*Acacia* species
Turkey	Near East	*Acacia* species
Verek	Sudan	*Acacia senegal*
Wattle	India, South Africa, Australia, New South Wales	*Acacia* species
White	West Africa	*Acacia senegal*
	Gum Arabic Substitutes	
Angico	South America	*Piptadenia* species
Cashew	India	*Anacardium* species
Goma de guanacaste	Central America, West Indies	*Enterolobium* species
Mesquite	U.S., Mexico, South America	*Prosopis* species
Mumuye	Nigeria	*Combretum* species
Siris	India	*Albizzia* species
Sonora	U.S., North and South America	*Prosopis* species

True gum arabic is a dried, gummy exudation obtained from trees belonging to various species of the genus *Acacia,* subfamily *Mimosoideae,* and family-*Leguminosae.* There are about 500 species of *Acacia* distributed over tropical and subtropical areas of Africa, India, Australia, Central America, and southwest North America. The gum is produced in many hot, dry, elevated locations, but only a few of the known *Acacia* species are commercially important. The effect of climatic conditions can be illustrated by the behavior of *Acacia decurrens* Willd. which produces a sparingly soluble gum in its native Australia, but when transplanted to Java, it yields a completely soluble gum.[1]

Gum arabic is unique in that it is produced by trees only when they are in an unhealthy condition. Healthy trees have not been observed to yield gum. Most authorities[8] believe that the formation of gum exudates is a pathological condition resulting from a microbial (fungal or bacterial) infection of the injured tree. Natural factors that tend to lessen the vitality of the trees, such as poor soil, lack of moisture, and hot weather, improve gum yields. Others,[9] however, consider the production of gum to be a normal metabolic process in the plant, with the quantity and quality produced being a function of environmental conditions.

Differences of opinion[10] also exist on whether gum exudates are formed at the site of the wound or generated internally and then transported to the site of the exudation. There have been many theories concerning the origin and functions of plant exudates, but to date, no one has proposed a universally accepted explanation.

The most important gum-yielding country is the Republic of the Sudan, followed by Senegal, Mauritania, and Mali. Nigeria, Tanzania, Morocco, Ethiopia, and the Republic of Somalia produce sizeable quantities of inferior gum.

World demand is increasing for gum arabic for use in glue, candy, bakery products, and as an adhesive for stamps and wrapping paper. Small quantities go to almost every country in the world where stamps are made.

Major markets[11] for gum arabic are the United States, the United Kingdom, Italy, West Germany, Japan, France, Belgium, and the Netherlands. These countries take about 90% of the Sudanese exports, or about 76% of the world's supply.

II. Sources and Production

Eighty-five percent of the world's supply of gum arabic, amounting to about 50,000 to 60,000 metric tons annually, is produced and exported by the Republic of the Sudan.[11] A peak harvest of 75,000 metric tons was reached in 1963 because of excellent rains and high prices for gum. The importance of gum arabic to the Sudanese economy is seen in the increase export value from about $7,000,000 in 1955 to $20,000,000 in 1965. Gum arabic accounts for 11% of the country's exports and is second only to cotton. About 20% of the gum is exported to the United States.[11]

More than two dozen distinct species of *Acacia* occur in the Sudan, but the great majority of the commercial gum comes from a single species, *Acacia senegal* (*Acacia verek*). This species of tree is about 15–20 ft (4.5–6 m) tall and has a life span of 25–30 years. It grows in poor, sandy, reddish soil, the color of which is caused by the presence of substantial amounts of iron. Orchards of *Acacia senegal* are found throughout the province of Kordofan in a large uninterrupted belt stretching westward through the province of Darfur and eastward to the Nile River. In some areas, the trees are cultivated and tended in small gardens. The gum collected from cultivated trees is colloquially known as *hashab geneina* (garden gum) as opposed to *hashab wady* (wild gum).

The best grade of gum comes from *Acacia senegal* and is commonly known as *hashab* in the Sudan and as *kordofan* gum in this country. One of the reasons for its excellent quality is that the limitation of botanical sources results in a uniform product. About 90% of the Sudan's production of gum arabic is of this high quality. Less than 10% of the Sudanese gum production comes from *Acacia seyal,* which is prevalent in the southwestern part of the country and in the Nile region. These trees are not tapped and only natural exudates are collected and sold as *tahl* gum or as *talca, talha, talki,* or *suakim* gum. The gum is inferior to *hashab* and does not find a ready market.

Several other species of *Acacia* occur in the Sudan and are known to exude gums. Gum from them is sometimes collected and mixed with other gums. Such *Acacia* trees are *A. abyssinica, A. albida, A. arabica, A. campylacantha, A. drepanolobium, A. farnesiana, A. flava, A. kirkii, A. laeta, A. melifera, A. orfota, A. sieberiana, A. spirocarpa, A. stenocarpa,* and *A. thunbergiana.*[5] The gum from *Acacia arabica* is sold as *sunt* gum.

Gum production starts in breaks or wounds in the trees that are sometimes caused naturally but, more commonly, artificially. Originally, all gum arabic marketed was the result of natural exudations, but about 60 years ago, tapping was introduced. A small axe is used to break the outer bark horizontally by making a shallow cut ~1.5 in (~3.8 cm) wide. The bark is then peeled one strip up and one strip down to form 2–3-ft. (0.6–1.0-m) wounds. Gum collects in the wound within 3–8 weeks, depending on weather conditions. Gum droplets formed are 0.75–3 in (1.9–7.5 cm) in diameter and gradually dry and harden on exposure to the atmosphere.

Each season the bark is torn from the tree in a different place, gradually working around the trunk. If this is done carefully, the tree is not harmed and the wounds eventually heal. A single tapper can tap about 100 trees a day. Usually the trees are tapped when they are 6–7 years old, at which age they yield gum freely. As the trees reach the latter part of their life, they yield less gum.[5] Annual yields from young trees have been reported to range from 188 to 2856 g, with an average of 900 g; older trees give 379 to 6754 g, with an average of 2000 g.[12] Gum is collected by hand about every 10 days during the dry season, which usually lasts from October to May or June, varying somewhat with the district. During the rainy season when the trees are in full bloom, no gum is formed.

Harvest of gum was previously restricted to areas within 2 to 3 days' travel from the natives' homes, because food and water had to be carried to the collecting areas. Recently the government has dug wells in some of the outlying districts to facilitate collection.[13] Because gum collecting is only a part-time occupation of the people, the amount collected depends on economic factors. The success or failure of the sorghum crop, *durra,* has the most direct effect on gum collections.

After collection, the gum is brought to villages and stored until there are sufficient quantities to transport to the marketing center. Upon arrival at market, the gum is auctioned under government supervision. The most important market is at El-Obeid, known as the Gum Market. This auction usually sets the price of gum arabic throughout the world. The Sudanese government derives important revenue from its levy upon the weight of gum sold. However, by the operation of a flexible royalty, market stability is effected.[13]

After auction, the gum is taken to cleaning sheds before shipment to Port Sudan for export. Various cleaning techniques are used. The best grades of gum are hand picked to insure uniformity of color and freedom from extraneous material.

Some gum is bleached by spreading it on sheets and exposing it to the sun for several weeks. This treatment develops many small cracks on the surfaces of the gum pieces so that the material is easily friable and has a uniform light color or frosted appearance. The gum is available in several grades: natural, cleaned and sifted, bleached, and dust.

The second important producing area of gum arabic is West Africa, comprising Senegal, Mali, and Mauritainia. Most of the industry, which has been in existence since the 17th century, is based in Senegal, which today exports one-sixth as much as the Sudan. During the Sudanese rebellion, 1881–1899, the normal flow of gum arabic was disrupted, stimulating gum arabic production in West Africa.

West African gum is collected from a single species, *Acacia senegal* (*Acacia verek*), and is commonly known as gum Senegal. There are two gum collection seasons, one from December to February and the other from April to July. The second season is the more important as it yields more gum of better quality.

Most of the gum is brought to ports on the Senegal river. Since a control service was established at the ports, the quality of the gums has definitely improved. At present there are two main gum qualities varying in coloring: *gum bas du fleuve* and *gum ferlo*. Smaller amounts of inferior gums are known as *gomme du haut de fleuve, gomme friable, salabreda,* and *sandra beida.*

Most of the gum is sent to Bordeaux, where it is sorted by hand into whole and broken tears. Whole tears are again sorted into color grades of decreasing quality: *gomme blanche* (almost colorless), *gomme petit blanche* (same, but smaller), *gomme blonde* (darker color), *gomme petit blonde* (same, but smaller), *gomme vermicelle* (almost white to pale yellow), and *gomme fabrique* (unusual shape or dark color). Broken tears are also sorted by sifting, giving material of uniform size but with colors varying from pale yellow to deep brown: these are called *gomme poussiere grabeaux, gomme menus grabeaux, gomme moyens grabeaux,* and *gomme gros grabeaux.*

In comparison with the Sudan or kordofan gum, gum Senegal is more yellow or reddish. It is not so clean and occurs in larger, more brittle tears than Sudan gum. Reported to be more viscous but much less adhesive, it is not used much in the U.S. but has an extensive European market.

Minor quantities of gum arabic are also produced by the African nations of

Nigeria, Tanzania, Morocco, Ethiopia, Somalia, and the Republic of South Africa. Gum arabic is found in India, which has about two dozen native species of *Acacia.* These gums are transported to Bombay, which has long been an important center for gum, although most of the gum arabic exported has been imported first from East Africa and then cleaned, sorted, and graded for reexport.

In Australia, over 300 species of *Acacia,* known locally as *wattles,* are found. The gums from wattle trees vary greatly. They are usually of low grade with poor solubility, a dark reddish-brown color, a strong taste caused by the presence of tannins, and an unusual tendency to form a gel rather than a mucilage with water.

In the western hemisphere, some gum arabic is collected in Central and South America, probably from *Acacia farnesiana,* and sold as *senaar* gum.

Several gum substitutes for gum arabic are also found in the western hemisphere. Sonora gum, mesquite gum, and cactus gum are found in the United States, Mexico, and South America. They are derived from the *Prosopis* genera of plants. *Goma de Guanacasta* from an *Enterolobium* species is produced in the West Indies and in Central America.

III. Types of Gum

Almost all the gum arabic used in the United States arrives in 100 kilo (220 lb) burlap bags. The following four grades of material are commonly imported:

1. Selected Sorts—These are the cleanest and largest pieces, having the lightest color and commanding the highest price.
2. Cleaned and Sifted Sorts—The color varies from pale amber to dark amber, and the gum contains a minimum of material smaller than 0.25 in (6.35 mm), that is, four-mesh size.
3. Clean Amber Sorts—The color varies from light to dark amber, and the gum contains various amounts of fines (chips finer than four mesh).
4. Siftings—This, the cheapest grade of gum, is the residue formed by sorting the above choicer grades. Siftings are fine chips containing considerable sand, dirt, and bark.

Upon arrival, the crude gum is reprocessed and sorted by form and quality. The basic types are whole sorts, granular, powder and spray-dried, and there are various grades of each. The most expensive grade is the spray-dried product, which is a filtered, heat-treated, purified product used mainly in pharmaceutical applications and sells for about $0.50 a pound. The most common grade is the #1 USP powder, which sells for about $0.40 per pound and meets vigorous specifications of quality. There are several lesser grades of gum of varying qualities of size, color, and viscosity. The lowest grade is technical grade, which, sells for about $0.24 a pound.

Imports of gum arabic into the United States have about tripled over the last 30 years (Table II) and are still are still increasing. World production has increased from about 20,000 tons in 1935 to about 45,000 tons in 1955 and about 60,000 tons in 1966. This has resulted in a fairly stable market in which a normal supply is available at reasonable prices.

TABLE II

U.S. Imports of Gum Arabic[14]

Year	Pounds	Value ($)
1937	9,494,117	770,820
1938	8,735,447	539,914
1939	9,199,623	634,610
1940	14,055,415	922,012
1941	16,088,213	1,104,358
1942	12,833,497	1,196,840
1943	7,008,744	660,566
1944	8,105,443	708,283
1945	16,812,983	1,389,437
1946	19,816,750	2,012,332
1947	11,723,403	1,215,770
1948	15,981,734	1,451,316
1949	16,011,034	1,435,660
1950	18,605,169	1,712,432
1951	14,075,598	1,640,344
1952	15,979,188	1,844,128
1953	16,694,649	1,635,622
1954	15,321,444	2,018,428
1955	17,261,736	2,706,927
1956	18,676,694	2,838,484
1957	19,151,298	2,988,536
1958	20,447,767	3,156,383
1959	23,361,004	3,885,204
1960	21,472,037	4,130,123
1961	20,580,944	3,534,895
1962	20,039,404	3,362,401
1963	22,160,315	3,815,042
1964	21,979,853	3,974,776
1965	25,084,576	4,652,114
1966	27,854,824	5,392,098
1967	32,268,480	7,622,563
1968	25,591,889	5,682,484
1969	26,748,986	6,677,815
1970	27,080,374	7,404,331

IV. Structure

Gum arabic as found in nature exists as a neutral or slightly acidic salt of a complex polysaccharide containing calcium, magnesium, and potassium cations. Its structure has been extensively examined[15–21] but is still not entirely clear.

1. *Molecular Weights*

Molecular weight determinations of gum arabic give variable results because of the heterogeneity of *Acacia* extracts,[22–30] the nature of the species extracted, its history, the method used, and the composition of the aqueous solvent.[31]

Molecular weights of *Acacia senegal* extracts can be obtained by the viscosity method using the Staudinger constants determined by Anderson and Rahman[32] using the equation $[\eta] = K\ (M_v)^a$, in which $[\eta]$ = instrinsic viscosity, M_v = viscosity average molecular weight, $K = 1.3 \times 10^{-2}$ and $a = 0.54$. The constants do not hold for acacia species other than *A. senegal*,[30] because of differences in molecular structure, it is necessary to determine individual sets of constants with each species.

Typical results of some recent studies show the variability obtained with a variety of conditions (Table III). Generally accepted values for gum arabic are $M_n = 250{,}000$ and $M_w = 580{,}000$.

Table III

Molecular Weights of Acacia Gums

Species	Molecular Weights	Method
A. senegal [33]	191,000	Osmometry
A. senegal [33]	217,000	Osmometry
A. senegal [28]	5000–125,000	Molecular sieve chromatography
A. senegal [34]	~250,000	Molecular sieve chromatography
A. senegal [30]	$0.26–1.16 \times 10^6$	Viscosity
A. senegal [35]	256,000–326,000	Ultracentrifuge
A. senegal [27]	$1.00 \pm 0.05 \times 10^6$	Light scattering
A. senegal [36]	580,000	Light scattering
A. senegal [29,32]	580,000	Light scattering
A. arabica [37]	2.3×10^6	Molecular sieve chromatography
A. arabica [32]	1.89×10^6	Light scattering
A. laeta [38]	250,000	Molecular sieve chromatography
A. nubica [32]	870,000	Light scattering
A. drepanolobium [39]	$> 200{,}000$	Molecular sieve chromatography

Precipitation with sodium sulfate showed gum arabic to have a very broad molecular weight distribution. Anderson and coworkers[30] reported weight-average molecular weights based on viscosity data ranging from 260,000 to 1,160,000. Wide variations were found from sample to sample with the *hennawi* variant having less uronic acid and much less L-rhamnose than other samples. Gum arabic is a mixture of many components, as indicated by its broad-skew molecular weight distribution.[30] It was described by Hirst[40] as a heterogeneous system consisting of a family of closely related polysaccharides.

2. *Molecular Shape*

Light-scattering data[27, 31] suggests that gum arabic molecules are in the shape of short, stiff spirals with numerous side chains. The length of the coil is 1050A at maximum molecular charge. Uncharged arabic acid molecules have an equivalent radius of 555A and an effective volume of 7.2×10^{-16} cm^3/mole. This concept of gum arabic as a flexible chain polyelectrolyte with a high degree of branching is supported by the viscometric studies of Basu, Dasgupta, and Sircar.[41] Overbeek[42] observed that the calculated change in the dissociation constant for gum arabic was smaller than that found experimentally, further indicating the presence of branched chains. Anderson and Rahman[32] picture the molecule in solution as a coil with a root-mean-square, end-to-end dimension varying from 0.06μ at zero charge to 0.24μ at maximum charge.

3. *Sugar Constituents*

Hydrolysis of gum arabic yields L-arabinose, L-rhamnose, D-galactose, and D-glucuronic acid. The glycosidic bonds vary in lability. Thus, autohydrolysis of arabic acid in water (pH $\sim$2) gave 34.4% of L-arabinose and 14.2% of L-rhamnose, plus 3-*O*-α-D-galactopyranosyl-L-arabinose and degraded gum arabic. Further hydrolysis with mineral acid produced 42.1% of D-galactose, 15.5% of D-glucuronic acid, and some 6-*O*-(β-D-glucopyranosyluronic acid)-D-galactose.[43–45]

Gum arabic from different sources contains the same sugars, but the proportions vary greatly (Table IV).[46–52]

4. *Structural Composition*

All gum arabic preparations appear to have common structural features. The main structural feature of gum arabic is a backbone chain of (1→3)-linked D-galactopyranose units, some of which are substituted at the C-6 position with various side-chains[40, 61] (Fig. 1). A complete structure, proposed on the basis of incomplete information, is shown in Figures 2 and 3. Three aspects of the molecular structure are the various acid-labile groups (R) located on the periphery of the molecule, the branched framework of D-galactopyranose residues,

TABLE IV

Structural Composition of Various Acacia *Gums*

Species	Calculated Equivalent Weight (from GlcA)	% N	Reported Equivalent Weight	$[\alpha]_D$	Molar Ratio					
					GlcA	Gal	Ara	Rha	MeO	MeOGlcA
A. senegal [30,53]	1493	—	1400	−28°	1.6	2.9	3.5	1.1	0.11	0.10
A. mollissima [53]	2157	—	1880	−49°	0.87	3.97	5.64	0.80	0.21	—
A. pycnantha [53,54]	9700	—	9000	−2°	0.93	32.5	16.2	0.9	—	—
A. cyanophylla [53,54]	809	—	730	−33°	0.91	1.9	0.39	0.95	—	—
A. karroo [53,54]	1617	—	1700	+54°	1.0	4.6	4.0	0.2	—	—
A. catechu [53]	1079	—	1025	−32°	0.95	2.8	1.2	0.87	—	—
A. sundra [53]	1079	—	980	−30°	0.91	2.3	1.8	0.84	—	—
A. seyal [52,53]	1403	0.1	1340	+58°	0.90	2.8	4.1	0.24	0.4	—
A. nilotica [53,55]	2110	0.02	1890	+108°	0.90	4.6	5.8	0.046	0.57	—
A. arabica [53,55]	2489	0.07	2370	+112°	0.95	4.7	8.5	0.058	0.65	—
A. campylacantha [53,56]	2368	0.28	2020	−3°	0.85	6.1	3.6	0.86	0.26	—
A. drepanolobium [53,56]	2157	1.12	2060	+75°	0.95	3.0	7.7	0.13	0.26	—
A. dealbata [53]	2489	0.61	1840	−25°	0.74	4.2	5.0	0.67	0.20	—
A. fistala [53]	2088	0.06	1530	+61°	0.73	3.1	5.1	0.037	0.43	—
A. nubica [53,57]	2659	0.21	3030	+100°	1.1	5.6	11.7	0.11	0.14	0.04
					(1.1)	(4.7)	(12.9)	(0.11)	(0.05)	(0.04)
A. laeta [38,58]	1407	0.65	1300	−42°	0.91	0.72	0.73	0.24	0.17	—
					(0.90)	(3.0)	(2.4)	(0.99)	(0.13)	—
A. mearnsii [54,59]	1607	—	1800	−55°	1.2	4.3	3.7	0.77	—	—
A. podalyriaefolia [54,60]	6450	—	3585	+5.3°	0.55	16.5	3.2	0.20	—	—
A. elata [54,60]	6450	—	6100	+4°	0.94	26.7	5.7	1.0	—	—

TABLE IV—*Continued*

Species	Calculated Equivalent Weight (from GlcA)	% N	Weight Reported Equivalent	$[\alpha]_D$	Molar Ratio					
					GlcA	Gal	Ara	Rha	MeO	MeOGlcA
A. adansoniana[34]	—	—	—	+83°	—	—	—	—	—	—
A. multijuga[34]	—	—	—	+40°	—	—	—	—	—	—
A. tortilis[34]	—	—	—	+96°	—	—	—	—	—	—
A. mellifera[34]	—	—	—	-51°	—	—	—	—	—	—
A. decurrens[54]	—	—	3300	-46°	—	—	—	—	—	—
A. giraffae[54]	—	—	730	+28°	—	—	—	—	—	—

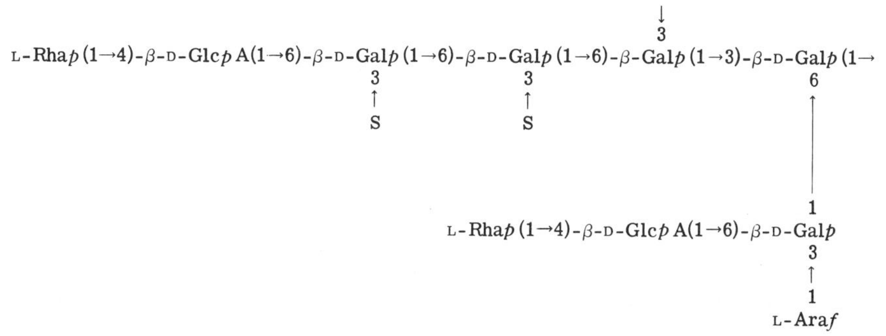

where S is a side chain of β-D-Gal$p$(1→3)-L-Ara$f$*(1→3)-L-Ara$f$(1→

or L-Ara$f$(1→3)-D-Gal$p$(1→
6
↑
1
L-Ara$f$

FIG. 1.—Structural features of gum arabic.[40, 61] *L-Arabinose is mainly present as L-arabinofuranosyl units, but some L-arabinofuranosyl units are present in the form → 3)-β-L-Ara$p$(1→ 3)-L-Ara$f$(1→.

and the portions of the molecule that give rise to acidic oligosaccharide fragments. A point of interest is that the galactan framework contains numerous chains with (1→3)-linked units bearing (1→6)-linked side chains. Location of the (1→6)-linkages is uncertain, but some of the L-rhamnopyranose

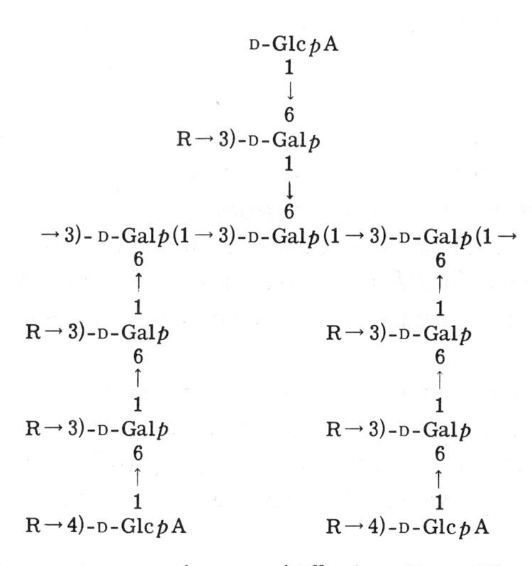

FIG. 2.—Possible core structure of gum arabic[62] where R = L-Rha$p$(1→, L-Ara$f$(1→, D-Gal$p$(1→ 3)-L-Ara$f$(1→, or L-Ara$p$(1→ 3)-L-Ara$f$(1→.

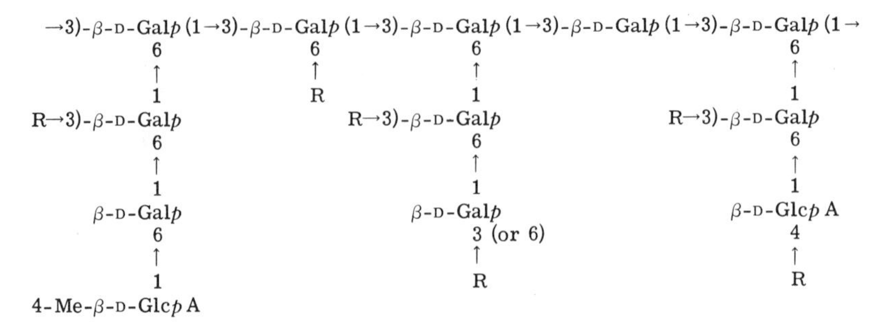

FIG. 3.—Proposed structure for *Acacia mearnsii* gum[63] where R = L-Ara*f*(1→, L-Ara*f*(1 → 3)-L-Ara*f*(1 →, L-Ara*f*(1 → 5)-L-Ara*f*(1 →, β-L-Ara*p*(1 → 3)-L-Ara*f*(1 →, or L-Rha*p*(1 →.

residues in gum arabic are joined (1→4) to D-glucopyranosyluronic acid residues.

The structure proposed by Hirst[61] includes most known features of fragments found to be present in gum arabic.

A very reasonable explanation for the biosynthesis of gum arabic has been proposed by Anderson and Dea.[64] They suggest that the gum has its precursor a highly branched arabinogalactan, to which is added L-rhamnose, D-glucuronic acid, and 4-*O*-methyl-D-glucuronic acid-terminated side chains.

5. *Components of Less Important Species*

Acacia sundra[63, 65] gum undergoes hydrolysis to yield L-rhamnose, L-arabinose, D-galactose, and D-glucuronic acid.

Acacia pycnantha[66] gum is a highly branched polysaccharide with a D-galactopyranose framework[62] of (1→3) and (1→6) linkages. Side chains joined by (1→6) linkages contain L-rhamnopyranose and L-arabinofuranose units.

Acacia laeta[38, 58] occurs in two varieties, *A. laeta* var. hashab (which resembles *A. senegal*) and *A. laeta* var. mellifera (which resembles *A. mellifera*). The structure of the gum of the *hashab* variant appears to consist of a main chain of D-galactopyranose units, mainly (1→3) linked but containing some (1→6) linkages.

Acacia arabica gum shows a high degree of branching.[37] It is interesting that, although *A. arabica* gum has a lower intrinsic viscosity than *A. senegal* gum, it has a higher molecular weight ($M_w = 2.3 \times 10^6$) than *A. senegal* ($M_w =$ 600,000). Important structural differences are responsible for this unusual phenomenon.

Acacia drepanolobium[67] gum is unusual in not being completely water soluble. It can be fractionated into a water-soluble fraction (A), a salt-soluble fraction

(B), and a gel fraction soluble in dilute alkali (C). Fraction A is very similar in structure and properties to *A. arabica* gum, but some differences do occur. Thus, gums of *A. drepanolobium* (fraction A), *A. arabica, A. nubica, A. nilotica,* and a few other species have remarkably low rhamnose contents.[53, 58]

Acacia seyal[68] has been found to contain four types of aldobiouronic acid units; two with α-D linkages and two with β-D linkages. These acids have been used as refence standards in a chromatographic examination of the aldobiouronic acids in the gums of seventeen other *Acacia* species. With the exception of *A. karroo* and *A. nubica* gums (which have low methoxyl contents), it appears that species having positive specific rotations contain the same four aldobiouronic acids types. *Acacia* species whose gums have negative specific rotations may contain the two aldobiouronic acids having β-D linkages. *Acacia pycnantha* gum is unusual in that it may contain only one aldobiouronic acid.

Acacia mearnali gum has a structure[59] reminiscent of *A. senegal* and *A. pycnantha* gums. It is a highly branched D-galactan linked principally (1→3) with branches linked mainly (1→6).

Acacia nilotica gum is unusual[56] for its low viscosity, low nitrogen content (0.08%), low L-rhamnose content (about 0.4%),[53] high methoxyl content (1.05%), and high specific rotation (+106° to + 108°).

Data pertaining to composition of various *Acacia* gums is summarized in Table IV.

V. Arabic Acid

1. Preparation

Arabic acid was prepared originally by the method described by Thomas and Murray[69] or a variation thereof. Commercial gum arabic was dissolved in ~0.1*N* hydrochloric acid and precipitated with four to five volumes of ethanol. Solution and precipitation was then repeated four times to give a pure product. Solutions of the purified gum after electrodialysis had a pH of 2.70. *Acacia senegal* gum has an equivalent weight of 1380.[63]

Moorjani and Narwani[70] prepared arabic acid by electrodialysis in Cellophane. This arabic acid, when dried at 110°, was insoluble.

Schleif and coworkers[71] prepared arabic acid by ion exchange and successfully dried it by spray-drying at 205° to yield a fluffy, white, finely subdivided powder. This procedure was extended by Swintosky and coworkers[72] to the preparation of other polysaccharidic acids, and a similar ion-exchange process for this purpose was patented by Wood.[73]

These methods were modified by Schweiger,[74] who recommends the following ion-exchange procedure for arabic acid and other polyuronic acids. Three lb (~1.5 kg) of Amberlite IR-120(H+) resin are packed into a glass column.

Then, ~3 gal (~11 liters) of a 7% aqueous gum arabic solution are passed through the column, followed by distilled water. All effluent with an acidic pH is collected. The solution is spray-dried at 200°, freeze-dried, or dried by solvent exchange to yield a fluffy, white, finely subdivided powder having an ash content of 0.05% or less.

2. *Properties*

Arabic acid is a moderately strong acid whose aqueous solutions have a pH of 2.2–2.7. Its $K_a = 2.0 \times 10^{-4}$ at 22° equating it to lactic acid. However, K_a varies with concentration; thus $K_a = 1 \times 10^{-3}$ for a 5% solution and 2×10^{-7} at infinite dilution.[4] Potentiometric titration of arabic acid[69] with strong alkali and alkaline earth bases gives titration curves typical of strong monobasic acids. It has an equivalent weight of about 1200.[24, 25, 41, 69, 70, 75–77] Arabic acid solutions show very little change in pH upon the addition of acid, indicating a buffering action. The addition of neutral salts lowers the pH of its solutions; barium chloride produces a greater depression of pH than sodium chloride.[78]

Arabic acid produces a higher solution viscosity than its salts. Thus, its viscosity can be lowered by the addition of sodium chloride and lowered further by addition of calcium chloride or other divalent salts. Emulsions prepared with arabic acid cream rapidly and are not as stable as those made with its salts. Also, the presence of other salts decreases the efficiency of the emulsifying action; divalent salts produce a greater decrease than monovalent salts.[79]

Pure arabic acid is nonreducing toward Fehling's solution and has a specific optical rotation of –27° to –30° in water.[43, 77]

3. *Degraded Gum Arabic*

Autohydrolysis of arabic acid releases L-arabinose, L-rhamnose, 3-*O*-α-D-galactopyranosyl-L-arabinose, and 3-*O*-β-D-galactopyranosyl-D-galactose from a somewhat hydrolysis-resistant core that contains only D-glucuronic acid and D-galactose.[41, 43, 80] Each species of *Acacia* may produce a gum that undergoes autohydrolysis to a unique product typical of the species extracted.

Acacia senegal gum has been subjected to autohydrolysis[28, 52] by heating a 2% solution to about 98° for 50 hr; the hydrolyzate was cooled, filtered to remove denatured protein, dialyzed, and freeze-dried to give a 50% yield of low-viscosity, degraded product of molecular weight[34] 4800; $[\alpha]_D$ –11°. The product contained 9.7% moisture, 19.2% uronic acid, 68% galactose and 2% arabinose. Further hydrolysis released two aldobiouronic acids, probably 6-*O*-(β-D-glucopyranosyluronic acid)-D-galactose and 6-*O*-(4-*O*-methyl-β-D-glucopyranosyluronic acid)-D-galactose. The dialyzate contained D-galactose, L-arabinose, L-rhamnose, three neutral disaccharides, traces of the two aldobiouronic acids, and some oligosaccharide.

Autohydrolysis of a 5% *Acacia nubica* solution yields a product of molecular weight 5730, DP 33, and $[\alpha]_D$ +44°, containing 12% glucuronic acid, 83% galactose, 5% arabinose, and 0.1% methoxyl. The reducing end-group was galactose.[57]

Degraded *Acacia nubica* resembles degraded gums of other *Acacia* species in that it consists mainly of (1→3)-linked β-D-galactopyranosyl units with some (1→6)-linked β-D-galactopyranosyl units. This gum, on autohydrolysis, produces a product of molecular weight 5200, $[\alpha]_D$ +31°, containing 17% uronic acid, 80% D-galactose, 3% L-arabinose, and 0.73 methoxyl.[39]

4. *Derivatives*

Several procedures have been reported for preparing sodium arabate from gum arabic, which occurs naturally as a mixed potassium, magnesium, and calcium salt. Krantz and Gordon[81] precipitated the calcium ions in a gum arabic solution by the addition of sodium carbonate and then filtered and evaporated the solution to yield sodium arabate. Briggs[82] directly neutralized an arabic acid solution with sodium hydroxide. Calcium arabate is prepared by washing gum arabic with calcium salts.[83]

Schleif, Higuchi and Busse[71] prepared sodium arabate by titrating arabic acid with sodium hydroxide solution. The solution was then spray dried at 205° to give a fluffy, fine, white powder. This method was extended to prepare spray-dried arabic acid salts of potassium, calcium, magnesium, zinc, iron(III), and aluminum.[79] Wood[73] prepared arabic acid salts of iron, copper, and silver by ion-exchange procedures. A more simplified, one-step ion-exchange process for preparing gum arabic salts was patented by Adams,[84] who passed a gum arabic solution through a cation-exchange resin in the salt form to prepare potassium, sodium, lithium, and ammonium salts.

Copper arabate, which is not stable, is made by treating Amberlite IRC-50-(H^+) cation-exchange resin with 10% w/v copper acetate solution, followed by treatment with a 5% w/v gum arabic solution. Another method is to allow a 5% arabic acid solution to stand over excess copper carbonate crystals. The copper arabate solution, which is pale bluish-green to dark olive green, is filtered and used without drying.[61] This copper salt has found some utility in copper plating on aluminum.

Organic salts and esters of arabic acid were prepared by Schleif and coworkers[79] by reacting a solution of arabic acid with the appropriate amine or alcohol and spray drying the neutral mixture. In this way, trimethylbenzyloxy arabate and salts, formed by neutralization or arabic acid with butylamine, ethylenediamine, diethylenetriamine and triethylenetetraamine, were prepared.

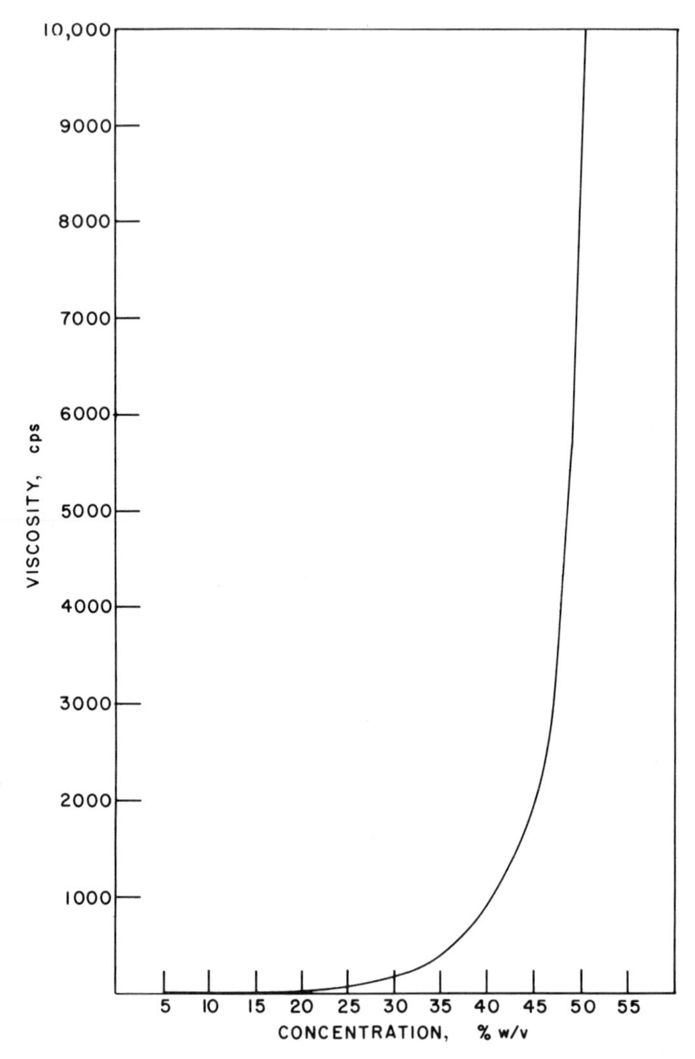

FIG. 4.—Viscosity as a function of concentration. Viscosities were determined with a Brookfield Synchro-Lectric Viscometer at 25.5° ±0.5° on solutions prepared with powdered gum.[85]

VI. VISCOSITY

Whereas most gums form very viscous solutions at low concentrations (1–5%) (Table V), gum arabic is unusual in that it is extremely soluble and can form solutions (Fig. 4) over a wide range of concentrations.[85] It is easy to pre-

TABLE V

Effect of Concentration[85] on the Viscosities (cps) of Gums[a]

%	Gum Arabic	Gum Traga-canth	Carra-geenan	Sodium Alginate	Methly-cellulose, 1500 cps	Locust Bean Gum	Guaran
0.5	—	—	23.5	86	16.75	20.0	1,389
1.0	—	54	57	214	38.9	58.5	3,025
1.5	—	—	—	1,102	—	—	—
2.0	—	906	397	3,760	512	1,114.3	25,060
2.5	—	—	—	8,300	—	—	—
3.0	—	10,605	4,411	29,400	3,850	8,260	111,150
3.5	—	—	—	—	—	—	—
4.0	—	44,275	25,356	—	12,750	39,660	302,500
5.0	7.25	111,000	51,425	—	67,575	121,000	510,000
6.0	—	183,500	—	—	—	—	—
10.0	16.50	—	—	—	—	—	—
20	40.50	—	—	—	—	—	—
30	200.00	—	—	—	—	—	—
35	423.75	—	—	—	—	—	—
40	936.25	—	—	—	—	—	—
50	4,162.5	—	—	—	—	—	—

[a] Measured with a Brookfield Synchro-Lectric viscometer at 25.5 ± 0.5°.

pare solutions containing up to 37% of gum arabic at 25° (Table V),[86, 87] and this solubility is responsible for its excellent stabilizing and emulsifying properties when it is incorporated with large amounts of insoluble materials.

The properties of the gum exudates are affected by the age of the parent tree, the amount of rainfall, the time of exudation, and the type of storage conditions. Thus, viscosities of solutions of gum arabic of similar grades can vary as much as 50% and are differently affected by pH, addition of salts or other electrolytes, and temperature.

The techniques and apparatus used in measuring gum viscosities may also give varying results. If the gum is dissolved in the total volume of water desired, a higher viscosity will sometimes result than if the gum is dissolved in a small amount of water and then diluted to the desired concentration. Higher viscosities are also obtained when powdered gums are used instead of chipped gums.[85] However, when carefully standardized techniques are used on pure gum arabic, its solutions have an intrinsic viscosity[25] of 0.20.

1. Concentration

Taft and Malm[87] determined the effect of concentration on the viscosity and density of purified gum arabic solutions (Table VI). The viscosities were

TABLE VI

Effect of Gum Arabic Concentration on Density and Viscosity[87]

Concentration, % by weight	Density, g/ml	Absolute Viscosity, poise	Relative Viscosity
1.22	1.000	0.0132	1.65
2.10	1.003	0.0160	2.00
2.70	1.006	0.0184	2.30
3.85	1.010	0.0234	2.92
3.95	1.011	0.0235	2.93
4.77	1.014	0.0259	3.23
5.54	1.017	0.0287	3.59
7.33	1.024	0.0390	4.87
8.55	1.029	0.0440	5.50
11.73	1.042	0.0707	8.84
15.68	1.050	0.0959	11.90
16.48	1.062	0.1400	17.50
18.69	1.071	0.190	23.70
34.92	1.141	2.410	300.00

measured with Ostwald pipettes at 30° (pH 7.14 ± 0.02). Kruyt and Tendeloo[88] found that the viscosity of gum arabic solutions at very low concentrations follows Poiseuille's law governing the flow of liquids through capillary tubes (Table VII).

At concentrations below 40%, gum arabic solutions exhibit Newtonian behavior as shown by the linearity in Figure 5. At concentrations of 40% and higher, pseudoplastic characteristics are observed as denoted by a decrease in viscosity with increasing shearing stress.[89]

TABLE VII

Relative Viscosity of Gum Arabic Solutions at Low Concentration[88] *and 25°*

Concentration, %	Relative Viscocity
0.0311	1.060
0.0622	1.104
0.1244	1.185
0.2490	1.321
0.4983	1.564
1.0000	2.015

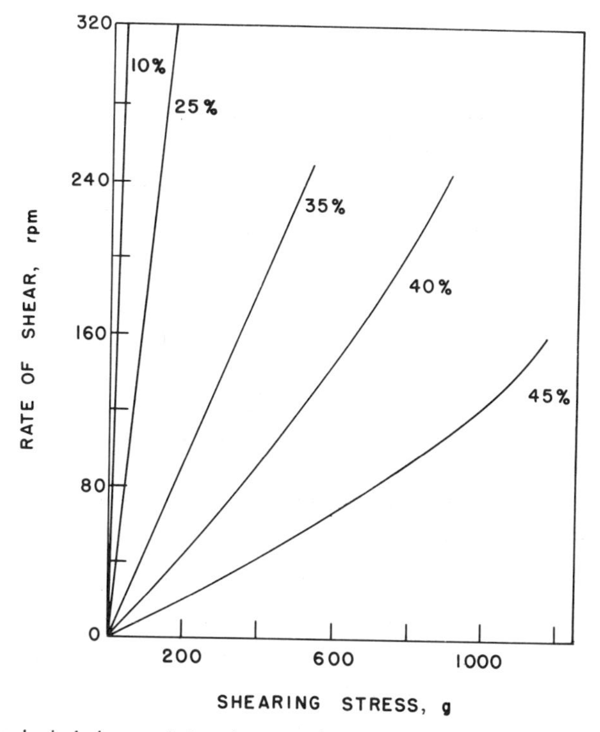

FIG. 5.—Rheological characteristics of gum arabic solutions. (redrawn from Aranjo.[89])

2. *Temperature*

Taft and Malm[87] showed that, as the temperature increased, the relative viscosity and density of gum arabic solutions decreased (Table VIII). Using higher concentrations (a 35% solution of *Acacia* mucilage, USP XV), Osborne and Lee[85] confirmed the fact that the viscosity is inversely proportional to temperature but varies to a smaller degree as the temperature is increased (Table IX).

It has been reported by Gabel[90] that the viscosity of gum arabic solutions can be increased by drying the gum over sulfuric acid or by heating the dry gum. Moorjani and Narwani[70] confirmed Gabel's results (Fig. 6). This was explained by an increase of imbibed water in gum micelles. When heated to 170° and then put into water, gum arabic does not dissolve, but swells to form a non-sticky gel.

Moorjani and Narwani[70] also examined the effect of sodium hydroxide on gum arabic that was heated to temperatures of 110°–170°. It was found that the quantity of sodium hydroxide required for neutralization increased with the rise in temperature. This was explained by the difference between the extent of

TABLE VIII

Effect of Temperature on Viscosity and Density of Gum Arabic Solutions[87]

Concentration, %	Temperature	Density, g/ml	Absolute Viscosity, Poise	Relative Viscosity
3.84	0°	1.019	0.0605	3.38
	15°	1.018	0.0357	3.13
	30°	1.101	0.0234	2.93
	45°	1.009	0.0170	2.74
9.09	0°	1.197	0.1280	7.17
	15°	1.034	0.0749	6.57
	30°	1.031	0.0478	5.97
	45°	1.025	0.0327	5.48

ionization of calcium and sodium arabates and the difference in hydration rates of calcium and sodium ions.

3. *Electrolytes*

Tendeloo[91] reported that the addition of electrolytes to 1% gum arabic solutions decreased the viscosity. The decrease in viscosity is proportional to the increase in the valence of the cation or the increase in the concentration of electrolyte. Addition of more than one electrolyte gives an additive effect.

Clark and Mann[92] found that all electrolytes, even in dilute solutions, lowered the viscosity of gum arabic. Lowering was more pronounced in the more concentrated solutions. In preparation of emulsions, the lowering of viscosity and interfacial tension produces favorable emulsifying conditions. Thus, whereas a very good benzene–water or kerosene–water emulsion can be produced with a 10%

TABLE IX

Effect of Temperature on Viscosity of Gum Arabic Solutions (35%)[85]

Temperature	Viscosity, cps
10°	344.5
20°	219.8
30°	136
40°	107.6
50°	79.9
60°	62.75
70°	52.75
80°	42.8
90°	39

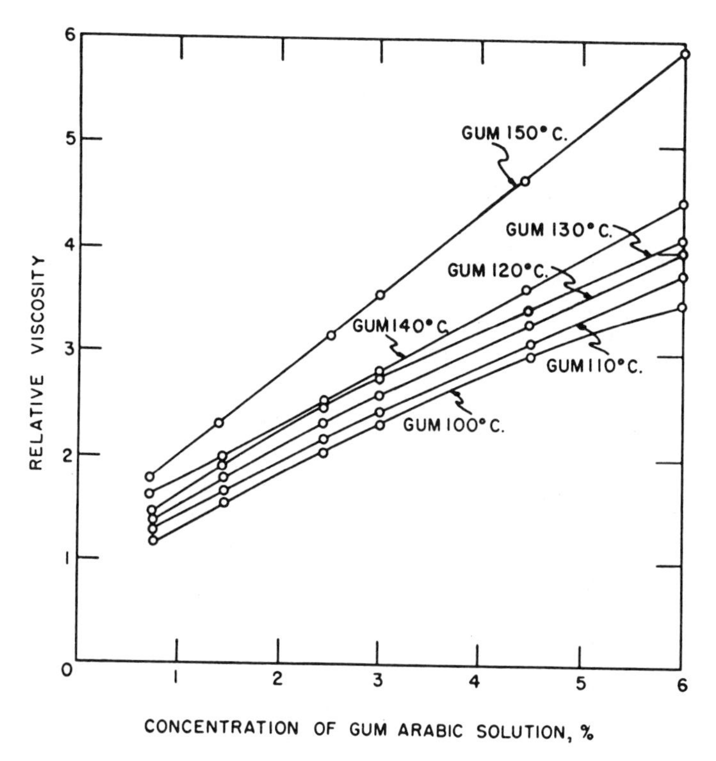

FIG. 6.—Effect of heating dry gum arabic on the relative viscosity of its solutions (redrawn from Moorjani and Narwani[70]).

gum arabic solution, equally good emulsions can be obtained with 0.5% gum arabic solutions in the presence of sodium hydroxide or sodium iodide for benzene and sodium sulfate or sodium bicarbonate for kerosene. The effect of salt additions depends upon the particular salt used. Taft and Malm[87] studied this effect by adding various sodium salts to a 5% solution of gum arabic at 30°; the concentration of the salt was 0.1 molal. They found that the addition of calcium chloride lowered the viscosity of gum arabic solutions, but as the concentration of calcium chloride increased, the relative viscosity increased. However, Borkowski and Olszak[93] showed that there was no interdependence between the calcium content of the gum and the viscosity of gum solutions. The calcium content of the gums investigated ranged from 0.6345 to 0.9511%.

Basu, Dasgupta, and Sircar[41] showed that the reduced viscosity, η_{sp}/C, of dilute sodium arabate solutions increases as the concentration of the gum decreases and decreases with addition of sodium chloride (Fig. 7). However, when very concentrated solutions of gum arabic ($\sim$35%) are mixed with aqueous

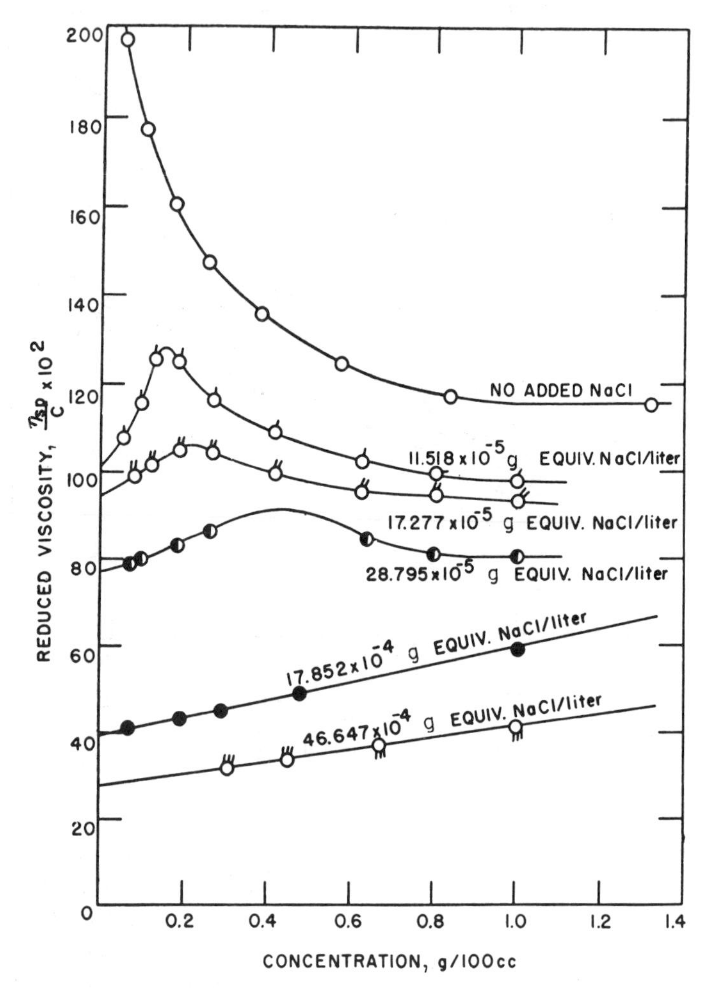

FIG. 7.—Reduced viscosity of sodium arabate as a function of sodium chloride concentration (redrawn from Basu, Dasgupta, and Sircar[41]).

solutions of salts, the effect is identical to that caused by dilution with distilled water; that is, the effect of the solute is negligible as compared with that of the solvent.[85]

At low concentrations, the electroviscous effect is graphically shown when the reduced viscosity is plotted against concentration. As the concentration of the gum decreases, the reduced viscosity decreases to a minimum and then rapidly rises. This rapid rise at low concentrations results from the charge on the molecule, which effects viscosity more at low concentrations. The addition of electrolytes will suppress this charge and give a lineal plot. The charge on the colloid is suppressed more by a divalent cation than by an equivalent concentration of a monovalent cation, and is still more depressed by a trivalent cation.

Thus, on gradual addition of potassium chloride, the viscosity curve straightens out and becomes linear as the electroviscous effect is neutralized (Fig. 8).[94] Bungenberg de Jong, Kruyt, and Lens[95] have also shown the effectiveness of calcium chloride in neutralizing the electroviscous effect of gum arabic solutions of increasing concentrations (Fig. 9). At higher concentrations, the electroviscous effect is self-suppressed; in gum arabic solutions, this occurs in the 0–10% concentration range.[96] The explanation for this is that the macromolecular electrolyte itself acts as an indifferent electrolyte and diminishes the interaction between its own charged areas, leading to a denser coil and a smaller, nonevident electroviscous effect.

4. pH

Thomas and Murray[69] found that the maximum viscosity of gum arabic solution occurs at pH 4.58–6.30 (Fig. 10). In the pH range from 5 to 10, there is a gradual change in viscosity with a marked change in pH. Taft and Malm[87]

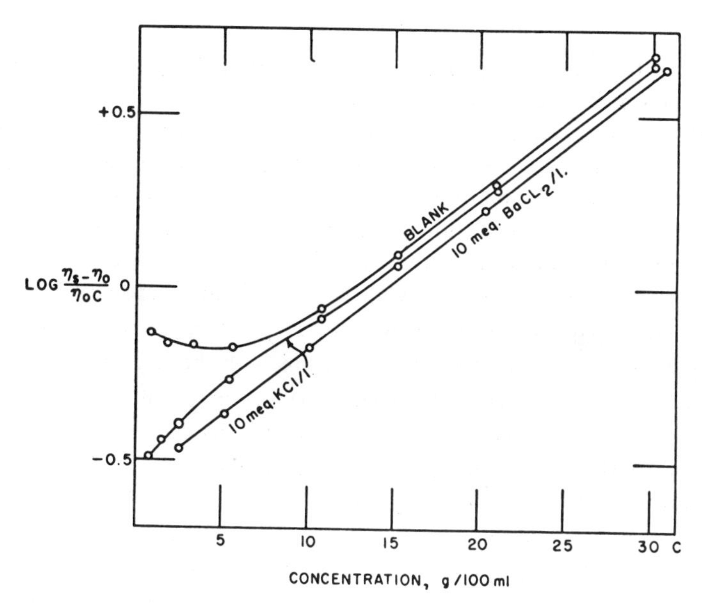

FIG. 8.—Comparative effect of monovalent and divalent salts on electroviscous phenomenon (redrawn from Bungenberg de Jong, Kruyt, and Lens[94]).

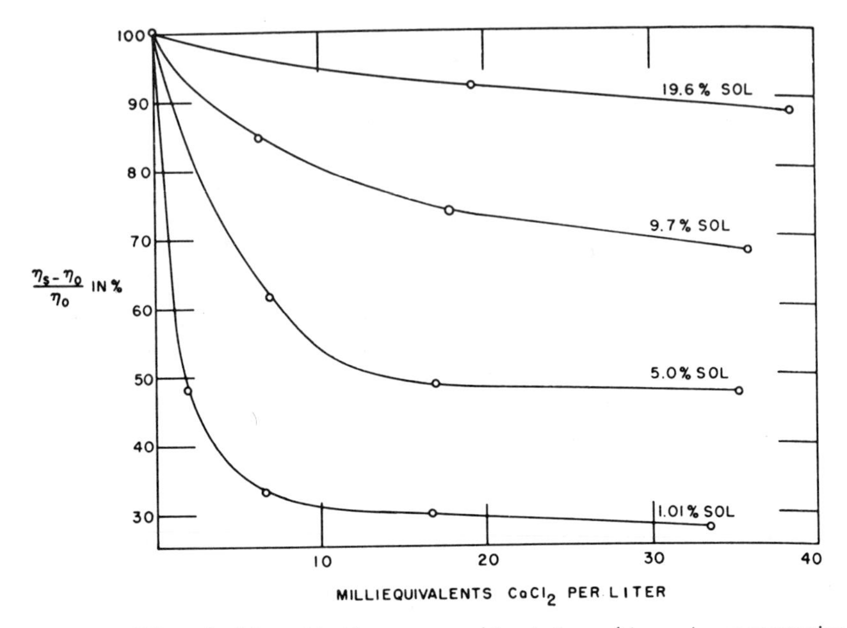

FIG. 9.—Effect of calcium chloride on gum arabic solutions of increasing concentration (redrawn from Bungenberg de Jong, Kruyt, and Lens[95]).

also found that the viscosity of gum arabic solutions changes with pH, but they found a maximum viscosity at pH 6–7.

The relative viscosity of dilute sodium arabate solutions falls slightly with increasing pH but decreases rapidly above pH 9. When the pH is reversed by the addition of acid, the viscosity remains practically constant from pH 11 to 6.8 Upon dialyzing the final solution, the viscosity gradually increases with time until the original value is almost regained after 8–10 hr (Fig. 11).[41]

A similar fall in viscosity is obtained by addition of sodium chloride, and this behavior is explained by changes in the shape of the gum arabic molecule.[41] If the gum arabic molecules are flexible chains with the carboxyl groups of the uronic acids distributed along the chain, changes in the degree of ionization, such as caused by the addition of an electrolyte, changes the average distance between the charge centers in the polymer chain. The polymer chain then either coils up or extends causing a decrease or increase in viscosity. With reversal of pH by the addition of hydrochloric acid, the excess hydroxide is simply neutralized to sodium chloride. There is no change in sodium ion concentration, and therefore the viscosity of the solution remains unchanged because the coiling of the chain remains constant. Upon dialysis, however, the sodium salt is lost,

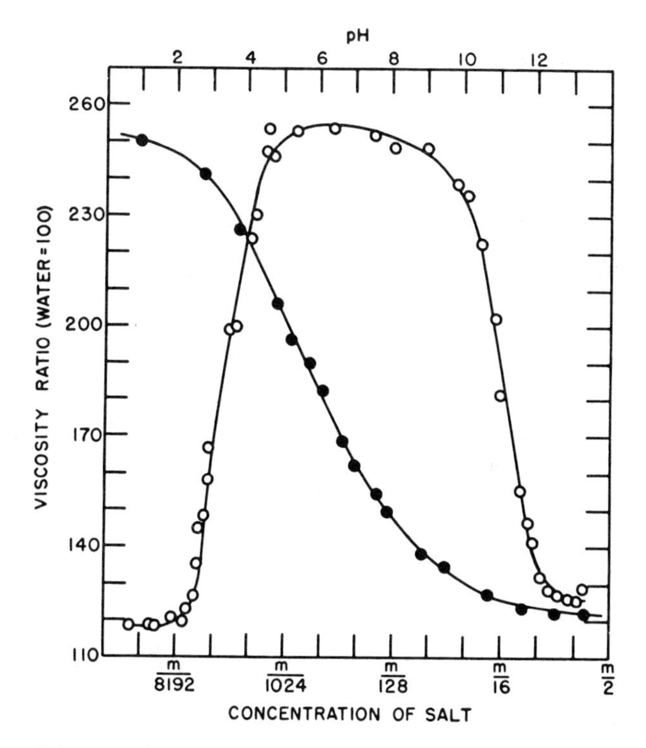

FIG. 10.—Viscosity of gum arabic solutions as a function of pH and salts (redrawn from Thomas and Murray[69]).

resulting in an increased viscosity caused by an increased extension of the polymer chain.

A detailed study of the effect of electrolytes that produce pH changes in gum arabic syrup was made by Woo and Huyck.[97] They prepared a gum arabic gum syrup containing sodium benzoate, tincture of vanillin, and soluble saccharin for use as a vehicle for potassium bromide, ammonium chloride, sodium citrate, or ferric ammonium citrate. The base syrup was stable for 2.5 months at room temperature and for 2 months at 37.5°, at which time a precipitate formed. There was little change in pH, and the viscosity increased, reaching a maximum in 4 weeks and then decreasing. Syrups with added neutral and acidic salts appeared to be stable for 6 months, but a syrup made with sodium citrate was stable for only 2 months.

In another series of experiments, *Acacia* syrup was prepared with glycerol as a preservative and evaluated under acidic (0.1% hydrochloric acid), neutral, and alkaline (0.1% dibasic sodium phosphate) conditions at 25° and 37.5°.

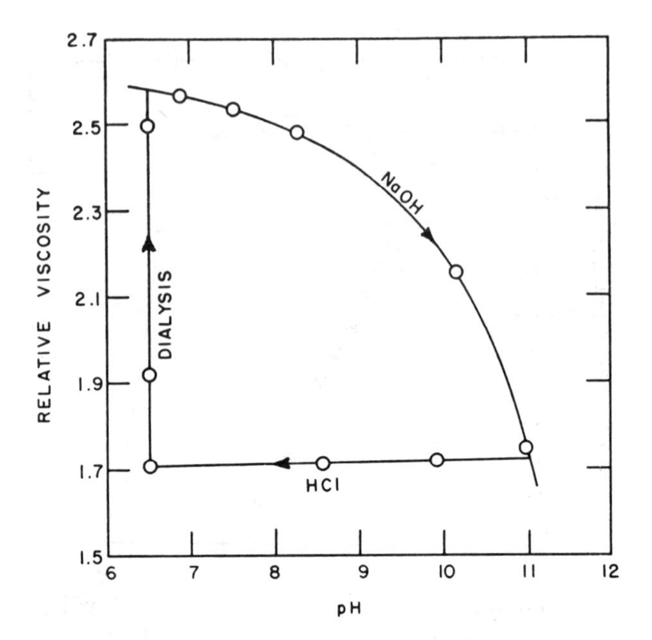

FIG. 11.—Effect of pH and sodium ions on viscosity of gum arabic solutions (redrawn from Basu, Dasgupta, and Sircar[41]).

All electrolytes lowered the viscosity of the solutions on standing except sodium citrate, which increased the viscosity. Neutral and acidic preparations stored at room temperature were stable at the end of 1 month, but alkaline syrups developed sediment after 3 weeks of storage.

5. *Organic Solvents*

The relative viscosity of gum arabic solutions decreases with the addition of ethanol.[98] At ~60% ethanol, gum arabic precipitates.

Miscible solvents such as glycerol act essentially as diluents when added to a 35% gum arabic mucilage, and the viscosity is a resultant of the combined effects of the components.[85]

6. *Aging*

The viscosity of gum arabic solutions decreases with time, but this decrease can be minimized by the addition of preservatives.[87] In a comparative study of several benzoic acid derivatives, Araujo[89] found the most effective preservative to be benzoic acid at 0.2% concentration.

As solutions of gum arabic age, the pH decreases slowly as does the viscosity. A 25% solution of gum arabic showed a decrease in pH from 4.8 to 3.5 in a

1 year period. This decrease in pH is minimized by the addition of preservatives.[89]

7. *Mechanical Treatment*

Normal mechanical treatment of gum arabic solutions does not affect their viscosities. A gum arabic solution drawn through a viscometer capillary about 20 times was found to be unchanged in viscosity.[87] However, mechanical passage through capillaries can be used to prepare emulsions using gum arabic as the stabilizer. Thus, emulsions of benzene or mixtures of benzene and liquid paraffin in water are made by passing the components, together with gum arabic, through capillaries under pressure.[99]

8. *Ultrasonic Vibrations and Ultraviolet Irradiation*

The maximum frequency of ultrasonic waves used for commercial applications[100] is of the order of 10^9 cycles per second, and the energy input for colloidal studies is somewhat greater than 10 W/cm^2. Industrially, ultrasonic vibration has been used for production of colloidal dispersions, but the structure of colloidal systems is easily broken down by these means; for example, vibration lowers the viscosity of gum solutions. Gum arabic solutions are depolymerized by ultrasonic waves.[2, 101] Piezoelectric vibrations[102] diminished the viscosity of gum arabic solutions by reducing the molecular weight rapidly to 150,000 and more slowly to 30,000. Continued exposure of gum arabic solutions to ultrasonic vibration releases monosaccharides.[103]

This lack of gum arabic stability to vibrational energy results in other emulsifiers being preferred for certain pharmaceutical preparations made by ultrasonic means. Hard soap, polyethylene glycol 400 monostearate, and polysorbate 80 were shown to be more stable emulsifiers than gum arabic, whereas sodium dodecyl sulfate and gum tragacanth were inferior.[104]

Ultraviolet radiation also reduces the viscosity of gum arabic mucilage irradiated for 2 hr in open dishes.[105] Gum arabic has an absorption maximum[106] at 265 nm.

VII. Physical Properties

1. *Solubility*

Gum arabic is insoluble in oils and in most organic solvents but usually dissolves completely in hot or cold water, forming a clear, mucilaginous solution. Solutions containing up to 50% of gum arabic can be prepared, and, as previously shown, the solubility in water increases as the temperature increases. There are certain samples of gum arabic, usually the exudates collected at the beginning of the dry season, that are not completely soluble. They form a stringy, mucus-like fluid that separates into the phases upon standing.

The solubility of gum arabic in organic solvents was examined by Taft and Malm.[107] Only hot ethylene glycol and glycerol were effective solvents for low-viscosity gum arabic. Gum arabic was soluble in aqueous ethanol up to about 60% of alcohol. Arabic acid behaves similarly. Very slight solubility was found with some acetate esters and acetate–alcohol mixtures.

Osborne and Lee[108] showed that gum arabic mucilage (35%) can be mixed with glycerol in all proportions to give a compatible solution with a viscosity equal to the resultant of the viscosities of both components. Evaporation forms a thick gel.

Although gum arabic is very soluble in water, finely divided colloid particles tend to clump when dispersed in water and do not dissolve readily unless vigorous agitation is used. This is caused by formation of a gel-like layer on the outside of particles as a result of rapid hydration, which prevents penetration of moisture to the inside. Many techniques have been developed for preparing gum arabic and other hydrophilic gum solutions. Two of the simpler methods are to mix the gum intimately with the other powdered ingredients before adding them to water, or to mix it thoroughly with ethanol or glycerol before dispersing it in water.

Christianson and Ramstad[109] improved the dispersibility of gums by bringing the moisture contents to 30–60%, mixing thoroughly, drying at room temperature, and then using the 40/100-mesh fraction. Coarser particles are ground to the desired mesh size and the fines reworked. This 40/100-mesh fraction is readily dispersible in water with a minimum of clumping or agglomeration. Henry's[110] dissolution process uses gum having a bulk density of at least 0.35 g/cm^3 and a particle size of 20–150 mesh. The gum is then treated with a water solution of any soluble alkali or alkaline earth metal salt of an inorganic or organic acid that has a pH of not less than 4.5. The concentration of the salt used is such that 5% by weight will be incorporated into the gum, which is then dried, ground, and sieved to 40/100-mesh.

Lachman and Chavkin[111] showed that freeze-dried gum arabic dissolved immediately to form a 10% solution, as compared to a solution time of 1 hr without freezing-drying. The viscosity of gum arabic is not affected by lyophilization nor is the pH of the solution significantly changed. Spray-dried gum arabic, also more rapidly soluble than powered material, can be purchased in commercial quantities at higher prices.

2. *Surface Tension*

The addition of gum arabic to water results in a lowering of the surface tension (Table X).[112] Temperature has an important effect on the surface tension

TABLE X

Effect of Gum Arabic Concentration on the Surface Tension of Water[112]

Gum Arabic Concentration, %	Surface Tension at 30°, dynes/cm
0.000	71.18
1.016	69.93
2.033	67.09
4.066	63.16

of a gum arabic solution. For a 4.066% solution of purified gum arabic at 30°, 40°, and 50°, the surface tension is 61.80, 60.46 and 59.03 dynes/cm, respectively. Surface tension also changes significantly with age of the solution.

Clark and Mann[92] showed that, as gum arabic concentration increases, there is a slight increase in surface tension, followed by a gradual decrease. Addition of electrolytes and especially mineral acid gives lower surface tensions. Sodium hydroxide produces a strong depression of surface tension in solutions containing up to 0.5% of gum arabic, but at higher gum concentrations, acid is most effective. This lowering of surface tension (or interfacial tension in mixtures) simultaneously with the lowering of viscosity by the addition of electrolytes produces favorable emulsifying conditions.[92, 112] Surface tension is lowered more by monovalent ions than by divalent ions.[112] The power of the different ions in lowering surface tension is in the order of $Li^+ > Na^+ > K^+ > Ca^{2+} > Ba^{2+}$.

Gum arabic is an effective emulsifying agent. The thickness and strength of the emulsifier film adsorbed at globule surfaces plays a more important role in the stability of emulsions than does size distribution.[113] The stabilization of the emulsion is dependant mainly on the coherence, rigidity, elasticity, and adsorption of the interfacial film.[114]

Warburton,[31] using a Wilhelm plate apparatus, measured surface tension of 0.1% w/v acacia solutions versus benzene at 25°. After 24 hr, the surface tension dropped 11.1 dynes/cm. A comparable experiment using 0.1% w/v copper arabate with 0.82% Cu dropped only about 4.5 dynes/cm in the same period, showing that the presence of copper profoundly affects the surface activity of acacia gum.

3. *Freezing Point*

The freezing point of gum arabic solutions decreases as the concentration of gum arabic increases (Fig. 12).[87] Mason[86] suggested that gum arabic dispersions should not be frozen, for when ice crystals are present, thawing does not regenerate the original viscosity.

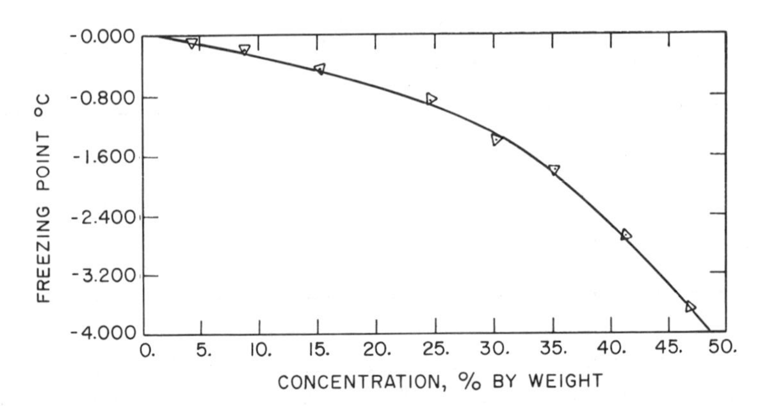

FIG. 12.—Freezing point of gum arabic solutions as function of concentration (redrawn from Taft and Malm[87]).

4. *Osmotic Pressure*

The osmotic pressure of sodium arabate solutions is at a maximum at a pH of 4.25, and a maximum at pH 3.35 was found for barium arabate (Fig. 13).[69]

Oakley[22] found that the osmotic pressure of boiled sodium arabate solutions is about 3% higher than that of unboiled solutions. There is a linear relationship between the osmotic pressure and the concentration at concentrations up to 1%. Above 1%, the plot becomes a gradual curve (Fig. 14). When sodium chloride is added, the osmotic pressure falls drastically; then the decline flattens out for concentrations above 0.2N sodium chloride. The limiting value of the osmotic pressure at any specific gum concentration is independent of the salt concentration.

5. *Conductance*

The effect of concentration of gum arabic solutions on conductance has been measured at 30°. When the equivalent conductances are plotted against the logarithms of volume, a curve (Fig. 15) is obtained that is typical of many electrolytes and that suggests that solutions of natural gum arabic are mixtures of calcium arabate and magnesium arabate.[87] Additional evidence for this hypothesis is seen when the equivalent conductance is plotted against the square root of the concentration. A straight line is obtained that can be extrapolated to zero concentration to obtain the equivalent conductance of the gum at infinite dilution. This value is 55.5 mho, whereas the values for the equivalent conductance at infinite dilution for calcium ion is 66.1 mho and for magnesium ion is 58.8 mho. Because the equivalent conductance of the gum is of the same order of

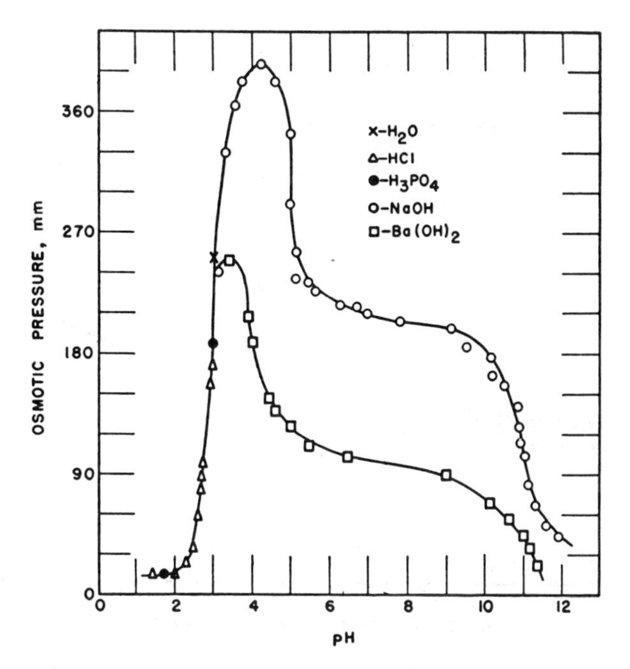

FIG. 13.—Effect of pH on osmotic pressure of gum arabic solutions (redrawn from Thomas and Murray[69]).

magnitude as that of its positive ions, it is assumed that the positive ions are the chief carriers of the current in these solutions.

VIII. Chemical Properties

1. *Chemical Reactivity and Compatibility*

Gum arabic as the calcium, magnesium, and potassium salt reacts with many reagents in a manner similar to that of other polysaccharidic acid salts. Solutions of gum arabic will produce precipitates or heavy gels on addition of borax, ferric chloride, basic lead acetate, mercuric nitrate, gelatin, potassium silicate, sodium silicate, Millon's reagent, and Stoke's acid mercuric nitrate reagent.[2–4] In general, trivalent metal ion salts will cause precipitation of gum arabic.[115] Dilute (1%) gum arabic solutions can be coagulated by the addition of Ruthenium red [$Ru_2(OH)_2Cl_4 \cdot 7NH_3 \cdot 2H_2O$], Hexol nitrate, or Desogen Geigy.[116]

However, in many applications, the flocculation or thickening of gum arabic can be prevented or retarded[117] by the addition of soluble alkali polyphosphates at concentrations of 0.1–1.0%.

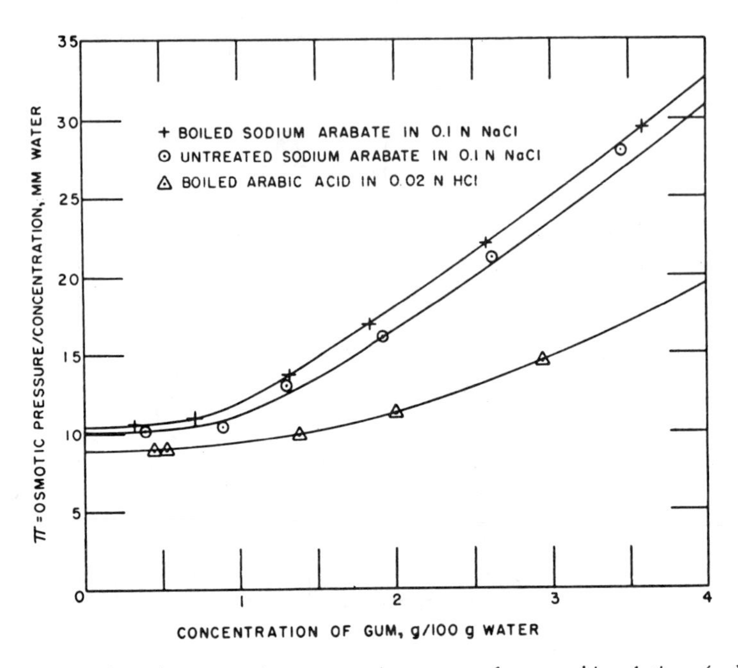

FIG. 14.—Effect of concentration on osmotic pressure of gum arabic solutions (redrawn from Oakley[22]).

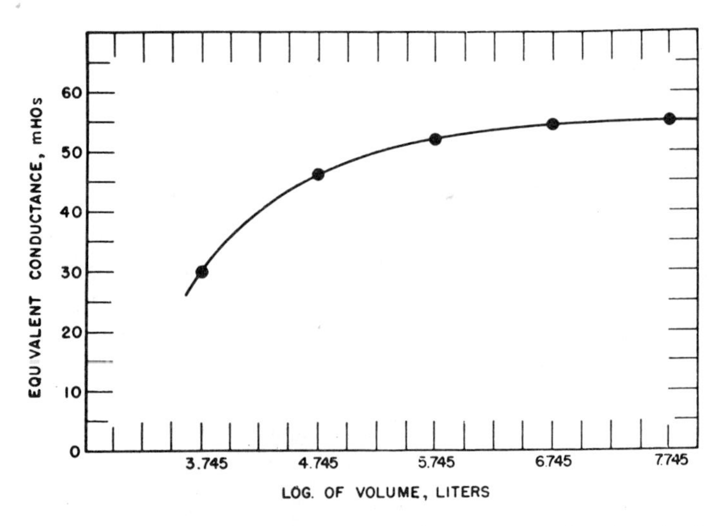

FIG. 15.—Equivalent conductance of a gum arabic solution as function of volume (redrawn from Taft and Malm[87]).

Upon treatment with dilute acids, gum arabic is hydrolyzed to yield a mixture of L-arabinose, L-rhamnose, D-galactose, and an aldobiouronic acid composed of D-glucuronic acid and D-galactose.[12] With hot nitric acid, gum arabic yields mucic (galactaric), saccharic (D-glucaric), and oxalic acids.[2]

The importance of *Acacia* mucilage in pharmaceutical preparations led to a study by Osborne and Lee[108] of its compatibility with various reagents and alkaloids. It is soluble in concentrated and dilute hydrochloric acid; concentrated and dilute acetic acid; concentrated and dilute ammonium hydroxide; dilute sodium hydroxide; and 10% solutions of sodium chloride, mercuric chloride, bismuth chloride, and silver nitrate, but insoluble in 10% ferric chloride and concentrated sodium hydroxide. *Acacia* gums have complete compatibility with 1 : 1000 dilutions of morphine sulfate, ephedrine hydrochloride, quinine sulfate, strychnine sulfate, and caffeine.

The "U.S. Dispensatory"[12] reports the following as incompatible with gum arabic: aminopyrine, pyrogallol, morphine, vanillin, phenol, thymol, carvol, α- and β-naphthol, pyrocatechol, guaiacol, cresols, eugenol, acetyleugenol, apormorphine, eserine, epinephrine, isobarbaloin, caffeotannic acid, gallic acid, and tannin.

Gum arabic is generally compatible with methycellulose derivatives. Tests run with methylcellulose showed that the upper limit of compatibility is a solution of 6% of methylcellulose (50 cps) and 4% of gum arabic, which is stable after standing for 3 days at room temperature. Carboxymethylcellulose (CMC) (medium viscosity) is compatible with gum arabic in both solutions and films. A 2% CMC solution and a 2% gum arabic solution are compatible when mixed in ratios[118] of 3 : 1, 1 : 1, and 1 : 3.

The gum arabic esters of acetic and propionic acids can be prepared by treating a gum arabic dispersion in formamide with the appropriate anhydride.[119]

Moe[120] prepared the sodium carboxymethyl ether of gum arabic. This material is readily dispersible in cold water and yields stable, low-viscosity solutions of good clarity. Two percent solutions are capable of forming clear films of good strength. The sodium carboxymethyl ether is useful as a sizing and finishing agent.

2. *Enzymes Present*

Gum arabic contains both oxidases and peroxidases that are inactivated by heating the gum solution to 80° or higher for 1 hr.[121] The peroxidase content varies but the enzyme can be inactivated without affecting the viscosity by heating to 100° for 30 min.[132]

Diastase[12] and pectinases have also been reported to be present in gum arabic. The pectinases will cause precipitation of pectin when both are present in the same medium.[123] This is undesirable in the preparation of citrus emulsions for flavoring beverages. The enzymes can be destroyed without materially affecting emulsifying properties by heat-treating gum arabic solutions of 20–40% concentration, at a pH ~5.0, for 15 min at 93°, followed by 1 hr at 77°. The pH, before treatment, is adjusted by addition of 0.1% of sodium benzoate, followed by a suitable amount of citric or tartaric acid. The heat-treated gum solution is cooled rapidly and then filtered. It can then be used for preparing citrus oil emulsions.[123] Gum arabic also has the ability to inhibit the action of certain esterases.[124]

IX. Coacervation

1. Introduction

Separation of colloid solutions into two phases, manifested by turbidity, droplet formation, or actual separation of a liquid layer, is termed coacervation.[96] The usual coacervate encountered is the single colloid type, as exemplified by the coacervate formed by addition of sodium sulfate or ethanol to a gelatin solution. However, many coacervates are known that contain two or more colloids.

Coacervates fall into two main classes, simple and complex. Although coacervates of different classes might appear similar, they are quite different in properties. The number of colloids present has no direct bearing on whether the coacervate is simple or complex. A simple coacervate is caused by partial miscibility. Thus, it is formed with the two colloids, gelatin and gum arabic, at pH values above the isoelectric point of gelatin. In this region, both colloids are negatively charged. If the solutions are sufficiently concentrated, there will be separation caused by mutual salting out. Simple coacervates occur only in concentrated solutions, because the main requisite is water deficiency. Addition of nonreacting salts to the coacervate might have no effect but often enhances simple coacervate formation. The effectiveness of a particular salt is largely dependent upon its position in the lyotropic series. No disintegration of a simple coacervate occurs in a direct-current electric field inasmuch as charged particles are not responsible for the phenomenon. Both layers of a simple coacervate are found to be rich in colloid, but each layer contains mainly one of them.

In contrast, complex coacervates occur only with two or more colloids having opposite charges. In the case of gelatin and gum arabic, a complex coacervate is formed with the negative gum arabic when the gelatin is positive (at pH values below its isoelectric point). Complex coacervates cannot occur in concentrated solutions, but only in dilute solutions, and may occur at concentrations as low as 0.001%. Nonreacting salts suppress complex coacervates with effectiveness depending upon valency. Position in the lyotropic series is important. Direct current electric fields cause disintegration of complex coacervates. Both layers in a complex coacervate contain colloid; one is fairly concentrated and the other relatively dilute. However, both layers contain the same proportion of the two colloids.

2. *Gum Arabic–Gelatin Coacervates*

Typical coacervates are formed with gum arabic and gelatin at certain optimum concentrations. They form clear, stable emulsions of concentrated microdroplets suspended in comparatively dilute solutions. Many factors, such as pH, salts, and other variables, affect coacervate formation.[125]

The nature of the materials forming the coacervate is important. Thus, an amphoteric substance, such as gelatin, can form complex coacervates with basic colloids at pH values above its isoelectric point and also with acidic substances at pH values below its isoelectric point. Nonamphoteric, acidic substances, such as gum arabic, can form complex coacervates only with basic materials.[126]

That a loose combination is formed in coacervation is indicated by the observation that a gelatin solution is capable of forming a film, whereas gum arabic does not.[127] Mixtures of gum arabic and gelatin, however, formed a double-strength film when mixed at pH 3.5 and 40° and allowed to set. This increase in strength is observed in the pH region 2–4.8, optimum for coacervate formation. The flow birefringence of gelatin changes in the presence of arabate ions,[128] additional evidence of complex coacervate formation.

Basu and Bhattacharya[129] discuss coacervation phenomena from the standpoint of modern polymer theory. Coacervates are considered as immense aggregates containing large amounts of trapped, immobilized solvent. Molecules coil in poor solvent. Consequently, no coacervation is expected in good solvents where molecules are extended. This theory was confirmed by precipitation of gelatin and gum arabic from aqueous solution with ethanol at both high and low pH. High pH causes extension of the gum arabic due to the repulsion of negative carboxyl groups, where low pH has a similar effect on the basic centers of gelatin. As expected, extended chains result in repression of the ability to form coacervates. Addition of salt has a suppressing effect and leads to precipitation of viscous liquids or gels at all pH values. This is attributed to a common ion effect of the salt.

Bungenberg de Jong and coworkers[130] consider the complex coacervate of gelatin and gum arabic to be a structure of gelatin cations to which portions of anion centers of gum arabic are randomly attached. Gum arabic does not contribute to the gel structure but is merely held by electrostatic attachment to the gelatin.[131]

3. Preparation of Coacervates

Coacervates of gelatin–gum arabic are easy to prepare. One mixes 3% solutions of gelatin and gum arabic and makes the solution weakly acidic with acetic acid. The optimum proportion[132, 133] of gum arabic is 59%. Coacervation does not occur below 19% nor above 81% of gum arabic.[134] When one or the other polymer is present in excess, asymmetrical behavior is found;[135] thus, when potassium arabate is present in excess, the weight value of the coacervate is higher than when gelatin is in excess.

A stable coacervate can be prepared by mixing solutions of 6 g of gum arabic in 10 ml of water and 5 g of gelatin in 190 ml of water and then adding a solution of 2 g resorcinol in 10 ml of 0.03*N* hydrochloric acid.[136] The resorcinol prevents gelation of the gelatin at room temperature, giving a coacervate that is stable at room temperature for long periods.[137] Hydrochloric acid is, perhaps, too strong an acid for coacervate formation; some workers prefer acetic acid or dilute acetate buffers, but buffers are more desirable than acids.[137]

When gelatin is partially heat-degraded by heating to 60° and the resulting product is reacted with gum arabic to form a coacervate, significantly less coacervate forms. Enzyme degradation has a similar effect.[138]

Coacervates in the form of hollow spheres with skinlike walls can be readily formed.[139] They are made by mixing gelatin and gum arabic solutions of the same pH, heating to 40°, allowing to cool, warming again to 40° and removing the supernatant solution by decantation. The coacervate is then added to a dilute hydrochloric acid solution at the same pH, and the mixture shaken. The slowly settling spheres have as their wall substances both gum arabic and gelatin.

4. General Properties, Physical Appearance, and Composition

The physical appearance of the gelatin–gum arabic complex depends on its history and mode of preparation.[130] Coacervates made by mixing and acidifying 3% solutions of gum arabic and gelatin may be homogeneous with interposed large vacuoles, network-like, or foamy and filled with many hollow spaces. If the material gels, it is a gelatin matrix in which gum arabic is ionically bound to the gelatin molecules.[130]

The composition of a coacervate system can be found by examining phase diagrams.[140, 141] Phase diagrams show that the excess components in coacervate formation go preferentially into the medium and not into the coacervate. The gelatin–gum arabic coacervate has a relatively low viscosity, about 20 times that of water.[96]

5. Effect of Temperature

Coacervates maintained above 30° do not change composition.[132] High temperatures, however, give low volumes of coacervate owing to increased solubility of the coacervate. Coacervates prepared at higher temperatures contain somewhat less water owing to lower hydration.[132] Vacuole formation does not seem to be related to temperature, but to the method of gel formation.[132] Cooling from 33°–40° to 28° often produces vacuoles.[142]

6. *Effect of pH*

Complex coacervate formation depends upon pH. Optimum pH is that at which one polymer is strongly negative and the other strongly positive. Thus, at pH 4.7, gelatin is only weakly positive, whereas gum arabic is strongly negative. As the pH is lowered, the situation is reversed, with the gelatin becoming strongly positive and the gum arabic weakly negative.[143] In the absence of salts,[144] the optimum acidity for gelatin–gum arabic coacervate formation is about pH 3.7. At this pH, coacervate droplets of $\sim 150\mu$ in diameter reach their equilibrium size within 10 min. This is their minimum diameter[145] and, hence, the greatest mutual bonding is at pH 3.7. At pH 1.2, the electric charge on gum arabic is so low that coacervate formation does not occur.[143]

7. *Reactions of Salts*

The effect of salts has been examined by sealing a mixture of 6 g of gum arabic and 5 g of gelatin in a colloidal membrane and causing coacervate formation by flowing dilute acetic acid down the membrane.[146] Low salt concentration has little effect, but increasing salt concentration caused first uniform vacuolization, then local vacuolization, and finally complete solution of the coacervate. Potency of a salt to react with the coacervate is as follows: $M^{+}{}_3A^{3-} < M^{+}{}_2A^{2-} < M^{+}\ A^{-} < M^{2+}\ A^{-}{}_2 < M^{3+}\ A^{-}{}_3$. Suppression of separation is favored by polyvalent anions and cations.

Mutivalent cation salts interact more strongly with the negative partner of the coacervate than do monovalent cation salts.[147] Hence, polycationic salts require a more alkaline medium to produce the additional negative charge needed to match the existing positive charge on the other member of the coacervate, for example gelatin. The polyvalent cation salt has the effect of reducing the negative charge on the gum arabic, whereas the higher pH tends to increase the negative charge, thus achieving balance. In a similar manner, a divalent anion tends to lower the optimum pH of coacervate formation by reacting primarily with the positive partner of the coacervate. With the omission of the $M^{+}{}_3A^{3-}$ salts, the effect of a series of salts of different valencies should have decreasing optimum pH values from the highest to the lowest as follows: $M^{3+}\ A^{-}{}_3 > M^{2+}\ A^{-}{}_2 > M^{+}\ A^{-} > M^{+}{}_2A^{2-}$.

When pH is held constant, addition of salts requires variation of the gum arabic : gelatin ratio for optimum coacervate formation as follows: (high gum arabic : gelatin ratio) $M^{3+}A^{-}_{3}$, $M^{+2}\ A^{-}_{2}$, $M^{+}\ A^{-}$, $M^{+}_{2}A^{2-}$, $M^{+}_{3}A^{3-}$ (low gum arabic : gelatin ratio[147]).

Distribution of salts between the coacervate and the equilibrium phase is rather complex. In one instance,[148] a mixture of 300 ml of 3% gelatin, 300 ml of 3% potassium arabate and 225 ml of water was brought to 40° and 75 ml of. 0.1*N* HCl was added, bringing the pH to 3.75. After 24 hr, the two clear liquid layers were analyzed. The lower layer had 14% coacervate and d_{40} = 1.057 and contained 9.0 meq/liter of potassium chloride. The upper equilibrium layer had a concentration of about 0.5% coacervate and d_{40} = 1.002 and contained 8.29 meq/liter of potassium chloride.

The counterion of the arabic acid salt plays an important part in the formation of gelatin–gum arabic complex coacervates. A study[149] of the effect of the alkali and alkaline earth cations of arabic acid resulted in the relationship:

$$\theta_{\max} = a\,(Ze/r^2) + b$$

in which θ is coacervation intensity, Ze is ionic charge, r is ionic radius, and a and b are constants.

Swelling of coacervate droplets is lowest at pH 3.7, where their water content is lowest. Reversible swelling at this pH caused by changes in salt concentration is lowest for monovalent salts.[145] The minimum swelling point can be raised as high as pH 4.81 by addition of potassium chloride.[144] Degree of swelling changes somewhat with time.[131]

Addition of sodium chloride to a buffered coacervate markedly lengthens equilibrium time.[150] Swollen gels follow typical polyelectrolyte behavior patterns with respect to effect of pH and neutral salt concentration.[151] For example, at pH 3.5 and 24°, complexes swell so slowly in many salt solutions that it is difficult to reach the equilibrium swelling point.

8. *Physical Phenomena*

Coacervate droplets can be obtained that move toward either the cathode or the anode or remain stationary in an electric field.[152] The excess component is decisive for lack of migration or its direction.[140, 153] In an individual gelatin–gum arabic coacervate droplet, the gum arabic accumulates at the anode and the gelatin accumulates at the cathode.

Lower interfacial tensions are obtained with potassium chloride, but the change in interfacial tension observed in electric fields is caused by migration of the components within single droplets.[153]

Coacervate droplets undergo motion in a diffusion field[92] owing to local concentration changes.[106]

It is found that coacervates with basic proteins[154] (for example, clupein) as a component form drops of 3.1–4.6μ in diameter, whereas coacervates containing gelatin are 7.3–7.7μ in diameter. Diameters of 1.5–3μ apparently were also found.[154] With basic proteins, no drops greater than 20μ were noted.

9. *Uses of Gum Arabic–Gelatin Coacervates*

Coacervates of gum arabic and gelatin have interesting properties that suggest many potential uses. By pouring some coacervates into water, immediate gels containing fine vaculoes are formed.[155] Coacervates have potential use in the food industry, such as in fixing flavors and colors.[156]

By adding a little salt to a gelatin–gum arabic coacervate, an edible viscous liquid or gel is formed that is stable over a wide pH range.[129]

Coacervates are used to concentrate amino acids, nucleotides, or polyelectrolytes of high molecular weight.[157] Gum arabic–gelatin coacervates have found wide application in microencapsulation of various materials for use in pressure-rupturable capsules, pressure-sensitive transfer record sheets, and water-insoluble capsules. A broad area of microençapsulation technology based on this type of coacervate formation has been developed by the National Cash Register Company and is covered by many patents.[158–162]

10. *Coexisting Coacervates*

Still more complicated coacervates can be prepared by combining three substances that can mutually react. Thus, one can form a coexisting complex coacervate by mixing solutions of gelatin, gum arabic, and "sodium nucleinate" at the proper pH.[163] Best coacervate formation is obtained with three parts of gelatin and one part each of gum arabic and "sodium nucleinate" at pH 3.6.

It is apparent that two types of complex coacervation can exist within a three-colloid system. In the above example, the gelatin is positively charged whereas the other two components are negatively charged. In such coexisting complex coacervates, each layer contains all three colloids, but one is rich in arabate and poor in "nucleinate" and the other is rich in "nucleinate" and poor in arabate. In this system, the two coacervates, gelatin–arabate and gelatin–"nucleinate", are only partly miscible in each other.

Another type of three-colloid system with two positive colloids can be formed from gelatin, arabate, and ichthyocol. Here, both proteins are positive and the arabate is negative. The coacervates, gelatin–arabate and ichthyocol–arabate, are mutually miscible in all proportions so that only two layers are formed. The similarity in properties of gelatin and ichthyocol is the determining factor that permits complete miscibility.

11. *Other Coacervates*

Gum arabic forms its best-known and most important coacervate with gelatin but it forms coacervates with many other substances, such as toluidine blue,[164] lecithin,[165] trypaflavin,[96] and myristylcholine.[133, 166]

Other proteins known to complex with gum arabic are hemoglobin, serum albumin, egg albumin, and casein.[133] Micheel and Heuer[167] formed complexes of gum arabic with gelatin and other substances. These products are not coacervates, as there is a covalent (amide) bond between the gelatin and gum arabic instead of the electrostatic linkage of the complex coacervate.

True coacervates are formed by gum arabic with serum albumin[168] at a pH as high as 6.9, but the maximum quantity is formed at about pH 4 using acetate buffer.[169] The formation of the coacervate can be used for separation of serum proteins as it is possible to react gum arabic with serum albumins while involving no reaction with globulins.[170] Some coacervation of α_1- and α_2-globulins was noted when coacervates were prepared from human serum, 10% gum arabic, and 10% acetic acid.[171] Addition of serum albumin, ovalbumin and hemoglobin to a gelatin–gum arabic mixture showed that serum albumin did not affect coacervate formation at pH 1.6–3.2, whereas the other proteins tended to lower the maximum of the formation curve in this region.[172] An electron microscopic examination gum arabic–serum albumin coacervates alone, with ribonucleic acid, and with both ribonucleic acid and ribonuclease in pH 4.1 acetate buffer gave similar droplets. Ribonuclease caused destruction of the droplets because of hydrolysis.[173]

X. Analysis of Gum Arabic

Standards for pharmaceutical grades of gum arabic have been established by the "U.S. Pharmacopeia"[174] and for food-grade gum arabic by the "Food Chemicals Codex."[175] Good quality gum arabic should have not more than 4% ash, more than 0.5% acid-insoluble ash, or more than 15% moisture. It should give no iodine color for starch nor a ferric chloride test for tannin. The heavy metal content should not exceed the following: arsenic, 0.5 ppm (mg/kg), lead, 1.7 ppm; zinc, 10 ppm; copper, 10 ppm.[176]

When found in nature, gum arabic is fairly free from contaminants and can be fairly easily identified by simple confirmatory tests.[174, 175] However, in commercial products, gum arabic occurs in combinations or mixtures with various materials, quite often other gums that have properties similar to those of gum arabic.

1. Isolation from Commercial Products

In simple cases, the mixture is heated with dilute acetic acid and filtered. The filtrate is cooled and treated with ethanol or acetone to precipitate the gum. Detailed methods for separating and identifying gum arabic[177] in complex food materials have been thoroughly reviewed by Glicksman[178] and are applicable to most other gums as well.

2. *Systematic Analytical Schemes*

When an unknown gum has been isolated from a complex mixture or product, it can be identified by systematic analytical schemes such as that of Ewart and Chapman[179] or Proszynski and coworkers.[180] These schemes and others are reviewed by Glicksman.[178] Positive tests for gum arabic or any other gum should be confirmed by specific confirmatory tests.

3. *Physical Confirmatory Tests*

Physical appearance.—The spheroidal tears (unground gum) of gum arabic are yellowish, white or translucent, almost odorless, and mucilaginous when wetted. Tears can be as large as 32 mm in diameter or can be broken into angular fragments. Gum arabic powder is nearly white. Microscopically, it is seen to consist of minute angular fragments.[177]

Infrared spectra.—Newberger and coworkers[181] determined the infrared spetra of gum arabic and many other common gums. The process can be used to identify gums isolated from foods, drugs, and cosmetics.

Solubility tests.—Gum arabic is practically unique among gums because of its very high water solubility. Although it is insoluble in ethanol and other organic solvents, its aqueous solutions will dissolve in glycerol.[177]

Optical activity.—A 10% aqueous gum arabic solution has almost no optical activity but might be slightly levorotatory.

4. *Chemical Confirmatory Tests*

Many confirmatory tests have been developed and reported for gum arabic.[2, 6, 12, 174–181] For a rapid test to distinguish gum arabic from gum tragacanth, iodine solution is added. Because gum arabic contains no starch, no color will be developed. Solutions of gum tragacanth, which is almost always accompanied by starch, turn blue.[2] Similarly addition of 0.02*N* iodine solution to powdered gum arabic causes no color change, while other gums such as agar and gum tragacanth, acquire a crimson or olive-green color.[12]

Gum arabic can be readily distinguished from quince or pectin gums by the procedure of Bryant.[182] To 10 ml of a 1% gum solution, 1 ml of 10% thorium nitrate is added. The solution is stirred and allowed to stand for 2 min. A gel indicates pectin or quince gums; gum arabic gives a negative test. Gum arabic can be distinguished from tragacanth, agar, karaya, carrageenan, quince, and locust bean gums by reacting 3 ml of gum solution with 0.5 ml of Millon reagent. A gel indicates carrageenan, agar, or locust bean gum. A flocculent, nonsettling precipitate indicates quince or tragacanth gum. A fine, powdery, or curdy precipitate shows gum arabic or gum karaya. If this precipitate dissolves in excess reagent, gum arabic is indicated.[177]

Inasmuch as gum arabic contains peroxidase, a drop of hydrogen peroxide added to a mixture of gum arabic solution and tincture of guaiacum produces a blue color. This serves as a useful identity test.[2] A deep blue color is also produced when 0.5 ml of hydrogen peroxide solution and 0.5 ml of a 1% benzidine solution (in 90% ethanol) are added to a solution of 0.25 g of gum arabic in 5 ml of water and allowed to stand.[12] Particles of gum arabic can be distinguished from agar and gum karaya by mounting the gums in solutions of ruthenium red examining them under a microscope. Gum arabic particles will not be stained red.[12]

5. *Direct Tests for Gum Arabic in Some Commercial Products*

Methods are available for detecting gum arabic in dairy products[183, 184] and in catsup.[185, 186] Detection of gum arabic in mural paintings by microchemical reactions has been investigated and reported by Bontinck.[187] Gum arabic is detected in blood by spectrophotometric examination[188] and in drugs by reaction with oxalic acid.[189] An excellent process by Redfern[190, 191] has been found effective for detecting the presence of gums in starchy foods, and a method for detecting gum arabic, carrageenan and other gums, excluding alginates, has been described by Mendelsohn.[192]

XI. Uses

1. *Foods*

The major use of gum arabic is in the food industry,[178] where it is used as a food additive to impart desirable properties through its influence on the viscosity, body, and texture of foods. In addition, it is nontoxic, odorless, colorless, tasteless, and completely water-soluble and does not affect the flavor, odor, or color of the food to which it is added.

Confectionery.—Gum arabic is used extensively in the confectionery industry, primarily because of its ability to prevent crystallization of sugar but also because of its thickening power.[132] It is used as a glaze in candy products and as a component of chewing gum, cough drops, and candy lozenges. Originally, gum drops were produced from gum arabic, but they are now made with thin-boiling starches or pectins.[193, 194]

In most confectionery products, gum arabic has two important functions.[195] The most important is to retard or prevent crystallization of sugar, and thus it finds its greatest application in confections where the sugar content is high and the moisture content comparatively low, as in jujubes and pastilles. Its second function is to act as an emulsifier, keeping the fat uniformly distributed throughout the product so as to prevent it from moving to the surface and forming an easily oxidizable greasy film.[196]

In the preparation of jujubes, pastilles, or gum drops, gum arabic is dissolved in water and the filtered solution is mixed with sugar and boiled. The desirable flavor is added with a minimum of stirring to prevent formation of bubbles or opaque spots.[197]

Lozenges are also prepared from gum arabic by mixing finely ground or powdered sugar with a thick mucilage of gum arabic. The flavors, usually peppermint or fruit, are added and thoroughly mixed to form a stiff dough, which is rolled into sheets and stamped out.[197]

Dairy products.—Gum arabic has been used as a stabilizer in frozen products, such as ice creams, ices and sherbets, because of its water-absorbing properties.[177, 198] Ice cream is prepared from a mixture of milk (skimmed, condensed, or powdered), butter, sugar, honey, eggs, flavoring agents, and a stabilizing colloid. Because ice cream is whipped during freezing, it contains numerous small ice crystals that tend to coalesce and form larger crystals, giving a coarse, gritty, and inferior product. Alternate melting and freezing caused by changes of temperature during storage and transportation would enable water crystals to grow. The addition of gum arabic prevents the formation of ice crystals by combining with large quantities of water and holding it as water of hydration, thus producing a finer texture in the ice cream. The chief objection to this use of gum arabic is that the ice cream does not melt readily.[199]

Scholz[200] patented a process using gum arabic for the preparation of packageable milk or cream. The proper amount of milk or cream is mixed with gum arabic; the mixture is heated mildly, poured into molds, cooled, and packaged. The product dissolves easily in hot beverages and keeps better under refrigeration than plain milk or cream. Walder[177] recommends gum arabic as a protective colloid in the preparation of processed baby food.

Bakery products.—Gum arabic is widely used in the baking industry for its viscosity and adhesive property. It is used in glazes and toppings. It also bestows smoothness when used as an emulsion stabilizer.

When used in a bun glaze, gum arabic imparts stability in conjunction with free-flowing and adhesive characteristics. The glaze is applied while still warm and adheres firmly to the bun upon cooling.[201] A stable icing base utilizing gum arabic has been patented by Wagner.[202]

Flavor fixative.—With the introduction of spray-dried flavors into foods,[203] an extensive use was found for gum arabic as a fixative for flavors. Spray drying is a fairly simple production operation producing desirable particle size, density, and solubility. By proper selection of spray-dryer nozzles and inlet pressures, droplets of various sizes can be formed and different drying times obtained.[204] Rapid drying at comparatively low temperatures is particularly advantageous for flavoring materials that are volatile or easily oxidized.

When used as a fixative, gum arabic forms a thin film around the flavor particle, protecting it from oxidation, evaporation, and absorption of moisture from the air. This is advantageous when hygroscopic materials are spray dried. Flavors spray dried with protective colloids last 10–20 times longer than the same flavors air dried on a solid. Tests made on spray-dried emulsions of hydroxycitronellal, benzaldehyde, and other easily oxidized aldehydes showed that the film protected the products from oxidation for years, whereas unprotected materials oxidized in seconds.[204]

Citrus oils and imitation flavors are prepared by emulsifying them in gum arabic solution before spray drying. The particle size can range from 1 to 50μ. Other colloids are also used for these purposes, but gum arabic is the most important and most widely used.

More recently, a new technique of microencapsulation has been developed and used for specialized flavor fixation.[205, 206] Gum arabic is reacted with gelatin in a medium containing the flavor to form an insoluble, protective film surrounding or encapsulating the sphere of flavor.

Flavor emulsifier.—Many flavor emulsions, such as orange, lemon, lime, cherry, root beer, and cola, are also prepared with gum arabic as the emulsifier.[207] Some flavor emulsions make use of the additive properties of a blend of gum arabic and gum tragacanth. Here gum tragacanth is the emulsion stabilizer, and gum arabic provides smoothness.[201]

In the preparation of bakers' citrus oil emulsions consisting of citric acid, lemon oil, glycerol, water, and coloring matter, the best emulsions are prepared with gum arabic–gum karaya mixtures.[208]

Beverages.—Gum arabic is an effective foam stabilizer in beverages and is largely responsible for the lace curtain effect on the sides of beer glasses.

Eye-appealing opacity in beverages and beverage dry mixes are produced by spray-dried combinations of vegetable oil and gum arabic sold commercially as a clouding agent.[209]

2. *Pharmaceuticals*

Probably no more than 5% of the gum arabic import is used for pharmaceutical purposes. Its inherent emulsifying and stabilizing properties plus its demulcent and emollient characteristics have led to a number of applications, ranging from the stabilization of emulsions to the preparation of tablets. Its applications are further extended because it retains its viscosity and stabilizing properties over a wide pH range.[210]

Suspending agent.—Gum arabic is listed in the "U.S. Pharmacopeia" as an effective suspending aid and has been employed to suspend insoluble drugs and to prevent the precipitation of heavy metals from solution through the formation of colloidal suspensions.

In a comparison of various other gums, gum arabic is one of the best emulsifying and suspending agents for calamine suspensions, kaolin suspensions, liquid petrolatum emulsions, and cod liver oil emulsions.[211] It is unsatisfactory for suspension of paraldehyde (2,4,6-trimethyl-1,3,5-trioxane).[212] It has been found to be excellent for preparing a stable, nonsettling magnesia suspension.[213, 214]

Intestinal absorption of poorly soluble medicinal substances, such as steroids, fat-soluble vitamins, and barbiturates, that are suspended in gum arabic can be facilitated by the incorporation of wetting-agents or other emulsifiers in the preparation.[215]

Demulcent syrup.—Gum arabic's demulcent or soothing characterstics have led to its use in many pharmaceutical syrups.[216] A gum arabic syrup that contains sodium benzoate, vanillin tincture, and sucrose is recommended for use as a flavored vehicle because of its demulcent effect. This syrup is often effective in masking the bitter or acid taste of medicaments by its protective colloid action. In recent years, the preparation of diabetic syrups has become increasingly important. Various nonsugar recipes have been developed, such as nonsugar cherry syrups, based on gum arabic.[217] A good syrup is made from saccharin, methyl *p*-hydroxybenzoate, water, and gum arabic.[79]

Emulsifying agent.—Gum arabic is superior to gum tragacanth for preparing cottonseed oil emulsions because the average diameter of the oil globules is much smaller.[218] With both vegetable and mineral oils, gum arabic emulsions are stable over the pH range 2–10.

The data are correlated with buffer capacities. Because gum arabic contains the potassium, calcium, and magnesium salts of weak acid groups, its buffer action is greater in the neutralization of acids than in the neutralization of alkalis.

Lotzkar and Maclay[219] compared pectin, gum tragacanth, gum karaya, and gum arabic as emulsifying agents for olive oil, cottonseed oil, and mineral oil in water while varying the pH, the oil-to-water ratio, and the gum concentration. They found that gum arabic stabilized emulsions over a wide pH range and that there was no relationship between initial viscosity and stability. Higher viscosity may prevent creaming but does not necessarily improve stability, and gum arabic emulsions are less viscous than those of other gum emulsions, although they are equally stable.

Paraffin oil and water emulsions for use as laxatives have been prepared with gum arabic, agar, and gum tragacanth.[220–222] Gum arabic has also been recommended for use in oral laxatives[223] and in laxative suppositories.[224]

One of the oldest applications of gum arabic has been in the preparation of cod liver oil emulsions.[210] Used in conjunction with gum tragacanth, stable emulsions with excellent shelf-life can be made of cod liver oil, linseed oil, and mineral oil. These emulsions are better than those prepared with either gum alone.[225]

The oxidase present in gum arabic may destroy 54% of the vitamin A content of cod liver oil emulsified with it.[226] Therefore, the oxidases must be destroyed by heating gum solutions before use. Gum arabic is effective in the preservation of vitamin A in vitamin-enriched margarine[227] and vitamin C in aqueous solutions.[228]

Antiseptic preparations.—Antiseptic preparations have been made with a mixture of colloidal silver bromide and gum arabic.[229] Silver arabate has antiseptic properties that make it suitable for use as a substitute for silver nitrate and organic silver compounds in the treatment of ophthalmic infections.[73] Silver compound preparations for the internal treatment of mucous membranes have been patented.[230]

It was reported that the use of gum tragacanth and other vegetable gums in pharmaceutical jellies can reduce the bactericidal action of the incorporated preservative.[231] It was found that gum tragacanth exerts a strong neutralizing effect upon the bactericidal activity of chlorbutanol (1,1,1-trichloro-2-methyl-2-propanol), the *p*-hydroxybenzoates, and benzalkonium chloride and, to a much lesser extent, a phenol, phenylmercuric acetate and Merthiolate, [(*O*-carboxyphenyl)thio]ethyl murcury. This depression of the bactericidal activity is a function of the concentration of gum tragacanth; the greater the concentration of gum, the less effective is the preservative agent. The reduction of bactericidal activity of quarternary ammonium salts by agar has been previously reported.[232, 233] Although no work of this nature has been done with gum arabic, it seems highly probable that it would also reduce the antibacterial effectiveness of preservatives.

Miscellaneous applications.—Gum arabic has been used as an adhesive or binder for pharmaceutical tablets, such as aluminum subacetate tablets,[234] and also as an excipient in the manufacture of pills and plasters. In additions, many types of coatings for pills employ gum arabic in their manufacture.[210]

Because of its emollient and demulcent properties, gum arabic is used in the production of cough drops and cough syrups. It produces a smooth viscous syrup and also prevents crystallization of sugar in both of these products.

Schwarzmann[235] has made a therapeutic gold–gum arabic–peptone complex, and another gum arabic–peptone complex without the addition of gold.[236] Mason[86] incorporated gum arabic with alkaloids for anesthetics. Iron arabate has been suggested for use as a medicinal hematinic preparation.[73]

3. *Medicine*

Gum arabic has been used for the treatment of low blood pressure caused by hemorrhage or surgical shock.[237–239] Intravenous saline injections alone were not successful because the salt escaped too rapidly from the blood vessels. The addition of a 7% gum arabic solution reduced the dissipation rate of the sodium chloride solution, and this treatment was successfully used in the 1920's. However, the subsequent development of such blood plasma extenders as dextran and polyvinylpyrolidone eliminated the use of gum arabic.

In 1933, intravenous injections of gum arabic solutions were recommended for the treatment of nephritic edema.[12, 237, 238] Some reports cited consequent liver and kidney damage;[12,237] whereas other reports, in which as much as 330 g of gum arabic were administered, presented no evidence of hepatic or renal damage but stated only that the treatment was successful in alleviating or eliminating the edema under treatment.[12, 238] It has been said that gum arabic, when used in infusion liquids, is harmful because it absorbs cations and causes an ionic unbalance.[240]

Some allergic reactions have occurred with injections of gum arabic.[237] The symptoms are flushing of the face, coldness of extremities, chills, nausea, vomiting, dyspnea, and uticaria. These reactions could be controlled or prevented by the administration of epinephrine.[12] Similar allergic reactions, manifested by asthmatic conditions, were caused by exposure to gum arabic used in offset printing.[12, 240] Most cases seem to have arisen as a result of inhaling the gum as dust.[242]

In plastic surgery, a 50% gum arabic adhesive has been used successfully in grafting destroyed peripheral nerves.[12]

4. *Cosmetics*

In lotions and protective creams, gum arabic stabilizes emulsions, increases the viscosity, assists in imparting spreading properties, adds a smooth feel to the skin, and forms a protective coating. It is also a binding agent in the formulation of compact cakes and rouges, and as an adhesive in the preparation of facial masks. A typical compact cake is composed chiefly of a color vehicle, mineral oil, and an aqueous solution of gum arabic. Gum arabic is also used as a foam stabilizer in the production of liquid soap.[243]

One of the advantages of gum arabic in cosmetics is its nontoxicity and its comparative freedom from dermatological and allergic reactions.[244, 245]

Face masks.—Several face mask formulas employing gum arabic have been developed.[246]

Hair creams and fixatives.—Nonoily hair fixatives based on vegetable gums have the advantage of vastly improving fixative properties without forming greasy stains on clothing. Gum arabic mucilages have been used in the preparation of these hair dressings.[246]

Face powder compact.—Cosmetic packs have been made using gum arabic as a binder.[247]

Protective cream.—Gum arabic is effectively used in protective creams as a stabilizer and film former.[248]

5. *Adhesives*

Gum arabic has been used widely in adhesives.[249] Powdered gum arabic is considered to be a safe, simple adhesive for miscellaneous paper products and is commonly used by dissolving it in two to three times its weight of water.[250, 251] A 40% solution makes an excellent mucilage for general office purposes.

Gum arabic glues are easy to prepare, light in color, odorless, and very stable. In addition, the adhesive strength of the gum can be improved by the addition of certain metal salts, such as calcium nitrate and aluminum sulfate. However, when these glues are heated too long or at too high a temperature, the gum arabic is degraded, resulting in a decrease in viscosity and a loss of adhesive strength. This degraded gum arabic will discolor bronze and colored papers. The problem can be minimized by using dilute solutions, which are then made more viscous by the addition of compatible thickening agents, such as gum tragacanth or methylcellulose. Gum arabic makes a brittle film, which is usually plasticized by the addition of 8–10% glycerol based on the weight of dry gum. Other suitable pasticizers are diethylene glycol, ethylene glycol, and sorbitol. Spoilage is prevented by glycerol or by the addition of a little calcium hydroxide.

Adhesives for envelopes, labels, and stamps are commonly dextrins or starches, but gum arabic finds some usage, particularly in the manufacture of adhesive for postage stamps. A typical formula contains 100 parts of gum arabic, 2.5 parts of sodium chloride, 2.0 parts of glycerol, 2.0 parts of starch, and 130 parts of water.[252]

A good transparent adhesive can be made with gum arabic, but to reduce costs, smaller amounts of gum arabic can be used, and the viscosity built up with small amounts of gum tragacanth.[253] A strongly caustic glue has been patented for use on greasy or waxed paper.[250] Wallpaper paste can be based on a mixture of gum arabic, bentonite, and starch.[254] Gum arabic is compatible with flour, starch, and dextrins and is used in conjunction with them in postage stamp adhesives,[252] mounting pastes,[255] extremely adhesive pastes,[251] and Lunel's paste for artificial flowers.[251] Gum arabic is sometimes used as a binder for water cements in foundry applications, such as gray[252] and iron[255] cements for castings. In the preparation of laminated papers, gum arabic has found specialized applications. The bonding of regenerated cellulose films can be accomplished by a special gum arabic paste.[256, 257] Some adhesives for bonding aluminum foil to paper have been prepared from gum arabic solutions modified with small amounts of tartaric or similar acids, but they are corrosive.[258] Gum arabic-based adhesives are frequently used for smooth, hard-surfaced glassine papers.[258]

6. *Paints*

The incorporation of protective colloids into a pigment-vehicle system, such as paint, is an old, established art.[259] This is usually done to obtain improved particle size during precipitation of pigments, to control pigment aggregates and wetting qualities, and to control consistency and setting.[100, 260, 261] Acheson[262] prepared graphite dispersions in water by use of tannic acid and gum arabic.

In paints and similar formulations, flocculation prevents hard settling of pigments. Water has been used for many years to prevent hard settling of the pigment in paint. Such controlled flocculation has been achieved by means of additives that presumably affect the coherence of particles in an immiscible liquid.[259] Gamble and Grady[263] treated pigments with water-soluble hydrophilic colloids such as gum arabic to give controllable thixotropic properties to paints. Less than 1%, based on the weight of the pigment, was recommended; but by varying the quantity, the thixotropic properties of the paint could be adjusted to any desired degree. Pigments that are readily wetted by water, such as titanium dioxide, zinc oxide, clays, and silica, require very small quantities. The amount of water required to produce suitable thixotropic properties in a paint is usually less than 1%. In water-emulsion paints, the use of gum arabic is limited owing to its unfavorable effect on the water resistance of the final paint film,[264] but it has found some use in the preparation of vinyl resin emulsions.[265]

7. *Inks*

Gum arabic is a constituent of many special purpose inks[266] because of its excellent protective colloid properties. The effectiveness of this protective colloidal action of gum arabic or other gums is designated by its gold number. The gold number is defined as the number of milligrams of gum that, when added to a standard gold dispersion, is just sufficient to prevent a change in color from red to blue when 1 ml of 10% sodium chloride solution is added.[266] The comparative effectiveness of several common colloidal materials can be seen from the following gold numbers:

Gum arabic	0.1–0.5
Gelatin	0.005–0.1
Casein	0.1
Gum tragacanth	0.5–2.0
Wheat starch	4–6
Potato starch	25

Early inks were simply dispersions of lampblack in water. Soon gum arabic was added routinely as a suspending agent or a protective colloid. Sometimes, the lampblack and gum were mixed into a thick paste and allowed to harden in molds. These ink sticks have been in continuous use with little or no change in processing for over 3000 years. To use an ink stick, the operator moistens a small brush with water and then rubs the brush on the end of the stick. To make liquid ink, he rubs the end of the stick in a little water in a shallow dish until the ink solution develops the proper color.[267, 268]

Record ink.—Government record ink had long used gum arabic in the formula as a protective colloid.[267] Some years ago, because of a clerical error, gum arabic was omitted from the formula when proposals for bids were being typewritten. This error was not discovered until after inks were prepared. However, tests made by the Department of Agriculture, Bureau of Chemistry, showed that the omission of gum arabic was a slight improvement, and it has been left out of the formula for standard record ink ever since.

Soluble inks.—Easily soluble inks are used by textile workers to mark cloth for cutting or sewing operations. The operator marks the position of buttonholes, hems, pleats, lapels, and pockets. When the garment is finished, a hot-water wash is used to remove the ink. A typical ink is composed of a mixture of dilute acetic acid, albumin, a basic dye, gum arabic, molasses, and triethanolamine.[269]

Water-color inks.—Inks suitable for water-color work are maintained in suspension by gum arabic. A typical ink base[270] is made by dissolving 1 part of gum arabic in 9 parts of glycerol and 1 part of water and then evaporating the mixture to a specific gravity of about 1.28. Pigment is subsequently added by grinding it into the base with a colloid mill.

Quick-drying inks.—Inks designed to dry faster than ordinary writing inks are usually based on solvents other than water. Ethanol is commonly used in conjunction with gum arabic. However, some fast-drying formulas using water as the solvent are known, such as one consisting of water, Lysol (as a preservative), sodium nitrate, gum arabic, and a water-soluble dye.[266, 271]

Fabric and laundry marking inks.—Many formulas for making inks for marking cloth are known.[266] Some of these are based on aniline derivatives. Gum arabic is frequently added to bring the ink to a desired viscosity for smooth writing on cloth. Other indelible inks make use of metallic pigments but also contain gum arabic as a thickener and protective colloid. Some of the metallic pigments used are silver, gold, platinum, manganese, chromium, and lead.[266, 267]

Pigmented inks (white, bronze).—The suspension of a pigment in a gum arabic solution makes a satisfactory ink that can be applied with a brush or pen. Pigments, as such as titanium dioxide or bronze powder used for gold inks, are preferably moistened first with a small amount of ethanol or acetone before mixing with the gum solution. A typical white ink is made by mixing a filtered solution of 20 parts of gum arabic in 160 parts of water with 30 parts of titanium dioxide paint pigment and 0.5 part of sodium salicylate preservative.[267]

Emulsion or typographic inks.—Some inks consisting of emulsions use gum arabic.[266] Many emulsion inks are oil-in-water emulsions with the pigment dissolved in the oil phase. A typical ink of this class consists of lampblack, mineral oil, rosin, catechu black, formalin, sodium silicate, sodium carbonate, gum arabic, aluminum resinate, coloring agent, and water.[272, 273]

Hectograph ink.—An ink that has been used for many years in hectograph duplicating equipment is prepared by mixing methyl violet dye with water or ethanol. Some formulas using hydrochloric acid, oxalic acid, lactic acid, tannic acid, or procatechol employ gum arabic to obtain the proper viscosity.[266]

Gloss-finish inks.—Decorative inks with a glossy finish for showcards or other display purposes can be prepared by incorporating a dye into the following formula: 25 lb (11.3 kg) of gum arabic, 1 gal (3.8 liters) of liquid soap, 100 lb (45.4 kg) of dextrin, 20 lb (9.1 kg) of starch, 40 lb (18.2 kg) of D-glucose, 0.5 pint (475 ml) of Lysol or other preserving agent, 10 gal (37.8 liters) of water.[274]

Electrically conductive inks.—Inks that conduct electricity are becoming increasingly important because of their use in the manufacture or repair of printed circuits in electronic applications. Activation of electronic calculators by electrically conductive areas on a card rather than by punched holes is also becoming of increased interest. One reason for this is the relative ease of putting a conductive ink mark on a card as compared to the work involved in punching a hole. Most conductive inks are made of carbon black, powdered graphite, and finely divided silver or powdered copper. The suspending agent can be a lacquer, but water-based inks function well when gum arabic, sodium chloride, and citric (or tartaric) acid are used.[275]

Wood-grain inks.—Inks devised to imitate expensive grains by printing a pattern of wood on metal, paper, stone, or inexpensive woods use gum arabic as a suspension aid. Similarly, wallpaper or metallic objects are frequently decorated with wood-grain patterns. A typical formula is 60% of ethylene glycol, 7.5% of water, 30% of pigment, and 2.5% of gum arabic.[276]

8. *Lithography*

Gums are important in lithographic applications, where they have many functions as sensitizers for lithographic plates, elements in the light-sensitive composition, ingredients of the fountain solution used to moisten the plates during pointing, and protectors during storage of the plate. Gum arabic has been used traditionally in this field, but it has been replaced to a small degree by cellulose and starch ethers[277] and to a large degree by arabinogalactan (larch gum),[278] which offers the benefits of constant uniformity, domestic availability, and greater solubility.

The desensitizing use of gum arabic takes advantage of the good wettability imparted to the solution by the gum and also of the viscosity control, which allows the wash solution to cling to the plate without running off or forming isolated droplets or pools on the plate. On metal plates, the desensitizing effect might be caused by the formation of an insoluble film (for example, aluminum or zinc arabate). A more plausible explanation is that a film of gum is absorbed by the plate. Studies have shown that such films occur on plates of zinc, aluminum, copper, silver, iron, tin, lead, glass, and fused silica. These films are not monomolecular but are composed of many molecular layers.[266] Measurement of the wettability of desensitizing solutions is conveniently evaluated by measurement and study of the contact angles. In this process, a section of a plate is partially immersed in water or in a solution of the gum to be tested. The plate is then turned at an angle to the surface of the liquid until the meniscus appears to be eliminated. The resulting angle of the plate to the surface of the liquid is known as the contact angle and is a measure of the wettability of that particular plate with the solution being tested.[279]

Mixtures of gum arabic and dichromates are tanned by the action of light to form water-insoluble substances. Certain other gums and proteins are reported to be superior to gum arabic and might be used in place of it, but gum arabic continues to be used to some extent as a constituent of the light-sensitive coating. Thus, a wettable, plastic-impregnated paper sheet, a metallic sheet, or even a stone surface can be made light-sensitive by application of a gum and a dichromate solution that is allowed to dry.[280] Exposure of a pattern through a negative will cause tanning or hardening of all areas that are struck by light, and the unexposed material can be removed by a water or dilute acid wash. The surface can be used either for planographic printing or for an etching process if the plate is used for direct printing. A related use is in the preparation of etched circuits and in the manufacture of small metallic parts. In the latter process, the part is sketched or photographed onto a transparent sheet, exposed onto a metal sheet made light-sensitive with gum arabic and a dichromate, washed in water, and etched to remove unwanted parts of the sheet. In this manner, intricate and delicate parts are readily produced.

Gum arabic used in lithography should be of the best quality and, preferably, should be converted to the sodium salt by ion exchange. Sodium arabate has increased stability, better viscosity characteristics, and greater resistance to bacterial and mold attack as compared to the starting material. The purified gum solution can be used directly for deep-etch sensitive coatings. To each 3.65 parts of gum solution, 1 part of ammonium dichromate is added; and the solution is made slightly alkaline (pH 8.8–10.9) with ammonium hydroxide. The final viscosity is preferably 26–29 cps. Coating, drying and exposure follow conventional procedures.

High-viscosity grades of gum arabic are not recommended for deep-etch coatings since they do not flow uniformly over the plate surface and can produce streaks. The strength of the solutions can be controlled by specific gravity readings with a Baumé hydrometer, and the viscosity can be adjusted by thinning with water.

Substitutes for gum arabic in offset copying have been used for many years.

They include dextrin, fish glue, methylcellulose, yeast protein, and now arabinogalactan (larch gum).

9. *Textiles*

Gums are widely used in the textile industry as sizing and finishing agents and in printing formulations for imparting designs or decorations to fabrics. Gum arabic is an efficient sizing agent for cloth. Gum arabic finds limited use in finishing silk and rayon by giving body to the fabric without interfering with the transparency. Although desirable effects can be created on cotton, the cost is prohibitively high.[281] However, sizings are sometimes added to cotton fabrics to give them a fuller finish and to make them feel more firm and compact. Most cloth sizings are lost in washing.[282]

Typical low-cost sizes are made of rye starch by simply heating a slurry of 90 kg of starch in 700 liters of water and then cooling. A considerable improvement can be made by substituting gum arabic for some of the rye starch. Thus, 45 kg of rye flour starch, 24 kg of gum arabic, and 700 liters of water is heated to boiling. A separate solution of 16.5 kg of castor oil, 2.5 kg of oleic acid, 7.5 kg of 15% sodium hydroxide, and 23.5 kg of water is added to the starch–gum arabic solution. After mixing, 1 liter of oleic acid is added; the mixture is stirred and boiled for 20 min, and acetic acid is added until the mixture is neutral.[283]

Acacia gum from *Acacia decurrens* has been found to be an effective thickening and binding agent for vat sizes.[284] The gum is light yellow to red in color and is free from tannins. It has excellent adhesive properties and will not interfere with the colors or dyes used in printing.

10. *Miscellaneous*

A coating of gum arabic on typing paper allows erasures to be made easily.[285] A double coating is preferred, with a short drying period between applications. Gum arabic based transfer inks can also be used for the preparation of good-grade carbon papers.[286]

Eichorn[287] described a novel photographic method for making templates utilizing gum arabic. A woven glass cloth is impregnated with a polyester or alkyl resin and an opaque drawing is made upon the surface, which has excellent heat and light dimensional stability. Conventional blueprint salts containing gum arabic (or other gums) and a tanning agent are poured over the work, which, for example, might be a sheet of metal being formed into an airplane wing. When the pattern has been printed photographically upon the metal sheet, the sheet can be cut, drilled or otherwise mechanically manipulated, according to the pattern printed on its surface.

Gum arabic phthalate and gum arabic succinate are used as overcoating and backing layers for photographic material.[288]

Gum arabic is used to inhibit metal corrosion in some applications. The addition of the gum, up to approximately 2% of the weight of the negative plate of a storage battery, increases the life of the battery by reducing the growth of surface projections. The gum arabic is intimately mixed with the active negative plate material, for example, zinc oxide, producing an increase in electrical resistance and reducing the rate of corrosion.[289]

The corrosion of iron is considerably accelerated by light and the presence of salts, but it is inhibited by the presence of gum arabic.[290] The gum also effectively controls the corrosion of aluminum.[291]

The gum is useful in the immersion plating of copper on aluminum.[292] A bright, dense, strongly adhering copper coating is formed on aluminum by dipping the aluminum with an electrically connected piece of iron into a copper sulfate bath containing 0.75–1.25% by weight of gum arabic. Gum arabic has been utilized in the preparation of drilling fluids[293] and oil-well cement.[294] It has been used as a glaze binder[295] in ceramics where it has had to meet rigid specifications pertaining to bonding power, absence of slip, and stability for long periods at elevated temperatures. A Japanese patent[296] recommends gum arabic as a binder for insecticides. Horn and Samko[297] have developed a nonglare coating for automobile windshields based on a water-soluble dye dissolved in a gum arabic solution. A typical formulation contains 0.25% of brilliant green dye, 0.25% of the dioctyl ester of sodium sulfosuccinate, 5.00% of gum arabic, and 94.50% of water.

XII. References

(1) J. F. Caius and K. S. Radha, *J. Bombay Nat. Hist. Soc.,* **41,** 261 (1939).

(2) C. L. Mantell, "The Water-Soluble Gums," Reinhold Publishing Corp., New York, 1947.

(3) C. L. Mantell, *in* "Natural Plant Hydrocolloids," *Advan. Chem. Ser.,* **11,** 20, (1954).

(4) R. L. Whistler and C. L. Smart, "Polysaccharide Chemistry," Academic Press, New York, 1953.

(5) F. N. Howes, "Vegetable Gums and Resins," Chronica Botanica, Waltham, Mass., 1949.

(6) F. Smith and R. Montgomery, "The Chemistry of Plant Gums and Mucilages," Reinhold Publishing Corp., New York, 1959.

(7) M. Glicksman, "Gum Technology in the Food Industry," Academic Press, New York, 1969.

(8) H. S. Blunt, "Gum Arabic, With Special Reference to Its Production in the Sudan," Oxford Univ. Press, New York, 1926.

(9) D. W. Malcolm, "Report on Gum and Gum Arabic," Government Printer, Dar-es-Salaam, Republic of Somalia, 1936.

(10) D. M. W. Anderson and I. C. M. Dea, *Carbohyd. Res.,* **6,** 109 (1968).

(11) "Chemicals," Quart. Ind. Rep., U.S. Dept. of Commerce, Vol. 14, No. 1, March 1967, pp. 30, 31.

(12) A. Osol and G. E. Farrar, "The Dispensatory of the United States," J. F. Lippincott, Philadelphia, 25th Ed., 1955, p. 1.

(13) *Chem. Drug.,* **161,** 702 (1954).

(14) J. S. Hopper, U. S. Dept. of Commerce, Washington, D.C., personal communication, 1969.

(15) E. L. Hirst, *J. Chem. Soc.,* 70 (1942).

(16) J. K. N. Jones and F. Smith, *Advan. Carbohyd. Chem.,* **4,** 246 (1949).

(17) E. G. V. Percival, "Structural Carbohydrate Chemistry," Muller, London, 1950, p. 169.

(18) E. L. Hirst and J. K. N. Jones, *Research (London),* **4,** 411 (1951).

(19) T. Dillon, D. F. O'Ceallachain, and P. O'Colla, *Proc. Roy. Irish Acad.,* **B55,** 331 (1953).

(20) D. R. Spriestersbach, "The Constitution of Plant Gums," Ph.D. Dissertation, University of Minnesota, Minneapolis, 1954.

(21) M. Heidelberger, J. Adams, and Z. Dische, *J. Amer. Chem. Soc.,* **78,** 2853 (1956).

(22) H. B. Oakley, *Trans. Faraday Soc.,* **31,** 136 (1935).

(23) H. B. Oakley, *Trans. Faraday Soc.,* **32,** 1360 (1936).

(24) H. B. Oakley, *Trans. Faraday Soc.,* **33,** 372 (1937).

(25) T. Saverborn, *in* "Svedberg Memorial Volume," Almquist and Wiksell, Uppsala, 1944, p. 508.

(26) O. Lamm, *Nova Acta Regiae Soc. Sci. Upsalienses,* **10,** 115 (1937); *Chem. Abstr.,* **31,** 7722 (1937).

(27) A. Veis and D. N. Eggenberger, *J. Amer. Chem. Soc.,* **76,** 1560 (1954).

(28) D. M. W. Anderson and J. F. Stoddart, *Carbohyd. Res.,* **2,** 104 (1966).

(29) D. M. W. Anderson, E. Hirst, and S. Rahman, *Carbohyd. Res.,* **3,** 308 (1967).

(30) D. M. W. Anderson, I. C. M. Dea, and K. A. Karamalla, *Carbohyd. Res.,* **6,** 97 (1968).

(31) B. Warburton, *in* "The Chemistry and Rheology of Water Soluble Gums and Colloids," *Soc. Chem. Ind. Monogr.,* **24,** 118 (1966).

(32) D. M. W. Anderson and S. Rahman, *Carbohyd. Res.,* **4,** 298 (1967).

(33) G. O. Aspinall, A. J. Charlson, E. L. Hirst, and R. Young, *J. Chem. Soc.,* 1696 (1963).

(34) D. M. W. Anderson, E. Hirst, and J. F. Stoddart, *J. Chem. Soc.,* 1959 (1966).

(35) S. Saverborn, "A Contribution to the Knowledge of Acid Polyuronaides" (Inaugural Dissertation), Almquist and Wiksells, Uppsala, 1945, 112 p.

(36) S. K. Deb and S. N. Mukherjee, *J. Indian Chem. Soc.,* **39,** 832 (1962).

(37) D. M. W. Anderson, E. Hirst, and J. F. Stoddart, *J. Chem. Soc.,* 1476 (1967).

(38) D. M. W. Anderson, I. C. M. Dea, and R. N. Smith, *Carbohyd. Res.,* **6,** 320 (1968).

(39) D. M. W. Anderson and I. C. M. Dea, *Carbohyd. Res.,* **6,** 104 (1968).

(40) E. L. Hirst, *in* "The Structure and Biosynthesis of Macromolecules," *Biochem. Soc. Symp.,* **21,** 57 (1962).

(41) S. Basu, P. C. Dasgupta, and A. K. Sircar, *J. Colloid Sci.,* **6,** 539 (1951).

(42) J. T. G. Overbeck, *Bull. Soc. Chim. Belges,* **57,** 252 (1948).

(43) F. Smith, *J. Chem. Soc.,* 744 (1939).
(44) F. Smith, *J. Chem. Soc.,* 1035 (1940).
(45) C. L. Butler and L. H. Cretcher, *J. Amer. Chem. Soc.,* **51,** 1519 (1929).
(46) E. L. Hirst and A. S. Perlin, *J. Chem. Soc.,* 2622 (1954).
(47) A. M. Stephen, *J. Chem. Soc.,* 646 (1951).
(48) A. J. Charlson, J. R. Nunn, and A. M. Stephen, *J. Chem. Soc.,* 269 (1955).
(49) A. J. Charlson, J. R. Nunn, and A. M. Stephen *J. Chem. Soc.,* 1428 (1955).
(50) R. K. Hulyalkar, T. R. Ingle, and B. V. Bhide, *J. Indian Chem. Soc.,* **36,** 31 (1959).
(51) S. Mukherjee and A. N. Shrivastava, *J. Amer. Chem. Soc.,* **80,** 2536 (1958).
(52) D. M. W. Anderson and M. A. Herbich, *J. Chem. Soc.,* 1 (1963).
(53) D. M. W. Anderson and K. A. Karamalla, *J. Chem. Soc.,* 762 (1966).
(54) M. Kaplan and A. M. Stephen, *Tetrahedron,* **23,** 193 (1967).
(55) D. M. W. Anderson and K. A. Karamalla, *Carbohyd. Res.,* **2,** 403 (1966).
(56) D. M. W. Anderson, I. C. M. Dea, P. A. Maggs, and A. C. Munro, *Carbohyd. Res.,* **5,** 489 (1967).
(57) D. M. W. Anderson and G. M. Cree, *Carbohyd. Res.,* **6,** 385 (1968).
(58) D. M. W. Anderson and R. N. Smith, *Carbohyd. Res.,* **4,** 55 (1967).
(59) G. O. Aspinall, J. J. Carlyle, and R. Young, *Carbohyd. Res.,* **7,** 421 (1968).
(60) P. I. Bekker, A. M. Stephen, and G. R. Woolard, *Tetrahedron,* **24,** 6967 (1968).
(61) E. Hirst, *in* "The Chemistry and Rheology of Water Soluble Gums," *Soc. Chem. Ind. Monogr.,* **24,** 3 (1966).
(62) G. O. Aspinall, *Recent Advan. Food Sci.,* **3,** 282 (1963).
(63) S. Mukherjee and A. N. Shrivastava, *J. Sci. Ind. Res., Sect. B,* **16,** 566 (1957); *Chem. Abstr.,* **52,** 10620 (1958).
(64) D. M. W. Anderson and I. C. M. Dea, *Carbohyd. Res.,* **6,** 104 (1968).
(65) S. Mukherjee and A. N. Shrivastava, *J. Amer. Chem. Soc.,* **80,** 2536 (1958).
(66) G. O. Aspinall, E. L. Hirst, and A. Nicolson, *J. Chem. Soc.,* 1697 (1959).
(67) D. M. W. Anderson and I. C. M. Dea, *Carbohyd. Res.,* **5,** 461 (1967).
(68) D. M. W. Anderson and G. M. Cree, *Carbohyd. Res.,* **6,** 214 (1968).
(69) A. W. Thomas and H. A. Murray, *J. Phys. Chem.,* **32,** 676 (1928).
(70) M. N. Moorjani and C. S. Narwani, *J. Indian Chem. Soc.,* **25,** 503 (1948).
(71) R. H. Schleif, T. Higuchi, and L. W. Busse, *J. Amer. Pharm. Ass., Sci. Ed.,* **40,** 98 (1951).
(72) J. V. Swintosky, L. Kennon, and J. Tingstad, *J. Amer. Pharm. Ass., Sci. Ed.,* **44,** 109 (1955).
(73) W. H. Wood, U.S. Patent 2,666,759 (1954); *Chem. Abstr.,* **48,** 3716 (1954).
(74) R. G. Schweiger, *Methods Carbohyd. Chem.,* **5,** 59 (1965).
(75) G. O. Aspinall, *Methods Carbohyd. Chem.,* **5,** 397 (1965).
(76) S. Mukherjee and K. B. Ghosh, *J. Indian Chem. Soc.,* **26,** 277 (1949).
(77) F. W. Tiebackx, *Pharm. Weekblad,* **59,** 574 (1922); Chem. Abstr., **16,** 2433 (1922).
(78) S. N. Mukherjee and A. C. Bhowmik, *J. Indian Chem. Soc.,* **26,** 313 (1949).
(79) R. H. Schleif, T. Higuchi, and L. W. Busse, *J. Amer. Pharm. Ass., Sci. Ed.,* **40,** 221 (1951).
(80) J. Jackson and F. Smith, *J. Chem. Soc.,* 120 (1948).
(81) J. C. Krantz and N. E. Gordon, *J. Amer. Pharm. Ass., Sci., Ed.,* **18,** 463 (1929).
(82) D. R. Briggs, *J. Phys. Chem.,* **38,** 867 (1934).

(83) K. S. Rao, *Curr. Sci.*, **9,** 19 (1940).
(84) D. N. Adams, U.S. Patent 2,694,057 (1954); *Chem. Abstr.*, **49,** 9310 (1955).
(85) G. E. Osborne and C. O. Lee, *Drug Stand.*, **19,** 13 (1951).
(86) C. F. Mason, *Chem. Ind. (New York)*, **53,** 630 (1943).
(87) R. Taft and L. E. Malm, *J. Phys. Chem.*, **35,** 874 (1931).
(88) H. R. Kruyt and H. J. C. Tendeloo, *Kolloid Chem. Beih.*, **29,** 396 (1929).
(89) O. E. Araujo, *J. Pharm. Sci.*, **55,** 636 (1966).
(90) L. F. Gabel, *J. Amer. Pharm. Ass., Sci. Ed.*, **19,** 838 (1930).
(91) H. J. C. Tendeloo, *Rec. Trav. Chim.*, **48,** 23 (1929).
(92) G. L. Clark and W. A. Mann, *J. Biol. Chem.*, **52,** 157 (1922).
(93) B. Borkowski and M. Olszak, *Acta Polon. Pharm.*, **13,** 53 (1956); *Chem. Abstr.*, **50,** 12516 (1956).
(94) H. G. Bungenberg de Jong, H. R. Kruyt, and J. Lens, *Kolloid-Beih.*, **36,** 429 (1932).
(95) H. G. Bungenberg de Jong, H. R. Kruyt, and J. Lens, *Kolloid-Beih.*, **37,** 395 (1933).
(96) H. R. Kruyt, "Colloid Science," Elsevier, Amsterdam, Vol. 2, 1949, p. 216.
(97) M. Woo and C. L. Huyck, *Bull. Nat. Formulary Comm.*, **16,** 140 (1948).
(98) A. Beutaric and M. Roy, *Bull. Soc. Chim. Fr.*, **6,** 316 (1939); *Chem. Abstr.*, **33,** 3658 (1939).
(99) L. Y. Kremnev and V. S. Kuprik, *Kolloid Zhur.*, **14,** 98 (1952).
(100) E. K. Fischer, "Colloidal Dispersions," Wiley, New York, 1950, p. 264.
(101) A. Szalay, *Z. Physik. Chem.*, **164A,** 234 (1933).
(102) G. Troger, *Angew. Chem.*, **62,** 42 (1950).
(103) J. Schlemmer, *Chemie* (Prague), **3,** 73 (1948); *Chem. Abstr.*, **46,** 2888 (1952).
(104) H. M. Beal and D. M. Skaven, *J. Amer. Pharm. Ass., Sci. Ed.*, **44,** 487 (1955).
(105) C. Zucca, *1st Botan. Univ., Lab. Crittogam., Pavia, Atti* (5) **10,** 85 (1935); *Chem. Abstr.*, **48,** 2399 (1954).
(106) T. C. Laurent and E. M. Wertheim, *Acta Chem. Scand.*, **6,** 678 (1952); *Chem. Abstr.*, **46,** 10885 (1952).
(107) R. Taft and L. E. Malm, *Trans. Kansas Acad. Sci.*, **34,** 116 (1931).
(108) G. E. Osborne and C. O. Lee, *Drug Stand.*, **19,** 4 (1951).
(109) G. Christianson and P. E. Ramstad, U.S. Patent 2,662,882 (1953).
(110) J. E. Henry, U.S. Patent 2,768,143 (1956).
(111) L. Lachman and L. Chavkin, *J. Amer. Pharm. Ass., Sci. Ed.*, **46,** 412 (1957).
(112) S. N. Banerji, *J. Indian Chem. Soc.*, **29,** 270 (1952).
(113) O. E. Araujo, *J. Pharm. Sci.*, **56,** 1141 (1967).
(114) E. Shotton and S. S. Davis, *J. Pharm. Pharmacol.*, **20,** 780 (1968).
(115) H. Deuel and J. Solms, *Kolloid-Z.*, **124,** 65 (1951).
(116) Henkel & Cie, G.m.b.H., Brit. Patent 449,428 (1936); *Chem. Abstr.*, **30,** 8434 (1936).
(117) R. Haller and B. Frankfurt, *Kolloid-Z.*, **80,** 68 (1937).
(118) Hercules, Inc., Wilmington, Delaware, "Cellulose Gum," 1966.
(119) J. F. Carson and W. D. Maclay, *J. Amer. Chem. Soc.*, **68,** 1015 (1946).
(120) O. A. Moe, U.S. Patent 2,520,161 (1950); *Chem. Abstr.*, **45,** 1365 (1951).
(121) J. P. Kieft, *Pharm. Weekblad,* **76,** 1133 (1939).

(122) A. B. Svendsen and E. Drottning, *Dansk. Tideskr. Farm.*, **27**, 129 (1953; *Chem. Abstr.*, **47**, 3131 (1953).

(123) H. T. Leo, C. C. Taylor, and J. W. Lindsey, U.S. Patent 2,380,115 (1945); *Chem. Abstr.*, **39**, 4168 (1945).

(124) P. I. Fodor, *Nature*, **158**, 375 (1956).

(125) D. Burgois and M. Joly, *J. Chim. Phys.*, **51**, 184 (1954).

(126) N. L. Feldman, *Doklady Akad. Nauk S.S.S.R.*, **89**, 343 (1953); *Chem. Abstr.*, **47**, 105077 (1953).

(127) D. Dervichian and C. Magnant, *Bull. Soc. Chim.*, **27**, 101 (1945).

(128) M. Joly, *Bull. Soc. Chim. Biol.*, **30**, 398 (1948).

(129) S. Basu and G. Bhattacharya, *Science*, **115**, 544 (1952).

(130) H. G. Bungenberg de Jong and O. Bank, *Protoplasma*, **33**, 321 (1939).

(131) H. G. Bungenberg de Jong and J. M. F. Landsmeer, *Proc. Kon. Ned. Akad. Wetensch., Ser. B.*, **51**, 295 (1948).

(132) H. G. Bungenberg de Jong, E. G. Hoskam and B. H. van den Brandhof-Schaegen, *Proc. Kon. Ned. Akad. Wetensch.*, **44**, 1104 (1941); *Chem. Zentr.*, **I**, 2969 (1942).

(133) H. G. Bungenberg de Jong and H. Sengers, *Rec. Trav. Chim.*, **53**, 171 (1934).

(134) L. de Ruiter and H. G. Bungenberg de Jong, *Proc. Kon. Ned. Akad. Wetensch., Ser. B.*, **50**, 836 (1947).

(135) R. Santamaria G. Hermann, and A. Loffredo, *Rend. Atti. Acad. Sci. Med. Chir.*, **117**, 151 (1963); *Chem. Abstr.*, **61**, 13911 (1964).

(136) H. G. Bungenberg de Jong and E. G. Hoskam, *Proc. Kon. Ned. Akad. Wetensch.*, **44**, 1099 (1941); *Chem. Zentr.*, **I**, 2969 (1942).

(137) H. G. Bungenberg de Jong and E. G. Hoskam, *Proc. Kon. Ned. Akad. Wetensch., Ser. B*, **45**, 585 (1942); *Chem. Zentr.*, **II**, 1771 (1942).

(138) H. G. Bungenberg de Jong and A. M. van Leeuwen, *Proc. Kon. Ned. Akad. Wetensch., Ser. B*, **54**, 91 (1951).

(139) H. G. Bungenberg de Jong and E. G. Hoskam, *Proc. Kon. Ned. Akad. Wetensch.*, **45**, 200 (1942); *Chem. Zentr.*, **II**, 16 (1942).

(140) H. G. Bungenberg de Jong, *Proc. Kon. Ned. Akad. Wetensch.*, **50**, 707 (1947).

(141) D. G. Dervichian, *Discuss. Faraday Soc.*, **18**, 231 (1954).

(142) H. G. Bungenberg de Jong and B. C. Bakhuizem van den Brink, *Proc. Kon. Ned. Akad. Wetensch., Ser. B*, **50**, 436 (1947).

(143) H. G. Bungenberg de Jong and W. A. L. Dekker, *Biochem. Z.*, **213**, 318 (1929).

(144) H. G. Bungenberg de Jong and J. M. F. Landsmeer, *Proc. Kon. Ned. Akad. Wetensch., Ser. B*, **51**, 137 (1948).

(145) H. G. Bungenberg de Jong and J. M. F. Landsmeer, *Rec. Trav. Chim.*, **65**, 606 (1946).

(146) H. G. Bungenberg de Jong and B. Kok, *Proc. Kon. Ned. Akad. Wetensch.*, **45**, 51 (1942); *Chem. Zentr.*, **II**, 16 (1942).

(147) H. G. Bungenberg de Jong and E. G. Hoskam, *Proc. Kon. Ned. Akad. Wetensch.*, **45**, 59 (1942); *Chem. Abstr.*, **37**, 4291 (1943).

(148) M. J. Voorn, *Rec. Trav. Chim.*, **75**, 1021 (1956); *Chem. Abstr.*, **51**, 4095 (1957).

(149) N. Tanaka, G. Hirata, T. Ida, and I. Utsumi, *Yakugaku Zasshi*, **85**, 799 (1965); *Chem. Abstr.*, **63**, 17803 (1965).

(150) N. Yoshida and C. Thies, *Amer. Chem. Soc., Div. Polymer Chem., Preprints,* **7,** 245 (1966); *Chem. Abstr.,* **66,** 69271u (1967).

(151) N. Yoshida and C. Thies, *J. Colloid Interface Sci.,* **24,** 29 (1967); *Chem. Abstr.,* **67,** 76549g (1967).

(152) H. G. Bungenberg de Jong and B. Kok, *Proc. Kon. Ned. Akad. Wetensch.,* **43,** 732 (1940).

(153) H. G. Bungenberg de Jong and L. de Ruiter, *Proc. Kon. Ned. Akad. Wetensch., Ser. B,* **50,** 1189 (1947).

(154) T. N. Evreinova and L. M. Galimova, *Biofiz. Kletki, Akad. Nauk S.S.S.R., Inst. Biol. Fiz., Sb. Statei,* 17 (1965); *Chem. Abstr.,* **65,** 2503 (1966).

(155) H. G. Bungenberg de Jong, *Proc. Kon. Ned. Akad. Wetensch.,* **45,** 393 (1942).

(156) H. G. Bungenberg and C. V. D. Meer, *Proc. Kon. Ned. Akad. Wetensch.,* **45,** 498 (1942); *Chem. Zentr.,* **II,** 1551 (1942).

(157) V. Liebl, J. Chaloupka, and D. Malek, *Probl. Evol. Tekh. Biokhim.,* 131 (1964); *Chem. Abstr.,* **61,** 10898 (1964).

(158) B. K. Green and L. Schleicher, U.S. Patent 2,800,457 (1957); *Chem. Abstr.,* **51,** 15842 (1957).

(159) B. K. Green, U.S. Patent 2,800,458 (1957); *Chem. Abstr.,* **51,** 15842 (1957).

(160) H. J. Eichel, U.S. Patent 2,907,682 (1959); *Chem. Abstr.,* **54,** 4062 (1960).

(161) H. S. Pesa and L. Schleicher, U.S. Patent 2,932,582 (1960); *Chem. Abstr.,* **54,** 14510 (1960).

(162) C. Brynko, J. A. Bakan, R. E. Miller, and J. A. Scarpelli, U.S. Patent 3,341,466 (1967); *Chem. Abstr.,* **68,** 99039h (1968).

(163) H. G. Bungenberg de Jong and E. G. Hoskam, *Proc. Kon. Ned. Akad. Wetensch.,* **45,** 387 (1942); *Chem. Zentr.,* **II,** 1551 (1942).

(164) H. G. Bungenberg de Jong and B. Kok, *Proc. Kon. Ned. Akad. Wetensch.,* **43,** 728 (1940).

(165) H. S. Reed, *Phytopathol.,* **33,** 739 (1943).

(166) D. G. Dervichian and C. Magnant, *Bull. Soc. Chim. Biol.,* **29,** 655 (1947).

(167) F. Micheel and U. Heuer, *Makromol. Chem.,* **3,** 217 (1949).

(168) A. I. Oparin, M. S. Bardinskaya, S. S. Melik-Sarkisyan, and K. B. Serebrovskaya, *Doklady Akad. Nauk S.S.S.R.,* **108,** 1125 (1956); *Chem. Abstr.,* **51,** 1336 (1957).

(169) E. Macovski and M. Cirsteanu, *Stud. Cercet. Biochim.,* **2,** 111 (1959); *Chem. Abstr.,* **53,** 22136 (1959).

(170) D. Motet-Grigoras, *Acad. Rep. Populare Romine, Inst. Biochim., Studii Cercetari Biochim.,* **2,** 181 (1959); **3,** 303 (1960); *Chem. Abstr.,* **54,** 2452 (1960); **55,** 9531 (1961).

(171) D. Motet-Grigoras, *Studi. Cercet. Biochim.,* **9,** 289 (1966); *Chem. Abstr.,* **66,** 9548h (1967).

(172) L. Arnet, *Stud. Cercet. Biochim.,* **1,** 363 (1958); *Chem. Abstr.,* **55,** 3671 (1961).

(173) A. I. Oparin, I. G. Stoyanova, K. B. Serebrovskaya, and T. A. Nekrasova, *Dokl. Akad. Nauk S.S.S.R.,* **150,** 684 (1963); *Chem. Abstr.,* **59,** 7781 (1963).

(174) "U.S. Pharmacopeia XVII," Mack Publishing Co., Easton, Pa., 1965, p. 13.

(175) "Food Chemicals Codex." Publ. 1406, Natl. Acad. Sci.–Natl. Res. Council, Washington, D.C., 1st Ed., 1966, p. 7, 8.

(176) J. C. Bartlett, E. List, M. Page, and R. A. Chapman, *Can. J. Technol.,* **31,** 146 (1953).

(177) W. O. Walder, *Food,* **18,** 86 (1949).

(178) M. Glicksman, "Gum Technology in the Food Industry," Academic Press, New York, 1969.

(179) M. H. Ewart and R. A. Chapman, *Anal. Chem.,* **24,** 1460 (1952).

(180) A. T. Proszynski, A. J. Mitchell, and C. M. Stewart, *Aust. Commonwealth Sci., Ind. Res. Organ., Div. Forest Prod. Technol. Paper* No. **38,** (1965).

(181) S. H. Newburger, J. H. Jones, and G. R. Clark, *Proc. Sci. Sect. Toilet Goods Ass.,* **19,** 25 (1935).

(182) E. F. Bryant, *Ind. Eng. Chem., Anal. Ed.,* **13,** 103 (1941).

(183) P. A. Racicot and C. S. Ferguson, *J. Ass. Offic. Agr. Chem.,* **21,** 110 (1938).

(184) F. L. Hart, *J. Ass. Offic. Agr. Chem.,* **23,** 597 (1940).

(185) T. E. Strange, *J. Ass. Offic. Agr. Chem.,* **35,** 354 (1952).

(186) J. D. Wildman, *J. Ass. Offic. Agr. Chem.,* **18,** 637 (1935).

(187) E. Bontinck, *Natuurwetensch. Tijdschr. (Ghent),* **23,** 97 (1949).

(188) S. A. Peoples and N. M. Phatak, *Proc. Soc. Exp. Biol. Med.,* **32,** 635 (1935).

(189) O. Frehden and L. Goldschmidt, *Mikrochim. Acta,* **2,** 184 (1937).

(190) S. Redfern, *J. Ass. Offic. Agr. Chem.,* **28,** 252 (1945).

(191) M. J. Gnagy, *J. Ass. Offic. Agr. Chem.,* **35,** 358 (1952).

(192) F. Y. Mendelsohn, *J. Ass. Offic. Agr. Chem.,* **36,** 599 (1953).

(193) L. F. Martin, *Advan. Food Res.,* **6,** 29 (1955).

(194) C. Carilli, *Candy Ind. Confect. J.,* **116,** No. 6, 7 (1961).

(195) K. E. Langwill, *Mfg. Confect.,* **19,** 37 (1939).

(196) C. T. Williams, *Confect. Mfg.,* **6,** 299 (1961).

(197) M. B. Jacobs, *Amer. Perfum. Essent. Oil Rev.,* **54,** 54 (1949).

(198) H. Pyenson and C. D. Dahle, *J. Dairy Sci.,* **21,** 169 (1938).

(199) M. J. Mack, *Ice Cream Trade J.,* **32,** 33 (1936).

(200) L. A. Scholz, U.S. Patent 2,568,369 (1951); *Chem. Abstr.,* **45,** 10433 (1951).

(201) S. J. Werbin, *Baker's Dig.,* **27,** 21 (1953).

(202) W. W. Wagner, U.S. Patent 2,682,472 (1954); *Chem. Abstr.,* **48,** 10256 (1954).

(203) C. N. Revie and B. R. J. Thomas, *Food Mfg.,* **37,** 40 (1962).

(204) J. M. Wenneis, *Proc. Flavor. Extract Mfr. Ass.,* **47,** 91 (1956).

(205) J. A. Herbig, *in* "Kirk–Othmer Encyclopedia of Chemical Technology," Interscience, 2nd Ed., Vol. 13, 1967, p. 436.

(206) L. L. Balassa and J. Brody, *Food Eng.,* **40** (11), (1968).

(207) C. Johnstone, *Mfg. Confect.,* **19,** 14 (1939).

(208) C. M. Ferri, *Food Ind.,* **19,** 784 (1947).

(209) D. E. Lakritz, *Food Process,* **24,** No. 6, 86 (1963).

(210) A. J. Andon, *Drug Cosmet. Ind.,* **79,** 762 (1956).

(211) G. E. Osborn and H. G. DeKay, *J. Amer. Pharm. Ass., Pract. Pharm. Ed.,* **2,** 420 (1941).

(212) C. C. Pfeiffer and H. L. Williams, *J. Amer. Pharm. Ass., Pract. Pharm. Ed.,* **8,** 572 (1947).

(213) Fr. Patent 1,002,540 (1952); *Chem. Abstr.,* **51,** 4606 (1957).

(214) Soc. pour l'ind. chim. à Bâle, P. W. Gerding, and G. J. Sperandio, *Drug Stand.,* **21,** 215 (1953).

(215) Société de recherches et d'entreprises industrielles et chimiques, Swiss Patent 225,886 (1943); *Chem. Abstr.,* **43,** 2375 (1949).

(216) "U.S. Pharmacopeia," Mack Publ. Co., Easton, Pennsylvania, 15th Ed., 1955.

(217) *J. Amer. Pharm. Ass., Pract. Pharm. Ed.,* **8,** 83, (1947).

(218) J. C. Krantz and N. E. Gordon, *J. Amer. Pharm. Ass.,* **15,** 93 (1926).

(219) H. Lotzkar and W. D. Maclay, *Ind. Eng. Chem.,* **35,** 1294 (1943).

(220) C. L. Brown and E. A. Lum, *Pharm. J.,* **131,** 341 (1933).

(221) F. W. Nitardy, F. F. Berg, and P. Georgi, U.S. Patent 1,913,561 (1933); *Chem. Abstr.,* **25,** 3130 (1931).

(222) F. W. Nitardy, F. F. Berg, and P. Georgi, U.S. Patent 1,913,561 (1933); *Chem. Abstr.,* **27,** 4348 (1933).

(223) W. A. Knight, *Pharm. J.,* **121,** 297 (1928).

(224) L. Baron, U.S. Patent 1,621,186 (1927); *Chem. Abstr.,* **21,** 1523 (1927).

(225) W. J. Husa and C. H. Becker, *J. Amer. Pharm. Ass., Sci. Ed.,* **30,** 171 (1941).

(226) G. Kedvessy, *Ber. Ungar. Pharm. Ges.,* **17,** 607 (1941); *Chem. Abstr.,* **37,** 4531 (1943).

(227) K. Naito and T. Mori, *Nippon Nogei Kagaku Kaishi,* **27,** 268 (1953); *Chem. Abstr.,* **49,** 8519 (1955).

(228) A. Ali, B. Kahn, and B. Ahmad, *Pakistan J. Sci. Res.,* **6,** 58 (1954).

(229) M. T. Leffler, Canad. Patent 401,754 (1941); *Chem. Abstr.,* **36,** 1741 (1942).

(230) K. von Neergaard, Brit. Patent 218,323 (1923); *Chem. Abstr.,* **19,** 560 (1925).

(231) P. C. Eisman, J. Cooper, and D. Jaconia, *J. Amer. Pharm. Ass., Sci. Ed.,* **46,** 144 (1957).

(232) M. B. Sherwood, *J. Bacteriol.,* **43,** 778 (1942).

(233) R. A. Quisno, I. W. Gibby, and M. J. Foter, *J. Amer. Pharm. Ass., Sci. Ed.,* **35,** 317 (1946).

(234) I. B. Wershaw, U.S. Patent 2,371,862 (1945); *Chem. Abstr.,* **39,** 3882 (1945).

(235) A. Schwarzmann, Swiss Patent 241,167 (1946); *Chem. Abstr.,* **43,** 8104 (1949).

(236) Chemische Fabrik Schweizerhall, Swiss Patent 307,867 (1954); *Chem. Abstr.,* **51,** 3937 (1957).

(237) "Merck Index," Merck and Co., Rahway, N.J., 8th Ed., 1968, pp. 2, 3.

(238) H. H. Anderson, F. Murayama, and B. E. Abreu, "Pharmacology and Experimental Therapeutics," Univ. California Press, Los Angeles, 1947, p. 1.

(239) C. K. Maytum and T. B. Magath, *J. Amer. Med. Ass.,* **99,** 2251 (1932).

(240) J. A. Maas, *Quart. J. Exp. Physiol.,* **28,** 315 (1938).

(241) C. B. Bohner, J. M. Sheldon, and J. W. Trenis, *J. Allergy,* **12,** 290 (1941).

(242) P. H. Sprague, *Can. Med. Ass. J.,* **47,** 253 (1942).

(243) P. I. Smith, *Amer. Perfum. Aromat.,* **67,** 67 (1956).

(244) L. A. Greenberg and D. Lester, "Handbook of Cosmetic Materials," Interscience, New York, 1954, p. 19.

(245) M. Feinberg and B. B. Schoenkerman, *Wisconsin Med. J.,* **39,** 734 (1940).

(246) R. G. Harry, "Modern Cosmeticology," Chemical Publishing Co., New York, 1947, pp. 56, 287, 314, 316.

(247) J. H. Helfrich, U.S. Patent 1,655,369 (1929); *Chem. Abstr.,* **22,** 1017 (1928).

(248) Compagnie Francaise pour l'exploitation des procédés Thomson-Houston, Fr. Patent 692,757 (1930); *Chem. Abstr.,* **25,** 1940 (1931).

(249) F. Braude, "Adhesives," Chemical Publishing Co., New York, 1943, pp. 68–73.

(250) J. Delmonte, "The Technology of Adhesives," Reinhold Publishing Corp., New York, 1947, p. 291.

(251) H. F. Wolfe, U.S. Patent 1,983,650 (1934); *Chem. Abstr.,* **29,** 857 (1935).

(252) "The Chemical Formulary," H. Bennett, ed., Chemical Publishing Co., New York, Vol. 4, 1939, pp. 19, 24.

(253) "The Chemical Formulary," H. Bennett, ed., Chemical Publishing Co., New York, Vol. 3, 1936, pp. 7, 9.

(254) J. Dumas, *Ann. Chim. Anal.,* **25,** 214 (1943).

(255) "The Chemical Formulary," H. Bennett, ed., Chemical Publishing Co., New York, Vol. 1, 1933, pp. 7, 10.

(256) "Handbook of Chemistry and Physics," C. D. Hodgman, ed., Chemical Rubber Co., Cleveland, Ohio, 35th Ed., 1953, p. 2984.

(257) "The Chemical Formulary," H. Bennett, ed., Chemical Publishing Co., New York, Vol. 5, 1941, p. 17.

(258) N. A. DeBruyne and R. Houwink, "Adhesion and Adhesives," Elsevier, New York, 1951.

(259) C. G. Sumner, "Clayton's Theory of Emulsions and Their Technical Treatment," Churchill, London, 5th Ed., 1954, p. 425.

(260) H. M. Johnson, U.S. Patent 99,907 (1870).

(261) J. Alexander, U.S. Patent 1,259,708 (1918; *Chem. Abstr.,* **12,** 1502 (1918).

(262) E. G. Acheson, U.S. Patent 964,478, (1910); *Chem. Abstr.,* **4,** 2717 (1910).

(263) L. D. Gamble and D. L. Grady, U.S. Patent 2,135,936 (1938); *Chem. Abstr.,* **33,** 1524 (1939).

(264) J. J. Mattiello, "Protective and Decorative Coatings," Wiley, New York, Vol. 4, 1944, p. 311.

(265) W. K. Wilson, U.S. Patent 2,508,343 (1950); *Chem. Abstr.,* **44,** 7583 (1950).

(266) C. Ellis, "Printing Inks," Reinhold Publishing Corp., New York, 1940, pp. 230, 334, 346, 398, 399, 417.

(267) C. E. Waters, *Nat. Bur. Stand., Circ.,* No. **C426,** 3, 33, 34, 45, 53 (1940).

(268) R. Neal, C. F. Bailey, and R. S. Casey, *J. Chem. Educ.,* **24,** 429 (1947).

(269) A. B. Poschel, Brit. Patent 393,132 (1933); *Chem. Abstr.,* **27,** 5553 (1933).

(270) J. Hoeffler, U.S. Patent 1,660,196 (1928); *Chem. Abstr.,* **22,** 1486 (1928).

(271) J. Rochat, Swiss Patent 153,496 (1931); *Chem. Abstr.,* **27,** 1216 (1933).

(272) K. Ishida, U.S. Patent 1,410,572 (1922); *Chem. Abstr.,* **16,** 1875 (1922).

(273) K. Ishida, Brit. Patent 194,156 (1922); *Chem. Abstr.,* **17,** 3615 (1923).

(274) C. T. Dean, U.S. Patent 1,607,060 (1926); *Chem. Abstr.,* **21,** 333 (1927).

(275) K. Suzuki, Jap. Patent 109,595 (1935); *Chem. Abstr.,* **29,** 4958 (1935).

(276) J. M. Verderosa, Brit. Patent 315,214 (1928); *Chem. Abstr.,* **24,** 1753 (1930).

(277) W. H. Wood, U.S. Patent 2,589,913 (1952); *Chem. Abstr.,* **46,** 6024 (1952).

(278) "Stractan (Arabinogalactan)," Stein, Hall and Co., Inc., New York, 1964.

(279) R. F. Reed, *Mod. Lithogr.,* **47,** 62 (1951).

(280) J. S. Mertle, *Graphic Arts Mon.,* **9,** 32 (1937).

(281) J. T. Marsh, "An Introduction to Textile Finishing," Wiley, New York, 1943, p. 23.

(282) E. Dyer, "Textile Fabrics," Houghton-Mufflin, Boston, Mass., 1923, p. 62.

(283) N. A. Mukoseev, *Tekstil Prom.,* 23 (1943).

(284) N. C. Gutheil and M. Formoso, *Anais. Ass. Quím. Brasil,* **10,** 335 (1951).

(285) J. A. Constance, U.S. Patent 2,676,119 (1954); *Chem. Abstr.,* **48,** 14208 (1954).

(286) H. W. A. Dixon and R. S. Moore, U.S. Patent 2,022,276 (1935); *Chem. Abstr.,* **30,** 792 (1936).

(287) A. Eichorn, U.S. Patent 2,801,919 (1957); *Chem. Abstr.,* **51,** 15317 (1957).

(288) C. R. Fordyce and J. Emerson, Brit. Patent 554,758 (1943); *Chem. Abstr.,* **39,** 467 (1945).

(289) H. J. Strauss, U.S. Patent 2,692,904 (1954); *Chem. Abstr.,* **49,** 2222 (1955).

(290) M. Charmandarian and N. M. Andronikova, *Bull. Soc. Chim. Fr.,* 99 (1952).

(291) G. T. Colegate, *Metallurgia,* **39,** 316 (1948).

(292) G. Norwitz, U.S. Patent 2,680,711 (1954); *Chem. Abstr.,* **48,** 9240 (1954).

(293) Badar-Ud-Din, *Pakistan J. Sci. Res.,* **2,** 28 (1950).

(294) C. T. Clark, U.S. Patent 2,620,279 (1952); *Chem. Abstr.,* **47,** 2454 (1953).

(295) T. Knapp, *Amer. Ceramic Soc. Bull.,* **33,** 11 (1954).

(296) T. Matsudaira, Jap. Patent 2997 (1953); *Chem. Abstr.,* **48,** 7840 (1954).

(297) E. N. Horn and J. Sanko, U.S. Patent 2,651,583 (1953); *Chem. Abstr.,* **48,** 339 (1954).

Chapter XI

GUM GHATTI

George Meer, William A. Meer and Thomas Gerard

Meer Corporation, North Bergen, N.J.

I. Introduction

Gum ghatti (Indian gum) is a complex water-soluble polysaccharide. It is a plant exudate that has been long in use and whose name is derived from the word *ghats,* which means passes, given to the gum because of its ancient mountain transportation routes. Gum ghatti is approved for food use and is in the GRAS (generally recognized as safe) list under the Federal Food, Drug and Cosmetic Act.

II. Source

Gum ghatti is an amorphous, translucent exudate of the *Anogeissus latifolia* tree of the Combretaceae family. The tree is quite large and is found abundantly in the dry, deciduous forests of India and, to a lesser extent, in Ceylon. It has a gray bark and red leaves during the dry season. In addition to its use as a source gum, the tree is widely used for timber, and tannin is extracted from its leaves. The gum has a glassy fracture and occurs in rounded tears, which are normally less than 1 cm in diameter, but it more often occurs in larger vermiform masses. The color of the exudate varies from very light to dark brown; the lighter the color, the better the quality and grade. Factors that determine the color are the proximity of the tear to the bark and the length of time it has remained on the tree before being picked, because tannins from the bark darken the gum. Airborne impurities also adhere to the gum while it is still plastic. Artificial incisions can be made in the tree bark to increase the yield; however, these incisions must be well planned so as not to destroy the tree. The gum is thought to act as a sealant when the bark is damaged.

Harvesting and grading of gum ghatti are done by methods similar to those used with gum karaya, near which it grows. The tonnage of gum karaya imported into the United States, however, far exceeds that of gum ghatti.

After picking, the gum is dried in the sun for several days. It is then taken to market and auctioned to exporters. Thereupon, the gum is moved to a classification and storage depot where it is hand sorted according to color and impurity. The grades are then placed in burlap bags and exported. Normally, 3 grades are imported to the United States (No. 1, No. 2, and unassorted). Number 1 grade is of lightest color (off-white to buff); No. 2 grade is light amber to brown, and the unassorted grade is dark brown to nearly black. The better the grade, the more effective is the gum and the less the amount of impurities. The best crops are obtained when the monsoon is not occurring, and normally the largest crop is picked in April. Table I shows average impurities (bark and sand), total ash, acid-insoluble ash, and moisture content of the three grades, determined according to methods described in the "National Formulary."[1]

Ghatti tears are processed in the United States. This processing is mainly a grinding operation by which the gum is pulverized to a fine powder. However, various other mesh separations can also be made to satisfy the demands of the consumer. During the process of particle breakdown, impurities are removed from the gum by sifting, aspiration, and density-table separation. From 1967 through 1971, between 450 and 600 tons of ghatti gum were imported to the United States per year. The price has varied from $0.35 to $0.68 a pound ($0.78 to $1.51 per kg), depending on the grade and market conditions at the time of purchase.[2] It is estimated that ~900–1150 tons are available on a worldwide basis. The amount of gum harvested depends on the economic conditions in the region in which the trees grow. If other crops pay a higher price for harvesting, the natives naturally prefer to harvest these, and ghatti is neglected. There are large areas in which the trees are as yet uncultivated but which could supply gum if the demand should justify it.

TABLE I

Ghatti Grades

Grade	Impurities, %	Total Ash, %	Acid-Insoluble Ash, %	Viscosity, cps (5% solution)	Moisture, %
1	0.9–1.6	1.4–1.9	0.02–0.2	30–400	12–15
2	1.4–3.6	2.2–3.9	0.2–1.0	30–350	12–15
Unassorted	11.0–15.0	6.0–10.0	3.7–5.8	30–300	12–15

It is well known to users of gum ghatti that gums of a different botanical origin are sometimes collected as ghatti and sold as such by exporters.

Ghatti has a bland taste and practically no odor. Only about 90% of the gum disperses in water, and this portion forms a colloidal dispersion.

III. Structure

Gum ghatti occurs as a calcium–magnesium salt. Aspinall and coworkers[3] have shown that it is composed of L-arabinose, D-galactose, D-mannose, D-xylose, and D-glucuronic acid in a molar ratio of 10 : 6 : 2 : 1 : 2 and traces (below 1%) of 6-deoxyhexose. Partial hydrolysis gives two aldobiouronic acids, namely 6-*O*-(β-D-glucopyranosyluronic acid)-D-galactose and 2-*O*-(β-D-glucopyranosyluronic acid)-D-mannose.[3] A previous investigation reported the presence of 50% pentose and 12% galactose or galacturonic acid.[4–6] Further work[7] has shown that the gum contains a backbone of (1→6)-linked β-D-galactopyranosyl units and that partial acid hydrolysis[8] affords two homologous series of oligosaccharides together with small amounts of 3-*O*-β-D-galactopyransoyl-D-galactose and 2-*O*-(β-D-glucopyranosyluronic acid)-D-mannose.[9] Acid-labile side chains are attached to the backbone through L-arabinofuranose resides.[10, 11]

Apparently, gum ghatti is similar in structure to *Anogeissus leiocarpus* gum, which contains two distinct polysaccharides, one a major component and one a minor component, that can be separated from each other.[12] Partial structures for the interior[13] and exterior[14] chains of the major fraction have been proposed.

In solution, the molecules may have an overall rod shape.[15] The heterogeneity of the polysaccharide is revealed by electrophoresis on glass-fiber paper[16] as well as on DEAE-cellulose.[17] A 1% dispersion of the gum has a pH of 2.63 and no buffering activity.[18] The equivalent weight[18] is 1340–1735. Riboflavin and thiamine are present in traces.[19]

IV. Properties

Gum ghatti forms viscous mixtures on dispersion in water to 5% concentration or greater. The dispersions are non-Newtonian in behavior, as is true with most of the water-soluble gums, and their viscosity increases geometrically with concentration (Table II). The dispersions are less viscous than those of

Table II

Viscosity of Gum Ghatti Dispersions Measured at 25° with a Brookfield Viscometer (Model LVF)

Concentration, %	1	2	5	7.5	10
Viscosity, cps	2	35	288	1012	2444

gum karaya, although they are more viscous than those of gum arabic. A higher apparent viscosity can be obtained by dispersing the gum in 25% ethanol or by increasing the pH to about 8.0 (Fig. 1). A dispersion in 50% ethanol does not hydrate completely. Gum ghatti is a good emulsifying agent and can emulsify more difficult systems than can gum arabic. Aging of gum ghatti dispersions increases their viscosity (Table III).

Gum ghatti solutions may be slightly colored because of traces of pigment remaining in the gum. The normal pH of the dispersion is 4.8. There may be more incompletely dissolved material than observed in dispersions of either karaya or arabic gums. Insoluble matter can be removed from gum ghatti by filtration of its aqueous dispersion, and spray-dried ghatti is currently available in which no insolubles are present. The viscosity of the spray-dried gum is somewhat lower than that of natural dry milled gum.

Table IV indicates the interaction of certain reagents with solutions of gum ghatti.

The adhesiveness of gum ghatti dispersions is similar to that of gum arabic. Because of its higher viscosity, it is not possible to prepare dispersions as concentrated as with gum arabic. Gum ghatti does not form a true gel. Films

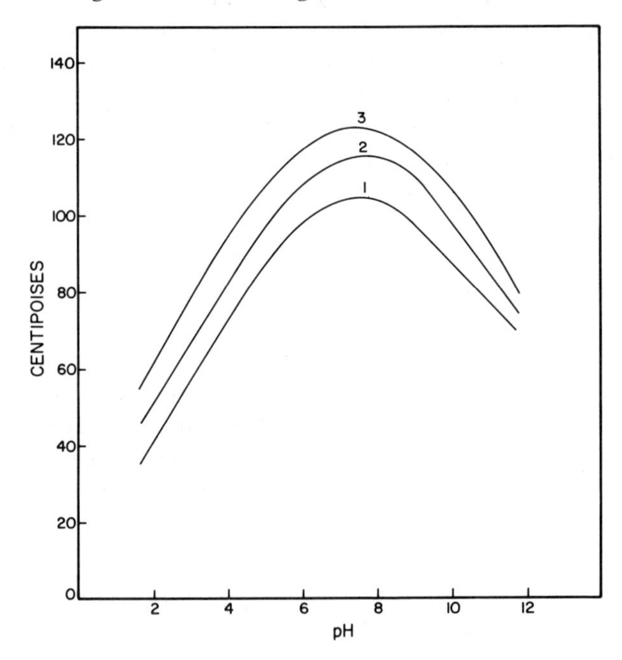

FIG. 1.—Viscosity as a function of pH and time. Viscosity of 5% gum ghatti solids measured at 25° with a Brookfield Viscometer (model LVF): curve 1, after 24 hours; curve 2, after 1 week; curve 3, after 2 weeks.

TABLE III

Viscosity of 5% Gum Ghatti Dispersions Measured at 25° with a Brookfield Viscometer (Model LVF)

pH	Cps		
	Aged Overnight	Aged 1 Week	Aged 2 Weeks
1.6	36	46	55
4.1	68	82	94
4.9[a]	86	96	105
5.6	93	104	112
7.8	105	115	122
8.9	95	112	116
10.8	77	90	95
11.8	72	74	78

[a] Control.

TABLE IV

Characteristics of the Precipitates of Gum Ghatti Dispersions

Reagent	Precipitate
Millon reagent	Fine precipitate
Lead acetate, 20% solution	None
Basic lead acetate (AOAC)	Translucent, flocculent precipitate
Potassium hydroxide, 10% solution	None
Ferric chloride, 5% solution	None
Ethanol, 4 volumes	Fine, flocculent precipitate, nonadherent
Sodium tetraborate, 4% solution	None
Stokes acidic mercuric nitrate reagent	None
Cationic soap, 1% solution	Fine precipitate
Saturated ammonium sulfate	None
Papain, 2% solution	None
Gelatin, 2% solution	Precipitate

prepared from gum ghatti dispersions are relatively soluble and brittle. Gum ghatti has good emulsifying properties, which serve as the basis for most of its applications.

V. APPLICATIONS

Gum ghatti is used in applications also served by gum arabic. It is often used in pharmaceutical preparations as an emulsifying agent. In the United States, ghatti is used in the preparation of stable, powdered, oil-soluble vitamins.[20] Gum ghatti has been used in combination with proteins as a means of stimulating the formation of eosinophils.[21]

Gum ghatti is used in table syrup emulsion containing about 2% butter to stabilize the emulsion. In such an application, about 0.4% ghatti is used in combination with 0.08% lecithin.[22, 23] The refractive index of table syrup containing emulsified butter may be adjusted by additional quantities of ghatti to produce clarity.

Gum ghatti prevents fluid loss in oil-well drilling muds when it is used at low concentrations in neutral mixtures and in high salt concentrations.[24] The gum also prevents fluid loss at elevated temperatures, which leads to its use in the maintenance of thin wall-cakes in oil wells where high temperatures occur.[25, 26] In dispersions at high pH, gum ghatti is not as effective as special starches.

Gum ghatti is used in the "acidizing" of oil wells. The gum is moistened with a water-insoluble nonaqueous liquid that is inert both to the gum and to the acid solution. Then acid is added with mixing to form a uniform dispersion, which is pumped under pressure to permeate the earth formation. This results in enlarged passageways or worm holes, which increase the productivity of the well. The drilling mud or other fracture-clogging deposits are washed out and the oil flows freely.[27]

Gum ghatti is used to emulsify petroleum and nonpetroleum waxes to form liquid and wax paste emulsions, which find wide uses in the paper industry as coatings and as barriers. Powdered gum ghatti is used in ammonium nitrate–semigelatin mixtures and powdered explosives to improve their resistance to water damage.[28] Gum ghatti is used in old and new varnishes[29] and is used for emulsifying oils, such as 40% kerosene oil[30] and other technical emulsions. It acts as a stabilizer in auto polishes and wallpaper gum sizings.

Gum ghatti has been used in combination with polyacrylamide to aid in the polymerization and formation of uniform and discrete prills of cross-linked polystyrene.[31] Salts of ghattic acid have reportedly been used for light-sensitive papers, pigments, and fungicides.[32] Because of its high L-arabinose content, gum ghatti is hydrolyzed to prepare pure L-arabinose on a commercial scale.[33] L-Arabinose is used as a flavor adjunct in food products and in the preparation of nucleosides used as antitumor drugs.

Gum ghatti is also used to stabilize the Prussian blue color in photoelectric determinations,[34, 35] and it is used in the polarographic determination of copper, lead, and iron.[36] It shows some promise in forming protective hydration layers around clay particles, thus maintaining a dispersion for the particle size analysis of soils.[37]

VI. References

(1) "National Formulary," American Pharmaceutical Association, Washington, D.C., 8th Ed., 1946.

(2) Meer Corporation, purchase and sales records.

(3) G. O. Aspinall, E. L. Hirst, and A. Wick-Strom, *J. Chem. Soc.,* 1160 (1955).

(4) D. Hanna, L. McReynolds, and E. H. Shaw, Jr., *Proc. S. Dak. Acad. Sci.,* **19,** 130 (1939).

(5) D. Hanna and E. H. Shaw, Jr., *Proc. S. Dak. Acad. Sci.,* **21,** 78 (1941).

(6) H. W. Carhart and E. H. Shaw, Jr., *Proc. S. Dak. Acad. Sci.,* **16,** 40 (1936); **19,** 130 (1939).

(7) G. O. Aspinall, B. J. Auret, and E. L. Hirst, *J. Chem. Soc.,* 221 (1958).

(8) G. O. Aspinall, B. J. Auret, and E. L. Hirst, *J. Chem. Soc.,* 4408 (1958).

(9) G. O. Aspinall and T. B. Christensen, *J. Chem. Soc.,* 2673 (1965).

(10) G. O. Aspinall and J. B. Christensen, *J. Chem. Soc.,* 3461 (1961).

(11) G. O. Aspinall, V. P. Bhavanadan, and J. B. Christensen, *J. Chem. Soc.,* 2677 (1965).

(12) G. O. Aspinall, J. J. Carlyle, J. M. McNab, and A. Rudowski, *J. Chem. Soc., C,* 840 (1969).

(13) G. O. Aspinall and J. M. McNab, *J. Chem. Soc., C,* 845 (1969).

(14) G. O. Aspinall and J. J. Carlyle, *J. Chem. Soc., C,* 851 (1969).

(15) P. H. Elworthy and T. M. George, *J. Pharm. Pharmacol.,* **15,** 781 (1963).

(16) B. A. Lewis and F. Smith *J. Amer. Chem. Soc.,* **79,** 3929 (1957).

(17) M. A. Jermyn, *Aust. J. Biol. Sci.,* **15,** 787 (1962).

(18) V. K. Srivastava and R. S. Rai, *Kolloid-Z.,* **190,** 140 (1963).

(19) R. Broker and J. V. Bhat, *Curr. Sci.,* **22,** 343 (1953).

(20) H. J. Dunn, U.S. Patent 2,897,119 (1959); *Chem. Abstr.,* **53,** 20709 (1959).

(21) J. S. Chapman, *Amer. J. Clin. Pathol.,* **40,** 357 (1963).

(22) H. Topalian and C. Elsesser, U.S. Patent 3,282,707 (1966); *Chem. Abstr.,* **66,** 1748 (1967).

(23) F. Smith, U.S. Patent 3,362,833 (1968); *Chem Abstr.,* **68,** 67920x (1968). 67920x (1968).

(24) G. R. Gray, *Tex., Eng. Exp. Sta. Bull.,* No. **96,** 63 (1946).

(25) G. Tchillingarian and C. M. Beeson, *Petrol. Engr.,* **24,** No. 4B, 45 (1952).

(26) W. L. Owen, *Sugar,* **45,** No. 11, 35 (1950).

(27) P. H. Cardwell and L. H. Eilers, U.S. Patents 2,824,833 and 2,824,834 (1958); *Chem. Abstr.,* **52,** 11400 (1958).

(28) S. H. Davidson and C. H. Rigby, U.S. Patent 2,680,068 (1954); See Brit. 645,039; *Chem. Abstr.,* **48,** 9694 (1954); S. H. Davidson, C. H. Rigby, and Imperial Chemical Industries Ltd., Brit. Patent 645,039; *Chem. Abstr.,* **45,** 4043 (1951).

(29) K. Letters, *Farbe Lack,* **58,** 293 (1952).

(30) L. N. Mukerjee and S. D. Shukla, *J. Indian Chem. Soc.,* **42,** 805 (1965).

(31) N. Chomitz, U.S. Patent 3,172,878 (1965); *Chem. Abstr.,* **62,** 13345 (1965).

(32) Harris-Seybold Co., Brit. Patent 689,623 (1953); *Chem Abstr.,* **47,** 10256 (1953).

(33) Pfanstiehl Laboratories, Inc., personal communication.

(34) S. Nussenbaum and W. Z. Hassid, *Anal. Chem.,* **24,** 501 (1952).

(35) M. E. Greig, *Science,* **105,** 665 (1947).

(36) C. A. Reynolds and L. B. Rogers, *Anal. Chem.,* **21,** 176 (1949).

(37) A. M. Wentermyer and E. B. Kinter, *Public Road,* **28,** No. 3, 55 (1954).

CHAPTER XII

GUM KARAYA

ARTHUR M. GOLDSTEIN AND EMIL N. ALTER

Stein, Hall & Co., Inc., Long Island City, New York

I. INTRODUCTION

Gum karaya is a partially acetylated polysaccharide that is characterized by a high degree of water absorption and a low solution pH. Among the exudation gums, it ranks second to gum arabic in commercial importance. Originally introduced as an inexpensive substitute for gum tragacanth, gum karaya has found many uses that depend on its unique properties.

II. Production

1. Source

Gum karaya is the name given to the dried exudation of the *Sterculia urens* tree. Almost all gum karaya comes from India. In the early days of its importation into the United States, many types of gums were introduced that had properties similar to those of gum karaya. In its monographs, "Drugs, Chemicals & Preparations," "The National Formulary" named it *Sterculia gum,* with the alternative name of *gum karaya.* It described Sterculia gum as the dried, gummy exudation from *Sterculia urens* (Roxburgh), *Sterculia villosa* (Roxburgh), *Sterculia tragacantha* (Lindley) or other species of *Sterculia* (Sterculiaceae family) or from *Cochlospermum gossypium* (De Candolle) or other species of *Cochlospermum* (Bixaceal family).[1] Howes established the fact that the gum shipped from India as gum karaya is obtained from the *Sterculia urens* tree.[2]

2. Producing Areas

Sterculia urens is widespread in India. It is a large bushy tree that grows to a height of 30 ft (9 m). It is found on the dry, rocky hills and plateaus of central and northern India.

3. Seasonal Effects

The best quality gum is collected during April, May, and June before the monsoon season. As the weather becomes warmer, the gum yield increases. The gum quality, as well as yield, improves during this period. Early rains reduce the size of the crop by washing away much of the exudation before it can dry. In addition, the gum collected during a rainy season has low viscosity. In September, after the monsoons, the collection cycle is repeated. The gum gathered in the fall has a grayish color and usually gives less viscous solutions than gum collected earlier in the year.

4. Collection

Gum exudes from *Sterculia* trees after they are tapped or blazed. Usually, only the younger trees are tapped. In trees over 3 ft (1 m) in girth, blazes or chars not exceeding 1 ft^2 (900 cm^2) are made with a maximum of two to a tree. The gum begins to exude immediately, and exudation continues for several days. The maximum amount of exudation occurs within the first 24 hr. The gum is in the form of large, irregular tears, some of which weigh several pounds. The dried tears are picked by natives and delivered to village collection points. Visiting dealers from Bombay buy the gum for shipment to central processing houses. Price is governed by both world market conditions and the gum quality. The yield of gum from mature trees is estimated at 2–10 lb (1–4.5 kg)/per tree per season. The average tree can be tapped about five times during its lifetime.[2]

5. *Purification*

The Indian gum merchants remove bark from the large tears of field-grade gums. The tears are then broken up, and the fragments are sorted into grades on the basis of color and amount of adhering bark. The gum is then sold to importers and processors in the United States and Europe.

The processor evaluates incoming raw materials according to impurity content, solution viscosity, and color. Purification is accomplished by size reduction and air flotation of loose bark. There are also devices, such as Sutton–Steele and Steele tables, that remove both bark and small pebbles by a combination aerodynamic–specific gravity principle.

6. *Grades*

Gum karaya can be separated into two classs, crude grades (the gum as it is imported) and processed grades (the gum as it is sold to the industrial user) (Table I). There is no uniformity in the nomenclature of commercial grades of gum karaya. However, in spite of the differences in nomenclature, many of the grades offered are based on similar processing techniques and assays.

Crude gum karaya is imported into the United States in grades that are based entirely on bark and foreign organic matter content and color. A processor usually offers three to five grades of gum karaya, varying in bark and foreign organic matter content from 0.1% to 3.0% and in color from white to tan. In order to be classified as NF-grade material and be acceptable for food and drug use, the gum must not contain more than 3% of bark and foreign organic

TABLE I

Gum Karaya Grade Standards

Grade[a]	Bark and Foreign Organic Matter, %	Color	Material Minus 10 Mesh, %
WSGA No. 1	0.5 (maximum)	White with slight gray cast	2
WSGA No. 2	1.5 (maximum)	Very light tan	3
WSGA No. 3	3.0	Tan	3
Technical siftings	6.0	Brown	—

[a] Water Soluble Gum Association Grades.[1]

matter.[1] Grades that contain more than 3% are offered for industrial uses. Gums having almost any content of bark and foreign organic matter from 3% to 10% can be purchased; the color of these gums varies from tan to dark brown.

Although designated as a water-soluble gum, gum karaya is one of the least soluble of the exudation gums. A gum particle placed in water does not dissolve but absorbs water and swells to many times its original size. Therefore, the particle size influences the type of dispersion obtainable. A coarse granulation yields a discontinuous, grainy dispersion; a finely powdered gum yields an apparently homogeneous dispersion.

Granulated or crystal gum karayas are usually processed so that the particle size is between 6 and 30 mesh. These granulated gum karayas are used principally as bulk laxatives.

Finely powdered grades are the most common types of processed gum karaya. The gum is prepared so that the powder completely passes through a 150-mesh screen and so that 80–90% also passes through a 200-mesh screen. The particle size of such a gum is so small that, in spite of individual particle swelling, an apparently homogeneous dispersion results.

Some grades of gum karaya have a particle size that lies between those of the granular and powdered types. These products are usually 80–200 mesh. They disperse in water with fewer lumps than do the finely powdered grades and are more dust-free. They represent a compromise between grades that are easily dispersed and those that form homogeneous dispersions.

There is no distinct correlation between viscosity and grade. Cost is determined on the basis of color and purity.

7. *Impurities*

The impurity in gum karaya is principally bark; there is also some wood fiber and a small amount of sand. The bark and wood fibers are present because they adhere to the undried exudation. "The National Formulary" (NF) specifications set a maximum of 3% on bark and foreign organic matter content. This includes sand. The NF specification also indicates a maximum of 1% of acid-insoluble ash, which is a measure of the sand content.[1]

8. *Quantities Marketed and Price Variation*

The poundage of gum karaya imported into the United States from 1951 to 1968 and its price are shown in Table II. Price variations of the various crude grades of gum are chiefly governed by demand and seasonal conditions. The table does not give quantities of each grade imported, nor are these figures available. However, it is known that grade 3 is imported in the largest volume and that grade 2 ranks second.

TABLE II

U.S. Imports and Prices

Year	Total Pounds[a] Imported into United States		Price of Gum Karaya, cents/lb[b]		
			WSGA No. 1	WSGA No. 2	WSGA No. 3
1951	7,696,225		35–47	26–27	19–27
1956	7,014,000		39	11–24	21–29
1961	7,345,125		31–33	27–30	24–26
1962	7,461,805		33–41	30–34	25–33
1963	7,132,397		41–46	35–38	33–35
1964	7,682,267		45–50	36–38	31–33
1965	9,061,322		45–47	36–38	32–33
1966	7,344,074	Before devaluation	45	36	34
1966		After devaluation (June 1966)	33	27	24
1967	7,201,505		35–36	26	23–24
1968	7,344,074		36–50	26–38	23–34
1969	8,281,000		62–70	53–58	46–49
1970	7,717,000		59–64	52–56	47–51
1971	7,650,000		58–62	48–52	41–43

[a] These figures were compiled from United Staees Department of Commerce Bulletins, entitled "U.S. Import of Merchandise for Consumption," from the Chemical and Rubber Industry Report of the U.S. Department of Commerce, As-OP, June 1955, and from the United States Department of Commerce Bulletin FT135 (Called FT125 through 1966).

[b] These figures include cost in India and freight charges. (Courtesy of Stein Hall & Co., Inc.)

9. *Potential Amount*

At the present time, it is estimated that the United States imports 75–80% of the available gum karaya. Any large increase in production could be achieved only by cultivation of *Sterculia urens* and would require 30–40 years to achieve.

III. USES

1. *History*

Gum karaya has been produced commercially for only a short time. There is evidence to indicate that, in the early part of the 20th century, the gum was sold as gum tragacanth or as an adulterant of gum tragacanth. However, since 1920, gum karaya itself has been imported in increasing amounts. The original incentive to its use was its price which was much lower than that of gum tragacanth, but recently, gum karaya has been found to be superior to other gums for some purposes.[3]

2. *Commercial Value*

Pharmaceuticals.—A large part of the gum karaya imported into this country is utilized in the pharmaceutical industry. As a bulk laxative, it is second to *Psyllium* seeds in importance. When used for this purpose, gum karaya is usually processed so that it is 8–30 mesh in size. These coarse gum particles absorb water and swell to 60–100 times their original volume, forming a discontinuous type of mucilage. This type of mucilage is very effective as a laxative. The gum is not digested nor is it absorbed by the body.

Another important pharmaceutical application of gum karaya is its use as a denture adhesive. The powdered gum is usually dusted on the dental plate, and it swells when it touches the moist surfaces of the mouth. This results in a more comfortable and tighter fit of the plate. The rapid swelling of the gum particles, their relative insolubility, and their unusual resistance to bacterial and enzymic breakdown make the gum suitable for this use.

Paper.—In the paper industry, gum karaya is largely confined to use with long-fibered, light-weight papers, for example, condenser tissues and fruit wrap tissues,[4, 5] in which it acts as a binder. The principle behind its use is the removal of the acetyl groups by treatment with alkali, thereby exposing more active carboxyl and hydroxyl groups. This increase in the polarity of the molecule makes the binding of the gum to hydrated cellulose fibers more efficient. Removal of acetyl groups also affords an internally cohesive or ropy solution. A method for the preparation of deacetylated solutions follows. A wooden or stainless steel tank is the recommended reaction vessel because of the acidic nature of the gum and the presence of a small percentage of tannins as impurities. Extreme agitation is needed to disperse the gum in water without lumping. Difficulty in dispersion can be overcome by the use of special dispersible gums. The gum is hydrated by allowing the dispersion to stand at room temperature for several hours or by heating it to 71°–82° for 15 min.

A weak alkali, such as ammonia or triethanolamine, is then added to raise the pH of the solution from its normal pH (4.5) to pH 8.0–8.5. The properties that are imparted to the gum by deacetylation are pH irreversible and are maintained even when the solution is added to an acidic pulp suspension.

The deacetylated gum solution can be added to the pulp suspension at the fan pump. Usual rates of addition are 10–15 lb of gum per ton of pulp (1 kg/200–300 kg of pulp). The gum deflocculates the fibers resulting in light-weight sheets of improved formation and strength. However, a deacetylated gum karaya solution has the disadvantage of decreasing the freeness and thereby slowing the sheet. In light-weight papers, this is not a very serious problem. In heavier papers, the slowing of the stock flow becomes sufficient to prohibit the use of deacetylated gum karaya. The use of alum or other cationic materials in the furnish impairs the efficiency of the gum as a deflocculant.

Foods.—Gum karaya is listed in the "Food Chemical Codex."[6] Powdered gum karaya is used in French dressing, ice pops, sherbets, cheese spreads, ground meat preparations, and meringue products. In French dressing, it is used as a stabilizer. It is used sometimes in conjunction with gum arabic, which acts as a protective colloid. Gum karaya is used in percentages of 0.2–0.4 in the manufacture of ice pops and sherbets. The gum prevents the bleeding of free water and the formation of large ice crystals. Its water-absorbing and water-holding capacities and its excellent acid compatibility make it suitable for this use. Concentrations of 0.8% or less of the gum are used in cheese spreads. Its acidic nature is not objectionable in this type of dairy product. It is added to prevent water separation and to increase the ease of spreading.

Gum karaya is also used as a meringue stabilizer because of its binding properties. In addition, its incorporation enables a greater volume of meringue to be prepared from a fixed amount of protein. Ground meat products, such as bologna, require an efficient water-holding substance that has a small amount of adhesiveness. Gum karaya, in concentrations of ~0.25% provides these characteristics and gives the product a smooth appearance.

Textiles.—In the textile industry, gum karaya is modified so that it can be used for printing operations in which it was considered unsatisfactory until a method of increasing its solubility was developed. This is done by cooking a water suspension of gum karaya under pressure. The rate of dissolution varies with the pressure.[7] In commercial practice, solutions containing 15–18% of solids are marketed as textile gum solutions. An alternative method of solubilization consists of treatment with sodium peroxide, persulfate, or persilicate. The textile gum solution is used principally as a thickening agent for the dye in direct color printing on cotton fabrics.

IV. Structure

Gum karaya contains approximately 8% acetyl groups and ~37% uronic acid residues. Partial acid hydrolysis yields D-galactose, L-rhamnose (6-deoxy-L-mannose), and D-galacturonic acid along with the aldobiouronic acids, 2-*O*-(α-D-galactopyranosyluronic acid)-L-rhamnose and 4-*O*-(α-D-galactopyranosyluronic acid)-D-galactose, and the acidic trisaccharide, *O*-(β-D-glucopyranosyluronic acid)-1→3-(α-D-galactopyranosyluronic acid)-(1→2)-L-rhamnose.

Methylation and hydrolysis before and after reduction of carboxyl groups furnishes information leading to the postulation of several structures. The isolation 3,4-di-*O*- and 3-*O*-methyl-L-rhanmnose, as well as 2,3-di-*O*-, 2-*O*-, and 3-*O*-methyl-D-galactose, following hydrolysis of the carboxyl reduced, methylated gum, shows that D-galacturonic acid and L-rhamnose units are branch points in the polysaccharide structure. Degradation of the gum yields a degraded polysaccharide, which, upon further hydrolysis, gives 2-*O*-(α-D-galactopyranosyluronic acid)-L-rhamnose and a trisaccharide believed to be β-D-glucopyranosyluronic acid-(1→3)-α-D-galactopyranosyluronic acid-(1→2)-L-rhamnose.[8]

Apparently, *S. caudata* gum[9] is structurally similar to *S. urens* gum. Partial hydrolysis of *S. caudata* gum yields, in addition to the acidic ogligosaccharides present in the hydrolyzate of *S. urens* gum, 2-*O*-acetyl-4-*O*-(β-D-galactopyranosyluronic acid)-D-galactose and 3-*O*-(β-D-glucopyranosyluronic acid)-D-galacturonic acid. Thus, D-glucuronic acid residues are present as nonreducing endgroups in both gums, whereas D-galacturonic acid residues are 4-*O*-substituted, with the majority being also 2- or 3-*O*-substituted.[10] It is not yet possible to propose a complete structure for the repeating unit of *Sterculia* gum, although portions of the gum may be represented by the partial structures shown in Fig. 1.

V. Properties

1. Dissolution and Viscosity Measurements

Gum karaya absorbs water very rapidly to form viscous mucilages at low concentrations. However, it is the least soluble of the commercial exudations; a powder that passes through a 150-mesh screen is used to produce visually homogeneous suspensions. In order to achieve a uniform dispersion with finely powdered gum karaya, one or more of the following procedures should be used: (*a*) applying vigorous agitation while adding the gum to water, (*b*) allowing two or more hours for hydration, (*c*) dispersing the gum into water as a mixture with another solid such as sugar, (*d*) premixing the gum with 1–5 times its weight of a water-miscible liquid, such as ethanol or glycerol, prior to the addition of water. The rapidity with which powdered gum karaya swells can be seen in the high initial rate of viscosity increase. The effect of particle size on viscosity development is shown by the fact that the rate of viscosity increase in a dispersion of 80–200 mesh gum karaya is slower than that in a dispersion of a more finely powdered gum of the same quality (Fig. 2).

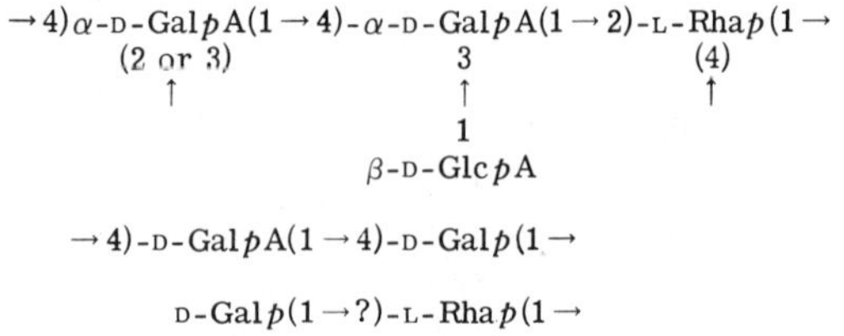

```
→4)α-D-GalpA(1→4)-α-D-GalpA(1→2)-L-Rhap(1→
     (2 or 3)            3              (4)
        ↑                ↑               ↑
                         1
                     β-D-GlcpA

     →4)-D-GalpA(1→4)-D-Galp(1→

         D-Galp(1→?)-L-Rhap(1→
```

FIG. 1.—Partial structure of *Sterculia* gum.[11]

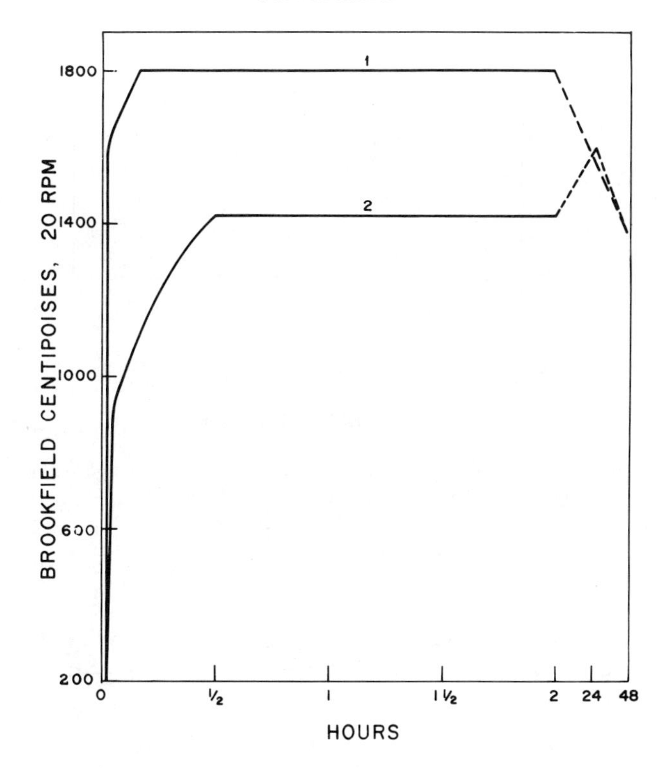

FIG. 2.—Effect of time on the viscosity of 1% dispersions of high grade gum karaya (25°). 1, minus 150-mesh; 2, 80/200-mesh.

Temperature.—When the temperature of a fully hydrated gum karaya solution is gradually raised from 20° to 85°, the viscosity decreases (Fig. 3). Boiling reduces the viscosity of gum karaya solutions, particularly when they are held at this temperature for more than 2 min. Higher maximum viscosity is obtained by cold hydration of gum karaya than is afforded by hot hydration (Fig. 4).

The reduction in viscosity that is obtained by cooking gum karaya suspensions, especially under pressure, is accompanied by an increase in the solubility of the gum. Under these conditions, it forms a smooth, homogeneous, translucent, colloidal dispersion. Concentrations as high as 15–18% can be prepared in this manner.

Concentration.—In dilute solutions of gum karaya, the viscosity increases linearly with concentration up to about 0.5%. Thereafter, gum karaya dispersions behave as non-Newtonian solutions (Fig. 5).

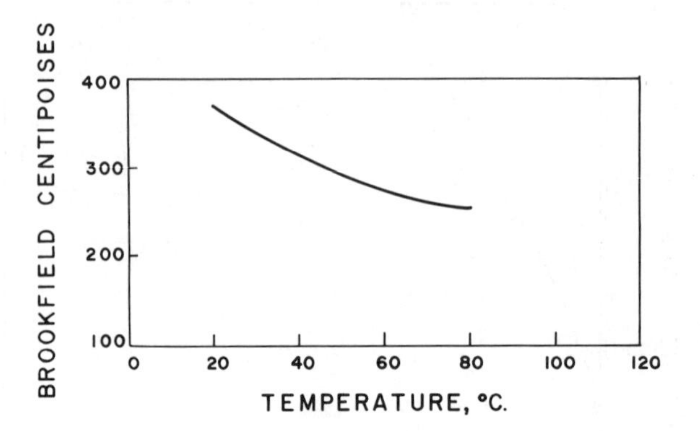

FIG. 3.—Effect of temperature on the viscosity of a fully hydrated 0.5% solution of high-grade gum karaya. Rate of temperature increase is 2°/min.

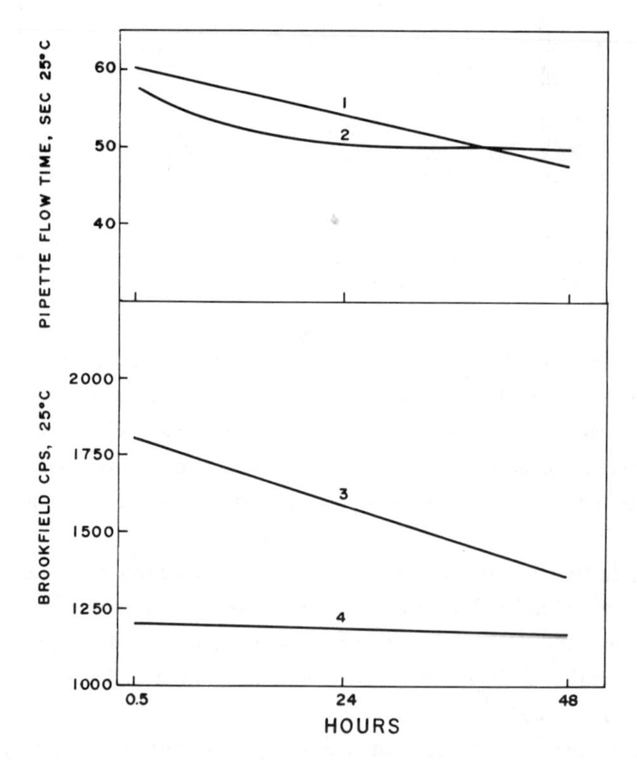

FIG. 4.—Viscosity development of 1% high-grade gum karaya dispersions versus time. Curves 1 and 3, gum karaya hydrated at 25°; 2 and 4, gum karaya hydrated initially at 80° for 10 min.

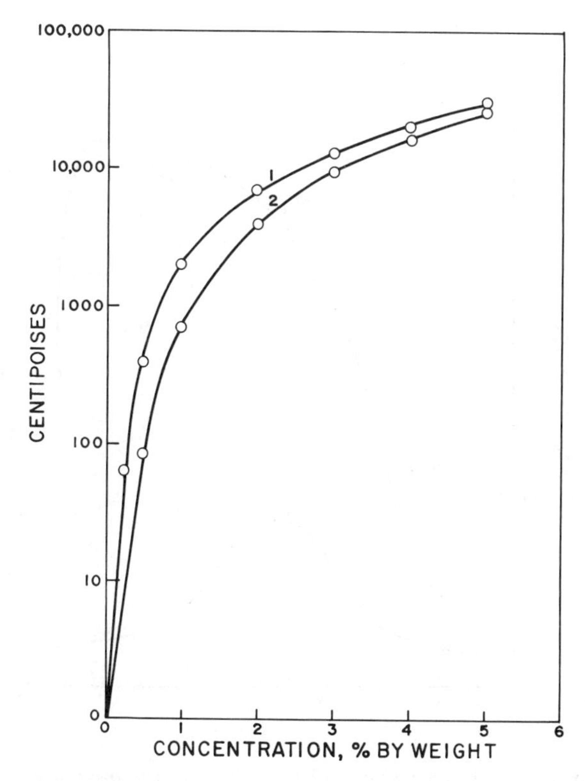

FIG. 5.—Effect of concentration on the viscosity of gum karaya dispersions as measured with a Brookfield viscometer at 20 rpm and 25°; curve 1, high-grade gum karaya; 2, average-grade gum karaya.

pH.—Gum karaya maintains its solubility with changes in pH; the viscosity decreases upon addition of acid or alkali (Fig. 6). Higher viscosities and pH stability over a wider range are obtainable when the gum is hydrated prior to pH adjustment. The solution color lightens in acidic media and darkens in alkaline solutions because of the presence of tannins. At pH 7 and above, the characteristic short-bodied gum karaya solutions become ropy mucilages. This irreversible transformation, which has been ascribed to deacetylation, is accompanied by an increase in viscosity.

Ionic strength.—The viscosity of gum karaya dispersions decreases when electrolytes, such as sodium, calcium, and aluminum chlorides and aluminum sulfate, are added. When gum karaya is hydrated in solutions containing as much as 25% of a strong electrolyte, the viscosity is stable, although there is an initial depression that is accompanied by a separation of solids from solution. The

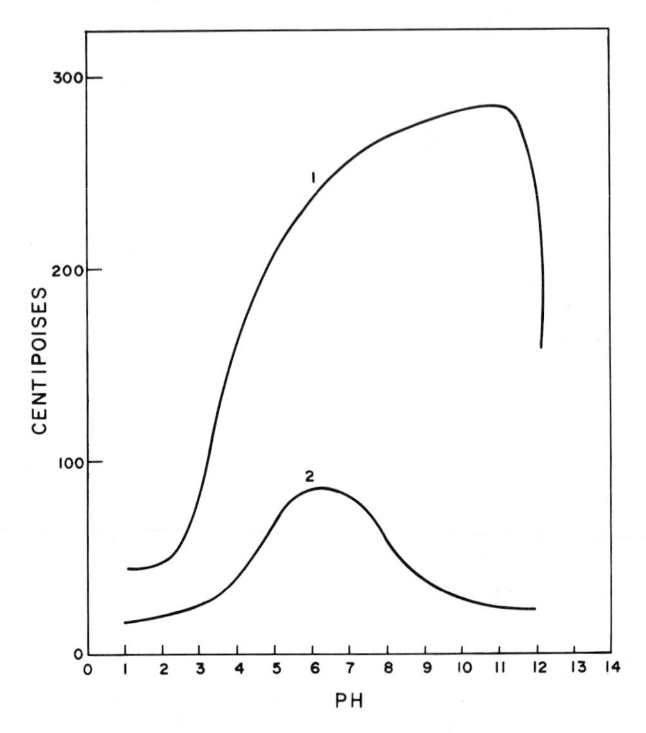

FIG. 6.—Effect of pH on viscosity of 0.5% gum karaya dispersions. Curve 1, gum karaya hydrated prior to pH adjustment; 2, unhydrated gum karaya added to acidic or alkaline solutions. Measurements made with a Brookfield viscometer at 25° and 20 rpm. The pH was adjusted with hydrochloric acid or sodium hydroxide.

normal viscosity of a 0.5% high-grade gum karaya solution (400 cps) drops below 100 cps when electrolytes are added (Fig. 7). Sensitivity of gum karaya to strong electrolytes begins at low salt concentrations; the gum is not as sensitive to solutions of weak electrolytes.

Age of solution.—The viscosity of gum karaya suspensions remain constant for several days. Increased stability can be provided by the addition of preservatives, such as chlorinated phenols, formaldehyde, mercuric salts, and benzoic or sorbic acid. Gum karaya loses viscosity forming ability when stored in the dry state; the loss is greater for a powdered material than for the crude gum.[12] This decrease is most noticeable in the first few weeks after the gum has been ground, especially if it is stored under conditions of high humidity and temperature. Cold storage inhibits this degradation. It has been suggested that the decrease in viscosity is related to the loss of acetic acid.[13]

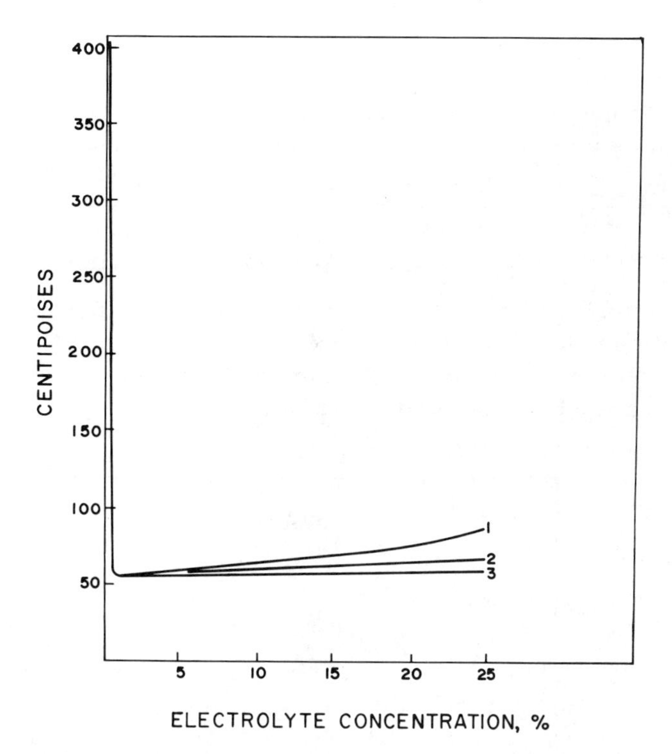

FIG. 7.—Effect of salt concentrations on the viscosity of 0.5% high-grade gum karaya dispersions. Curve 1, aluminum chloride or sulfate; 2, calcium chloride; 3, sodium chloride. Measurements made with a Brookfield viscometer at 25° and 20 rpm.

2. *Gels*

Gum karaya forms heavy, nonflowing pastes at concentrations above 2–3%. Its dispersions do not pass through a noticeable gel stage although they exhibit the absorption characteristics of gels, especially at high concentrations. Its compatibility with a wide variety of other gums and proteins indicates that it could be useful in cosmetic, pharmaceutical, and food products.

3. *Films*

Gum karaya forms smooth films. When plasticized with compounds such as glycols, it finds use in hair-setting preparations. However, practically all other applications of the gum depend on its viscosity and water-absorbing characteristics.

4. *Adhesiveness*

At concentrations of 20–50% in water, gum karaya exhibits strong wet-adhesive properties. Its use as a binder in denture powders is due to it's good gel strength and the strong bonds produced. Its ability to absorb water enables the gel to resist loss of strength when diluted. Blending the powdered gum with a mild alkali improves adhesiveness.

5. *Acid Resistance*

Gum karaya undergoes hydrolysis slowly and resists hydrolysis in 10% hydrochloric acid solutions at room temperature for at least 8 hr (Fig. 8). Acid concentrations of 4% at 50° for 10–24 hr or 90% for 1–2 hr are required to hydrolyze a gum karaya solution to water thinness.

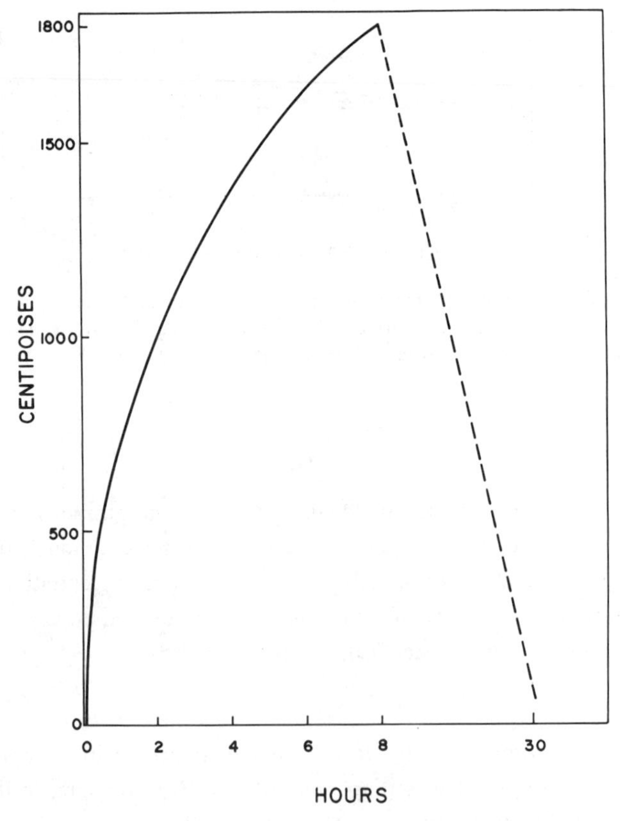

FIG. 8—Viscosity of 1.5% gum karaya dispersion in 10% hydrochloric acid as measured with a Brookfield viscometer at 25° and 12 rpm.

6. *Ropiness*

The short, gel-like body of a gum karaya mucilage can be transformed by alkaline treatment into a coherent, ropy paste. An empirical measurement of a solution's ropiness is obtained by determining the volume of a dilute gum karaya solution that pours into a graduate cylinder as an unbroken string. Viscosity is, to a considerable degree, correlated with degree of ropiness.

7. *Water Retention*

Gum karaya is available in coarse grades, commonly 8–30 mesh. A coarse-mesh product has a slower viscosity buildup than a finer mesh grade and produces a discontinuous mucilage. A modified osmometric procedure has been used to determine the water retention properties of a coarse gum karaya which is used as a bulking agent.[14]

VI. References

(1) "The National Formulary," American Pharmaceutical Association, Washington, D.C., 8th Ed., 1956, p. 500.

(2) F. N. Howes, "Vegetable Gums and Resins," Chronical Botanica Co., Waltham, Mass., 1949, pp. 41–44.

(3) C. L. Mantell, "The Water-Soluble Gums," Rheinhold Publishing Corp., New York, 1947, p. 50.

(4) T. R. Le Compte, U.S. Patent 2,069,766 (1937); *Chem. Abstr.*, **31,** 2434 (1937).

(5) T. R. Le Compte, U.S. Patent 2,088,479 (1937); *Chem. Abstr.*, **31,** 6881 (1937).

(6) "Food Chemicals Codex," Publ. 1406, Natl Acad. Sci.–Natl. Res. Council, Washington, D.C., 1st Ed., 1966, p. 359.

(7) E. Knecht and J. B. Fothergill, "The Principles and Practice of Textile Printing," Griffin and Co., London, 2nd Ed., 1924, pp. 123, 124.

(8) G. O. Aspinall and Nasir-ud-din, *J. Chem. Soc.*, 2710 (1965).

(9) E. L. Hirst, *Verhandlungsber. Kolloid-Ges.*, **18,** 104 (1958).

(10) G. O. Aspinall and R. N. Fraser, *J. Chem. Soc.*, 4318 (1965).

(11) G. O. Aspinall, *Pure Appl. Chem.*, **14,** 43 (1967).

(12) R. W. Money, J. *Sci. Food Agr.*, **2,** 385 (1951).

(13) A. M. Goldstein, *in* "Natural Plant Hydrocolloids," *Advan. Chem. Ser.*, **11,** 35 (1954).

(14) A. L. Monaco and E. J. Dehner, *J. Amer. Pharm. Ass., Sci. Ed.*, **44,** 237 (1955).

CHAPTER XIII

GUM TRAGACANTH

GEORGE MEER, WILLIAM A. MEER, AND THOMAS GERARD

Meer Corporation, North Bergen, New Jersey

I. INTRODUCTION

Gum tragacanth is one of the most widely used natural emulsifiers and thickeners available to the food, drug, and allied industries. The high viscosity imparted to water by the gum makes it useful for preparing aqueous suspensions of insoluble substances.

Gum tragacanth was known and used in the days of Theophrastus, who described it in the 3rd century BC.[1] It is a water-soluble plant exudate, which places it in the general classification of plant hydrocolloids used as thickeners. It is listed in the "United States Pharmacopeia"[2] and in the "Food Chemicals Codex."[3] Gum tragacanth is approved for food use and is on the GRAS list (generally recognized as safe for its intended use) under the Federal Food, Drug, and Cosmetic Act.

II. Source

Gum tragacanth is the dried gummy exudation of several species of *Astragalus* (family Leguminosae), a small, low, bushy perennial shrub characterized by a relatively large tap root, which, along with the branches, is tapped for the gum. The name tragacanth is derived from the Greek *tragos* (goat) and *akantha* (horn) and probably refers to the curved shape of the ribbons, the best grade of commercial gum. It also occurs in flake form. The two types are obtained from different shrubs. A plant will produce only ribbons or flakes, and generally it is unusual for the same locality to produce high-quality gum of both types. The lower grades of gum may also result from several tappings from one bush.

The plants are common in certain sections of Asia Minor and in the semidesert and mountainous regions of Iran, Syria, and Turkey. The best ribbon areas, in the order of quality produced, are Shamiran, West of Teheran, Hamadan, Teheran, Isfahan, and Kerman. The quality of ribbons decreases as the collection area goes further south of Teheran. The best flake districts, in order of quality produced, are Isfahan, Shiraz, Teheran, Hamadan, Kermanshahan, and Kerman.

III. Collection

The plants are longtiudinally incised with a sharp blade and the gum exudes spontaneously from these wounds or breaks and dries or hardens as ribbons or flakes. The bushes grow rather sparsely, and it is a time-consuming task to tap the roots and branches and then return later to collect the dried gum.

The ribbons occur in flattened, lamellated, frequently curved fragments or in straight, spirally twisted, linear pieces. They are white to off-white in color, and 2–4 in (5–10 cm) in length. The flakes are oval, thick, and brittle, varying in size from 0.5–2 in (1.25–5 cm) in diameter. The better gum is milky white, whereas low-viscosity gum is translucent and somewhat glassy in appearance and fracture. Gum tragacanth is ordorless and tasteless.

The plants require an abundance of water during the growing season but need a relatively dry climate during collecting time. Ribbon tragacanth is collected between April and September, whereas the flake varieties are collected during August to November. Excessive rain and wind storms during the collection season will cause discoloration of the gum and reduced solution viscosity.

Collected gum is brought by mule to the cities, where it is sorted and transhipped. The principal cities are Hamadan, Teheran, and Isfahan. An experienced overseer is responsible for initial separation and sorting of the gum. Women sitting at long, low tables sort the gum into five grades each of ribbons and flakes. The lowest quality of gum makes the technical grade. The original sorting is primarily visual, the lightest colored and longest ribbons or largest flakes comprising the best grades. Viscosity measurements have been introduced and are now being employed with varying degrees of success. After sorting, the ribbons are packed in lined boxes and the flakes in double burlap bags. Almost the entire production is exported.

The United States and England are normally the largest importers of gum tragacanth, accounting for much of the production, with the remainder going to Russia, West Germany, France, Italy, and Japan. The importation of gum into the United States has increased in the past few years, except for 1971. In 1968, it amounted to 1,560,000 lb (700,000 kg); in 1969, 1,820,000 lb (820,000 kg); in 1970, 2,100,000 lb (945,000 kg); and in 1971, 1,600,000 lb (720,000 kg).

The gum is available to United States manufacturers both as whole ribbon and powdered to standard specifications. Gum tragacanth is processed in the United States by selectively sifting, aspirating, and grinding the ribbons or flakes to a fine powder.

The preferred medium for conveying the gum is air, which acts to cool the powder and prevents viscosity loss. The final product is blended to insure unformity and standardized for viscosity.

IV. PROPERTIES

1. *Structure and Analysis*

When tragacanth is mixed with water, only the soluble fraction, called *tragacanthin,* dissolves to give a colloidal hydrosol whereas the insoluble fraction, consisting of 60–70% bassorin,[4] swells to a gel. Chemically, tragacanthin is a complex mixture of acidic polysaccharides containing D-galacturonic acid. The other sugars produced on hydrolysis are D-galactose,[5] L-fucose (6-deoxy-L-galactose),[5] D-xylose,[6] and L-arabinose.[6,7] Bassorin appears to be a methylated acidic polysaccharide. Demethoxylation of bassorin probably gives tragacanthin.[8] Cellulose and starch are present in very small amounts. The "Food Chemicals Codex"[3] has established standards for food-grade gum tragacanth.

Gum tragacanth has a high molecular weight (840,000), and the molecules have an elongated shape, 4500A by 19A, which accounts for its high viscosity.[9] When a solution of the gum is boiled with a few drops of 10% aqueous ferric chloride solution, a deep yellow, stringy precipitate is formed. A stringy precipitate is also formed on heating the gum solution with Schweitzer reagent, made by dissolving freshly precipitated copper oxide in concentrated ammonium hydroxide.[10]

2. *Viscosity*

Gum tragacanth solutions, at 0.5% and higher, exhibit structure viscosity, a phenomenon displayed by elongated molecules and recognized by the fact that the rate of flow of the solution in a capillary tube is not proportional to the pressure.[11] Mechanical grinding of dry gums leads to molecular cleavage. Thus, the same concentration of gum solutions made from whole gum is more viscous than that from powdered gum.[12] Gum tragacanth solutions reach maximum viscosity after being held for 8 hr at 40°, and the viscosity is proportional to the methoxyl content of the gum.[4]

Viscosity versus gum concentration curves all show a sharp, almost logarithmic rise with increasing gum content (Fig. 1). Viscosity decreases with increasing rate of shear (Fig. 2). Increasing concentration of alcohol and most organic solvents depress solution viscosity. The very high shear rates used in dissolving gum tragacanth greatly reduce the viscosity.

An important property of gum tragacanth is its ability to produce solutions with high viscosity. A 1% solution of high-grade gum has a viscosity of 3600 cps at 60 rpm using a Brookfield viscometer. At 25°, the solution viscosity reaches a maximum at about 24 hr.[8] This maximum may be obtained in about 2 hr at 50°. A thick gel is produced at 2–4% concentration. Gum tragacanth is an effective emulsifying agent because it delays coalescence of oil globules by increasing the viscosity of the external phase and thus slows down movement of the dispersed oil phase. It forms stable solutions as seen in Figure 3.

3. *Viscosity and pH*

The maximum initial viscosity of gum tragacanth solutions occurs at pH 8, but maximum stable viscosity is near pH 5.[13] Compared to other gums, traga-

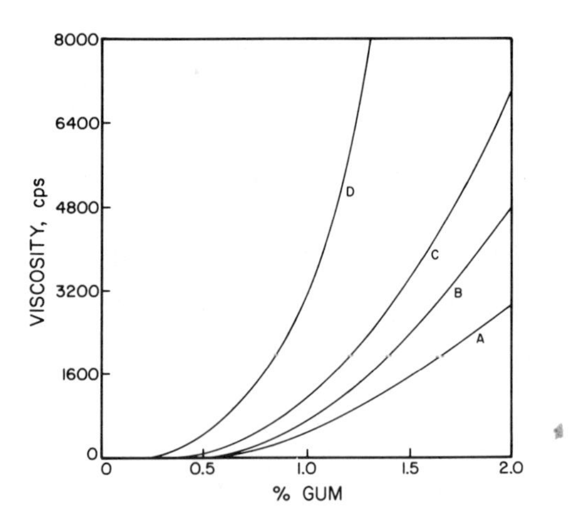

FIG. 1.—Viscosity versus gum tragacanth concentration for several gum types indicated by the letters A–D.

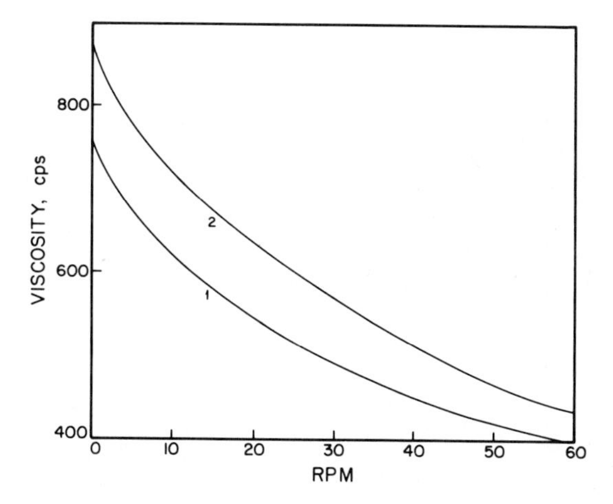

FIG. 2.—Effect of shear rate on viscosity. Curve 1, 1% flake gum tragacanth; 2, 0.5% ribbon gum tragacanth.

canth is quite stable over a wide pH range and is stable at about pH 2.[14, 15] The viscosities of different concentrations of gum in a 1% acetic acid solution are shown in Figures 4 and 5.

4. Preservatives

Preservatives are necessary for most gum solutions, and the choice will depend on the finished product and the formulation. Glycerol or propylene glycol, at 12 oz/gal (94 ml/liter) serve as excellent preservatives in many emulsions. Benzoic acid, chlorobutanol and a combination of the methyl and propyl esters of *p*-hydroxybenzoic acid are effective preservatives when used under the proper conditions.[16] A combination of 0.17% methyl and 0.03% propyl *p*-hydroxybenzoates give best preservation.

5. Infrared Spectra

Films suitable for infrared analysis may be made from all water-soluble gums. Gum tragacanth exhibits a strong carbonyl absorption at about 5.75μ.[17] Spectra of gum tragacanth, gum karaya, and pectin appear fairly similar, with slight but definite variations occurring at 7.5–8μ, 9–10μ, and 12μ. These differences are definite enough to serve as a means of identification and classification.

V. APPLICATIONS

Gum tragacanth is widely used in many industries because of its stability to

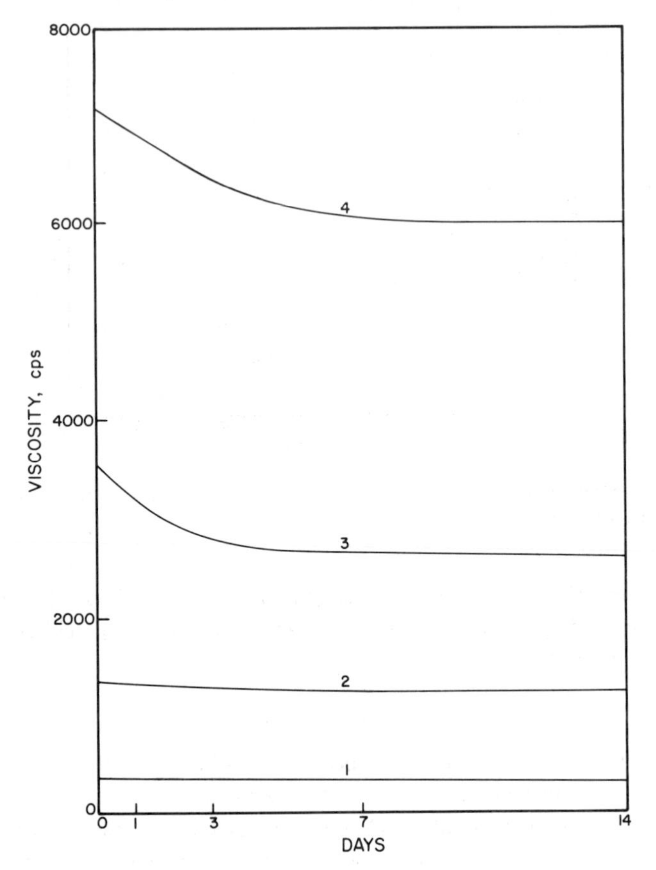

FIG. 3.—Viscosity stability of flake gum tragacanth in water. Curve 1, 0.75%; 2, 1%; 3, 1.5%; 4, 2%.

heat and acidity and because it is an effective emulsifying agent with an extremely long shelf life.

1. Pharmaceuticals and Medicinals

Gum tragacanth is an effective suspending agent for many pharmaceutical products. It generally acts to suspend the active ingredient by increasing the viscosity of the external phase and thus prevents undissolved material from settling out. Tragacanth is used in practice as a suspending agent in aqueous mixtures containing resinous tinctures and heavy insoluble powders. Glycerite of tragacanth is a useful excipient to bind tablet masses.

Mucilage of tragacanth is used in lotions for external applications. Gum tragacanth is also used at higher concentrations as a base for jelly lubricants.

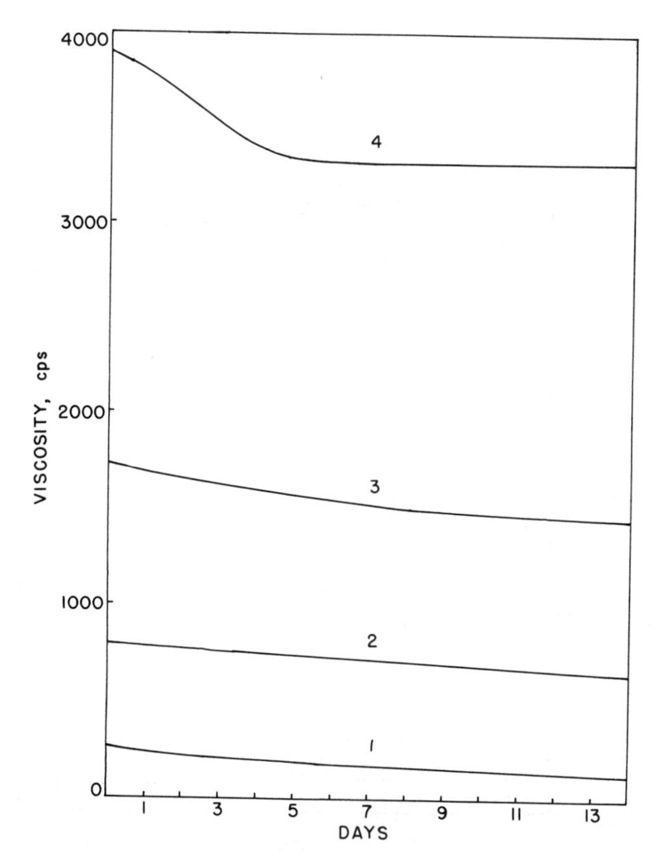

FIG. 4.—Viscosity stability of flake gum tragacanth, type I (curve B in Fig. 1), in 1% acetic acid; curve 1, 0.75% gum; curve 2, 1% gum; curve 3, 1.5% gum; curve 4, 2% gum.

One of the more important uses of gum tragacanth is in spermicidal jellies. Spermicidal jellies and creams act as chemical agents immobilizing spermatozoa with which they come into contact. Their thickening action also enables them to have an obstructive function.

In formulating a stable oral suspension of procaine penicillin, 0.2% w/v gum tragacanth has the best suspending qualities.[18] Gum tragacanth is used in medicinal oil emulsions, such as cod liver oil. As an emulsifier, it facilitates the absorption of poorly soluble substances, such as steroid glycosides and fat-soluble vitamins. It is also used in various types of elixirs and syrups where low-calorie intake is required.

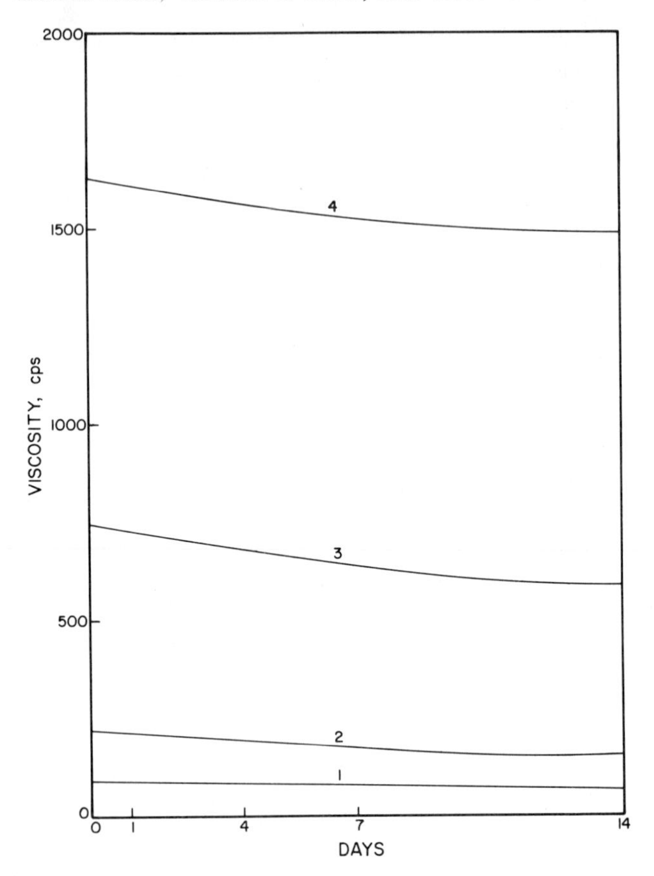

FIG. 5.—Viscosity stability of ribbon gum tragacanth, type E–1 (curve D in Fig. 1), in 1% acetic acid: curve 1, 0.5% gum; curve 2, 0.75% gum; curve 3, 1% gum; curve 4, 1.25% gum.

2. Foods

Gum tragacanth is widely used in the preparation of salad dressings; relishes; sauces; condiment bases; sweet pickle liquors; soft jellied products, such as gefilte fish; thick broths; beverage and bakery emulsions; ices and sherbets; bakery toppings and fillings; and confectionary cream centers.[19]

Salad dressings.—Because of its acid resistance and its long shelf life, gum tragacanth is useful in preparing stabilized French, Italian, Roquefort, and other creamy dressings. These are generally considered to be the pourable type and a few, such as the French and Italian types, are covered by standards of identity and must contain not less than 35% vegetable oil. The gum acts to thicken the water phase and prevents the oil phase from coalescing. Generally 0.4–0.75%

gum, based on the total weight of dressing, is used.[20] The preferred procedure is to wet the gum with a small amount of oil to inhibit lumping and then to disperse the mixture in water with rapid agitation. After all the ingredients have been added, the mixture is heated to approximately 72° for 30 min. It is then homogenized in a colloid mill or another type of homogenizer.

Recently, there has been a shift to the preparation of these pourable dressings by a continuous process, using rigid process tolerances and techniques.[21] Low calorie dressings have become an important part of the pourable dressings market. In these, the oil content is 1–5% and the gum content is 0.5–1.2% of the total dressing weight. The action of the gum is similar to that in a standard dressing, but a higher use level is required because of the larger aqueous phase.

The advantage of gum tragacanth in dressings is that it forms a creamier, more natural looking dressing, has excellent shelf life, and has good refrigerator stability in the home after the dressing has been opened by the homemaker.

Condiments and sauces.—Condiments and sauces are important product groups in which acid stability and long shelf life are important. Vinegar is usually an important ingredient in them, and gum tragacanth acts both as an emulsion stabilizer and thickener of the aqueous phase for such ingredients as spice flavorings and natural flavor extracts. Condiments are made[22] by heating a smooth mixture of ingredients to boiling. The gum may be added to the condiment at the end of the boiling period, and the mixture may be cooled in a heat exchanger. Generally between 0.4 and 0.8% of gum based on total weight of sauce, barbecue sauce, or condiment is used.

Bakery emulsions and toppings.—Gum tragacanth is used to stabilize bakery emulsions and fillings in which suspended fruit, fruit purees, natural flavor extracts and other flavors are used. The gum forms a creamy filling with good shine and transparency and gives a long shelf life in conjunction with the fruit acids in the product. Recently,[23] gum tragacanth has been used in a fruit topping for frozen cheese cakes in which the whole fruit is suspended in a thick jelly giving clarity, brilliance, and improved texture. It is important to use a high-grade gum for such an application to give the fruit a natural and rich appearance. Gum tragacanth is used in frozen pie-fillings and has been suggested as a cold-process stabilizer for meringues.

Miscellaneous.—Gum tragacanth has been used as a stabilizer in ice cream mixes at concentrations of 0.2–0.35%,[24] giving smooth body and texture. A favorable characteristic in this application is the ability of tragacanth to maintain and increase its viscosity during heat processing. It is also used with frozen fruits that are to be suspended in the ice cream. It has been used as a stabilizer for water ices, ice pops, and sherbets, in concentrations of approximately 0.5%.

In ice pops, gum tragacanth prevents the syrup from separating from the ice matrix.

Gum tragacanth is used in cream centers of candies that contain natural fruit acids.[25] It has been found useful for the stabilization of vitamin C in aqueous solutions.[26] A patent has been issued using gum tragacanth for preserving milk.[27]

3. *Cosmetics*

Gum tragacanth acts as the suspending agent in various types of toothpastes with a humectant, such as glycerol or propylene glycol. It forms a creamy and brilliant product. Its long shelf life and its film-forming properties make it useful in hair lotions and hand creams and lotions.[28] Generally, 0.4–0.8% based on total weight is used in these applications.

4. *Polishes*

Gum tragacanth is useful in various types of polishes, such as furniture, floor, and auto polishes.

5. *Textiles*

Gum tragacanth is used in print pastes and sizes because of its good release properties. It is used for stiffening silks and crepes. The gum is also used in the dressing of leather and in the preparation of leather polishes.

6. *Ceramics*

Certain grades of gum tragacanth are useful as binding agents in ceramics because they contain a low ash content, and the gum acts to suspend the various materials in a mass prior to the firing of the ceramic in the furnace.

7. *Insecticides*

Gum tragacanth forms stable emulsions containing 50% insect repellant. They are as effective as pure repellant compounds against mosquitoes, mites, chiggers, ants, and certain fleas.

VI. Price

Prices of gum tragacanth vary over a rather large range because of the grades available. The cost of labor for harvesting the gum is an important part of its total cost. The price range per lb for 1968–1970 has been as follows:

Ribbon #1	$5.50–$6.50
Ribbon #2	5.00– 6.25
Ribbon #3	3.75– 5.00
Powdered high viscosity	3.50– 5.85
Powdered medium viscosity	2.00– 3.00
Powdered low viscosity	0.85– 2.00

VII. References

(1) C L. Mantell, "The Water-Soluble Gums," Reinhold Publishing Corp., New York, 1947.

(2) "The United States Pharmacopeia," Mack Publishing Co., Easton, Pa., 18th Ed., 1970.

(3) "Food Chemicals Codex," Pub. 1406, Natl. Acad. Sci.–Natl. Res. Council, Washington, D.C., 1st. Ed., 1966.

(4) J. M. Rowson, *J. Pharm. Pharmacol.,* **10,** 161 (1937).

(5) A. Hilger and W. E. Dreyfus, *Ber.,* **33,** 1178 (1900).

(6) J. A. Widtsoe and B. Tollens, *Ber.,* **33,** 132 (1900).

(7) J. K. N. Jones and F. Smith, *Advan. Carbohyd. Chem.,* **4,** 243 (1949).

(8) Meer Corporation, brochure on water-soluble gums, 1958.

(9) N. Graten and M. Karrholm, *J. Colloid Sci.,* **5,** 21 (1950).

(10) H. N. Bundesen and M. J. Martinels, *Milk Food Technol.,* **17,** 79 (1954).

(11) E. M. Frith and R. F. Tuckett, "Linear Polymers," Longmans Green, London, 1951.

(12) W. P. Chambers, *J. Pharm. Pharmacol.,* **1,** 103 (1949).

(13) T. W. Schwarz, G. Levy, and H. H. Kawagoe, *J. Amer. Pharm. Ass., Sci. Ed.,* **47,** 695 (1958).

(14) G. Levy and T. W. Schwarz, *J. Amer. Pharm. Ass., Sci. Ed.,* **47,** 451 (1958).

(15) G. Levy and T. W. Schwarz, *Drug Stand.,* **26,** 153 (1958).

(16) A. Taub, W. A. Meer, and L. W. Clausen, *J. Amer. Pharm. Ass., Sci. Ed.,* **47,** 235 (1958).

(17) Sadtler Research Laboratories, Inc., Philadelphia, Pa., Infrared Spectrogram #D958.

(18) G. R. Sabatini and J. J. Gulesich, *J. Amer. Pharm. Ass., Pract. Pharm. Ed.,* **17,** 806 (1956).

(19) M. Glicksman, "Gum Technology in the Food Industry," Academic Press, New York, 1969.

(20) Meer Corporation, Product Information Bulletin #10R.

(21) Anderson Clayton Foods, personal communication.

(22) J. R. Burrell, *Food Mfr.,* **33,** 10 (1958).

(23) Sara Lee and Co., Inc., personal communication.

(24) F. E. Potter and D. H. Williams, *Milk Plant Mon.,* **39,** No. 4, 76 (1950).

(25) C. M. Ferri, *Mfg. Confect.,* **39,** 37 (1959).

(26) A. Ali, B. Khan, and B. Ahmad, *Pakistan J. Sci. Res.,* **6,** 58 (1954); *Chem. Abstr.,* **49,** 3425 (1955).

(27) P. H. Perrin, Fr. Patent 860,210 (1941); *Chem. Abstr.,* **42,** 6471 (1948).

(28) G. Meer, W. A. Meer, and T. Gerard, *Amer. Perfum.,* **77,** No. 2, 34 (1962); No. 4, 49 (1962); No. 5, 49 (1962).

Seed Gums

CHAPTER XIV

GUAR GUM

ARTHUR M. GOLDSTEIN, EMIL N. ALTER, AND JAMES K. SEAMAN

Stein, Hall and Co., Inc., Long Island City, New York

I. INTRODUCTION

The consumption of guar gum has grown rapidly since its commercial introduction in 1953. This rapid growth is largely because of guaran's function not only in the traditional role of a viscosity builder for water systems, but also as a hydrogen-bonding, reagent-type chemical for such industries as mining and paper making. The gum, a galactomannan, is a nonionic polysaccharide.

II. Production

1. Source and Producing Areas

Guar gum is derived from the seed of the guar plant, *Cyanaposis tetragonolobus,* family Leguminosae. The guar plant is a pod-bearing, nitrogen-fixing legume. It has been grown for centuries in India and Pakistan, where it is one of the principal crops and is used as a food for both humans and animals. Guar was introduced into this country in the early 1900's. However, not until 1946 did its potential as a cash crop and its value in contributing to soil fertility arouse interest here.[1] Guar has been steadily increasing in popularity as both a rotation and a strip crop in the Vernon and Kenedy areas of Texas.

2. Agronomics

The guar plant needs very little surface water during its growing season and is, therefore, adaptable for growth in those semiarid regions in which less hardy crops perish. In India, the crop is harvested with hand labor, whereas in Texas, grain combines are used. It is essential that the crop be harvested before the first rain following the first frost. If it is not, many of the seeds wither and turn black, the yield of gum is lowered, and the gum is contaminated with black specks.

3. Purification

The guar seed is dicotyledonous, having a diameter of approximately ⅛ in (8 mm). The average composition of the various seed components is given in Table I. It is apparent from an examination of the data that to obtain relatively pure galactomannan, the endosperm must be separated from the hull and germ.

Hull removal.—The hull is loosened by water soaking and removed by grinding. The difference in hardness of the various seed components is utilized, and a purification is accomplished by multistage grinding and sifting.

Germ removal.—Differential grinding is used for germ separation. There

Table I

Composition of the Components of Guar Seed

Seed Part	Protein (N × 6.25), %	Ether Extract, %	Ash, %	Moisture, %	Crude Fiber, %	Type of Carbohydrate
Hull (14–17%)	5	0.3	4	10	36.0	D-Glucose
Endosperm (35–42%)	5	0.6	0.6	10	1.5	Galactomannan
Germ (43–47%)	55.3	5.2	4.6	10	18.0	D-Glucose

are many types of grinders, such as attrition mills and special types of hammer and roller mills, that can be employed because of the difference in hardness of the endosperm and germ.

Endosperm grinding.—After the endosperm is separated from the hull and germ, it is ground to a fine particle size and marketed as guar gum.

4. Grades

Food-grade guar gum is substantially pure endosperm. It usually has a small residue of hull and germ owing to imperfect purification. However, because the entire seed is edible, this contamination only dilutes the amount of available galactomannan. Guar gum is included in the Food and Drug Administration's standards of identity for cheeses and cheese products,[2] frozen desserts,[3] and salad dressings.[4]

A typical analysis of the impurities in food-grade guar gum is as follows: crude fiber, 2.5%; moisture, 10–15%; protein (nitrogen $\times$ 6.25), 5–6%; ash, 0.5–0.8%. A comparison of this analysis with that of pure endosperm (Table I) shows the effectiveness of the separation as done by the usual commercial processing techniques.

Most guar gums marketed to the food industry produce viscosities of 3000–5000 cps in a 1% solution. Viscosity is not only influenced by purity but by processing techniques.

Industrial grades of guar gum are compounded with chemicals that aid in dispersion, control viscosity, cause gelling, act as preservatives, or in any other way aid in achieving specific results for a particular industrial use. In addition, such properties as rate of hydration, controlled gelation, and viscosity breakdown can be varied. These properties can be controlled by processing techniques and, therefore, specific grades of guar gum can be tailor-made for special applications. Many grades and modifications are commercially available.

5. *Quantities Marketed and Price Variation*

Guar gum has been in commercial production since 1953. An estimate of the United States' consumption since 1953 is shown in Table II. These figures do not include the amounts processed and sold abroad. The price is between $0.30 and $0.46 a pound, depending on grade, modification and formulation and has been stable over the past 6 years.

6. *Potential Amount*

There is no practical limit to the potential amount of guar gum that could be made available if the demand rises. The guar plant is an annual plant that is hardy and can be easily grown in many areas. An increasing market could

TABLE II

Estimated Consumption of Guar Gum in the United States

Year	lb
1954	2,500,000
1955	5,500,000
1956	12,000,000
1957	15,000,000
1960	20,000,000
1965	25,000,000
1969	30,000,000
1970	33,000,000
1971	33,000,000

even be accommodated by existing crops as the percentage of the total crops of India and Pakistan that are used for gum production is still very small.

III. USES

1. History

In this country, guar was first investigated as a source of gum in 1945. General Mills undertook agricultural production of the plant in the Southwest, and by 1949, 1700 acres had been planted in Texas. At the same time the Institute of Paper Chemistry in Appleton, Wisconsin, was comparing the use of guar gum with that of locust bean gum and other beater and size press additives in paper manufacturing.[1] However, it was not until 1953 that the gum was produced in commercial quantities. During this intervening period, the demand for locust bean gum grew rapidly because of the food, paper, and textile industries. By 1951, the demand for locust bean gum exceeded its limited supply, forcing its price to new highs. (Locust bean gum is harvested from perennial trees. See Chapt. XV.) Two American companies, Stein, Hall & Co., Inc. and General Mills, Inc., built plants in the United States and began commercial production of guar gum in the middle of 1953. Some time later, the European processors of locust bean gum also began to produce guar gum.

Guar gum was originally developed as a replacement for locust bean gum. However, it soon became obvious that, although both gums are galactomannans, there are significant differences in their chemical composition and behavior. Locust bean gum requires cooking at elevated temperatures to achieve its maximum viscosity, whereas guar gum will hydrate in cold water. Locust bean gum has an average of only one D-galactopyranosyl unit as a branch on every fourth D-mannopyranosyl unit. Guar gum has a D-galactopyranosyl branch on every other D-mannopyranosyl unit.[5,6] The greater number of branches in guar gum is responsible for both its easier hydration and its different hydrogen-bonding activity as compared to that of locust bean gum. Some industrial applications will be described in order to indicate both the conventional uses of the hydrophilic colloid as a viscosity builder and water binder and also its specialized uses, which depend on chemical composition and molecular configuration.

2. *Mining Industry*

Guar gum is used in froth flotation of potash as an auxiliary reagent, depressing the gangue minerals, which might be clay, talc, or shale.[7,8]

This action of guar gum results from the hydrogen-bonding of the gum molecule with the hydrated mineral surfaces of the clays. The over-all straight-chain configuration of the molecule with its regularly occurring D-galactopyranosyl branches makes it well suited to this type of reaction. The fact that both the D-mannopyranosyl and D-galactopyranosyl units of guaran have cis hydroxyl groups is also thought to promote more efficient hydrogen bonding.

Settling agent.—Guar gum is also used as a flocculant or settling agent to concentrate ores and tailings in the mining industry. In practice, slimes and tailings from ore beneficiation processes must be concentrated so that the water can be reused and the solids easily disposed of or reprocessed. Normally, large tanks and thickeners are used to effect this settling. By adding guar gum at 0.1–0.5 lb/ton of dry solids (1 kg/ 4100–20,500 kg of dry solids) faster settling of the suspended solids can be achieved. This reduces the amount of thickener required to settle a given amount of pulp solids.

Filter aid.—In mining, filtration operations are designed to remove suspended slimes or clay particles. These small particles have a tendency to form a very compact filter cake that traps water within itself and plugs the filter surface. The addition of guar gum to a pulp, by methods similar to those described above, results in the flocculation of these small particles. Twice as much guar gum is used in this connection as is required for settling operations. These large flocs no longer have a tendency to blind the filter screen and, at the same time, they allow a quicker channeling of liquid through the filter cake. This action increases filtration efficiency and results in a drier cake.

Water treatment.—Guar gum is nontoxic and approved by the U.S. Public Health Service for use in potable water treatment as a coagulant aid in conjunction with such coagulants as alum (potassium aluminum sulfate), iron(III) sulfate, and lime (calcium oxide). It increases the size of floc initially formed by the coagulant, thereby increasing the rate of settling of solid impurities, reducing solids carryover to the filters, and increasing periods between backwashes. In industrial waters, guar gum flocculates clays, carbonates, hydroxides and silica when used alone or in conjunction with inorganic coagulants.

3. Foods

Guar gum is often used in foods as a thickener and a binder of free water in sauces and salad dressings.

Guar gum is also used as a binder of free water and as a stabilizer in ice cream.[9] Free water in an ice cream mix causes a grainy texture, ice crystals, poor meltdown properties, and poor heat-shock resistance in the finished ice cream. The incorporation of a stabilizer containing guar gum in quantities of $\sim 0.3\%$ of the ice cream mix yields a smooth-textured, chewy product with slow meltdown properties and good heat-shock resistance. Guar gum binds the free water without disturbing the viscosity characteristics of the mix. It is also particularly suitable for use in flash pasteurization because of its rapid hydration properties.[10] It is included in the ice cream standards[3] and the Food Chemicals Codex.[11]

This gum is used in the stabilization of ice pops and sherbets. The principle of use here is the same as that which applies to ice cream. The retention of maximum hydration properties in acidic media is an important factor in this use of guar gum. Other foods that can be stabilized with guar gum because of its ability to bind water are frozen foods, cheeses, pie fillings, icings, and dog foods.

4. Cosmetics and Pharmaceuticals

Guar gum is used to thicken various cosmetics and pharmaceuticals.[12] It is also used as a binder and distintegrator for compressed tablets.[13] Coarse granulations of guar gum are ordinarily used for this purpose. Guar gum in coarse granulations is being investigated as a bulk laxative.

5. Paper Industry

The paper industry has become the largest user of galactomannans. It is estimated that about 22 million pounds (10 million kilograms) of galactomannans (locust bean gum and guar gum) are used annually by this industry, with guar accounting for a major portion.

Wet end additive.—The major use of galactomannan in paper making is in the wet end of the process. This means that the gum is added to the pulp suspension just before the sheet is formed on either a Fourdrinier machine or a cylinder machine. The pulping process, which is designed to remove lignin and thereby produce a fibrous cellulosic pulp, also removes a large part of the hemicelluloses normally present in the wood. These hemicelluloses, which are mostly mannans and xylans, could contribute greatly to the hydration properties of the pulp and the strength of the paper formed from the pulp. Galactomannans replace or supplement the natural hemicelluloses in paper bonding. Advantages gained by addition of galactomannans to pulp are: (*a*) improved sheet formation with a more regular distribution of pulp fibers (less fiber bundles); (*b*) increased Mullen bursting strength; (*c*) increased fold strength; (*d*) increased tensile strength; (*e*) increased pick (in printing grades, pick is a measure of the force required to pull a fiber from the surface of a sheet; (*f*) easier pulp hydration (the pulp is normally passed through jordans or wet mills which increase the surface of the fiber, thereby allowing it to bind more water; galactomannans added to a pulp aid in the binding of water and decrease the amount of jordaning necessary, consequently lowering power consumption); (*g*) improved finish (as measured by the smoothness and a count of protruding fibers); (*h*) decreased porosity; (*i*) increased flat crush of corrugating medium (this refers to the pressure required to crush a corrugated flute); (*j*) increased machine speed with maintenance of test results; (*k*) increased retention of fines.

It is generally agreed that the hydrogen-bonding effect is one of the major factors affecting fiber–fiber bonding.[14] An examination of the molecular structure of galactomannans reveals a rigid molecule with primary and secondary hydroxyl groups. It is difficult to compare the efficiency of locust bean gum and guar gum in paper products because too many variables exist, such as type of gum formulation, pulp type, beating time, and machine conditions. However, many in the paper industry believe that guar gum is the more efficient additive. One other property of guar gum makes its practical application in paper more acceptable. The grades of guar and locust bean gums used contain a small amount of hull as a contaminant. Locust bean hulls are red-brown, and in white sheets, these particles appear as small but visible specks. However, guar hull is almost white and, when present in the paper sheet, is indistinguishable from the bleached pulp.

Gum solutions are prepared by batch or continuous cooking. After preparation, the solution is usually pumped to the suction side of the fan pump. For mills that do not wish to install cooking equipment, there are specially processed guar gums that can be added directly to the beater.

6. *Explosives*

In the production of a water-resistant ammonium nitrate stick explosive,[15] guar gum is added as a waterproofing agent. The ability of guar gum to hydrate in saturated solutions of ammonium nitrate has resulted in the large scale use of guar as a thickener and gelling agent for slurry explosives. Many patents describing the use of guar gum for this application have been granted.[16–22]

7. *Tobacco*

Guar gum has found use as a binder for fragmented tobacco fines in the production of reconstituted tobacco for smoking products.[23] In this process, a moistened blend of high-quality tobacco fines, guar gum, and humectant is passed between closely spaced steel rollers revolving at different peripheral speeds. The sheet is then partially dried. These flexible sheets with the tensile strength and thickness of leaf tobacco can be blended with leaf tobacco and still retain the taste, flavor and aroma of an all-leaf product.

8. *Derivatives*

Many derivatives of guar gum have been prepared and reported in the literature. These include oxidized guar gum, carboxymethylated guar gum, hydroxyalkylated guar gums, and many of the other typical carbohydrate modifications. Some of these derivatives have found important uses in the paper and textile industries.

In the last few years hydroxyalkylguar has been assuming commercial importance. Hydroxyethyl- and hydroxypropyl-guar materials varying in viscosity and in molar substitution have become available. Derivatization through the use of alkylene oxides has modified the properties of guar for certain useful areas.

Regular guar, for example, is rapidly adsorbed onto hydrated mineral surfaces by hydrogen bonding to cause flocculation. By the introduction of hydroxyalkyl side chains, adsorption rates are modified. In this way, a slightly reduced flocculation rate to almost no flocculation can be achieved for a wide range of mining applications.

Similarly, guar interacts with hydrated cellulosic surfaces to affect the formation and strength of a sheet during paper production. By hydroxyalkylation of guar, adsorption rates onto hydrated cellulose surfaces are altered for certain useful paper making applications.

One of the novel improvements attainable through hydroxyalkylation of guar is solubility in water-miscible solvents such as glycols and alcohol. Solvent miscibility enables hydroxyalkylguar derivatives to be used as thickeners for certain slurry explosives.

Electrolyte compatibility, a valuable property of guar, has been extended by hydroxyalkylation to cover saturated calcium salt solutions. This is particularly useful in applications such as oil well fracturing and aqueous slurry explosives.

Hydroxyalkylguar derivatives have a significantly lower BOD than modified guar, yet are more easily biogradable than cellulose ethers. This intermediate degree of biodegradability suggests that hydroxyalkyl-guar is suited to textile thickening and sizing operations.

IV. Structure

Methylation, fragmentation, and periodate oxidation show that guar gum (guaran) consists of linear chains of (1→4)-β-D-mannopyranosyl units with α-D-galactopyranosyl units attached by (1→6) linkages.[24–26] Enzymic hydrolysis of guaran gives mannobiose (4-*O*-β-D-mannopyranosyl-D-mannose), mannotriose, and 6-*O*-α-D-galactopyranosyl-D-mannopyranose, confirming earlier methylation results and showing that β-D-(1→4) linkages are present in the mannan and that side chain D-galactose units are attached by α-D-(1→6) linkages.[27-29]

The ratio of D-galactose to D-mannose in guar gum is 1 : 2, with single D-galactopyranosyl unit side chains attached to every other D-mannopyranosyl unit[30] (Fig. 1). The molecular weight has been reported[31-32] as 220,000.

V. Properties

1. *Viscosity*

Guar gum forms viscous, colloidal dispersions (solutions) when hydrated in cold water; viscosities can be measured with a rotational, shear-type viscometer, such as the Brookfield Syncro-Lectric or the Haake Rotovisco. Its solutions exhibit the variance of viscosity with shear rate which is typical of non-Newtonian fluids. Figures 2 and 3 illustrate the change in viscosity of a 1% food-grade guar gum solution with shear rate at 25° as measured with a Brookfield Synchro-Lectric and Haake Rotovisco, respectively. More dilute solutions of guar gum show less pseudoplastic response. Viscosities of 0.3% solutions change only slightly with increasing shear up to 60 rpm (Fig. 4).

Viscosity of guar gum solutions can also be measured with pipette-type instruments, such as the Dudley, Engler, and Ostwald viscometers. These instruments are most useful for measuring gum solutions at concentrations below 1000 ppm (mg/liter). Extrusion rheometers are suitable for measuring high viscosity ranges at concentrations above 2%.

To prepare guar gum solutions, the dry gum is rapidly sifted into a vigorously agitated tank of water. When dispersion is complete, the agitation is reduced, but stirring is continued until the solution becomes viscous. The guar gum is

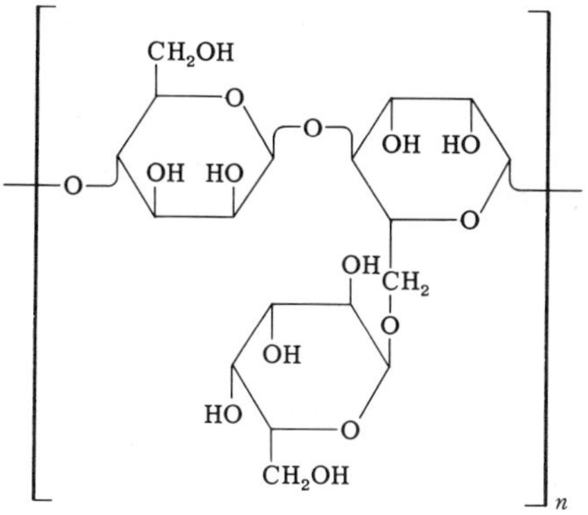

FIG. 1.—Structure proposed for guaran.[6]

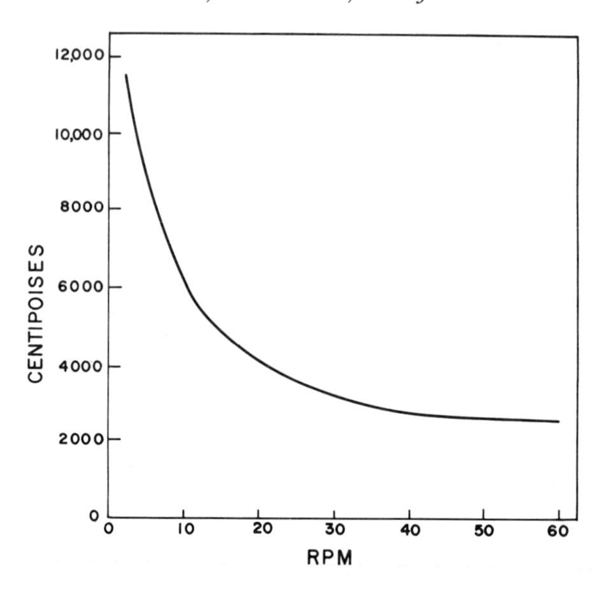

FIG. 2.—Viscosity versus shear rate in a 1% food-grade guar gum solution at 25° as measured with a Brookfield Synchro-Lectric viscometer.

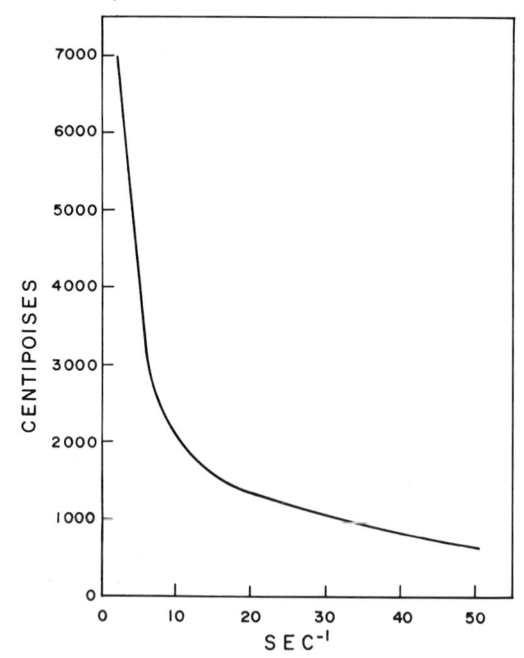

FIG. 3.—Viscosity versus shear rate in a 1% food-grade guar gum solution at 25° as measured with a Haake Rotovisco viscometer.

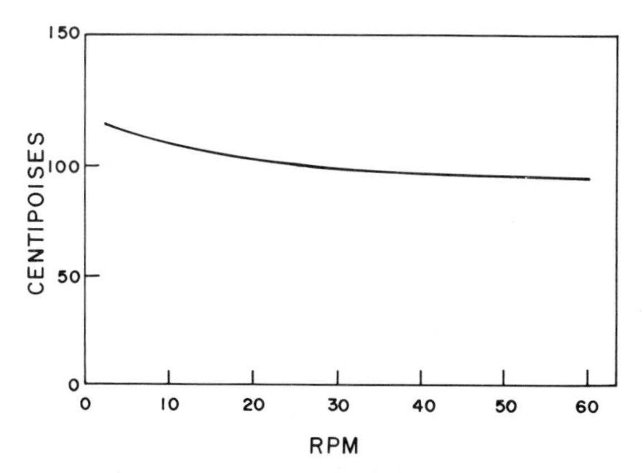

FIG. 4.—Viscosity versus shear rate in a 0.3% food-grade guar gum solution at 25° as measured with a Brookfield Synchro-Lectric viscometer.

permitted to hydrate and thicken for at least 5 min and preferably for 30 min. (Some grades of gum may continue to thicken for as much as 24 hr.) Viscosity is dependent on time, temperature, concentration, pH, ionic strength, and type of agitation.

Time.—The viscosity, in centipoises, of two different grades of guar gum in 1% solutions at 25°, measured as a function of time, is shown in Figure 5.

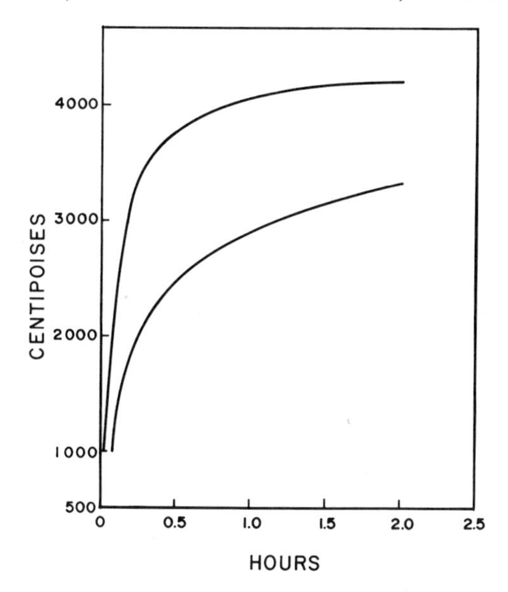

FIG. 5.—Change in viscosity with time of two different grades of guar gum. 1% dispersions at 25°. Brookfield Synchro-Lectric viscometer at 20 rpm.

To maintain viscosity over a period of time, preservatives are added to the solution in sufficient concentration to provide protection. Among the more effective preservatives are formaldehyde and chlorinated phenolics. Food preservatives, such as benzoic and sorbic acids, are also used where appropriate.

Temperature.—The viscosity of a fully hydrated 1% guar gum solution varies almost directly with changes in temperature over the range 20°–80° (Fig. 6). The maximum viscosities that can be obtained when dry guar gum is dispersed in water at various temperatures for periods up to 2 hr is shown in Figure 7. The factors influencing the rate of viscosity development are the increase in viscosity caused by short-term heating and the decrease in viscosity caused by the degradative effect of prolonged heat. After maintaining the desired temperatures for 2 hr, the solutions are immediatly brought to 25°. Viscosities are again measured at the latter temperature after 24 hr. Guar gum exhibits good stability toward prolonged heat for periods up to 2 hr at 80°. Maximum viscosity of the guar gum dispersion is achieved with temperatures of 25°–40°. The lower the temperature, the lower the rate at which the viscosity increases.

Concentration.—In dilute solutions, the viscosity of guar gum increases linearly with concentration up to about 0.5%. Thereafter, guar gum solutions

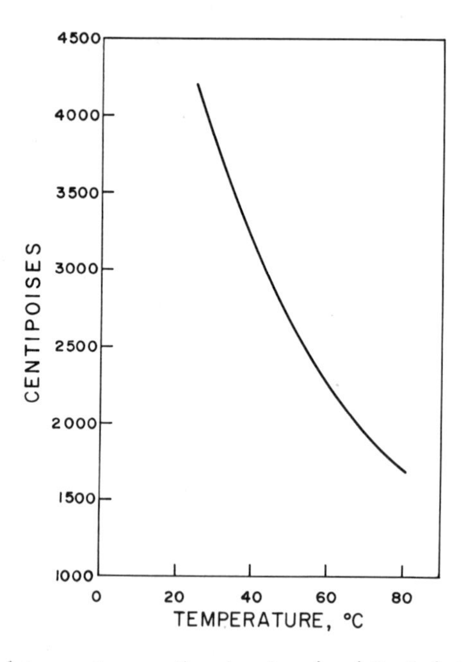

FIG. 6.—Effect of temperature on the viscosity of a fully hydrated, 1% guar gum solution as measured with a Brookfield Synchro-Lectric viscometer at 20 rpm. The rate of temperature increase is 2°/min.

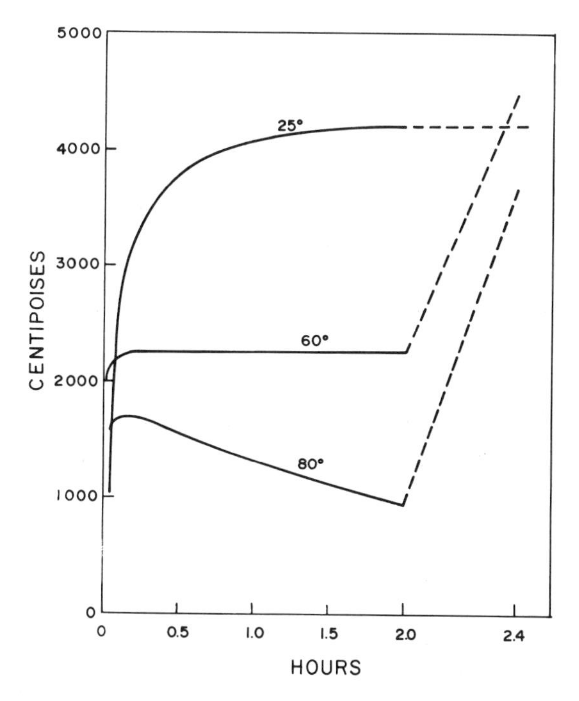

FIG. 7.—Viscosity variation with time of a 1% food-grade guar gum dispersion maintained at constant temperature as measured with a Brookfield Synchro-Lectric viscometer at 20 rpm. After 2 hr, all dispersions were cooled to 25°.

behave as non-Newtonian solutions mainly as a result of the complex surface attractions at higher concentation (Fig. 8). Table III compares the viscosity-imparting ability of various colloids that were measured in a similar manner.

pH.—Guar gum is stable over a wide pH range. The nonionic nature of the molecule is responsible for the almost constant viscosity of solutions in the pH range 1–10.5. However, previously hydrated guar gum is compatible with high alkalinity; and modified guar gums that develop high viscosity in concentrated caustic solutions have recently become available. The optimum rate of hydration for guar gum occurs between pH 7.5 and 9. Despite the rate differences, maximum viscosities are the same in both acidic and alkaline media.

Ionic strength.—Because guar gum is nonionic, compatibility with salts is exhibited over a wide range of electrolyte concentration (Fig. 9). High concentrations of multivalent salts affect hydration and produce gels.

2. *Friction Reduction*

Guar gum, with its long, straight-chain configuration, reduces and minimizes

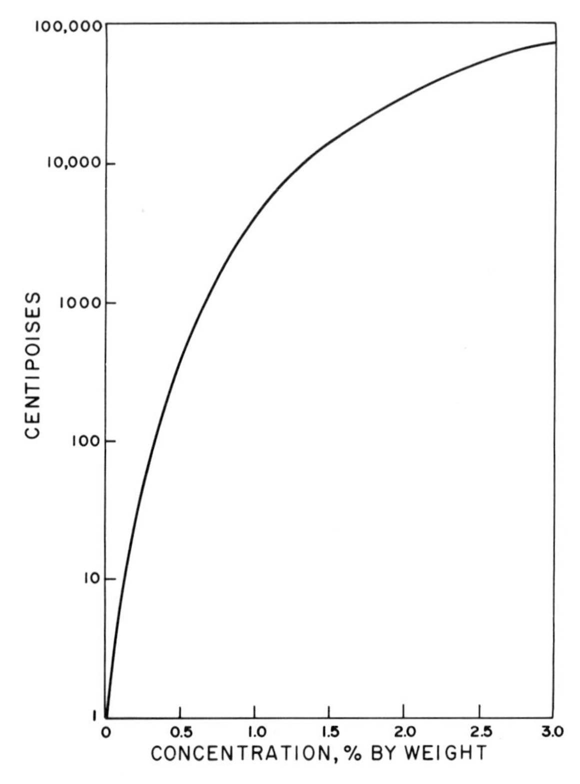

FIG. 8.—Viscosity as a function of concentration of food-grade guar gum hydrated for 24 hr at 25° as measured with a Brookfield Synchro-Lectric viscometer at 20 rpm.

TABLE III

Comparative Viscosities of 1% Gum Solutions Hydrated at 25°

Gum	cps
Gum arabic (20% by wt)	50
Locust bean gum	100
Methylcellulose	150
Gum tragacanth	200
Carrageenan	300
High-viscosity sodium carboxymethylcellulose	1200
Gum karaya	1500
Sodium alginate	2000
Guar gum	4200

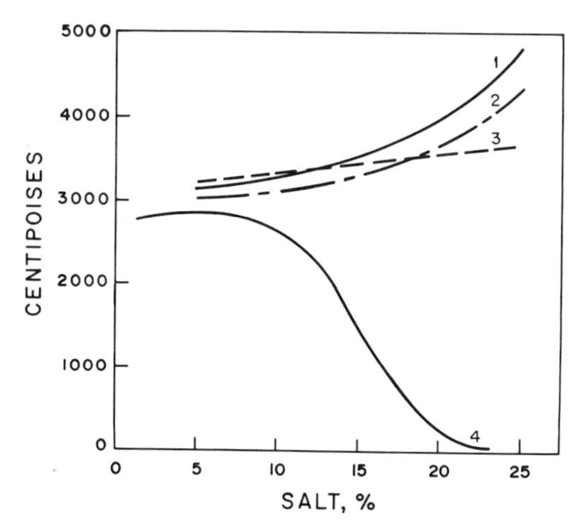

FIG. 9.—Effect of salts on the viscosity of 1% food-grade guar gum solutions held for 24 hr at 25° as measured with a Brookfield Synchro-Lectric viscometer at 20 rpm. Curve 1, aluminum chloride (pH 3); 2, calcium chloride (pH 5); 3, sodium chloride (pH 6); 4, aluminum sulfate (pH 3–4)

the frictional pressure losses of water when in turbulent flow (Fig. 10). Figure 11 shows the great resistance of guar to shear degradation when constantly recycled.

Friction reduction efficiency is conveniently expressed as the percentage of pressure loss that has been eliminated by the presence of the gum.

$$\%\ \text{friction reduction} = \frac{\Delta P_{H_2O} - \Delta P_{\text{gum soln.}}}{\Delta P_{H_2O}} \times 100$$

3. *Gels*

Borate ion acts as a crosslinking agent with hydrated guar gum to form cohesive structural gels. The formation and strength of these gels are dependent on the pH, temperature, and concentrations of reactants. The optimum pH range for gel formation is 7.5–10.5. The solution–gel transformation is reversible; the gel can be liquified by dropping the pH below 7 or by heating. By the former method, a normal solution is obtained and by the latter, a stringy solution. Polysaccharides with numerous adjacent hydroxyl groups in the cis position can form these three-dimensional borated gels.[33] Borated gels can also be liquified by the addition of a low-molecular-weight polyol, such as glycerol or mannitol, capable of reaction with the borate ion.

Borate ion will inhibit the hydration of guar gum if it is present at the time the powdered gum is added to water. The minimum concentrations that

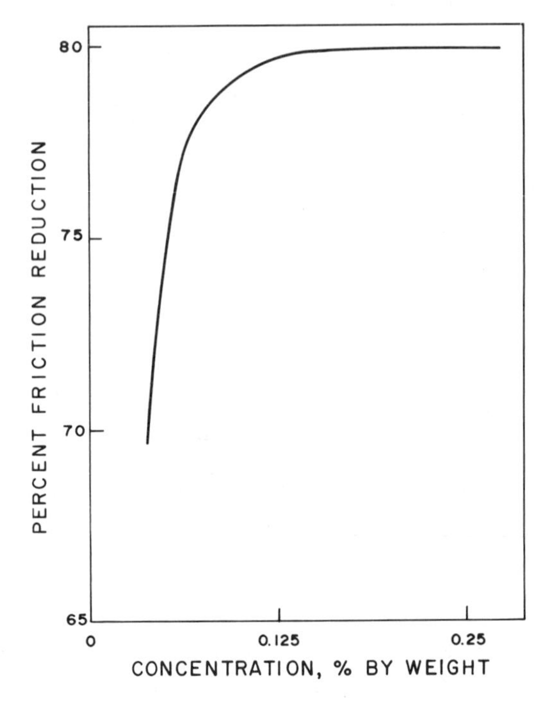

FIG. 10.—Effect of guar gum concentration on the percent friction reduction in a 0.269 in (6.83 mm) id test pipe at a linear velocity of 50–60 ft/sec (15–18 m/sec).

are necessary to inhibit hydration are dependent on pH. For example, with 1% of guar gum, 0.25–0.5% (based on the guar gum weight) of borax (sodium tetraborate) is needed at pH 10–10.5; at pH 7.5–8.0, 1.5–2.0% of borax is required. The complexing reaction is reversible because decreasing the pH below 7 permits the gum to hydrate normally. This technique is often used for better mixing and easier dispersion.

Guar gum can be insolubilized or gelled by transition metal ions to form commercially useful gels.[34] These gels are pH dependent and not reversible, unlike the borate ion gel. Gel characteristics vary widely with the specific cation being used.

4. *Films*

Guar gum forms tough, pliable films that are sensitive to water. Palmer and Ballantyne[35] described a method for producing highly oriented films that are elongated 150%. Smart and Whistler[36] examined the triacetate derivative. The films produced from the triacetate are colorless and transparent and have a

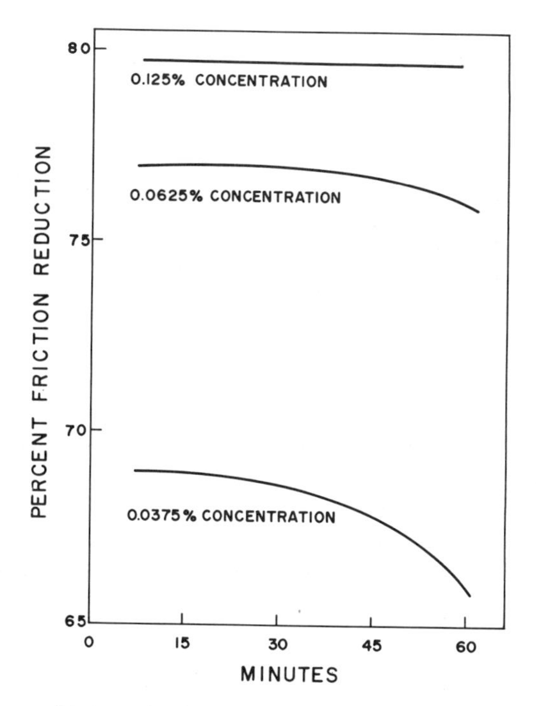

FIG. 11.—Percent friction reduction versus time in a closed circulating system in a 0.269 in (6.83 mm) id test pipe at a linear velocity of 50–60 ft/sec (16–18 m/sec).

tensile strength that is slightly lower than that of films of cellulose triacetate or potato amylose triacetate. Its films can be plasticized to make them soft and pliable.

5. *Miscellaneous*

Some industrial grades of guar gum are prepared so that they have a temporarily inhibited hydration in acidic media (Fig. 12). Viscosity of the dispersion begins to develop after 10–15 min. This time lag enables the gum to disperse uniformly. The viscosity then increases in a normal manner. Food-grade products can be prepared with improved dispersibility by agglomerating fine-mesh powder into a coarse, granular form.

VI. REFERENCES

(1) R. L. Whistler, *Chem. Ind.* (New York), **62,** 60 (1948).

(2) "Service and Regulatory Announcements, Food and Drug Administration, Foods, Drugs and Cosmetics," Title No. 2, Part 19, Federal Security Agency (June 1952).

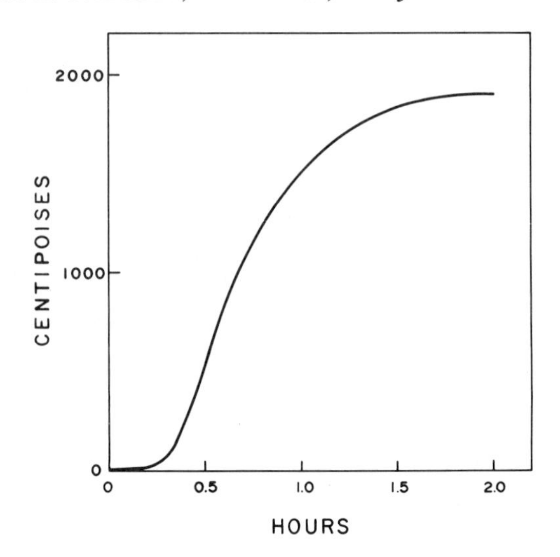

FIG. 12.—Viscosity versus time for a 1% dispersion of a special grade of guar gum that has temporarily inhibited hydration as measured with a Brookfield Synchro-Lectric viscometer at 20 rpm and 25°.

(3) "Service and Regulatory Announcements, Food and Drug Administration, Foods, Drugs and Cosmetics," Title No. 2, Part 20, Federal Security Agency (March 1958).

(4) "Service and Regulatory Announcements, Food and Drug Administration, Foods, Drugs and Cosmetics," Title No. 2, Part 25, Federal Security Agency (June 1952).

(5) Z. F. Ahmed and R. L. Whistler, *J. Amer. Chem. Soc.*, **72,** 2524 (1950).

(6) R. L. Whistler and C. L. Smart, "Polysaccharide Chemistry," Academic Press, New York, 1953, pp. 296–299.

(7) G. E. Atwood and D. J. Bourne, U.S. Patents 2,696,912 (1954); *Chem. Abstr.*, **49,** 4246 (1955).

(8) G. E. Atwood and D. J. Bourne, *Mining Eng.*, **5,** 1099 (1953).

(9) S. J. Werbin, U.S. Patent 2,502,397 (1950); *Chem. Abstr.*, **44,** 5496 (1950).

(10) R. J. McKiernan, paper presented at the Michigan Dairy Manufacturers Annual Conference, East Lansing, Mich., Nov. 7, 1957.

(11) "Food Chemicals Codex," Publ. 1406, Natl. Acad. Sci.–Natl. Res. Council, Washington, D.C., 1st Ed., 1966, p. 315.

(12) H. H. Hutchins and R. E. Singiser, *J. Amer. Pharm. Ass., Pract. Pharm. Ed.*, **16,** 226 (1955).

(13) L. E. Eatherton, P. E. Platz, and F. P. Cosgrove, *Drug Stand.*, **23,** 42 (1955).

(14) J. W. Swanson, *Tappi,* **39,** 257 (1956).

(15) W. J. Taylor, U.S. Patent 2,654,666 (1953); *Chem. Abstr.*, **48,** 3692 (1954).

(16) M. A. Cook and H. Farnam, U.S. Patent 2,930,685 (1960); *Chem. Abstr.*, **54,** 14689 (1960).

(17) R. Barnhart and F. Sawyer, U.S. Patent 3,072,509 (1963); *Chem. Abstr.*, **58,** 6641 (1963).

(18) J. McIrvin, U.S. Patent 3,108,917 (1963); *Chem. Abstr.*, **60,** 1533 (1964).

(19) H. Clarlevado, U.S. Patent 3,129,126 (1964); *Chem. Abstr.,* **61,** 4143 (1964).
(20) J. S. Logan and J. A. Zalowsky, U.S. Patent 3,214,307 (1965); *Chem. Abstr.,* **63,** 17791 (1965).
(21) J. D. Ferguson, U.S. Patent 3,235,423 (1966); *Chem. Abstr.,* **64,** 12455 (1966).
(22) W. Lyerly, U. S. Patent 3,355,336 (1967).
(23) M. M. Samfield, B. A. Brock, and E. E. Locklair, U.S. Patent 2,708,175 (1955); *Chem. Abstr.,* **49,** 11248 (1955).
(24) E. L. Hirst and J. K. N. Jones, *J. Chem. Soc.,* 1278 (1948).
(25) R. L. Whistler and D. F. Durso, *J. Amer. Chem. Soc.,* **73,** 4189 (1951).
(26) R. L. Whistler and D. F. Durso, *J. Amer. Chem. Soc.,* **74,** 5140 (1952).
(27) R. L. Whistler, T. K. Li, and W. Dvonch, *J. Amer. Chem. Soc.,* **70,** 3144 (1948).
(28) R. L. Whistler and J. S. Stein, *J. Amer. Chem. Soc.,* **73,** 4187 (1951).
(29) R. L. Wistler and C. G. Smith, *J. Amer Chem. Soc.,* **74,** 3795 (1952).
(30) E. Heyne and R. L. Whistler, *J. Amer. Chem. Soc.,* **70,** 2249 (1948).
(31) A. D. Boggs, M. S. Thesis, Purdue University, Lafayette, Indiana, 1949.
(32) J. W. Hoyt, *J. Polymer Sci., Part B,* **4,** 713 (1966).
(33) H. Deuel, H. Neukom, and F. Weber, *Nature,* **161,** 96 (1948).
(34) J. D. Chrisp, U.S. Patent 3,301,723 (1967); *Chem. Abstr.,* **67,** 92485w (1967).
(35) K. J. Palmer and M. Ballantyne, *J. Amer. Chem. Soc.,* **72,** 736 (1950).
(36) C. L. Smart and R. L. Whistler, *J. Polymer Sci.,* **4,** 87 (1949).

CHAPTER XV

LOCUST BEAN GUM

FRANCO ROL

Cesalpinia S. p. A., Bergamo, Italy
(Subsidiary of Hercules, Inc., Wilmington, Delaware)

I. INTRODUCTION

Locust bean or carob, known botanically as *Ceratonia siliqua,* belongs to the family of Leguminosae, subfamily Caesalpiniaceae. The carob tree takes root and is cultivated easily but is sensitive to low temperatures. A typical fruit of the Mediterranean area, the carob is not found elsewhere.

The name *carob* probably originates from the Hebrew, in which language it is called *kharuv,* a noun that may derive from *kherev* meaning *saber.* The carob seed was used by the Arabs as a unit of weight. They called the seed *karat,* and the standard weight of the carob seed became the unit of weight for precious stones, that is, *carat.* The carob fruit, which has even been immortalized in the Bible as the food of St. John the Baptist, is still called *Johannisbrot* in the German language and is known also in English as St. John's Bread.

II. Production

1. Source

The carob is an evergreen tree growing up to 30 ft (9 m) in height and has luxuriant perennial foliage. The tree grows on rocky soil unfit for most other trees, has very long roots that can penerate 60–90 ft (18–27 m), and can thrive even in regions with very little rainfall. It takes 15 years before the tree starts producing the fruit regularly and 50 years before the tree is fully grown.

The flowers, which have petals and stamens with red-orange anthers, grow in clusters on the branches in the midst of foliage. Pollination is done by insects; therefore, it is advisable to have beehives on the carob plantation. Flowering begins in the period from January to March, depending on the area, while the last fruits of the previous season are still on the tree.

The fruit of the locust bean tree, the very well-known locust bean or carob, is a dark, chocolate colored pod 4–8 in (10–20 cm) long and 0.75–2 in (1.9–5 cm) wide, which has a thickness of 0.25–0.375 in (65–95 mm) and a weight of 0.75–1.5 oz (20–40 g). In April or May, the fruit starts to grow, and in June or July, it reaches its maximum size but is still green. By September, October, or November, the fruit is ripe and has a dark-brown color. About 3 or 4 years after grafting, the locust bean tree begins to produce fruit. Sometimes there are gigantic trees that produce a ton of fruit. The locust bean tree can be planted, although it usually grows naturally.

The carob tree is subject to diseases such as those caused by the fungi *Oidium ceratoniae* and *Cercospora ceratoniae,* the latter being a leaf disease that occurs when there is high humidity.

2. Producing Areas

Spain: Catalonia (Tarragona), Valencia (Castellon Valencia), Murcia (Cartagena), Andaulsia (Malaga), Balearic Islands (Majorca, Menorca, Ibiza)
Italy: Sicily, Apulia, Calabria, Sardinia
Greece: Crete, Peleponnesus, Phthiotis
Cyprus: The whole island
Algeria: Oran, Bougie, Constantine, Algiers
Portugal: Algarve
Morocco: Casablanca, Essaouira, Agadir
Israel: Mostly in the area around Jerusalem
Turkey: Izmir, Mersin

The best quality locust bean kernels (seeds) come from Sicily, where the carob trees were probably planted about the 16th–17th century.

The world harvest for 1970 was estimated at between 32,000 and 35,000 tons.

3. *Seasonal Effects*

The tree grows well in hot, dry weather. Temperatures below freezing may seriously compromise the yield of the tree and can even destroy the crop. If the weather is cold and the sun insufficient, the fruit remains small. If rainfall is too abundant, the fruit grows large in size but contains only a small number of kernels. If the fruit does not ripen fully, it is difficult to kibble the pod.

4. *Collection*

Pods are harvested by shaking the twigs with poles and are then picked up from the earth by hand or with rakes. Farmers bring the fruit to the kibblers in bulk.

5. *Purification*

The pods are cleaned, then crushed and kibbled in kibbling machines. A pod consists of pulp, 90%; kernels, 8%; and other materials, 2%. The kernel is composed of husk, 30–33%; germ 23–25%; and endosperm, 42–46%.

The percentage of seeds contained in the pod as well as the gum portion (endosperm) contained in a seed depends upon the quality and origin of the crop. The dicotyledonous seed contains a central hard, yellow embryo surrounded by a corneous and comparatively large layer of white, semitransparent endosperm. This is the desired gum. The endosperm in turn is surrounded by a tenacious, dark brown husk.

Successful production of the gum (especially high-grade locust bean gum) involves separation of the endosperm from the embryo. The color of the gum (white and with no brown specks) depends on efficient removal of the outer husk either by chemical attack or by mechanical abrasion. In 1954, Cesalpinia invented an improved system for the decortication and degermination of seeds.

The dehusked or decorticated seeds are split lengthwise, and the embryo is separated from the gum. The presence of yellow germ (embryo) in the gum increases the rate of fermentation of gum solutions. It is, therefore, important to have very clean endosperm. Finally, the gum is ground.

6. *Grades*

There are four main grades of locust bean gum: top high-grade, normal high-grade, industrial grade, and technical grade.

Italian manufacturers have placed on the market a new type of top high-grade gum of 3600–4000 cps. High-grade locust bean gum is produced in four standard meshes, whereas industrial and technical grades are produced in only one standard mesh. Mesh sizes of top high-grade and normal high-grade locust bean gum are 100 (rarely used), 150, 175, 200 and 250 (rarely used). The color of high-grade gums is determined by eye because there are no standards for whiteness of the gum.

7. *Impurities*

Commercial high-grade locust bean gums contain virtually no specks. Natural moisture of locust bean gum is about 14%. Other substances, such as germ particles, should not exceed 8–10% in top high-grade locust bean gum.

8. *Quantities Marketed*

In 1970, the world production of locust bean gum was approximately 15,000 tons.

9. *Price Variation*

Average prices for high-grade locust bean gum in the United States have varied from $0.28 to $0.50 a pound ($0.62 to $1.11 per kg) in the years 1961 through 1970.

10. *Potential Amount*

The crop of carob pods depends upon weather conditions, and the availability of seeds depends upon the quality of the carob kibbled. A great problem exists in the industrial utilization of the pods. If there is no market for the kibbled carob, the farmers are not willing to break the pods. In fact, unkibbled carobs can be stored for one or even two years whereas kibbled carobs cannot be stored long because fermentation occurs if they become moist.

Under normal circumstances about 15,000 tons of locust bean gum can be produced in Europe now. There is still a considerable growth of wild carob trees. If a part of these wild trees were grafted, the availability of kernels could presumably be increased in some of the growing areas.

III. Uses

1. *History*

The history of the carob tree is lost in remote ages. Its cultivation was known many centuries before the Christian era. The carob fruit was used as a feed for horses, cattle, and swine and as food for poor people. In Palestine, carob wood was employed to make furniture and idols and was exported to Egypt as material for the construction of temples.

The ancient Egyptians prepared the strips with which they bound their mummies with carob paste (locust bean gum). Dioscorides referred to the curative properties of the fruit in the first century AD. Greek and Roman pharmacy made use of it in decoctions, syrupy extracts, laxatives, and similar remedial substances.

Much later on, the beans were used extensively for tobacco curing and were bought in Germany for roasting as a coffee substitute. Poor people in North Africa and other carob-growing areas still eat the pulp, which, because of its high sugar content (30–40%), is liked by children. During the revolution and civil war in Spain, 1936–1939, people living in the carob-growing areas of Southern Spain ate the fruit and overcame the difficult years in better health than the people of Northern Spain.

There has been industrial utilization of locust bean kernels for more than half a century. It was first started by two English firms, Tragasol of Hooton and Ellis Jones of Stockport, and an Italian firm, Cesalpinia of Milano. Locust bean gum is also marketed under other names, such as carob gum, gum *gatto,* gum *hevo, jandagum, lakee* gum, Rubigum, Lupogum, Luposol, gum Tragon, Tragarab, and Tragasol.

2. *Applications*

The important derivatives of locust bean gum are the water-soluble hydroxyethyl and carboxymethyl ethers. Locust bean gum and its derivatives are used today in the following industries.

Food industry.—Locust bean gum derivatives especially designed as ice cream stabilizers have stabilizing and swelling power and great ability to absorb water. They impart smooth meltdown and excellent heat-shock resistance. The derivatives are not affected by lactic acid or calcium salts. Therefore, mixes having a high content of lactic acid need not be neutralized before stabilization. Mixes stabilized with locust bean gum derivatives can be easily cooled because no increase in viscosity occurs on cooling coils. Mixes do not whey off, and they yield excellent overrun. The cost of stabilization is low, and ice cream stabilized with locust bean gum derivatives has smooth texture and slow, creamy meltdown with no masking of flavor.

Locust bean gum derivatives are used in the manufacture of pork specialties, such as salami, bologna, *salchichon,* and *chorizo.* Locust bean gum acts as a binding and stabilizing agent in sausages so that sausages can be made more homogeneous and have greater stability and smoother texture. Locust bean gum has a lubricating effect on meat mix. As a consequence of its addition, the mix is easier to handle, and thus the extruding and stuffing operations are facilitated. Because of the highly hydrophilic properties of the gum, losses in weight of sausages during storage are noticeably decreased. The use of locust bean gum and its derivatives in sausage manufacturing to improve quality provides both technical and economic advantages.

In soft-cheese manufacture, locust bean gum and its derivatives promote an increase in yield of curd solids. In many cases, it has been possible to increase yield by more than 10%. The resulting curd is as soft and compact as curd obtained by the usual process and has an excellent texture. The separated whey is limpid. Improvements in characteristics of cheese are obtained because of the buffer action shown by acidified locust bean gum solutions, which act as colloid-protecting agents. The finished cheese shows constancy in pH. From a merely technical point of view, the addition of locust bean gum and its derivatives is particularly advantageous because it speeds up coagulation and makes separation and removal of curd easier. The finished cheese has excellent body and structure, is smoother, more resilient, and more homogeneous. Exudation of fresh cheese is considerably reduced.

Melted cheese and cheese spreads can be prepared from very soft cheese with a high water content by the addition of locust bean gum. If no stabilizer were added, the finished products would be excessively soft and therefore very difficult to refine. The addition of locust bean gum and its derivatives produces finished cheese of better consistency, easier to spread. Moreover, higher quantities of water can be incorporated into the cheese, which is homogeneous and has a very fine texture.

Locust bean gum is used as a thickener for canned meat and fish and as a stabilizer and thickener for sauces and salad dressings. Considerable locust bean gum is used in pie filling preparations for thickening purposes, particularly in fruit-type fillings. While preventing the filling from flowing from the shell, locust bean gum also yields a clearer, more fruit-like filling, which is more palatable and does not mask the flavor. The quantity of locust bean gum to be used for this purpose ranges between 1 and 2% of the weight of the fruit juice and water of the filling.

Locust bean gum is occasionally used to thicken soups. The gum concentration usually ranges between 0.2 and 0.6%. In many food applications, vegetable gums are often preferred to starches because they do not tend to retrograde at low temperatures. Consequently, because of the increasing consumption of frozen food products, the demand for water-soluble gums is growing progressively.

There is today widespread use of specialty products, such as those based on locust bean and guar gums plus other natural colloids. These are utilized as stabilizers and binding agents in such up-to-date prepared foods as instant dry sauces and soups, frozen concentrated soups, frozen butter and cheese sauces for table-ready vegetable and fish dishes, and a host of other food products. Locust bean gum has been successfully used to stabilize natural whipped cream, as well as imitation whipped cream. Other important applications are in mayonnaise and tomato ketchup.

Bakery products.—The manufacture of bread and other leavened bakery products is sometimes difficult because of differences in quality of flour. Flours may vary in gluten content, and very often a flour which is deficient in gluten makes an unsatisfactory product because of poor water-holding properties of the dough. The use of high-grade locust bean gum and its derivatives produces doughs with constant properties and prevents the dough from becoming flabby. Because of the increased water-holding properties of the dough, higher yields can be obtained. Moreover, resulting doughs have greater resiliency and drier appearance, and baked products have better texture, are much softer, have greater oven outturn, and can be preserved for a longer time.

The use of high-grade locust bean gum and its derivatives in cakes and biscuits also offers many advantages. Higher yields can be obtained along with a considerable saving of eggs. Finished cakes and biscuits are softer and maintain their freshness for a longer time. Cakes can be easily cut into slices without crumbling and are easily removed from the moulds.

Textile industry.—Locust bean gum derivatives can be used either alone or in combination with starch and synthetics as sizing agents for cotton and staple fiber yarns. Its derivatives bring about uniform penetration, constant moisture content, elasticity, and smoothness. They have outstanding film-forming and emulsifying powers and allow a simple desizing operation because it is sufficient to wash in hot water without preliminary treatment with enzymes. In particular, new formulations foreseen for modern sizing products are based on locust bean gum as the film-forming agent and synthetic resins as the cohesion agent so as to abate oxygen consumption in the biodegradation of desizing waters (BOD).

Locust bean gum derivatives can be used as print-paste thickeners that are resistant to most chemicals and are suitable for the preparation of all kinds of dyestuffs. Locust bean gum derivatives can be considered universal print-paste thickeners for roller printing as well as for screen printing, both flat and rotary. Because they possess a higher thickening power than most other print-paste thickeners, their use is economical. For example, 2 oz (57 g) of treated locust bean gum gives the same results as 18 oz (510 g) of British gum, 8 oz (230 g) of starch, 2.5 oz (70 g) of gum tragacanth, or 18 oz (510 g) of gum arabic. Locust bean gum printing pastes are homogeneous, transparent, and free-flowing and have high stability. In addition, they can easily be washed off the printed cloth to impart a soft and smooth touch even if the prints are large. When used in roller or screen printing, locust bean gum derivatives help to obtain purity and uniformity of shades, neatness of outlines, and deeper penetration of dyes. Locust bean gum derivatives can be used in conjunction with starch, British gum, gum tragacanth, and dextrins to improve the homogeneity and flow properties of the paste. Locust bean gum is preferred for the Colloresin and flash-ageing process.

Locust bean gum derivatives used as finishing agents impart body and smooth touch, attractive appearance, and bright colors. Cloth and fabrics finished with locust bean gum derivatives have a distinctive look. As finishes, locust bean gums are not evident on the fabric and the transparency of the finsh does not lower the brightness of either color or print; on the contrary, it improves the appearance of the finished fabric. Finish with locust bean gum derivatives remains clean without dusting of the weighing material, because both loading and weighing materials are firmly held to the fabric by the thin and transparent film.

Velvet and plush finished with locust bean gum derivatives are more durable. It is sufficient to size the back of such fabrics, thus avoiding finishing of the surface which has a negative effect on the piles or naps.

Paper industry.—This is the widest and most important field for locust bean gum and its derivatives. Derivatives are used as fiber bonding and beater additives for the paper making industry. Locust bean gum derivatives have several advantages. They improve the physical characteristics of all types of paper and can be used even for high-grade white paper. The preparation of gum solutions is easy, and the moderate cost of the added solution is rewarded by technical and commercial advantages. The addition of locust bean gum derivatives to the pulp allows a reduction of the beating time, giving an improvement of the drainage rate of the pulp, and therefore of the machine speed.

Oil drilling industry.—Locust bean gum derivatives are effective as drilling-mud additives. Oil well drilling mud represents a complex colloidal system based mainly on a dispersion of clay in water. The clay base of the system is partly provided by the drilled soil and partly by a direct addition of clay to the mud in the course of its preparation. It is well known that a dispersion of clay in water possesses a certain stability, which means that the solid phase of the system does not easily separate from the liquid phase as would, for example, occur with a water–sand mixture. However, the stability of the system is insufficient under certain operating conditions, and in such cases, the dispersion must be integrated by protective colloids.

Clay-base drilling muds generally possess a remarkable sensitivity to the action of electrolytes. An addition of a small quantity of chlorides to the mud is sufficient to bring about marked changes in its colloidal structure. Chlorides are often encountered in drilling operations, mainly in the form of saturated aqueous solutions. Their presence can cause serious alterations in the mud's physical properties, such as an excessive increase in viscosity. Through the action of chlorides, the mud might become so viscous that the pressure of the pumps will be inadequate to cause the mud to circulate satisfactorily within the drilling system. The action of chlorides can also cause a separation and subsequent sedimentation of the sand or barium oxide that is often added to the mud to increase its density. The separated minerals sometimes cause blocking of the bit.

The negative influence of chlorides on clay-base drilling muds can be greatly reduced by the addition of colloid protecting agents. Those substances which, when added to a colloidal suspension, prevent or delay coagulation are called *protectors*. Electrolytes and colloids are generally used as protecting agents. As drilling-mud additives, modified starches and locust bean gums have proved to be the most efficient agents. Locust bean gum derivatives have been developed to control muds used when perforating deep salt layers or when working with salt water. Under these extreme conditions, locust bean gum derivatives satisfactorily control water losses and viscosity of the mud.[1]

Pharmaceutical and cosmetic industries.—Locust bean gum is used as an excipient for tablets, a thickener for tooth paste, and a thickener and stabilizer for lotions and creams.[2–5].

Others.—Locust bean gum is used as an auxiliary for the manufacture of hydroresistant photographic papers, as well as an auxiliary for explosives. In the latter case, the gum acts by absorbing the moisture of the other ingredients and by forming a gelatinous coating impermeable to water.[6]

Locust bean gum can be used in the manufacture of smoking tobacco. If the gum is properly mixed and kneaded with the purest parts of tobacco, it is possible to obtain a sheet having flexibility and strength characteristics comparable with those of the tobacco leaf.[7] In this same field, locust bean gum can also be employed in the preparation of slow combustion cigarette paper.[8]

Locust bean gum finds application in the field of flocculations as an agent that accelerates the settling rate.[9] Locust bean gum can also be used in the preparation of adhesives.[10, 11]

IV. Structure

Commercial locust bean gum contains about 88% of D-galacto-D-mannoglycan, 4% of pentan, 6% of protein, 1% of cellulose, and 1% of ash.[12] Ratios of D-galactose to D-mannose in the galactomannan have been reported as 27 : 73,[13] 20 : 80,[14, 15] 18 : 82,[16] 16 : 84,[12] and 14 : 89.[17] Whistler and Smart[18] state that "while these differences may be due in part to variation in the analytical method employed, it is also quite probable that the D-galactose to D-mannose ratio may vary with the origin of the gum or the growth conditions of the plant at the time of production." Although methylation studies[14, 17] have given slightly different molar ratios of the hydrolysis components, it is evident that the locust bean gum molecule is a linear chain of β-D-mannopyranosyl units linked (1→4) with every fourth or fifth D-mannopyranosyl unit substituted on C-6 with an α-D-galactopyranosyl unit. The structure of locust bean gum probably differs from the structure of guaran only in the smaller number of D-galactosyl units as side chains.

Chemical reactions of locust bean gum are the same as those of other neutral, water-soluble polysaccharides. Derivatives can be made by esterification or etherification. Among those which have been produced are carboxyalkyl ethers,[19, 20] the hydroxyethyl ether[21] and cross-linked derivatives from the use of bifunctional reagents.[22] Locust bean gum will complex with borate.[23–29]

V. Properties[12]

Locust bean gum, which has a molecular weight[30] of 310,000, is incompletely dispersed in water at room temperature. Most of the gum is dispersed in concentrations up to 5%, but part remains as insoluble flocs that settle upon standing. Incomplete dispersibility has been attributed to some crystalline regions in the dry gum that disintegrate as the temperature of the dispersion is raised.[31] If locust bean gum is dispersed in cold water, then as the temperature of the dispersion is raised, a decrease in viscosity is followed by a sharp increase (Fig. 1). The increase that begins at a temperature of about 45° continues as insoluble aggregates are more completely dispersed.[32]

When preheated dispersions are cooled, a further increase in viscosity is noted.[31] Dispersions that are first heated to ~95° and then cooled have a much higher viscosity than dispersions made in cold water; no flocculent material settles out from preheated dispersions. Therefore, to obtain the greatest efficiency as a thickener, dispersions should be made in hot water and then cooled. For example, when the gum is dispersed in cold water at a concentration of 5%, a heavy, opaque paste that is practically nonflowing is formed; but when it is dispersed in hot water, a concentration of only 2% produces the same type of paste. The effects of temperature and concentration on locust bean gum dispersions made in both hot and cold water are shown in Figures 1 and 2, respectively.[32]

Locust bean gum dispersions are pseudoplastic (Fig. 3).[32] Because locust bean gum is a neutral polysaccharide, pH has little effect on viscosity in the range pH 3–11.[32]

Neutral salts will not precipitate the gum,[33] but certain basic salts, such as basic lead acetate, will cause precipitation.[23] Locust bean gum is also precipitated by tannin and is therefore used in leather manufacture as a restraining agent in the tanning process. Through precipitation with quartenary ammonium salts, it is possible to obtain the separation of locust bean gum from other neutral polysaccharides.[34] The effect of ionic strength on viscosity is given in Figure 4.[32]

Disperions of locust bean gum do not gel, but as with other galactomannoglyclans, the addition of borate in neutral or alkaline solutions will gel dispersions of less than 0.3% of polysaccharide.[35] These firm, plastic, ropy, transparent gels, which show no syneresis and are cohesive but will not stick to glass or other substances, can be liquefied by the addition of a low-molecular-weight polyhydroxy compound, such as mannitol or glycerol. Locust bean gum can also

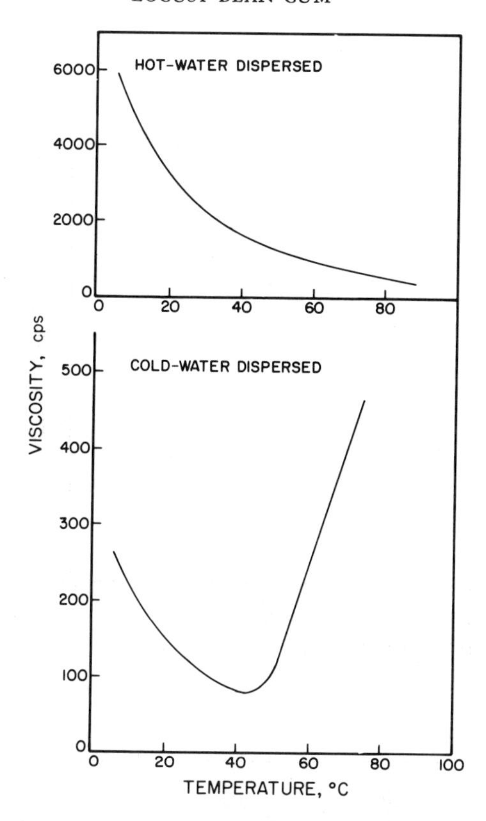

FIG. 1.—Effect of temperature on viscosity of 1% w/v locust bean gum dispersions as measured with a Brookfield Synchro-Lectric viscometer with spindle No. 1 at 20 rpm for the cold-water dispersion and spindle No. 4 at 20 rpm for the hot-water dispesion.

influence the gelling of other polysaccharides. Gels prepared from locust bean gum, carrageenan, and potassium chloride are resilient and form retaining, are not tough or fragile, are edible and heat-reversible, and can be made in a wide range of solids content and pH.[36–38] Gels of locust bean gum and agar mixtures have an increased elastic deformability and breaking strength over those prepared from agar alone and show no syneresis.[39]

Viscosities of locust bean gum dispersions can be increased by cross-linking molecules.[22] Viscosity, on the contrary, decreases by depolymerization through exposure to ultrasounds.[40]

When dispersions are evaporated on a plate, tough, pliable films are formed.

In general, the physical properties of locust bean gum are similar to those of guaran, and the two gums can be used interchangeably in such things as pharmaceutical preparations.[41]

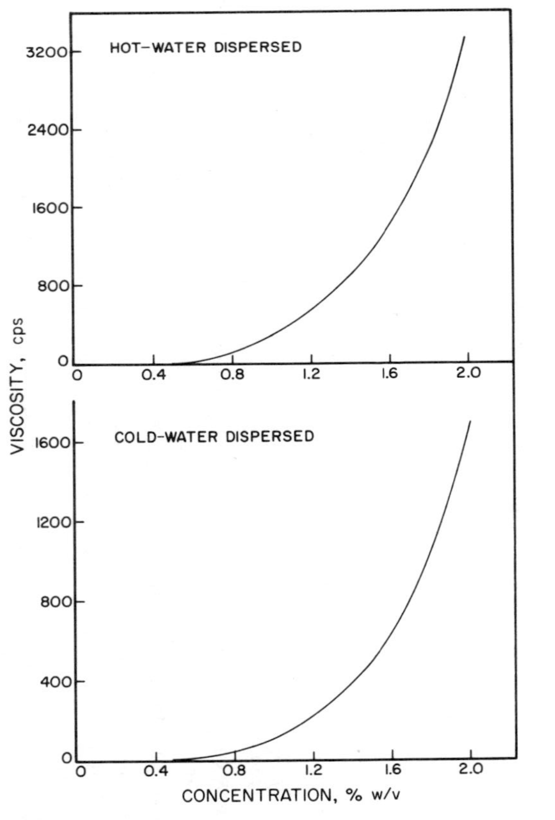

FIG. 2.—Effect of concentration on viscosity of locust bean gum dispersions at 25° as measured with a Brookfield Synchro-Lectric viscometer with spindle No. 1 at 5 rpm for the cold-water dispersed gum and spindle No. 5 at 10 rpm for the hot-water dispersed gum.

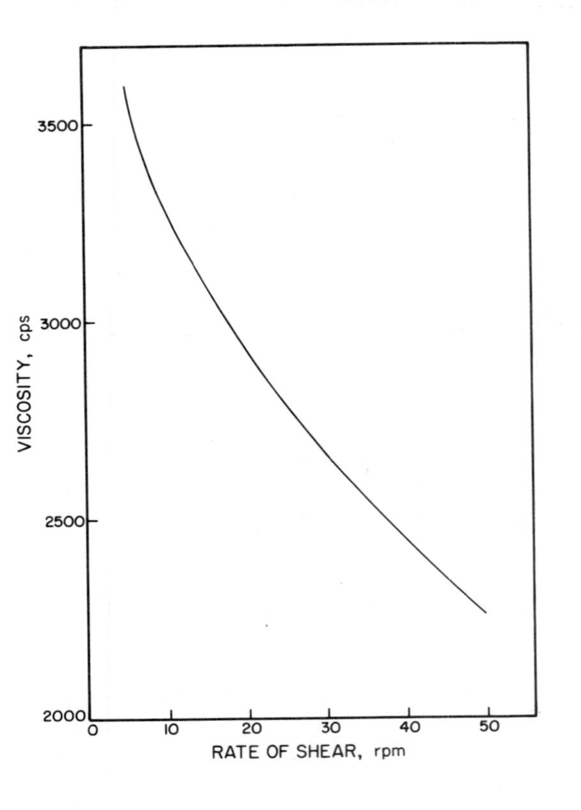

FIG. 3.—Viscosity versus rate of shear for a 1% w/v dispersion of locust bean gum made in hot water and measured with a Brookfield Synchro-Lectric viscometer with spindle No. 4 at 25°.

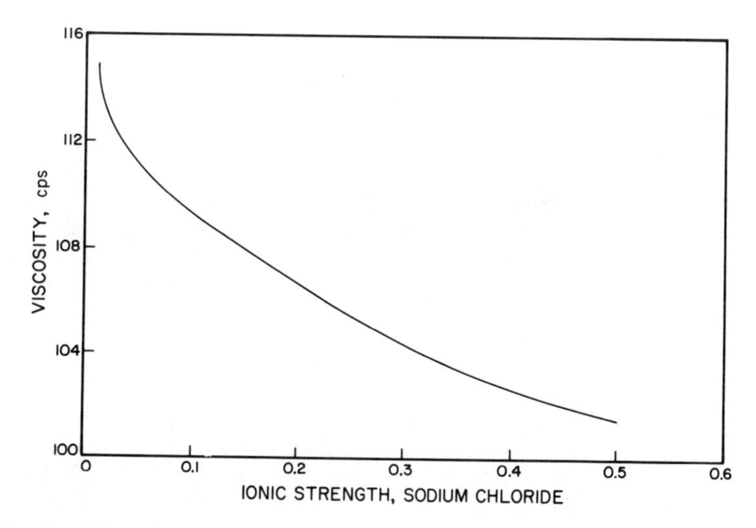

FIG. 4.—Effect of ionic strength on the viscosity of 1% w/v locust bean gum dispersions made in cold water as measured with a Brookfield Synchro-Lectric viscometer with spindle No. 1 at 50 rpm and 25°.

VI. REFERENCES

(1) Wintershall Akt.-Ges. and Kalle & Co. Akt.-Ges., West Ger. Patent 1,017,560 (1957); *Chem. Abstr.,* **54,** 15921 (1960).

(2) H. H. Hutchins and R. E. Singiser, *J. Amer. Pharm. Ass., Pract. Pharm. Ed.,* **16,** 226 (1955); *Chem. Abstr.,* **49,** 10582 (1955).

(3) C. H. Becker, *Amer. Prof. Pharm.,* **20,** 939, 987 (1954); *Chem. Abstr.,* **49,** 1279 (1955).

(4) P. I. Smith, *Amer. Perfum. Aromat.,* **67,** No 6, 67 (1956); *Chem. Abstr.,* **50,** 11688 (1956).

(5) B. N. Patel, *Drug Cosmet. Ind.,* **95,** 337 (1964); *Chem. Abstr.,* **61,** 14473 (1964).

(6) G. L. Griffith, Jr., and D. G. Samuel, Jr., U.S. Patent 2,860,041 (1958); *Chem. Abstr.,* **53,** 3697 (1959).

(7) M. M. Samfield, B. A. Brock, and E. E. Locklair, U.S. Patent 2,708,175; *Chem. Abstr.,* **49,** 11248 (1955).

(8) A. W. Schoenbaum, U.S. Patent 2,775,970 (1957); *Chem. Abstr.,* **51,** 6156 (1957).

(9) G. Hulot, U.S. Patent 2,771,996 (1956); *Chem. Abstr.,* **51,** 9106 (1957).

(10) W. R. Moffitt and M. B. Berkey, U.S. Patent 2,967,836 (1961); *Chem. Abstr.,* **55,** 9966 (1961).

(11) C. Love and Co., Ltd., Austral. Patent 232,807 (1961); *Chem. Abstr.,* **57,** 2484 (1962).

(12) C. Griffiths, *Mfg. Chem.,* **20,** 321 (1949); *Food,* **21,** 58 (1952).

(13) A. Spada, *Atti Soc. Nat. Mat. Modena,* **70,** 20 (1939).

(14) F. Smith, *J. Amer. Chem. Soc.,* **70,** 3249 (1948).

(15) L. E. Wise and J. W. Appling, *Ind. Eng. Chem., Anal. Ed.,* **16,** 28 (1944).

(16) B. Lew and R. A. Gortner, *Arch. Biochem.,* **1,** 325 (1943).

(17) E. L. Hirst and J. K. N. Jones, *J. Chem. Soc.,* 1278 (1948).

(18) R. L. Whistler and C. L. Smart, "Polysaccharide Chemistry," Academic Press, New York, 1953, p. 300.

(19) O. A. Moe, U.S. Patent 2,477,544 (1949); *Chem. Abstr.,* **44,** 360 (1950).

(20) O. A. Moe, U.S. Patent 2,520,161 (1950); *Chem. Abstr.,* **45,** 1365 (1951).

(21) K. M. Gaver, E. P. Lasure, and D. V. Tieszen, U.S. Patent 2,572,923 (1951); *Chem. Abstr.,* **46,** 1786 (1952).

(22) O. A. Moe, U.S. Patent 2,650,917 (1953); *Chem. Abstr.,* **48,** 394 (1954).

(23) A. L. Williams, *Analyst,* **53,** 411 (1928).

(24) R. Hart, *Ind. Eng. Chem., Anal. Ed.,* **2,** 329 (1930).

(25) O. A. Moe, S. E. Miller, and M. H. Iwen, *J. Amer. Chem. Soc.,* **69,** 2621 (1947).

(26) E. Anderson, *Ind. Eng. Chem.,* **41,** 2887 (1949).

(27) H. Deuel and H. Neukom, *Makromol. Chem.,* **3,** 13 (1949).

(28) H. Deuel, H. Neukom, and F. Weber, *Nature,* **161,** 96 (1949).

(29) J. W. Swanson, *Tappi,* **33,** 77 (1950).

(30) J. V. Kubal and N. Gralen, *J. Colloid Sci.,* **3,** 457 (1948).

(31) H. Deuel and H. Neukom, *in* "Natural Plant Hydrocolloids," *Advan. Chem. Ser.,* **11,** 51 (1954).

(32) R. L. Whistler and R. J. McCredie, data published *in* "Industrial Gums," R. L. Whistler, ed., Academic Press, New York, 1st Ed., 1959, p. 370. Measurements were made with a Brookfield Synchro-Lectric viscometer (multispeed model RVT) on commercial high-grade locust bean gum.

(33) F. F. Bryant, *Ind. Eng. Chem., Anal. Ed.,* **13,** 103 (1941).

(34) S. A. Barker, M. Stacey, and G. Zweifel, *Chem. Ind. (London),* 330 (1957); *Chem. Abstr.,* **51,** 10099 (1957).

(35) R. E. Schachat and L. Z. Raymond, *in* "Physical Functions of Hydrocolloids," *Advan. Chem.. Ser.,* **25,** 11 (1960); *Chem. Abstr.* **54,** 19097 (1960).

(36) G. L. Baker, J. W. Carrow, and C. W. Woodmansee, *Food Ind.,* **21,** 617 (1949).

(37) G. L. Baker, U.S. Patent 2,466,146 (1949); *Chem. Abstr.,* **43,** 5132 (1949).

(38) G. L. Baker, U.S. Patent 2,669,519 (1954); *Chem. Abstr.,* **48,** 6051 (1954).

(39) H. Deuel, G. Huber, and J. Solms, *Experientia,* **6,** 138 (1950).

(40) W. F. Head, Jr., and W. M. Lauter, *J. Amer. Pharm. Ass.,* **46,** 617 (1957); *Chem. Abstr.,* **52,** 1660 (1958).

(41) G. E. Osborne and C. O. Lee, *Bull. Nat. Formulary Comm.,* **19,** 4 (1951).

CHAPTER XVI

QUINCE SEED, PSYLLIUM SEED, FLAX SEED, AND OKRA GUMS

J. N. BEMILLER

Department of Chemistry and Biochemistry, Southern Illinois University at Carbondale, Carbondale, Illinois

I. QUINCE SEED GUM[1]

1. *Source of Quince Seed*[1–3]

Quince, *Cydonia oblonga,* is a small, deciduous tree (15–20 ft, 4.5–6 m) or large deciduous bush, of the family Rosaceae, that is grown for its fruit. Its leaves, flowers, and fruit closely resemble those of apple and pear trees. The fruit, a yellow pome, is covered with woolly hairs, which fall off as the fruit matures; it is used to make preserves, marmalades, and confections. It can be used alone or to enhance the flavor of fruit mixtures. Each fruit possesses five ripened carpels, which constitute its core. Each of these carpels contains 6–15 seeds arranged in two rows. The ovate, angled, inodorous seeds, 6 mm (0.25 in) in length, are reddish-brown, have an insipid, slightly bitter taste, and have a coriaceous envelope covered with a mucilaginous epithelium that causes the seeds in each carpel to adhere to each other.

Quince, native to central Asia and found abundantly in northern Iran, grows extensively throughout temperate regions of the world and is widely cultivated for its fruit over much of the apple-growing areas of the United States, mainly California, Ohio, Michigan, Pennsylvania, and New York. Iran is the main source of commercial quince seed, supplying ~75% of the total world production. The chief producing areas in Iran are Teheran and Hamadan. Other countries that have furnished quince seed to world markets are Spain, India, the Union of South Africa, Iraq, and Portugal. It is estimated that 11,000 tons of the fruit are produced and consumed annually in Syria and Lebanon, but very little seed is reclaimed in these countries. Because the fruit ripens in late summer or early fall, the seed is usually ready in export centers in December or January.

To produce quince seed, often called *gum quince seed, semen cydonia, golden apple seed,* or *cydonia seed,* the fruit is allowed to rot or air dry. The reclaimed seeds, which are likely to cluster, have fruit pulp, dirt, and other foreign matter adhering to them.

The value of quince seed is determined by three factors. In order of importance, they are gum content (about 20% of the seed by weight),[4] color of solution after extraction with water, and freedom from adulteration with dirt, insects, and other inert material. Seed free from foreign matter can be utilized in special applications and is more valuable; unclean seed is less expensive.

The seed, collected in small lots, mainly by nomads, is sold to collectors or traders, who in turn sell it, usually uncleaned, to exporters. Exporters usually clean and grade the seed before resale. The seed is graded according to size and often to country of origin. Of the three sizes, small, mixed, and large, the latter brings the highest price.

Quince seed can be packed in barrels, boxes or tins but is usually packed for export in bags of 1 quintal (220 lb). The seed, when stored or shipped, should be dried and placed away from moist or odorous goods to prevent damage. The stored seed should be protected from insects, particularly those that bore.

Quince trees grow abundantly, but no official statistics of production or exports of quince seed are available. Collection or production of seed varies from 300,000 to 500,000 lb (135,000 to 225,000 kg) annually. Prices paid to collectors determine to a large extent the actual quantity produced.

At the present time, the United States imports the largest part of the world production (>300,000 lb, >135,000 kg).[1] Other countries that import the seed in large quantities are France, England, and Canada. As no restrictions are placed on the import of quince seed, it is imported duty free. There has been no effort made to grow and collect seed in the United States. Although quince is cultivated in several states, the high cost of labor has precluded any attempt at commercial production of the seed.

The price, which varies with grade, has been rather stable in recent years (Table I).

2. *Properties of the Gum*

Quince seed gum hydrates slowly to form slightly colored, smooth-flowing, highly viscous dispersions at concentrations up to 1.5% (Figs. 1 and 2). It is most difficult to disperse quince seed gum at concentrations exceeding 2% (Fig. 2); at these concentrations, a slimy, mucilaginous mass forms. Quince seed gum dispersions contain fibrillar particles (30–35%) of cellulose.[6] Their viscosities can be estimated from a nomograph.[7]

Quince seed gum dispersions are thixotropic; that is, they show a decrease in viscosity as shear rate is increased, a decrease in viscosity with time at a constant rate of shear, a recovery of viscosity when the shearing stress ceases, and a yield value or force required to produce movement (Fig. 3). They are relatively little affected by temperature at temperatures between 15° and 50° (Fig. 4). There is no change in viscosity with pH in the pH range 4 to 10, but a rapid increase in viscosity results above and below this range (Fig. 5). The viscosity is unaffected by sodium chloride at concentrations up to 0.1M (Fig. 6).

3. *Preparation, Use, and Development of the Gum*

Mucilage is prepared from the seed by the user by methods that differ slightly depending upon the expected application. In the commercial method of preparation, the seed is extracted with water (2 parts of seed per 100 parts of hot or cold water) for about 30 min, and the mixture is filtered through muslin without pressure.

Quince seed mucilages are used primarily by the cosmetic industry and to a lesser extent by the pharmaceutical industry. They could also be used in many

TABLE I

Price[5] of Quince Seed in the United States in June of Each Year, 1965–1971 (dollars/pound)

1965	1966	1967	1968	1969	1970	1971
1.50–1.75	1.50–1.65	1.50–1.65	1.65–2.25	1.65–2.25	1.65–2.25	1.65–2.25

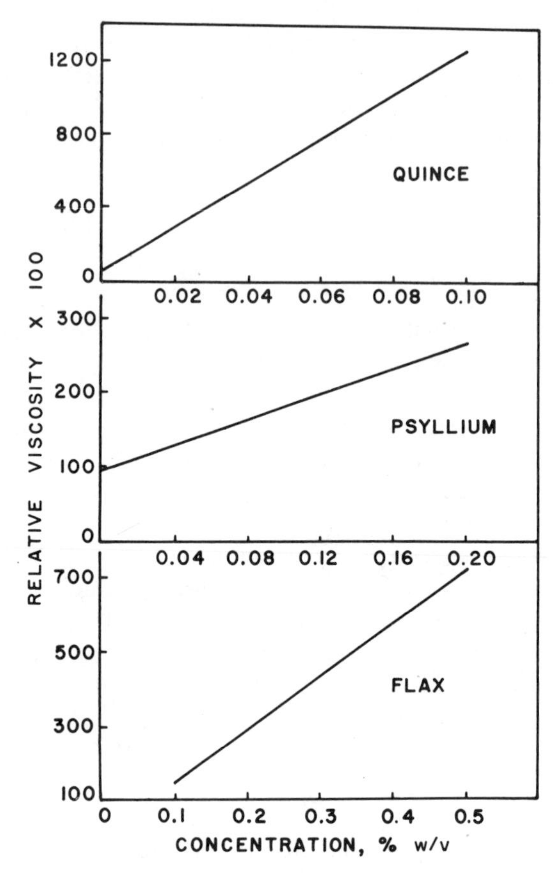

FIG. 1.—Viscosity vs. dilute concentrations of quince, psyllium, and flaxseed gums measured in a Ubbelohde viscometer.[1]

other industrial operations where a mucilaginous liquid or suspending agent is needed. However, their high cost does not seem to destine them for large industrial growth. As examples of its potential use, quince seed gum could be useful in drilling muds in the salt dome area of the Near East and other parts of the world.[8] Quince seed gum can be used as a stabilizer in chocolate milk and ice cream.[2] In ice cream, quince seed gum compares flavorably with gelatin except for the price, which is prohibitive.[9] A nonsettling, chocolate-flavored drink that contains 0.30–0.10% of quince seed gum and the process of production have been patented.[10] Quince seed gum can be used on natural, synthetic, or mineral fibers as a deflocculant that makes it possible to water-lay all lengths of fibers to produce paper or fabric with good tensile and flexural (tear) strength.[11]

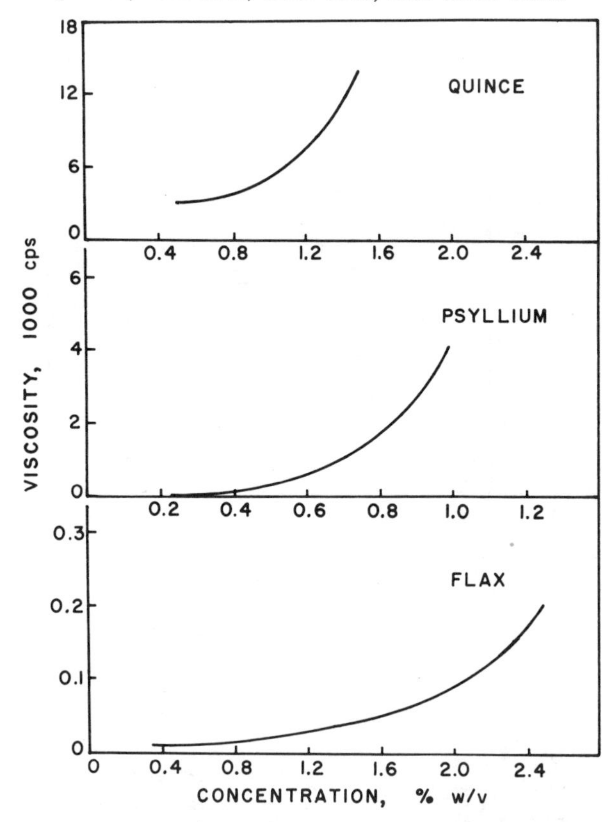

FIG. 2.—Viscosity vs. concentration measured with Brookfield viscometer: quince seed gum, spindle No. 5 at 20 r.p.m.; psyllium seed gum, spindle No. 3 at 30 r.p.m.; flaxseed gum, spindle No. 1 at 20 r.p.m.[1]

Quince seed gum can be used in wave-setting lotions. The viscous and rather slimy solutions of gum bind hair together so that it can be shaped readily into waves and curls that will stay in position until dry. The lotion on evaporation leaves very little residue that would make the hair appear dull or dusty. In any wave-setting fluid, the solids content must be as low as possible, and therefore, only those gums that have excellent thickening properties are used. Among the natural, water-soluble gums, quince seed gum is particularly suitable, for it yields highly viscous dispersions at low concentrations and has the property of glairiness or stringiness desired in certain cosmetics.[12] Quince seed gum has also been found to do a better job of keeping hair in place during rainy weather. To prepare a satisfactory waving lotion, it is only necessary to mix quince seed gum, water, 5–10% ethanol to reduce drying time, a preservative, and various other substances such as *p*-ethylphenethyl alcohol, coloring, and perfumes.[13, 14]

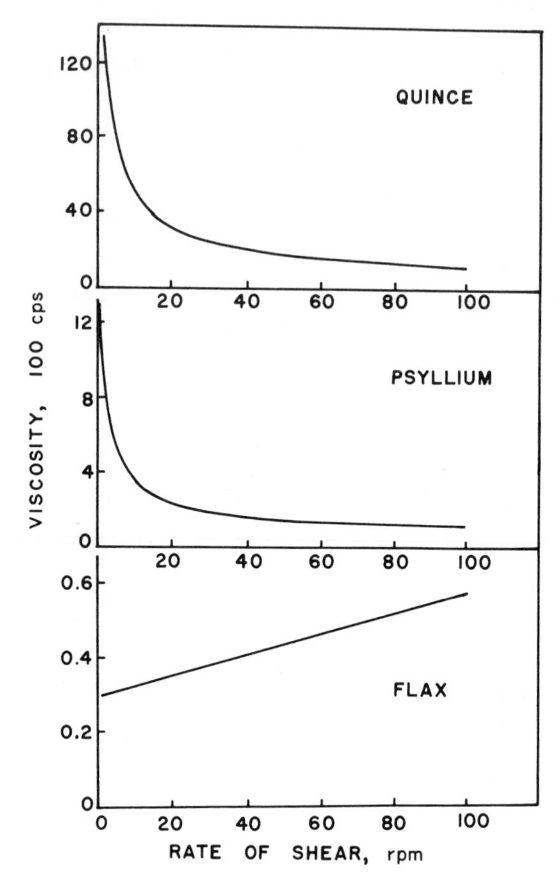

FIG. 3.—Viscosity vs. rate of shear: quince seed mucilage (0.5%), spindle No. 4; psyllium seed mucilage (0.5%), spindle No. 3; flaxseed mucilage (1.0%), spindle No. 1.[1]

A cream mascara has been made that contains quince seed gum and that is reliable because it does not dry in the tube and stays on the eyelashes well.[15] Both a liquid cleansing milk[16] and a liquid cleansing cream[17] that contain quince seed gum have been described.

A 10% hydrogel that is useful for the treatment of rhagades of female breasts was prepared by digesting quince seeds for 15 min with 1% boric acid at 50°–60°, boiling the solution, cooling it to ~70°, filtering it, and holding it at 56° for 48 hr.[18]

4. *Structure*

Quince seed gum contains cellulose microfibrils suspended in an amorphous

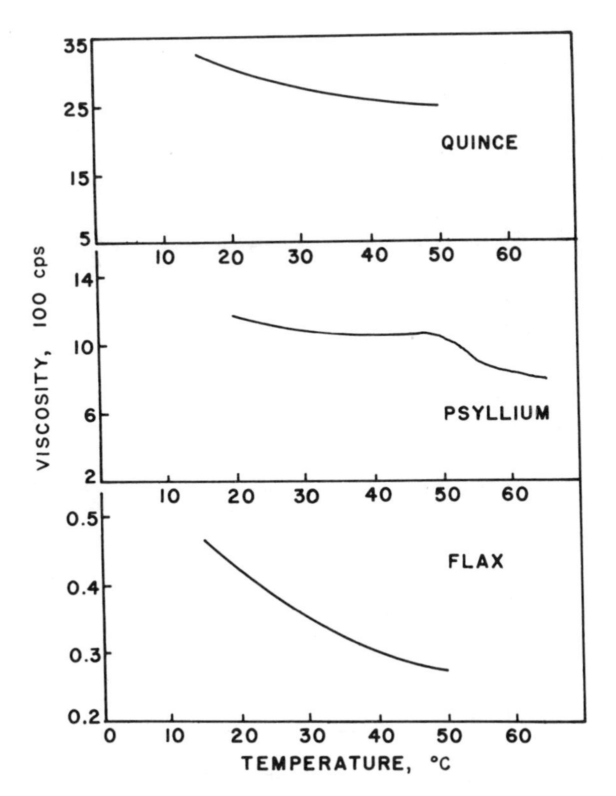

FIG. 4.—Viscosity vs. temperature: quince seed mucilage (1%), spindle No. 4 at 50 r. p. m.; psyllium seed mucilage (0.75%), spindle No. 2 at 20 r.p.m.; flaxseed mucilage (1.0%), spindle No. 1 at 100 r.p.m.[1]

polysaccharide(s).[6, 19–22] Mild, acid-catalyzed hydrolysis produces cellulose (30–35%), a small amount of L-arabinose, and a mixture of aldobiouronic acids apparently composed of D-xylose combined with mono-*O*-methylhexuronic acid (72%) and hexuronic acid (28%).[23, 24] Further hydrolysis releases D-glucose, galactose, L-arabinose, D-xylose, and hexuronic acid.[18]

II. Psyllium Seed Gum[1]

1. *Source of Psyllium Seed*[1, 25, 26]

Psyllium seed comes from plants of the *Plantago* genus, which is comprised of a large number of acaulescent or short-stemmed herbs of the Plantaginaceae family, having very small, greenish flowers in close bracketed spikes or heads. Distribution of this genus is quite wide; several species, such as *Plantago lan-*

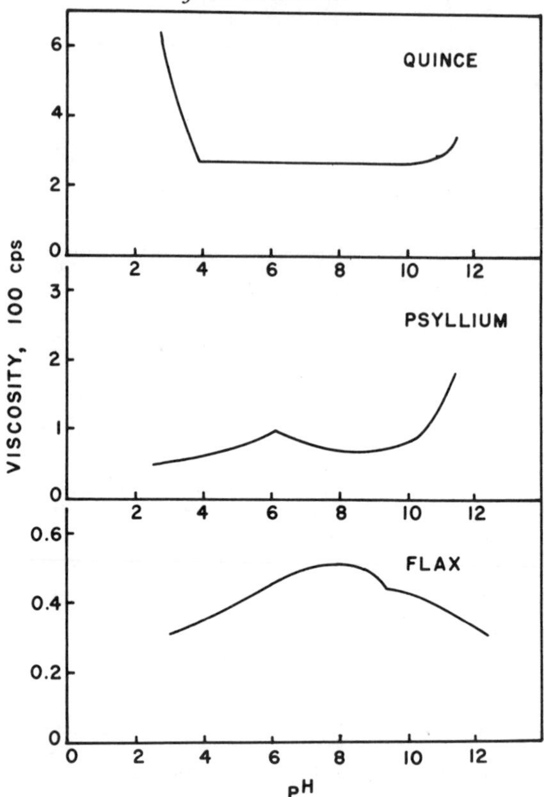

FIG. 5.—Viscosity vs. *p*H: quince seed mucilage (0.25%), spindle No. 1 at 20 r.p.m.; psyllium seed mucilage (0.5%), spindle No. 1 at 50 r.p.m.; flaxseed mucilage (1.0%), spindle No. 1 at 50 r.p.m.

ceolata, commonly known as buckhorn, *P. rugelii,* and *P. major,* or plantain, are common weeds.

Psyllium is cultivated chiefly in the Mediterranean region and in India. Commercial psyllium is the seed of several European and Asiatic species of *Plantago,* which are cultivated in France, Spain, and India. French (black) psyllium comes from *Plantago indica* (L.) also referred to as *P. arenaria* (Waldstein and Kitaibel), Spanish psyllium from *P. psyllium* (L.), and Indian (blonde) psyllium from *P. ovata* (Forskal).

At one time, France was the main supplier of psyllium to the U.S., but now this role is held by India. This change has occurred for several reasons: (1) World War II shut off the supply from France; (2) Indian seed is available at a lower price; (3) the gum content of the Indian species is greater; (4) the blonde seed yields a practically colorless mucilage; (5) the seed coat of Indian psyllium cracks off under slight mechanical pressure and can be readily separated

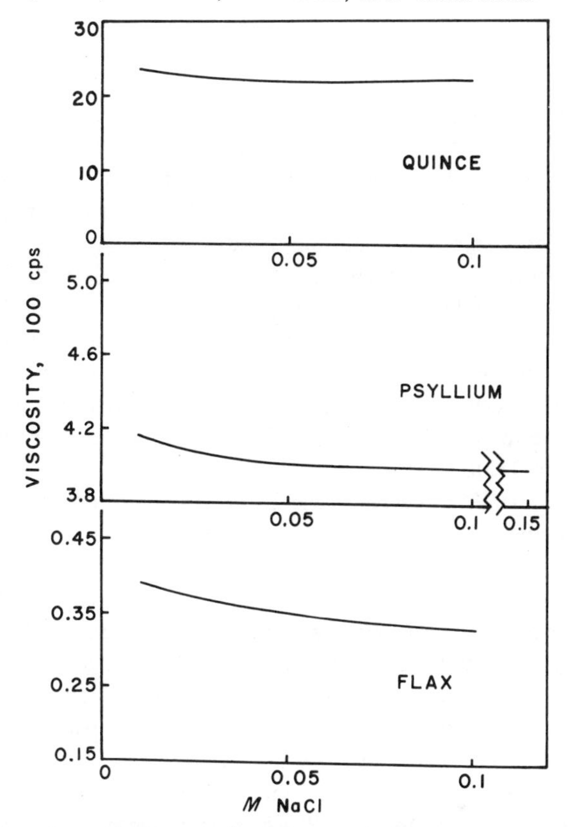

FIG. 6.—Viscosity vs. ionic strength: quince seed mucilage (0.5%), spindle No. 3 at 20 r.p.m.; psyllium seed mucilage (0.75%), spindle No. 2 at 50 r.p.m.; flaxseed mucilage (1.0%), spindle No. 1 at 100 r.p.m.[1]

from the rest of the seed. Output in India has been increasing as a result of an increasing demand for the husk (seed coat) in the United States, which consumes almost the entire production. *Plantago ovata* seed is commonly known in India by the Persian word *ispaghula.*

About two thirds of the Indian seed (*P. ovata*) is produced in the State of Baroda and the remainder in the State of Palanpur. The plant grows on most types of soil, but preferably on well-drained, sandy loam along the rivers. Seeds are sown 8–10 lb/acre (9–11 kg/hectare) in October and November when the temperature ranges from 17° to 40° (63° to 104° F). In recent years, the yield per acre has varied between 575 and 750 lb (260–340 kg), depending on weather conditions at the time of flowering and harvesting. Cool, dry weather assures a good crop. Frost at flowering time, or rain or cloudy weather during the time of maturation, reduces the yield.

Spanish psyllium, known locally as *zaragatona,* grows wild in the region of Cadiz and is gathered in June or July. The seed is collected by workers who gather about 5 kg (11 lb) daily, but only when there is a demand.

Psyllium seed and seed husks are packed for export in bags. Standard size of Indian bag is 168 lb (76 kg) and a French bag is 198 lb (90 kg). For proper conservation, it is necessary that the seed and husk be packed and kept in a dry place away from moist and oily goods.

Grades of psyllium are: fleaseed, domestic, French; fleaseed, French, black; fleaseed, Indian, blonde. Seed is also graded as either technical or National Formulary. The price of psyllium is determined by grade and the amount of foreign organic matter contained in the products. Black seed commands a higher price than blonde seed; and Indian seed, although having a greater gum content, is less expensive than the French seed. The price structure over the past several years is given in Table II.

The gum content[26] of several varieties of psyllium seed on the U.S. market are: French psyllium, 11.8%; Indian psyllium, 30.9%; German psyllium, 11.5%.

There are no over-all data on exports of psyllium from important world sources, but imports into the United States demonstrate that psyllium is a stable commodity.[1] Because of the number of different countries that can grow and export it as a commercial crop, it seems that the available amount could be substantially increased if there were a greater demand. Moreover, two native unofficial *Plantago* seeds have been described and compared to those in the National Formulary.[27] These two species are *Plantago rhodosperma* (Decne), which is distributed throughout the sandy soil of Missouri and Oklahoma to Louisiana, Texas, and Arizona, and *P. wrightiana* (Decne), native to Texas and Arizona and found in dry, sandy soil. The latter grows well under cultivation on marginal and submarginal soil, favors a deep sand, and requires little attention and no irrigation. When planted in October, the mature seeds can be harvested in May. The fruiting spikes extend over the tops of the leaves with sufficient clearance for easy harvesting with a modified combine. *Plantago wrightiana* is a hardy, high-seed-yielding plant. The seed is relatively large, con-

TABLE II

Price[5] of Psyllium Seed and Husks in the United States in June of Each Year, 1965–1971 (cents/pound)

Grade	1965	1966	1967	1968	1969	1970	1971
Black	40–42	48–50	48–50	55–60	55–60	55–60	55–65
Blonde	32–35	28–30	25–30	26–30	26	32–33	32–35
Husks	80	65	70–80	65–70	50	75	85

tains a high percentage (23%) of gum and, in all respects, compares favorably with official seeds.

Several additional native species of *Plantago,* as well as a homegrown, cultivated *P. ovata,* have been investigated.[28] The gum content (17.5%) of each of three native Texas species, *P. inflexa* (Morris), *P. helleri* (Small), and *P. rhodosperma* (Decne), is slightly less than that of commercial *P. ovata.* Homegrown *P. ovata* produces three types of seeds, the majority of which differ in appearance from the imported seed. The yield of gum from the homegrown variety equals that of the official varieties. The mucilage-forming ability of two additional natives species, *P. purshii* (R. and S.), and *P. aristata* (Michx), compares well with official seeds. The mucilage-forming capacity of psyllium seeds increases in the order *P. ovata* < *P. purshii* < *P. indica* < *P. aristata* < *P. psyllium.*[29]

2. *Properties of the Gum*

Purified psyllium seed gum is a white, fibrous material that hydrates slowly to form viscous dispersions at concentrations up to 1% (Figs. 1 and 2). At 2% solids, a clear, gelatinous mass is formed. The dispersions are clear, even though solution is incomplete. Portions of the gum that have hydrated but not dissolved cannot be removed by filtration in the usual manner but can be by centrifugation. When seeds are placed in 25 times their weight of water and left overnight, a dense gel settles out with the seeds and the clear solution can be removed by decantation or centrifugation.[30]

Psyllium seed gum dispersions are thixotropic; that is, they show a decrease in viscosity as shear rate is increased, a decrease in viscosity with time at a constant rate of shear, a recovery of viscosity when the shearing stress ceases, and a yield value or force required to produce movement (Fig. 3). They are relatively little affected by temperature at temperatures between 20° and 50° (Fig. 4) and are unaffected by sodium chloride at concentrations up to 0.15*M* (Fig. 6). There is little change in viscosity with pH in the pH range 2 to 10 (Fig. 5). No increase in viscosity of *P. ovata* gum dispersions occurs at low pH values because intramolecular hydrogen bonding, which persists up to the point at which ionization of 20% of the carboxyl groups is suppressed, prevents expansion of the molecular conformation.[31, 32]

Of 31 organic and inorganic substances tested for compatibility with *P. ovata* gum, only basic lead acetate was found to be incompatible.[33]

3. *Preparation, Use, and Development of the Gum*

Several methods are reported for extraction of psyllium seed gum. Often, the mucilage-forming capacity of the seed is such that it is not necessary to extract the gum; instead, intact or macerated seeds are used. However, if it is necessary to purify the gum, it can be readily extracted from the seed with hot water.

Because the gum is located in the seed coat (husk, hull), it is advantageous to separate the seed coat from the rest of the seed before extraction. Seed coats are removed by application of slight mechanical pressure, which cracks them. An alternative process uses freezing to break the hull loose.[34]

Purification of the gum from *Plantago* seed is suggested in a process in which it is first extracted with boiling water until the seed has swelled.[35] The dispersed gum is then separated from cellulose and other insoluble portions of the material by gravity under conditions of low viscosity by dilution with a dilute agar dispersion. Agar reduces the stickiness of the dispersed gum and facilitates separation so that a clean top layer of psyllium seed gum and agar is formed. Alternatively, centrifugation can be used without addition of agar.[30] Varying conditions of preparation give products with different properties.[36, 37]

Mucilages formed from psyllium seed have been used since ancient times, both in home remedies and in the pratice of medicine as a demulcent in dysentery, erosion of intestines, dry coughs, hoarseness, burns, excoriations, and inflammations of the eye.[1, 38]

The Pharmacopoeia of India officially recognizes psyllium seed[38] and its husk.[39] Various uses of psyllium seed are described in ancient Indian Ayurvedic books. The "Indian Materia Medica" of Nadkarni, published in 1927, reports that the seed is recommended as a demulcent and is given in urinary disorders and dysentery as well as to arrest flux and griping pains in the bowels.

Psyllium has had a rather spectacular career in markets throughout the world. In the early part of the 1920's, it was regarded as an agricultural nuisance, and its commercial value was little suspected. By 1930, psyllium had taken its place in the market as a mechanical laxative. Over 2,000,000 lb (900,000 kg) of psyllium were imported into the United States during that year, and there were no less than 117 different brands of psyllium on the American market.

The most extensive use of psyllium is as a bulk laxative,[38] and it was because of this use that psyllium rose rapidly in the world market. There was a tendency at the time psyllium was introduced into the United States to restrict its use to the mitigation of constipation. When ingested with the proper amount of water, it not only swells and increases the size of the fecal mass, but its mucilaginous dispersions have a lubricant action equal to that of oil without any of oil's disadvantages. Psyllium has no nutritive value; it becomes an integral part of the fecal mass instead of surrounding it as does oil. In addition, psyllium seeds are practically tasteless and preferred by persons who have an aversion to oily medicaments.[39]

In 1932, it was observed that kidneys of rats fed ground *P. psyllium* became brown to black in color.[40] This change in color of the kidney was caused by a pigment material in the cortex of the kidneys that was observed as fine brown granules in the epithelial cells of the collecting tubules, particularly in the proximal convoluted tubules, and in Henle's loop. The nature of the pigment was not determined. Rats fed whole psyllium seed did not show any changes in kidney color.

In a later study, four preparations of good psyllium seed were fed to rats.[41] The body weight gain of the rats was normal. No pigmentation of the epithelial cells of the collecting tubules of the kidneys was observed. However, the kidneys of a group of rats fed tailings from *P. ovata* were slightly darker brown than normal after 6 weeks of continuous feeding. This pigmentation is probably related to a fraction of the seed hull.

In a study of the swelling properties and gel textures of 24 commercially available bulk-type laxatives, it was observed that the swelling of psyllium seed gum was not influenced to any extent by distilled water, 2% sodium chloride, 0.5% hydrochloric acid, 1% sodium hydrogen carbonate, artificial gastric juice, and artificial intestinal juice.[42] In artificial intestinal juice, psyllium seed gum increased in volume 5–14 times, locust bean gum 5–10 times, and methylcellulose 16–30 times in 24 hr. Compared to flaxseed gum, psyllium seed gum was found to be better as a laxative because it remained in gel form, whereas flax seed gum did not.[43]

From an *in vivo* evaluation of hydrophilic properties of bulk laxatives, it was claimed that the physiochemical properties of methylcellulose and carboxymethylcellulose give them two advantages over the natural gums.[44] Methylcellulose is a more efficient bulk laxative because of its greater water-retentive capacity, whereas carboxymethylcellulose gives uniform distribution through the intestinal contents and eliminates any tendency to produce intestinal blockage. A modified osmometric procedure used to compare the water-retentive properties of some hydrophilic colloids, including psyllium seed gum and several grades of methylcellulose, showed that dispersions of methylcellulose of relatively low viscosity have greater water-retentive capacity than equal concentrations of natural gum dispersions of greater viscosity.[45] Even so, many of the better-known bulk laxatives contain psyllium seed gum or psyllium husk (hull) powder.[46]

A preliminary investigation on the use of hydrogels to delay allergic reactions shows that there is evidence that many food substances do not become offending allergens until they reach the colon and become intermediate or end-products of digestion, bacterial degradation, or both.[47] Because psyllium seed gum gels and holds water and possibly soluble toxins and allergens against the absorptive capacity of the colon, its clinical value in certain diseases of allergenic origin is suggested. Psyllium seed gum also causes a decrease in skeletal deposition of strontium fed as strontium chloride.[48]

Preparations suitable for the treatment of intestinal diseases by oral administration can be prepared with an activated natural silicate (attapulgite) as the active component and a mild laxative such as psyllium seed gum.[49]

In addition, sterile dispersions of purified psyllium seed gum with 2-phenylethanol, invert sugar, and sodium chloride have been given intravenously as sclerosing agents in the treatment of varicose veins.[50]

Oral intake of psyllium seed hull powder lowered the serum cholesterol level in humans.[50]

Removal of cations from psyllium seed gum solutions with a sulfonic acid cation-exchange resin in the acid form, followed by spray drying, produces the free acid form of the polysaccharide.[51] Some possible pharmaceutical applications of this and other gum acids, which are innocuous when taken internally, include removal of exogenous sodium from the intestinal tract, treatment of gastric hypoacidity, use as enteric coating materials, use as a tablet disintegrant or retarder of tablet disintegration, use in sustain-release drug therapy, inhibition of bacterial growth by the pH effect, and treatment of local infections occurring in certain types of vaginitis.

Psyllium seed mucilage, like quince seed and flax seed mucilages, has the property of glairiness or stringiness so much desired in the preparation of cosmetics. It has been used successfully in hair-setting lotions with small amounts of ethanol.[16] Although its mucilage is stringy and has a fast drying time, five to ten times as much psyllium seed gum is required to give the same viscosity as a 2% quince seed gum dispersion.[11]

Water-resistant explosive compositions can be prepared with psyllium seed gum alone or in admixture with other gums.[52, 53] Psyllium seed gum has also been proposed for using in acidizing oil wells to remove drilling mud or other fracture-clogging deposits and to give enlarged passageways.[54] It can also be used on natural, synthetic, or mineral fibers as a deflocculant that makes it possible to water-lay all lengths of fibers to produce paper or fabric with good tensile and flexural (tear) strength.[11]

Emulsions prepared with the gum compared favorably with those prepared with gum arabic.[55]

4. *Structure*

Erskine and Jones[30] fractionated the polysaccharides from *Plantago ovata* and *P. arenaria* with copper(II) acetate and found that each contained at least two components, at least one of which was a neutral polysaccharide. Earlier, Laidlaw and Percival[56] had obtained an acidic polysaccharide fraction from *P. ovata* by cold water extraction and a neutral polysaccharide fraction by subsequent hot water extraction. Erskine and Jones[29] suggest that there might also be a third fraction. The gum from *P. lanceolata* was also shown to contain two components by electrophoresis on glass fiber paper.[57]

Table III gives the reported monosaccharide composition of psyllium seed

TABLE III

Composition of Plantago *Polysaccharides*

Source	Monosaccharide Components	Aldobiouronic Acid Isolated	References
Plantago arenaria	D-xylose (62%) D-galactose (6%) L-arabinose (17%) L-rhamnose D-galacturonic acid	2-*O*-(α-D-galacto-pyranosyluronic acid)-L-rhamnose (13%)	58–61
Plantago arenaria Fraction A	D-xylose L-arabinose D-galactose		30
Fraction B	L-rhamnose D-xylose L-arabinose D-galactose D-galacturonic acid		30
Plantago lanceolata	D-xylose (72%) D-galactose L-rhamnose (11%) D-galacturonic acid (15%)		62, 63
Plantago ovata Cold water extract	D-xylose (46%) L-arabinose (7%) L-rhamnose D-galacturonic acid	2-*O*-(D-galacto-pyranosyluronic acid)-L-rhamnose (40%)	37
Hot water extract	D-xylose (80%) L-arabinose (14%)	None	56
Plantago ovata Fraction S1	L-rhamnose D-xylose L-arabinose D-galactose D-galacturonic acid		30
Fraction S3	D-xylose L-arabinose		30
Fraction G	L-rhamnose 4-*O*-methyl-D-glucuronic acid D-xylose L-arabinose D-galactose D-galacturonic acid		30

TABLE III—*Continued*

Source	Monosaccharide Components	Aldobiouronic Acid Isolated	References
Plantago psyllium	L-arabinose D-xylose D-galacturonic acid		36
Plantago fastigiata	L-arabinose D-galactose D-galacturonic acid		64

gums. There would seem to be species differences in detailed composition. However, exact analysis of such differences awaits purification of each polysaccharide found in the husk of each species.

III. Flax Seed Gum[1]

1. *Source of Flax Seed*[1, 65, 66]

Linum usitatissimum, of the family Linaceae, the only species of the flax family that has any commercial importance, produces fiber from which linen is made and seed from which linseed oil and linseed oil meal are derived. Flax grown to produce fiber for making linen and flax grown for its oilseed belong to the same species but are different varieties, and generally speaking, the two products are not obtained from the same crop. Varieties of flax grown for fiber are tall, little branched, and early maturing, whereas those grown for seed are shorter, more highly branched, and produce a high yield of seed. Ordinary fiber flax is harvested before the seed is mature when the quality of the fiber is best, but the seed at this stage does produce an oil of good quality. Seed not required for replanting is usually sold in the oilseed market.

Flax, an annual plant that grows to a height of 12–40 in (30–100 cm), has a distinct main stem and a short taproot. The slender root branches can extend to a depth of 3–4 ft (90–120 cm) in light soil. In thick seeding, as for fiber flax, only the main stem develops; whereas in thin stands, two or more branches might develop from the base of the plant. The flax fruit is a five-celled boll, or capsule, which when filled contains ten seeds, except in abnormal double bolls. Selective breeding has led to shorter, more highly branched varieties with higher seed yields.

Flax is grown in many parts of the world, principally in temperate regions. Seed varieties are grown largely in North and South Dakota, western Minnesota, and eastern Montana. In this area, nearly 95% of the flax produced in the U.S.

is grown. More recently, it has been introduced as a cash crop in California. Other major producers of seed flax are Argentina, India, Russia, and Canada. European countries that cultivate flax are, in order of importance, Czechoslovakia, France, Germany, Italy, Belgium, and Ireland.

Quantities of seed flax grown in the various countries of the world have fluctuated greatly. In recent years, the United States has supplied most of its own requirements from domestic production, the small amount imported coming from Canada. The production, imports, exports, and average price of linseed meal, the primary source of flax seed gum,[67] are given in Table IV.

Results from a comprehensive study of four varieties of flax grown for several years at 54 different field stations in North America representing a wide range of latitudes, longitudes, altitudes, climates and soils, indicate, without much doubt, that drought and high temperature during the seed-filling period (that is, after 25 days past blossoming) reduce the oil content of the seed as well as the iodine number of the oil in the seed.[68] The yield per acre and the seed size decrease under these conditions, but the crude protein content of the meal increases. Fats and proteins increase remarkably in flax seed with increasing water availability, whereas the amounts of cellulose and other non-nitrogenous components increase in the seed of plants grown in a water deficiency.[69]

In the U.S., flax is grown principally to obtain linseed oil. Methods used for milling flax seed include the principal methods employed for the milling of any

TABLE IV

Linseed (Flax Seed) Meal in the United States (1958–1967): Production, Imports, Exports, and Average Price[66]

Year	Production, 1000 tons	Imports, 1000 tons	Exports, 1000 tons	Price, $/ton[a]	
				New York[b]	Minneapolis[c]
1958	409	4	14	—	—
1959	426	2	67	81.25	62.95
1960	378	2	39	71.35	53.35
1961	358	2	16	80.45	62.70
1962	388	1	54	85.60	67.50
1963	360	1	33	79.90	61.20
1964	381	1	80	75.10	59.10
1965	398	1	115	86.80	70.55
1966	363	6	111	96.05	76.40
1967	293	<0.25	92	94.35	75.90

[a]Price for extracted meal was usually $5–20/ton lower.
[b]Average price, bulk, carlots, 34% protein.
[c]Average price, bulk, carlots, 32–37% protein.

oil seed, that is, hydraulic pressing in box or cage presses, pressing in continuous screw presses, extraction with solvent, and a combination of pressing and extraction. Because flax seed gum is commercially feasible only as a by-product of the linseed oil industry, methods of gum preparation are based on oil meal or oil meal cake.

2. *Properties of the Gum*

Flax seed gum, also called *linseed gum,* is a white, powdery material that hydrates slowly to form a milky dispersion of relatively low viscosity (Fig. 1). It may dissolve completely when the concentration is kept below 0.1–0.2%. At constant concentration, viscosity decreases logarithmically with increasing temperature (Fig. 4), more rapidly at higher concentrations.[70] At constant shear, viscosity increases logarithmically with concentration[70] (Fig. 2). Flax seed gum forms non-Newtonian dispersions that are not thixotropic (Fig. 3). Mechanical mixing for 1 min reduced the viscosity of 0.5–1.0% dispersions by ~50%; very little was recovered upon standing, and the loss of viscosity was progressively greater with longer mixing times.[70] Therefore, the gum can only be used where mild mixing procedures are employed or where it is desirable to thin solutions after some time period.

Dispersions of flax seed gum show a slight pH optimum at about pH 8 (Fig. 5). An increase in ionic strength slightly decreases the viscosity (Fig. 6).

3. *Preparation, Use, and Development of the Gum*

A mucilaginous material can be extracted from flaxseed meal.[71] Solvent-extracted meal is air separated and screened to obtain kernel and hull fractions. The kernel is high in protein, and the hull is high in gum. The hull concentrate, which contains 11% of protein, is then mixed with water in the ratio of 1 part of concentrate to 30 parts of water; the pH is adjusted to 4.5, and the mass is heated to 60°–80° and agitated 1 hr. The liquid is centrifuged, adjusted to pH 7, concentrated by evaporation, and spray dried. The dried product is substantially protein- and fiber-free.

Another procedure obtains the gum by water extraction of oil meal cake, followed by centrifugation and evaporation under diminished pressure.[72] The concentrated liquid is treated with clarifying agents (charcoal, activated carbon, or infusorial earth) and then added to ethanol or another organic solvent miscible with water. The precipitate is collected and spray or drum dried.

Pressure steam extraction of the gum from flax seed has also been suggested.[73] In this process, seeds are extracted with wet steam under pressure, and the resulting liquid extract is passed into a receptacle under a substantially lower pressure. In still another method, seed or meal cake is immersed in cold or warm water containing an iron salt to prevent extraction of tannin pigments; the gum is separated from other impurities by centrifugation.[74]

The processing methods now in use for linseed oil have no significant effect on the properties of the gum,[75] although the mucilage-forming capacity of linseed tailings is reduced by heating.[76]

Repeated dissolution in water, centrifugation, and precipitation do not remove protein or minerals from the gum.[77] Washing with hydrochloric acid in ethanol or cation-exchange treatment effects hydrolysis,[77] although flax seed gum acid can be prepared under controlled conditions.[78]

Flax seed gum has not been used to any degree as an industrial gum, although large amounts of linseed meal are available and simplicity of extraction should allow a comparatively low price. It has some potential commercial importance in cosmetics and pharmaceuticals because of its action as a demulcent and emollient, and it has been stated that it can replace gum arabic, gum tragacanth, and gum karaya in certain applications; however, no information is available as to any definite uses of this gum in the U.S. today. There seems to be more interest in its applications in Eastern Europe.

It has been suggested that it can be used in medicinal preparations.[70] However, as the protein cannot be easily removed, it is possible that flax seed gum contains allergens.[1] Ointments and pastes containing flax seed gum are effective in the treatment of furunculosis, carbunculosis, impetigo, and ecthyma.[79] Flax seed gum can also be used as a bulk laxative (although it is not as effective as psyllium seed gum) and as a cough emollient agent.[80] It has also found to be useful for the stabilization of barium sulfate suspensions for x-ray diagnostic preparations.[70, 81] Flax seed gum also has lipotropic activity.[82] The free acid form of flax seed gum has been suggested for removal of exogenous sodium from the intestinal tract, for treatment of gastric hypoacidity, as an enteric coating material, as a retarder of tablet disintegration, for use in sustain-release drug therapy, for inhibition of bacterial growth in pharmaceutical preparations, and to produce more palatable fluid preparations of alkaloids or other drugs by removal of all or a portion of the distasteful materials from solution.[51]

Flax seed gum can replace gum arabic in stabilizing emulsions.[83] The concentration necessary to stabilize an emulsion is from 0.5–1.5%, depending on the quality. At 2.5% concentration, it is a good base for an eye ointment. It has also been suggested for use in hair-dressing preparations.[10] The mucilage is stringy and has a fast drying time, but it takes five to ten times as much gum to give the same vicosity as a 2% quince seed dispersion. A hand cream formulation contains flax seed gum.[83]

Egg white substitutes for use in bakery products and ice cream can be made from flax seed gum.[84] Its strong buffering action also makes it useful in the manufacture of fruit drinks.[85]

Flax seed gum was found to be one of the two most promising materials for reducing pressure losses in the Mohole drill string design, that is, it was one of two of a large number of natural and synthetic polymers that had a high degree of viscoelasticity at various shear rates.[86] At a concentration of ~6 lb/barrel (~23 g/liter), it reduces by 50% the pressure loss incurred when pumping sea water. It is recommended for drilling where heat stability and resistance to bacterial action are not required characteristics.

Flax seed gum has also been found useful in the water flooding process for recovering petroleum.[87] It has less tendency to be adsorbed on clay surfaces than some other polymers that have been examined. A gum dispersion can be deactivated with clay to remove the adsorbable fraction. The remaining solution is then used for the pusher flood to avoid decreasing the permeability around the well bore and to keep the solution in contact with the oil at essentially the same viscosity as the oil.

Flax seed gum can be used as a deflocculant to water-lay all lengths of fibers to produce paper or fabric with good tensile and flexural (tear) strength.[11] It can also be used to prepare a glue.[88] Other suggested potential uses are in water paints; in the manufacture of soluble fibers;[71] in food; and in the printing, textile, and cigar industries.[72]

The gum is a source of L-galactose.[89]

4. *Structure*

Acid-catalyzed hydrolysis of the gum releases L-galactose, D-xylose, L-arabinose, L-rhamnose (6-deoxy-L-mannose), D-galacturonic acid,[85, 90–93] 2-*O*-(D-galactopyranosyluronic acid)-L-rhamnose, and, perhaps, a trace of D-glucose.[93] L-Arabinose is the first component to be released, suggesting that it is in the furanose form. D-Xylose is released next and is found in eight times the amount of L-arabinose. The components of the aldobiouronic acid are only slowly released.[93–96]

Partial hydrolysis of the free acid form of flax seed gum yields an aldotriouronic acid composed of D-galacturonic acid, L-rhamnose, and L-galactose.[97]

It is clear that flax seed gum is a mixture of polysaccharides. It was first fractionated into fractions of different compositions with barium hydroxide.[98] Later, copper(II) acetate and ethanol were used for fractionation.[99] Using a modification of the latter procedure, it was shown that flax seed gum is a mixture of not less than three polysaccharides, one of which was an arabinoxylan (~25 : 75), containing only traces of uronic acid and D-galactose.[100, 101]

Partial acid-catalyzed hydrolysis of the purified arabinoxylan gave two disaccharides, tentatively identified as 4-*O*-β-D-xylopyranosyl-D-xylose (xylobiose) and 5-*O*-β-D-xylopyranosyl-L-arabinose. Mild partial hydrolysis resulted in the preferential removal of L-arabinose, suggesting the presence of the furanose ring form. Periodate oxidation indicated a branched structure. Periodate oxidation, after removal of L-arabinofuranosyl units, suggested that the L-arabinofurnosyl units protect some D-xylopyranosyl units from oxidation. These results indicate that D-xylopyranosyl end-units are present in the polysaccharide, that the proportion of these end-units is not greatly increased on elimination of some of the L-arabinofuranosyl units, and that the L-arabinofuranosyl units are not attached to D-xylopyranosyl end-units but to a main chain of β-D-(1→4)-linked xylopyranosyl units.[101]

Methylation analysis yielded 2,3,4-tri-*O*-methyl-, 2, 3-di-*O*-methyl-, and 3-*O*-methyl-D-xylose; D-xylose; 2,3-di-*O*-methyl- and 2,4-di-*O*-methyl-L-arabinose; and 2,3,4,6-tetra-*O*-methylgalactose. The optical rotation of the methylated galactose indicated that it was composed mostly of the D form. Erskine and Jones[100] interpret the evidence as indicating a main chain of β-D-(1→4)-linked xylopyranosyl units to which are attached at the *O*-2 position side chains of (1→5)- or (1→3)-linked D-xylopyranosyl-L-arabinopyranosyl and -L-arabinofuranosyl units.[101] They compare it to the neutral polysaccharide of psyllium seed gum.[37]

Using a combination of borate and cetyltrimethylammonium bromide precipitation, Hunt and Jones[102] separated flax seed gum into acidic and neutral fractions, then further separated the acidic mixture into copper(II) acetate solution-insoluble and -soluble fractions.

The copper(II) acetate solution-insoluble fraction contained L-rhamnose, L-galactose, and D-galacturonic acid in the approximate molar ratio of 2 : 1 : 2. Methylation analysis after reduction of the uronic acid gave 2,3,4-tri-*O*-methyl-, 3,4,-di-*O*-methyl-, and 4-*O*-methyl-L-rhamnose and 2,3,4,6-tetra-*O*-methyl- and 2,3,6-tri-*O*-methyl-D-galactose. Another derivative tentatively identified as 2,3-di-*O*-methyl-D-galactose was obtained; it probably arises from the reduction of D-galacturonic acid. L-Galactose was lost during methylation. Hunt and Jones[102] suggest that the polysaccharide is composed of a main chain of (1→2)-linked L-rhamnopyranosyl units with some branching at *O*-3, with most of the L-galactopyranosyl units attached as nonreducing end-units and with the D-galactopyranosyluronic acid units in the main chain or as internal units (substituted at *O*-4) of side chains. In support of this structure, autohydrolysis releases L-galactose first.

The copper(II) acetate solution-soluble fraction contained L-rhamnose (6-deoxy-L-mannose), L-fucose (6-deoxy-L-galactose), L-galactose, and D-galacturonic acid in the approximate molar ratio of 4 : 1 : 2 : 2. Methylation analysis of the reduced polysaccharide gave 2,3,4-tri-*O*-methyl-L-fucose, 2,3,4,6-tetra-*O*-methyl-L- and 2,3,6-tri-*O*-methyl-D-galactose, 4-*O*-methyl-L-rhamnose, L-rhamnose, and possibly 2,3-di-*O*-methyl- and 3-*O*-methyl-D-galactose. Periodate oxidation, in which about two thirds of the units were unoxidized, and Smith degradation[103] indicated the presence of a poly-L-rhamnosyl backbone with L-fucopyranosyl and L-galactopyranosyl units attached as nonreducing end-units. Again, the D-galactopyranosyluronic acid units appeared to be in the main chain or as internal units of side chains (substituted at *O*-4).[102]

The neutral material was composed of two polysaccharides in almost equal amounts and one polysaccharide in very small amounts. This neutral material seems to be similar, if not identical, to the arabinoxylan isolated by Erskine and Jones.[100, 101] A Smith degradation[103] yielded a polysaccharide in addition to low-molecular-weight products. Some D-xylose remained unoxidized after a second oxidation, indicating a high degree of branching. Methylation analysis of the once degraded material yielded 2,3,4-tri-*O*-methyl-, 2,3,-di-*O*-, and 4-*O*-methyl-D-xylose; D-xylose; and traces of either 2,3,4-tri-*O*-methyl- or 2,5-di-*O*-methyl-L-arabinose and 2,4-di-*O*-methyl- and 3-*O*-methyl-D-xylose. The available evidence indicates that the polysaccharide has a backbone of β-D-(1→4)-linked xylopyranosyl units, that there are some (1→3)-linked D-xylopyranosyl units, that some of the L-arabinosyl units are in the pyranose ring form, and that the polysaccharide is highly branched.[102]

Kalac[104] fractionated flax seed gum by a procedure almost identical to that used by Hunt and Jones.[102]

An enzyme from *Aspergillus niger* rapidly decreases the viscosity of flax seed gum solutions[105] and may be useful in further structural studies of the fractions.

IV. Okra Gum

1. *Source of Okra*

Okra, *Hibiscus esculentus,* family Malvaceae, is cultivated for its immature pods, which contain a gum that makes a thick, slimy mucilage. The pods are, therefore, used to thicken soups and stews and as a vegetable.

The plant, also called *gumbo* in the United States, *bhindi* in India, and *bamia* in the Arabic countries, is a native of Africa and is grown chiefly in the southern part of the United States. It is an annual plant that grows to a height of 2–8 ft (0.6–2.4 m). The many-ribbed fruit pods are picked for cooking, canning, and freezing when they are young and tender, 2–4 in (5–10 cm) in length.

The production of commercial pack frozen okra[106] in the U.S. varied between 30,000,000 and 47,500,000 lb (between 13,500,000 and 21,500,000 kg) a year during the period 1964–1968. Okra is also imported to the U.S. from Mexico, Panama, and other Central American countries.

Okra pods for the commercial production of okra gum could be shipped economically only in a dry, ground form.

2. Properties of the Gum

An investigation of the properties of okra gum was made by Walter E. Castro of the Water Resources Research Institute of Clemson University, Clemson, South Carolina.[107, 108] His main objective was to evaluate it as a friction reducing additive in pipe flow.

Dilute solutions of okra gum at concentrations $<0.001\%$ exhibit Newtonian behavior when viscosity is measured with a capillary tube viscometer, as do other polymers, such as guar gum.[107] However, at slightly higher concentrations, okra gum solutions have significant viscosity, with the viscosity increasing rapidly with increasing concentration.[108] There is little loss of viscosity at constant shear.[108]

Dilute aqueous solutions at concentrations $>0.001\%$ have significantly less friction in turbulent flow than water. A concentration of 0.0025% gives optimum friction reduction for Reynolds numbers less than 30,000 in a 0.666 in (1.69 cm) pipe and gives optimum friction reduction for Reynolds numbers less than 10,000 in a 0.186 in (4.72 mm) pipe. At higher Reynolds numbers, greater concentrations are required for optimum results. In general, maximum friction reduction is a function of pipe diameter, concentration, Reynolds number and, perhaps, other factors.[109]

The viscous, slightly cloudy okra gum solutions exhibit pituity or thread formation, a property characteristic of most friction-reducing solutions.[109]

In a product information sheet published by Morningstar-Paisley, Inc. about 1960, it is stated that the powdered gum may be dispersed directly in water by gentle agitation and that the tendency for lumping may be avoided by first wetting the powder with three times its weight of alcohol and adding the water with rapid stirring.

The rheological properties of okra dispersions are relatively little affected by the presence of salts, such as sodium chloride and sodium sulfate, much more so by divalent salts, such as calcium chloride, and are coagulated by trivalent ions, such as aluminum. Organic acids, such as citric and lactic acids, have little effect.

Anionic and nonionic surface-active agents appear to have no appreciable effect on the dispersions, but cationic agents bring about coagulation.

One of the unique features of okra gum is its remarkable ability to impart ropiness or stringiness to water solutions; a 0.25% dispersion is far more stringy than uncooked egg white. It is believed that this property is caused by huge molecular aggregates. Electron photomicrographs of okra dispersions show branched-chain structures $>4\mu$ long.

Okra dispersions are also unusual in their lubricity properties. They tend to form tenacious coatings on skin that are difficult to remove by washing.

Extensive investigation of the properties of okra gum remains to be done.

3. *Preparation, Use and Development of the Gum*

Okra gum is not now commercially available. For the studies of Castro and Neuwirth,[107, 108] dried okra was obtained from the John A. Manning Paper Co., Troy, N.Y., which has investigated the use of okra gum in the formation of long-fibered papers. This material was prepared by countercurrent extraction of ground, dried okra pods with 2-propanol to remove lipids. The extracted material was redried to remove the alcohol and reground. The resulting stable powder, which contained about 90% of water-insoluble substances, was stirred with water for 24 hr. The insoluble material was removed from the mixture by filtration through fine-mesh nylon screens under diminished pressure, centrifugation and, sometimes, filtration through carbon. The resulting viscous solution was used as prepared. However, it is assumed that okra gum could be recovered from this solution and purified by the usual procedures used for polysacccharide gums.

Because the gum is not commercially available, it has no regular uses in the U.S. However, it has been commonly used in the clarification of sugar cane juice in *gur* manufacture in India.[109] Castor seed and peanut extracts have now been found to be better clarifying agents in the open-pan boiling of sugar cane juice, making a *gur* of superior quality.[110] Okra gum appears to inhibit sucrose crystal formation. When added to a supersaturated sucrose solution at a concentration of 0.2% based on sucrose, very fine crystals are formed upon cooling.

A whipping composition for reconstituted dried egg white can be made with okra gum alone or in combination with a second whipping agent.[111]

An adhesive based on corn flour that gives viscosity and tack without gelling can be made with methylcellulose, hydroxyethylcellulose, okra gum, or sodium carboxymethylcellulose.[112] Okra gum can also be used on natural, synthetic, or mineral fibers as a deflocculant that makes it possible to water-lay all lengths of fibers to produce paper or fabric with good tensile and flexural (tear) strength.[11]

Okra gum can be used as a brightening agent in the electrodeposition of such metals as silver, nickel, copper, and cadmium;[113] it can be used in a quenching medium.[114] The most marked brightening effect is seen with nickel.

The use of okra gum in the preparation of a plasma replacement has been suggested.[115] It is claimed to be pyrogen-free, even though it contains a glycopeptide. It does inhibit prothrombin activation (threone activity).[116] This suggested use is interesting when considered with the ability of okra gum to reduce friction in turbulent fluid flow.[107, 108] Although flow in the cardiovascular system is primarily laminar, some turbulence does occur normally around vessel side branches such as intercostal and renal arteries. Also, in certain pathologic conditions, such as atheroscelerosis, the constricted vessels produce turbulence and high blood pressures with Reynolds numbers as high as 10,000.[108]

The Morningstar-Paisley, Inc. product information sheet also states that the protective colloid and emulsifying properties of okra gum make it possible to add it to cheese so that water can be added to make a smooth, creamy spread. It further states that okra gum exhibits antioxidant properties when incorporated into lard and peppermint oil. It presumably should be effective in protecting other animal and vegetable fats and oils from becoming rancid.

Because okra gum is obtained from an edible material, okra itself is used as a flavoring material and bodying agent in foodstuffs, and okra gum has some unique properties, the use of okra gum in the food and pharmaceutical industries warrants further consideration.

4. *Structure*

Whistler and Conrad[117] macerated fresh okra pods with 95% ethanol, filtered the mixture, and washed the residue with ethanol and acetone. The dried residue was extracted with water, and after the residue was collected by centrifugation, it was reextracted with water. The gum was precipitated from the aqueous extract with ethanol; collected by filtration; washed with 95% ethanol, absolute ethanol, and ether; and dried. The yield was 1.5% of the dry weight of the fresh pods.

From a partial acid-catalyzed hydrolysis of this preparation, they isolated two disaccharides, 4-*O*-D-galactopyranosyl-D-gallactose[117] and 2-*O*-(D-galactopyranosyluronic acid)-L-rhamnose,[118] and two trisaccharides,[118] D-galactopyranosyl-(D-galactopyranosyluronic acid)-L-rhamnose and (D-galactopyranosyluronic acid)-L-rhamnopyranosyl-D-galactose.

Amin[119] reprecipitated the gum as a copper(II) complex which was not easily soluble in water but was soluble in alkaline solution. He found D-galactose, L-arabinose, L-rhamnose, and D-galacturonic acid in a molar ratio of 80 : 3 : 10 : 6. A methylation analysis yielded 2,3,4,6-tetra-*O*-methyl-D-galactose (12 moles), 2,3,6-tri-*O*-methyl-D-galactose (60 moles), 2,6-di-*O*-methyl-D-galactose (6 moles), 3,4-di-*O*-methyl-L-rhamnose (8 moles), and ?-*O*-methyl-L-rhamnose (9 moles).

Kelkar and coworkers[120] disagreed with Whistler and Conrad[118] and Amin[119] on the gum's composition. They found 11.9% ash and 3.26% nitrogen that could not be removed using acidic ethanol. The ash contained iron(III), calcium(II), and phosphate. Hydrolysis with 8*N* hydrochloric acid at 90° for 24 hr yielded L-histidine (1.1%), L-arginine (0.6%), L-lysine (4.7%), L-glutamic acid (0.8%), L-serine (7.2%), and L-alanine (0.6%). Hydrolysis with 2*N* hydrochloric acid at 90° for 24 hr yielded D-glucose and D-glucosamine (10.1%). Hydrolysis with 0.5*N* barium hydroxide at 90°–100° for 10 hr yielded a tetrasaccharide.

Kishida and Fukui[121] found that okra gum is composed of 14 amino acids, D-glucose, D-fructose, possibly some uronic acid, and some protein. Kishida[122] later characterized a glycopeptide as being L-glycyl-(L-aspartyl or L-asparaginyl)-L-threonyl-L-seryl-(L-glutamyl or L-glutaminyl)-L-serine to which is attached D-glucose and D-fructose.

V. References

(1) R. J. McCredie and R. L. Whistler, *in* "Industrial Gums," R. L. Whistler, ed., Academic Press, 1st Ed., 1959, p. 433.

(2) G. Hadary, *Food Ind.,* **15,** 76 (1943).

(3) "U.S. Dispensatory," J. B. Lippincott Co., Philadelphia, Pa., 24th Ed., 1947.

(4) R. K. Aliev, A. A. Namazova, and A. K. Rakhimova, *Materialy Nauch.-Prakt. Konf. Azerbaidzhan. Nauch. Farm. Obshchestva Posvyashchen. Dostizhen. Aptechn. Dela Izuch. Lekarstv.-Syr'evykh Resursov Respublike Gody Sov. Vlasti, Baku, Sbornik,* 59 (1960); *Chem. Abstr.,* **56,** 6090 (1962).

(5) *Oil, Paint, Drug Rep.*

(6) E. Husemann and G. Kelich, *Cellul. Chem. Technol.,* **3,** 445 (1969); *Chem. Abstr.,* **72,** 123086b (1970).

(7) D. S. Davis, *Chem. Process. (Chicago),* **20,** 208 (1957).

(8) G. Tchillingarian and C. M. Beeson, *Petrol. Eng.,* **24,** No. 4, 45 (1952); *Chem. Abstr.,* **46,** 5299 (1952).

(9) G. Hadary, *Ice Cream Rev.,* **26,** No. 11, 22 (1943).

(10) E. A. Siehrs, U.S. Patent 2,267,624 (1941).

(11) L. R. B. Hervey, U.S. Patent 3,102,838 (1963); *Chem. Abstr.,* **60,** 5745 (1964).

(12) M. G. DeNavarre, "The Chemistry and Manufacture of Cosmetics," D. Van Nostrand, New York, 1941, pp. 289–290.

(13) *Schimmel Briefs,* **1,** No. 181 (1950).

(14) R. S. Manly, U.S. Patent 2,850,410 (1958); *Chem. Abstr.,* **53,** 2657 (1959).

(15) K. N. Richardson, *Soap, Perfum. Cosmet.,* **18,** 286 (1945).

(16) R. G. Harry, "Modern Cosmeticology," Chemical Pub. Co., New York, 1956, p. 133.

(17) H. Bennett, "The New Cosmetic Formulary," Chemical Publ. Co., New York, 1970, p. 20.

(18) J. Morvay and K. Szendrei, *Gyogyszereszet,* **11,** 178 (1967); *Chem. Abstr.,* **67,** 84794f (1967).

(19) C. Schmidt, *Ann.,* **51,** 29 (1844).

(20) W. L. Kirchner and B. Tollens, *Ann.,* **175,** 205 (1875).

(21) K. Muhlethaler, *Makroml. Chem.,* **2,** 143 (1948).

(22) K. Muhlethaler, *Exp. Cell. Res.,* **1,** 341 (1950).

(23) A. G. Renfrew and L. T. Cretcher, *J. Biol. Chem.,* **97,** 503 (1932).

(24) K. Bailey and F. W. Norris, *Biochem. J.,* **26,** 1609 (1932).

(25) H. W. Youngken, *J. Amer. Pharm. Ass.,* **21,** 1265 (1932).

(26) "The National Formulary," Amer. Pharm. Ass., Washington, D.C., 13th Ed., 1970, p. 560.

(27) D. Greenberg, *J. Amer. Pharm. Ass.*, **37,** 139 (1948).

(28) M. J. Jones and C. C. Albers, *J. Amer. Pharm. Ass., Sci. Ed.*, **44,** 100 (1955).

(29) A. Neva and E. B. Fischer, *J. Amer. Pharm. Ass., Sci. Ed.*, **38,** 34 (1949).

(30) A. J. Erskine and J. K. N. Jones, *Can. J. Chem.*, **35,** 1174 (1957).

(31) P. C. Bandyopadhyay, *J. Sci. Ind. Res., Sect. B,* **19,** 378 (1960); *Chem. Abstr.*, **55,** 10003 (1961).

(32) C. Pranab, *J. Sci. Ind. Res., Sect. B,* **20,** 101 (1961); *Chem. Abstr.*, **55,** 22503 (1961).

(33) B. M. Mithal and V. D. Gupta, *J. Birla Inst. Technol. Sci.*, **1,** 162 (1967); *Chem. Abstr.*, **71,** 128648a (1969).

(34) M. Hefti and Co., Ital. Patent 498,739 (1954); *Chem. Abstr.*, **51,** 8459 (1957).

(35) H. B. Near, A. J. Pacini, R. W. Crosley, M. M. Gerth, F. T. Breidigam, and J. D. Kelly, U.S. Patent 2,010,880 (1935); *Chem. Abstr.*, **29,** 6674 (1935).

(36) E. Anderson and M. Fireman, *J. Biol. Chem.*, **109,** 437 (1935).

(37) R. A. Laidlaw and E. G. V. Percival, *J. Chem. Soc.*, 1600 (1949).

(38) J. F. Montague, "Psyllium Seed: The Latest Laxative," Montague Hospital for Intestinal Ailments, New York, 1932.

(39) C. K. Atal and K. K. Kapur, *Indian J. Pharm.*, **25,** 376 (1963); *Chem. Abstr.*, **60,** 7870 (1964).

(40) E. M. MacKay, E. M. Hall, and F. M. Smith, *Proc. Soc. Exp. Biol. Med.*, **30,** 152 (1932).

(41) F. Coulston and J. C. Seed, *J. Amer. Pharm. Ass., Sci. Ed.*, **45,** 716 (1956).

(42) J. N. Bone and L. W. Rising, *J. Amer. Pharm. Ass., Sci. Ed.*, **43,** 102 (1954).

(43) R. Wasicky, *Planta Med.*, **9,** 232 (1961); *Chem. Abstr.*, **56,** 1526 (1962).

(44) R. H. Blythe, J. J. Gulesich, and H. L. Tuthill, *J. Amer. Pharm. Ass., Sci. Ed.*, **38,** 59 (1949).

(45) A. L. Monaco and E. J. Dehner, *J. Amer. Pharm. Ass., Sci. Ed.*, **44,** 237 (1955).

(46) R. S. Goodhart, "Modern Drug Encyclopedia and Therapeutic Index," Reuben H. Donnelley Corp., New York, 9th Ed., 1963.

(47) A. E. Man, *Clin. Med.*, **49,** 188 (1942).

(48) N. S. MacDonald, R. E. Nusbaum, F. Ezmirlian, R. C. Barbera, G. V. Alexander, P. Spain, and D. E. Rounds, *J. Pharmacol. Exp. Ther.*, **104,** 348 (1952).

(49) A. H. Beaufour and G. H. Beaufour, Fr. Patent CAM87 (1965); *Chem. Abstr.*, **64,** 534 (1966).

(50) J. E. Garvin, D. T. Forman, W. R. Eiseman, and C. R. Phillips, *Proc. Soc. Exp. Biol. Med.*, **120,** 744 (1965).

(51) J. V. Swintosky, L. Kennon, and J. Tingstad, *J. Amer. Pharm. Ass., Sci. Ed.*, **44,** 109, 112 (1955).

(52) D. Davidson, U.S. Patent 2,768,073 (1956); *Chem. Abstr.*, **51,** 720 (1957).

(53) J. F. M. Craig and E. L. Falconer, Ger. Offen. Patent 1,950,580 (1970); *Chem. Abstr.*, **73,** 5606u (1970).

(54) P. H. Cardwell and L. H. Eilers, U.S. Patents 2,824,833 and 2,824,834 (1958); *Chem. Abstr.*, **52,** 11400 (1958).

(55) B. M. Mithal and J. L. Kasid, *Indian J. Pharm.*, **26,** 316 (1964); *Chem. Abstr.*, **62,** 14427 (1965).

(56) R. A. Laidlaw and E. G. V. Percival, *J. Chem. Soc.*, 528 (1950).

(57) B. A. Lewis and F. Smith, *J. Amer. Chem. Soc.*, **79,** 3929 (1957).

(58) W. A. G. Nelson and E. G. V. Percival, *J. Chem. Soc.*, 58 (1942).

(59) F. Hostettler and H. Deuel, *Helv. Chim. Acta,* **34,** 2440 (1951).

(60) E. L. Hirst, E. G. V. Percival, and C. B. Wylam, *J. Chem. Soc.*, 189 (1954).

(61) A. J. Erskine and J. K. N. Jones, *Can. J. Chem.*, **34,** 821 (1956).

(62) J. Mullan and E. G. V. Percival, *J. Chem. Soc.*, 1501 (1940).

(63) E. G. V. Percival and L. C. Willox, *J. Chem. Soc.*, 1608 (1949).

(64) E. Anderson, L. A. Gillette, and M. G. Seeley, *J. Biol. Chem.*, **140,** 569 (1941).

(65) A. C. Dillman and L. G. Goar, *U.S. Dept. Agr. Farmer's Bull.* No. 1793 (1937).

(66) A. C. Dillman, *U.S. Dept. Agr. Farmer's Bull.* No. 1328 (1924); No. 2036 (1952); *U.S. Dept. Agr. Tech. Bull.* No. 1064 (1953).

(67) "Agricultural Statistics," U.S. Dept. Agr., U.S. Govt. Printing Office, 1969, pp. 122, 123.

(68) A. C. Dillman and T. H. Hopper, *U.S. Dept. Agr. Tech. Bull.* No. 844 (1943).

(69) A. D. D. Rose and L. Tombesi, *Ann. Sper. Agr.*, **4,** 15 (1950); *Chem. Abstr.*, **44,** 6493 (1950).

(70) J. Zemanova-Simaljakova and J. Kalac, *Cesk. Farm.*, **15,** 142 (1966); *Chem. Abstr.*, **65,** 2071 (1966).

(71) D. H. Bolley and R. H. McCormack, U.S. Patent 2,593,528 (1952); *Chem. Abstr.*, **46,** 9333 (1952).

(72) C. T. Mason and L. A. Hall, *Food Ind.*, **20,** 382 (1948).

(73) J. E. Sanftleben, U.S. Patent 1,841,763 (1932); *Chem. Abstr.*, **26,** 1734 (1932).

(74) G. Tomoda and Y. Asami, Jap. Patent 3359 (1950); *Chem. Abstr.*, **46,** 10654 (1952).

(75) A. M. Goldovskii and S. S. Berestovskaya, *Masloboino-Zhirovaya Prom.*, **27,** No. 3, 21 (1961); *Chem. Abstr.*, **55,** 15953 (1961).

(76) W. Huss, *Landwirt Forsch.*, **22,** 196 (1969); *Chem. Abstr.*, **72,** 62814p (1970).

(77) J. Kalac, E. Hornakova, and J. Zemanova-Simaljakova, *Cesk. Farm.*, **17,** 423 (1968); *Chem. Abstr.*, **70,** 84393m (1969).

(78) R. G. Schweiger, *Methods Carbohyd. Chem.*, **5,** 59 (1965).

(79) R. K. Aliev, *Amer. J. Pharm.*, **118,** 439 (1944).

(80) A. Boichinov, K. Akhtardzhiev, and D. Kolev, *Farmatsiya (Sofia),* **16,** 21 (1966); *Chem. Abstr.*, **66,** 68861z (1967).

(81) N. Tufegdzic, E. Tufegdzic, and A. Georgijevic, *Acta Pharm. Jugoslav.*, **12,** 121 (1962); *Chem. Abstr.*, **62,** 8944f (1965).

(82) V. Hrstka and L. Hronec, *Arch. Int. Pharmacodyn.*, **132,** 197 (1961); *Chem. Abstr.*, **56,** 847 (1962).

(83) "Drug and Cosmetic Catalog," Drug and Cosmetic Industry, New York, 12th Ed., 1956–1957.

(84) O. E. Nikkila, Finn. Patent 34,558 (1965); *Chem. Abstr.*, **65,** 11250h (1966).

(85) J. Schormuller and H. Winter, *Nahrung,* **2,** 83 (1958).

(86) J. L. Lummus and B. V. Randall, PB 169605, 1964, 38 pp; from *U.S. Govt. Res. Develop. Rep.*, **41,** 519 (1966); *Chem. Abstr.*, **67,** 4914g (1967).

(87) J. L. Lummus, U.S. Patent 3,338,304 (1967); *Chem. Abstr.*, **67,** 110310s (1967).

(88) G. Maton and M. Maton, Fr. Patent 1,272,979 (1962); *Chem. Abstr.*, **57,** 1139 (1962).

(89) C. Araki and K. Arai, *Methods Carbohyd. Chem.*, **1,** 122 (1962).

(90) A. Hilger, *Ber.*, **36,** 3197 (1903).

(91) A. Neville, *J. Agr. Sci.*, **5,** 113 (1913).

(92) H. Akhtardzhiev and D. Kolev, *C.R. Acad. Bulgare Sci.*, **10,** 387 (1957); *Chem. Abstr.*, **52,** 14203 (1958).

(93) K. Zagorski, *Ann. Univ. Mariae Curie-Sklodowska, Lublin-Polonia, Sect. AA,* **14,** 59 (1961); *Chem. Abstr.*, **57,** 8877 (1962).

(94) E. Anderson and J. A. Crowder, *J. Amer. Chem. Soc.*, **52,** 3711 (1930).

(95) C. Niemann and K. P. Link, *J. Biol. Chem.*, **104,** 205 (1934).

(96) R. S. Tipson, C. C. Christman, and P. S. Levene, *J. Biol. Chem.*, **128,** 609 (1939).

(97) E. Anderson and H. J. Lowe, *J. Biol. Chem.*, **168,** 289 (1947).

(98) K. Bailey, *Biochem. J.*, **29,** 2477 (1935).

(99) D. G. Easterby and J. K. N. Jones, *Nature,* **165,** 614 (1950).

(100) A. J. Erskine and J. K. N. Jones, *Can. J. Chem.*, **34,** 821 (1956).

(101) A. J. Erskine and J. K. N. Jones, *Can. J. Chem.*, **35,** 1174 (1957).

(102) K. Hunt and J. K. N. Jones, *Can. J. Chem.*, **40,** 1266 (1962).

(103) I. J. Goldstein, G. W. Hay, B. A. Lewis, and F. Smith, *Methods Carbohyd. Chem.*, **5,** 361 (1965).

(104) J. Kalac, *Cesk. Farm.*, **19,** 143 (1970); *Chem. Abstr.*, **73,** 80438u (1970).

(105) J. Kalac and L. Rexova, *Biochim. Biophys. Acta,* **167,** 590 (1968).

(106) Ref. 66, p. 203.

(107) W. E. Castro, "Reduction of Flow Friction with Polymer Additives," Completion Report OWRR Project No. A-009-SC (1972).

(108) W. E. Castro and J. G. Neuwirth, Jr., *Chem. Tech.*, **1,** 697 (1971).

(109) K. L. Khanna and A. S. Chacravarti, *Indian J. Agr. Sci.*, **19,** 137 (1949); *Chem. Abstr.*, **45,** 5430 (1951).

(110) A. S. Chacravati, K. Prasad, and K. L. Khanna, *J. Sci. Ind. Res., Sect. B,* **11,** 305 (1952); *Chem. Abstr.*, **47,** 4637 (1953).

(111) H. F. Ziegler, Jr., and H. J. Buehler, U.S. Patent 3,219,457 (1965); *Chem. Abstr.*, **64,** 4177 (1966).

(112) C. G. Fortney, Jr., and K. R. Hunt, U.S. Patent 3,251,703 (1966); *Chem. Abstr.*, **65,** 7439 (1966).

(113) R. S. Robinson, U.S. Patent 2,932,610 (1960); *Chem. Abstr.*, **54,** 13909 (1960).

(114) R. Viala, Fr. Patent 891,204 (1944); *Chem. Abstr.*, **47,** 3209 (1953).

(115) H. B. Benjamin, H. K. Ihrig, and D. A. Roth, *Rev. Can. Biol.*, **10,** 215 (1951); *Chem. Abstr.*, **45,** 9805 (1951).

(116) W. H. Seegers, W. G. Levine, and S. A. Johnson, *J. Appl. Physiol.*, **7,** 617 (1955).

(117) R. L. Whistler and H. E. Conrad, *J. Amer. Chem. Soc.*, **76,** 1673 (1954).

(118) R. L. Whistler and H. E. Conrad, *J. Amer. Chem. Soc.*, **76,** 3544 (1954).

(119) El S. Amin, *J. Chem. Soc.*, 828 (1956).

(120) G. M. Kelkar, T. R. Ingle, and B. V. Bhide, *J. Indian Chem. Soc.*, **39,** 557 (1962).

(121) T. Kishida and H. Fukui, *Kaseigaku Zasshi,* **18,** 71 (1967); *Chem. Abstr.*, **67,** 70558v (1967).

(122) T. Kishida, *Kaseigaku Zasshi,* **20,** 397 (1969); *Chem. Abstr.*, **72,** 107134k (1970).

CHAPTER XVII

TAMARIND

P. S. RAO

Forest Research Laboratory, Bangalore, India

AND

H. C. SRIVASTAVA

Ahmedabad Textile Industry's Research Association, Ahmedabad, India

I. Introduction

A recent addition to industrial gums is the polysaccharide obtained from the seed kernels of the tamarind tree, *Tamarindus indica* (L). Coming into regular use in 1943, this material was one of the direct discoveries made because of the exigencies of World War II. In a search for new sizing materials, the Forest Research Institute, Dehra Dun, discovered that the polysaccharide from the seed kernels of the tamarind tree possesses excellent sizing properties.

1. Tree

Tamarindus indica is one of the most important and common trees of India. In addition to being a forest tree, it is cultivated throughout India, Bangladesh, Ceylon, and Burma. It is also found in many other parts of the world: Florida, Egypt, Sudan, Formosa, Malaya, and other Southeast Asian countries. It is probably indigenous to Africa and possibly to some parts of South India.[1,2]

The tamarind tree is a large evergreen tree that grows rapidly and lives for more than 100 years. A full-grown tree is ~24 m (~80 ft) high, with a circumferance of ~7.5 m (~25 ft). It usually begins to bear fruit after 13–14 years and continues to give abundant crops for more than 60 years. The tree flowers in spring and its fruit ripens in winter. A full-grown tree yields 180–225 kg (400–500 lb) of fruit.[2, 3]

Almost every part of the tamarind tree finds some use in India.[1, 4–6]

2. Fruit

The fruit is a large flat pod, 10–15 cm (4–6 in) long. The ripe pods are gathered from the tree by merely shaking the branches. The pods that remain on the tree are gradually blown off and are picked up.[7]

The fruit contains seed, stringy fibrous matter, and acidic pulp. Its analysis has been reported by Hooper[3] as follows: pulp, 55.0%; seed, 33.9%; shell and fiber, 11.1%. Analysis of the edible portion of the ripe pod is as follows:[8] water, 63.3–68.6%; total solids, 33.6–31.3%; proteins, 3.1–1.6%; fat, 0.27–0.69%; total sugars, 22.0–30.4%; sucrose, 0.1–0.8%; cellulose, 3.4–2.0%; ash, 1.6–1.2%. It contains 16–18% of acids; the pH is 3.15. The ash contains 45–49% of potassium and appreciable amounts of silicon, sodium, phosphorus, and calcium. Similar data have been reported for the Florida variety.[9]

3. Pulp

Tamarind pulp is the chief souring agent in Indian curries, *chatnis* (chutneys), and sauces. In the West Indies, after hand removal of the seed and the epicarp, the pulp is mixed with sugar and exported to Europe in casks to be used by confectioners.[4] In India, for preservation it is mixed with about 10% of salt, trodden into a mass, and made into balls. Subsequently, the latter are exposed to sun and dew for about a week. In another process, the pulp, freed from fiber and seeds, is packed into stout gunny bags lined with matting made from date-palm or pamyra leaves.[7]

Because of the presence of sugars and acid, the pulp is largely employed in India in the preparation of congies, syrups, *chatnis* (chutneys), and beverages. It has also long been used in the preparation of laxative drinks.[7, 15]

4. Seed

Tamarind seed, which is a by-product of the tamarind pulp industry, is a valuable raw material for certain industries in India. The seed is flat and of irregular shape, being round, ovate, or obtusely four sided. The length of the side is 1.5 cm (0.6 in), whereas the thickness is 0.75 cm (0.3 in), with the edge broadly keeled and more often slightly furrowed. A kernel (~70%) is covered by a testa (~30%) which is of a rich brown color.[11] For a pharmacognostic description in comparison with other leguminous seeds, reference can be made to the work of Trease.[12] In the powder form, it can be distinguished from locust bean gum and guar gum under a microscope or by determination of the proportions of sugars released by hydrolysis, although it is not possible to do so by infrared spectroscopy.[13]

It is estimated that about 140,000 metric tons of the seed are available annually in India, and the chief collection centers are Calcutta and Ghatal in Bengal; Cuttack, Jeypore, Parlakimidi, Berhampore, and Rayghada in Orissa; Madura and Pollachi in Tamil Nadu; Hindpur, Bobbili, Hyderabad, and Warangal in Andhra; Thoduvatti in Kerala; and Bombay, Ahmednagar, Satara, and Nasik in Maharashtra. Their current price is Rs.22–25 a quintal (220 lb), although it rose in 1965 to as high as Rs.45/- a quintal (Rs.7.50 = $1.00 at the present rate of exchange). Since the kernels form nearly 70% of the whole seed, their potential availability can be estimated at nearly 98,000 metric tons a year.

When roasted, the testa becomes brittle and, if gently pounded with a wooden mallet, is detached from the kernel. When soaked, the seed coat becomes loose and is easily removed as in the blanching of almonds.[3] The kernels are then boiled or fried before they are eaten. They are said to be tolerably palatable. Alternatively, they are dried and ground into flour from which cakes or bread can be prepared after admixture with other cereal flours.[1–3] In certain areas, the kernels are mixed with rice, fried, powdered, made into balls with sugar and ghee, and eaten by the poorer classes. Certain hill tribes eat the kernels combined with mahwa flowers (*Madhuca latifolia*). In all these preparations, it is essential that the outer testa be completely removed; otherwise some untoward effects, such as depression, constipation, diarrhea, and inflammation might result. However, the analysis of the kernels does not indicate any toxic principles; rather, they are regarded as having medicinal effects. They are said to function as a stomachic.[3] The kernel powder is also recommended as a valuable remedy in diarrhea and dysentery.

Proteins of tamarind seed kernels are of a higher biological value than those of wheat and corn, and lower only than those of millet; but their net utilization is lower than that of the others.[14] Recently, this problem has been further examined by the rat-growth method at a 10% level of protein in diets supplemented with other dietary essentials.[15] It is reported that the proteins have a nutritive value that compares well with that of cereal proteins. It is also suggested that the isolated proteins can be used either as an adhesive or as a feed for cattle. A method has been developed for isolation of the proteins from the kernel powder by suspending it in a nonpolar, nonhydrolyzing solvent, such as trichloroethylene, toluene, or benzene, and then passing the suspension through a 200-mesh sieve, which retains the protein portion.[16] The nutritive value of the carbohydrate portion has also been studied by the rat-growth method.[17] Replacement of 25% or less of the rice by tamarind kernel powder produces a significant improvement in the over-all nutritive value of a rice diet; but when rice is completely replaced by tamarind kernel powder, the diet cannot support life in rats. Nutritional experiments show that tamarind seed, although rich in proteins of a high biological value, is lacking in certain dietary essentials and that supplementation with available carbohydrates and mineral mixtures is necessary to offset this deficiency.

In composition, tamarind kernels resemble the cereals:[3, 14, 15, 18–20] protein, 15.4–22.7%; oil 3.9–7.4%; crude fiber, 0.7–8.2%; nonfiber carbohydrates, 65.1–72.2%; ash 2.45–3.3% on a dry basis. The oil present in the tamarind kernels resembles peanut oil. It is obtained in a yield of 3.89–6.38% of the weight of the dry seed[3, 18, 20–23] and 6–8% of the weight of the kernels.[3, 15, 19, 24, 25]

Tamarind seed is promising as a cattle feed. Apart from its consumption in India to some extent for the purpose mentioned, it is reported to be exported to America and other countries to serve as feed for pigs. The amount of total carbohydrates and minerals compares favorably with those of oats, whereas the protein content is much higher than that of cereals.

II. Tamarind Kernel Powder (TKP)

Tamarind seed is most valued for sizing in the textile industry. When its kernel is boiled, a sort of gruel is obtained that, on drying in thin layers, produces strong, smooth, continuous, and elastic films — properties useful in sizing. Therefore, its powder, called TKP (tamarind kernel powder) commercially, finds extensive use as a textile size.

1. Preparation

Preparation of TKP is a relatively simple operation and consists of decortication of the seeds and pulverization of the creamy white kernels. First, the seeds are thoroughly washed with water in order to free them from attached pulp. Hollow seeds, which float, are discarded. Decortication of the seeds is the most difficult step because the testae are tenaciously held to the kernels and cannot be removed easily. According to one patented process for removal of the testae, the seeds are parched in an oven or on hot sand at 150° for 10–15 min so that the inner kernels are not charred.[26, 27] The testae, which become brittle by this treatment, are removed by light pounding and winnowing. In another process,[28, 29] the seeds are immersed for about 20 sec in molten paraffin kept at 200°. Prior to this treatment, the seeds are soaked in cold 10–25% sodium chloride solution for ~15 hr, then washed, and finally boiled in water for about 5 min. This process, it is claimed, not only readily separates the testae from the kernels, but also removes the radicles of the seeds. Two other methods,[30, 31] similar to the one adopted by Krishna,[26] involve heating the seeds either alone or mixed with fine sand and/or powdered inorganic salts, such as sodium chloride, sodium sulfate, magnesium sulfate, potassium chloride, and potassium sulfate, and/or filings of metals such as iron, aluminum, copper, and tin, for 15–30 min at temperatures not exceeding 120°. In one method, the heating is done under pressure.[30] In still another process, the seeds are hammered, scratched, filed, pierced, or treated with concentrated sulfuric acid so as to crack the outermost layers without exposing the kernels. The seeds are then steeped for 8–15 hr in water of dilute acid.[32] In all the above processes, with the exception of the last one, the testae become brittle and friable and are removed by a decortication process. After their separation by winnowing, the remaining creamy white kernels are ground into powder.

For simultaneous decortication of seeds and pulverization of kernels, the seeds are ground directly. Here, advantage is taken of the fact that the testae and kernels have different degrees of pulverizability. The process, therefore, involves (*a*) breaking the seeds in a disintegrator running at such a speed (700–1100 rpm) that the testae alone are pulverized while the kernels remain in a crushed condition, (*b*) removing the testae, and (*c*) grinding the crushed kernels to a fine powder in a disintegrator running at a high speed (2000–2500 rpm).[33, 34] In the process of Dutt and Chatterji,[33] an impact-*cum*-attrition type of disintegrator is used for the initial grinding of the seeds and also for the final pulverization of the crushed cotyledons. An air blower is suggested for removal of the finely divided testae; they are blown off in the air stream, while the broken cotyledons fall out or remain undischarged. Alternatively, they can be separated either by winnowing or by flotation.

Of the two general processes, the cold process has the advantage of lower costs, higher rates of production, and less chances of degradation of the sizing component through heat, but in this method, a higher capital outlay and greater supervision are required. The parching process, on the other hand, can be done on a cottage-industry basis, requires a lower initial capital, and enables the testae to be removed more easily. However, unless proper controls are introduced at the parching stage, there is the likelihood of heat degradation of the polysaccharide. Overroasting of the seed results in deterioration of the kernel powder. When overroasted, the color of the flour becomes darker and the consistency of the paste decreases, whereas the percentage of insoluble matter increases. Comparison of unroasted TKP and TKP roasted at temperatures of 100°–200° for different periods, shows that, as judged by the copper number and the alkali number, no appreciable heat degradation takes place up to 150°. In fact, samples roasted at 100°–125° have much lower copper numbers and alkali numbers than the unroasted samples, and do not develop marked undesirable color. This strange behavior is caused by the presence of the non-polysaccharide fraction of TKP. In both corn starch and the tamarind seed polysaccharide, the copper number and the alkali number increase gradually on roasting up to a temperature of 150° and then rise very rapidly, as is also the case with tamarind kernel powder.[35]

Combining the principles of the two processes, an integrated scheme has been suggested for the production of TKP so that the quality of the flour is assured and, at the same time, production costs are not high.[36] The seed is first parched in a roaster at a definite temperature for a measured period and is then fed into a grain cleaner provided with beaters and an exhaust fan. The dirt and husk are removed here, and the decorticated seed is then ground in two stages. It is first crushed in an impact grinder and then pulverized to the required fineness in a stone grinder. In a typical experiment, the yield of TKP is 51.5% of the weight of the seed; the yield of testa is 33.1%, and the remaining percentage is moisture lost during parching. The powder, on long storage, might deteriorate through the action of enzymes, particularly in moist areas. Therefore, it is desirable to mix the powder with 0.5% by weight of sodium bisulfite and pack it in moistureproof containers[37, 38] or to store it in a dry place and to fumigate it with sulfur dioxide at frequent and regular intervals.[39]

At present, about 50,000 metric tons are produced annually in India.

2. *Properties*

The powder, which is creamy white, is a mixture of substances.[3, 14, 15, 19, 20, 25, 40] In addition to polysaccharides, proteins, fiber, fat, and inorganic salts, it contains free sugars and tannins.[25, 40] The free sugars are D-xylose, D-galactose, and D-glucose. Evidence for the presence of free uronic acid is conflicting,[41–43] but the presence of D-galacturonic acid in the water-insoluble polysaccharide fraction of TKP has been demonstrated.[44, 45] No detailed work has been done on the proteins, except to show that they contain prolamins, albumins, and glutelin.[48] although some work was reported on seedlings.[54, 55] The oil contains 28% of saturated and 72% of unsaturated (oleic and linoleic) fatty acids.[56]

3. *Use as a Size*

The main function of sizing in yarns is to give them a protective coating so that they withstand the stresses, strains, and abrasions that take place on the looms during weaving. In fabrics, the size adds weight and imparts smoothness and stiffness to low-quality materials. A good sizing paste should have viscosity, fluidity, and adhesiveness. During sizing, the size, if it has good fluidity, penetrates the yarn; if it has good adhesiveness, it glues the fibers together so that the strength of the yarn increases. At the same time, it forms a smooth, strong film on the outer surface of the yarn, which cements together all the protruding fibers. Because of this smooth film, the friction between the warps and wefts and between the yarn and machinery is minimized so that weaving is facilitated. The best sizes are those which produce strong films on the warp yarn. The strength of the film is dependent on the viscosity of the size solutions; in general, the higher the viscosity, the greater is the strength of the film.

In addition to being strong, the surface films should be elastic; otherwise the yarn becomes stiff. A similar result ensues if the size has too much penetrating power, in which case it goes into the center of the yarn and glues all the fibers together. If this happens, the tensile strength of the yarn increases, but the flexibility decreases, having an adverse effect on the efficiency of weaving. Therefore, it is essential to maintain a cushioning effect in the center of the yarn in order to increase the shock resistance.[29]

Tamarind kernel powder, when heated with water, forms thick, viscous, colloidal dispersions, that yield strong, transparent, elastic films on evaporation. Its dispersions also have sufficient gluing power, but their fluidity and consequently their power of penetration into the yarn is rather poor when compared with thin-boiling starches.

Viscosity.—When TKP is treated with water, it disperses instantly even in the cold, but a uniform solution is obtained only on heating the mixture. The maximum viscosity is attained after boiling for 20–30 min.[49, 50] It is to be noted that these solutions exhibit a non-Newtonian flow, that is, the rate of flow in the viscometer increases more rapidly than the pressure applied.[51] Hence, solutions of TKP are pseudoplastic and, in this respect, they resemble those of gum tragacanth but differ from those of gum arabic.

As with starches, the viscosity of TKP dispersions is dependent on a number of factors, such as the purity of the material, the concentration of the solution, the conditions of its preparation, the pH of the medium, and the addition of other chemicals. For instance, the product from seeds that have been overroasted or from those that have undergone microbiological degradation yield solutions of lower viscosity. Similarly, products prepared from washed and unwashed kernels differ. The viscosity of the dispersions or pastes prepared from ordinary TKP decreases rapidly on prolonged boiling, but if the product is prepared from thoroughly washed kernels, the viscosity does not fall even on boiling for 4 hr.[51] Apparently the hydrolyzing agents are eliminated during washing.

In general, the viscosity of TKP dispersions is much higher than that of starch solutions of equal strength. A 5% (w/v) paste of TKP has a peak viscosity of more than 1600 Brabender Units (BU), whereas a paste of corn starch of the same concentration has a viscosity of only 300 BU (Fig. 1).[50] As with starches, the viscosity of TKP rises with the concentration. It increases gradually in concentrations of up to 0.5% and more rapidly at higher concentrations. Above 1%, a small increase in concentration results in a considerable increase in viscosity, and a very rapid rise is noticed between 3 and 4% concentrations; above 5% concentration a paste without any free flow is formed (Fig. 2). These results have a great practical bearing on the loom performance of the sized yarn. Bulk trials with jute yarns have shown that the incidence of warp

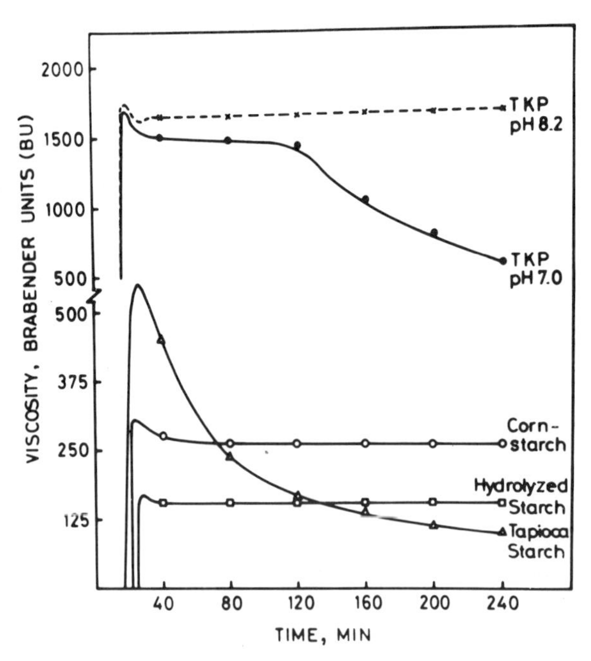

FIG. 1.—Viscosity as a function of heating of solutions of TKP and other common sizing materials[50] at 95°.

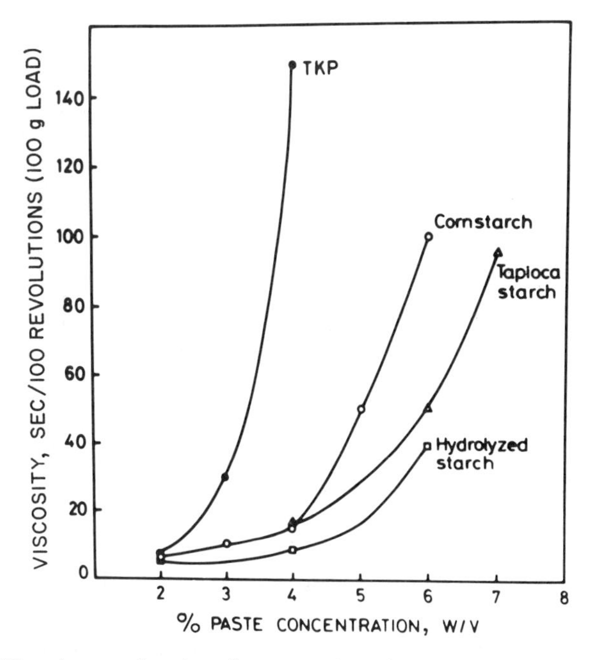

FIG. 2.—Viscosity as a function of concentration of solutions of TKP and other common sizing materials.[50] Viscosity measurements were made with a Stormer viscometer at 70°.

breakage during weaving tends to decrease when the concentration of the TKP size paste is above 1%; in fact, a 1.2% TKP paste is as effective in reducing loom stoppage caused by warp breakages as a 4.2% paste of wheat flour. On further increasing the concentration up to 2%, the number of warp breakages progressively diminishes; but any additional increase has very little effect.[51]

Boiling has a profound effect on the viscosity of TKP dispersions or pastes. As with starches the viscosity first reaches a maximum and then begins to fall. As stated earlier, the maximum viscosity is attained on boiling for 20–30 min and falls to nearly one-half of that value in about 5 hr.[49] The change in the viscosities of TKP, sago, and corn starch dispersions on boiling are presented in Table I. Even prolonged heating at a lower temperature (95°) brings about similar changes (FIG. 1). However, if the paste is made slightly alkaline, the viscosity is stabilized.

A TKP solution attains its maximum viscosity when its pH lies between 7.0 and 7.5, and beyond these limits, whether on the acid or alkaline side, the viscosity falls. This decrease is more pronounced in acid media. Hence, strong acids and strong alkalis should not be used in conjunction with TKP solutions. Boiling mineral acids, even in low concentrations, bring about a rapid

TABLE I

Effect of Continued Boiling on Viscosity of TKP and Starch Dispersions[49]

Boiling Time, min	Relative Viscosity		
	TKP (0.5%)	Sago Starch (1.0%)	Corn starch (1.5%)
10	4.7	5.2	4.9
20	5.0	4.0	4.6
30	4.3	3.3	4.5
60	3.8	2.9	4.0
180	2.9	2.8	2.7
300	2.5	2.5	2.4

fall in the viscosity and a consequent thinning of the paste, whereas organic acids do not have such a marked effect.[51] The fall in the viscosity above and below pH 7.0–7.5 is caused by depolymerization, whereas the maximum viscosity in that range is attributed to uncoiling of the molecules. To attain pH 7.0–7.5, sodium hydroxide, sodium carbonate, sodium bicarbonate, or even sodium tetraborate (borax) can be used with advantage. These substances are added so as to form 0.001*M* solutions in the final size pastes, which do not subsequently thin, even after boiling for as long as 5 hr.[49]

The action of sodium tetraborate in increasing concentration is somewhat different from that of the other alkaline salts. Unlike the others, it does not bring about molecular degradation, even on boiling, so that there is no fall in the viscosity; instead, it promotes an increase in the viscosity.[51] At higher concentrations of borax, the TKP solution is converted into an elastic, self-adhesive, rubberlike mass through the formation of cross linkages. In fact, a 2% solution, when treated with sodium tetraborate in the amount of 7.5% of the weight of the gum, changes into a semisolid gel that thins down on warming and sets again on cooling. The dried and powdered material of this complex behaves very much like gum tragacanth or locust bean gum and has been named *gum taminda*.[52, 53] It is recommended for the finishing of textiles and for preparation of printing pastes for calico.

Vigorous stirring has an adverse effect on the viscosity of starch solutions. However, TKP solutions do not thin down on agitation. This dissimilarity might be attributed to a basic difference in their mechanisms of dispersion. TKP possibly undergoes a straightforward dispersion rather than an initial swelling of particles followed by disintegration and dispersion.[51] Measurement of the particle size of TKP supensions at various temperatures has given support to this view. There is no evidence of extreme distension of the particles at any stage when TKP is heated in water from 30° to 98°; the mean diameter, taken at the long and short axes of a number of random particles in a sample, varies from 44μ at 33° to 25μ at 92°. Furthermore, TKP particles appear to have no characteristic shape or size as starches have; the size of the particle depends largely on the extent of grinding. The absence of particle swelling in TKP is reflected in the viscous properties of its paste, and hence under suitable conditions, a constant viscosity can be maintained throughout the sizing operation.

The viscosity of a TKP paste increases as the temperature decreases; but in contrast with most starches, the pastes do not gel on cooling and are, therefore, easily redispersed even after storage for several days. However, precautions, such as the addition of an antiseptic, must be taken to prevent microbial degradation.

Variation of the viscosity of TKP solutions with temperature is not completely reversible. This behavior cannot be explained by a change in molecular structure with the change in temperature. The decrease in viscosity at higher temperatures is attributed by Das and Basak[49] mainly to the molecules' coiling, although depolymerization may make a contribution.

The use of antiseptics, such as copperII sulfate and zincII chloride, in the preparation of sizings is not recommended because their addition adversely affects the viscosity. However, soluble calcium and magnesium salts, which are often used in size pastes, can be added because they have no appreciable effect on the viscosity of TKP pastes. Macmillan and Chakraverti[51] reported that a slight initial drop in the consistency of TKP paste is caused by the addition of these salts but the viscosity does not further decrease significantly with their increasing concentration. However, Das and Basak[49] have recorded an increase in viscosity caused by the addition of calcium chloride. As mentioned, these calcium and magnesium salts are used in size mixings because they enable the yarn to retain the required amount of moisture through their deliquescence. However, they should not be used in excess as the damp conditions they create might induce mildew growth.

Film-forming properties.—Films can be prepared from TKP by pouring the paste onto a glass plate and spreading it uniformly into a thin layer by means of a doctor blade. The water from the layer is removed gradually by passing over it air having a relative humidity of 60–80%. The films from TKP are smooth, strong, continuous, and extensible. The properties of films prepared from size mixings of corn starch, TKP, hydrolyzed starch, and thin-boiling TKP have been studied.[54] The film properties of TKP compare well with those of starch and those of thin-boiling TKP, compare well with those of hydrolyzed starch. Some investigations of TKP films have also been done by Ghose and Krishna[27] and by Macmillan and Chakraverti.[51] Although properties such as tensile strength and extensibility of the film give some idea of the quality of the sizing agent, a direct correlation of the film properties and the performance of the sizing agent during weaving on the loom is not possible. For testing purposes, the films are prepared on a smooth surface, such as glass or polished metal, while in actual working the size film is deposited on a rough surface, that is, yarn, and the thickness of the film varies from place to place. Another factor of importance is the adhesion of the size film to the yarn, which is subjected to forces of flexing and abrasion in addition to stretching. The size film should, therefore, not only have high tensile strength and extensibility but also good adhesion to yarn, and should not crack or crumble under flexing and abrasion.

Adhesiveness and penetration.—Adhesive characteristics of TKP, acid- and enzyme-modified TKP, corn starch, and hydrolyzed corn starch have been studied[55] by the procedure of Mehta and coworkers.[56] It has been found that TKP and its products have greater adhesiveness than starch and hydrolyzed starch.

TKP dispersions are not sufficiently fluid and consequently they do not penetrate deep into the yarn during sizing. Photomicrographic studies of cross-sections of sized yarns reveal that the size does not penetrate between bundles of fibers. It only forms a coating around the surface of the yarn, as does corn starch.[57] However, homogenization of TKP pastes improves the size penetration so that size films adhere well to the yarn.[58] At the same time, the size does not penetrate so deep into the core that the dynamic resilience of the yarn is lost.

4. *Specifications*

As with starches, the performance of TKP depends on its purity and on its availability in a suitable form. The Indian Standards Institution has, therefore, prescribed specifications for the material for use in the cotton textile industry.[37] According to these specifications, TKP should be creamy white, not exceeding 0.3 red and 0.5 yellow against the standard white of the Lovibond Tintometer; it should not have any disagreeable odor; it should all pass through 85-mesh and, at least to an extent of 80%, through 100-mesh British Standard test sieves; it should be free from testae; it should have a relative viscosity of 4.5–5.0 at 35° in 0.5% solution; it should not have more than 3.0% ash; it should contain not less than 50% of polysaccharide and its D-xylose content should be between 19 and 21%. The Standard also specifies the test methods. TKP conforming to the above specifications gives excellent results.[57, 58]

For use in the jute industry the powder need not meet such rigid specifications, at least in certain respects, such as color, fineness, and testa content. The testae may be present in amounts up to 3%. However, the color limits should not exceed 0.8 yellow and 0.8 red in the Lovibond Tintometer, and the powder should all pass through Indian Standard Sieve 40 and, at least to an extent of 80%, through Indian Standard Sieve.[38]

5. *Physical Properties of Yarn Sized with TKP*

Jute yarn.—A systematic study was made by Macmillan and Chakraverti[51] of the physical properties of jute yarn sized with TKP, wheat flour and corn starch. Yarns sized with TKP show an average increase of 12% in strength, but their elongation at rupture is decreased by about 10%. Thus, TKP improves one desirable property at the expense of another. However, ballistic work of rupture, which takes into account both the strength and the extensibility of the yarn, increases by 10–12%, and resistance to abrasion, indicated by the number of strokes required to break a yarn abrading itself, increases by an average of 40%. These figures compare very well with similar data obtained by employing wheat flour and corn starch.

Cotton yarn.—Mehta and Shah[57] systematically examined different kinds of cotton yarn sized with TKP, starches and cereal flours. They have found that the residual elongation and the increase in tensile strength of yarns sized with TKP are more or less similar in value to those obtained using starches. Similar results were reported by Bhat.[59]

Abrasion resistance increases with increase in the concentration of TKP paste and even more with addition of sodium silicate. Addition of tallow adversely affects abrasion resistance because the natural fat present in TKP acts as a softener, and addition of extraneous fat only makes the yarn excessively soft.[58]

Warp breakages and weaving efficiency.—Results of warp breakage give a fairly comprehensive idea of the relative efficacies of the different sizes. The results obtained by Macmillan and Chakraverti[51] on jute yarn using TKP, corn starch, and wheat flour indicate that the incidence of warp breakage during weaving of TKP-sized yarn is much less than that occurring when wheat flour is used as the size. Doubtless, a decrease in the concentration of TKP results in increased breakage, but even with quantities that are only 30% of the weight of wheat flour, these breakages are less than with the latter.

The weaving efficiency of yarns sized with TKP is slightly lower than that of those sized with corn starch. However, it is suggested by Macmillan and Chakraverti[51] that this reduced efficiency might not be caused by the effect of the TKP itself but to the difference in humidity at the time of weaving. Furthermore, the quantity of TKP used is only 40% of that of corn starch. Nevertheless, TKP paste is more viscous and tacky than corn starch paste, and yarn sized with it compares very favorably in smoothness and pliability to that sized with corn starch and also stands the mechanical operations of weaving equally well.

Working with cotton warp covering a wide range of counts (from 18's to 70's) and of fabric constructions from 22 to 45 ends per cm, Mehta and Shah[57] came to the conclusion that yarns sized with tamarind kernel powder mixings have approximately the same breakage rates as those sized with the commonly used starches. TKP mixings have given slightly better results than those based on wheat and tapioca starches, and a 60 : 40 mixture of TKP and corn starch. However, when TKP mixings are used, there is a slightly higher breakage rate than when sago or thin-boiling corn starch is used as a size; the maximum difference is about 0.5 break per 10,000 picks. On the whole, TKP of a standard quality compares well with any of the starches generally used for sizing, but it should be used in lower concentration in the size mixing in order to give comparable results in weaving.

6. *Fungal and Bacterial Decomposition*

Because TKP that contains a high percentage of proteins and a large amount of carbohydrates offers a good food for microorganisms, there is a great danger that yarn sized with TKP could be attacked. If mildew growth starts, it rapidly attacks first the size and then the fiber. However, yarn sized with TKP is only slightly more susceptible to mildew attack than that sized with wheat flour.[60]

It is essential that suitable antiseptics are used in size mixings based on TKP. The relative effect of a number of preservatives, such as β-naphthol, shirlan, salicylic acid, hexamine, formaldehyde, formic acid, propionic acid, and furfural, has been studied, and formaldehyde in about 0.05% concentration was found to be a satisfactory preservative.[61] The problem was examined in greater detail by evaluating the bacteriostatic and fungistatic potencies of some of the antiseptics that are readily or potentially available in India.[62] In these studies pentachlorophenol, β-naphthol, *p*-chloro-*m*-cresol, and salicylic acid were found to be capable of inhibiting the growth of both bacteria and fungi. The first three compounds effectively prevent deterioration of size pastes during overnight or weekend storage when used in concentrations of 0.04%, whereas salicylic acid requires a slightly higher concentration. Salicylanilide, copper(II) sulfate, and sodium tetraborate possess good fungistatic potency, but their bacteriostatic power is relatively low.

Mixed antiseptics that are compatible with one another and with the size medium are generally more effective. For example, salicylanilide possesses excellent fungistatic action but a poor bacteriostatic value; alum, on the other hand, is by itself a poor fungistat. However, a combination of the two inhibits the growth of both bacteria and fungi. The same results are obtained with a mixture of alum and sodium pentachlorophenate. The pH of the size medium also has a pronounced effect on the potency of antiseptics; alkalinity generally tends to decrease this potency, and acidity, to increase it. This is particularly so with salicylic acid because in an alkaline medium it is converted into its sodium salt, which possesses a markedly lower potency than the free acid.

Tamarind kernel powder is more susceptible to decomposition by bacteria and fungi than is corn starch, but only a partial reduction of the protein content, for example, from the original 19% to 13%, reduces its mildew susceptibility to the order of that of corn starch. Elimination of water-solubles, as occurs, for example, in the case of washed TKP, also reduces the mildew liability, but nôt to the same extent as does the removal of proteins. The fat in TKP does not appear to be responsible for this microbial activity because its removal does not produce any change in the susceptibility of the powder to mildew attack. Sizes can be ranked in increasing order of mildew susceptibility as follows: corn starch, low-protein TKP, washed TKP, wheat flour, defatted TKP, commercial-grade TKP.[63]

An investigation of steeping as a preventive of bacterial decomposition or mildew attack showed that it did not appreciably lower the susceptibility of the product. Although a better dispersion of TKP in water can be achieved by steeping, the polysaccharide portion is hydrolyzed during this treatment so that the viscosity of the dispersion, and hence its sizing value, falls at a very rapid rate and no improvement is obtained.[63]

7. *Cost*

In general, a good-quality TKP costs nearly one-half as much as corn starch, and its current price is Rs.60–70/per quintal, that is, $8.00–9.30 for 220 lb at the present rate of exchange. In 1965, when there was a scarcity of food grains in India and consequently no starch could be freely diverted for use in the textile industry, the price of TKP rose to nearly Rs.90/- a quintal, and even up to Rs.110/- a quintal for one or two months.

8. *Typical Size Mixings*

Tamarind kernel powder has completely replaced starches in the sizing of jute yarns, because no blending is necessary. It is also being extensively used as size for cotton warps. Typical recipes of size mixes for jute and cotton warps are given in Tables II and III.

TABLE II

Size Mixings of TKP for Jute Yarn[38]

Ingredients	Size Mixings (1.5–2% w/v)
TKP	30–40 kg
Sodium carbonate	1 kg
Salicylic acid or	1 kg
β-naphthol	0.5 kg
Sizing auxiliaries	As required
Water to make	2000 liters

TABLE III

Size Mixings for Cotton Warps[64, 65]

Ingredients	Size Mixing for Bleached Goods (7% w/v)	Heavy Size Mixing for Grey Goods (14% w/v)
TKP	67.0 kg	50.5 kg
Guar gum	—	1.0 kg
Sodium silicate	6.7 kg	—
China clay	—	63.0 kg
Magnesium chloride	—	12.6 kg
Mutton tallow	1.3 kg	15.2 kg
Paraffin wax	—	5.0 kg
Sodium silicofluoride	0.2 kg	0.3 kg
Water to make	1000 liters	1000 liters

The warp breakages during weaving of jute and cotton yarns are of the same order with size mixings containing TKP as with those containing starch or thin-boiling starch. Because TKP contains 17–20% proteins in addition to carbohydrates and fatty matter, it is susceptible to microbiological degradation in solution as well as under high humidity conditions, which are present in the weaving departments of textile mills. In order to prevent microbiological attack on TKP pastes and size films on yarns, it is essential that an antiseptic or a mixture of antiseptics is included in the size mixing. It is desirable to have an antiseptic that has both antifungal and antibacterial properties. Pentachlorophenol, β-naphthol, *p*-chloro-*m*-cresol, and salicylic acid are reported to belong to this class of antiseptics.[62] In the cotton textile industry in India, sodium silicofluoride and zinc chloride are commonly used as antiseptics in TKP size mixings. It is a general practice to use 5–8% of a softener such as mutton tallow in starch-based size mixings meant for cotton warp. The purpose of the softener is to make the starch film smooth, soft, and pliable, and thereby to prevent excessive abrasion against the loom parts. Inasmuch as TKP contains 6–8% of fatty material, in size recipes based on TKP, little or no softener need be added. Furthermore, to prevent the sized yarn from getting excessively soft, as well as to prevent the peeling of the size film, it has been found that addition of sodium silicate (10% on the weight of TKP) improves the weaving efficiency. Furthermore, addition of sodium silicate makes the size paste slightly alkaline, thus stabilizing the paste viscosity during sizing operations. It has been reported that by homogenization of TKP, its paste viscosity can be reduced somewhat and, therefore, slightly more size can be put on the yarn.[58]

9. Desizing

Desizing is an important operation in the textile industry when bleached sorts and printed or dyed goods are produced, because the size has to be removed before bleaching, printing, or dyeing. When starch is used, the common desizing practice is to use enzyme preparations containing α- and/or β-amylases as desizing agents. These enzyme preparations, however, do not work satisfactorily in the desizing of yarns sized with TKP, although the British Tentative Textile Standard No. 26 (1956)[66] recommended treatment with malt extract for the removal of TKP size both on cotton and viscose.

Dried films of TKP can be easily redissolved.[27] For this reason, scouring the sized goods once or twice with boiling water containing small amounts of sodium carbonate is a simple, inexpensive desizing procedure and is the method used by many mills. The different methods, that is, acid steeping, enzyme treatment and boiling with alkali, were studied systematically by Mehta and Shah.[57] Of these, boiling with 1% sodium carbonate for 3 hr was found to be an efficient and industrially practicable method for white goods. In the case of colored goods, the same treatment at a lower temperature (70°) was recommended.[58] For the desizing of grey yarn, Bhat[59] recommended boiling for 3 hr with a solution containing 0.15% of ammonium sulfate and 0.3% of sodium chlorite which gives simultaneous bleaching.

It is reported that enzymes from garden soil bacteria and several common fungi liquefy tamarind kernel paste with little production of reducing sugars when treated for 1 hr at a pH of 5.5–6.0 and a temperature of 55°–60°. By this treatment, the size can be quantitatively removed from cotton or rayon fabrics.[67]

10. Defects and Disadvantages

Although TKP is a good sizing material, it has certain defects that must be eliminated before it can fully replace cereal starches in the sizing industry. These defects, along with suggestions for their rectification,[29, 58] are given below.

Inherent defects.—Decortication is not an easy process, and the kernel is very hard, making pulverization to the required fineness a difficult job.

Even the best-quality TKP is only creamy white, not pure white as are the cereal starches. Many methods have been suggested for improving the color.[16, 68] However, attempts to bleach TKP by a practical process have been so far unsuccessful. Still, its appearance can be improved by admixture with a small amount of soluble blue or a suitable commercial whitening agent.

Tamarind kernel powder has a sticky appearance and forms lumps. These properties are attributed to the presence of fat. This fat makes fine pulverization of the kernels very difficult and causes the powder to look dull and dirty.

Defatting the kernel by solvent extraction in the preparation of TKP has been suggested,[28, 69] but the desirability of defatting is doubtful, because the elimination of the natural fat very seriously affects the weaving efficiency. The fat present in TKP acts, as already stated, as a softener and eliminates the need for the addition of any extraneous material, such as mutton tallow.[58, 59]

Excessive frothing is commonly encountered in the sow box. The entrapped air in the structure of the yarn and the presence of alkaline surface-active agents, such as sodium silicate, in the size paste are mainly responsible. By adding alum or acetic acid to adjust the pH to only slightly above 7.0 and by adding antifoaming agents, the frothing can be controlled.

Because of the very high viscosity of TKP pastes, high amounts of size ($>10\%$) cannot be put on the yarn as is necessary for high-count yarns and yarns means for high-reed-pick fabrics.

Imposed defects.—When TKP is not prepared properly, difficulties arise that are more serious than the inherent defects.[39, 58]

Testae, if not entirely removed, impart a reddish tinge to the powder, which cannot then be used for sizing grey calendar sorts sold in the loom state. In general, oxidizing agents do not improve the color; reducing agents do better. However, most of the reducing agents, with the exception of sodium sulfoxylate formaldehyde, which gives satisfactory results, decompose rapidly at the high temperatures involved in size preparation.

Accumulation of foreign matter on the fents arises from the presence of large quantities of fibrous and foreign matter in TKP, but does not occur if a standard product is used. Furthermore, the trouble can be eliminated by washing the fent of the squeeze roller with hot water once every shift.

Sticking of fibers on the drying cylinders occurs because of the high temperature of the drying cylinder and is not a defect of TKP itself. Reduction in alkalinity and maintenance of a proper drying temperature remove this difficulty. When drying cylinders have to be kept at higher temperatures for heavy sorts, the first drying cylinder should be held at a slightly lower temperature than the second.

Tamarind kernel powder forms more viscous pastes than does starch and has, therefore, to be used in lower concentrations to give pastes with viscosity in a workable range. Hence, to obtain the same amount of dried size on the yarn, a larger quantity of water must be evaporated when TKP size is used. This is a drawback but it is partially offset by the fact that smaller quantities of TKP size are necessary for a comparable weaving performance. It is also possible to prepare more concentrated pastes with moderate viscosities by homogenization. Another compensating factor is that TKP solutions do not show retrogradation or thinning down. It is possible to use the pastes after storing even for more than 24 hr.

Tamarind kernel powder gives best performance only at about 75% relative humidity and excessive humidity, which occurs, for example, in India during monsoons, leads to a poorer weaving efficiency. For this reason, proper humidity conditions should be maintained in the loom shed.

11. TKP as a Thickener for Textile Printing and Finishing

Tamarind kernel powder, along with other thickeners, has been evaluated for printing of cellulosic materials by vat dyes.[70] It has been found that TKP has a considerably higher affinity for vat dyes than cellulose has. This limits the transfer of vat dyes from the TKP phase to cellulose to a very large extent. Whatever little dye transfer takes place is through a dye–TKP complex.

Keller[71] reported that, for finishing of fabrics with TKP, it is necessary that fatty substances be destroyed by the addition of alkali, that the greyish yellow color be brightened by optical brightening agents, and that the unpleasant odor of TKP be suppressed by the use of a perfume. However, until a practical method of bleaching TKP is discovered, it is not likely to find favor as a finishing agent.

12. Other Uses

Because of its hydrophilic character and its ability to form solutions of high viscosity even in low concentrations, TKP serves as a good creaming agent for the concentration of rubber latex. It has been found that the addition of a 3% solution of TKP to the ammoniated latex in an amount calculated to give 0.3% of the weight of the water phase brings about an effective concentration. After creaming for nearly 38 hr, the dry rubber content of the latex increases to 58.3%.[72] In another patent,[73] a borated solution of the tamarind seed polysaccharide itself was used instead of a solution of TKP. A 1.5–2.0% solution of the polysaccharide containing sodium tetraborate in small quantities, when added to four times its volume of rubber latex, effects the concentration. It is said that this process reduces the time of creaming to less than 30 hr and that the concentration period is reduced to less than 6 hr if the latex mixed with a borated tamarind seed polysaccharide solution is put through a low-speed centrifuge of the cream-separator type.

It is also reported that TKP acts as a soil stabilizer.[74–76] When 2–4 parts of it are thoroughly mixed with 94–97 parts of dry sand, soil, clay or earth along with 1–4 parts of calcium hydroxide and 10–25 parts of water, stabilization occurs; and this stabilized material can be molded into strong and compact bricks for building purposes. A chlorinated mixture of TKP and magnesium oxide yields hard, insoluble, and infusible compositions.[77]

Sawdust briquettes for fuel have also been made by employing 10% TKP as the binder.[78]

Paper adhesives of good strength have been prepared from TKP and boric acid;[79] label pastes, which are suitable not only for paper but also for glass and metal, have also been produced.[80]

Prevention of the flocculation of the fiber suspension is important in paper and rayon manufacture. Tamarind kernel powder has been found to be a good dispersing agent for fiber suspensions. In this respect, it is superior to starch ethers and pectins but inferior to locust bean gum, guar gum, and cellulose ethers.[81]

Inclusion of TKP in compositions for blasting explosives has been suggested. It is reported that addition of 0.5–15% of TKP to blasting explosives containing water-soluble salts makes the explosives water resistant.[82] The powder is also recommended as a thickener for aqueous slurry explosives.[83]

Tamarind kernel powder is used in the plywood industry as an extender for urea–formaldehyde resin adhesives. Quality plywood for tea chests is produced by making use of TKP-extended urea–formaldehyde resin glue in which there is 50% TKP.[84] In lower-grade commercial plywood, only 40% of TKP is used. It is estimated that about 15–20 million m^2 of plywood are manufactured annually in India using TKP-extended glue.

III. Modification of TKP

Tamarind kernel powder has established itself as a sizing agent for jute and cotton warps. Being cheap and available in fairly large amounts, it may also find use as an adhesive, a thickener, or a dispersant. For each of these end uses, TKP in its native form may not be entirely suitable. Similarly, for removing the TKP size from yarns, it may be necessary to depolymerize it to such an extent that it is removed easily by water or weakly alkaline solutions. There is, therefore, the need to modify TKP to suit specific end uses. Modification of TKP polysaccharide has been brought about by the following methods: (*a*) enzymic degradation, (*b*) acid hydrolysis and dextrinization, and (*c*) derivatization.

1. Enzymic Degradation

Savur[85] isolated from tamarind leaves two enzyme systems; one which liquefies TKP paste without extensive hydrolysis of glycosidic bonds and the other increases the reducing power of the polysaccharide without appreciable decrease in its viscosity. The first enzyme appears to be an endoenzyme that breaks a small number of inner β-D-(1→4)-glucosidic linkages of the polymer, thereby bringing down its viscosity without the generation of small fragments. The enzyme merits further examination inasmuch as TKP pastes of lower viscosity have great application potential in textile sizing.

Chakraverti and Nag[67] have examined the degradation of pastes of TKP, as well as those prepared from the pure polysaccharide, by bacterial and fungal enzymes in connection with the problem of drop in viscosity of TKP paste when the latter is transferred from the size-preparation kettle to the sow box during sizing of jute yarns. Bacterial enzymes were found to be more potent than fungal enzymes. The enzyme preparation from *Aspergillus flavus* was the most active among those from the many fungi tested, the optimum activity of the enzyme being at 55°–60° and pH 5.5–6.0. Interestingly, the pure polysaccharide was found to be less susceptible to enzymic degradation than TKP prepared from raw or roasted tamarind seeds. The authors also showed that the bacterial and fungal enzymes could be used for desizing of fabrics woven from TKP-sized yarns.

Srivastava and coworkers[86] have treated dispersions of TKP in water with a fungal cellulase to reduce the viscosity of TKP pastes to that of corn starch and hydrolyzed starches. The optimum activity of the enzyme, as measured by the fall in viscosity, was found to be at pH 4.4–4.5 and 55°. Slurries of high concentration of TKP (up to 25%) can be degraded to the required level of viscosity by cellulase preparations at 0.1% of the dry weight of TKP. Brabender viscosity curves of corn starch, TKP, TKP hydrolyzed to the viscosity levels of corn starch (MTKP-1), acid-hydrolyzed thin-boiling TKP (MTKP-2), and thin-boiling starch (acid-hydrolyzed corn starch) are given in Figures 3 and 4. It is clear that the heating, cooking, and cooling curves of cornstarch are very similar to those of MTKP-1. Although hydrolyzed starch and MTKP-2 display similar behavior on heating and cooking, the increment in viscosity on cooling of the pastes is much more in the case of hydrolyzed starch than in MTKP-2 (Fig. 4). It is noteworthy that the well-known property of setback (retrogradation) of corn starch and hydrolyzed starches is not shown by MTKP-1 and MTKP-2, the pastes of which can be consequently redispersed easily even after storage at room temperature for several days.

Film properties of sizing materials give, at least in part, an indication of their performance as a protective coating on yarn during weaving operations on the loom. Film characteristics of corn starch, a commercial sample of thin-boiling starch, TKP, MTKP-1, and MTKP-2 as constituents of sizing recipes were, therefore, examined (Table IV).[86] Films from both TKP and MTKP-1 have higher tensile strength than those prepared from starches. Similarly, films prepared from MTKP-2 have a higher tensile strength than those from thin-boiling starch even though their viscosities are similar. However, so far as elongation at break is concerned, starch shows the maximum value.

2. *Acid-Catalyzed Hydrolysis and Dextrinization*

Reduction of viscosity, which is the main objective of modification of TKP for use in textile sizing, can be achieved by hydrolysis of TKP in the form of a

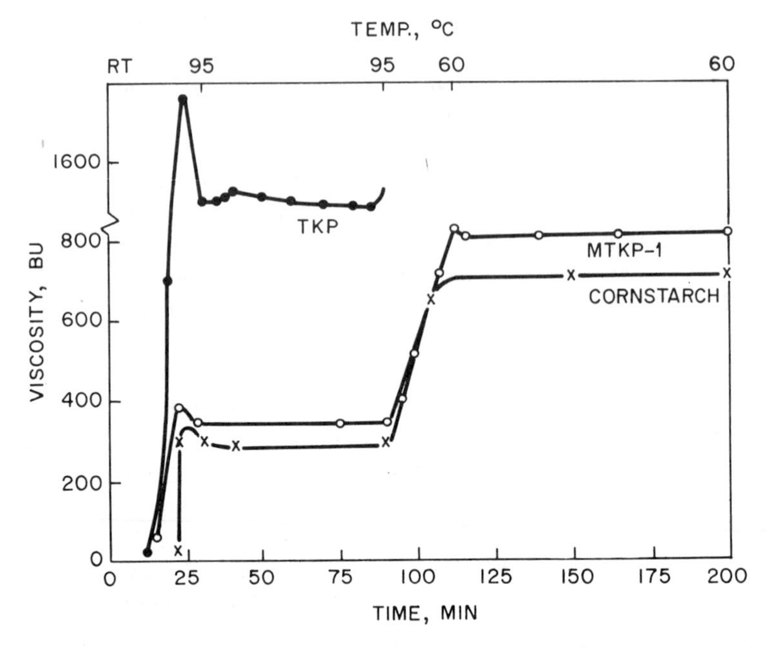

FIG. 3.—Brabender viscosity curves of 5% (w/v) pastes of corn starch, TKP and MTKP-1.[86] See text for explanation of abbreviations.

TABLE IV

Recipes for Film Preparation and Properties of Films Prepared from Corn Starch, Hydrolyzed Starch, TKP and Modified TKP[86]

Adhesive	Ingredients, g				Water	Tensile strength, kg/cm²	Elongation at break, %
	Adhesive	Mutton tallow	Sodium silicate	Sodium silicofluoride			
Corn starch	6.0	0.60	—	0.01	100	267.87	5.7
Thin-boiling starch	6.0	0.42	—	0.01	100	235.53	4.8
TKP	6.0	0.15	0.72	0.016	100	283.34	4.2
MTKP-1	6.0	0.12	0.72	0.016	100	274.41	4.3
MTKP-2	6.0	0.12	0.72	0.016	100	266.46	4.9

slurry with acid or by dry roasting in the presence of acid (dextrinization). In the wet process, a TKP slurry of the required concentration is treated with acid, the concentration of which is chosen so that, by heating the slurry at 95°

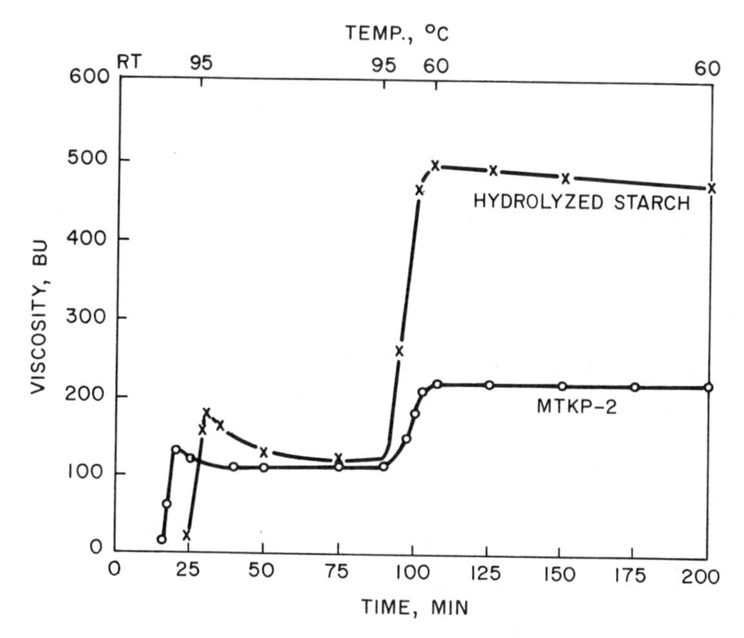

FIG. 4.—Brabender viscosity curves of 5% (w/v) pastes of thin-boiling starch (acid-hydrolyzed corn starch) and MTKP-2[86] See text for explanation of abbreviations.

for 1.5 hr, the desired level of viscosity is obtained.[87] To the resulting paste, 10–12% of sodium silicate is added not only to neutralize the residual acidity but also to improve the quality of the size paste.

In the dextrinization process,[50] TKP is heated in a relatively dry state (15% moisture or less) in the presence of 0.3–3.6% of hydrochloric acid at temperatures of 110°–115° for 0.5–5.0 hr. After the reaction is over, the product is cooled and the requisite amount of an alkaline substance is added to neutralize the residual acid in the powder. By a proper selection of the reaction conditions, modified TKP of the desired properties can be obtained.

Acid-modified TKP of the viscosity levels of thin-boiling corn starch, prepared either by the wet process or by dextrinization, displays Brabender viscosity curves similar to those given for enzyme-modified products (MTKP-1 and MTKP-2).[50, 86–88] It is so be noted that, whereas the enzyme process and wet-acid method give low viscosity pastes that have to be used soon after their preparation, by the dextrinization process dry powders of thin-boiling TKP that can be transported and stored indefinitely are obtained. The performance of thin-boiling TKP as a sizing adhesive for cotton warps has been evaluated in mill-scale trials.[89] The results (Table V) show that thin-boiling

TKP can be used for sizing of medium-count cotton yarns meant for medium-construction fabrics for which starches are usually employed. It is also to be noted that, although with sizes based upon starches, gums and softeners (generally a tallow) are used, with thin-boiling TKP no such additives are necessary in the size mix. Considering the low cost of TKP and the fact that no expensive additives, such as softeners and gums, need be added in the size mix, sizing with thin-boiling TKP is more economical than sizing with starches.

Working with the dextrins produced by heating dry TKP with stirring at 150° for varying periods up to 8 hr with and without a catalyst (0.5% hydrochloric acid on the weight of dry TKP), Srivastava and Patel[90] noticed that dextrinization brings about marked changes in the solubility, reducing power (ferricyanide number), and viscosity of TKP (Table VI).

Whereas increase in reducing value and decrease in viscosity are recognized changes associated with the dextrinization process, decrease in solubility of TKP on progressive dextrinization is contrary to what happens in starch dextrinization. The loss in solubility may be caused either by insolubilization of protein or protein–polysaccharide complexes or by the formation of a more linear polysaccharide, the branches having been cleaved preferentially. Similarly, it has been demonstrated that highly branched polysaccharides, such as gum karaya[91] and guar gum,[92] become less soluble in water when subjected to dextrinization. Another point of note is that catalytic and noncatalytic dextrinizations give more or less similar products at any particular time of reaction. This

TABLE V

Textile Sizes Based on Starch and Thin-boiling TKP and Their Performance on the Loom[89]

Sizing Conditions			Loom Shed Performance	
Size mix	Size add-on	Fabric construction	Duration of observation, hr	Average breaks/loom hr
Based on starch[a]	16.55	30ˢ/36ˢ;88/64	96	1.44(0.07)[c]
	12.20	30ˢ/36ˢ;88/64	135	1.51(0.37)
Based on thin-boiling TKP[b]	10.25	30ˢ/36ˢ;88/64	60	1.19(0.31)
	8.4	30ˢ/36ˢ;88/64	143	1.41(0.43)

[a] Tapioca starch (43.75 kg), guar gum (1.88 kg), mutton tallow (3.13 kg), ethylene glycol (0.13 kg), and French chalk (0.28 kg) made up to 450 liters.

[b] Thin-boiling TKP (33.33 kg), French chalk (0.38 kg), sodium silicate (2.67 kg), and sodium silicofluoride (0.43 kg) made up to 450 liters.

[c] The figures in parentheses give multiple breaks.

TABLE VI

Properties of TKP Dextrins[90]

Time, hr	Solubility		Ferricyanide Number		Intrinsic Viscosity	
	A[a]	B	A	B	A	B
0	1.70	1.70	5.5	5.50	2.25	2.25
2	1.04	1.58	7.0	8.25	1.80	1.70
4	0.94	1.20	7.6	8.80	1.45	1.44
6	0.90	1.16	8.0	9.37	1.15	1.10
8	0.82	1.03	9.3	10.20	0.95	0.90

[a] A, without catalyst; B, with catalyst.

may be because of the acid-binding power of the protein component of TKP. A similar observation has been made in the case of dextrinization of guar gum.[92]

Dextrinization of a purified sample of tamarind seed polysaccharide at pH 3.1 and at 153° for 5 hr has been examined.[93] Extraction of the dextrin with methanol afforded L-arabinose, D-xylose, D-galactose, D-glucose, and 1,6-anhydro-β-D-glucose (levoglucosan), the pentoses being in the largest amount. The residual dextrin, after purification, was devoid of L-arabinose residues, and had the following analysis: D-glucose (10 moles), D-galactose (1 mole), and D-xylose (2 moles). Methylation and hydrolysis of the dextrin produced 2,3,4-tri-*O*-methyl-D-xylose (2 moles), 2,3,4,6-tetra-*O*-methyl-D-glucose (1 mole), 2,3,4,6-tetra-*O*-methyl-D-galactose (1 mole), 2,3,4-tri-*O*-methyl-D-glucose (1 mole), 2,3,6-tri-*O*-methyl-D-glucose (4 moles), and 2,3-di-*O*-methyl-D-glucose (4 moles). Upon oxidation of the dextrin with periodate, 1.4 moles of oxidant were consumed and 0.38 mole of formic acid was liberated. Smith degradation of the dextrin produced erythritol, glycerol, and ethylene glycol in the molar ratio of 8 : 3 : 2. These results show that the main reactions during dextrinization of tamarind seed polysaccharide are that of debranching and chain depolymerization. However, the presence of levoglucosan in the methanol extract of dextrin and 2,3,4-tri-*O*-methyl-D-glucose in the hydrolyzate of methylated dextrin indicates that some transglycosidation also takes place during dextrinization.

3. *Derivatization*

As already stated, the main component of TKP is a polysaccharide that can give rise to a number of derivatives possessing useful properties which will be presented later with the discussion of the properties of the polysaccharide. Though possessing useful properties, the preparation of these derivatives from the isolated polysaccharide for sizing or similar purposes will not be feasible for reasons of economy. Hence, attempts have been made to produce and utilize derivatives of TKP.

By treating TKP with an alkylating or hydroxyalkylating agent, Schlageter[94] produced a product having excellent swelling properties in water, which, when added to paper pulp, gave paper of high strength and excellent resistance to water. Sized papers with high alkali resistance are produced from pulp sized in an alkaline medium with TKP and saturated fatty acids having 12–20 carbon atoms. The sizing composition also includes a water-soluble precipitating agent to promote the formation of water-insoluble soaps throughout the fiber mass.[95]

Schiavio and Maderno[96] treated alkali suspensions of TKP with alkyl chlorides or carboxyalkyl chlorides and produced alkyl or carboxyalkyl ethers that were reported to be useful as emulsifiers and in textile printing pastes.

The products obtained by the alkoxylation of TKP, when added to gypsum, are reported to yield highly viscous dispersions that are especially useful for vertical walls and ceilings. Only very small quantities (0.1–1.0% on the weight of the gypsum) are needed for the purpose.[97]

IV. Tamarind Seed Polysaccharide

The creaming and sizing properties of TKP have been shown by Krishna[73] and by Ghose and Krishna[27] to be due to the polysaccharide present. The latter was originally thought to be either starch or related to it because of the blue color imparted to its aqueous solutions by iodine.[19, 98, 99] It was, however, shown by Rao and coworkers,[100] from a study of its hydrolytic products and the characteristically different iodine stain, that the polysaccharide was not starch. It forms mucilaginous solutions with water and possesses the characteristic property of forming gels with sugar concentrates as does pectin, but it differs from pectin in that it forms gels over a wide pH range.[101, 102] There are differences in other respects also. For example, it sets to a thick gel when its solutions are treated with small amounts of sodium tetraborate,[53] whereas those of the fruit pectins do not. Furthermore, it does not contain any uronic acid or methyl uronate groups, nor does it precipitate as a calcium salt,[103] when treated according to the method of Carre and Haynes.[104]

The property of forming gels with sugar concentrates under diverse conditions of pH is peculiar to tamarind seed polysaccharide and is not possessed by any other class of neutral polysaccharides,[105] although coffee-seed "pectin" might be in the same group.[106] For this reason, the name *jellose* was suggested by Rao and Krishna[107] as being descriptive of both its gel-forming property and its carbohydrate character. Others have called the polysaccharide *polyose*,[108] *tikernose*,[109] *and Tamarindus amyloid*.[110]

1. Preparation

Large scale.—For the preparation of tamarind seed polysaccharide, tamarind kernel powder is a convenient starting material. According to the original method of Ghose and Krishna,[25] a paste of the powder in ten times its weight of cold water is added to 30–40 times its weight of boiling water, boiled for 20–30 min, strained through cloth, and allowed to stand (preferably overnight) after the addition of a little sulfur dioxide to effect bleaching. The solution is then centrifuged, concentrated under reduced pressure to the consistency of a viscous liquid, and dried in a drum drier. During this process, the sulfur dioxide degrades the polysaccharide so that it loses some of its original gelling properties. However, if the polysaccharide is prepared without this treatment, it forms very tough gels. It is suggested that a weaker acid, such as citric or tartaric acid, brings about only the required degree of degradation so that softer, but still firm, gels are produced. According to this method,[111] TKP is gradually added to 30–40 times its weight of boiling water containing the organic acid in a concentration of 0.2%. The mixture is stirred vigorously and the boiling is continued for 30–40 min. The resulting thin, whitish liquid is then discharged into a tall settling tank and allowed to remain overnight so that most of the proteins and fibers settle out. The next day the supernatant liquor is siphoned into a steam-heated concentration pan and reduced to nearly one half of the original volume. It is then mixed with infusorial earth (0.5% of the volume of the liquor), stirred well, and passed through a filter-press. The resulting translucent solution is then evaporated to dryness in a drum drier, producing a solid that is pulverized in a ball mill to yield a powder of the required mesh. The yield of the product is a little more than 50% of the starting material. As an alternative to drum drying, the solution can be poured into twice its volume of ethanol to precipitate the polysaccharide as a fibrous mass. In this case, the precipitate is collected in bag filters of heavy canvas, felt, or thick cloth; dried in an oven at 50°–60°; and finally powdered in a ball mill. If this method is used, an ethanol-recovery plant is necessary. Tamarind seed polysaccharide obtained by the above-mentioned procedures is a pale, creamy white substance without taste or odor. The air-dried sample contains 15.3% of moisture, 0.5% of nitrogen, 0.11% of phosphorus, 1% of fibrous matter, and a small amount of oil. The material readily dissolves in water to give a translucent solution (1% concentration) that can pass through filter paper under suction. The aqueous solution is slightly acidic, is nonreducing toward Fehling solution, and has an $[\alpha]_D$ +71.4° (*c* 0.25, water).

A method of isolating the polysaccharide without dissolution has been reported.[112] This process involves the dispersion of the nonpolysaccharidic materials into a homogeneous aqueous solution at room temperature, while the polysaccharide fraction remains undissolved. The two components are then separated by dilution and dialysis. The product obtained in this process is reported to precipitate almost completely upon addition of ethanol to give a fine white powder; it is also reported to contain less than 1% of proteins. Rao[113] has examined this product and has not been able to substantiate the claims.

Recently, Gordon[114] reported a method for the recovery of the polysaccharide. The process comprises extracting TKP with a polar organic solvent, such as isopropyl alcohol, at a temperature of 78°, recovering the extracted powder, and extracting it with an excess of water. The product recovered from this water extract, that is, the polysaccharide, is reported to be precipitated[115] by the addition of sulfates of Na^+, NH_4^+, Mg^{2+}, and Al^{3+}.

Laboratory scale.—In the laboratory, tamarind seed polysaccharide is usually prepared by the alcohol-precipitation method, and the product normally contains about 14% of associated proteins.[103, 116] Its purification has been attempted in a number of ways. Repeated precipitation from dilute aqueous solutions by the addition of ethanol, as suggested by Ghose and Krishna,[116] is tedious and difficult, whereas mild acid-catalyzed hydrolysis results in simultaneous degradation to varying degrees of both the proteins and the polysaccharide. Preferential precipitation of the proteins by phosphotungstic or tannic acids or by saturation with ammonium or sodium sulfate is not possible because the polysaccharide is also precipitated; and enzymes, such as pepsin, remove only 30–35% of the nitrogenous fraction. Allowing the extract to stand for several hours and using centrifugation for removal of coagulated proteins has also been only partially successful. However, when the extract is diluted with five to six times its original volume of water and allowed to stand overnight, most of the proteins settle out. The supernatant liquor is then centrifuged, concentrated under reduced pressure, and treated with an excess of ethanol to precipitate the polysaccharide. It has been claimed by Savur and Sreenivasan[40] that, by this method, the protein content of the polysaccharide is reduced to less than 2%; but Macmillan and coworkers[63] have shown that this is not the case.

Ghose and Krishna[116] prefer using crushed seeds, instead of seed flour, as the starting material for the extraction. In this process, most of the proteins and fibers remain with the swollen seeds while the polysaccharide dissolves, giving a purer material. Further purification is effected by dissolving the polysaccharide in water to form a thin solution, centrifuging the solution, and then precipitating the polysaccharide with ethanol. The precipitate thus obtained usually contains less than 1.5% of proteins. Final purification is accomplished by preparing a suitable metal complex, purifying it, and regenerating the polysaccharide.[100]

2. Metal Complexes and Purification

Metal complexes are obtained when an aqueous solution of the polysaccharide is treated with a solution of a metal hydroxide or an alkaline salt. Thus, the copperII compound is formed by the addition of either Fehling solution or cooper ammonium sulfate; the lead complex, by the addition of potassium plumbate; the barium complex, by the addition of a saturated solution of barium hydroxide; the strontium complex, by the addition of strontium hydroxide; and the calcium complex, by the addition of calcium hydroxide. These compounds separate as gels, and their purification is difficult because they have a tendency to adsorb some of the unreacted polysaccharide. Boiling water cannot be used to remove the adsorbed polysaccharide because the compounds decompose when heated. Therefore, they are purified by a thorough wash with a dilute solution of the complexing agent.[117] Other metal salts have also been reported.[108] The polysaccharide also forms an insoluble complex with tannic acid.[100]

The polysaccharide can be regenerated with ease from both the tannic acid and metal complexes, using alkali in the first case and acid in the second. A fine powder of these compounds is agitated with water to form a uniform suspension, which is treated with a slight excess of the acid or alkali. The mixture is then filtered through silk. Addition of ethanol to the filtrate precipitates the polysaccharide as a gelatinous mass that, on maceration with warm ethanol, breaks up into a crisp granular powder. This is subsequently filtered and dried at 50°–60°. The copper and barium compounds provide two convenient materials for purifying the polysaccharide.[117, 120] Samples of the polysaccharide thus purified are free from fiber and fat but still contain nitrogen (0.2%) and phosphorus (0.04%).[100]

3. *Properties*

Tamarind seed polysaccharide, as prepared and purified by any of the methods described in the previous section, does not show any heterogeneity on repeated precipitations from dilute aqueous solutions. It retains its original specific rotation and a pentose content of 30.5% even after the eighth precipitation.[121] Further, the polysaccharide purified via copper and barium complexes contained a single component when examined by boundary and zone electrophoresis and by ultracentrifugation.[117, 120] Although the claims of Savur[122] that tamarind kernels contain a mixture of three polysaccharides with different solubilities and powers of gelation have not been fully substantiated, a re-examination of the fractionation of TKP has shown that it contains a water-insoluble, protein-rich fraction (I, 27%), a water-soluble fraction whose copper complex is insoluble (II, 60%), and a third fraction whose copper complex is soluble in Fehling solution (Table VII).[93]

The molecular weight of tamarind seed polysaccharide based on viscosity,[109] osmometry,[118] copper number,[118] and reduction of 3,5-dinitrosalicylate[120] has been

TABLE VII

Properties of TKP Fractions[93]

Fraction	Nitrogen Content, %	$[\alpha]_D$ (water)	Molar Ratios of Sugars			
			Glc	Xyl	Gal	Ara
I	7.8	—	3	1	1	1[a]
II	0.4	+75°	8	4	2	1
III	2.6	+60°	4	2	1	1

[a] Fraction I also contained D-galacturonic acid, but its molar ratio was not determined.

calculated as 52,350, 54,600, 55,600, and 115,000, respectively. It is rather surprising that the determination of the molecular weight based on the reducing power of the polysaccharide by two different methods has given widely differing values.

The polysaccharide disperses easily in cold water, forming viscous, mucilaginous solutions even in low concentrations. The viscosity increases so rapidly with concentration that it is difficult to prepare a mobile solution of more than 2%. The viscosity is not affected by pH of the solution or by the addition of sodium, calcium, or iron salts. Addition of sucrose, D-glucose, starch syrup, and other oligosaccharides increases viscosity, whereas hydrogen peroxide decreases it.[123]

A 2% solution, treated with a small quantity of sodium tetraborate (7.5% of the weight of the polysaccharide) and then stirred, yields a semisolid gel that becomes thin on warming but sets again when cooled. Both the gel and the solutions become hard when dried on hot rollers (drum-dried). This material, which has been named *gum taminda,* looks and behaves like powdered gum tragacanth when crushed and can be used as a substitute for it. It has been recommended for use in cotton fabric finishings and in calico printing pastes.[53]

A bright yellow color is formed when a few drops of iodine are added to a 0.5% solution, but with an excess of the reagent, a greenish blue color, which rapidly changes to yellow if diluted with water, is produced. In a more concentrated solution of the polysaccharide, a greenish blue gel appears, which changes to an orange solution on the addition of water. With excess iodine, the aqueous solutions are greenish blue, but on dilution with water the color disappears. A microscopic examination of an iodine-treated section of the seed kernel shows that the polysaccharide is present only in the cell wall and not within the cell.[100, 124]

The polysaccharide gives a hexasaccharide *tamarindose*, composed of D-glucose (3 moles), D-xylose (1 mole) and D-galactose (1 mole), on fermentative degradation with *Cladosporium herbarum* (Pers.) Link.[125] On enzymic degradation with heated cellulase of *Myrothecium verrucaria,* the polysaccharide yields a series of oligosaccharides. The three main products of this degradation are a hepta-, an octa- and a nonasaccharide. The heptasaccharide is composed of D-glucose (4 moles) and D-xylose (3 moles), whereas the octa- and nonsaccharides contain one and two additional units of D-galactose, respectively. On partial acid hydrolysis, the polysaccharide produces an alcohol-insoluble product consisting almost exclusively of D-glucopyranosyl units, whose x-ray diffraction diagram is almost identical with that of cellohexaose.[126] Further degradation of this hexaose by heated *Myrothecium verrucaria* cellulase, as well as by emulsin, results in the liberation of D-glucose and a number of oligosaccharides.

The polysaccharide undergoes progressive acetylation, depending on the conditions of the reaction, yielding a number of acetyl derivatives, some of which give fairly strong, flexible, glossy, transparent films that adhere to glass, metallic, and wooden surfaces. They might be useful as thermoplastic resins because of their wide melting range.[127]

The polysaccharide has also been subjected to the action of dinitrogen pentaoxide, epibromohydrin, diazomethane, and allyl halide, and products containing small amounts of nitro, halo, methyl, and allyl groups, respectively, have been obtained. These are reported to give good gels with strengths varying from 57 to 71% of that of the parent polysaccharide.[128] The low-DS allyl ether is reported to be soluble in organic solvents, giving hard films on drying. Savur[128] has also reported that the condensation of the polysaccharide with formaldehyde in the presence of a suitable catalyst, such as phosphoric, hydrochloric, or sulfuric acid, yields a product containing 2.8% of formyl groups. The viscosity and gel strength of this substance are said to be higher than those of the polysaccharide itself by 33% and 26%, respectively. Similarly, if the polysaccharide is suspended in a mixture of water, ethanol, and 2-propanol and then treated with ammonia, products containing some ammonium ions are formed.[129] These, like the partially acetylated derivatives, are said to give good gels that are more transparent but less strong than those obtained from the underivatized polysaccharide.

4. *Structure*

In early studies, tamarind seed polysaccharide, as obtained by the method of Rao and coworkers,[100] was shown to be composed of D-galactose, 4-xylose, and D-glucose in the molar ratio of 1 : 2 : 3.[40, 125, 130]

Whether or not tamarind seed polysaccharide contains L-arabinose has been a subject of controversy. According to Srivastava and Singh[117] and Chakraverti and coworkers,[118] L-arabinose is an integral part of TKP polysaccharides, even after they have been rigorously purified.[93, 118] Earlier, Savur and Sreenivasan,[40, 130] Rao and coworkers,[100] Kahn and Mukherjee,[109] and Kooiman[120] had not been able to detect its presence, but Damodaran and Rangachari[41] did. Indeed, it was found that if tamarind seed polysaccharide is hydrolyzed under mild conditions (0.01 *N* HCl), the first sugar released is L-arabinose.[131] It may be worthwhile to investigate whether locality, season, or age are factors controlling the occurrence of L-arabinose as a constituent of tamarind seed polysaccharide.

When subjected to methylation and subsequent methanolysis, tamarind seed polysaccharide gave rise to methyl glycosides of 2,3,4-tri-*O*-methyl-D-xylose, 3,4-di-*O*-methyl-D-xylose, 2,3,4,6-tetra-*O*-methyl-D-galactose, 2,3,6-tri-*O*-methyl-D-glucose, and 2,3-di-*O*-methyl-D-glucose in the molar ratio of 1 : 1 : 1 : 1 : 2.[132] The repeating unit of the polysaccharide molecule was suggested by White and Rao[132] to be three (1→4)-linked D-glucopyranosyl units, with branches on *O*-6 of two of the three units, one of which might be a single D-galactopyranosyl unit and the other two (1→2)-linked D-xylopyranosyl units. The authors also pointed out that some of the D-glucopyranosyl units might be in the side chain. Later, Kahn and Mukherjee,[109] working independently on more or less the same lines, proposed a similar structure; obviously, the work of White and Rao[132] escaped their notice. Based on enzymic hydrolysis and infrared spectral data, they concluded that the polysaccharide contains only β-D linkages. That the polysaccharide has a backbone of (1→4)-linked β-D-glucopyranosyl units was demonstrated by Kooiman and Kreger[126] from a study of the x-ray diffraction of the partially hydrolyzed polysaccharide; the x-ray diagram was almost identical to that of cellohexaose. In further support of β-D-(1→4) linkages, celloboise was identified as one of the main products when the product of partial hydrolysis was further hydrolyzed with heated *Myrothecium* cellulase. Furthermore, acetolysis of the polysaccharide acetate gave cellobiose octaacetate as one of the products.

In a fragmentation analysis, Kooiman[120] isolated and identified a number of oligosaccharides. Partial hydrolysis with acid yielded 2-*O*-β-D-galactopyranosyl-D-xylopyranose, whereas with the fungal enzyme *Luizym,* 6-*O*-α-D-xylopyranosyl-D-glucopyranose was produced along with other products. As the molar ratio of D-glucose to D-xylose in the polysaccharide was 4 : 3, Kooiman[120] concluded that approximately three out of every four D-glucopyranosyl units are substituted with D-xylopyranosyl units. From these data, he concluded that the polysaccharide consists of a main chain of β-D-(1→4)-linked glucopyranosyl units and that a side chain consisting of a single xylopyranosyl unit is attached to every second, third, and fourth D-glucopyranosyl unit through a α-D-(1→6) linkage. One D-galactopyranosyl unit is attached to one of the xylopyranosyl units through a β-D-(1→2) linkage. To which D-xylopyranosyl unit the D-galactopyranosyl unit is attached is not known.[120] The finding by Kooiman[120] that the D-galactopyranosyl units are attached to D-xylopyranosyl units is in contrast to the suggestion of White and Rao[132] that these residues are connected directly to D-glucopyranosyl units.

Srivastava and Singh[117] also studied the structure of tamarind seed polysaccharide, employing fragmentation, methylation, and periodate oxidation. They reported the isolation of not only cellobiose but also 2-*O*-β-D-galactopyranosyl-D-xylopyranose upon acetolysis of the polysaccharide and of 4-*O*-β-D-galactopyranosyl-D-glucose (lactose) and 6-*O*-α-D-xylopyranosyl-D-glucopyranose on hydrolysis with an enzyme preparation (hemicellulase). Methylation of the polysaccharide followed by hydrolysis yielded 2,3,5-tri-*O*-methyl-L-arabinose (1 mole), 2,3,4-tri-*O*-methyl-D-xylose (3 moles), 2,3,4,6-tetra-*O*-methyl-D-galactose (2 moles), 3,4-di-*O*-methyl-D-xylose (1 mole), 2,3,6-tri-*O*-methyl-D-glucopyranose (2 moles), and 2,3-di-*O*-methyl-D-glucopyranose (6 moles). Tamarind seed polysaccharide was also partially hydrolyzed to give a degraded polymer, composed essentially of D-glucopyranosyl units, which upon methylation and hydrolysis did not produce 2,3,4-tri-*O*-methyl-D-glucose. From this finding, coupled with the fact that the D-glucopyranosyl-(1→6)-D-glucopyranose bond is more stable to acid hydrolysis than the corresponding (1→4) linkage,[133] the authors concluded that the lactose unit constitutes one of the side chains. On the basis of the foregoing evidence, Srivastava and Singh[117] proposed the structure in Figure 5, although the sequential distribution of branches along the main chain is uncertain.

Oxidation of tamarind seed polysaccharide with sodium metaperiodate consumed 1.35 moles of the oxidant with the liberation of 0.34 mole of formic acid. The oxopolysaccharide, upon reduction and hydrolysis (Smith degradation[134]), yielded erythritol, glycerol, and glyceraldehyde in molar ratios of 8 : 2.5 : 1.[117] These results are in agreement with the proposed structure (Fig. 5).

Recently, the structure of fraction III of tamarind seed polysaccharide, the fraction which is soluble in Fehling solution, has been examined by methylation analysis.[93] Methylation and hydrolysis of the polysaccharide produced 2,3,5-tri-*O*-methyl-L-arabinose (1 mole), 2,3,4-tri-*O*-methyl-D-xylose (1 mole) 2,3,4,6-tetra-*O*-methyl-D-galactose (1 mole), 2,3,6-tri-*O*-methyl-D-glucose (1 mole), 3,4-

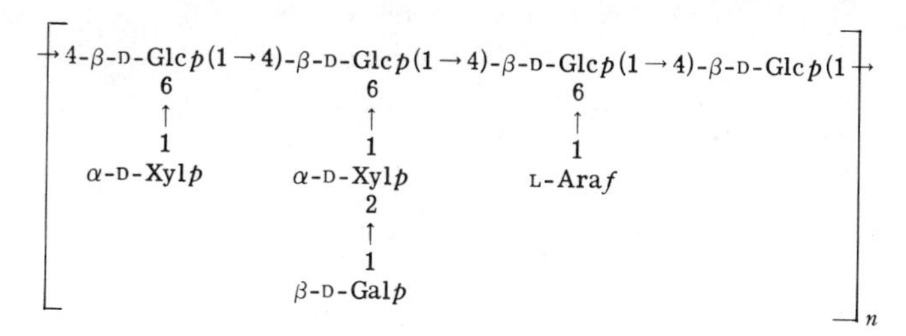

FIG. 5.—Structure proposed for tamarind seed polysaccharide.[117]

di-*O*-methyl-D-xylose (1 mole), and 2,3-di-*O*-methyl-D-glucose (3 moles). These results show that the linkages in fraction III are essentially the same as those in the main polysaccharide, fraction II. Thus, the two polysaccharides are quite similar structurally, although the percentage of L-arabinosyl units is higher in fraction III.

It is to be noted again that the finding by Srivastava and Singh[117] that the polysaccharide contains L-arabinosyl units is contrary to findings by some of the earlier workers, and this has made the determination of structure more difficult. As mentioned earlier, it is to be seen whether these authors were dealing with a somewhat different product, the differences arising from season, locality, and age factors.

5. *Uses*

Confectionery.—The most important and characteristic property of tamarind seed polysaccharide is its ability to form jellies with sugar concentrates over a wide pH range. Because of this property, it is an excellent substitute for fruit pectins in jams, jellies, and marmalades.[135] With TKP selling at Rs. 60/- –70/- a quintal, the estimated current cost of production of tamarind seed polysaccharide is Rs. 2.50/kg, that is, $0.15/lb at the present rate of exchange. In gel strength, as judged by the weight required to break the surface, the gels prepared from tamarind seed polysaccharide compare very favorably with those obtained from the fruit pectins (Table VIII).[136] The strength is not influenced by the kind of sugar used. In this study, the tamarind seed polysaccharide gels were prepared with only one-half the concentration used to make the pectin gels, yet they were much stronger than those of the pectins. In fact, the best tamarind seed polysaccharide gels are formed when its concentration is between 0.7 and 0.9% of the weight of the final gel, whereas the required concentration of fruit pectins is nearly double that amount. Figure 6 shows the variation of gel strength with concentration of tamarind seed polysaccharide and lemon pectin. Differences are also noticed in other respects. During tamarind seed polysaccharide gel formation, the gel strength increases tremendously with boiling time. Thus, the gel strengths on boiling for 5, 7, and 10 min are, respectively, 420, 540, and 650,[136] whereas the strength of fruit pectin gels is not appreciably affected[137] Furthermore, tamarind seed polysaccharide can form gels over a wide pH range, including pH 7,[102, 136] whereas fruit pectins set only in acid media. It may be noted, however, that the highest gel strength is obtained at a pH of 2.7–2.8,[136, 138] (cf. Yoshida and coworkers[139] according to whom the highest strength is recorded at pH 3.8–3.5). Figure 7 shows the influence of pH on the strength of tamarind seed polysaccharide and lemon pectin gels. It appears as though a pH of 2.8 is optimum in all cases, but in the case of tamarind seed polysaccharide, the gel strength does not fall to zero as the pH rises. The gels

TABLE VIII

Comparative Gel Strengths of Tamarind Seed Polysaccharide and Fruit Pectins[136]

Gelling Principle	Gel Strength, g
Lemon pectin	340
Orange pectin	361
Apple pectin	253
Wood apple pectin	346
Plantain pectin	239
Carrot pectin	327
Beet pectin	0
Tamarind seed polysaccharide	630

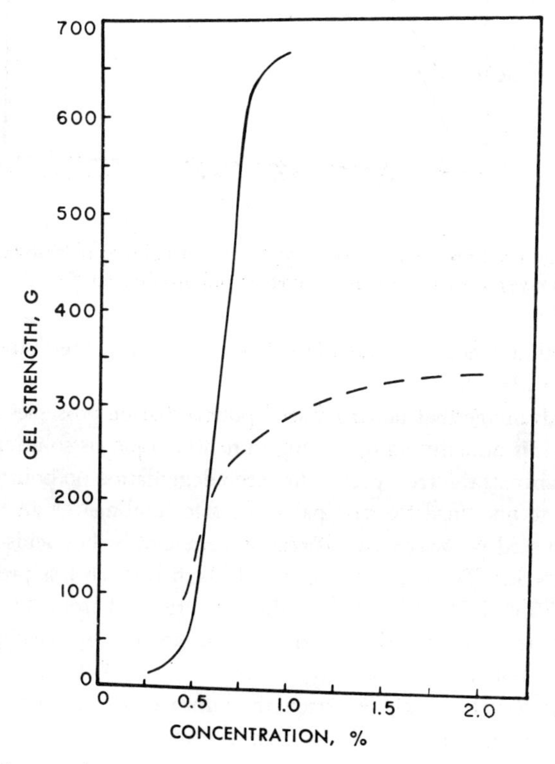

FIG. 6.—Influence of concentration of tamarind seed polysaccharide and lemon pectin (broken curve) on the gel strength. (Redrawn from Savur and Sreenivasan[136].)

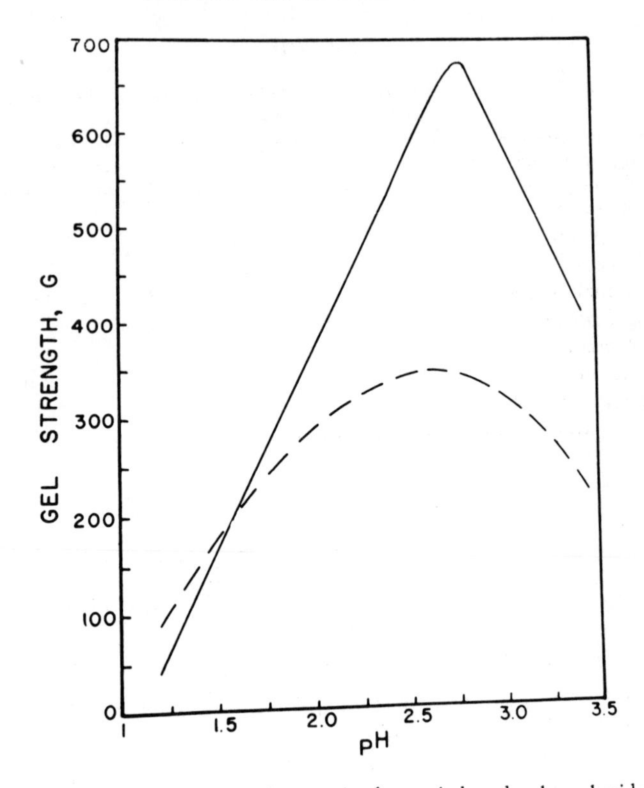

FIG. 7.—Influence of pH on the gel strength of tamarind seed polysaccharide and lemon pectin (broken curve) gels. (Redrawn from Savur and Sreenivasan.[136])

formed in alkaline media, even at pH 9.8, are quite firm; their strength is about 60% of that at pH 2.8.[140]

Another advantage that tamarind seed polysaccharide possesses over the fruit pectins is that it is not affected by boiling in neutral aqueous solutions even for as long as 2 hr, whereas the fruit pectins undergo degradation on boiling, the gelling power falling to one-third the original value after boiling for an hour.[141] However, tamarind seed polysaccharide is readily degraded by hot acids and alkalis as are the fruit pectins. The degrading effect of alkali is much less pronounced than that of acid (Table IX). It is said that the gels prepared from the original crude polysaccharide are tough in bite, opaque in appearance, and lacking in sheeting qualities and spoon feel.[142] However, a pretreatment with mild acid or alkali reduces the undesirably high gel strength, improves the transparency, and deceases the toughness of the gel; in fact, treatment with citric acid has been patented.[143]

Maximum gel strength is obtained when the sucrose concentration is 65–

TABLE IX

Effect of Acid and Alkali on the Gel Strength of Tamarind Seed Polysaccharide[136]

Reagent	Boiling Time, hr	Yield of Ethanol Precipitate, %	Gel Strength, g
0.1*N* Sulfuric acid	0	100	684
	1	76	216
	2	43.5	73
0.1*N* Sodium hydroxide	0	100.0	683
	1	90	580
	2	80.4	302
	3	72.3	241
	4	65.3	115
	5	58.2	76
	6	50.8	71

72% (Fig. 8). Polyvalent cations, such as calcium, have little effect on the strength of low-sugar tamarind seed polysaccharide gels, although they greatly increase the strength of low-sugar pectin gels. Addition of 0.075% of calcium hydrogen phosphate to a gel with a 0.75% of tamarind seed polysaccharide, 50% of sucrose, and 1% of citric acid only increased the gel strength from 92 to 124 g.

Tamarind seed polysaccharide jellies compare favorably in strength and transparency with, and show less syneresis than, fruit pectin jellies.[144] In view of its potential use as a substitute for fruit pectins, methods have been developed for direct utilization of the tamarind seed kernels without resorting to the isolation of tamarind seed polysaccharide.[145] In this process, TKP or crushed seed kernels are extracted with boiling water as previously described. The extract is allowed to settle in a settling tank, mixed with infusorial earth, and passed through a filter-press, whereby a clear solution is obtained. It is then heated in a steam-jacketed stainless steel open kettle until the concentration reaches 1%. The required amounts of sugar, citric acid (if acid taste is desired), and color are added, and the mixture is boiled until a gel consistency is attained (65°-70° Brix). The jelly solution, after skimming, is mixed with a flavor, transferred into jelly glasses, and allowed to set in a cool place. If intended for storage, the jelly mixture is somewhat underboiled and, while still hot, is poured into enamel-lined cans, which are sterilized and sealed. From 1 kg of kernels, about 58 kg of jelly are obtained.

In addition to its important use in the jelly industry, tamarind seed polysaccharide also finds an application in preparing other items of confectionery, such as jujubes. The set jelly is cut with a jelly knife first into slabs of 1–1.5 cm (0.4–0.6

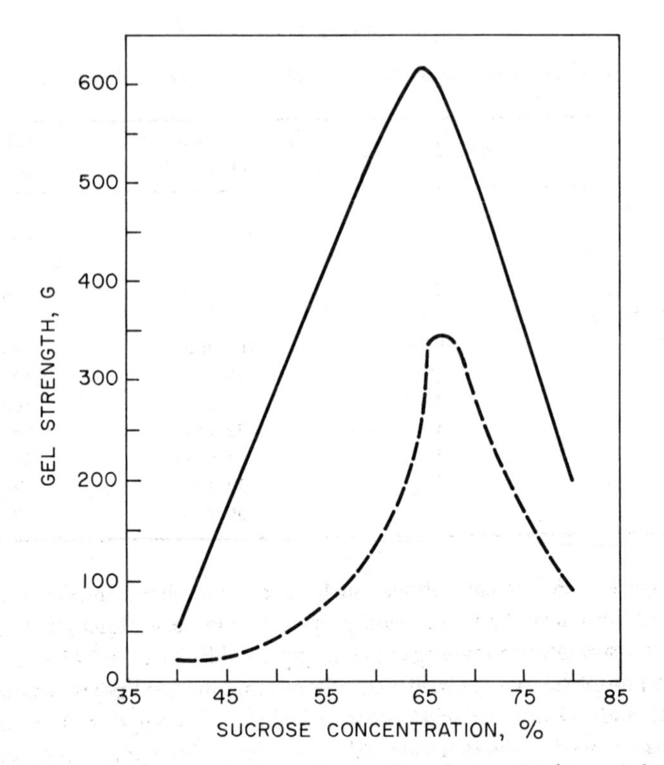

FIG. 8.—Influence of sucrose concentration on the gel strength of tamarind seed polysaccharide and lemon pectin (broken curve) jellies. (Redrawn from Savur and Sreenivasan.)[136]

in) thickness and finally into cubes or blocks of any suitable shape and size. Alternatively, the jelly solution can be poured into molds prepared in a starch bed and allowed to set. Incidentally, this starch also absorbs the excess moisture present in the jelly pieces. The latter are then dipped into a 1% solution of *babul* gum an inferior kind of gum arabic from *Acacia arabica* (Willd), dried either in the sun or in an air-oven at 40°–45°, during the course of which the jelly pieces lose a further quantity of water, and finally rolled in crystalline sugar, which forms a coating on the surface. After about a week, they are rolled once again in sugar and also in corn flour and packed in polyethylene bags. The cost of production usually works out to double the cost of sugar.[146]

In view of the colloidal nature of tamarind seed polysaccharide, its use is further advocated as a stabilizer in ice creams and mayonnaise.[147] These and other aspects of utilization of tamarind seed polysaccharide in food industries have been discussed in detail by Savur.[142, 148]

Textiles.—Savur recommends the use of the isolated polysaccharide in the textile industry for sizing, finishing, and printing cotton and artificial silk.[122, 149] The results of his investigations suggest that tamarind seed polysaccharide, either alone or in combination with gum tragacanth, can be used economically in printing pastes with most colors. With some dyestuffs, it does not produce a suitable paste because, upon addition of the dye, it gels and then becomes ropy and insoluble. The formulations and lists of dyes which can be used with the polysaccharide, as well as those which cannot, were also indicated.[150] These and other aspects of the industrial utilization of tamarind seed polysaccharide have been reviewed.[149, 151, 152]

Other possible uses.—Use of tamarind seed polysaccharide is indicated in a number of industries in addition to those of food and textiles.[24, 122] Its use depends upon its ability to serve as an adhesive, as an emulsifying agent, as a thickener, and as an inert, tasteless, colorless, and odorless substance. In cosmetics, it can be used for preparing emulsions of essential and fatty oils, shaving creams, and dentifrices. In pharmaceuticals, it can be used as a binder in the manufacture of pills and tablets, as an excipient in making greaseless ointments, and as a gelling agent in the preparation of colloidal iodine jelly. It can be used in insecticidal preparations for emulsifying the active principles and mineral oils. Measurements of change of the interfacial surface tension of the emulsions with time have shown that tamarind seed polysaccharide is comparable to other gums, such as tragacanth, arabic, and karaya, in stabilizing oil emulsions.[153] A 2–3% solution of the substance has good adhesive properties and can, therefore, be used for bookbinding and for cardboard manufacture.

Because tamarind seed polysaccharide has many properties similar to those of fruit pectins, it might find use in the treatment of colitis, diarrhea, dysentery, and other intestinal disorders.

In the leather industry, tamarind seed polysaccharide can be used in sizing and weighting compositions.

V. References

(1) G. Watt, "A Dictionary of the Economic Products of India," W. H. Allen & Co., London, 1893.

(2) *Indian Agr.*, **23,** 241 (1908).

(3) D. Hooper, *Agr. Ledger*, No. **2,** 13 (1907).

(4) L. H. Burkill, "A Dictionary of the Economic Products of the Malay Peninsula," Crown Agents for the Colonies, London, 1935.

(5) C. D. Mell, *Textile Colorist*, **51,** 111 (1929).

(6) S. Krishna and S. Ramaswami, *Indian Forest Bull.*, No. **79** (1932).

(7) *Indian Agr.*, **13,** 92 (1898).

(8) J. R. de Almeida and O. Valsecchi, *Brasil Acucareiro*, **38,** 365 (1951); *Chem. Abstr.*, **46,** 5255 (1952).

(9) A. L. Stahl, *Florida Agr. Expt. Sta. Bull.*, No. **283** (1935); *Chem. Abstr.*, **30,** 529 (1936).

(10) W. C. Taber, *J. Ind. Eng. Chem.*, **7,** 607 (1915).

(11) "Pharmacographia Indica," 1890.

(12) G. E. Trease, *Pharm. J.*, **126,** 211 (1931).

(13) N. Tseuneo, T. Itiro, and T. Shintaro, *Shoyakugaku Zasshi,* **21,** 47 (1967); *Chem. Abstr.*, **69,** 53042w (1968).

(14) G. C. Esh and K. N. Bagchi, *J. Proc. Inst. Chem. Calcutta,* **24,** 11 (1952).

(15) S. M. Bose and N. Subramanian, *Bull. Central Food Technol. Res. Inst.*, **3,** 66 (1954).

(16) A. N. Puri, Indian Patent 44,427 (1953); *Chem. Abstr.*, **47,** 9640 (1953).

(17) S. M. Bose, M. Swaminathan, and V. Subrahmanyan, *Bull. Central Food Technol. Res. Inst.*, **3,** 67 (1954).

(18) N. D. Kehar and B. Sahai, *Sci. Cult.*, **14,** 534 (1949).

(19) Rajnarian and S. Dutt, *Indian J. Agr. Sci.*, **15,** 209 (1945).

(20) P. Padilla and F. A. Soliven, *Philippine Agr.*, **22,** 408 (1933).

(21) *Olien en Vetten,* No. **30,** 387 (1920); *Chem. Abstr.*, **14,** 1230 (1920).

(22) K. Kafuku and C. Hata, *Nippon Kagaku Kaishi,* **55,** 369 (1934).

(23) D. N. Grindley, *J. Soc. Chem. Ind., London,* **65,** 118 (1946).

(24) M. S. Tawakley and R. K. Bhatnagar, *Indian Soap J.*, **19,** 113 (1953).

(25) T. P. Ghose and S. Krishna, *J. Indian Chem. Soc., Ind. News Ed.*, **5,** 114 (1942).

(26) S. Krishna, Indian Patent 29,620 (1943).

(27) T. P. Ghose and S. Krishna, *Indian Text. J.*, **53,** 236 (1943).

(28) N. S. Bhathena, Indian Patent 38,758 (1947).

(29) N. S. Bhathena, "Tamarind Powder Industry," Din Products, Ltd., Bombay, India, 1953.

(30) G. W. Deodhar, Indian Patent 48,164 (1953); *Chem. Abstr.*, **48,** 3702 (1954).

(31) G. W. Deodhar, Indian Patent 49,313 (1954); *Chem. Abstr.*, **48,** 7314 (1954).

(32) G. W. Deodhar, Indian Patent 49,314 (1954); *Chem. Abstr.*, **48,** 7315 (1954).

(33) K. G. Dutt and U. Chatterji, Indian Patent 30,321 (1943).

(34) M. L. Schroff, Indian Patent 30,487 (1944).

(35) W. G. Macmillan, S. Nag, and I. B. Chakraverti, *J. Sci. Ind. Res., Sec. B,* **16,** 418 (1957).

(36) W. G. Macmillan and I. B. Chakraverti, *J. Sci. Ind. Res., Sect. B,* **10,** 13 (1951).

(37) Indian Standard: 189 (1951), Indian Standards Institution, New Delhi.

(38) Indian Standard: 511 (1954), Indian Standards Institution, New Delhi.

(39) P. S. Rao and S. Krishna, *Indian Text. J.*, **59,** 314 (1949).

(40) G. R. Savur and A. Sreenivasan, *J. Biol. Chem.*, **172,** 501 (1948).

(41) M. Damodaran and P. N. Rangachari, *Curr. Sci.*, **14,** 203 (1945).

(42) D. B. Das, P. K. R. Chaudhuri, and J. F. Wareham, *J. Text. Inst.*, **44,** T402 (1953).

(43) P. B. Sarkar and A. K. Mazumdar, *J. Text. Inst.*, **43,** T453 (1952).

(44) P. S. Rao and E. E. Dickey, *Sci. Cult.* **21,** 102 (1955).

(45) H. C. Srivastava and T. N. Krishnamurthy, in press.

(46) K. K. Patnaik and M. M. Laloraya, *Curr. Sci.*, **32,** 279 (1963).

(47) R. N. Pal and M. M. Laloraya, *Physiol. Plant.*, **20,** 789 (1967).

(48) S. G. Bhat, *Indian Oil Soap J.*, **32,** 53 (1966).

(49) D. B. Das and K. K. Basak, *J. Indian Chem. Soc.*, **27,** 115 (1950).

(50) K. F. Patel, M. M. Patel, S. N. Harshe, J. R. Modi, and H. C. Srivastava, *Proc. Technol. Conf., A.T.I.R.A., B.T.R.A. and S.I.T.R.A., 9th, 1967,* 219 (1968).

(51) W. G. Macmillan and I. B. Chakraverti, *J. Sci. Ind. Res., Sect. B,* **10,** 270 (1951).

(52) S. Krishna, Indian Patent 29,228 (1942).

(53) T. P. Ghose and S. Krishna, *Indian Text. J.,* **56,** 506 (1946).

(54) H. U. Mehta, P. K. Ghouthoy, and P. C. Mehta, *Proc. Technol. Conf., A.T.I.R.A., B.T.R.A., S.I.T.R.A., 7th, 1965,* **A,** 78 (1966).

(55) H. C. Srivastava and S. N. Harshe, unpublished data.

(56) H. U. Mehta, R. S. Parmar, R. S. Patel, and J. K. Patel, *Tappi,* **52,** 664 (1969).

(57) P. C. Mehta and C. C. Shah, *A.T.I.R.A. Res. Notes,* **3,** No. 2 (1953).

(58) P. C. Mehta, P. N. Bhatt, S. S. Trivedi, and C. C. Shah, *J. Sci. Ind. Res., Sect. A,* **16,** 194 (1957).

(59) S. G. Bhat, *J. Sci. Ind. Res., Sect. A,* **16,** 563 (1957).

(60) N. F. Desai, A. Sreenivasan, and K. Venkataraman, *J. Sci. Ind. Res., Sect. B,* **44,** (1946).

(61) P. K. Saha, *Text. Mfr.,* **78,** 625 (1952).

(62) W. G. Macmillan, I. B. Chakraverti, and P. N. Pal, *J. Sci. Ind. Res., Sect. B,* **11,** 438 (1952).

(63) W. G. Macmillan, I. B. Chakraverti, and P. N. Pal, *J. Sci. Ind. Res., Sect. B,* **13,** 66 (1954).

(64) J. R. Modi, personal communication.

(65) A.T.I.R.A., *Tech. Dig.,* **2,** No. 2, 19 (1968).

(66) *J. Text. Inst.,* **47,** P375 (1956).

(67) I. B. Chakraverti and S. Nag, *J. Sci. Ind. Res., Sect. C,* **20,** 113 (1961).

(68) G. W. Deodhar, Indian Patent 47,457 (1954); *Chem. Abstr.,* **48,** 7816 (1954).

(69) S. V. Puntambekar, *Bull. Natl. Inst. Sci. India,* No. **6** (1955).

(70) P. C. Mehta and J. Singh, *Indian J. Technol.,* **4,** 296 (1966).

(71) G. W. Keller, *Melliand Textilber.,* **40,** 423 (1959).

(72) E. Rhodes and K. C. Sekar, Brit. Patent 474,651 (1937); *Chem. Abstr.,* **32,** 3665 (1938).

(73) S. Krishna, Indian Patent 29,870 (1943).

(74) S. Krishna, Indian Patent 32,035 (1947); *Chem. Abstr.,* **42,** 7506 (1948).

(75) President, Forest Research Institute and Colleges, Dehra Dun, Indian Patent 31,362 (1944).

(76) S. V. Puntambekar, *Bull. Natl. Inst. Sci. India,* No. **6** (1955).

(77) N. S. Bhathena, Indian Patent 44,751 (1953); *Chem. Abstr.,* **47,** 6630 (1953).

(78) A. C. Shekar, P. L. Taneja, V. K. Sood, and M. G. Karnik, *Res. Ind.,* **8,** 291 (1963).

(79) D. Narayanamurti and P. R. Rao, *Res. Ind.,* **6,** 127 (1961).

(80) P. S. Rao and A. K. Misra, *Indian Forest.,* **89,** 686 (1963).

(81) A. J. de Roos, *Tappi,* **41,** 354 (1958).

(82) G. Boyd, Australian Patent 229,190 (1960); *Chem. Abstr.,* **55,** 25255 (1961).

(83) H. R. Fee, J. D. Ferguson, and R. W. Lawrence, U.S. Patent 3,350,246 (1967); *Chem. Abstr.,* **68,** 14630w (1968).

(84) P. Maity, *Indian Wood Panels,* **2,** 30 (1969).

(85) G. R. Savur, *J. Indian Chem. Soc., Ind. News Ed.,* **20,** 55 (1957).

(86) H. C. Srivastava, S. N. Harshe, and G. P. Nudia, *Indian J. Technol.,* **8,** 347 (1970).

(87) H. C. Srivastava, S. N. Harshe, and M. M. Gharia, unpublished data.

(88) H. C. Srivastava and S. N. Harshe, Indian Patent 112,845 (1969).

(89) M. M. Gharia, S. N. Harshe, J. R. Modi, M. C. Paliwal, and H. C. Srivastava, *Proc. Technol. Seminar, Textile Ass., 3rd, Delhi Branch,* 43 (1971); H. C. Srivastava and S. N. Harshe, *J. Text. Ass.,* **33,** 148 (1972).

(90) H. C. Srivastava and M. M. Patel, unpublished data.

(91) R. S. Parmar, Ph.D. Thesis, (H. C. Srivastava, Preceptor), University of Gujarat, Ahmedabad, India, 1967.

(92) G. B. Dave, Ph.D. Thesis, (H. C. Srivastava, Preceptor), University of Gujarat, Ahmedabad, India, 1969.

(93) T. N. Krishnamurthy, Ph.D. Thesis (H. C. Srivastava, Preceptor), University of Gujarat, Ahmedabad, India, 1971.

(94) R. Schlageter, U.S. Patent 3,223,699 (1965); *Chem. Abstr.,* **64,** 17857 (1966).

(95) M. L. Oliver, U.S. Patent 3,392,085 (1968); *Chem. Abstr.,* **69,** 60174u (1968).

(96) A. Schiavio and C. Maderno, Italian Patent 574,928 (1958); *Chem. Abstr.,* **54,** 25911 (1960).

(97) Chemische Fabrik Gruenau G.m.b.H., Germ. Patent 1,206,777 (1965); *Chem. Abstr.,* **64,** 6290 (1966).

(98) M. J. Schleiden, *Ann. Phys. Chem.,* **43,** 391 (1838).

(99) M. J. Schleiden and T. Vogel, *Ann. Phys. Chem.,* **46,** 327 (1839).

(100) P. S. Rao, T. P. Ghose, and S. Krishna, *J. Sci. Ind. Res.,* **4,** 705 (1945–1946).

(101) R. H. McDowell, *Chem. Ind. (London),* 198 (1946).

(102) P. S. Rao, *J. Sci. Ind. Res., Sect. B,* **7,** 89 (1948).

(103) H. R. Nanji, G. R. Savur, and A. Sreenivasan, *Curr. Sci.,* **14,** 129 (1945).

(104) M. H. Carre and D. Haynes, *Biochem. J.,* **16,** 60 (1922).

(105) P. S. Rao, *J. Sci. Ind. Res., Sect. A,* **16,** 138 (1957).

(106) K. Gorter, *Ann.,* **359,** 217 (1908).

(107) P. S. Rao and S. Krishna, *Curr. Sci.,* **16,** 256 (1947).

(108) G. R. Savur, *J. Indian Chem. Soc., Ind. News Ed.,* **19,** 67 (1956).

(109) N. A. Khan and B. D. Mukherjee, *Chem. Ind.* (*London*), 1413 (1959).

(110) P. Kooiman, *Nature,* **180,** 201 (1957).

(111) P. S. Rao, *Indian Export Trade J..* **10,** 120 (1956).

(112) G. R. Savur, Indian Patent 53,429 (1954).

(113) P. S. Rao, unpublished data.

(114) A. L. Gordon, U.S. Patent 3,399,189 (1968); *Chem. Abstr.,* **69,** 97856u (1968).

(115) Dainippon Pharmaceutical Co. Ltd., Brit. Patent 1,007,303 (1965); *Chem. Abstr.,* **63,** 17809 (1965).

(116) T. P. Ghose and S. Krishna, *Curr. Sci.,* **14,** 299 (1945).

(117) H. C. Srivastava and P. P. Singh, *Carbohyd. Res.,* **4,** 326 (1967).

(118) I. B. Chakraverti, S. Nag, and W. G. Macmillan, *J. Sci. Ind. Res., Sect. D,* **20,** 380 (1961).

(119) I. B. Chakraverti, S. Nag, and W. G. Macmillan, *Ind. J. Technol.,* **1,** 216 (1963).

(120) P. Kooiman, *Rec. Trav. Chim.,* **80,** 849 (1961).

(121) P. S. Rao and S. Krishna, *Curr. Sci.,* **15,** 133 (1946).

(122) G. R. Savur, *Curr. Sci.*, **24,** 235 (1955); *Chem. Ind. (London)*, 212 (1956); *Sci. Cult.*, **21,** 744 (1956); *J. Chem. Soc.*, 2600 (1956).

(123) A. Tamura, K. Wada, and O. Shoji, *Nippon Nogei Kogaku Kaishi,* **38,** 300 (1964); *Chem. Abstr.*, **63,** 6247 (1965).

(124) T. P. Ghose and S. Krishna, *Indian J. Agr. Sci.*, **16,** 410 (1946).

(125) P. S. Rao and R. M. Beri, *Proc. Indian Acad. Sci., Sect. A,* **35,** 1 (1952).

(126) P. Kooiman and D. R. Kreger, *Biochim. Biophys. Acta,* **26,** 207 (1957).

(127) P. S. Rao and R. M. Beri, *Proc. Indian Acad. Sci.*, **42A,** 199 (1955).

(128) G. R. Savur, *Indian Food Packer,* **10,** No. 3, 7 (1956).

(129) G. R. Savur, *Indian Food Packer,* **10,** No. 2, 11 (1956).

(130) G. R. Savur and A. Sreenivasan, *Curr. Sci.*, **15,** 43 (1946).

(131) P. P. Singh, Ph.D. Thesis, (H. C. Srivastava, Preceptor), University of Gujarat, Ahmedabad, India, 1966.

(132) E. V. White and P. S. Rao, *J. Amer. Chem. Soc.*, **75,** 2617 (1953).

(133) M. L. Wolfrom, A. Thompson, and C. E. Timberlake, *Cereal Chem.*, **40,** 82 (1963).

(134) F. Smith, M. Abdel-Akher, J. K. Hamilton, and R. Montgomery, *J. Amer. Chem. Soc.*, **74,** 4970 (1950).

(135) Daurala Sugar Works, Indian Patent 28,409 (1941).

(136) G. R. Savur and A. Sreenivasan, *J. Soc. Chem. Ind., London,* **67,** 190 (1948).

(137) G. L. Baker, *Ind. Eng. Chem.*, **18,** 89 (1926).

(138) K. Wada, A. Tamura, and O. Shoji, *Nippon Nogei Kagaku Kaishi,* **38,** 347 (1964); *Chem. Abstr.*, **63,** 6247 (1965).

(139) T. Yoshida, Y. Yokoo, and Y. Koyama, *Nippon Shokuhin Kogye Gakkaishi,* **11,** 200 (1964); *Chem. Abstr.*, **64,** 14858 (1966).

(140) P. S. Rao, unpublished data.

(141) C. L. Hinton, "Fruit Pectins: Their Chemical Behaviour and Jellying Properties," Food Investigation Special Report No. 48, His Majesty's Stationery Office, London, 1939.

(142) G. R. Savur, *Indian Food Packer,* **9,** No. 7, 15 (1955).

(143) H. R. Pithawala and A. Sreenivasan, Indian Patent 40,835 (1950); *Chem. Abstr.*, **44,** 9588 (1950).

(144) G. R. Savur, *Indian Food Packer,* **9,** No. 3, 31 (1955).

(145) P. S. Rao, *J. Sci. Ind. Res., Sect. A,* **8,** 354 (1949).

(146) P. S. Rao, *Res. Ind.*, **4,** 173 (1959).

(147) O. Shojo, K. Wada, A. Tamura, and K. Wada, U.S. Patent 3,342,608 (1967); *Chem. Abstr.*, **68,** 2121 (1968).

(148) G. R. Savur, *Indian Food Packer,* **9,** No. 2, 13 (1955).

(149) G. R. Savur, *Indian Text. J.*, **65,** 418, 547 (1955).

(150) G. R. Savur, *Indian Text. J.*, **66,** 309 (1956).

(151) G. R. Savur, *Indian Text. J.*, **66,** 33 (1956).

(152) G. R. Savur, *Melliand Textilber.*, **37,** 588 (1956); *Chem. Abstr.*, **50,** 15091 (1956).

(153) R. P. Patel and Y. Raghunathan, *Indian J. Pharm.*, **21,** 159 (1959).

Plant Extracts

CHAPTER XVIII

LARCH ARABINOGALACTAN

MARK F. ADAMS AND BRUCE V. ETTLING

College of Engineering Research Division, Washington State University, Pullman, Washington

I. INTRODUCTION

Arabinogalactan is a water-soluble gum found in unusually high concentration (up to 35%) in the heartwood of all species of larch. Arabinogalactans are found in small quantities in a number of other trees, plants, and plant seeds and in cell walls of certain bacteria. Even though a report of this polysaccharide first appeared in the literature in 1898, arabinogalactan did not appear on the commercial market until 1964. It has many unique characteristics, such as complete miscibility with water and low viscosity at high dissolved solids content (23.5 cps at 40% concentration and 20°). These unique characteristics have made the gum useful in such industries as the printing, mining, carbon black, and food industries; particularly in the latter are new and exciting uses being generated. Several good review articles are available.[1–4]

II. PRODUCTION

1. *Source and Producing Areas*

Arabinogalactan is contained in high concentration (5–35% on a dry wood basis) in the heartwood of the many species of larch.[5] It is unevenly distributed throughout the tree from the root system to the ends of the branches. In western larch, *Larix occidentalis* (Nutt.), the highest concentration is in the lower portion of the tree, with increasing concentration from the center of the tree to the sapwood.[6] Depending on its location in the tree, the gum varies in composition and molecular weight.[7]

Tamarack (*Larix laricina*) also contains a high concentration of arabinogalactan; however, the concentration in the heartwood increases with increased height in the tree. The highest concentrations (21%) appear in the branches. The concentration is lowest at the pith and increases in radial direction to reach a maximum at the heartwood–sapwood boundary, as it does in all the industrially important larch species. The arabinogalactan is contained in the lumen of the tracheides and also in the ray cells. There is some evidence that the polysaccharide is an amorphous substance with little or no connection to the cell wall.[8]

The genus *Larix* (Larches) is scattered throughout the world: Dahurian larch (*Larix dahurica*) in Russia, eastern larch (*Larix laricina*) in eastern North America, European larch (*Larix decidua,*) *Japanese larch* (*Larix leptolepsis*), Siberian larch (*Larix siberica*), western larch (*Larix occidentalis*) in western North America, and others. The only commercial production of arabinogalactan is from western larch by the St. Regis Company in northwestern Montana, USA. Western larch stands are found along the western slopes of the Rocky Mountains as far south as Utah, with heavy stands extending into British Columbia.[9]

The commercial forests of western larch (*Larix occidentalis*) within the states of Montana, Idaho, Washington, and Oregon comprise ~29,000,000,000 board-feet of standing timber. The average annual production of larch from this area is ~200,000,000 board-feet of lumber. Extraction of the bark-free wood residue from the production of 1000 board-feet of lumber will yield approximately 100 pounds (45 kg) of arabinogalactan. Utilization of larch wood residues currently available at the saw mills in this region would yield ~20,-000,000 pounds of the gum a year.[10]

Arabinogalactan is found in small amounts in other genera of trees. It is found in the cambium layer and sapwood of hemlock.[11] Black spruce, parana pine, mugo pine, douglas fir, incense cedar, and juniper contain minor amounts,[12–14] and a small amount of arabinogalactan (L-arabinose to D-galactose ratio 1 : 1.7) is contained in the sapwood of the sugar maple.[15]

An arabinogalactan with an L-arabinose to D-galactose ratio of 2 : 5 is present in small amounts in the green coffee bean,[16–18] and one with a ratio of 1 : 13 is found in centrosema seeds.[19] Wheat flour contains a small amount of arabinogalactan[20] with a sugar ratio of 7 : 3. The major component (70%) of the water-soluble fraction of soybean flour is an arabinogalactan[21, 22] with a sugar ratio of 1 : 2. The water-soluble part of gum tragacanth contains an arabinogalactan[23] of sugar ratio 6 : 1.

2. *Method of Production*

Arabinogalactan can be extracted from larch by chipping or grinding the heartwood and extracting it with water.[27] The extraction of finely divided larch wood can also be accomplished using dilute sulfuric acid at temperatures up to 160° in an autoclave. The acid extraction gives a hydrolyzed product.[28] The rate of extraction of gum from larch chips is dependent upon the particle size of the chip and the extraction temperature. The concentration of the extract solution is dependent upon the amount of gum in the larch chip.[29]

Arabinogalactan is produced by the St. Regis Company at Libby, Montana,[30] and marketed under the trade-name Stractan by the Stein, Hall Company.[31]

The present commercial plant uses a countercurrent extraction process[32] and has a production capacity in excess of 400,000 pounds (180,000 kg) a year. Waste wood from the plywood mill, the stud mill, and the lumber mill is chipped into normal pulp wood chips. The fresh chips are placed in a tank of a countercurrent extraction system. Each time (~1 hr) that a tank of fresh chips is put into the system, a tank of depleted chips is removed. Fresh water is introduced into the tank battery and flows from the freshest toward the most exhausted chips. The final extract contains 8–10% of arabinogalactan. This solution is dried on a drum dryer to produce technical-grade Stractan.

In another procedure described in the same patent, finely divided larch wood is mixed with sufficient water to give a moisture content of 200%. The water-saturated, finely divided wood is pressed to give approximately 70% recovery of water extract. The time for this cycle is approximately 10 min. Another pressing cycle can be used to recover 70% of the remaining gum. Thus, in two pressings, 90% of the original gum can be extracted.

Both processes can be operated at ambient temperature or higher, but preferably not above the boiling point of water, for phenolic impurities and water-soluble substances are extracted in amounts that increase with temperature.

The phenolic impurities (expressed as tannic acid) constitute about 3% of the gum. Almost all the phenolic impurities are contained in a polymer of higher molecular weight than that of the gum.

These two production procedures are conducted in a manner that minimizes hydrolysis. The content of reducing sugars in the final product is one-third less than that obtained by extractions done at the boiling point.

The commercial product has a very slight buff color, a slightly woody odor, and a bland taste reminiscent of fine sawdust, and contains phenolic compounds, oligosaccharides and trace amounts of terpenes as impurities. The material is soluble in water and has a viscosity of 23.5 cps at 40% (w/v) concentration[32] and 20°. Arabinogalactan is noncrystalline. In its production, some fine powder is produced which could be an explosive hazard. The most sensitive dust (minus 325 mesh) has a higher minimum ignition temperature (MIT) (511°) than that of other common powders such as cornstarch, sucrose, D-glucose, and wood flour; a much higher minimum ignition concentration (MIC) (0.22 oz/ft^3, 220 g/m^3) and one of the lowest maximum explosive pressures (MEP) (52 psi, 3.66 kg/cm^2) of any the common dusts.[33]

3. *Purification*

For certain uses, particularly in the food industry, it is desirable to have an arabinogalactan that is colorless and without taste. Consequently, attempts have been made to remove impurities from the gum. These efforts have included the treatment of gum solutions with various types of carbons,[34] multiple precipitation with alcohol,[35] use of various ion-exchange columns,[36] and electrodialysis. No purification step has been fully successful. Some of the processes did not adequately remove impurities, and others were uneconomical.

One of the better procedures for removing iron and phenolic compounds is treatment of a gum solution with active magnesium oxide.[37] The active magnesium oxide (5–10% on the basis of the gum) is rapidly mixed with an almost boiling solution of the gum. A flocculent, brownish precipitate forms almost immediately, and the slurry is filtered quickly to prevent formation of a slimy precipitate. The final solution, which is essentially colorless, is dried on a drum dryer to produce a white powder that is odorless but has a very slight aromatic taste.

In some food uses, even this slight taste is objectionable and can be removed by treating a solution of the purified gum with a decolorizing carbon. The final, highly purified arabinogalactan has the same solubility, viscosity density, and refractive index as the technical-grade gum, but it has none of the surface active properties characteristic of the technical product.

4. *Grades*

Stractan is available in various grades, ranging from those of high phenolic content for use in the printing, pelletizing, and mining industries to those that are highly purified for use in foods and certain pharmaceuticals.

III. Characterization

Arabinogalactan from western larch has been the most extensively examined.[32] Its physical and chemical properties depend upon the method used for its isolation and purification.

1. Physical Properties

Purified arabinogalactan is completely colorless, odorless, and tasteless,[36] but the gum as extracted from western larch and from other plant sources is tan when dried and powdered. Solutions in water are tan to brown and usually are moderately turbid above 10% concentration. The gum from western larch has a mild, woody flavor and odor when wet.[38]

Water solutions of arabinogalactan are fluid to 60% concentration. Above this concentration, they form a thick paste and, finally, a glass that is friable when the moisture content is below about 10%.

The viscosity of a water solution of arabinogalactan is less than that of most other natural gums. A 10% w/v solution of pure or technical gum has a viscosity of 1.74 cps at 20°. Neither electrolytes at 0.1*M* concentration nor a pH of 1.0 have any effect on the solution viscosity. At a pH 11.5, the viscosity of a 30% w/v solution at 20° is 8.15 cps as compared with a viscosity of 7.8 cps at pH 4, a 7% increase.[38] The Brookfield viscosity of a 40% solution of arabinogalactan in water is a linear function of the spindle speed up to 60 rpm, thus showing Newtonian flow properties up to that speed.[39]

The refractive index of arabinogalactan solutions is a linear function of the concentration and can be used to determine gum concentrations in a production plant.[39] The refractive index is essentially the same for pure and technical gum. The n_D^{20} of a 10% solution at 20° is 1.3480. The change in refractive index per 1% change in concentration is 0.0015.

The surface tension[38, 39] of the gum in water depends greatly upon the purity. Pure arabinogalactan has a surface tension of 72.3 dynes/cm at 20° and 10% concentration by the du Nuoy method. Technical gum under the same conditions has a value of 65.1 dynes/cm. At 30% concentration, the value of pure polymer is 71.7 dynes/cm, whereas for technical polymer it is 59.9 dynes/cm. The impurities in the technical grade lower the surface tension and similarly the interfacial tension with both mineral oil and cottonseed oil.[38, 39] These same impurities are presumably responsible for the dispersant properties of technical gum. The surface tension of the technical product is lower at low pH values.

The optical rotation of both technical-grade arabinogalactan and the purified polysaccharide is $[\alpha]^{20}\text{D}$ $+12°$ (water). [36, 40] The high-molecular-weight component of the gum has $[\alpha]^{20}\text{D}$ $+7°$ and the low-molecular-weight component has $[\alpha]^{20}\text{D}$ $+10°$ (water).[41] The optical rotation of arabinogalactan from eastern larch[42] is $+13°$ and from European larch[43] is also $+13°$, with one report[44] of $+20°$.

The ultraviolet spectrum of a 0.1% solution of purified gum shows virtually no absorption above 225 nm (Fig. 1). A similar spectrum of commercially

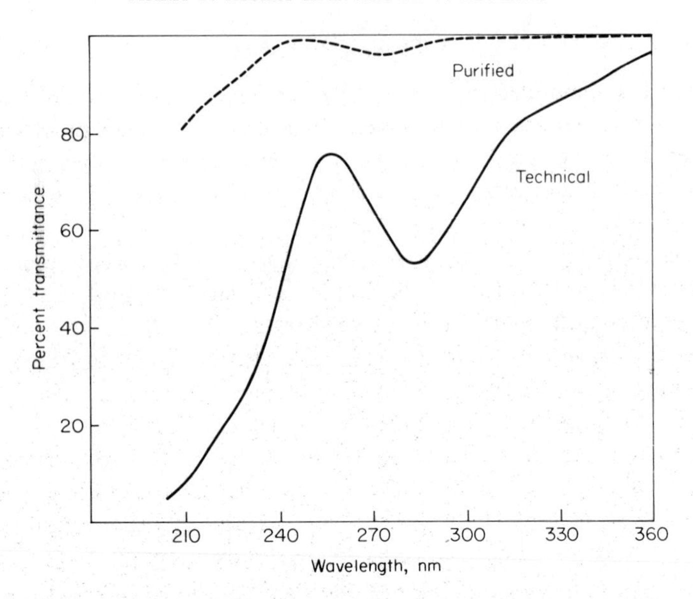

FIG. 1.—Ultraviolet spectra of technical arabinogalactan (0.10% solution) and purified arabinogalactan (1.0% solution).

produced arabinogalactan shows some absorption in the blue and violet end of the visible range (to give the amber to tan color) that continues to increase to a peak at 280–285 nm. The maximum absorbance varies greatly with the method of producing the polysaccharide but is generally in the range of 0.26–0.31 absorbance units per 0.10% concentration for the commercial product. This absorption apparently results from the phenolic components of the technical-grade gum.

The infrared spectrum of technical gum (Fig. 2) is virtually identical to that of purified arabinogalactan.[32] The relatively small amount of phenolic im-

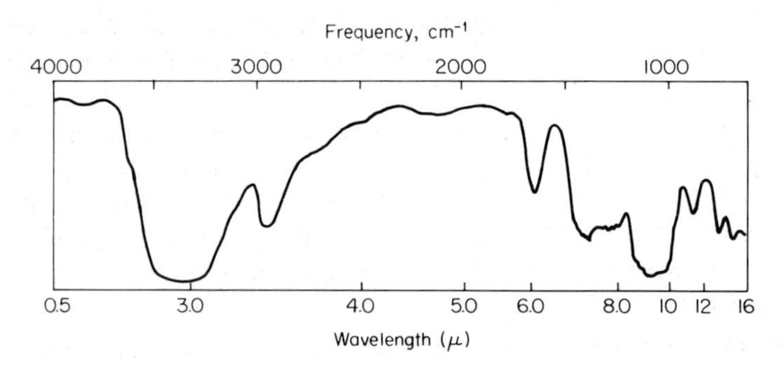

FIG. 2.—Infrared spectrum of technical arabinogalactan.

purities are not indicated in the infrared spectrum except by differential techniques using purified polysaccharide as a reference.

2. *Chemical Properties*

Purified arabinogalactan is a neutral polysaccharide, whereas the technical gum is slightly acidic because of the phenolic components. The pH of technical-grade gum solutions[38, 39] is 4.0 to 4.5. The pH of purified gum solutions will vary depending upon the treatment, because pH changes are involved in some methods of purification.[37]

Arabinogalactan reacts with acid anhydrides or acyl halides to produce esters. Acetate,[39, 40, 45] sulfate,[47] propionate,[46] and benzoate[46] esters have been made. Methyl[39, 45, 48] and benzyl[48] ethers have also been prepared. The esters and ethers are not water soluble but become soluble in moderately polar solvents, such as acetone and chloroform.

The polysaccharide can be oxidized to give oxidation products in which the D-galactopyranosyl units are oxidized. Thus, D-galactonic acid and its lactones are produced by limited oxidation of a crude larch extract with chlorine[49] at 50°–70°. More vigorous oxidation of larch extracts with nitric acid give mucic acid (D-galactaric acid) plus oxalic acid and other oxidation products.[50, 51]

The arabinofuranose units in the gum are labile to acid and can be removed under mild hydrolytic conditions.[52] Therefore, boiling solutions of the gum in the dilute acid produces furfural by dehydration of the L-arabinose released by hydrolysis. Dulcitol (galactitol) can be obtained from hydrogenolysis of the gum.[53]

The purified gum is free of phenolic, terpenoid, or other extraneous chemicals but contains any oligosaccharides that are part of the larch extract. The technical gum without purification contains small amounts of volatile terpenoid oils, low-molecular-weight phenolics, polymeric phenolics, and simple sugars.[54] The major impurity of commercial gum was shown by gel filtration to be a material with higher molecular weight than the high-molecular-weight fraction of western larch gum.[54] When recovered from the magnesium oxide purification process,[37] this material was found to have an infrared spectrum similar to that of arabinogalactan, but it had an ultraviolet spectrum similar to that of the lignin-like phenolics.

3. *Structure*

Arabinogalactan from western larch is composed of D-galactopyranosyl and L-arabinofuranosyl units with only a trace of a uronic acid.[55] The ratio of D-galactose to L-arabinose has been reported as 6 : 1.[35] The ratio was shown to increase from 8 : 1 in the outer heartwood to 23 : 1 near the pith for a 700 year old tree,[7] or from 5 : 1 to 10 : 1 in a 320 year old tree.[56] The ratio was found to vary from 7.2 : 1 for a high-molecular-weight fraction to 4.3 : 1 for a lower-molecular-weight fraction.[55]

The first conclusive evidence of separate components of larch arabinogalactan was provided by ultracentrifugation, in which a fast sedimenting component was separated from a slower and more diffuse component.[57] Later, fractional precipitation techniques allowed separation of working quantities of the two components whose identity was confirmed by sedimentation studies.[58] These fractions are referred to as A (faster sedimenting) and B (slower sedimenting). The A component also moves faster in electrophoresis in a borate buffer than does the B component.

The most used method of fractionating arabinogalactan involves cetyltrimethylammonium hydroxide precipitation of borate complexes.[58] The gum has also been fractionated by gel filtration, which involves fewer chemical transformations and adds no extraneous degradations or changes in the gum (Fig. 3).[7, 54, 55]

Unfractionated polymer and its two major components have been studied by methylation and hydrolysis. The Barry degradation[59] (periodate oxidation and formation of derivatives with phenylhydrazine) and Smith degradation[60, 61] (periodate oxidation, reduction and hydrolysis) have also been used.

From present information[52, 58, 62–68] it can be concluded that the backbone chain is composed of β-D-galactopyranosyl units linked (1→3) with branches occurring at some *O*-6 positions. The A and B components are generally similar except for somewhat less branching or shorter branches in the B fraction. The

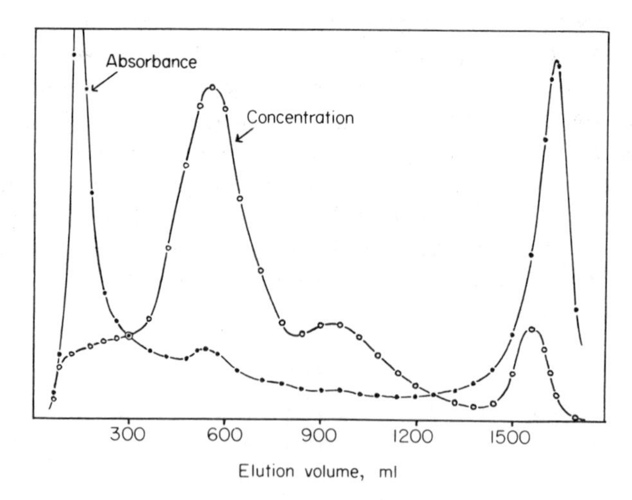

FIG. 3.—Elution pattern from the gel filtration of technical arabinogalactan on Sephadex G–100.

nonreducing ends of the branches are terminated with either D-galactopyranosyl or L-arabinopyranosyl units. Much of the L-arabinose occurs in side chains as 3-*O*-β-L-arabinopyranosyl-L-arabinofuranosyl units. Some D-galactose units seem to bear two side chains, the second being at the *O*-4 position. A likely structure of a portion of an arabinogalactan molecule from western larch is shown in Figure 4.

The structures of arabinogalactans from European, eastern, Japanese, and mountain larches have been investigated. They are generally similar to the arabinogalactan of western larch, but some differences have been noted.

The backbone chain of arabinogalactan from tamarack is thought to contain β-D-(1→6)- as well as β-D-(1→3)-linked galactopyranosyl units.[69] D-Galacturonic acid and an aldobiouronic acid have been isolated from a hydrolyzate of tamarack arabinogalactan.

European larch arabinogalactan is structurally similar to that from western larch,[70] with a β-D-(1→3)-linked galactopyranosyl backbone and side chains attached at the *O*-6 position.[71]

Arabinogalactan from mountain larch also has a β-D-(1→3)-linked galactopyranosyl backbone [including possibly some β-D-(1→6) linkages] with side chains at the *O*-6 positions.[72] The side chains are composed of (1→6)-linked D-galactopyranosyl units, probably in the β configuration. L-Arabinopyranosyl and L-arabinofuranosyl end-units are present. D-Glucopyranosyluronic acid (6–8%) may be present also as nonreducing end units.

The arabinogalactan from Japanese larch is very similar to that from western larch.[73]

The D-galactose to L-arabinose ratios for the A and B components of several species are given[62] as: western A, 4.3; western B, 3.7; mountain A, 4.9; mountain B, 2.6; Japanese A, 7.8; tamarack A, 5.4; and European A, 5.7. These values, as expected, do not necessarily match the values given for unfractionated arabinogalactan. Although it was once thought that some species of larch were devoid of component B, a later study showed that, whereas mountain and western larch had 25–30% of component B, six other species had only 5–10%.[63]

It has been suggested that the arabinogalactans from all larch species are

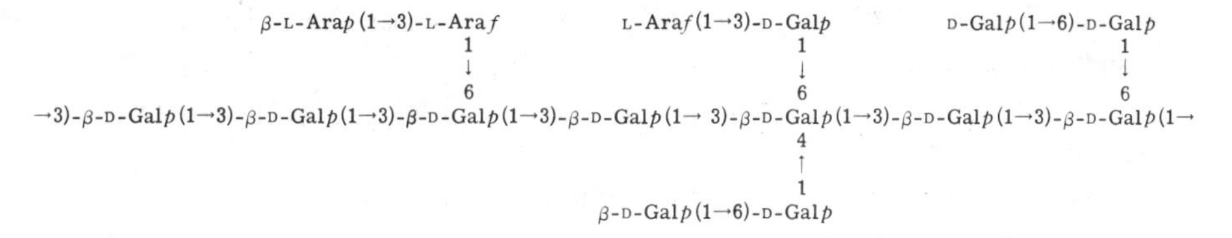

FIG. 4.—A likely structure of a portion of a molecule of arabinogalactan.

very similar and that observed differences might be caused by different analytical techniques,[72] as well as varying methods of isolation.

Viscosity studies[74, 75] have suggested that the gum molecules are spherical or highly branched. The relative viscosity of solutions of component A was reported to be the same as that of solutions of component B,[76] which might imply a spherical shape for component A and a more nearly linear shape for B.

4. *Molecular Weight*

The weight-average molecular weight of arabinogalactan was determined by ultracentrifugation to be 100,000 for the high-molecular-weight component and 16,000 for the low-molecular-weight component.[41] Osmotic pressure measurements on the unfractionated polymer have given number-average molecular weights of 13,000 to 65,500[7] and 46,300 to 70,000,[56] depending on the age of the heartwood. The number-average molecular weights of gum fractions prepared by gel filtration were 6300 to 51,000. The unfractionated gum had a molecular weight of 36,000 by osmotic pressure measurements.[55] The same material showed molecular weights of 6700 to 61,000 by rapid sedimentation equilibrium. Number-average molecular weights of the components prepared by gel filtration from magnesium oxide purified gum were reported as 55,000 for component A and 20,000 for component B, using light scattering measurements.[76]

IV. Uses

Arabinogalactan, although commercially marketed only within the past 5 years, has already acquired many and varied uses. It may be used under the Federal Food, Drug and Cosmetic Act as an emulsifier, stabilizer, binder or bodying agent in essential oils, non-nutritive sweetners, flavor bases, non-standardized dressing, and pudding mixes in the minimum quantity required to produce its intended effect.[77]

A solution containing arabinogalactan, ammonium dichromate, and phosphoric acid can be used as a fountain solution in the lithographic industry. A 20% solution of the gum can be used to protect the printing plates while in storage.[78] A solution of arabinogalactan and a light sensitizing chromic acid compound can be applied to a metal base plate for the preparation of photographic images for printing.[79]

Arabinoglactan does not make a good beater additive in the manufacture of paper.[80]

There is some question as to the usefulness of arabinogalactan solutions as blood plasma extenders.[81] However, the gum has many unique and interesting properties when compared to other hydrocolloids used in cosmetics and pharmaceutical dispersions.[82] Arabinogalactan can be used as a tablet binder and an emulsifier for pharmaceutical oil-in-water or water-in-oil emulsions.[39]

The gum can be hydrolyzed and oxidized with nitric acid to form mucic acid.[50, 51, 83] It improves beer foam adhesion.[84]

The commercial gum has found some use in the mining industry for the reverse flotation of iron ores[85] and in the copper mining industry.[86]

An improvement in detention of flavorants, such as ethyl acetate, in solid form is made by fixing the unstable flavor material in a covering consisting of at least 50% by weight of the gum.[87] An artificial sweetener–gum composition, such as that with maple syrup, can be formulated using arabinogalactan for the bulking and bodying agent (usually provided by sucrose), an artificial sweetener and other ingredients, including maple flavoring and water.[88]

Arabinogalactan can be mixed with silver halides and gelatin emulsions to form radiological or lithographic films.[89]

V. References

(1) M. F. Adams and C. Douglas, *Tappi,* **46,** 544 (1963).

(2) G. O. Aspinall, *Ann. Rev. Biochem.,* **31,** 79 (1962).

(3) H. O. Bouveng, *Sv. Kem. Tidskr.,* **73,** 113 (1961).

(4) T. E. Timell, *Advan. Carbohyd. Chem.,* **20,** 409 (1965).

(5) W. A. Côté, Jr., A. C. Day, B. W. Simpson, and T. E. Timell, *Holzforschung,* **20,** No. 6, 178 (1966).

(6) R. L. Mitchell and G. J. Ritter, *J. Forest.,* **49,** 112 (1951).

(7) D. C. Jones, B. W. Simson, and T. E. Timell, *Cellul. Chem. Technol.,* **2,** 391 (1968).

(8) W. A. Côté, Jr., and T. E. Timell, *Tappi,* **50,** 285 (1967).

(9) J. D. Stevens and M. F. Adams, *Washington State Inst. Technol. Circ.,* **9** (1964).

(10) A. W. Stout, *in* "Industrial Gums," R. L. Whistler, ed., Academic Press, New York, 1st Ed., 1959, p. 307.

(11) O. Goldschmid and H. L. Hergert, *Tappi,* **44,** 858 (1961).

(12) N. S. Thompson and O. A. Kaustinen, *Tappi,* **49,** 83 (1966).

(13) A. Roudier, *Ass. Tech. Ind. Papet., Bull.,* **16,** 343 (1962); *Chem. Abstr.,* **60,** 1928 (1964).

(14) R. A. Laidlaw and G. A. Smith, *Chem. Ind. (London),* 462 (1962).

(15) G. A. Adams, *Sv. Papperstidn.,* **67,** 82 (1964); *Chem. Abstr.,* **61,** 2053 (1964).

(16) M. L. Wolfrom and L. E. Anderson, *J. Agr. Food Chem.,* **15,** 685 (1967).

(17) M. L. Wolfrom and D. L. Patin, *J. Org. Chem.,* **30,** 4060 (1965).

(18) J. F. Courtois, F. Percheron, and J. C. Glomaud, *Cafe, Cacao, The,* **7,** 231 (1963); *Chem. Abstr.,* **60,** 9589 (1964).

(19) A. M. Unrau, *Can. J. Chem.,* **42,** 916 (1964).

(20) H. Neukom, L. Providoli, H. Gremli, and P. A. Hui, *Cereal Chem.,* **44,** 238 (1967).

(21) M. Morita, *Agr. Biol. Chem.,* **29,** 626 (1965).

(22) M. Morita, *Agr. Biol. Chem.,* **29,** 564 (1965).

(23) G. O. Aspinall and J. Baillie, *J. Chem. Soc.,* 1702 (1963).

(24) K. Tsuchiya and A. Misaki, *Takeda Kenkyusho Nempo,* **24,** 130 (1965); *Chem. Abstr.,* **67,** 1785e (1967).

(25) A. Misaki and S. Yukawa, *J. Biochem.* (*Tokyo*), **59,** 511 (1966).

(26) I. Azuma, H. Kimura, and Y. Yamamura, *Amer. Rev. Resp. Dis.,* **96,** 536 (1967).

(27) H. F. Weiss, U. S. Patent 1,358,129 (1920); *Chem. Abstr.,* **15,** 374 (1921).

(28) S. F. Acree, U.S. Patent 2,073, 616 (1937); *Chem. Abstr.,* **31,** 3322 (1937).

(29) G. T. Austin, *J. Forest Prod. Res. Soc.,* **4,** 7 (1954).

(30) A. B. Anderson, *Econ. Bot.,* **21,** 15 (1967).

(31) *Chem. Eng. News,* **41,** No. 19, 58 (1963).

(32) M. F. Adams, U.S. Patent 3,337,526 (1967); *Chem. Abstr.,* **67,** 83185w (1967).

(33) B. E. Ettling, *Northwest Sci.,* **38,** No. 3, 88 (1964).

(34) K. H. Ekman, *Tappi,* **44,** 762 (1961).

(35) L. E. Wise and F. C. Peterson, *Ind. Eng. Chem.,* **22,** 362 (1930).

(36) K. H. Ekman, *J. Chromatog.,* **7,** 419 (1962).

(37) I. Herrick, M. F. Adams, and E. M. Huffaker, U.S. Patent 3,325,473 (1967); *Chem. Abstr.,* **67,** 91878q (1967).

(38) K. H. Ekman and C. Douglas, *Tappi,* **45,** 477 (1962).

(39) M. R. Nazareth, C. E. Kennedy, and V. N. Bhatia, *J. Pharm. Sci.,* **50,** 560 (1961).

(40) A. W. Schorger and D. F. Smith, *Ind. Eng. Chem.,* **8,** 494 (1916).

(41) G. Lystad-Borgin, *J. Amer. Chem. Soc.,* **71,** 2247 (1949).

(42) L. E. Wise, P. L. Hamer, and F. C. Peterson, *Ind. Eng. Chem.,* **25,** 184 (1933).

(43) F. C. Peterson, M. Maughan, and L. E. Wise, *Cellul.-Chem.,* **15,** 109 (1934).

(44) W. G. Campbell, E. L. Hirst, and J. K. N. Jones, *J. Chem. Soc.,* 774 (1948).

(45) L. E. Wise, F. C. Peterson, A. J. Barry, and H. Unkauf, *J. Amer. Chem. Soc.,* **62,** 2361 (1940).

(46) Industrie Biologique Francaise S.A., Fr. Patent M4431 (1966); *Chem. Abstr.,* **68,** 6181k (1968).

(47) Industrie Biologique Francaise S.A., Fr. Patent M4874 (1966); *Chem. Abstr.,* **69,** 44182q (1968).

(48) N. I. Nikitin and I. A. Soloviev, *Zh. Prikl. Khim.,* **8,** 1016 (1935); *Chem. Abstr.,* **30,** 5563 (1936).

(49) R. S. Black, U.S. Patent 1,864,229 (1932); *Chem. Abstr.,* **26,** 4346 (1932).

(50) S. F. Acree, U.S. Patent 1,816,137 (1931); *Chem. Abstr.,* **25,** 5554 (1931).

(51) S. F. Acree, Brit. Patent 160,777 (1921); *Chem. Abstr.,* **15,** 2545 (1921).

(52) E. V. White, *J. Amer. Chem. Soc.,* **63,** 2871 (1941).

(53) A. F. Zaitseva, A. Z. Karpov, S. Z. Levin, and S. D. Antonovskii, *Zh. Prikl. Khim.,* **32,** 690 (1959); *J. Appl. Chem. USSR,* **32,** 720 (1959).

(54) B. V. Ettling and M. F. Adams, *Tappi,* **51,** 116 (1968).

(55) H. A. Swenson, H. M. Kaustinen, J. J. Bachhuber, and J. A. Carlson, *Macromolecules,* **2,** 142 (1969).

(56) W. A. Côté, Jr., B. W. Simson, and T. E. Timell, *Holzforschung,* **21,** No. 3, 85 (1967).

(57) H. Mosimann and T. Svedberg, *Kolloid-Z.,* **100,** 99 (1942).

(58) H. O. Bouveng and B. Lindberg, *Acta Chem. Scand.,* **12,** 1977 (1958).

(59) V. C. Barry, *Nature,* **152,** 538 (1943).

(60) M. Abdel-Akher, J. H. Hamilton, R. Montogmery, and F. Smith, *J. Amer. Chem. Soc.,* **74,** 4970 (1952).

(61) J. K. Hamilton, G. W. Huffman, and F. Smith, *J. Amer. Chem. Soc.,* **81,** 2176 (1959).

(62) E. V. White, *J. Amer. Chem. Soc.,* **64,** 302 (1942).

(63) E. V. White, *J. Amer. Chem. Soc.,* **64,** 1507 (1942).

(64) E. V. White, *J. Amer. Chem. Soc.,* **64,** 2838 (1942).

(65) H. Bouveng and B. Lindberg, *Acta Chem. Scand.,* **10,** 1515 (1956).

(66) H. Bouveng, *Acta Chem. Scand.,* **13,** 1869 (1959).

(67) H. Bouveng, *Acta Chem. Scand.,* **13,** 1877 (1959).

(68) H. Bouveng, *Acta Chem. Scand.,* **15,** 78 (1961).

(69) S. Haq and G. A. Adams, *Can. J. Chem.,* **39,** 1563 (1961).

(70) G. O. Aspinall, E. Hirst, and E. Ramstad, *J. Chem. Soc.,* 593 (1958).

(71) G. O. Aspinall and A. Nicholson, *J. Chem. Soc.,* 2503 (1960).

(72) J. K. N. Jones and P. E. Reid, *J. Polym. Sci., Part C,* 63 (1963).

(73) G. O. Aspinall, R. M. Fairweather, and T. M. Wood, *J. Chem. Soc., C,* 2174 (1968).

(74) E. Husemann, *Naturwissenschaften,* **27,** 595 (1939).

(75) H. Owens, *J. Amer. Chem. Soc.,* **62,** 930 (1940).

(76) B. V. Ettling and M. F. Adams, paper presented at the Northwest Regional Meeting, Amer. Chem. Soc., 23rd, 1968.

(77) *Federal Register,* **30,** 2430 (Feb. 25, 1965).

(78) W. H. Wood, U.S. Patent 2,250,516 (1914).

(79) W. H. Wood, U.S. Patent 2,297,932 (1942); *Chem. Abstr.,* **37,** 1535 (1943).

(80) J. O. Thompson, J. W. Swanson, and L. E. Wise, *Tappi,* **36,** 534 (1953).

(81) R. E. Semple, *J. Polym. Sci., Part C,* 73 (1963).

(82) B. N. Patel, *Drug Cosmet. Ind.,* **95,** 337 (1964).

(83) A. W. Schorger, U.S. Patent 1,718,837 (1929); *Chem. Abstr.,* **23,** 3932 (1929),

(84) E. Segel and P. R. Glenister, U. S. Patent 3,443,957 (1969); *Chem. Abstr.,* **71,** 37497j (1969).

(85) S. I. Gorlovskii, *Obogashch. Rud.,* **6,** No. 5, 18 (1961); *Chem. Abstr.,* **57,** 476 (1962).

(86) M. R. Knudson, St. Regis Paper Company, personal communication, 1965.

(87) M. Glicksman and R. E. Schachat, U.S. Patent 3,264,114 (1966); *Chem. Abstr.,* **65,** 12786f (1966).

(88) G. L. Stanko, U.S. Patent 3,294,544 (1966); *Chem. Abstr.,* **66,** 45627p (1967).

(89) V. C. Chambers, Belg. Patent 611,621 (1961); *Chem. Abstr.,* **57,** 12005 (1962).

CHAPTER XIX

PECTIN

GORDON A. TOWLE AND OTTO CHRISTENSEN

The Copenhagen Pectin Factory Ltd., Lille Skensved, Denmark
(Subsidiary of Hercules, Inc., Wilmington, Delaware)

I. INTRODUCTION

Pectin is a general term for an interesting group of polysaccharide substances present in the cell walls of all plant tissues which functions, in combination with cellulosic material, as an intercellular cementing material. The main component of this group is a linear D-galacturonan, many of the carboxyl groups of which are esterified with methanol. Other members include L-arabinan and D-galactan components, whose combined presence may account for up to 20% of the group.

As a result of its properties and ready availability, pectin has assumed great importance as an industrial gum. Its main role continues to center around its use as a component of the classic sugar–acid–pectin gel important to jam and jelly production. Other uses account for a portion of the production uptake, however, and new uses for pectin continue to be discovered.

Pectin has been known for some time[1–3] and has been the object of much study brought on in large part by its remarkable chemical and physical properties. Several recent review articles concerning the chemistry and biochemistry[4–9] as well as industrial uses of pectin[10–12] have appeared. A comprehensive treatment of the pectic substances can also be found in a book by Kertesz.[13]

II. Production

1. *Raw Material and Supply*

The main raw material for pectin today is the peel of citrus fruits, preferably lemon peel, but peel of lime, orange, and grapefruit are also used. Pectin is a constituent of the albedo, the white, spongy, inner part of the peel, and of the internal membranes within the fruit. The peel used in pectin production is obtained from juice industries, is void of juice and most of the essential oils, and has normally been washed to remove excess soluble materials and dried, especially if long storage or transportation is expected. It is imperative for the quality of the pectin that the peel is used when fresh or that it be suitably treated to deactivate pectic enzymes that, within a relatively short time, cause undesirable changes in the pectin. The peel after juice extraction contains 2–4% pectin and in dried form 20–40% pectin.

Another important source of pectin is apple pommace, the residue remaining from pressing of apples to obtain apple juice. As with cirtus peel, the pommace can be used either fresh or dried, but the short apple season makes it almost imperative to dry some of the pommace to extend pectin production. The pectin content of dried apple pommace varies from 10 to 20%.

Other sources of pectin have been suggested, some of which have been tried on an industrial scale. During World War II, there was some production in Europe of pectin from sugar beet waste, and in Sweden this production extended into the peacetime years following, but apparently without success as production has now stopped. Pectin from this source differed from citrus or apple pectin in having a higher amount of acetyl-ester groups and was mainly useful for the production of low-methoxyl pectin. There has been some interest, particularly in eastern European countries, in pectin from sunflower bottoms,[14] although it is unknown at present if commercial production based on this raw material has been attempted.

The potential world supply of citrus peel and apple pommace far exceeds possible demand, but a lower price limit is set by the alternative use of the peel as cattle feed, for which most of the peel production goes. The drying of peel for this use is less expensive, gives higher yield, and demands less rigid control than the careful washing and drying needed for pectin raw materials. Although the supply is seemingly unlimited, there is some competition among the largest pectin producers to obtain raw materials of top quality. The price range for pectin quality citrus peel in 1970 was about $100–150 a metric ton (2205 lb) and for pectin quality apple pommace $40–55.

2. *Purification*

The production of pectin is, in principle, a very simple process, consisting of extraction, purification of extract, and precipitation. In practice, however, a number of technical refinements and control measures have come into use, and the relatively few producers are very secretive about process details. Some fairly detailed descriptions are to be found in the literature,[15, 16] including one rather detailed description of equipment design.[17] The general steps in pectin production for four commonly used processes are represented schematically by the flow-sheet shown in Figure 1.

Extraction.—The first step in pectin production, aside from the various steps necessary in readying the raw material, consists of freeing the pectin from its loosely bound position in the fruit tissue. To accomplish this, the raw material, fresh or dried, is added to water; and a mineral acid, usually nitric acid or hydrochloric acid, is added to lower the pH and bring about some hydrolysis of the native pectin (protopectin) so that it is more easily dissolved. The temperature is somewhat elevated to aid diffusion and lower the viscosity of the solution.

Great care must be taken, particularly during the production of high-methoxyl pectin, in controlling time, temperature, and acidity; as the pectin, if treated too harshly, will be hydrolyzed or deesterified excessively. The practical pH range during extraction is 1.5–3.0, at temperatures of 60°–100° and for times ranging from less than one to several hours. Along with extraction, a certain amount of deesterification invariably occurs, and it is necessary to adjust extraction conditions relative to the raw material being used in order to arrive at a finished product having the desired methyl ester content. By allowing deesterification to proceed for longer periods, pectins of considerably lower methoxyl content are obtained, and this procedure is presently used by some manufacturers in the production of low-methoxyl pectins.

Purification and recovery.—After extraction, the extract is separated from the residue by centrifugation, filtration, or a combination thereof. The residue has some feed value and is often used by nearby farmers as a supplementary

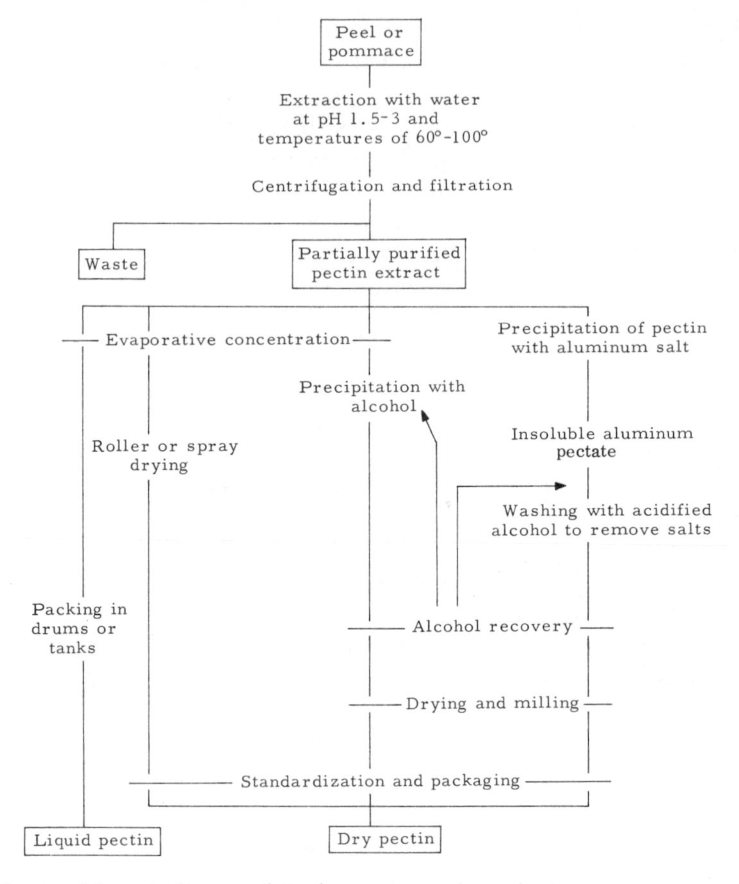

FIG. 1.—Schematic diagram of the four major pectin production processes.

cattle feed. Pectin extracts from some raw materials, in particular apple pommace, may contain small amounts of starch. This starch, where objectionable, can be hydrolyzed by treatment of the extract with an amylase preparation, free of pectin-degrading enzymes. This treatment is usually followed by a polishing filtration with diatomaceous earth as filter aid.

Further treatment of the extract depends on which type of pectin is to be made. The simplest process consists of an evaporative concentration leading to a viscous, usually more or less brownish fluid which is sometimes sold as *liquid pectin.* Liquid pectin (with sulfur dioxide added as a preservative) is sold packed in metal drums or delivered in tank cars. Trade in this product is decreasing in volume because of high shipping costs, limited shelf life, and demand for more highly purified products. Thus most of the pectin produced at present is sold in dry form obtained by (*a*) spray drying or roller drying the extract, (*b*) precipitation as an insoluble salt followed by washing with acidified alcohol and drying, or (*c*) precipitation with alcohol and drying. Spray or roller dried pectin contains all the soluble impurities present in the original extract, and as the drying process does not show significant economic advantage over precipitation methods, these alternative recovery methods are by far the most utilized.

Salt precipitated pectin is obtained by adjusting the pH of the extract and adding certain precipitation salts, normally soluble salts of aluminum, and to a lesser extent copper, after which the precipitate is washed with acidified aqueous alcohol to effect salt removal. The pH of the precipitate is adjusted with alkali or ammonia, and the pressed precipitate is dried and milled.

Alcohol-precipitated pectin is obtained by adding to a suitably concentrated pectin extract sufficient alcohol to cause insolubilization of the pectin. Isopropanol (2-propanol) is often employed as a precipitant although methanol and ethanol are also used. The gelatinous precipitate of pectin is then washed with additional alcohol, pressed to remove much of the entrained alcohol and solutes, and dried. The alcohol is recovered by distillation and reused. A final milling of the dry pectin completes the operation.

Many variations to the simple process illustrated above are in use. For example a pickling of the peel[18] is employed by some producers to ease extraction and to control the degree of esterification. Sulfur dioxide can be used for acidification and offers the advantage of ease of removal by evaporation rather than by neutralization. Some manufacturers deesterify at an early stage of processing before extraction; others, as mentioned earlier, deesterify the extract itself. It is difficult to avoid excessive depolymerization in this process, but with careful control, both acid- and base (ammonia)-catalyzed deesterification is possible.[19]

Deesterification leading to the production of low-methoxyl pectin is often done in connection with the precipitation process and even as late as the drying and milling stages. Depolymerization is usually reduced by the presence of alcohol. Hydrochloric or nitric acid is commonly used as an acid catalyst, whereas for base-catalyzed deesterification, ammonia is usually the reagent of choice. When ammonia is used, it has been shown that, commensurate with deesterification, a certain amount of amidation takes place.[20, 21] A consequence of the variations employed with regard to methods of deesterification is that low-methoxyl pectins from different manufacturers tend to show greater differences in properties than do high-methoxyl pectins.

3. *Grades and Standardization*

The finished pectin is standardized according to procedures developed by the more advanced producers so that the customer has a choice of a few uniform product groups. The most important factors in establishing product grades are gel strength and gelling time, which are related to some degree, making it difficult to speak in absolute terms of either property.[22]

Gel strength.—Gel strength is a measure of the amount of pectin required to assure a certain stiffness of a pectin gel. The concept of a gel grade was early agreed upon, and the grade of a pectin was expressed as the number of grams of sugar one gram of pectin could hold in a gel of standard texture.[23] At that early state of technology, a pectin grade of 100 was taken as the standard, but technology has since improved, and today's standard grade in 150. Even this grade has normally been brought artificially down from higher values through standardization procedures. The steady increasing economic importance of the pectin trade made a more concise definition of the jelly grade necessary, and in 1947, the Institute of Food Technologists established a committee to devise a standard method. In 1948 this committee began work on this problem in several American and European laboratories with the result that in 1959 a standard method was adopted and published.[24] This procedure, developed from the method of Cox and Higby,[25] is referred to in the trade as method 5-54 or the sag method of pectin standardization. It is based on the principle of measuring the sag of a freely standing, unmolded gel. The sag expressed as the percentage of the original height of the jelly is then used to calculate the true grade of the pectin.

Since its introduction, the sag method has gradually been adopted by the pectin producer and user alike as the standard of gel strength measurement. Other methods are still in use, however, often advantageously as supplementary controls on pectin quality. Some alternative procedures are similar in principle to the sag method while others differ markedly. The correlation between results obtained by applying different methods to a given gel is often poor.[26–28]

Procedures for determination of the gel strength of low-methoxyl pectins have not been developed as extensively as those for high-methoxyl pectins, and there is no generally agreed upon method. Some producers rely upon a modified sag method, in which the amount of calcium ion in the gel is carefully controlled.[29] However, sag readings for low-methoxyl pectin vary with the degree of esterification, reflecting a calcium ion requirement within rather strictly defined limits.

Grades in excess of 200 are commonly obtained for raw high-methoxyl pectin as a result of today's improved manufacturing procedures. Standardization downward to an industry norm grade of 150 is usually achieved by mixing the raw pectin with an inert diluent such as sucrose or dextrose in finely powered form.

Gelling time.—A second factor important in determining grade, that of setting time, is defined as the time between the moment when setting conditions are present and the moment when setting of the jelly begins. This interval is of considerable practical importance for several reasons. Most importantly, it represents the time available for packing of jellies into containers. Too short a setting time might result in gel formation in the filling equipment, whereas too long a setting time might lead to delays in packaging. When fruit particles, for example strawberries, are present, setting conditions are timed to give a slight set just prior to filling and thus prevent berries from partitioning unevenly in the container. Hence, the customer needs a pectin with a well-defined and uniform gelling time to establish his process conditions.

As with gel strength, the gelling time for a particular pectin, under set conditions of pH, solutes concentration, and rate of temperature decrease, is an intrinsic property of that pectin, which can be established easily and with adequate precision by several methods. The method that is most widely accepted today is the Joseph and Baier method,[30] which has the advantage that it can be incorporated with the gel strength measurement according to the now accepted sag method. The procedure which has been adopted as a standard method by the National Preservers Association in the United States is as follows: Ground pepper is added to freshly prepared jelly and the jelly is poured into a jar placed in a constant temperature bath. The jar is turned at short intervals, and with a aid of a light beam, the time is noted at which the pepper particles return elastically after a sharp turn. This time, usually expressed in seconds, is a measure of the gelling time.

That point at which gelation begins could, of course, be specified more exactly in terms of temperature, and in England, a method that expresses setting temperature rather than setting time is widely used.[31] For practical purposes, however, the setting time under defined conditions has been found to be a useful parameter on which to base pectin grades. Commercial pectins are generally grouped into rapid-, medium-, and slow-set types according to their gelling time. The range of gelling times ordinarily encountered for each of the three types and the relation of gelling time to degree of esterification is presented in Table I.

Pectins with degrees of esterification outside these limits are available for

TABLE I

Type of Pectin	Degree of Esterification, %	Gelling Time, sec
Rapid-set	72–75	20– 70
Medium-set	68–71	100–135
Slow-set	62–66	180–250

special uses. Through methylation with methanol in the presence of dilute hydrochloric acid, pectins having degrees of esterification of up to 85% are obtained; these are useful in certain household pectin preparations.[32] Pectins with degrees of esterification slightly below 62%, the lower limit for a normal slow set type, are produced for special purposes, such as use in jelly candies with high solids (>75%) content.

From the foregoing considerations, it can be seen that adjustment of gelling time should be made at some stage of production by employing conditions that result in the correct degree of esterification in the finished product. It is possible, however, to adjust gelling time by blending individual lots and to effect small changes in gelling time by incorporation of certain buffering salts. The pectin user can also control gelling time somewhat by adjusting the pH and solids content of the jelly mix. Within certain limits, the gelling time is reduced by decreases in pH and by increases in total solids content.

4. *Production Geography and Statistics*

It is natural to expect pectin production to follow raw material availability, and with few exceptions, production is geographically connected with apple or citrus growing and processing areas. The pectin industries have tried to avoid too narrow a dependence on season and crops, however, and many original apple pectin producers have included citrus pectin based on dried citrus peel in their production.

California has several factories producing citrus pectin. There are large factories producing pectin from apple pommace in England, France, Germany, and Switzerland. In Denmark, a pectin industry has grown into one of the largest in the world without any link to a national raw material source.

Estimates based on production statistics from the major producers place world production of pectin in 1971 at about 8000 metric tons of standardized high-methoxyl pectin (150 grade) and a further 1000 metric tons of low-methoxyl pectin. Of this, approximately 3000–3500 metric tons were produced in the United States, with the bulk of the remainder being produced in the aforementioned west European countries. Based on past trends, production in the immediate future is expected to show a modest but steady growth rate, with low-methoxyl pectin making disproportionate gains as a result of the increasing trend to jams and jellies of low sugar content. The price of 150 grade, high-methoxyl pectin to large scale consumers is presently in the range of $1.40 to $1.50 a pound, while low-methoxyl pectin sells for approximately $2.00 a pound.

III. Applications

Pectin has been produced commercially since the beginning of this century, at first as a crude liquid extract, and now almost entirely as refined powdered pectins of high purity. From its beginnings as a manufactured gum, pectin has always been looked upon as a natural constituent of human foods and has not been questioned as a safe food additive. Its use is allowed in all countries, and the joint FAO/WHO Food Standards Program Codex Committee on Food Additives at its October Meeting[33] in Arnhem in 1969 recommended pectin as a safe food additive with no limit on acceptable daily intake except as dictated by good manufacturing practice.

Pectin enjoys a unique position in its main market, that of sale for use in jams, jellies, and preserves, and is almost free of competition from other gums. Reasons for its market strength include a complete absence of off-taste, nearly perfect function, and the fact that food standards seem to protect pectin as a natural component in these products. Its usage is not limited strictly to foods, however; there are various other applications for pectin, particularly in the pharmaceutical industry.

1. *Foods*

Jams and jellies.—Food gels based on pectin have been known and developed as a home art for over 500 years. Only within the present century, however, has the role of pectin interacting with sugar, acid, and water to yield a gel been established and this understanding been used in the development of technology. Present manufacturing procedures represent a rather high degree of refinement over simple cooking-until-thick methods used in the past. The various steps in use are essentially the same for jellies made from fruit juices or for jams and marmalades made from whole fruit or from juice and shredded peel.

The first step in jam making consists of a brief cooking of the fruit to liberate juice and pectin through conversion to protopectin to soluble pectin. Additional pectin, as required, may be added at any point in this initial process, either as a dry powder mixed with 5–10 parts of sugar as a dispersing aid or as a solution. When a dry powder is used, it is important that complete dissolution of the pectin be achieved before the addition of sugar, as sugar in excess of 20% strongly retards hydration of the pectin. Sugar is then added, and a boiling process in which solubilization of the sugar is achieved ensues, usually with some removal of water through evaporation. The evaporative process has a great effect on the quality of the jam. Most manufacturers today use a vacuum concentration process whereby the mixture is stirred and heated in a closed container under reduced pressure, thus allowing the process to be effected at temperatures of 75°–80° and above. The lower range represents the minimum temperature at which boiling can be conducted with complete solubilization of the added sugar. Thus, relatively low vacuum equipment is adequate. The required pectin, if not added initially, may be added just prior to or at some point during the second boiling cycle, the only requirement being that the pectin be in a solubilized form because of the high sugar content of the mixture. In a final step, the pH is adjusted by adding a concentrated solution of an organic acid, usually citric acid. Acid addition must be made as late in the process as possible and preferably just prior to filling to avoid excessive amounts of sugar inversion, which can lead to syneresis in the finished jam.

Jelly production requires an additional operation, that of a straining or pressing step to remove insoluble materials after a boiling of the fruit. The extracted juice, containing soluble sugars, flavor components, solubilized pectin, and other soluble components, is usually filtered to improve clarity and then used directly for jelly manufacture. In both jams and jellies, the proportion of fruit derived component to sugar is 45 : 55 by weight and the final product contains 65% or more soluble solids. The combination of high solids content and low pH of such products prevents the development of spore-forming bacteria, and to assure a long keeping time, it is only necessary that containers be filled while hot and sealed tightly to exclude molds and yeast.

The time and manner of addition of pectin constitute important process variables in jam and jelly manufacture. Most manufacturers prefer to add the necessary pectin as a concentrated aqueous solution. Its advantages, besides those of convenience, are that complete dissolution of the pectin is assured and that the pectin can be added late in the process and is subjected to less heating. Losses of pectin from heat degradation can be appreciable, particularly if the pH of the solution during boiling is above 4.2. Pectin solutions of concentrations ranging from 4 to 8% may be prepared by adding pectin, mixed with 5 parts sugar, to water in a high-speed mixer. The upper concentration limit is determined by viscosity; thus water low in calcium ion content is used.

Other variables to consider are the setting time of the added pectin, the cooking temperature, the soluble solids content, and the pH. As mentioned above, an increase in soluble solids content or decrease in pH can shorten setting time. In addition, a lower cooking temperature through use of vacuum boiling equipment lowers appreciably the temperature differential between the boiling and setting temperatures. Thus, rapid set pectins, having a high setting temperature, cannot be used in vacuum processes, as gelation ensues almost immediately after boiling, giving insufficient time for filling.

A final variable is the natural pectin content of fruits and juices. In general, this pectin which may exist in rapid-set, slow-set, or even low-methoxyl forms must be accepted as is, and adjustments must be made through addition of well-defined pectin. Occasionally it is preferable to depectinize the fruit or juice enzymically and to add sufficient pectin of desired type for gelation.

Various analytical checks are necessary during the cooking operation, particularly in regard to soluble solids content and pH. The soluble solids content at various stages in the cooking operation is usually determined by measuring the refractive index of the mixture. Determination of the percentage of invert sugar is usually not necessary, but when required, may be made through use of Fehling solution or by polarimetry. A critical relationship exists between pH, soluble solids content and pectin amount and type, as well as other variables; and for this reason, it is usual for small test batches to be prepared as a means of control on production. Most manufacturers of pectin supply detailed handbooks on the use of pectin in jams and jellies, which provide information on formulas, selection of pectin types and manufacturing techniques. Information on the selection and use of pectins can also be found in a review by Ehrlich.[34]

Low-sugar or sugar-free jams and jellies.—There is a growing demand among consumers of jams and jellies for products with less sugar and even without sugar, partly for calorie-consciousness reasons and partly to fill the need for sugar-free products for diabetics. Sweetness in such products can be provided by synthetic sweetners, but as the sugar content is reduced, high-methoxyl pectin is no longer able to produce a gel. Thus, quite naturally, the preserves industry has been one of the first to explore the possibility of using low-methoxyl pectin in such products and the formation of a low-methoxyl pectin–calcium gel.

At present, this application provides the most important market for low-methoxyl pectin. Very acceptable products are possible, but great care is necessary to control process conditions because the concentration of calcium ion optimum for gelation is often near the amount of calcium provided by many fruits. It is, therefore, difficult to control the concentration of calcium ion in the finished product to a level that is sufficient for thickening or gelation and yet not so high that immediate gelling leading to a broken gel with excess bleeding occurs. Even through analysis, it is difficult to establish the content of ionizable calcium in fruit or other foods, as calcium may be bound in reversible fashion and behave quite differently at varying temperatures and pH.

In preparing gels with low-methoxyl pectin, it is usually desirable to begin with a mixture low in calcium content and to add calcium in the form of a soluble salt, such as calcium chloride, later in the process. There is also some advantage to be gained through addition of the pectin as late in the process as possible. The requirement for pectin and calcium ion to maintain desirable gel characteristics is related to the pH and soluble solids content and, in general, increases as the soluble solids content is reduced and as the pH increases.[35] By adherence to recently developed methods,[35] it is possible, using low-methoxyl pectin, to prepare low-sugar, syneresis-free gels comparing favorably in quality with jams and jellies prepared with high-methoxyl pectin.

Whereas high-methoxyl pectin is almost without competition when used in high-sugar jams and jellies, low-methoxyl pectin meets high competition from other natural gums such as agar and carrageenan in low-sugar and sugar-free products. The most important advantage of low-methoxyl pectin over its competitors is its greater stability under acid conditions, although the difficulty of controlling the setting time of low-methoxyl pectin gels may be counted as a disadvantage.

Confectioner's jellies.—Candies, made of flavored high-methoxyl pectin jellies containing a higher than usual ($\sim$80%) content of soluble solids, are a common commodity in most countries. The candies, usually molded into shapes simulating fruits or in gumdrop shape, are usually finished with a coating of sugar or other dusting material to prevent stickiness. The high soluble solids content of such jellies imparts resistance to drying and improves shape retention, leading to products with very good shelf-life characteristics.

Slow-set (degree of methylation of $\sim$60) pectins are well suited for this product. The preparation of confectionary jellies is analogous to the preparation of ordinary 60–65% solids jellies, with the exception that a slightly higher final pH (3.5–3.8) is required for optimal gelation. Citric acid is commonly used to lower the pH, often in combination with buffering salts to assist in controling the final pH.

Bakers' instant jelly.—A popular product in many European countries is an instant jelly preparation applied to fruit tarts and similar bakery items. The jelly is prepared by mixing a cold precooked pectin–sugar syrup with a pH above the setting pH and a solution of citric acid. After mixing, the solution gels in about 2 min, allowing time for application to baked goods.

Reversible bakers' jellies.—Synthetic or semisynthetic jellies prepared from low-methoxyl pectin having heat reversibility are used by bakers to cover certain baked goods with a jelly layer.

Frozen berries.—Incorporation of a small amount (generally 0.05–0.2%) of pectin in the syrup of frozen berries results in improved shape and reduced loss of syrup through drainage during thawing of the frozen product. The quality of frozen strawberries in particular are improved by use of pectin.[36–39] Both high- and low-methoxyl pectins may be used, with the latter showing some advantage because of their somewhat better performance under freeze–thaw conditions. Coating compositions containing low-methoxyl pectin applied to fruits for use in ice cream improve fruit texture and quality.[40, 41]

Beverages.—Pectin finds some application in the beverage industry, particularly for the stabilization of soft drinks. Pectin has several functions, of which the most important is as an emulsion and suspension stabilizer for essential oils and suspended fruit particles. In addition, a slight increase in viscosity is obtained, which gives a pleasantly rich mouthfeel. For this application, however, pectin has many competitors among other gums. One of the advantages of pectin is that it can be considered as a natural constituent, as pectin is always present in fruit juices.

Household pectin.—As low-methoxyl pectin does not require the same rigid conditions for gelling as high-methoxyl pectin, it is sometimes preferred for use in household pectin preparations for homemade preserves, even when sugar contents are sufficiently high for use of high-methoxyl pectin. A special preparation called *gelierzucker* which is a mixture of low-methoxyl pectin and sugar, is sold in Germany.[42]

Artificial cherries.—One of the most successful applications of low-methoxyl pectin has been in the production of artificial cherries,[43] where the completely synthetic medium makes it possible to control setting conditions.

Milk products.—High-methoxyl pectin has been suggested for use as a stabilizer in certain sour milk products.[44, 45] Low-methoxyl pectin reacts with milk to produce a gel that is suitable for a pudding base. Other gums such as agar, furcellaran, and carrageenan, however, have proved more economical for this purpose and supply the largest part of this market. Low-methoxyl pectin in combination with gelatin has been suggested for use in the preparation of a sour cream mix to prevent wheying off and provide body.[46]

Stabilizer for frozen desserts.—Incorporation of pectin, either alone or in combination with other gums, in water ices and sherbets reduces crystal growth and produces a rich mouth feel.[47] Pectin is one of the best stabilizers known for this application, but competition from other natural gums and carboxymethylcellulose limits its use.

Miscellaneous food uses.—Foods coated with a transparent edible coating of low-methoxyl pectin may be dehydrated osmotically without impregnation of the tissue with dehydrating media.[48] In addition, the film coating serves to protect the dehydrated material during storage and to minimize loss of nutrients and flavor during rehydration.

Low-methoxyl pectin amide has been suggested for use in the preparation of canned, storage-stable, ready to eat, gelled food products.[49, 50]

In an application related to its use in low-sugar gels, low-methoxyl pectins have been found to function well as a component of both hot and cold-process mixes for dessert gels and pudding. In hot-process formulations, a mixture consisting of low-methoxyl pectin along with a calcium salt and other ingredients is simply dispersed in water and heated to boiling to dissolve the pectin. After cooling, a gel forms. Slightly or slowly soluble calcium salts may be used to improve the dissolving characteristics of the pectin. Cold-process desserts using low-methoxyl pectins pose a special problem because of the difficulty of dissolving the pectin in the presence of calcium ion. Two approaches to this problem have been successful, one in which the dissolving properties of the pectin are improved by use of a modified pectin[51, 52] and another in which the calcium salts used are carefully chosen for their limited or slow solubility.[52, 53]

2. *Pharmaceuticals*

Pectin shows potential for use in a variety of pharmaceutical applications, as seen by the rather extensive literature in this area.[54] In spite of its applicability, however, very little pectin finds its way into pharmaceutical use, perhaps partly because of a somewhat higher cost in comparison with competing gums. The specifications for pectins for pharmaceutical uses are defined by the "U.S. Pharmacopoeia."

Diarrhea treatment.—The classic pharmaceutical application for pectin is in diarrhea treatment, and several products incorporating pectin as the main ingredient are available commercially. Indeed, the apple diet long prescribed and used for the treatment of diarrhea, particularly in infants, owes its effectiveness to the high content of pectin in apples.

Miscellaneous pharmaceutical uses.—Pectin administered intravenously shortens the coagulation time of drawn blood and is useful in controlling hemorrhage or local bleeding.[10, 54] Pectin sulfates, on the other hand, prolong the clotting time of blood and have been suggested for use as replacements for heparin.[55–57] The relatively effective anticoagulant action of pectin sulfates, however, is offset by their toxicity, which although low, limits their use over a long time period or in a large dose.

Recent work has shown that pectin administered orally is somewhat effective in reducing cholesterol levels in the blood.[58–60]

Pectin in combination with gelatin has been suggested for use as an encapsulating agent for medicinals to promote sustained release.[61]

A complex of degraded pectin and iron has been found to be useful for the intramuscular administration of iron for treatment of iron deficiency anemia,[62] whereas a bismuth–D-galacturonan mixture has been claimed to be an effective means for administration of bismuth in medicinal preparations.[63]

High-methoxyl pectin has been claimed to promote sustained release of aspirin and to act as a demulcent in minimizing the gastrointestinal irritation sometimes noted during its administration.[64] The pectin–aspirin mixture may be prepared in tablet form or as a powder mixture to be added to water. In the latter case, a special dispersible pectin is used.

A preparation consisting of low-methoxyl pectin (degree of methylation 32–37), aluminum hydroxide, and magnesium oxide has been suggested for use in the treatment of gastric and duodenal ulcers.[65]

3. *Technical Uses*

Pectin is well suited for use in several industrial applications. In particular, pectin functions well as an emulsion stabilizer for water–oil emulsions,[66, 67] and because of its excellent film-forming properties, it has been shown to be useful as a sizing agent for paper and textiles. Extensive use in this area, however, is again limited by a price disadvantage to lower cost natural gums as well as various starch and cellulose derivatives.

Pectin has been shown to be useful for the preparation of membranes for ultracentrifugation and electrodialysis[68] and may find use in this area where price is less of a factor. A recently suggested use of pectin is that of a component of sulfuric acid sols for use in lead accumulators.[69] Sols free of air bubbles are prepared by blending pectin at a level of approximately 1% into sulfuric acid.

IV. Structure

Pectin is a complex polysaccharide, the composition of which varies with the source and conditions used in its isolation. D-Galacturonic acid is always the principal sugar component, but varying amounts of D-galactose, L-arabinose, L-rhamnose, and occasionally trace amounts of other sugars are usually present also. The relationship of the sugar units in pectin has been the object of much study with the result that a number of structural details are now known.

The D-galactopyranosyluronic acid units in pectin make up a polysaccharide that on the basis of optical rotation and hydrolysis,[70–76] methylation,[70, 72, 75, 77] and periodate oxidation[78] analyses has been shown to consist of a linear (1→4)-linked α-D-galacturonan (Fig. 2). Recent evidence on the basis of work conducted on pectins from several sources[70, 79–82] suggests that this galacturonan is not a homopolymer but usually contains L-rhamnose as an integral part of the main chain. The frequency of its occurrence remains to be established, although it has been suggested that L-rhamnosyl units may be concentrated in rhamnose-rich areas interposing relatively long galacturonan segments.[83]

D-Galactose and L-arabinose are probably present in pectin as constituents of galactan and arabinan molecules. Their relation to the galacturonan main chain has been the object of much controversy because of the difficulty of separating these two neutral components from the main acidic component. It now appears that, whereas neutral polymers can be separated by fractionation procedures,[71, 72, 84–88] portions of the D-galactopyranosyl and L-arabinofuranosyl

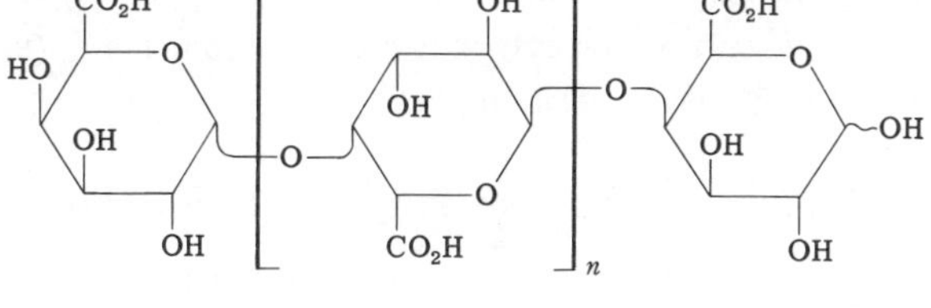

Fig. 2.—Pectin D-galacturonan.

units exist as integral constituents of the galacturonan[89] and are removable only under conditions leading to chemical modification.

Structures for pectic L-arabinans[70–72, 84, 85, 90] and a D-galactan[91] have been elucidated from fragmentation analysis. The D-galactan associated with pectin appears to be made up of an essentially linear chain of (1→4)-linked β-D-galactopyranosyl units. The L-arabinan consists of a (1→5)-linked main chain of L-arabinofuranosyl units to which is attached single (1→3)-linked α-L-arabinofuranosyl units as single side chains. The amounts of neutral glycans in commercial pectin preparations are usually of a low order because of their breakdown during extraction and demethylation steps.

Commercial pectin can, for practical purposes, be regarded as a homopolymer preparation of a D-galacturonan. Most of the D-galactopyranosyluronic acid units of this macromolecule do not occur free however, but are esterified with methanol to varying degrees, and some of the hydroxyl groups at positions C-2 and C-3 may be esterified by acetic acid. The number and distribution of respective ester groupings in the molecule has a great deal of bearing on physical properties and will be discussed in a later section.

The range of polymers possible through variation in methoxyl content has made it necessary to adopt a standard nomenclature defined by certain arbitrary limits in methoxyl group content. Although a comprehensive classification exists,[92] pectic materials of commerce are generally differentiated into two classes, with *pectins* designating those materials having appreciable amounts of methoxyl groups, the unesterified carboxyl groups of which may exist in various salt forms, and with *pectic acids* designating those substances that are essentially void of methoxyl groups. Only the former types are capable of forming gels with sugar and acid under suitable conditions.

An extensive range is thus possible within the pectin group, depending upon the degree of methylation (DM), and it is common to classify pectins somewhat arbitrarily according to their ester content. Pectins having a DM of 70 or higher (70% or more of carboxyl groups esterified) are classified as rapid-set or high-methoxyl (HM) pectins. Those of DM 60–65 are termed slow-set pectins; pectins of DM less than 50 fall into the category of low-methoxyl (LM) pectins. It has been recently suggested[12] that pectins be classified as (*a*) regular types of methoxyl content >7%, which require sugar and acid for gelling and (*b*) low-methoxyl types having methoxyl contents <7% (~43 DM) and requiring calcium ion for gelation.

V. Biosynthesis and Enzymic Degradation

There is evidence[93] to suggest that the main component of the pectic group, the D-galacturonan, is generated from uridine diphosphate D-galacturonic acid through the action of a synthetase enzyme. Uridine diphosphate D-galacturonic acid in turn may be formed enzymically from uridine diphosphate D-glucose[94–96] or from D-galacturonic acid *via* phosphorylation and a transferase mediated reaction involving uridine triphosphate.[97] Methylation probably occurs at some point after formation of the glycosidic bond, the methyl group being donated by S-adenosyl-L-methionine.[98, 99]

Less is known concerning the biosynthesis of neutral components of the triad, although it is likely that the L-arabinan and D-galactan polymers are both built up *via* enzyme-catalyzed condensation of respective uridine diphosphate sugar precursors.[7]

Several types of enzymes are known that degrade the pectin molecule. These may be catagorized according to their mode of action as (*a*) pectin esterases, (*b*) trans-eliminases or lyases, and (*c*) polygalacturonases.

Pectin esterases are found in most plants and are synthesized by a number of microorganisms. Their action on pectin results in the hydrolysis of the methyl ester group with formation of methanol and free D-galacturonic acid units and most probably account in part for the decline in methyl ester content of pectins with increasing age of the plant. Pectin esterase preparations free of other pectin degrading enzymes are available and in theory could be useful for production of low-methoxyl pectins. Hydrolysis, however, has been found to commence at a terminal end or some other point on the pectin molecule and to proceed sequentially along the molecule. Thus, partially enzymically demethylated pectins have properties different from those of acid or alkali demethylated pectins, in which a random demethylation has occurred,[100] and are not attractive for commercial use.

A number of microorganisms produce lyases that breakdown pectic substances.[101–109] This deploymerization proceeds by the same type of β-elimination mechanism that occurs during the base-catalyzed depolymerization of pectin yielding an unsaturated nonreducing end-group[110] (see Section VI. 3). Lyases from various organisms differ in regard to their specificity and action and are classified according to their activity toward pectins of different methoxyl content and their *endo* or *exo* mode of attack.[111, 112]

Deploymerizing enzymes are also known that cleave the glycosidic bond in a normal hydrolytic manner. Termed polygalacturonases, they are produced by a number of microorganisms and are found in small amounts in some fruits and vegetables, but are essentially absent in citrus.[113, 114]

There are indications[75, 115] from work conducted with impure enzyme preparations that there are enzymes capable of splitting the galactan and arabinan components of pectin materials.

VI. Properties

1. Solubility and Dispersibility

Pectins are, in general, soluble in water and insoluble in organic solvents, although they swell markedly and dissolve to some degree in some of the more polar organic solvents, such as formamide, *N,N*-dimethylformamide, and methyl sulfoxide. Their water solubility is determined by the number of methoxyl groups, their distribution, and the molecular weight. Generally, solubility increases with decreasing molecular weight and with increasing methyl ester content, although solution pH, temperature, and the type and concentration of salts present, as well as the presence of other organic substances such as sugar, have a marked effect on solubility. Divalent salts of pectic and pectinic acids are only slightly soluble in water and must be converted to their sodium or potassium forms for dissolution.

The ease of solubilization of commercial pectins is usually a more important consideration than absolute solubility and this, in turn, is determined largely by dispersibility. Dry, powdered pectins, when added to water, have a tendency to hydrate very rapidly with the result that lumps consisting of semidry packets of pectin contained in an envelope of more highly hydrated material form. Further dissolution of such lumps is very slow. Their formation can be prevented by improving dispersibility through mixing the pectin with a small quantity of a water-soluble carrier material or by use of pectins having improved dispersibility characteristics imparted through special treatment during manufacture.

Finely powdered sucrose or D-glucose (dextrose) are common dispersing agents. They are mixed with the pectin in amounts of 5–10 parts by weight either during manufacture or by the user. The association of the sugar and pectin may be improved by exposing the mixture briefly to superheated steam, followed by drying and screening.[116] Sugars, in addition to serving as dispersing aids, also are used by the manufacturer as a convenient diluent for standardization purposes.

Several processes have been described for the preparation of dispersible pectins. One of the earliest methods described, though not in us today, consisted of incorporating aluminum or calcium pectate as a coating or integral part of the pectin particle.[117–119] The less soluble pectate material tends to retard hydration sufficiently to allow dispersal and uniform wetting of the particles.

A novel approach to the problem of dispersibility has been to incorporate into the dry pectin powder mixtures of an organic acid and an inorganic carbonate. The effervescence resulting during addition of such a mixture to water serves to agitate and thus disperse the particles throughout the medium. A similar effect results from the incorporation of sodium bicarbonate in dry pectin,[120] the pectin itself serving as the organic acid.

The use of specially prepared dispersible pectins is often limited by economic and taste considerations. Thus, powdered pectins used in the jam industry are usually added in dry form along with 5–10 parts by weight of sugar or are first prepared as a solution by adding the pectin to warm water in a specially constructed high-speed mixer.[121] Standard pectin solutions for use the same day are prepared in less than 10 min by this latter method and can be added at any time to a batch or continuous process operation.

In the laboratory, pectin solutions are easily prepared by either mixing the pectin with a water-soluble dry dispersant such as sugar or by first wetting the powder with approximately 2 parts v/w of a water-miscible solvent such as alcohol or glycerol. Alternatively, the pectin may be added slowly to water agitated vigorously in a Waring Blendor. Dissolution is greatly expedited by use of water that has been warmed to approximately 55.°

2. *Solution Properties*

Solutions of pectin are viscous, the viscosity being related to the molecular weight and degree of esterification of the pectin and to the concentration, pH, and composition of the solution. As with solubility, the viscosity is proportional to the degree of esterification, although an apparent increase in viscosity results when the methoxyl content is sufficiently low because of an insolubilization of the molecule.

Addition of salts of monovalent cations, such as sodium chloride, to pectin solutions causes a reduction in viscosity; this effect is more pronounced with decreasing methoxyl content and is apparently caused by suppression of charges on carboxyl groups, thus lessening repulsion between adjacent chains and allowing closer association. Soluble salts of di- and trivalent cations show an opposite effect,[122] the viscosity increasing in a regular fashion with the addition of salts of calcium, barium, strontium, and similar cations. This effect has been ascribed to a bridging effect between suitably positioned carboxyl groups, although it is likely that other effects as well are involved.[123]

Solutions of pectic materials, as is common with solutions of most linear polymers, display non-Newtonian flow characteristics seen mainly as a decrease in viscosity with increasing rates of shear. This decrease is most probably the result of an orientation effect.[124] As the rate of shear is increased, the pectin molecules, which are long, randomly positioned chains, become increasingly aligned in the direction of flow with a lessening of interaction between adjacent polymer chains.

Solutions of pectin are most stable at a pH range of 3–4 and at higher and lower pH values exhibit a decline in viscosity and gel strength caused by depolymerization. This breakdown is most severe at higher pH values and is greatly accelerated at temperatures above 50°. For this reason, it is important that conditions during industrial use of pectins be carefully chosen and regulated. Pectins, for reasons discussed in the next section, are much more susceptible to base-catalyzed deploymerization than are pectic acids.

3. Chemical Properties

Pectins are stable under mildly acidic conditions at normal temperatures but undergo depolymerization under more strongly acidic conditions, particularly at elevated temperatures. A mechanism involving first a protonation of the glycosidic oxygen atom followed by a cleavage of the glycosidic bond and addition of water to the resulting resonance-stabilized carbonium ion is envisioned.[125] Pectic acid is much more resistant to acid-catalyzed deploymerization and requires more rigorous conditions for complete hydrolysis.

Pectins, in addition to undergoing methoxyl group saponification under basic conditions, are also deploymerized. The rate of depolymerization at any given time is proportional to the amount of remaining methoxyl groups. Hence, anaerobic alkaline saponification of pectin to produce pectate frequently results in depolymerization, which is complete when deesterification is complete.[126–131] Kenner,[132] Whistler and BeMiller,[133] and Neukom and Deuel[134] explained the observed depolymerization as a beta-elimination reaction (Fig. 3). Hence, the products are the same as those produced by the action of trans-eliminase on pectin. Saponification proceeds simultaneously with deploymerization and at a faster rate. Launer and Tomimatsu[135] reported that all reactions of pectin in a pH 10.2 buffer at 25° are complete in 30 min and, on the average, about one glycosidic bond is cleaved for every 80 ester groups hydrolyzed. Pectin is also unstable in neutral solutions, being degraded in a temperature-dependent reaction.[136–138] Again, the rapid decrease in viscosity and increase in reducing end-groups is indicative of splitting of glycosidic bonds.[138] However, by controlling temperature, pH, and time, it is possible to produce pectic acids by base-catalyzed deesterification during which only a small amount of depolymerization takes place.[139, 140]

Pectic acids, because they lack a methoxyl substituent at C-6, have a much less acidic proton at C-5 and cannot be resonance stabilized in the transition state; they are, therefore, much more resistant to base-catalyzed depolymerization. In addition, the charge on the carboxyl group is thought to repel the approaching hydroxyl anion. Elimination does not occur in residues where an amide function replaces the methoxyl group, suggesting that amidated pectins may be somewhat more resistant to depolymerization under basic conditions.[141]

Pectin and pectic acid are reactive toward a number of reagents and have been used as starting materials for the preparation of a number of interesting and potentially useful derivatives.

Pectin derivatives in which the hydroxyl groups at positions C-2 and C-3 have reacted include the pectin nitrates,[142–145] sulfates,[55, 57] and acylates,[146–149]

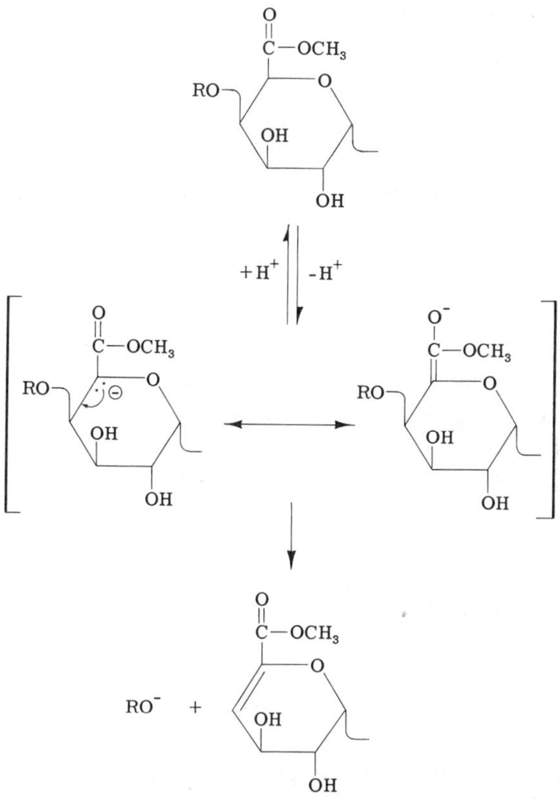

FIG. 3.—Mechanism of base-catalyzed depolymerization of pectin.

and the formaldehyde cross-linked pectins.[148] Ethylene oxide and propylene oxide[149] react with the carboxyl groups of pectin and pectic acid to yield the respective hydroxyethyl and hydroxypropyl esters. Primary alcohol groups are formed either by the action of sodium or potassium borohydride on pectin or hydroxyethylpectin[150] or by the action of diborane[151] on pectic acid. The latter reduction is conducted in an ether solvent, the pectic acid being first acylated to render it soluble.

A number of substituted and unsubstituted amides of pectin can be prepared by reaction of pectin with ammonia[152—156] or with primary and secondary alkylamines.[157—159] Low-methoxyl pectins prepared by treatment of pectin with ammonia emerge as low DS pectin amides. The amide group in low amounts does not appear to detract from the gelling properties of the pectin and may reduce calcium sensitivity.

4. *Gels and Their Characterization*

The importance of pectin to industry and more specifically to the food industry lies in the somewhat unique ability of its solutions to yield thermoreversible gels in the presence of a dehydrating agent at a pH at or near 3 or in the presence of calcium ion. The normal high-methoxyl pectin gel, as established through years of development and use, is considered for practical purposes as being composed of a dilute (<1%) solution of pectin containing from 58–75% sugar as a dehydrating agent and adjusted to a pH of 2.8–3.5 by addition of a suitable acid. Citric, malic, and tartaric acids are commonly used.

The mechanism of formation of this classic gel is regarded[123, 160–162] as one in which the pectin chains, normally strongly hydrated by water molecules, become less hydrated through replacement of water molecules with solute molecules. The result is a greater contact between pectin chains, which yields a complex network of polysaccharide molecules, parts of which are associated through hydrogen bonding and other forces into junction zones. Trapped in the interstices of this network are water and solute molecules.

Gel formation and the nature of the gel formed are governed by several factors, chief of which are the degree of polymerization and degree of methylation. Under similar conditions, the degree of gelation of a pectin gel is generally proportional to the molecular weight and is inversely proportional to the degree of esterification.[35, 163, 164] Extrinsic factors, such as sugar content, pH, and concentration of calcium ion, are also important; their requirements are established by the degree of methylation of the pectin. Generally, as the degree of methylation of a normal high-methoxyl pectin increases, so also does the requirement for dehydrating agent necessary to bring about gelation. Pectins with increasingly greater degree of methylation gel at a somewhat higher pH, perhaps because they have fewer free carboxyl groups, which in ionized form cause repulsion of polymer chains.

Low-methoxyl pectins,* pectinic acids,* and pectic acids* show a diminishing requirement for dehydrating solute and an increasing requirement for protons or calcium ions for gelation, both of which can act to remove charges from carboxyl

---

*The term *pectins* designates those water-soluble pectinic acids of varying methyl ester content and degree of neutralization that are capable of forming gels with sugar and acid under suitable conditions. The term *pectinic acids* designates colloidal galacturonoglycans containing more than a negligible content of methyl ester groups with varying degrees of neutralization. The term *pectic acids* designates colloidal galacturonoglycan, essentially free of methyl ester groups, with varying degrees of neutralization. The term *protopectin* designates the water-insoluble parent pectic substance that occurs in plants and which, upon restricted hydrolysis, yields pectin or pectinic acids. The term *pectic substances* designates complex colloidal polysaccharides that occur in plants and which are primarily derivatives of galacturonoglycans.[92]

groups. Pectinic and pectic acids gel only in the presence of calcium or other divalent cations. This effect has been attributed to a bridging effect of calcium ions between suitably positioned carboxyl groups, although as pointed out Rees,[123] the energy of such simple electrostatic attractions is probably too low to be regarded as an important factor in gel formation.

The distribution of methoxyl groups on the polymer as well as the presence of other substituents are also important in affecting gelling properties. Thus, the presence of as little as one *O*-acetyl group on position C-2 or C-3 of one in every eight D-galactopyranosyluronic acid units is sufficient to prevent gelation, presumably by interferring with crystallite formation. It is for this reason that beet and potato pectins, which contain unusually large amounts of *O*-acetyl residues, are not regarded as being good sources of pectins for commercial use.

The effect of methoxyl group distribution on gelling properties can be seen through a comparison of acid or base demethylated pectins with those demethylated through enzyme action.[160, 165, 166] The former types, which would be expected to have a random distribution of methoxyl groups along the polymer chain, show normal variation in gel properties with decreasing methoxyl content, whereas the latter show anomalities seen mainly as a lower than expected gel strength with increasing deesterification. This is attributable to nonrandom removal of methoxyl groups to give both pectin- and pectic acid-like chain segments, analogous to a mixing of pectin and pectic acid molecules.

A variety of methods have been used to characterize pectin gels. A number have been designed for use on pectin gels, whereas others, have been adapted for use on pectin. Such methods can be grouped[27] according to their action on the gel into (*a*) those which provide a measure of gel strength through controlled deformation of the gel within the elastic limits and (*b*) those which measure gel strength by rupturing the gel.

One of the simplest and most widely used methods of determining gel strength, which relies on the deformation within elastic limits principle, is the sag method of Cox and Higby[25] that measures the rigidity of a standard gel after the supporting vessel is removed. A modification of this method has been adopted by the Institute of Food Technologists Committee on Pectin Standardization.[24] The heart of the sag method is the Ridgelimeter[167] which consists essentially of a vertically mounted micrometer that is used to measure the amount of sag of an unsupported gel, prepared under standard conditions, in a given time period (usually 2 min). The percentage of sag as compared with sag values of standard gels serves as an accurate and reproducible measure of gel strength.

Still in limited use today are several instruments that measure gel strength in terms of the deformation resulting from an applied force. One such instrument, the Bloom gelometer,[168] originally developed for use in the gelatin industry, measures gel strength as the force required to depress a plunger of set diameter a distance (usually 4 cm) against the gel surface. The force required for the deformation is registered either mechanically or in more recent instruments, electronically. Another instrument, developed by the British Association of Research[169] and known in its various modifications as the BAR or FIRA jelly tester, measures the force required to cause torsional deflection of a vane inserted in a pectin gel through a set number of degrees.

Methods within the second group, those measuring the force required to cause a rupture of the gel surface, although not used as widely, often serve as valuable supplementary aids to gel characterization and are useful for quality control purposes. Foremost of these is the Delaware jelly tester of Tarr[170] and Baker[171] and its various modifications,[172, 173] in which a plunger of set diameter is driven down upon the gel surface until rupture occurs. A disadvantage of this method is that nonuniformity of gel surfaces can lead to appreciable imprecision in results.

An inherent drawback of gel strength measurement methods presently in use is that usually only one deformation or break value is provided. Although sufficient for many purposes, it does not provide an overall picture of the characteristics of the gel system. Attempts have been made to overcome this deficiency through the design of more sophisticated instruments that provide a graphical representation of deformation and rupture in relation to force. One such instrument, termed the Gel Characterization Apparatus[174] utilizes the plunger principle of Bloom and Delaware designs coupled to sensitive measuring and recording devices. In addition to providing a measurement of gel strength information, measurements of cohesiveness, rigidity, and adhesiveness are also provided. Other interesting instruments able to simulate chewing action have been described;[175–178] they are able to furnish information on several textural properties and should prove useful for the characterization of pectin gels.

5. *Analytical Methods and Calculations*

Methods for qualitative analysis.—Pectin in pure form is easily identified on the basis of its physical characteristics and reactivity toward specific reagents. In admixture with other gums, however, or in foods, special tests or separation procedures are usually necessary to minimize interference from other components. Such procedures are made difficult by the fact that gums in general are quite similar with regard to physical properties and chemical reactivity.

The dissolving characteristics, solution appearance, and appearance of the precipitate obtained by addition of alcohol are useful preliminary tests for identification of pectin.[179] Pectic acid and calcium pectate are insoluble in water, whereas monovalent salts of pectin and pectic acid dissolve on heating to yield a clear solution that separates into a stringy white precipitate on addition of alcohol. Several other hydrocolloids display similar characteristics, however, and are easily confused with pectin.

Infrared spectroscopy has been investigated[180] as a possible method for gum indentification, but again structural differences are not sufficient to allow a sharp distinction. The majority of gums can be placed in one of four groups on the basis of spectral features in the 7–15μ range with pectins occupying a group distinct from those of other gums. Other identification methods suggested, but upon which only limited work has been done, include electrophoresis[181] and differential thermal analysis.

Both pectin and its hydrolysis products react in a specific manner with basic lead acetate solution. Addition of this reagent to a solution of pectin results in the formation of a flocculent precipitate which, upon warming, exhibits a yellow coloration in the upper layers. This test is effective at pectin concentrations as low as 0.2% and in the presence of a number of other gums.[10] Pectin, following hydrolysis with pectic enzymes, treatment with basic lead acetate solution, and warming yields a red precipitate. This reaction, based on the presence of D-galacturonic acid in the hydrolyzate, is used as a specific test for pectic materials.[182]

Pectin reacts with alkaline hydroxylamine to yield a hydroxamic acid, which, in turn, gives a red percipitate with ferric chloride.[183] Pectic acid, being void of methoxyl groups, does not show this reaction, and pectinic acids give only a faint reaction. Treatment of pectin or its hydrolysis products with sulfuric acid followed by L-cysteine (Dische test) results in the slow development of a blue color.[184] Both of the above tests show good specificity and sensitivity.

Group identification schemes whereby commonly used gums, including pectin, can be routinely identified have been described.[6, 179, 180] Such schemes rely upon the precipitation reaction of gums with various reagents as preliminary identification followed by confirmatory tests. The quaternary ammonium salts, cetylpyridinium chloride and cetyltrimethylammonium bromide (Cetavlon), have been used successfully for the separation of neutral and acidic polysaccharides. Pectin can be separated from neutral polysaccharides by reaction with Cetavlon and its identity confirmed by specific tests.

Methods for quantitative analysis.—Analyses for anhydrouronic acid (AUA) and ester methoxyl content are generally done on pectins that have been acid washed to convert the free carboxyl groups to the protonated form. This conversion is achieved by washing the pectin thoroughly with 60% alcohol containing 5% of hydrochloric acid and then by alcohol and drying.[185]

The ester methoxyl content is usually determined using a modification[186] of the saponification method of Hinton:[187] A sample (0.5 g) of acid-washed pectin is dissolved in water and titrated with standard base to neutrality. The base required represents the neutralization equivalent (A) of the carboxyl groups and may be used later in a calculation to determine the anhydrouronic acid content. Twenty-five ml of $0.25N$ sodium hydroxide solution are added, and the solution is allowed to stand for 0.5 hr. Twenty-five ml of $0.25N$ hydrochloric acid are then added, and the excess acid is back-titrated with standard base. This value (B) as corrected with a suitable blank, is used to calculate the methoxyl group content according to the relation: % methoxyl $= 3.1N \times B/W$, where N is the normality of standard sodium hydroxide solution used in the final titration and W is the weight of pectin. When necessary a second term may be added to correct for acetyl group content as determined in a separate analysis.[188]

The anhydrouronic acid content can be found by addition of the ester content to the number of free carboxyl groups corresponding to the amount of standard base (A) used in the first titration. Considerable error may be introduced, however, by the presence of mineral constituents, including ammonia, or amide groups in the pectin, and for this reason, determination based on decarboxylation methods or colorimetric methods are preferred. Decarboxylation is achieved by treatment of the pectin with 19% hydrochloric acid, the evolved carbon dioxide being collected in standard sodium hydroxide solution and back titrated with standard acid.[186] The decarboxylation method, although considered to be the most reliable, is time consuming and, as an alternative, a colormetric method[189, 190] in which pectin is treated with sulfuric acid followed by carbazole to yield a colored reaction compound is sometimes used.

The degree of esterification, defined as the percentage of esterified uronic acid carboxyl groups, can be calculated according to the relation: degree of esterification $= 176 \times$ % methoxyl $\times 100/31 \times$ % anhydrouronic acid.

The degree of esterification of pectins may also be found by measuring the methanol liberated by base treatment.[191] The methanol is oxidized to formaldehyde and reacted with pentane-2,4-dione in the presence of ammonium ion to give a colored condensation product that is determined spectrophotometrically. This procedure has the advantages of being specific for methyl ester and having high sensitivity and is suggested for studies where semimicro assay is required. In addition to methyl ester determination, its use in measuring pectin esterase activity has been suggested.

6. *Physiological Properties*

Pectin, as found naturally in fruit bodies and other plant parts, has constituted a normal part of the diet of plant-eating animals since the emergence of higher plant and animal life forms. Although this does not automatically insure that pectic substances are void of any unfavorable physiological effects, it does suggest that pectin has been shown to be safe as a food additive by many centuries of normal use. That it is completely safe has been substantiated by the feeding of pectin in large doses to experimental animals and human subjects[192, 193] with no untoward effects.

Pectin in its passage through the animal gastrointestinal tract is extensively broken down but only slighty assimilated. This breakdown, at first believed caused by the action of pectinolytic enzymes produced by the animal, has been shown[194, 195] to be caused by the action of the microbial flora of the large intestine. The products of this breakdown do not appear to enter into the metabolism of the organism to an appreciable extent because pectin has been shown to provide negligible energy value.[196, 197] On the other hand, a hemostatic effect, [198] characterized as a shortening of the coagulation time of drawn blood samples, is observed after ingestion of pectin. Thus, a slight absorption of pectin, most probably degraded, is suggested.

Many species of bacteria and fungi are able to degrade and utilize pectin as a carbon source. Certain fungi are especially efficient in degrading the pectin molecule. Intestinal organisms chiefly responsible for the breakdown of pectin have been shown to belong to *Lactobacillus, Enterococcus, Aerobacillus,* and *Micrococcus* genera.[193] *Escherichia coli,* however, does not appear to be a contributor to pectin degradation in the gastrointestinal tract and may show diminished growth in the presence of pectin. Recent evidence[199] suggests that under certain conditions, pectins may show a slight antimicrobial action toward *E. coli.*

VII. References

(1) M. Vauquelin, *Ann. Chim.,* **5,** 92 (1790).

(2) A. Payen, *Ann. Chim. Phys.,* **26,** 329 (1824).

(3) B. Braconnot, *Ann. Chim. Phys.,* **28,** 173 (1825).

(4) E. L. Hirst and J. K. N. Jones, *Advan. Carbohyd. Chem.* **2,** 235 (1946).

(5) R. L. Whistler and C. L. Smart, "Polysaccharide Chemistry," Academic Press, New York, 1953, p. 161.

(6) R. M. McCready and H. S. Owens, *Econ. Bot.,* **8,** 29 (1954).

(7) H. G. J. Worth, *Chem. Rev.,* **67,** 465 (1967).

(8) H. Deuel and E. Stutz, *Advan. Enzymol.,* **20,** 341 (1958).

(9) W. Pilnik and P. Zwiker, *Gordian,* 202 (1970).

(10) W. A. Bender, in "Industrial Gums," R. L. Whistler, ed., Academic Press, New York, 1st Ed., 1959, p. 377.

(11) J. J. Doesburg, "Pectic Substances in Fresh and Preserved Fruits and Vegetables," I.B.V.T. Commun. No. 25, Institute for Research on Storage and Processing of Horticulture Produce, Wageningen, The Netherlands, 1965.

(12) M. Glicksman, "Gum Technology in the Food Industry," Academic Press, New York, 1969, p. 159.

(13) Z. I. Kertesz, "The Pectic Substances," Interscience, New York, 1951.

(14) S. A. Stoĭkov, *Nahrung,* **2,** 287 (1958).

(15) G. H. Joseph and C. R. Havighorst, *Food Eng.,* **24,** 87 (1952).

(16) W. G. Hull, C. W. Lindsay, and W. E. Baier, *Ind. Eng. Chem.,* **45,** 876 (1953).

(17) *Food Mfr.,* **42,** 37 (1967).

(18) Pomosin Werke Grossenbrode G.m.b.H., Germ. Patent 1,224,134 (1966); *Chem. Abstr.*, **65,** 20760 (1966).

(19) W. Bock and D. Lange, *Nahrung,* **7,** 71 (1963).

(20) E. F. Bryant, U.S. Patent 2,480,710 (1949); *Chem. Abstr.,* **44,** 2018 (1950).

(21) G. H. Joseph, A. H. Kieser, and E. F. Bryant, *Food Technol.,* **3,** 85 (1949).

(22) C. J. B. Smit and E. F. Bryant, *J. Food Sci.,* **33,** 262 (1968).

(23) C. P. Wilson, *Amer. Food J.,* **21,** 279 (1926).

(24) Institute of Food Technologists, Committee on Pectin Standardization, *Food Technol.,* **13,** 496 (1959).

(25) R. E. Cox and R. H. Higby, *Food Ind.,* **16,** 441 (1944).

(26) M. Olliver, *Food Technol.,* **4,** 370 (1950).

(27) P. E. Christensen, *Food Res.,* **19,** 163 (1954).

(28) M. Olliver, P. Wade, and K. P. Dent, *J. Sci. Food Agr.,* **8,** 188 (1957).

(29) "Pectin L. M." Sunkist Growers, Inc., Ontario, California, 1953, p. 10.

(30) G. H. Joseph and W. E. Baier, *Food Technol.,* **3,** 18 (1949).

(31) C. L. Hinton, *J. Sci. Food Agr.,* **1,** 300 (1950).

(32) W. A. Bender, R. V. MacAllister, and B. Sienkiewicz, U.S. Patent 3,133,053 (1964); *Chem. Abstr.,* **61,** 2407 (1964).

(33) FAO Nutritions Meetings, *Rep. Ser.,* No. **46A,** 133 (1969).

(34) R. M. Ehrlich, *Food Prod. Develop.,* **2,** 36 (1968).

(35) A. Lopez and L.-H. Li, *Food Technol.,* **22,** 91 (1968).

(36) G. H. Joseph, *Food Eng.,* **25,** 71, 114 (1953).

(37) J. B. Wegener, B. Baer, and P. D. Rogers, *Food Technol.,* **5,** 76 (1951).

(38) R. R. Barton, *Proc. Amer. Soc. Hort. Sci.,* **58,** 95 (1951); *Chem. Abstr.,* **46,** 8286 (1952).

(39) M. W. Hoover and R. A. Dennison, *Proc. Amer. Soc. Hort. Sci.,* **65,** 188 (1955); *Food Sci. Abstr.,* **28,** 1511 (1956).

(40) C. W. Decker, *Can. Dairy Ice Cream, J.,* **30,** 46 (1951); *Chem. Abstr.,* **47,** 227 (1953).

(41) Sunkist Growers Bull. No. 384, Sunkist Growers, Inc., Ontario, California.

(42) Pfeifer and Langen K.-G., Neth. Patent 6,607,121 (1966); *Chem. Abstr.,* **66,** 94091t (1967).

(43) W. J. S. Peschardt, Canad. Patent 520,983 (1956).

(44) J. J. Doesburg and L. De Vos, *Int. Fruit Juice Congr., 5th, Vienna, 1959,* 32 (1959).

(45) H. Exler, Auslegeschrift (Germ. Patent Appl.) 1,270,938 (1968).

(46) National Dairy Products Corp., Brit. Patent 1,124,238 (1968).

(47) H. T. Leo and C. C. Taylor, U.S. Patent 2,754,214 (1956); *Chem. Abstr.,* **51,** 3868 (1957).

(48) W. M. Camirand and R. R. Forrey, U.S. Patent 3,425,848 (1969); *Chem. Abstr.,* **70,** 86417w (1969).

(49) R. H. Waitman and J. W. Hoos, U.S. Patent 3,367,784 (1968); *Chem. Abstr.,* **68,** 94761c (1968).

(50) E. L. Moore, A. H. Rouse, C. D. Atkins, and E. C. Hill, *Proc. Fla. State Hort. Soc.,* **82,** 224 (1969).

(51) A. D. Shepherd, R. M. McCready, and H. S. Owens, U.S. Patent 2,673,157 (1954); *Chem. Abstr.,* **48,** 8444 (1954).

(52) G. W. Twieg and A. E. Poarch, U.S. Patent 2,701,767 (1955); *Chem. Abstr.,*

49, 13552 (1955).

(53) A. E. Poarch and G. W. Twieg, U.S. Patent 2,809,893 (1957); *Chem. Abstr.,* **52,** 1507 (1958).

(54) G. H. Joseph, "Pectin Bibliography of Pharmaceutical Literature," Sunkist Growers, Inc., Ontario, California, 1956.

(55) A. Cannava and B. Chiarlo, *Med. Sper.,* **26,** 114 (1955).

(56) H. E. Alburn and J. Seifter, U.S. Patent 2,729,633 (1956); *Chem. Abstr.,* **50,** 8144 (1956).

(57) J. R. Geigy, Swiss Patent 305,888 (1955); *Chem. Abstr.,* **51,** 6097 (1957).

(58) A. Keys, F. Grande, and J. T. Anderson, *Soc. Exp. Biol. Med.,* **106,** 555 (1961).

(59) E. H. Groot, *Voeding,* **27,** 549 (1966).

(60) T. A. Anderson and R. D. Bowman, *Proc. Soc. Exp. Biol. Med.,* **130,** 665 (1969).

(61) CIBA Ltd., Fr. Patent 1,462,506 (1966); *Chem. Abstr.,* **67,** 67587y (1967).

(62) H. J. Eichel, U.S. Patent 3,324,109 (1967).

(63) Artha Company, Brit. Patent 1,196,158 (1970); *Chem. Abstr.,* **73,** 59297n (1970).

(64) W. A. Bender, U.S. Patent 3,485,920 (1969); *Chem. Abstr.,* **72,** 59069u (1970).

(65) Sigma-Tau Industrie Farmaceutiche Riunite, S.p.A., Brit. Patent 1,206,870 (1970); *Chem. Abstr.,* **73,** 133985d (1970).

(66) F. F. Hansen, U.S. Patent 3,310,408 (1967); *Chem. Abstr.,* **67,** 2270v (1967).

(67) L. N. Mukherjee and S. D. Shukla, *Indian J. Appl. Chem.,* **28,** 177 (1965); *Chem. Abstr.,* **66,** 41006n (1967).

(68) H. Thiele, Germ. Patent 1,249,517 (1967); *Chem. Abstr.,* **67,** 109390m (1967).

(69) Varta Pertrix-Union G.m.b.H., Brit. Patent 1,025,975 (1966); *Chem. Abstr.,* **64,** 18969 (1966).

(70) G. O. Aspinall and R. S. Fanshawe, *J. Chem. Soc.,* 4215 (1961).

(71) A. J. Barret and D. H. Northcote, *Biochem. J.,* **94,** 617 (1965).

(72) S. S. Bhattacharjee and T. E. Timell, *Can. J. Chem.,* **43,** 758 (1965).

(73) M. A. Jermyn and R. G. Tomkins, *Biochem. J.,* **47,** 437 (1950).

(74) J. K. N. Jones and W. W. Reid, *J. Chem. Soc.,* 1361 (1954).

(75) E. J. Bourne, J. B. Pridham, and H. G. J. Worth, *Phytochem.,* **6,** 423 (1967).

(76) V. Zitko and C. T. Bishop, *Can. J. Chem.,* **44,** 127 (1966).

(77) S. Luckett and F. Smith, *J. Chem. Soc.,* 1106 (1940).

(78) P. A. Levene and L. C. Kreider, *J. Biol. Chem.,* **120,** 591 (1937).

(79) G. O. Aspinall, I. W. Cottrell, S. V. Egan, I. M. Morrison, and J. N. C. Whyte, *J. Chem. Soc., C,* 1071 (1967).

(80) G. O. Aspinall, K. Hunt, and I. M. Morrison, *J. Chem. Soc., C,* 1080 (1967).

(81) G. O. Aspinall, B. Gestetner, J. A. Molloy, and M. Uddin, *J. Chem. Soc., C,* 2554 (1968).

(82) G. O. Aspinall, J. W. T. Craig, and J. L. Whyte, *Carbohyd. Res.,* **7,** 442 (1968).

(83) G. O. Aspinall, "Polysaccharides," Pergamon Press, Oxford, 1970, p. 119.

(84) E. L. Hirst and J. K. N. Jones, *J. Chem. Soc.,* 1221 (1947).

(85) E. L. Hirst and J. K. N. Jones, *J. Chem. Soc.,* 2311 (1948).

(86) E. L. Hirst, J. K. N. Jones, and W. O. Walder, *J. Chem. Soc,* 1225 (1947).

(87) H. Neukom, H. Deuel, W. J. Heri, and W. Kündig, *Helv Chim. Acta,* **43,** 46 (1960).

(88) J. Rosík, V. Zitko, and J. Vašátko, *Coll. Czech. Chem. Commun.*, **27,** 1346 (1962).

(89) G. O. Aspinall, *in* "The Carbohydrates," W. Pigman and D. Horton, ed., Academic Press, New York, Vol. IIB, 1970, p. 516.

(90) E. L. Hirst and J. K. N. Jones, *J. Chem. Soc.*, 496 (1938).

(91) E. L. Hirst, *J. Chem. Soc.*, 70 (1942).

(92) Committee for the Revision of the Nomenclature of Pectic Substances, *Chem. Eng. News,* **22,** 105 (1944).

(93) C. L. Villemez, Jr., T. S. Lin, and W. Z. Hassid, *Proc. Nat. Acad. Sci. U.S.*, **54,** 1626 (1965).

(94) R. C. Bean and W. Z. Hassid, *J. Biol. Chem.*, **218,** 425 (1956).

(95) J. L. Strominger and L. W. Mapson, *Biochem. J.*, **66,** 567 (1957).

(96) E. E. B. Smith, G. T. Mills, H. P. Bernheimer, and R. Austrian, *Biochim. Biophys. Acta,* **28,** 211 (1958).

(97) E. F. Neufeld, D. S. Feingold, S. M. Llves, G. Kessler, and W. Z. Hassid, *J. Biol. Chem.*, **236,** 3102 (1961).

(98) H. Kauss, A. L. Swanson, and W. Z. Hassid, *Biochem. Biophys. Res. Commun.*, **26,** 234 (1967).

(99) H. Kauss and A. L. Swanson, *Z. Naturforsch. B.*, **24,** 28 (1969).

(100) Ref. 5, pp. 179–182.

(101) C. W. Nagel and R. H. Vaughn, *Arch. Biochem. Biophys.*, **94,** 328 (1961).

(102) P. Albersheim and U. Killias, *Arch. Biochem. Biophys.*, **97,** 107 (1962).

(103) J. D. Macmillan, H. J. Phaff, and R. H. Vaughn, *Biochemistry,* **3,** 572 (1964).

(104) R. D. Edstrom and H. J. Phaff, *J. Biol. Chem.*, **239,** 2409 (1964).

(105) C. W. Nagel and M. M. Anderson, *Arch. Biochem. Biophys.*, **112,** 322 (1965).

(106) S. Nasumo and M. P. Starr, *J. Biol. Chem.*, **241,** 5298 (1966).

(107) C. W. Nagel and S. Hasegawa, *Arch. Biochem. Biophys.*, **118,** 590 (1967).

(108) F. M. Rombouts and W. Pilnik, *Antonie van Leeuwenhoek; J. Microbiol. Serol.*, **37,** 247 (1971).

(109) A. F. Perley and O. T. Page, *Can. J. Microbiol.*, **17,** 415 (1971).

(110) S. Hasegawa and C. W. Nagel, *J. Biol. Chem.*, **237,** 619 (1962).

(111) A. L. Demain and H. J. Phaff, *Wallerstein Lab. Commun.*, **20,** 119 (1957).

(112) D. F. Bateman and R. L. Miller, *Ann. Rev. Phytopath.*, **4,** 119 (1966).

(113) G. E. Hobson, *Nature,* **212,** 717 (1962).

(114) E. J. Gizes, Ph.D. Thesis, Michigan State University, East Lansing, Michigan, 1964.

(115) W. W. Reid, *J. Sci. Food Agr.*, **1,** 234 (1950).

(116) Sueddeutsche Zucker A.-G., Fr. Patent 1,565,127 (1969); *Chem. Abstr.*, **72,** 53910g (1970).

(117) A. G. Olsen, U.S. Patent 2,261,858 (1941); *Chem. Abstr.*, **36,** 1106 (1942).

(118) E. F. Bryant, U.S. Patents 2,418,865 and 2,418,866 (1947); *Chem. Abstr.*, **41,** 4895 (1947).

(119) H. T. Leo and C. C. Taylor, U.S. Patents 2,703,758 and 2,703,759 (1955); *Chem. Abstr.*, **49,** 7152 (1955).

(120) R. E. Cox, U.S. Patent 3,236,657 (1966); *Chem. Abstr.*, **64,** 16539 (1966).

(121) For further information see, "Genu Pectin Handbook," The Copenhagen Pectin Factory, Lille Skensved, Denmark, 3rd Ed., p. 12.

(122) H. Lotzkar, T. H. Schultz, H. S. Owens, and W. D. Maclay, *J. Phys. Chem.*, **50,** 200 (1946).

(123) D. A. Rees, "The Shapes of Molecules," Oliver and Boyd, Edinburgh, 1967, p. 123.

(124) R. W. Stoddart, I. P. C. Spires, and K. F. Tipton, *Biochem. J.*, **114,** 863 (1969).

(125) J. N. BeMiller, *Advan. Carbohyd. Chem.*, **22,** 25 (1967).

(126) C. L. Hinton, *Food Invest. Special Rep.*, 48 (1939); *Chem. Abstr.*, **34,** 1090 (1940).

(127) G. Schneider and U. Fritschi, *Ber.*, **69,** 2537 (1936).

(128) H. P. Bulmer and Co., Ltd., and P. Garrick, Brit. Patent 625,675 (1949); *Chem. Abstr.*, **44,** 245 (1950).

(129) B. Vollmert, *Angew. Chem.*, **61,** 329 (1949).

(130) B. Vollmert, *Makromol. Chem.*, **5,** 110 (1950).

(131) H. Neukom, *Deut. Lebensm.-Reundschaw.*, **61,** 35 (1965); *Chem. Abstr.*, **62,** 16543 (1965).

(132) J. Kenner, *Chem. Ind. (London)*, 727 (1955).

(133) R. L. Whistler and J. N. BeMiller, *Advan. Carbohyd. Chem.*, **13,** 289 (1958).

(134) H. Neukom and H. Deuel, *Chem. Ind. (London)*, 683 (1958).

(135) H. Launer and Y. Tomimatsu, *J. Org. Chem.*, **26,** 541 (1961).

(136) P. Albersheim, H. Neukom, and H. Deuel, *Arch. Biochem. Biophys.*, **90,** 46 (1960).

(137) P. Albersheim, *Biochem. Biophys. Res. Commun.*, **1,** 253 (1959).

(138) G. Kvachanov and D. Dimitrov, *Nach. Tr., Visshiva Inst. Khanitelna Vkusova Prom.-Plovdia.*, **12,** 301 (1959); *Chem. Abstr.*, **64,** 16194 (1966).

(139) R. G. Schweiger, *Methods Carbohyd. Chem.*, **5,** 60 (1965).

(140) R. M. McCready, *Methods Carbohyd. Chem.*, **5,** 167 (1965).

(141) J. Kiss, *Tetrahedron Lett.*, **23,** 1983 (1970).

(142) H. Bock, J. Simmerl, and M. Josten, *J. Prakt. Chem.*, **158,** 8 (1941).

(143) R. Speiser and C. R. Eddy, *J. Amer. Chem. Soc.*, **68,** 287 (1946).

(144) L. H. Lampitt, R. W. Money, A. Uric, and B. E. Judge, *J. Soc. Chem. Ind.*, **67,** 101 (1948).

(145) H. H. Schlubach and H. P. Hoffman-Walbeck, *Makromol. Chem.*, **4,** 5 (1949).

(146) J. F. Carson and W. D. Maclay, *J. Amer. Chem. Soc.*, **68,** 1015 (1946).

(147) J. F. Carson and W. D. Maclay, *J. Amer. Chem. Soc.*, **67,** 787 (1945).

(148) H. Deuel, *Helv. Chim. Acta,* **30,** 1269 (1947).

(149) H. Deuel, *Helv. Chim. Acta,* **30,** 1523 (1947).

(150) G. O. Aspinall and A. Canas-Rodrigues, *J. Chem. Soc.*, 4020 (1958).

(151) F. Smith and A. M. Stephen. *Tetrahedron Lett.*, **7,** 17 (1960).

(152) A. H. Kieser and E. F. Bryant, *Food Technol.*, **3,** 85 (1949).

(153) W. A. Mitchell, *Food Technol.*, **4,** 135 (1950).

(154) J. Sohns, A. Denzler, and H. Deuel, *Helv. Chim. Acta,* **37,** 2153 (1954).

(155) J. Sohns and H. Deuel, *J. Polym. Sci.*, **12,** 551 (1954).

(156) D. Dargel and A. Hock, *Arch. Tierernaehr.*, **17,** 197 (1967).

(157) J. F. Carson, *J. Amer. Chem. Soc.*, **68,** 2723 (1946).

(158) J. F. Carson, U.S. Patent 2,520,123 (1946); *Chem. Abstr.*, **44,** 10733 (1950).

(159) H. Deuel and H. Neukom, *Nature,* **159,** 882 (1947).

(160) A. G. Olsen, *J. Phys. Chem.,* **38,** 919 (1934).

(161) G. H. Joseph, *J. Phys. Chem.,* **44,** 409 (1940).

(162) R. Speiser, M. J. Copley, and G. C. Nutting, *J. Phys. Colloid Chem.,* **51,** 117 (1947).

(163) D. Sulc. D. Ciric, and J. Helman, *Tehnika* (Belgrade), **22,** 858 (1967); *Chem. Abstr.,* **68,** 70370y (1967).

(164) C. J. B. Smit and E. F. Bryant, *J. Food Sci.,* **33,** 262 (1968).

(165) E. F. Jansen and L. R. MacDonnell, *Arch. Biochem.,* **8,** 97 (1945).

(166) C. H. Hills, H. H. Mattern, G. C. Nutting, and R. Speiser, *Food Technol.,* **3,** 90 (1949).

(167) H. C. Lockwood and R. S. Hayes, *J. Soc. Chem. Ind.,* **50,** 145T (1931).

(168) O. T. Bloom, U.S. Patent 1,540,979 (1925); *Chem. Abstr.,* **19,** 2280 (1925).

(169) L. E. Campbell, *J. Soc. Chem. Ind.,* **57,** 413 (1938).

(170) L. W. Tarr, *Del. Agr. Expt. Sta. Bull.,* 142 (1926).

(171) G. L. Baker, *Ind. Eng. Chem.,* **18,** 89 (1926).

(172) E. E. Meschter and L. I. Lataillade, *Food Technol.,* **3,** 28 (1949).

(173) W. A. Bender, *Anal. Chem.,* **21,** 408 (1949).

(174) R. V. MacAllister and C. J. Reichenwallner, U.S. Patent 2,912,855 (1959).

(175) H. H. Friedman, J. Whitney, and A. S. Szczesniak, *J. Food Sci.,* **28,** 390 (1963).

(176) A. S. Szczesniak, *J. Food Sci.,* **28,** 410 (1963).

(177) A. S. Szczesniak, *Candy Ind. Conf. J.,* **127,** No. 9, 22, 43 (1966).

(178) A. S. Szczesniak, *Candy Ind. Conf. J.,* **127,** No. 10, 45, 49, 51 (1966).

(179) M. H. Ewart and R. A. Chapman, *Anal. Chem.,* **24,** 1460 (1952).

(180) A. T. Proszynski, A. J. Michell, and C. M. Stewart, *Austr. Commonw. Sci. Ind. Res. Organ., Div. Forest Prod., Technol. Pap.* No. **38** (1965).

(181) M. T. Cuzzoni and T. P. Lissi, *Farmaco, Ed. Prat.,* **16,** 416 (1961); *Chem. Abstr.,* **56,** 7564 (1962).

(182) H. S. Owens, R. M. McCready, A. D. Shepherd, T. H. Schultz, E. L. Pippen, H. A. Swenson, J. C. Miers, R. F. Erlandsen, and W. D. Maclay, "Methods Used at Western Regional Research Laboratory for Extraction Analysis of Pectic Materials," U. S. Dept. Agr. Bull. AIC-340 (1952).

(183) R. M. McCready and R. M. Reeve, *J. Agr. Food Chem.,* **3,** 260 (1955).

(184) Z. Dische, *Arch. Biochem.,* **16,** 409 (1947).

(185) "The National Formulary, "American Pharmaceutical Association, Washington, D. C., 12th Ed., 1965, p. 291.

(186) T. H. Schultz, *Methods Carbohyd. Chem.,* **5,** 189 (1965).

(187) C. L. Hinton, "Fruit Pectins," Chemical Publishing Co., New York, 1940, pp. 27, 32, 69.

(188) T. H. Schultz, *Methods Carbohyd. Chem.,* **5,** 187 (1965).

(189) E. A. McComb and R. M. McCready, *Anal. Chem.,* **24,** 1630 (1952).

(190) S. M. Stark, Jr., *Anal. Chem.,* **22,** 1158 (1950).

(191) P. J. Wood and I. R. Siddiqui, *Anal. Biochem.,* **39,** 418 (1971).

(192) K. Voit and H. Friedrich, *Klin. Wochschr.,* **14,** 1792 (1935); *Chem. Abstr.,* **30,** 3030 (1936).

(193) S. C. Werch, R. W. Jung, A. A. Day, T. E. Friedmann, and A. C. Ivy, *J. Infec. Dis.*, **70,** 231 (1942).
(194) Z. I. Kertesz, *J. Nutr.*, **20,** 289 (1940).
(195) S. C. Werch and A. C. Ivy, *Amer. J. Dig. Dis.*, **8,** 101 (1941).
(196) B. H. Ershoff and H. B. McWilliams, *Amer. J. Dig. Dis.*, **12,** 21 (1945).
(197) S. Viola, G. Zimmerman, and S. Mokady, *Nutr. Rep. Int.*, **1,** 367 (1970).
(198) H. Violle and L. Saint-Rat, *C. R. H. Acad. Sci.*, **180,** 603 (1925).
(199) M. A. El-Nakeeb and R. T. Yousef, *Planta Med.*, **18,** 295 (1970).

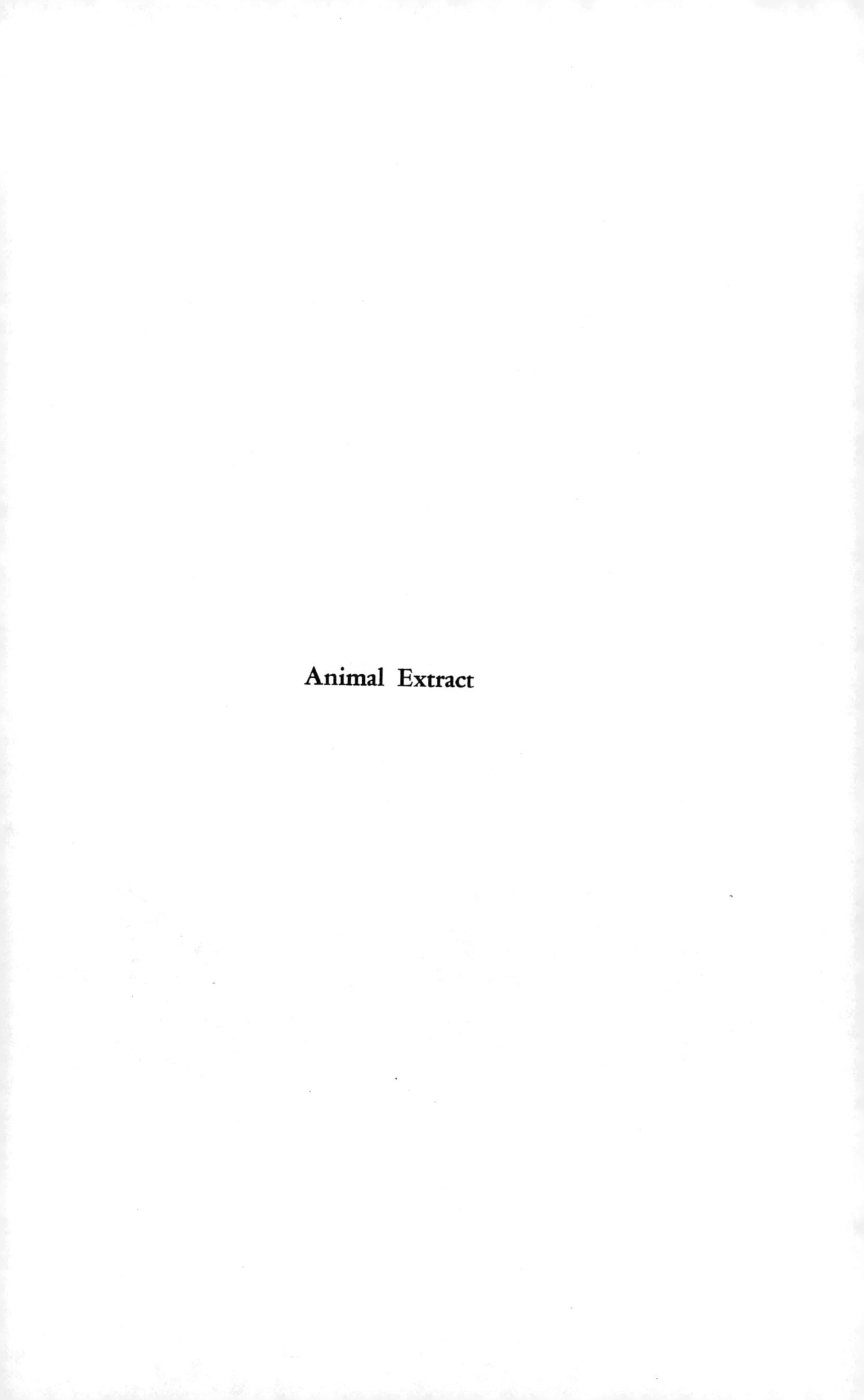

Animal Extract

CHAPTER XX

CHITIN

ROY L. WHISTLER

Department of Biochemistry, Purdue University, Lafayette, Indiana

I. INTRODUCTION

Chitin, somewhat like cellulose, is not an industrial gum because it is insoluble in water. However, many of its derivatives and many of its chemical modifications dissolve in water to produce solutions of high viscosity and sometimes are capable of gelation. Although chitin is not now used commercially, it is expected that future demands will bring it into industrial use. Lobster, crab, and shrimp shells, a prime source of chitin, are collected in large quantities at numerous processing sites and it would be logical to extract from this huge amount of refuse a by-product of commercial utility.

Chitin is a high-molecular-weight linear polymer of 2-acetamido-2-deoxy-D-glucopyranosyl (*N*-acetyl-D-glucosaminyl) units. As in cellulose, the monomer units are linked by β-D-(1→4) bonds. Its name comes from the Greek word *chiton,* meaning coat of mail, because of its function as a protective coating for invertebrates. It also occurs in significant amounts in the cell walls of many of the lower plants.

II. SOURCE

Chitin is the most abundant organic constituent in the skeletal material of invertebrates. It is found in arthropods, annelides, and mollusks, where it provides skeletal support and body armor. Its primary economic source is crustaceans, which are gathered on a large scale for food use. Modern freezing and canning operations with lobster, crab, and shrimp result in the availability of substantial quantities of crustacean waste materials. These wastes consist mainly of shells and heads that can easily be processed to yield chitin. Hard crustacean shells contain 15–20% chitin and as much as 75% calcium carbonate along with skeletal protein. Softer shelled crustaceans, such as shrimp, contain 15–30% chitin and 13–40% calcium carbonate plus skeletal protein. Lobster shells contain less calcium carbonate than do hard shelled crabs, but more than do the shells of shrimp.

Fungi are a potential source of commercial chitin. It is well known that chitin occurs as a constituent of the mycelia and spores of many fungi. Its presence has been demonstrated in the Chytridiaceae, Blastocladiaceae, and Ascomydes. Mycelia from some species of *Penicillium* may contain up to 20% chitin.

Chitin is found throughout the exoskeletons of most insects, where it may be present in amounts ranging up to 60% in special parts such as the flexible portions. The average chitin content in the cuticle of a number of different species is reported to be 33%. The cuticle consists of alternate layers of protein and chitin impregnated with calcium carbonate and pigments and interspersed with polyphenols.

III. Preparation

Shells of lobster, crab, or shrimp may be extracted directly but it is preferable to grind the shells to a size that will pass a 6-mm screen and be retained on a 1-mm screen. The first treatment is one of demineralization, wherein the ground shells are mixed with 5% hydrochloric acid solution. Commonly, this is done in a countercurrent extraction, flowing the acid solution from tank to tank. Because of the hardness of the shells, approximately 24 hr are required for the demineralization, which reduces the ash content of the shell to 0.4–0.5%.

The next processing step may be one of deproteinization, accomplished by treating the demineralized shells with pepsin or trypsin.

Alkali deprotcinization may be used, and is preferred when the final product is to be deacetylated chitin (chitosan). Alkali deproteinization is accomplished by stirring the demineralized ground shells with three successive quantities of 5% sodium hydroxide solution at 85°–90°. Each treatment is 30–45 min in duration and is followed by a brief water wash to remove partially solubilized residues.

Chitin at this stage may be light pink in color, caused by remaining pigments. These are solubilized and removed by a mild oxidation with acidified hydrogen peroxide solution for 6–7 hr at room temperature.

Removal of *N*-acetyl groups from chitin to produce chitosan requires concentrated alkali. A typical deacetylation bath may consist of two parts of potassium hydroxide to one of 95% ethanol and one of ethylene glycol. The mixture is usually heated to 120°, at which temperature the alcohol gently refluxes. During the deacetylation reaction, some alkaline cleavage of the polysaccharide occurs with accompanying decrease in viscosity. Consequently, the reaction is terminated as soon as the acetyl content has been diminished to low levels. The over-all yield of chitosan from raw crab shells is approximately 7%.

Good quality deacetylated chitin at a concentration of 1.25% in dilute acetic acid has a viscosity of ~1200 cps, medium-grade a viscosity of ~160 cps, and low-grade a viscosity of ~15 cps.

Commercial deacetylated chitin is approximately 80–85% deacetylated. It contains about 6.5% of nitrogen rather than the theoretical nitrogen value of 8.7% calculated for the completely deacetylated polysaccharide.

IV. Amount Available

Approximately 150,000 tons of shrimps, 25,000 tons of lobsters, and 85,000 tons of crabs are processed in the United States. Some of the catch is shipped alive and other portions are partially cleaned at sea, but most is processed in land installations. Other large sources of shells are available at processing centers in many parts of the world.

V. Structure

Chitin may be looked upon as a 2-acetamido-2-deoxycellulose because it is composed of *N*-acetyl-D-glucosaminyl units linked together by β-D-(1→4)-linkages in the same way D-glucopyranosyl units are linked in cellulose molecules. The molecular weight of 143,000 to 210,000 suggests that the molecule is a highly linear array of monomer units. This regularity of structure and the high hydrogen bonding and dipole interactions possibly account for the insolubility of natural chitin. Deacetylated chitin is a cationic polysaccharide of fascinating physical and chemical properties. For example, hydroxyethyl and other soluble derivatives make useful wet-end additives in papermaking and are useful precipitants for anionic waste waters such as those from meat-packing plants.

VI. Derivatives

A variety of derivatives of chitin can be made in the same way that derivatives can be produced from cellulose. These derivatives have unusual properties that extend its characteristics and potential usefulness.

VII. Applications

A variety of uses of chitin, deacetylated chitin, and the product of complete hydrolysis, D-glucosamine, are possible.

Deacetylated chitin has been recommended as a sizing agent for rayon, cotton, synthetic fibers, and wool. It has been recommended as a beater stock additive for paper and as a spinning bath additive for cellophone. It has great adhesivetivity to glass and plastics. Both *N*-acetyl-D-glucosamine and D-glucosamine have numerous pharmaceutical uses and potentialities.

Part B: Prepared Gums

Biosynthetic Gums (Chapters XXI–XXIII)

Starch Fractions and Derivatives (Chapters XXIV–XXVII)

Cellulose Derivatives (Chapters XXVIII–XXXI)

Biosynthetic Gums

CHAPTER XXI

XANTHAN AND SOME OTHER BIOSYNTHETIC GUMS

WILLIAM H. MCNEELY AND KENNETH S. KANG

Kelco Company, San Diego, California

I. Introduction

Among the candidates for commerialization as new water-soluble gums are extracellular polysaccharides produced by microorganisms. Production of such extracellular polysaccharides is frequently encountered where carbohydrates are present. Particularly in food and beverage processing, the holding of solutions under nonsterile conditions for excessive periods of time has led to frequent reports of the development of thickened or gummy solutions.

Although many microorganisms growing on solid media produce gummy colonies, only a fraction of these colonies produce extracellular polysaccharides in broth culture. Few of the microorganisms that produce extracellular polysaccharides convert sufficient substrate to polysaccharide to be of economic interest. Moreover, only a fraction of the microorganisms that meet the requirements of extracellular polysaccharide production in broth culture and economic conversions of substrate produce water-soluble gums having properties of possible commercial interest. Nevertheless, the number of microbial species is large and the degree of polysaccharide production by a given organism usually can be improved by variations in cultural conditions. Thus, a substantial number of biosynthetic polysaccharides are candidates for commercialization. Among these candidates, xanthan gum has met the greatest commercial success. Dextrans have been extensively investigated and enjoy some specialty uses, but no large-scale uses have been developed (Chapter XXIII).

II. Biology of Microbial Polysaccharides

1. *Relation to Cell Structure*

Microorganisms produce polysaccharides of three distinct types: extracellular polysaccharides, structural components, and intracellular storage polysaccharides. The extracellular polysaccharides, the main subject of this chapter, can be further classified into two forms: (*a*) capsules that are integral with the cell wall, as well as structurally demonstrable microcapsules, and (*b*) slimes that accumulate outside of the cell wall and diffuse constantly into the culture medium. These extracellular polysaccharides contribute a gummy texture to bacterial colonies on a solid medium and an increased viscosity during growth in a liquid medium.

The capsular components can be separated from the amorphous loose slime by centrifugation. The slime formers may produce slime in large quantities and, in some cases, the viscosity becomes so great a liquid culture remains in place when the culture flask is inverted. The polysaccharides produced by slime formers have the greatest potential for industrial development because the gums can be recovered in large quantities from the culture fluids.

2. *Natural Functions of Polysaccharides*

Extracellular polysaccharides probably protect microorganisms against various adverse environmental factors.[1] For example, the remarkably high moisture-holding capacity of the polysaccharide enables bacteria to maintain at least a minimum of moisture in their immediate environment even after prolonged exposure to low humidity.[2] Also, the thick capsule associated with extracellular polysaccharide production by bacteria provides definite protection against bacteriophages, simply by serving as a physical barrier against the agent.[1] This fact was substantiated by Maxted for group A streptococci.[3] The capsules also provide a partial protection against phagocytosis and amebic attack.[1]

In general, extracellular polysaccharides do not appear to function as a reserve energy source. Thus, slime-producing bacteria normally are not capable of catabolizing their own extracellular polysaccharides.[1] There are indications, however, that fungi degrade and reutilize their own capsular polysaccharides,[4] and the relatively large amount of starch stored by *Polytomella coeca* during growth serves as an energy reserve.[5]

3. *Environmental Influences*

Many polysaccharide producers respond to environmental factors in a clear-cut way. A specific carbohydrate is required for the microbial synthesis of most homopolysaccharides because homopolysaccharides, such as dextrans and levans, are generally produced by a single or simple enzyme system. Because of the simplicity of these systems, they are easily adaptable to cell-free, enzymic synthesis. In contrast, heteropolysaccharides are formed from any utilizable carbon source, although the synthetic efficiency varies to a considerable degree because complex enzyme systems that can handle various carbon sources are involved. For these systems, a practical, cell-free, enzymic synthesis has not yet been attained.

Ordinarily, the conversion efficiency of a carbon source into polysaccharides is inversely related to the carbon substrate concentration. For example, it has been reported that the conversion efficiency of D-glucose to polymer by *Xanthomonas campestris* decreased markedly with an increase in D-glucose concentration.[6]

Although a nitrogen source is necessary for both cell growth and product formation, an excess of nitrogen, in general, reduces conversion of the carbohydrate substrate to extracellular polysaccharide. Usually, the effects of aeration can be related to the relative efficiencies of energy production during aerobic and anaerobic culture. Under anaerobic conditions, the "Pasteur effect" takes place because of a low efficiency of energy production. Therefore, it is expected that commercial heteropolysaccharide production requires an aerobic process. With aerobic fermentations, a major problem is a rapidly increasing media viscosity caused by an increasing heteropolysaccharide concentration. The increasing fermentation beer viscosity in turn causes a markedly decreased oxygen transfer rate between the air bubbles and the culture fluid.

Temperature is often critical in polysaccharide synthesis. In general, the optimum temperature for cell growth is also optimum for product formation. With this optimum temperature, the Q_{10}, or temperature coefficient, for most polysaccharide synthesis is greater than 1, as it is in enzyme-catalyzed reactions.

Another important factor affecting polysaccharide production is pH. During the latter stages of acidic polysaccharide fermentations, such as those involving *Xanthomonas*[7] or *Azotobacter* gums,[8] an increase in acidity occurs because of formation of the uronic acid-containing polymers and of organic acids. Nevertheless, the bacterial polysaccharides of possible commercial significance appear to have an optimum pH for synthesis between 6.0 and 7.5. For fungal gum production, the optimum pH lies between 4.0 and 5.5.

Many microorganisms have a strict requirement for certain mineral elements. Among these elements are K, P, Mg, Mn, and Ca. Other elements, such as Mo, Fe, Cu, and Zn, may also be required. However, in some cases, certain of the minerals that might be added can inhibit product formation. Thus, the mineral requirement for polysaccharide synthesis varies from species to species.

III. Mechanism of Synthesis

1. *Homopolysaccharides*

The mechanism of microbial biosynthesis of such extracellular homopolysaccharides as starch- and glycogen-like products along with the polysaccharides from disaccharide donors has been reviewed.[9–12] Therefore, the mechanism of homopolysaccharide synthesis in microorganisms is discussed here only briefly.

In general, the synthesis of most polysaccharides involves the addition of a single monomeric unit to the nonreducing end of an existing primer polysaccharide molecule. The general reaction for homopolysaccharide synthesis may be represented by the following simple equation:

$$\text{G—O—X} + (\text{G—O})_n\text{—G} \xrightarrow{\text{E}} (\text{G—O})_{n+1}\text{—G} + \text{X}$$

In this equation, G—O—X designates a donor system in which G is the carbohydrate unit of a glycoside and X is a nonpolymeric product, such as pyrophosphate, UDP, or sugar. The symbol E denotes the transglycosidation enzyme, and n is the molar quantity.

In glycogen synthesis, this addition reaction begins with the formation of D-glucose 1-phosphate (G-1-P) from D-glucose 6-phosphate (G-6-P) by the action of the enzyme phosphoglucomutase. The glycosyl phosphate ester reacts with uridine triphosphate (UTP) to yield uridine diphosphate glucose (UDPG) in which the phosphoryl groups of G-1-P and of uridine monophosphate (UMP) are linked by a pyrophosphate (PPi) bond (releasing pyrophosphate).

$$\text{G-1-P} + \text{UTP} \rightarrow \text{UDPG} + \text{PPi}$$

This reaction is catalyzed by a specific enzyme, UDPG pyrophosphorylase. By the action of the enzyme UDPG transglucosylase in the presence of a primer polysaccharide molecule, the activated D-glucopyranosyl units of UDPG are transferred to the nonreducing termini of the primer in new glycosidic linkages. The energy required for the synthesis of these glycosidic linkages is directly supplied by UTP, and the UTP is resynthesized from UDP by a direct reaction with adenosine triphosphate (ATP).

$$\text{UDP} + \text{ATP} \rightarrow \text{ADP} + \text{UTP}$$

Robbins and his coworkers[13] demonstrated, based on the thermodynamics and equilibria of these enzyme reactions, that in animal tissue systems, glycogen synthesis from UDPG is much more efficient than that from G-1-P.

Where X in the above equation is phosphate, as in the case of the starchlike polysaccharides synthesized by a strain of *Neisseria perflava,*[14, 15] *Polytomella coeca*[16] and by *Corynebacterium* and certain streptococci,[17] the original bond energy is conserved in the 4-glucosyl–enzyme complex.

$$\text{G-1-P} + \text{enzyme} \rightarrow \text{D-glucosyl—enzyme} + \text{Pi}$$

Calculations have revealed that the energy of the C—O—P linkage of G-1-P is 4800 cal and is comparable to that of the glycosidic linkage of the polysaccharides.[18, 19]

For many polysaccharides, such as dextran and levan synthesized from disaccharide donors, X in the equation is a single monomeric sugar unit D-fructofuranosyl for the former and D-glucopyranosyl for the latter. It was clearly demonstrated that the enzymes dextransucrase and levansucrase function in the absence of any detectable amount of inorganic phosphate. It was also found that dextransucrase has no detectable catalytic action on inorganic phosphate and G-1-P,[20] or on G-6-P, D-fructose 6-phosphate (F-6-P), D-fructose, 1,6-diphosphate (FDP), and ATP.[21] Likewise, the levansucrases produced by *Aerobacter levanicum,*[22] *Bacillus polymyxa,* and *B. subtilis* are not affected by inorganic phosphate, F-6-P, or FDP.[23, 24] Most *Neisseria* strains, except the one mentioned above, have an amylosucrase that catalyzes the polymerization of the D-glucopyranosyl unit of sucrose to yield a starchlike polysaccharide. Again, in this pathway, the intermediary formation of G-1-P is not involved.[25, 26]

The evidence indicates that dextransucrase, levansucrase, and amylosucrase directly transfer a single monomeric sugar unit that originated from the disaccharide donors.

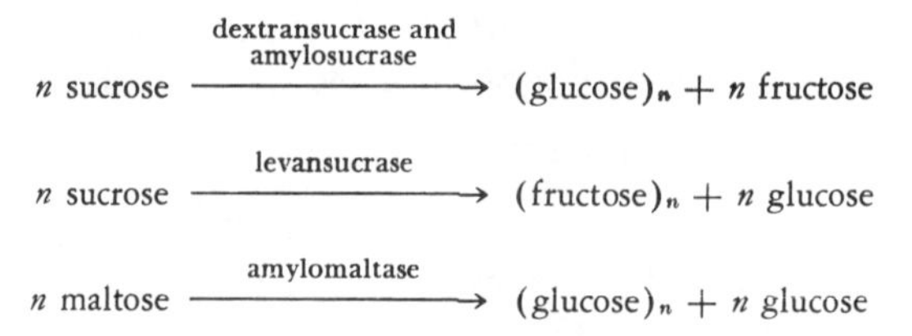

The production of a starchlike polysaccharide by an amylomaltase obtained from mutants of *Escherichia coli*[27, 28] is another example of transglycosylation. In this case, maltose serves as the donor substrate whereas sucrose and raffinose are substrates for levansucrase. It should again be noted that the monomeric units from the donor molecules are transferred to a suitable acceptor through the mediation of a monomer–enzyme complex, and in this complex, the energy of the glycosidic bond is maintained.

In the biosynthesis of branched polysaccharides, such as glycogen- and starch-like substances as well as levan and dextran, it is believed that the branching results from additional enzymes in the reaction mixture. Although dextran synthesized by *Leuconostoc dextranicum* appears not to be branched, those dextrans synthesized by other organisms, such as many strains of *L. mesenteroides*[29] and *Betacoccus arabinosaceus*,[30, 31] which is a strain of *L. mesenteroides*, are highly branched. The Q enzyme, which is capable of converting linear α-D-(1→4)-linked glucans to branched glucans, has been found in *Neisseria perflava*[14] and *Polytomella coeca*.[32] Therefore, it is concluded that branching enzymes introduce the branch points into a growing polysaccharide chain.

2. *Heteropolysaccharides*

Attempts at complete elucidation of the mechanism of microbial heteropolysaccharide synthesis for *in vitro* systems have not been completely successful. Consequently, there is no direct experimental evidence by which heteropolysaccharide synthesis can be completely explained.

The ability to synthesize a particular heteropolysaccharide appears usually to be a stable genetic character as was demonstrated by a genetic transformation reaction mediated by DNA.[33] In support of the conclusion, it was reasoned that the heteropolysaccharides normally are produced within a quite narrow range of molecular weights.[1] Furthermore, *Klebsiella aerogenes* type 54 produces an extracellular polysaccharide that is specific as to composition and characteristic of the generating bacterial strain, regardless of the variety of carbohydrates used as the sole carbon source.[34]

It appears that changes in composition or molecular size of most heteropolysaccharides can be brought about only by mutation in the generating bacteria. In fact, Wilkinson[1] questioned whether heteropolysaccharide biosynthesis is enzymic or "whether the failure to find enzymes is because of their nonexistence." In this regard, he proposed a template hypothesis in which DNA provides the initial replica for synthesis. The hypothesis states that only one gene could control the template mechanism. A similar hypothesis was advanced more recently by McMullen.[35] However, Anderson[9] pointed out that "Wilkinson's hypothesis does not obviate the need for several enzymes in the preparation of various glycosyl donors from a single monosaccharide source." Nucleotide sugars have been demonstrated to be glycosyl donors in heteropolysaccharide synthesis. The associated reactions, particularly for capsular polysaccharides, have been reviewed recently by Neufeld and Hassid[36] and by Horecker.[37]

It becomes apparent from several lines of experimental evidence that heteropolysaccharide synthesis is catalyzed by a multienzyme system and that each enzyme assumes a high degree of specificity. Type III pneumococcal capsular polysaccharide can be synthesized in a cell-free, crude enzyme preparation from UDP-glucose and UDP-glucuronic acid but not from G-1-P or D-glucose.[38] It was shown also that synthesis of capsular polysaccharide by some pneumococcal strains is closely linked to an intact pathway of UDP-glycosyl metabolism and that the lack of a single enzyme in the pathway results in a block in the synthesis of such polysaccharides.[39] Similarly, UDP-glucuronic acid[40] and UDP-*N*-acetylglucosamine (UDP-2-acetamido-2-deoxy-D-glucose)[41] have been established as the precursors of hyaluronic acid. Bernstein and Robbins[42] demonstrated that the formation of the nucleotide sugar is mediated by a specific enzyme. Furthermore, the enzymes that catalyze the transfer of glycosyl units from nucleotide diphosphate sugars to acceptor molecules exhibit absolute specificity with respect to the linkage position and anomeric configuration.[43]

Wright and coworkers[44] reported the synthesis of long sequences of mannosyl–rhamnosyl–galactosyl units by cell-free systems of *Salmonella newington*. In the presence of UDP-galactose and TDP-rhamnose, but in the absence of GDP-mannose, D-galactose and L-rhamnose were incorporated into a L-rhamnosyl-(1→3)-D-galactosyl phosphate. Addition of UDP-mannose to the cell-free system containing the derivative resulted in the incorporation of mannosyl-rhamnosyl–galactosyl repeating units into a polysaccharide. These results suggest a mechanism of biosynthesis of heteropolysaccharides based upon synthesis of subassembly units by a group of enzymes. Other enzymes then assemble these units into the heteropolysaccharide.

Primer molecules play an important role in polysaccharide biosynthesis. As far as it is known, the primer cannot be completely excluded from the microbial system, although hyaluronic acid synthesis by the protoplast membrane of a Group A *Streptococcus* appears to be an exception.[45]

Recent work on cell-wall polysaccharide synthesis in cell-free systems has revealed that elements of the cytoplasmic membrane are involved in the synthetic process.[43] However, it apparently is not known whether the membrane is also involved in the synthesis of extracellular slimes and, if it is, how it functions.

IV. Bacterial Polysaccharides

1. *Arthrobacter Polysaccharides*

Arthrobacter sp. NRRL B-1973 produces a high-viscosity, extracellular heteropolysaccharide during fermentation with D-glucose as the carbon source.[17] The medium contains 2–4% D-glucose, 0.3% hydrolyzed casein, dipotassium hydrogen phosphate, and traces of magnesium and manganese salts and is adjusted to a pH of 7.0. Yields of 40–45% based on dextrose have been obtained in a 4 day aerobic fermentation at 25°. After clarification, the native polysaccharide B-1973 is recovered by precipitation from a minimum addition of 70% ethanol by volume.

This polysaccharide contains D-glucose, D-galactose, and D-mannuronic acid (as a salt) in the molar ratio of 1 : 1 : 1.[46] In addition, the linear polysaccharide is highly acetylated with 25% by weight of *O*-acetyl groups present. The structure is of particular interest in that, aside from its presence in algin, D-mannuronic acid is rare in polysaccharides.

The high proportion of acetylated groups in the native polysaccharide interferes with hydration and molecular alignment to the extent that poor films are formed when aqueous solutions are dried. However, flexible films are formed after deacetylation.

Native polysaccharide B-1973 solutions are quite pseudoplastic. Also, they will not tolerate as high concentrations of some salts as will solutions of xanthan gum. Deacetylation of polysaccharide B-1973 increases its sensitivity to precipitation by salts and water-miscible solvents. Heating or autoclaving the native polysaccharide causes first an increase in viscosity caused by cross-linking and then degradation.

Polysaccharide B-1973 has not been produced commercially. Although the high viscosity and pseudoplasticity of its solutions are similar to those of xanthan gum, its potential market would be much smaller because of its greater salt sensitivity, its cross-linking tendencies on heating, and the lack of toxicity studies.

2. *Starchlike Polymers*

Many strains of *Neisseria* produce starchlike extracellular polysaccharides from sucrose.[15] The product is a mixture of a linear α-D-(1$\rightarrow$4)-linked fraction termed amylose and a branched chain portion resembling amylopectin or glycogen. In these biosyntheses, the enzyme amylosucrase transfers D-glucopyranosyl units from the sucrose molecule to the end of the growing polymer chain.

A mutant strain of *E. coli* produces an amylose from a specific substrate, maltose. This biosynthesis, catalyzed by amylomaltase, is reversible. If the freed nonreducing glucose moiety in maltose is allowed to accumulate, oligosaccharides having a chain length of approximately 10 units predominate. If the D-glucose is removed as it is formed, the reaction proceeds to completion, and high-molecular-weight amylose is formed.

Starch is available from plant sources at low prices and in very large amounts. It is unlikely, therefore, that the microbial starchlike polymer could develop into any significant commercial production.

3. *Bacterial Cellulose*

In 1886, Brown[47] reported that *Acetobacter xylinium* elaborated tough membranes of cellulose when grown in a suitable nutrient solution. Cellulose formation also has been reported for *Acetobacter pasteurian, A. rancens, Sarcina ventriculi, and Bacterium xyloides.*[48] In a study of this reaction, Tarr and Hibbert[48] established the preferred substrate for *A. xylinium* as 5–10% of a hexose sugar, 0.1% L-asparagine, 0.5% potassium dihydrogen phosphate, 0.1% sodium chloride, and 0.5% ethanol. Aerobic fermentation for 10 days at 30° formed extracellular cellulose as a meshwork of intertwined microfibrils;[49] the product is identical with plant cellulose.[50]

A highly specialized use for bacterial cellulose in membranes for osmometry has been reported.[51] Nevertheless, because cellulose is available in great quantities from plant sources at very low prices, it is unlikely that bacterial cellulose will attain any appreciable production volume.

4. *Levan*

Bacterial fructan or levan is produced by a number of microorganisms. These include species of *Bacillus, Pseudomonas, Acetobacter,* and *Corynebacterium.* Levans are polymers of β-D-fructofuranosyl units containing mainly β-D-(2→6)-glycosidic linkages.

The mechanism of levan biosynthesis from sucrose is similar to that of dextran. *Bacillus subtilis,* as an example, produces an extracellular inducible enzyme, levansucrase, that transfers the D-fructofuranosyl unit of sucrose to the growing levan chain. D-Glucose accumulates in the solution along with some free D-fructose. Biosynthesis of levan is readily accomplished with cell-free extracts of *Bacillus subtilis* by the addition of sucrose.

Like most dextrans, levans are branched homopolysaccharides having molecular weights in the range[12] of $1–100 \times 10^6$. In spite of the high molecular weight, the viscosities of aqueous solutions of levans are considerably lower than those of linear polysaccharides, such as algin. Thus, levans appear to be unlikely prospects for development as commercial water-soluble gums.

5. Bacillus polymyxa *Polysaccharide*

As the name of the organism indicates, *B. polymyxa* has long been known as a polysaccharide producer. Recently, Cadmus and coworkers[52] reported the production of its polysaccharide by fermentation. Under the authors' conditions, the organism could utilize 3% D-glucose within 72–96 hr with 30–35% conversion. An apparently different strain of *B. polymyxa,* isolated by Ninomiya and Kizaki,[53] converts 5% D-glucose into polysaccharide with an efficiency of 36% after a 62 hr fermentation. In our laboratory, fermentation conditions have been devised in which 3% D-glucose can be completely utilized within 20–24 hr with a satisfactory conversion efficiency. Under these rapid fermentation conditions, microbial contamination is not a major problem because the native bacteria outgrow most possible contaminants.

Bacillus polymyxa slime is a heteropolysaccharide composed of D-glucose, D-mannose, D-galactose, and D-glucuronic acid in a molar ratio[53] of 3 : 3 : 1 : 2. Solutions of the gum exhibit relatively high viscosity and high pseudoplasticity. The viscosity increases sharply in aqueous solution when various salts are added, but it decreases drastically with a small increase in temperature. As the organism is a spore former, the gum should be further processed in most cases in order to eliminate spores. This has discouraged commercial development of *B. polymyxa* polysaccharide.

6. Azotobacter *Polysaccharides*

A number of strains of *A. indicus* and *A. vinelandii* produce copious amounts of extracellular acidic slimes. It is interesting that *A. agilis* does not produce such slimes, but the polysaccharide is confined to the bacterial capsule.[54]

In Burk nitrogen-free broth, a strain of *A. vinelandii* produces an extracellular polysaccharide consisting of D-mannuronic acid accompanied by small amounts of L-glucuronic acid.[55] Cohen and Johnstone[8] showed a relation between pH and slime formation in Burk broth by several strains of *A. vinelandii.* They found that all slime-forming strains lowered the pH of the medium, whereas non-slime formers did not.

Patrikh and Jones[56] isolated an extracellular polysaccharide from *A. indicus* and found it to be a repeating unit of D-glucuronic acid, D-glucose, and D-*glycero*-D-*manno*-heptose.

Others[57] reported that three strains of *A. vinelandii* produce extracellular polysaccharides that consist of galacturonic acid, D-glucose, and rhamnose in an approximate molar ratio of 43 : 2 : 1. Only a minor difference was observed among the strains.

Little work has been done on the properties of the *Azotobacter* polysaccharides, but the efficiency of *Azotobacter* gum production, under the present experimental conditions, is too low to be of any commercial significance.

7. Alcaligenes *Polysaccharides*

Harada[58] isolated *A. faecalis* var. *myxogenes* 10C3 from soil and found that it produces a large amount of extracellular polysaccharide from D-glucose or ethylene glycol. The polymer was named *succinoglucan 10C3,* and it is composed of 70–80% D-glucose, approximately 10% succinic acid, and small amounts of galactose and mannose. The glycosidic linkages appear to be β-D-(1→2) and (1→3). This polymer obtained with D-glucose as the carbon source has a rather high viscosity.

A mutant derived from *A. faecalis* var. *myxogenes* 10C3 (mutant K) produces a homopolysaccharide known as *curdlan.* This polymer consists of a nonbranched linear chain of D-glucopyranosyl units joined entirely by β-D-(1→3) glycosidic linkages.[59] In a defined or in a semisynthetic medium, the organism converted 50% of added D-glucose into curdlan after 5 days incubation[60] at 32°. This polysaccharide is neutral and is neither soluble in water nor in an acidic solution, but it dissolves in alkaline solutions, such as 0.5*N* potassium hydroxide. Curdlan forms a resilient gel when its water suspension or alkaline solution is heated to ~80° for a few minutes followed by cooling to ~25°.

8. *Miscellaneous Bacterial Polysaccharides*

Several species of *Rhizobium* are slime producers, notably *R. trifolii, R. leguminosarium,* and *R. meliloti.*

Early work for component analysis of the extracellular polysaccharide produced by *R. radicicolum* (clover strain) revealed 67% D-glucose and 23% aldobiouronic acid.[61]

Corpe[62] described the formation of an extracellular polysaccharide by *Chromobacterium violaceum.* The organism in peptone broth converted D-glucose into a polysaccharide that consisted of D-glucose, uronic acid, and aminosugar in an approximate molar ratio of 5 : 1 : 1. It was found that growth and polysaccharide synthesis in a defined medium were markedly enhanced when amino acids were substituted for ammonia as a nitrogen source, and the best polysaccharide yield occurred when the ratio[63] of carbohydrate to nitrogen was 10 : 1. The polysaccharide exhibited an unusual resistance to degradation by soil microorganisms[62] and some interesting rheological properties, but the product yield was much lower than in *Xanthomonas* polysaccharide fermentations.

V. Yeast Polysaccharides

1. *Phosphomannan*

Yeasts of the genus *Hansenula* and related genera produce extracellular phosphomannans from D-glucose.[64] These polysaccharides yield on hydrolysis only D-mannose and D-mannose 6-phosphate. Developmental work was done at the

Northern Utilization Research and Development Division of the U.S. Department of Agriculture on the phosphomannan produced by *Hansenula holstii* NRRL Y-2448.[65] This phosphomannan was produced in an aerobic 4 day fermentation[66] at 28° in a medium containing 6% D-glucose, organic nitrogen, potassium dihydrogen phosphate, and trace elements at pH 5.0. The product was isolated by alcohol precipitation from the fermentation beer. Yields were 40–55% based on D-glucose. *H. capsulata* Y-1842 also produces a phosphomannan.[67]

Phosphomannan from *Hansenula holstii* Y-2448 contains a repeating unit of five D-mannopyranosyl units connected (1→6) by an orthophosphate unit.[65] The molecular weight by light scattering is $\sim 16 \times 10^6$.

Soft gel-like solutions are formed by the phosphomannan at concentrations of >1% in water. The viscosity versus concentration curve is unusual for a water-soluble gum in that the viscosity reaches a maximum of about 2000 cps at a 1.25% concentration with no further increase in viscosity up to a concentration of 5%. Solutions are thixotropic and quite sensitive to salts. Weak acids split the diester cross-link and result in loss of viscosity.[68]

Development of phosphomannan as a commercial water-soluble gum would be difficult because of its sensitivity to salts, instability in acid solutions, and relatively slow formation.

2. *Polysaccharide Y-1401*

Various species of *Cryptococcus* and *Tremella* produce extracellular heteropolysaccharides. Like extracellular polysaccharides in general, biosynthesis of these polysaccharides is a stable genetic property. As a result, the individual polysaccharides are unique in structure and are characteristic of the generating species.[69]

A *Cryptococcus* polysaccharide was produced by *C. laurentii* var. *flavescens* NRRL Y-1401 in a 6 day aerobic fermentation at 25° with 5% D-glucose substrate in the presence of autolyzed brewer's yeast and inorganic salts.[70] The yield was 30–35% based on D-glucose. Polysaccharide Y-1401 contains D-mannose, D-xylose, D-glucuronate, and acetyl in the molar ratio 8 : 2 : 2 : 3. Certain *Tremella* heteropolysaccharides contain the same components found in the *Cryptococcus* polymers.[69] These polysaccharides have not been commercialized.

VI. Fungal Polysaccharides

1. Plectania *and* Helotium *Polysaccharides*

Some species of fleshy fungi produce extracellular β-D-glucans from D-glucose and in certain cases from sucrose. *Plectania occidentalis* NRRL 3137 and *Helotium* sp. NRRL 3129 utilize both D-glucose and sucrose.[71] The mycelium from *P. occidentalis* NRRL 3137 gives 30% conversion of D-glucose to the biosynthetic polysaccharide in 10 days of submerged aerobic fermentation. The pH is maintained at 6.0 and the temperature at 25°; adjuncts are hydrolyzed soy protein, salts, and thiamine. *Helotium* sp. NRRL 3129 gives a 50% conversion of D-glucose to the extracellular β-D-glucan in 3 days with the same medium and conditions as for *P. occidentalis* NRRL 3137. Recovery of the polysaccharide is by alcohol precipitation after removal of the mycelium.

Both polysaccharides NRRL 3137 and 3129 yield only D-glucose on hydrolysis. They are mainly β-D-(1→3)-glucans with side chains attached by β-D-(1→6) bonds. Polysaccharide NRRL 3129 is somewhat more complex than NRRL 3137 and may contain some (1→4) or (1→2) glucosidic bonds interspersed with regions of (1→3) bonds.

Solutions of polysaccharides NRRL 3137 and 3129 at concentrations of 1% or greater behave as thixotropic gels. These solutions have relatively good salt compatibility and heat stability.

The extracellular β-glucans differ sharply from most extracellular heteropolysaccharides in that the glucans serve as reserve carbohydrates. If the fermentation is not stopped at the right time, little or no yield of β-glucan is obtained.

2. Sclerotium *Polysaccharides*

Polysaccharides are produced by many species of *Sclerotium* in media containing sucrose as carbon source, autolyzed yeast and salts.[72, 73] Various carbohydrate sources are utilized by these organisms (see also Chapter XXII.)

With *S. glucanicum* NRRL 3006, approximately 0.6% of polysaccharide is obtained from 3% of added sucrose or D-glucose at 28° in a 5-day fermentation period. The finished fermentation beer is diluted with water and then filtered to remove fungal cells.

The polysaccharide scleroglucan produces a viscous gel when dispersed in water at a concentration of 1.5% and is pseudoplastic.

3. *Miscellaneous Fungal Polysaccharides*

Bouveng and coworkers[74] reported the formation of a water-soluble polysaccharide in an *Armillaria mellae* culture grown in a broth containing D-glucitol as the carbon source. They were able to fractionate this gum into four fractions. Two fractions consisted of a glycogen-type glucan and a β-D-(1→3)-glucan, respectively. The third fraction contained a glucan and a xylomannan, and the fourth was a heteropolysaccharide from which was released upon hydrolysis D-galactose, D-mannose, L-fucose, and 3-*O*-methyl-D-galactose in a molar ratio of 6 : 1 : 2 : 2.

Szaniszlo and coworkers[4] reported the isolation of a capsular polysaccharide from *Leptosphaeria albopunctata,* a marine fungus. It is interesting that polysaccharide synthesis by this organism requires concentrations of salts similar to those in seawater and, moreover, that the synthesis is enhanced by sodium chloride concentrations above that of artificial seawater. Use of an extracellular, slime-forming, marine fungus such as this would appear to be desirable, because the use of seawater as a fermentation medium should result in a lower production cost. However, organisms of this type that produce sufficient gum to have industrial significance have not been reported.

Buck and coworkers[75] reported an extracellular glucan produced by a *Claviceps* species. The polysaccharide was a branched glucan consisting of a β-D-(1→3)-linked main chain with single D-glucopyranosyl units as branches linked β-D-(1→6) at intervals along the main chain.

An excellent review of the structural chemistry of yeast and fungal polysaccharides has been published.[76]

VII. Xanthan Gum

1. History

It has long been known that many species of *Xanthomonas* produce viscous gummy colonies. More than 100 species of *Xanthomonas* have been reported in the last 80 years, but these publications contain a large number of duplications.[77] Of the remaining distinct species, *X. campestris, X. phaseoli, X. malvacearum,* and *X. carotae* are reported to be the best producers of extracellular polysaccharides.[78]

The Northern Utilization Research and Development Division of the U.S. Department of Agriculture conducted an extensive evaluation of their culture collection for water-soluble gum producers of possible commercial importance. Of the various biosynthetic polysaccharides suggested by them, polysaccharide B-1459 (xanthan gum) produced by *Xanthomonas campestris* NRRL B-1459 appeared to have the most interesting properties, properties that would allow this biosynthetic polysaccharide to supplement rather than compete with natural and synthetic water-soluble gums.

The Kelco Company, San Diego, California placed xanthan gum under the trade name, *Kelzan,* in pilot production in 1960, in semicommercial production in 1961, and in substantial commercial production in early 1964.

2. Production

Literature methods.—A study[79] of the nutritional requirements of 30 species of the genus *Xanthomonas,* including *X. campestris,* was reported in 1946. A simple medium including D-glucose, ammonium chloride, phosphate buffer, magnesium sulfate, and trace salts met the minimal growth requirements. The addition to 0.5% casein hydrolyzate, however, gave much better growth than was given by the simple medium. A mixture of amino acids can replace the casein hydrolyzate (unpublished data).

Lilly and coworkers[78] studied the conditions providing the best yields of extracellular polysaccharides by *X. phaseoli* and *X. campestris.* In an aerobic fermentation at 28°, concentrations of D-glucose ranging from 1% to 5% were suitable. The higher concentration of D-glucose, however, gave poor conversions to polysaccharide. D-Glucose, sucrose starch, and hydrolyzed starch were found to be about equally useful as substrates. Either hydrolyzed casein or soy bean protein was suitable as an organic nitrogen source. Various antifoam agents also were evaluated.

A kinetic examination of xanthan gum fermentation was made by Moraine and Rogovin.[7] Growth of *X. campestris* took place during the early phase of the fermentation period, whereas polymer synthesis occurred throughout the period. The rate of polymer formation increased with increasing cell concentration; and the cell concentration, in turn, was doubled by doubling the concentration of distillers solubles in the medium. Because certain assimilable nitrogen compounds in the distillers solubles were consumed exclusively for growth, it was suggested that polymer formation might be separated from cell growth by means of multistage continuous fermentation.

Rogovin and coworkers[6] reported on conditions for xanthan gum production by *X. campestris* using conventional aerated and agitated pilot plant fermentors. For inoculum buildup, a vigorously growing slant culture was transferred to YM broth in flasks and incubated for 24 hr on a shaker. This culture was used to inoculate seed flasks containing the final medium composition. Inoculum levels for intermediate stages and the final fermentation were suggested as 10% and 5%, respectively. The final medium contained 2.5–3.0% D-glucose, 0.4% dried distillers solubles, 0.5% dipotassium hydrogen phosphate, and 0.01% magnesium sulfate heptahydrate. Oxygen absorption levels were adjusted to 1.0 mmole oxygen/liter/min when measured according to the sulfite oxidation method in water.[80] With the temperature held at 28°, approximately 50% conversion of D-glucose to xanthan gum was obtained in 96 hr. The final fermentation beer was diluted with water to lower the viscosity and centrifuged to partially remove cells; the gum was precipitated at a concentration of 50% (w/w) of methanol in the presence of 2% (w/w) of potassium chloride.

Alternative recovery methods have been proposed. These include drum drying or spray drying for a technical product.[81] With drum drying, difficulty is encountered because of the tenacious adherence of xanthan gum to the metal drying surface.

Long-chain quaternary ammonium salts have been proposed as precipitants.[82] This process however, requires, essentially complete recovery of the relatively expensive quaternary ammonium salt. In addition, it is doubtful that such a process could be used in producing a food-grade product. Precipitation of the xanthan gum as a calicum salt above pH 10 and subsequent recovery of the polysaccharide have been patented by McNeely and O'Connell.[83, 84]

Variations of continuous fermentation processes for xanthan gum have been suggested.[85, 86] These processes work well in laboratory equipment, and they will offer economic advantages if they can be scaled up. However, the overriding problem of maintaining sterility in commercial batch fermentors may make a commercial continuous fermentor installation a high risk with the medium used in xanthan gum fermentation.

Commercial methods.—For commercial production, a mutant strain of *Xanthomonas campestris,* maintained as a lyophile in a nutrient medium, is transferred to agar slants. After passing through several serial transfers with associated identification tests, the inoculum is increased in several steps to a large shake flask inoculum. This seed is used to inoculate an aerated and agitated seed fermentor. At the proper time, the seed is forced by sterile air into the sterile medium in the final fermentor.

Raw materials.—D-Glucose, sucrose, and other forms of carbohydrates may be used as substrates, depending upon the grade desired. A protein supplement and an inorganic nitrogen source are necessary for efficient xanthan gum production. Small amounts of phosphate and magnesium ions are required; necessary trace minerals are usually furnished in the water supply.

Fermentation.—Because the pH and medium employed are also suitable for the growth of a number of microorganisms other than *X. campestris* and large amounts of sterile air are required, extreme precautions must be taken to avoid contamination. Precise control of the oxygen transfer rate is critical during the 2 day fermentation. This involves proper design of the agitation system to achieve good oxygen transfer at the high viscosities encountered. The fermentation is done at pH 6.0–7.5 and at 28°–31°.

Controls.—The carbohydrate content of the fermentation beer is followed by standard methods. Xanthan gum content is determined by precipitation and colorimetric methods. Careful microbiological control is maintained throughout the process.

Recovery.—Several methods of recovery are useful, depending on the type of product and its end use. Each process starts with pasteurization of the fermentation beer in a continuous heat exchanger to kill the bacterial cells. Because xanthomonads are nonspore formers and the vegetative cells are sensitive to high temperatures, a complete kill is readily obtained. For clear products, the beer is diluted and clarified by filtration or other methods before recovery. For standard products, the xanthan gum content of the beer may be recovered by precipitation with 2-propanol or as an insoluble salt. In the isopropyl alcohol process, practically complete recovery of the alcohol is essential to the economics of the process. Recovered xanthan gum is dried, milled, tested, and blended to specifications.

Grades available.—The biosynthesis of xanthan gum is genetically controlled both as to composition and molecular weight. Hence, a product of outstanding uniformity for a water-soluble gum is obtained. Xanthan gum is now available in a number of commercial grades for industrial applications.

The Kelco Company's food-grade xanthan gum was the subject of extensive feeding studies, which established its safety for use as a food ingredient. The Food and Drug Administration on March 19, 1969, issued a food additive regulation authorizing its unrestricted use in foods for which standards of identity did not preclude such use.

3. *Derivatives*

A number of derivatives have been reported for xanthan gum. These include deacetylated xanthan gum,[87] the carboxymethyl ether,[88] the propylene glycol ester,[89] cationic derivatives,[90] and formalydehyde cross-linked products.[91]

Very few of the many patented derivatives of water-soluble gums have been put into production. Commercialization of these few, such as propylene glycol alginate,[92] has been dependent on developing an inexpensive manufacturing process that is compatible with the main production process and on unique properties of the derivative.[93] The deacetylated derivative meets these requirements, and a partially deacetylated xanthan gum is available commercially.

4. *Structure*

Xanthan gum is a heteropolysaccharide with a molecular weight[94] of $>10^6$. It contains D-glucose, D-mannose, and D-glucuronate[95] in the molar ratio of 2.8 : 2.0 : 2.0. The polysaccharide is partially acetylated with $\sim$4.7% acetyl.

Xanthan gum is unusual in that it contains pyruvate. The pyruvate, amounting to $\sim$3% of the total weight, is attached to a single unit D-glucopyranosyl side chain as a ketal.

Xanthan gum contains a basic repeating unit of 16 units, of which 13 are present in a main linear chain and 3 are attached as single-unit side chains. The D-glucopyranosyluronic acid units are linked β-D-(1→2) to D-mannopyranosyl units. The D-glucopyranosyl and D-mannopyranosyl units are joined to the next unit in the linear main chain by (1→4) linkages; all these linkages appear to have the β-D-configuration. D-Mannopyranosyl single-unit side chains are attached to the main chain by (1→2) or (1→3) glycosidic bonds. A proposed structure of xanthan gum has been given (Fig. 1). In addition to sugar units, acetyl groups, pyruvate acetal (1-carboxyethylidene) groups, and monovalent cations are present.

[→4)-β-D-Glc*p* A(1→2)-D-Man *p*(1→4)-D-Glc*p* (1→4)-β-D-Glc*p*A(1→2)-D-

Man*p*(1→4)-β-D-Glc*p*(1→4)-D-Glc*p*(1→4)-β-D-Glc*p*A(1→2)-D-Man(1→4)-β-D-Glc*p*A(1→
3 or 2
↑
1
D-Man*p*

2)-D-Man*p*(1→4)-β-D-Glc*p*(1→4)-D-Glc*p*(1→]$_n$
3 or 2
↑
1
D-Man*p*

Fig. 1.—Proposed basic repeating unit of xanthan gum. Also present in the 16 unit repeating unit are 3.4 moles of *O*-acetyl groups, 5.0 moles of potassium and sodium ions, and 1.0 mole of 4,6-*O*-(1-carboxyethylidene)-D-glucopyranose as a single unit side chain.[95]

5. *Properties*

Viscosity.—Xanthan gum dissolves readily in either hot or cold water upon mechanical stirring to give high viscosity solutions at low concentrations. A concentration of 1% gives a viscosity of approximately 1000 cps when measured at 60 rpm with a Brookfield Model LVF viscometer at ~25° (Fig. 2).

Pseudoplasticity.—Solutions of xanthan gum in water are highly pseudoplastic; thus, the measured viscosity decreases rapidly as the shear rate is increased. This is an instantaneous and reversible process. Solutions containing 0.75% or more of xanthan gum have a rheological yield point. The pseudoplastic nature of

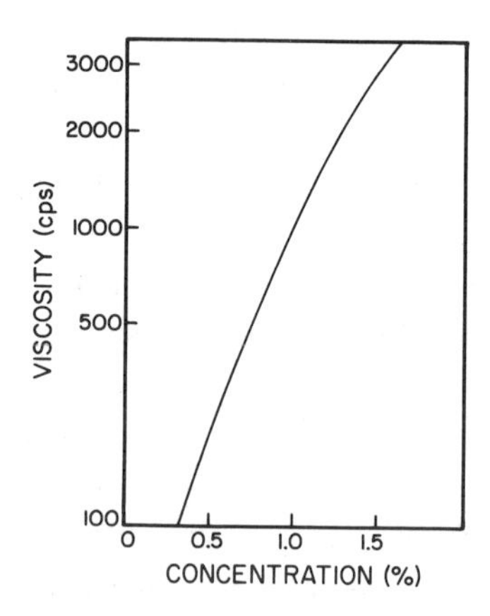

Fig. 2.—Viscosity of aqueous solutions of xanthan gum measured at 60 rpm with a Brookfield Model LVF viscometer.

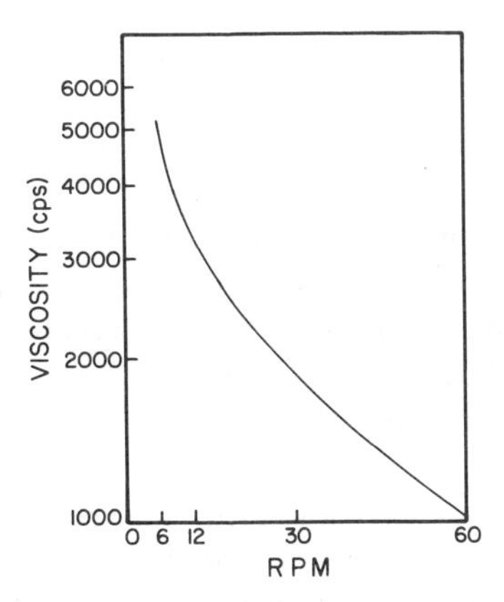

Fig. 3.—Change in viscosity of xanthan gum with shear rate as measured with a Brookfield Model LVF viscometer.

xanthan gum in solution is illustrated by the dramatic reversible decrease in viscosity of the solutions with increasing rotational speed (Fig. 3).

Temperature effect on viscosity.—A remarkable property of xanthan gum is that the viscosity of its aqueous solutions is nearly independent of temperature over a wide temperature range. Between 10° and 70°, a xanthan gum solution having a viscosity of 1000 cps will change in viscosity by not more than 100 cps.

pH Effect on viscosity.—The viscosity of aqueous solutions of xanthan gum is nearly independent of pH between pH 6 and 9 and shows only minor variations in viscosity over a pH range of 1 to 11. At pH 9 or above, xanthan gum is gradually deacetylated, but this deacetylation has little effect on its solution properties.

Compatibility.—Xanthan gum has unusually good compatibility with high concentrations of many salts, and it will dissolve directly in moderate concentrations of a wide range of salts. When the salt or salt solution is added to a predissolved xanthan gum, the gum is compatible; for example, it is compatible with 25% solutions of aluminum sulfate, calcium chloride, or zinc chloride and with 15% solutions of sodium chloride. Incompatibilities are few; calcium salts at pH >10 will precipitate xanthan gum,[84] and long-chain quaternary ammonium salts or amines with more than eight carbon atoms in the main chain may precipitate it.[82]

Xanthan gum will dissolve directly in 5% solutions of sulfuric acid, nitric acid, and acetic acid, in 10% solutions of hydrochloric acid, and in 25% solutions of phosphoric acid. Such solutions are reasonably stable for several months unless the temperature is elevated. Xanthan gum will dissolve directly in 5%

sodium hydroxide solutions. Thickened 10–15% sodium hydroxide solutions can be obtained by adding 50% sodium hydroxide solutions to predissolved xanthan gum. Such highly alkaline solutions have surprising viscosity stability.

Concentrations of most water-miscible solvents of up to 40–45% are compatible with xanthan gum.

Stability to heat.—Xanthan gum is remarkably resistant to degradation by heat. Long exposures to temperatures as high as 80° appear to have little effect on xanthan gum solutions. The resistance to degradation by heat is improved by the presence of salts. Thus, gum solutions in the presence of a small amount of a salt such as potassium chloride can be autoclaved at 121° for 15–30 min with only minor viscosity changes.

Film formation.—Xanthan gum forms strong films on evaporation of its aqueous solutions. These films adhere strongly to glass and many metal surfaces if formed by evaporation in contact with the surface. Deacetylated xanthan gum forms stronger and more flexible films than the native polysaccharide.[87]

Properties of particular interest in foods.—Solutions, emulsions and gels containing xanthan gum have excellent freeze-thaw resistance. Compatibility of xanthan gum with food-grade salts and acids is unusually good. Obtained are exceptionally stable xanthan gum suspensions and emulsions having extreme resistance to heat and viscosities almost independent of temperature.

Other properties.—The equivalent weight of xanthan gum, on a dry basis, after transformation to the free acid is ~725. The specific optical rotation is close to zero. Xanthan gum cross-links with trivalent ions, such as chromium ions, at an alkaline pH.

6. *Industrial Applications*

Textile printing and dyeing.—The excellent compatibility and pseudoplastic nature of xanthan gum in solution have made this gum useful as a textile print paste thickener. It has found exceptionally good reception as a thickener in space dyeing of yarn to achieve a nonpatterning effect in rugs. The low viscosity at high shear rates provides color migration control in the yarn. The concentrations of xanthan gum in the solutions used are ~0.5% in space dyeing and ~1% in print pastes.

Oil well drilling fluid additive.—The excellent compatibility of xanthan gum with salts together with its unequaled resistance to degradation by temperature have made it useful as an additive to oil well drilling fluids. Xanthan gum at low concentrations gives the drilling fluid low water-loss properties. These drilling fluids then have high carrying capacity, and they reduce friction losses in the drill string. Adding a trivalent chromium salt to the drilling fluid containing the xanthan gum and adjusting the pH to >7 improves the efficiency of the xanthan gum and further reduces water loss to the strata.[96] Concentrations of xanthan gum used range from 0.5 to 1.5 lbs/bbl (140 to 430 g per 100 liters) of drilling fluid. The chromium salt is added at about one-third the xanthan gum level.

Viscous water flooding.—Maximum secondary recovery of oil is achieved in viscous water flooding by matching the viscosity of the crude oil with that of the viscous water flood.[97] Such thickened water may encounter salts and may be subjected to elevated temperatures in the underground formation for periods of a year or more. At the same time, the additive must not plug the small pores of a tight formation. Xanthan gum with its exceptional resistance to heat degradation, its excellent compatibility, its high viscosity, and its pseudoplastic flow properties is the only polysaccharide that promises to meet these severe requirements. The use level of xanthan gum ranges from about 0.03% to 0.15%.[98]

Ceramic glaze and related suspensions.—Xanthan gum is an excellent suspending agent in ceramic glazes and many hard-to-suspend mixtures. It is compatible with all ingredients in ceramic glazes, promotes grinding, improves wet adhesion, controls the drying time, and reduces imperfections.[99] It is used at a concentration of about 0.5% of the total weight of the glaze.

Cleaners and polishes.—The excellent compatability of xanthan gum with ingredients together with its exceptional suspending and emulsifying properties make it the ideal stabilizer for difficult-to-suspend cleaners and polishes. Insolubles such as abrasives are suspended indefinitely, yet the product is easy to pour or spread because of the pseudoplastic properties of xanthan gum solution. The concentrations used range from 0.1% to 1.0%.

Latex emulsion paints.—Xanthan gum is used in both conventional flat latex paints and dripless paints because of its good wettability, easy dispersion, and high pseudoplasticity. Pigments in paints containing xanthan gum at levels of 3–6 lb per 100 gal (360–720 g per 100 liters) of paint are suspended perfectly in the can, yet the paints have excellent workability combined with dripless properties.

7. *Food Applications*

Although a food additive regulation has been issued,[100] it does not authorize the use of xanthan gum in foods that are the subject of federal definitions and standards of identity. Discussion here of the possible usefulness of xanthan gum in standardized foods does not imply that the standards of identity of such foods have been amended to permit the use of xanthan gum.

Beverages.—Xanthan gum gives citrus and fruit-flavored beverages enhanced mouth feel with full-bodied taste and good flavor release.[101] At the higher use levels, xanthan gum is an excellent suspending agent for the desirable cloud and pulp in some of these drinks. Xanthan gum also is promising as a stabilizer for the flavor oil emulsions employed in certain of these beverages. Concentrations of xanthan gum in the finished drink may range from 0.001% to 0.15%.

Canning.—Because of its unique heat stability combined with excellent emulsion stabilizing and suspending properties, xanthan gum is quite useful in canned food systems. Excellent quality canned tuna, chicken, ham, potato, and macaroni salads have been formulated with heat-stable salad dressings containing xanthan gum.

Frozen foods.—Frozen starch-based pie fillings and puddings may show syneresis and retrogradation after one or two freeze–thaw cycles. The addition of 0.1% of xanthan gum gives a dramatic improvement in freeze–thaw stability.

Relishes.—Pickle, tomato, onion, and mustard relishes impose severe requirements on the stabilizer used. These systems may contain high salt, high acidity, or a combination of the two. Xanthan gum not only is compatible with these systems, but it is an excellent stabilizer, suspending agent, and thickening agent under these adverse conditions. Concentrations of ~0.1% xanthan gum are effective in relishes.

Nonstandardized salad dressings.—The pseudoplastic properties of xanthan gum give thousand island, Russian and green goddess dressings that are exceptionally stable because of the apparent high viscosity within the container.[102] Yet the dressings are easily pumped and flow readily into and out of the bottles. Viscosity and pourability changes little from refrigerator temperature to maximum storage temperatures. Concentrations of xanthan required are only about 0.25%. Flavor release and mouth feel are excellent.

Xanthan gum–locust bean gum gels and puddings.—Excellent instant puddings are made from the gelling reaction between xanthan gum and locust bean gum.[103] Combinations of about 0.1% xanthan gum, 0.25% locust bean gum, and 0.4% tetrasodium pyrophosphate when stirred with cold milk and then refrigerated for a few minutes set to a delightful pudding body. Excellent mouth feel and flavor release is obtained. There is a complete absence of gummy body or starch flavor.

8. *Pharmaceutical and Cosmetic Applications*

Experimental tests have indicated that xanthan gum has excellent properties as a bodying agent for toothpaste and ointments. It is an excellent suspending and emulsifying agent and gives high viscosities at low concentrations. Because of the pseudoplastic properties of xanthan gum, it enables toothpaste and ointments to hold their shape. Nevertheless, these formulations spread with extreme ease. The addition of a new ingredient to a formulation for which drug claims are made requires the filing of a New Drug Application with the Food and Drug Administration. Xanthan gum is a permitted ingredient in all nonmedical cosmetics for which no drug claims are made.

VIII. References

(1) J. F. Wilkinson, *Bacteriol. Rev.,* **22,** 46 (1958).

(2) H. R. Morgan and T. D. Backwith, *J. Infect. Dis.,* **65,** 113 (1939).

(3) W. R. Maxted, *Nature,* **170,** 1020 (1952).

(4) P. J. Szaniszlo, C. Wirsen, Jr., and R. Mitchell, *J. Bacteriol.,* **96,** 1474 (1968).

(5) E. J. Bourne, M. Stacey, and I. A. Wilkinson, *J. Chem. Soc.,* 2694 (1950).

(6) S. P. Rogovin, R. F. Anderson, and M. C. Cadmus, *J. Biochem. Microbiol. Technol. Eng.,* **3,** 51 (1961).

(7) R. A. Moraine and P. Rogovin, *Biotechnol. Bioeng.,* **8,** 511 (1966).

(8) G. H. Cohen and D. B. Johnstone, *Nature,* **198,** 211 (1963).

(9) R. F. Anderson, *in* "Biochemistry of Industrial Microorganisms," Academic Press, New York, 1963, Chapt. 8.

(10) S. A. Barker and E. J. Bourne, *Quart. Rev., Chem. Soc.,* **7,** 56 (1953).

(11) T. H. Evans and H. Hibbert, *Advan. Carbohyd. Chem.,* **2,** 203 (1946).

(12) S. Hestrin, *in* "The Bacteria," Academic Press, New York, Vol. 3, 1962, Chapt. 8.

(13) P. W. Robbins, R. R. Traut, and F. Lipmann, *Proc. Nat. Acad. Sci. U.S.,* **45,** 6 (1959).

(14) E. J. Hehre, A. S. Carlson, and J. M. Neill, *Science,* **106,** 523 (1947).

(15) E. J. Hehre and D. M. Hamilton, *J. Bacteriol.,* **55,** 197 (1948).

(16) A. Lwoff, H. Ionesco, and A. Gutman, *Biochem. Biophys. Acta,* **4,** 270 (1950).

(17) A. S. Carlson and E. J. Hehre, *J. Biol. Chem.,* **177,** 281 (1949).

(18) F. Lipmann, *Advan. Enzymol.,* **1,** 99 (1941).

(19) W. E. Trevelyan, P. F. E. Mann, and J. S. Harrison, *Arch. Biochem. Biophys.,* **39,** 419 (1952).

(20) E. J. Hehre, *Proc. Soc. Exp. Biol. Med.,* **54,** 240 (1943).

(21) W. W. Carlson, C. L. Rosano, and V. Whiteside-Carlson, *J. Bacteriol.,* **65,** 136 (1953).

(22) S. Hestrin, D. S. Feingold, and G. Avigad, *J. Amer. Chem. Soc.,* **77,** 6710 (1955).

(23) S. Hestrin, *Nature,* **154,** 581 (1944).

(24) S. Hestrin and S. Aineri-Shapiro, *Biochem. J.,* **38,** 2 (1944).

(25) E. J. Hehre, *J. Biol. Chem.,* **177,** 267 (1949).

(26) E. J. Hehre and D. M. Hamilton, *J. Biol. Chem.,* **166,** 777 (1946).

(27) M. Doudoroff, W. A. Hassid, E. W. Putnam, A. L. Potter, and J. Lederberg, *J. Biol. Chem.,* **179,** 921 (1949).

(28) J. Monod and A. M. Torriani, *Ann. Inst. Pasteur,* **78,** 65 (1950).

(29) A. Jeanes and C. A. Wilham, *J. Amer. Chem. Soc.,* **72,** 2655 (1950).

(30) W. Z. Hassid and H. A. Barker, *J. Biol. Chem.,* **130,** 163 (1940).

(31) M. Stacey and G. Swift, *J. Chem. Soc.,* 1555 (1948).

(32) A. Bebbington, E. J. Bourne, and J. A. Wilkinson, *J. Chem. Soc.,* 246 (1952).

(33) R. Austrian, *Bacteriol. Rev.,* **16,** 31 (1952).

(34) J. F. Wilkinson, W. F. Dudman, and G. O. Aspinall, *Biochem. J.,* **59,** 446 (1955).

(35) A. I. McMullen, *Int. Congr. Cell Biol., 11th, 1964 (Excerpta Med. Found. Int. Congr. Ser., n***77**), 32 (1964).

(36) E. Neufeld and W. Z. Hassid, *Advan. Carbohyd. Chem.,* **18,** 309 (1963).

(37) B. L. Horecker, *Ann. Rev. Microbiol.,* **20,** 253 (1968).
(38) E. E. B. Smith, G. T. Mills, H. P. Bernheimer, and R. Austrain, *J. Biol. Chem.,* **235,** 1876 (1960).
(39) R. Austrian, H. P. Bernheimer, E. E. B. Smith, and G. T. Mills, *J. Exp. Med.,* **110,** 585 (1959).
(40) L. Glaser and D. H. Brown, *J. Biol. Chem.,* **228,** 729 (1957).
(41) A. Markovitz, J. A. Cifonelli, and A. J. Dorfman, *J. Biol. Chem.,* **234,** 2343 (1959).
(42) R. L. Bernstein and P. W. Robbins, *J. Biol. Chem.,* **240,** 391 (1965).
(43) P. W. Robbins, A. Wright, and M. Dankert, *J. Gen. Physiol.,* **49** (No. 6, Pt. 2), 331 (1966).
(44) A. Wright, M. Dankert, and P. W. Robbins, *Proc. Nat. Acad. Sci. U.S.,* **54,** 235 (1965).
(45) A. Markovitz and A. Dorfman, *J. Biol. Chem.,* **237,** 273 (1962).
(46) A. Jeanes, C. A. Knutson, J. F. Pittsley, and P. R. Watson. *J. Appl. Polym. Sci.,* **9,** 627 (1965).
(47) A. J. Brown, *J. Chem. Soc.,* **49,** 172, 432 (1886).
(48) H. L. A. Tarr and H. Hibbert, *Can. J. Res.,* **4,** 372 (1931).
(49) K. Muhlethaler, *Biochem. Biophys. Acta,* **3,** 527 (1949).
(50) S. Hestrin and M. Schramm, *Biochem. J.,* **58,** 345 (1954).
(51) C. R. Masson, R. F. Menzies, and J. Cruikshank, *Nature,* **157,** 74 (1946).
(52) M. C. Cadmus, K. A. Burton, A. A. Lagoda, and K. L. Smiley, *Bacteriol. Proc.,* A102 (1967).
(53) E. Ninomiya and T. Kizaki, *Angew. Makromol. Chem.,* **6,** 179 (1969).
(54) G. H. Cohen and D. B. Johnstone, *J. Bacteriol.,* **88,** 1695 (1964).
(55) P. A. J. Gorin and J. F. T. Spencer, *Can. J. Chem.,* **44,** 993 (1966).
(56) V. M. Patrikh and J. K. N. Jones, *Can. J. Chem.,* **41,** 2826 (1963).
(57) G. H. Cohen and D. B. Johnstone, *J. Bacteriol.,* **88,** 329 (1964).
(58) T. Harada, *Arch. Biochem. Biophys.,* **112,** 65 (1965).
(59) T. Harada, A. Misaki, and H. Saito, *Arch. Biochem. Biophys.,* **124,** 292 (1968).
(60) I. Maeda, H. Saito, M. Masada, A. Misaki, and T. Harada, *Agr. Biol. Chem.,* **31,** 1184 (1967).
(61) E. A. Cooper, W. D. Daker, and M. Stacey, *Biochem. J.,* **32,** 1792 (1938).
(62) W. Corpe, *Can. J. Microbiol.,* **6,** 153 (1960).
(63) W. Corpe, *J. Bacteriol.,* **88,** 1433 (1964).
(64) M. E. Slodki, L. J. Wickerham, and M. C. Cadmus, *J. Bacteriol.,* **82,** 269 (1961).
(65) A. Jeanes, J. E. Pittsley, P. R. Watson, and R. J. Dimler, *Arch. Biochem. Biophys.,* **92,** 343 (1961).
(66) R. F. Anderson, M. C. Cadmus, R. G. Benedict, and M. E. Slodki, *Arch. Biochem. Biophys.,* **89,** 289 (1960).
(67) M. E. Slodki, *Biochem. Biophys. Acta,* **69,** 96 (1963).
(68) "Information on Phosphomannan Y-2448," Northern Utilization Research and Development Division, U.S. Dept. Agr., Peoria, Ill., 1958.
(69) M. E. Slodki, L. J. Wickerham, and R. J. Bandoni, *Can. J. Microbiol.,* **12,** 489 (1966).
(70) M. C. Cadmus, A. A. Logoda, and R. F. Anderson, *Appl. Microbiol.,* **10,** 153 (1962).
(71) E. N. Davis, R. A. Rhodes, and H. R. Schulk, *Appl. Microbiol.,* **13,** 267 (1966).

(72) J. J. Johnson, Jr., S. Kirkwood, A. Misaki, T. E. Nelson, J. V. Scaletti, and F. Smith, *Chem. Ind.* (*London*), 820 (1963).

(73) F. E. Halleck, U.S. Patent 3,301,848 (1967); *Chem. Abstr.*, **66**, 84772b (1967).

(74) H. O. Bouveng, R. L. Fraser, and B. Lindberg, *Carbohyd. Res.*, **4**, 20 (1967).

(75) K. W. Buck, A. W. Chen, A. G. Dickerson, and E. B. Chain, *J. Gen. Microbiol.*, **51**, 377 (1968).

(76) P. A. J. Gorin and J. F. T. Spencer, *Advan. Carbohyd. Chem.*, **23**, 367 (1968).

(77) H. Stolp and M. P. Starr. *Phytopathol. Z.*, **51**, 442 (1964).

(78) V. G. Lilly, H. A. Watson, and J. G. Leach, *Appl. Microbiol.*, **6**, 105 (1958).

(79) M. P. Starr, *J. Bacteriol.*, **51**, 131 (1946).

(80) C. M. Cooper, G. A. Fernstrom, and S. A. Miller, *Ind. Eng. Chem.*, **36**, 504 (1944).

(81) S. P. Rogovin, W. J. Albrecht, and V. Sohns, *Biotechnol. Bioeng.*, **7**, 161 (1965).

(82) S. P. Rogovin and W. J. Albrecht, U.S. Patent 3,119,812 (1964); *Chem. Abstr.*, **60**, 11348 (1964).

(83) W. H. McNeely and J. J. O'Connell, Can. Patent 681,525 (1964).

(84) W. H. McNeely and J. J. O'Connell, U.S. Patent 3,232,929 (1966); *Chem. Abstr.*, **64**, 14878 (1966).

(85) B. J. Lipps, Jr., U.S. Patent 3,251,749 (1966); *Chem. Abstr.*, **65**, 2969 (1966).

(86) J. T. Patton and G. P. Lindblom, U.S. Patent 3,020,206 (1962); *Chem. Abstr.*, **56**, 12094 (1962).

(87) A. Jeanes and J. H. Sloneker, U.S. Patent 3,000,790 (1961); *Chem. Abstr.*, **56**, 2625 (1962).

(88) R. G. Schweiger, U.S. Patent 3,236,831 (1966); *Chem. Abstr.*, **64**, 14402 (1966).

(89) R. G. Schweiger, U.S. Patent 3,256,271 (1966); *Chem. Abstr.*, **65**, 10423 (1966).

(90) R. G. Schweiger, U.S. Patent 3,244,695 (1966); *Chem. Abstr.*, **65**, 15490 (1966).

(91) J. T. Patton, U.S. Patent 3,020,207 (1962); *Chem. Abstr.*, **56**, 13150 (1962).

(92) A. B. Steiner and W. H. McNeely, U.S. Patent 2,453,824 (1949); *Chem. Abstr.*, **43**, 4291 (1949).

(93) A. B. Steiner and W. H. McNeely, *Ind. Eng. Chem.*, **43**, 2073 (1951).

(94) J. G. Leach, V. G. Lilly, H. A. Watson, and M. R. Purvis, Jr., *Phytopathol.*, **47**, 113 (1957).

(95) J. H. Sloneker, D. G. Orentas, and A. Jeanes, *Can. J. Chem.*, **42**, 1261 (1964).

(96) J. T. Patton and W. E. Holman, U.S. Patent 3,243,000 (1966); *Chem. Abstr.*, **64**, 17322 (1966).

(97) K. D. Detting, U.S. Patent 2,341 500 (1944); *Chem. Abstr.*, **38**, 4785 (1944).

(98) G. P. Lindblom, G. D. Ortloff, and J. T. Patton, Can. Patent 654,809 (1962).

(99) W. H. McNeely, W. P. Fairchild, and A. R. Hunter, Can. Patent 727,071 (1966).

(100) *Fed. Reg.* 34,5376 (March 19, 1969); 21CFR 121.1224.

(101) H. R. Schuppner, U.S. Patent 3,413,125 (1968); *Chem. Abstr.*, **70**, 46304a (1969).

(102) J. J. O'Connell, U.S. Patent 3,067,038 (1962); *Chem. Abstr.*, **58**, 8360 (1963).

(103) H. R. Schuppner, Can. Patent 797,202 (1968).

CHAPTER XXII

SCLEROGLUCAN

NELSON E. RODGERS

The Pillsbury Company, Minneapolis, Minnesota

I. INTRODUCTION

Scleroglucan is a capsular polysaccharide secreted by the mycelia of certain imperfect fungi, notably by species of the genus *Sclerotium*. This uniquely structured polysaccharide was extracted from *Sclerotium glucanicum* by Halleck,[1] who recognized in its physical properties a potential significance as an industrial gum. Subsequently, gums similarly structured but with higher molecular weights have been detected in species of the genus *Corticum*, as well as in other species of *Sclerotium*. Gums differing slightly in structure from scleroglucan have been noted in cultures representing the genera *Sclerotium, Sclerotinia, Corticum*, and *Stromatinia*.

Scleroglucan is a neutral glucan. Its solutions are highly pseudoplastic, notably stable to high temperature and a broad range of pH, and tolerant to a variety of electrolytes. Viscous flow is relatively insensitive to change in temperature. This array of useful properties has stimulated commercialization.[a]

[a]Marketed by The Pillsbury Company, Minneapolis, under the trade name Polytran®.

II. Production

Scleroglucan is produced by submerged aerobic fermentation of D-glucose by a selected species of *Sclerotium*. The process is conducted in conventional deep fermentation equipment, using a sterile medium and aseptic management of the culture. The medium contains D-glucose, corn steep liquor, nitrate as the principal nitrogen source, and mineral salts.

Synthesis of scleroglucan proceeds concomitantly with mycelial growth and the medium develops a gel-like consistency. The pH of the fermentation, initially 4.5, drops to about 2 with accumulation of oxalic acid. At 30°, the fermentation is complete in about 60 hr.

Growth is pelletlike. Hyphyl filaments radiate outward from a central mycelial mass into an enshrouding capsule of extracellular polysaccharide. In its native association with the mycelium, the polysaccharide is a gel phase largely immiscible with water and the capsular structure tends to retain its identity through the fermentation. Transport of oxygen and nutrients through this gel layer to the mycelial core is rather remarkable.

The mature fermentation is heated to inactivate glucan-degrading enzymes and to kill the fungus, which is a plant pathogen. Then, the fermentation liquor is homogenized to disengage the gum from the mycelium and to disorient the molecular association of the polysaccharide so that it is completely dispersible in water.[2] Homogenization predisposes a molecular state in the subsequently dried product that favors rapid development of viscosity when rehydrated.

For some industrial applications that can tolerate the residual mycelium, nutrients, and metabolites, spray-drying the homogenized fermentation liquor provides a satisfactory product. Usually, an agent is added to the liquor to suppress molecular association in the dried product and insure rapid rehydration.

For food, cosmetic, and drug applications, scleroglucan is refined by filtration and precipitation with alcohol. The homogenized fermentation liquor is extremely viscous and must be diluted for efficient filtration. Filtration at elevated pressure and temperature, using diatomaceous filter aids, removes the mycelium and other particulate matter. The clear filtrate is concentrated in a swept film evaporator to a level compromised to a manageable viscosity and economical use of alcohol. Precipitation of the gum is conducted continuously in 50 volume percent 2-propanol or methanol. The alcohol extract is drained from the fibrous precipitate, which is then, successively treated with higher concentrations of alcohol to effect further extraction and dehydration. After the solvent has been drained off in a screen centrifuge, the fibrous mass is dried in trays by hot air and is ground to a powder. The fiinished product contains not more than 6% moisture and not less than 85% scleroglucan. The impurities are largely accounted for by $<5\%$ ash and nitrogenous materials equivalent to $<0.4\%$ nitrogen.

III. Structure

The structure of scleroglucan produced by *Sclerotium glucanicum* has been characterized as a linear chain of glucopyranosyl units linked β-D-(1→3) with single glucopyranosyl units linked β-D-(1→6) to every third unit of the chain.[3] Only D-glucose is produced on acid-catalyzed hydrolysis. A low specific optical rotation of $[\alpha]_D$ $-1°$ (1 *N* sodium hydroxide) indicates mainly a β-D anomeric configuration. Periodate oxidation at ~25° consumes 2 moles of periodate for every 4 D-glucosyl units with formation of 1 mole of formic acid. Acid-catalyzed hydrolysis, following borohydride reduction of the periodate-oxidized polysaccharide, liberates 1 mole of glycerol per 4 D-glucosyl units and precipitates a degraded glucan that is insoluble in water and alkali. These reactions presumably establish that 1 of every 4 D-glucosyl units is a terminal nonreducing pyranosyl unit.

Selective hydrolyses, involving two types of β-D-(1→3) glucanases, have been used to confirm the structure. An exo-type enzyme derived from a basidiomycete specifically releases D-glucose units one at a time from the end of the chain. An endo-type enzyme, isolated from *Rhizopus arrhizus,* attacks the chain randomly. Hydrolysis of the glucan by the exo-enzyme yields 1 mole of gentiobiose per 2 moles of D-glucose, this porportion remaining constant through the hydrolysis. The isolation of gentiobiose confirms the identity of the side branch. The same enzyme produces only D-glucose from the insoluble glucan degraded by periodate. The endo-type enzyme yields D-glucose, gentiobiose, and laminaribiose but no laminaritriose from the intact glucan. This is consistent with the chemical evidence for a single D-glucopyranosyl side branch for each 3 linear D-glucopyranosyl units. Treatment of the degraded glucan with the endo-enzyme yields D-glucose, laminaribiose, and laminaritriose. Collectively, this evidence convincingly indicates the following structure for the repeating unit of the scleroglucan derived from *Sclerotium glucanicum.*

```
                                   β-D-Glcp
                                      1
                                      ↓
                                      6
—[→3)-β-D-Glcp(1→3)-β-D-Glcp(1→3)-β-D-Glcp(1—]→n
```

Examination of selected cultures representing the genera mentioned above yielded gums with physical properties and structures quite similar to the glucan from *Sclerotium glucanicum.* These particular gums present a closely graded series in which single glucopyranosyl units linked β-D-(1→6) as side branches occur on from about 30 to 43% of the β-D-(1→3)-linked glucopyranosyl units in the linear chain. It seems appropriate to refer to these and related gums as the scleroglucan group.

The degree of polymerization (DP) of the scleroglucan from *Sclerotium glucanicum* has been estimated to be ~110 by end-group analysis.[3] The related scleroglucans are considerably larger. Several have degrees of polymerization in the range 200–500 and most in the range 500–1600. The commercial polysaccharide has a DP of ~800 and about 34% of the D-glucopyranosyl units in the chain bear branches.

β-D-(1→3) Glucans have been variously reported in fungi, algae, and higher plants, some of which contain β-D-(1→6)-linked side branches.[4, 5] These appear to be largely structural elements of cell walls. In the native state, they are insoluble in water, probably because of either association with other constitutents of the cell wall or restriction of solubility by a limited degree of branching. *Claviceps purpurea* produces a gum with 25% branching similar to the scleroglucans.[6] The culture is viscous, but the alcohol-precipitated glucan is insoluble in water until dissolved in alkali and neutralized. Recently, a water-soluble glucan with a structure apparently similar to that of scleroglucan has been isolated from a culture of *Stereum sanguinolentum*.[7]

Although possibly not a homoglucan, the principal structure of the gum produced by the fleshy fungus *Plectania occidentalis* NRRL 3137 is similar to the scleroglucans and is notable because of an apparent 50% side branching and similarity of certain rheological properties.[8]

IV. Properties

The properties of scleroglucan will be described in terms of two products. One is a commercially available dried fermentation residue, containing about 30% scleroglucan. The second is an experimental refined product with a polysaccharide content of 85–90%. Aside from the content of scleroglucan, the rheological and other functional properties are similar.

The flow properties of scleroglucan described here are in terms of a Brookfield LVT viscometer using standard spindles with a wide cup or the UL configuration. Measurements with a Rotovisco instrument yield comparable indications.

1. Dissolution

The gum products disperse easily in water, provided reasonable precautions are taken to prevent clumping of the freshly wetted powder. The rate at which viscosity develops is influenced by mixing, temperature, pH, and concentration.

Intense agitation of a dispersion of the commercial gum at 21°, such as in an Osterizer blender at high speed, can yield 80% of equilibrium viscosity within 1 min. A batch of the same size stirred with a 2-in (5-cm) propeller at 700 rpm reaches 80% of equilibrium viscosity in 50 min. Equilibrium viscosity is defined as the viscosity attained after holding the dispersion for 24 hr, heating

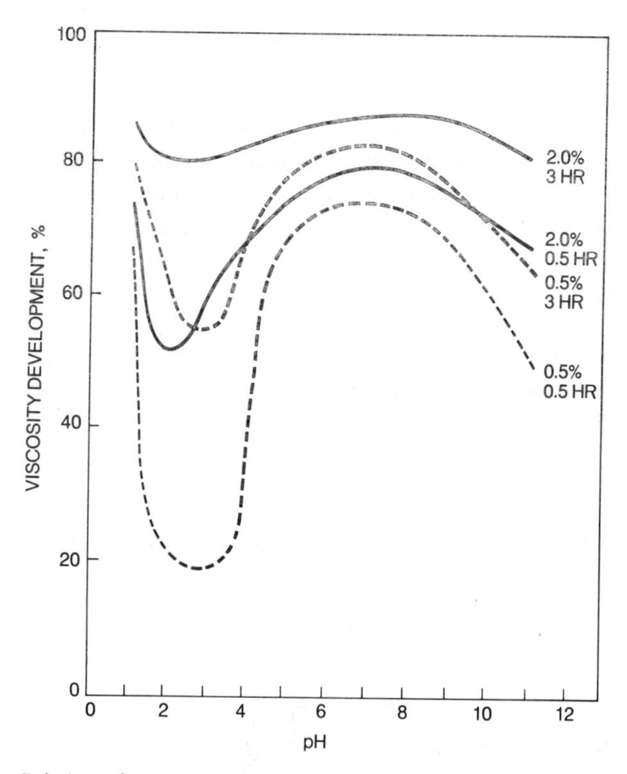

FIG. 1.—Relation of rate of development of viscosity to pH in relatively high and low concentrations of commercial scleroglucan. Viscosity development at 0.5 and 3.0 hr is expressed as a percent of the equilibrium viscosity after heating to 90° at pH 7.0. The final viscosity is similar at all levels of pH.

to 90° and cooling Once a substantial viscosity has been reached, continued mixing has little effect on the course of viscous development. Because of the pseudoplastic character of the solution, agitation becomes closely localized around the impeller.

Curiously, concentrated dispersions of scleroglucan develop viscosity proportionately faster than dilute dispersions. This effect is particularly noticeable at pH levels above and below pH 5.0–8.0 (Fig. 1). Seemingly, the associative forces of polysaccharide molecules in solution attract molecules in the solvated particulate phase and draw them into solution in a manner quite inverse to simple solute systems.

Viscosity evolves rapidly between pH 5.0 and 8.0, but the rate of development is depressed above pH 8.0 and particularly between pH 1.5 and 4.0 (Fig. 1). At pH 1.0, viscosity develops nearly as fast as at pH 7.0. The pH effects are largely dissipated in 4–5 hr. By 24 hr, the viscosity is quite uniform from pH 1 to 11.

In practical dispersion systems, raising the temperature accelerates development of viscosity in a peculiar manner. Shortly following initial hydration of the refined scleroglucan, the temperature effect is quite marked up to ~40°, but above this temperature the rate of acceleration tends to diminish or even reverse at very high temperatures; the effect is transient. Thus, although viscous development at 80° with mild agitation may initially lag behind that at 25°, the effect is soon reversed and near equilibrium viscosity is attained at 80° within 30 min. It may be that the thermal activity of water at high temperature delays accumulation of stable hydration layers around the polysaccharide molecule until the system becomes molecularly oriented.

2. *Rheology*

Pseudoplasticity or shear thinning is a salient characteristic of scleroglucan solutions (Fig. 2). Thus, it is necessary to designate a measured viscosity as that

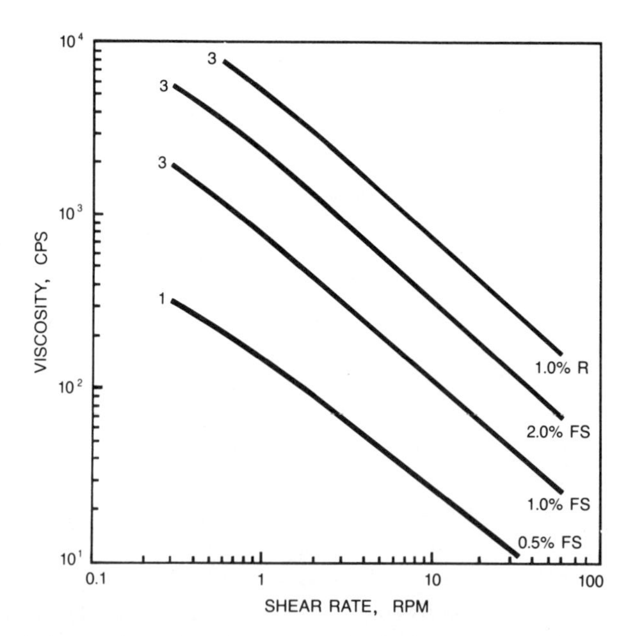

FIG. 2.—Relation of viscosity of solutions of refined (R) and commercial (FS) gums to shear rate. Brookfield LVT viscometer, spindles 1 and 3.

apparent at a specified rate of shear. Pseudoplasticity is evident in gum solutions containing the equivalent of 0.02% or less of scleroglucan, but the flow becomes progressively more Newtonian as the concentration decreases below 0.2%. It is doubtful that this viscous system displays a definable yield value, in view of the slight downward bend of the curves at low shear rates evident in Figure 2.

Solutions containing less than 0.8% of scleroglucan are not significantly thixotropic, except at temperatures approaching 10° and below. Above 20°, as sensed by the Brookfield-type viscometer, the hysteresis loops traced by the response of shear stress to increasing and decreasing shear rate are of negligible proportions. Somewhat more prominent but shallow loops are displayed by the Rotovisco viscometer.

Because of the high degree of pseudoplasticity, gel states are not always clearly defined. Thus, 1.2–1.5% solutions of purified gum form self-supporting sliceable gels at ~25° but at temperatures below 10°, even very dilute solutions form diffusely structured gels that tend to shrink and undergo syneresis when left undisturbed for long periods of time. Such diffuse gels disperse quickly with mild agitation.

At a given rate of shear, the viscosity of a scleroglucan solution varies with concentration according to $V = bC^n$, where V is the measured viscosity, C is the concentration and b and n are constants. A plot of this relation in the form $\log V = \log b + n \log C$ yields a straight line, with b equal to the viscosity at 1.0% concentration and n as the slope. The slopes in Figure 3 indicate that, in the form of either commercial or refined gums, a doubling of the concentration of scleroglucan yields approximately a threefold increase in viscosity.

The viscosity of scleroglucan solutions is relatively insensitive to temperature in the range 10° to >90° (Fig. 4). Below 10°, and particularly around 5°, gel-like transformations occur. However, the viscosity of a 2.0% solution of commercial scleroglucan, measured at 30 rpm with a Brookfield viscometer, declines only 13% between 10° and 90°. Temperature dependence increases with decreasing concentration and decreasing rate of shear. Compared to most viscous systems, the thermal effects are remarkably small. It appears that solutions of scleroglucan have a quasistructure of some sort and that the thermal activity is characteristic of that structure as distinguished from that in an unassociated molecular dispersion.

3. *Suspending Properties*

A pseudoplastic flow system inherently combines a capacity for suspending fine particles with good pourability of the suspension. A fine particle with a small mass generates a low gravitational rate of shear. Consequently, it exists with relation to an apparent high viscosity and settles slowly. The mass of a

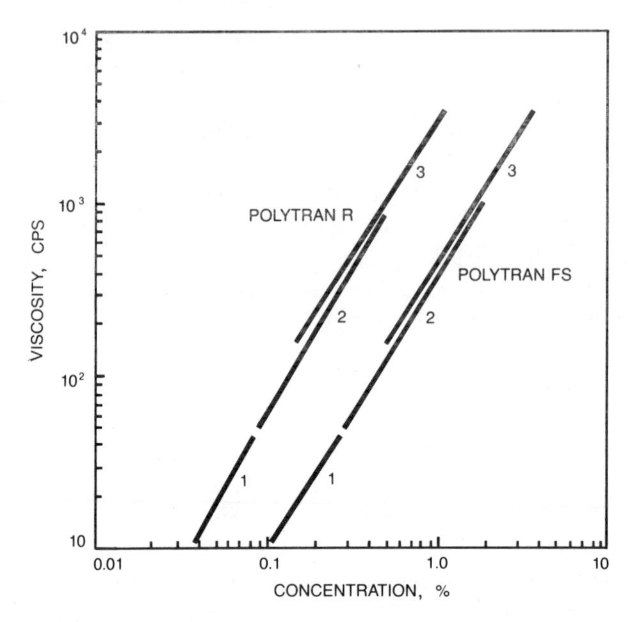

FIG. 3.—Relation of viscosity of solutions of refined (Polytran R) and commercial (Polytran FS) scleroglucans to concentration. Brookfield LVT viscometer, spindles 1, 2, and 3; 30 rpm.

suspension creates a relatively high shear rate when disturbed. In this state, the viscosity is apparently low and the suspension pours easily. This aspect is important in many applications. Gums with Newtonian flow characteristics usually require some compromise of suspending function and mobility of the suspension.

Purified scleroglucans in concentrations of 0.1–0.2% effectively stabilize 5–10% aqueous suspensions of fine powders, such as zinc oxide, reprecipitated calcium carbonate, and sulfamerazine. The minimum effective concentration of gum varies somewhat with the substance and its concentration. It is important to recognize that very low concentrations of scleroglucan destabilize and flocculate suspended materials. Thus, a curve relating stabilization of a suspension to concentration usually is in the form of a U, the high concentration leg of which terminates abruptly in a plateau of stabilization. In some cases, a 10% change in concentration marks the boundary between flocculation and complete stabilization.

The viscosity of combinations of scleroglucan with bentonite suspensions reflects a marked synergism. Thus, whereas the apparent viscosities of 0.15%

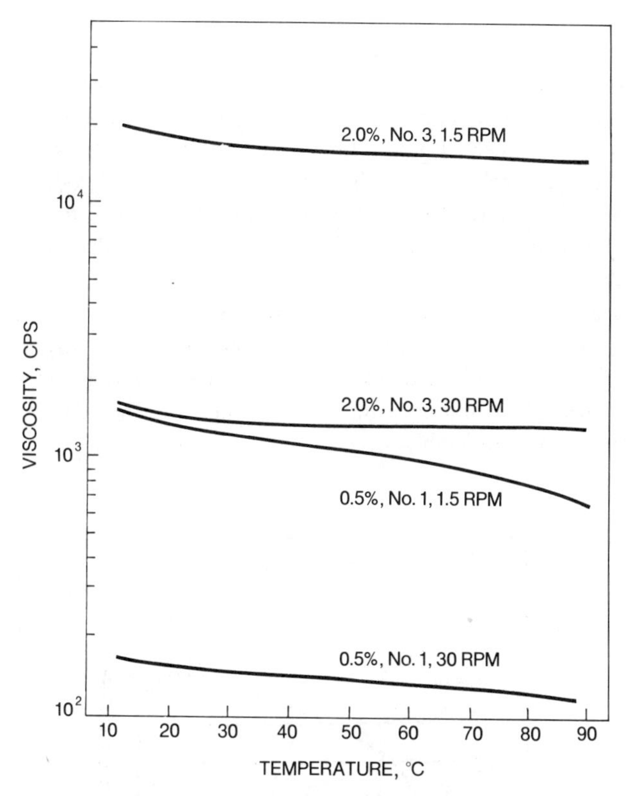

FIG. 4.—Relation of viscosity of solutions of commercial scleroglucan to temperature at varied concentration and shear rate. Brookfield LVT viscometer, spindles 1 and 3.

purified gum and 5.0% bentonite are around 200 and 300 cps, respectively, a combination yields a viscosity of >4000 cps (Brookfield LVT, spindle 3, 30 rpm). This and the foregoing observations on suspensions suggest that adsorption phenomena in addition to viscosity are involved in fluocculation and stabilization reactions.

Although not a primary emulsifier in the sense of a surfactant, scleroglucan enables very low energy dispersion in creating stable oil-in-water emulsions. For example, equal parts of a 0.25% solution of purified gum and a vegetable oil briefly shaken by hand effects an emulsion that does not separate after several months. Such emulsions pour easily. In addition to the suspending action of the pseudoplastic system, prevention of coalescence seems to underlie this kind of stabilization.

4. pH and Temperature Stability

Although the initial development of viscosity is retarded by mildly acidic and alkaline conditions, the viscosity of equilibrated solutions is remarkably tolerant to extreme acidity and alkalinity. In Figure 5, the zero hour curve reflects the viscosity of a commercial gum solution aged 24 hr, heated to 90°, cooled, and adjusted to varied levels of pH with hydrochloric acid and sodium hydroxide. Aside from a minor elevation at pH 0 and a notable increase around pH 12, the viscosity between pH 1 and 10 is uniform. Incipient degradation of the system is evident above pH 12.

Heating at 121° for 15 min increases the general level of viscosity somewhat but causes degradation below pH 1. The system is substantially stable for 20 hr at 121° from initial pH 3 to 11. The pH of these solutions does not change importantly through 5 hr but, after 20 hr of heating, the pH of the alkaline solutions decreases considerably.

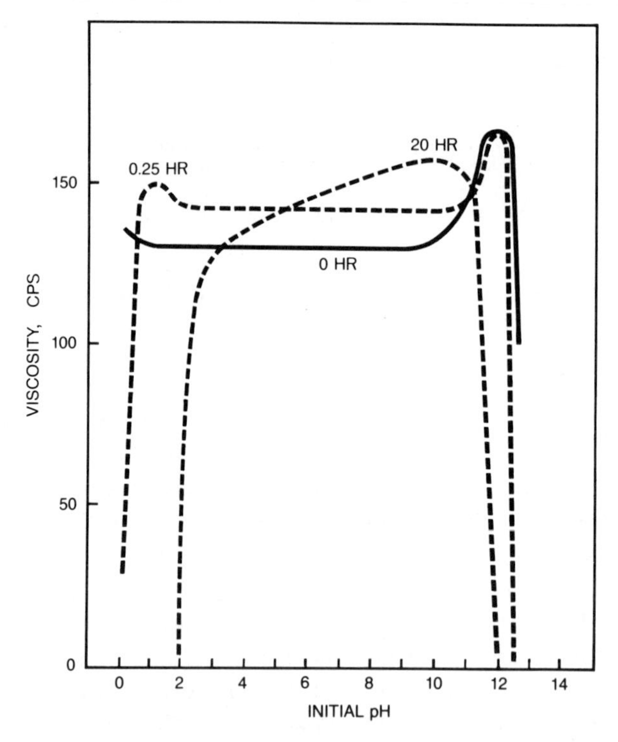

FIG. 5.—Effect of pH on stability of 0.5% solutions of commercial scleroglucan heated at 121° for 0.25 and 20 hr. Brookfield LVT viscometer, spindle 1, 30 rpm.

5. *Compatibilities*

Scleroglucan is compatible with a variety of electrolytes in moderate to high concentrations. Table I depicts the behavior of a 0.5% commercial gum[a] in the presence of several mono- and polyvalent cations and anions. The salts were added as solutions to a 2.0% gum solution, previously equilibrated for 24 hr, to provide the indicated concentrations. Viscosity was measured after aging the resulting solution 24 hr; gelation was observed after heating at 121° for 15 min.

With the exception of a moderate depression in viscosity by 5.0% aluminum sulfate, none of the salts in 5% concentration affect vicosity beyond an increase attributable to the increased ratio of polysaccharide to water. Scleroglucan is insensitive to all chloride salts with the exception of sodium chloride. In the latter case, viscosity apparently declines at concentrations above 10% salt, and the solutions gel upon heating. It is possible that the apparent decline in viscosity is caused by incipient gelation, making the system unresponsive to rotational viscometry. Ten percent or more of sulfate salts cause precipitation, the extent of which is approximately proportional to the sulfate content of the salt. Disodium phosphate is inactive at 10%, the limit of solubility.

Initial development of viscosity while hydrating commercial gum is not significantly influenced by 1 and 5% of the salts listed in Table I with the exceptions of calcium chloride, aluminum sulfate, and ferric chloride; 5% calcium chloride slows development of viscosity slightly, whereas aluminum sulfate and ferric chloride are strongly retardant, probably because of the low pH of their solutions.

TABLE I

Effect of Electrolytes on Performance of 0.5% Prehydrated Commercial Scleroglucan[a]

Salt	Salt Concentration, %					
	1	3	5	10	15	20
NaCl	I	I	I	IG	DG	
$MgCl_2$	I	I	I	I	I	I
$CaCl_2$	I	I	I	I	I	I
$FeCl_3$	I	I	I	I	I	I
Na_2SO_4	I	I	I	D	P	
$MgSO_4$	I	I	I	P		
$Al_2(SO_4)_3$	I	I	D	P		
Na_2HPO_4	I	I	I	I		

[a]Conditioned for 24 hr: I, viscosity increase; D, viscosity decrease; P, precipitate; G, gels upon heating 15 min at 121° and cooling.

Commercial scleroglucan in 0.5% concentration is compatible with 15 volume percent ethanol but precipitates in 20 volume percent. Ethylene glycol, propylene glycol, and glycerol in concentrations up to 50 volume percent cause no precipitation. Viscosities of polyalcohol solutions rise to peaks at 20 to 30 volume percent, followed by decline to levels equal to or not less than one-half the viscosity of a water solution. When heated at 121° for 15 min, the polyalcohol solutions at 20 to 30 volume percent and higher are fluid while hot but gel on cooling to room temperature. The gels disperse easily when agitated. Similarly, the precipitates formed in 20 volume percent and higher concentrations of ethanol dissolve while hot and gel upon cooling.

6. *Other Properties*

Scleroglucan complexes slowly with borate in alkaline solution to form a stable gel that does not revert to a sol on aging or acidifying. It is precipitated in alkaline solution by quarternary ammonium compounds. Derivatives have been prepared; the following are examples: methyl, carboxymethyl, acetate, benzyl, benzoate, carbanilate, cyanoethyl, sulfate,[1] and sulfoalkyl.[9]

Dried films cast from solutions of refined scleroglucan tend to be somewhat brittle but remain tough and pliable indefinitely when plasticized with glycerol. Matured films resist dispersion in water.

Scleroglucan solutions and gels are tactily distinctive. They are neither slimy nor sticky. In dropwise flow, solutions do not string out. Gels exhibit a short break. Solutions do not rub out on the hands or leave a sticky residue. Solutions clear the mouth quickly without leaving a slimy, mucilaginous sensation.

V. Physiology

Rats assimilate purified scleroglucan as the caloric equivalent of starch. Dogs initially absorb limited amounts of the polysaccharide but later appear to utilize it well. Short- and long-term feeding studies with rats and dogs show no toxicity, blood abnormalities, or significant tissue pathology. Eye and skin tests, involving guinea pigs, rabbits, and humans, demonstrate no significantly adverse reactions or sensitization. Ducks exhibit no effects symptomatic of aflatoxin.

Scleroglucan in the diet of chicks reduces dietary cholesterolemia and increases excretion of lipid and cholesterol.[10] Similarily, dogs fed a standard diet supplemented with scleroglucan exhibit a reduced level of cholesterol in the blood and increased excretion of lipid.

VI. Uses

The scleroglucan gums appear to be generally useful in many of the areas occupied by other industrial gums. However, the presently early stage of commercialization has sought to identify those uses in which the pseudoplastic properties and adaptability to adverse conditions are particularly advantageous. The following are some examples of recognized uses.

The gum has been suggested for use in porcelain and ceramic glazes, extruded refractory products, integrated circuit chips, insulators, water-base paints,[11] paper coatings, printing inks, pesticide and defoliant sprays, agricultural seed coatings, secondary oil recovery,[9, 12, 13] drilling muds, and liquid animal feed concentrates,[14] as source of gentiobiose,[15] and as a ceramic binder.[11]

Suggested cosmetic applications are in formulations for hair sprays, hand lotions, and antisoilant skin coatings.

Pharmaceutical applications include use in table coatings,[16] ophthalmic solutions,[17] injectable antibiotic suspensions, anti-acid suspensions, and calomine lotion.

Refined scleroglucan is potentially adapted to the many-faceted requirements of the food industry for bodying, suspending, coating, and gelling agents. The tendency of scleroglucan solutions to form diffuse gels and undergo syneresis in liquid systems at temperatures below 10° may be limiting in some circumstances. However, this condition is unobjectionable in situations where it can be dissipated by mild agitation.

VII. References

(1) F. E. Halleck, U.S. Patent 3,301, 848 (1967); *Chem. Abstr.*, **66,** 84772b (1967).

(2) R. B. Ferguson and J. D. Westover, U.S. Patent 3,436,311 (1969); *Chem. Abstr.*, **71,** 2147t (1969).

(3) J. Johnson, Jr., S. Kirkwood, A. Misaki, T. E. Nelson, J. V. Scaletti, and F. Smith, *Chem. Ind.* (*London*), 820 (1963).

(4) A. E. Clarke and B. A. Stone, *Pure Appl. Chem.*, **13,** 139 (1963).

(5) M. Kitahara and Y. Takeuchi, *Nippon Nogei Kagaku Kaishi,* **35,** 474 (1961).

(6) A. S. Perlin and W. A. Taber, *Can. J. Chem.*, **41,** 2278 (1963).

(7) K. Axelsson, H. Bjorndal, and K. E. Eriksson, *Acta Chem. Scand.*, **22,** 1363 (1968).

(8) L. L. Wallen, R. A. Rhodes, and H. R. Shulke, *Appl. Microbiol.*, **13,** 272 (1965).

(9) S. A. Williams, U.S. Patent 3,373,810 (1968).

(10) P. Griminger and H. Fisher, *Proc. Soc. Exp. Biol. Med.*, **122,** 551 (1966).

(11) F. Halleck, U.S. Patent 3,447,940 (1969); *Chem. Abstr.*, **71,** 51304a (1969).

(12) S. A. Williams, U.S. Patent 3,372,749 (1968).

(13) J. D. Westover and R. B. Ferguson, U.S. Patent 3,436,346 (1969); *Chem. Abstr.*, **71,** 37510h (1969).

(14) *Fed. Reg.*, **34,** 13162 (1969).

(15) F. E. Halleck and F. Smith, U.S. Patent 3,423,288 (1969); *Chem. Abstr.*, **70,** 86288e (1969).

(16) P. Sheth and L. Lachman, U.S. Patent 3,421,920 (1969).

(17) L. Lachman and P. Sheth, U.S. Patent 3,415,929 (1968); *Chem. Abstr.*, **70,** 50467y (1969).

CHAPTER XXIII

DEXTRANS

P. T. MURPHY AND ROY L. WHISTLER

Department of Biochemistry, Purdue University, Lafayette, Indiana

I. INTRODUCTION

A multitude of dextrans with structures varying from slightly branched to highly branched have been obtained from a number of microorganisms and by enzymic synthesis. Most have been examined and proposed for industrial uses. However, few actual uses for dextrans have developed. Several thousand papers have been devoted to dextran synthesis, modification, and recommended applications. No other group of polysaccharides has been so extensively investigated and has developed so few actual applications. The field is almost overworked and enthusiasm is floundering. It may be expected, however, that one or more structures will eventually prove useful and will be made available at low cost.

The term *dextran* was coined by Scheibler in 1869. Evans and Hibbert[1] review the early literature (to 1946) and several reviews present recent chemistry.[2–5] Jeanes[6] has given an extensive bibliography of patents on the production and use of native dextrans, partially degraded dextrans, and their derivatives. Reviews dealing with the production, chemistry, and uses of dextrans have been published by Aiyar[7] and Fujiwara and Shinoda.[8]

Sometimes dextran synthesis by microorganisms is undesirable. Dextrans produced in wine processing create unpleasant thickening, and in the sugar industry, they may induce sugar to crystallize as impure, elongated crystals.[9] Dextrans are often deposited on teeth as dental plaques.[10, 11] To remove these plaques and assist with oral hygiene, dentifrices containing dextranases have been developed.[12–14]

Commercial production is achieved by fermenting sucrose by either of two methods, whole-culture or cell-free enzymic fermentation. The basic processes are still the same as outlined in the first edition of this book.[15]

II. Production

The enzyme or group of enzymes (glucosyltransferases) that produce dextrans from sucrose are termed dextransucrases. Ebert and coworkers[16–20] propose a two-step mechanism for the enzymic propogation of the dextran chain.

$$EP_n + S \rightleftarrows SEP_n$$
$$SEP_n \rightleftarrows EP_{n+1} + F$$

Where E is the enzyme, S is sucrose, P_n is a dextran molecule containing n D-glucopyranosyl units, F is fructose, and SEP_n is a complex.

SEP_n is relatively stable and its dissociation controls the rate of the overall reaction. Building of the dextran chain stops only when the molecule is detached from the enzyme complex. The action of various acceptors (A) detaches dextran molecules from the enzyme and leads to termination of polymer growth.

$$SEP_n + A \rightleftarrows ES + AP_n$$

Sucrose is an acceptor as well as a substrate and, in high concentrations, has an inhibitory action. Branches in dextran molecules are produced by already formed dextrans acting as acceptors. Increasing reaction time or temperature, or decreasing enzyme or sucrose concentration, produces dextrans with higher molecular weights and lower intrinsic viscosities.[19, 21] This indicates an increase in the degree of branching. The acceptor reaction of sucrose becomes less important as the acceptor reaction of the dextran molecules becomes correspondingly more important. When tritium-labeled dextran of molecular weight 30,000 is added during dextran synthesis, the branched polymer formed has a labeled main chain and unlabeled side chains.[20] The addition of methyl α-D-glucopyranoside increases the rate of formation of dextrans but diverts the reaction from formation of high-molecular-weight to low-molecular-weight dextrans.[22]

Hehre and Suzuki[23] theorize that family of dextransucrases, differing in affinity for the various D-glucose hydroxyl groups of dextran chains, may account for the structural diversity of dextrans. They also report that a trisaccharide having a β-D-galactopyranosyl unit linked to the O-4 position of the fructosyl unit of sucrose can be utilized as a substrate by dextransucrase.

1. *Production of Native Dextran*

All commercial dextrans are biosynthetic products but a dextran can be prepared by polymerizing 1,6-anhydro-2,3,4-tri-*O*-benzyl-β-D-glucopyranose using phosphorus pentachloride as catalyst and subsequently removing benzyl groups.[24,25] The product has a molecular weight of 32,400–36,500 and $[\alpha]^{25}D$ $+ 196°–200°$. Infrared and NMR spectra indicate the absence of β-D-glucosidic linkages. The product undergoes enzymic hydrolysis similar to natural dextran.[26]

Several organisms produce dextrans but only *Leuconostoc mesenteroides* and *L. dextranicum* have been used commercially. *Betacoccus arabinosaceous* (Birmingham strain) is considered to be the same as *Leuconostoc*. Reviews on the manufacture and uses of dextrans have been given by Behrens and Ringpfeil[27] and Foster.[28]

Whole-culture process.—Specific procedures for the preparation of two water-soluble dextrans (from strains NRRL B-512F and B-1146) and one water-insoluble dextran (from B-523) are given by Jeanes.[29] The composition of the medium for the buildup of inoculum and for production of B-512F dextran is, in grams per liter of distilled water: sucrose, 100.0; Difco yeast extract, 2.5; magnesium sulfate heptahydrate, 0.20; dipotassium hydrogen phosphate, 5.0. The initial pH is adjusted to 7.0. The culture is incubated at 25° until the pH reaches 4.8, which should also be the point of maximum viscosity.

The preculture in which the bacteria are produced may be supplied with some carbon source other than sucrose, although sucrose is always used for the production of dextran. For example, four pure strains of *Leuconostoc mesenteroides* were cultivated on a substrate of 10% L-arabinose or D-xylose,[30] and preliminary cultures of *L. mesenteroides* were prepared utilizing 0.01–2.0% of a carbon source of sodium acetate, D-glucose, D-fructose, or sucrose in low concentrations.[31] Strains of *L. mesenteroides* have also been cultivated on low-molecular-weight fatty acids, di- and tricarboxylic acids, and low-molecular-weight alcohols. However, low concentrations of sucrose, which are not conducive to the formation of dextran, are most commonly used. A preculture used by Farbwerke Hoescht Akt.[23] for multiplying *L. mesenteroides* consists of yeast extract, 0.15%; meat extract, 0.15%; sodium chloride, 0.35%; potassium dihydrogen phosphate, 0.13%; dipotassium hydrogen phosphate, 0.1%; and sucrose, 1%.

The methods for preparing native dextrans have not changed recently but several refinements have been patented. In a semicontinuous process,[31] 80–90% of a preculture of *L. mesenteroides* is used as an inoculum for a nutrient solution containing a large amount of sucrose. Fresh nutrient is added to the remaining preculture, and incubation is repeated in the same vessel; again, 80–90%

is used for inoculating a new culture. This procedure can be repeated ten times without affecting the quality of the dextran. Malek and Lacko[33] reserve a portion of the dextran-producing medium as inoculum for the next fermentation. This procedure can also be repeated ten times without adversely affecting dextran quality. The rate of production of dextran is increased by separating 40–80% of the medium at the onset of decline in maximum production rate. This is worked up to the final product while fresh medium is added to the residual to make up the original volume.

An impure dextran, which is nevertheless suitable for technical use, is obtained using final molasses as substrate.[34] Jiricek[35] developed a procedure using, in the fermentation medium, 0.5–0.7% of calcium carbonate, which acts as a buffer and stimulator of dextransucrase, allows higher concentrations of sucrose to be used, and consequently gives higher yields of dextran.[35]

Enzyme synthesis.—Jeanes[36] states that, among the numerous advantages of this method over production in growing cultures, are greater uniformity of product, resulting from synthesis under more constant and favorable conditions, and simpler purification of the product.

Ebert and Schenk[37] purified and characterized dextransucrase from *Leuconostoc mesenteroides* NRRL B-512F. Jeanes[36] gives a method of synthesis of dextran using this dextransucrase. The medium for the first stage of inoculum buildup contains in grams per 100 ml of distilled water: sucrose, 2.0; Difco yeast extract, 0.5; Difco Tryptone, 0.25; and dipotassium hydrogen phosphate, 0.5. The medium for successive stages includes small amounts of metal salts and omits the Difco Tryptone. Control of pH is important because the maximum rate of enzyme production is at about pH 6.7; however, inactivation is also rapid at this pH, the enzyme being more stable at pH 5.2–5.0. Bacterial cells are removed by centrifugation. Dextran is produced simply by combining an appropriate dilution of cell-free culture medium at pH 5.0–5.2 and sucrose and allowing the reaction to proceed at 25°–30°.

At low temperatures,[38] high-viscosity dextrans are produced in a cell-free enzymic synthesis by enzymes from *L. dextranicum* and *L. mesenteroides* strains NRRL B-512, B-1146M, B-1298, and B-1149. A review of dextransucrase is given by Tsuchiya.[39]

Production of dextrans from Streptococcus strains.—In 1956, Bailey and Oxford[40] reported that many strains of *Streptococcus bovis* of rumen origin give good yields of dextran when incubated anaerobically for one day at 38° in a liquid sucrose medium. Dextran is produced from *Streptococcus* NRRL B-1351 using a whole-culture process.[41] The culture medium, composed of sucrose, tryptose (a protein hydrolyzate product of the Fischer Scientific Co.), yeast extract, and dipotassium hydrogen phosphate, is adjusted to pH 7.6 and inoculated with the bacteria. Dextrans with an average molecular weight range of 25,000–200,000 can be produced by changing the culture medium and by fractionation. Rogovin and coworkers[42] have produced clinical dextran (m.w. 40,000-70,000) by direct fermentation with *Streptococcus* NRRL B-1351 strain DS-50. Their medium consists of 15% sucrose, dipotassium hydrogen phosphate, and yeast hydrolyzate. The fermentation is complete in 48 hr and a 43% yield of dextran is obtained. Cybulska and Pakula[43] describe Streptococcal dextransucrases and state that the optimum conditions for the enzymic synthesis of dextran require a temperature of 37°–45° and a pH of 7. Using enzymes from two streptococci, *Streptococcus sanguis* type I/II and *Streptococcus* strain Challis, dextrans containing about 51% of (1→6) linkages are produced.

2. *Production of Clinical Dextran*

Directed synthesis of clinical dextrans.—Native dextran normally has a high molecular weight, which makes it unsuitable as a blood plasma substitute. Clinical dextrans have previously been prepared by controlled depolymerization of native dextrans. Consequently, an intensive research effort, especially by Behrens and coworkers at VEB Serum-Werk Bernburg, has been directed to the one-step synthesis of clinical-size dextran from sucrose.

Dextrans with molecular weights less than 60,000 are prepared from nutrient solutions with high concentrations of sucrose and small amounts of D-glucose, D-fructose, or maltose.[44] Low-molecular-weight dextran with DP 50–400 can be prepared by inoculating *L. mesenteroides* into a liquor containing, besides the conventional nutrients, 20–80% of sucrose and 0.1–10% of a low-molecular-weight sugar, such as D-glucose, D-fructose, maltose, and oligosaccharides of the maltose or isomaltose series.[45] The synthesis of low-molecular-weight product is further controlled by the addition of a primer of low-molecular-weight dextran. Dextrans with a molecular weights of 10,000–50,000 are produced in cultures containing 20–40% of sucrose, 1–6% of D-glucose, and 0.5–4% of a low-molecular-weight dextran.[46, 47] Dextrans with molecular weights controlled in the range of 40,000–300,000 are obtained in 85–90% yield by culturing *L. mesenteroides* in a medium containing 10% of sucrose and 2% of a primary dextran (m.w. 20,000), as well as other nutrients.[48] A further control is obtained by using partial hydrolyzates of animal and plant proteins as a carbon source in the inoculum.[49] Using this procedure, the molecular weight of product dextran (M_1) can be calculated.[50, 51]

$$M_1 = M_2\,(1 + C_1/C_2)$$

where M_2 is the molecular weight of primer dextran, C_2 is its concentration, and C_1 is the concentration of the glucopyranosyl units of sucrose.

Novak and Witt[52] produced dextran of desired molecular weight by concurrent synthesis from a sucrose-containing medium by *L. mesenteroides* enzyme and degradation of the dextran by a dextranase from *Aspergillus wentii.*

Chemical methods of depolymerization of native dextrans.—The classical method of producing dextrans suitable for clinical use is to partially depolymerize native dextrans by acid-catalyzed hydrolysis.[53, 54] The desired molecular weight is obtained by limiting reaction time, temperature, and acid concentration. This basic process is used by U.S. commercial firms[55–57] and the USDA.[58, 59] The fermentation mixture may even be hydrolyzed directly.[60, 61] Increased yields are obtained by restricting the extent of hydrolysis and rehydrolyzing the higher-molecular-weight fractions recovered.[62] Hydrolysis in the presence of a water-miscible alcohol or ketone apparently inhibits cleavage and permits more uniform, higher viscosity products.[63, 64] A process involving hydrolysis in alternate steps with solvent extraction increases the yield of dextran in the clinical molecular weight range;[65] low-molecular-weight dextrans are extracted into the solvent layer.

The kinetics of the acid-catalyzed depolymerization reaction have been examined.[59, 66, 67] The activation energy for the reaction is 33.0 kcal/mole. Antonini and coworkers[68] show that random splitting of a statistically branched polymer occurs.

Dextran solutions are simultaneously depolymerized and depyrogenized by heating in the presence of hydrogen peroxide.[69, 70] Iron(II) salts[71] and calcium carbonate,[72] which maintains the pH at 7.0 in the depolymerization mixture, may be used in conjunction with hydrogen peroxide.

Partial degradation of dextran is effected by treatment with alkali metal hydroxides.[73, 74] Treatment with potassium hydroxide[75] at 26° reduces the molecular weight of dextran from 5×10^6 to $1.5–5.0 \times 10^5$.

Dextran is depolymerized by alcoholysis in polyhydric alcohols, which, on heating, are both solvent and reactant.[76] Small amounts of acid catalyze the reaction. Heating in methyl sulfoxide at 100°–150° also depolymerizes dextran.[77] Depolymerization can also be achieved by heating in the presence of water and neutral salts at 120°–160° under pressure.[78]

Enzymic methods of depolymerization of native dextrans.—Enzymes from molds, such as *Penicillium funiculosum* and *Verticillium coccorum,* are capable of degrading dextran.[79] Potent dextranase products are elaborated by four strains of *Penicillium funiculosum,* four strains of *P. lilacinum,* a *P. verrucolosum* strain, and a *Spicaria violacea* strain.[80] Dextranases from a strain of *P. lilacinum* and a strain of *P. funiculosum* hydrolyze *Leuconostoc mesenteroides* (Birmingham) dextran to complex mixtures of straight-chain and branched oligosaccharides.[81] In the vicinity of α-D-(1→3) branch points, several α-D-(1→6) linkages of the main chain are resistant to hydrolysis by these dextranases.[82] The monoenzymic nature of the *P. funiculosum* enzyme has been shown, and the activation energy of the hydrolysis reaction is calculated as 17.4 kcal/mole, which is about 54% of that for acid-catalyzed hydrolysis.[83] The incomplete hydrolysis of dextran by this enzyme can be explained by the decrease in affinity of dextranase toward substrate with decreasing molecular weight.[84] Excellent yields of dextranase are obtained from *P. funiculosum, P. lilacinum,* and *Spicaria violacea* grown on oxidized dextran that has most of the carbonyl groups at C-3 of the D-glucopyranosyl units as the carbon source.[85]

Lactobacillus bifidus dextranase hydrolyzes *L. mesenteroides* (Birmingham) dextran to a complex mixture of oligosaccharides.[86] Dextranases can be isolated from species of *Chaetomium, Streptomyces, Humicola, Sporotrichum, Anixella, Macrosporium, Gibberella, Gloeosporium,* and *Glomerella.*[87] A dextranse from *Chaetomium gracile* F-218-8 lowers the viscosity of a dextran solution to half its initial value in 13 min.[87]

Aspergillus wentii elaborates an endodextranase that degrades dextran to relatively low molecular weights.[88, 89] Clinical dextrans are produced by adding the dextranase to the dextran culture solution after boiling to inactivate any bacterial enzymes and cooling.[90] A refinement of this process concurrently produces dextran with a *L. mesenteroides* enzyme and degrades it with *Aspergillus wentii* dextranase.[52] An glucoamylase concentrate from *Asperigillus niger* degrades dextrans completely to D-glucose.[91]

Dextran $(1\rightarrow6)$-glucosidase isolated from human and rabbit livers cleaves not only α-D-$(1\rightarrow6)$-glucosidic bonds but also α-D-$(1\rightarrow3)$-bonds.[92] Normally animal dextran glucosidase hydrolyzes α-D-$(1\rightarrow3)$ bonds more readily than α-D-$(1\rightarrow2)$ or α-D-$(1\rightarrow6)$ bonds, but if the dextran is preincubated with concanavalin A, this order is reversed, apparently because the concanavalin A complexes with the terminal nonreducing glucose units in dextrans with α-D-$(1\rightarrow3)$ bonds and retards enzyme activity.[93]

Physical methods of depolymerization of native dextrans.—Lockwood and coworkers[94] and Stacey[95] used ultrasonic vibrations to depolymerize dextrans. In an aqueous dextran solution exposed to 840 kc/sec ultrasonic waves, the molecular weight decreased from 35,000 to a minimum of $\sim$20,000 after 3 hr.[96] Ultrasonic degradation occurred more slowly when carbon dioxide instead of air was bubbled through the solution.[97] Hawley and Dunn[98, 99] measured the ultrasonic absorption of dextran in aqueous solutions as a function of molecular weight and frequency. Sonic vibrations showed a tendency to cleave $(1\rightarrow6)$ linkages, whereas ultraviolet irradiation tended to cleave non-$(1\rightarrow6)$ linkages in solutions of B-512 dextran.[100]

Partial degradation of dextran is effected by subjecting it to an alternating current.[101, 102] Degradation by gamma irradiation has been extensively studied; the main effects observed are depolymerization, a change in branching patterns, and oxidative reactions.[103–105] Whistler and Ingle[106] review the effect of high-energy radiation on some polysaccharides, including dextrans. Cathode rays also have been shown to depolymerize dextran.[107]

3. *Recovery*

Native dextran is normally recovered from culture media by precipitation with one volume of a water-miscible organic solvent, such as methanol, ethanol, 2-propanol, or acetone.[55, 58, 108, 109] Some nondextran impurities may be separated initially by the addition of 15–30% alcohol.[110] Precipitated dextran is dissolved in warm water and reprecipitated to obtain a dextran of acceptable purity.[109] Dextran can also be purified by continuously putting small increments of a 5% solution of dextran and methanol in contact in a colloid mill to obtain a suspension that is then centrifuged.[111] Contaminating D-fructose can be removed by passing a solution through a column of either sulfonic acid cation-exchange or quaternary ammonium anion-exchange resin.[112]

Alkaline earth hydroxides may also be employed to precipitate dextran from culture media.[113, 114] D-Fructose in culture media can be consumed by fermention with yeast before precipitation with alkaline earth hydroxides.[115]

Dextran hydrolyzates are also usually precipitated with organic solvents, but having lower molecular weights, they require higher concentrations of organic solvent. To reduce the consumption of solvent, Malek[116] proposed reducing the dielectric constant of the partial hydrolyzate with an polar organic solvent and precipitating the desired fractions by adding a concentrated alkaline hydroxide solution to raise the pH to 12–14. In another process, high-molecular-weight components are removed with ethanol or acetone, and then phosphate buffer is added; this reduces solvent consumption by 12–14.4 times.[117] A different approach is to add the organic solvent at room temperature and cool to precipitate the fractions.[118]

III. Structure

One constant feature of dextrans is their almost exclusive content of α-D-glucopyranosyl units. The other chemical and physical characteristics of the dextrans vary considerably. More than 50% of the linkages are (1→6) and the remainder are (1→3), (1→4), and/or (1→2).[119]

Periodate oxidation of *Leuconostoc mesenteroides* NRRL B-512 dextran shows that there are 95.6% of (1→6) linkages and that the remaining linkages are equally divided between (1→3) and (1→4) linkages.[120] Sloan and coworkers[121] reports 95% (1→6) linkages and 5% (1→3) linkages for a B-512 dextran. Moyer and Isbell[122] report that this dextran has 93.2% (1→6), 4.4% (1→3), and 1.3% (1→4) linkages. Lindberg and coworkers,[123] investigating B-512 dextran, find that all main chain residues are α-D-(1→6) linked and that all branches originate at C-3.

On evidence from periodate oxidation, infrared spectometry, and optical rotation, dextrans from six strains of *Streptococcus bovis* have been purported to possess only α-D-(1→6) linkages and, therefore, to be unbranched.[124] However, although synthetic linear dextran does not precipitate with concanavalin A, *S. bovis* dextran produces a typical precipitation curve and gives nigerose on acetolysis, indicating that the molecule is branched.[125] Four clinical dextrans examined by methylation and periodate oxidation show branch points at C-3, and two of them are also branched at C-4.[126] A Cuban dextran examined by periodate oxidation, Smith degradation, and methylation seems to be similar to a B-512 dextran with estimated 93% α-D-(1→6), 6.1% α-D-(1→3), and 1.1% α-D-(1→4) linkages.[127] Misake and coworkers[128] have produced a dextran from *Leuconostoc* 117 that has a structure similar to that of B-512 dextran, with 94.5% α-D-(1→6) and 5.5% α-D-(1→3) linkages.

On methylation and hydrolysis, an *Acetobacterum capsulatum* dextran yields 2,3,4,6-tetra-, 2,3,4-tri-, and 2,3-di-*O*-methyl-D-glucose in the percentage molecular ratio 7.6 : 83.2 2 : 9.2, showing that it is a (1→6)-linked dextran with (1→4) branches.[129] Methylation analysis and partial acid-catalyzed hydrolysis have led one group[130–132] to propose a repeating structure for B-512 dextran of approximately 23 D-glucopyranosyl units, of which 21 are (1→6) linked, one is an end-group, and one is a branch point linked (1→3). Methylation analysis of *L. mesenteroides* strain C (NRRL B-1298) suggests that the molecule has an average chain length of 4–5 units with most branches at C-2, a smaller number at C-3, and some branching at both C-2 and C-3 of a single unit.[133] The small amount of 2,4,6-tri-*O*-methyl-D-glucose isolated may not be structurally significant but may be the result of undermethylation or demethylation during hydrolysis. The relative rate constants for the methylation of hydroxyl groups at C-2, C-3, and C-4 in dextran are calculated to be 8 : 1 : 3.5.[134] The strain C dextran gives a strong precipitin reaction with type-12 *Pneumococcus* antiserum and with concanavalin A.[133] 2,4,6-Tri-*O*-methyl-D-glucose is also isolated from the methylation analysis of dextran from Tibi grains, which are a symbiotic association of *Lactobacillus brevis, Streptococcus lactis,* and *Saccharomyces cerevisiae.*[135] The dextran is mainly (1→6) linked with branches at C-3. From periodate oxidation and methylation studies, the dextran produced by the cariogenic bacterium *Streptococcus mutans* E49 is found to be highly branched with 18–20% of (1→3) linkages.[136] About 8 mole percent of 2,3,6-tri-*O*-methyl-D-glucose is isolated from methanolysis of the methylated dextran but this is ascribed to contaminating amylose that is produced at the same time.

Fragmentation analysis has been very useful in the structural analysis of dextrans. Glycosidic bonds to primary and secondary hydroxyls have different reactivities toward cleavage by acid-catalyzed hydrolysis and acetolysis. For example, partially hydrolyzed *L. mesenteroides* strain C (NRRL B-1298) dextran yields the homologous series of isomaltodextrins, whereas acetolysis gives kojibiose, nigerose, isomaltose, and trisaccharides derived from the branch points.[137] A method for the production of isomaltodextrins by acid hydrolysis is given by Whelan.[138] It is postulated that approximately 80% of the branches in B-512 dextran consist of single D-glucopyranosyl units.[130–132] A carboxylated dextran prepared by catalytic oxidation of B-512 dextran, on subsequent partial hydrolysis, gives as the predominant aldobiouronic acid, 6-*O*-(α-D-glucopyranosyluronic acid)-D-glucose, indicating that most of the side chains are longer than one D-glucose residue.[123] However, similarly oxidized dextrans of *L. mesenteroides* NRRL B-1375 and NRRL B-1415, yield 3-*O*- and 4-*O*-(α-D-glucopyransyluronic acid)-D-glucose, respectively, showing that branches in these dextrans consist mainly, if not exclusively of single D-glucopyranosyl units.[139]

Acetolysis is important in establishing branching linkages. Yamauchi and Matsuda[140] have isolated from *L. mesenteroides* B (NRRL B-421) dextran, isomaltotriose, 3-*O*-α-isomaltosyl-D-glucose, 3,6-di-*O*-α-D-glucosyl-D-glucose, and nigerotriose, the isolation of nigerotriose suggesting that some (1→3) linkages are contiguous. Sixteen bacterial dextrans subjected to acetolysis have yielded the disaccharide fragments: maltose from 8 dextrans, kojibiose from 12, and nigerose from all 16. Type-12 cross reactivity can be correlated with the amount of kojibiose isolated.[141] The main products on acetolysis of NRRL 1355-S dextran are D-glucose, 21%; isomaltose, 2%; nigerose, 20%; and 3-*O*-α-isomaltosyl-D-glucose, 2%.[142] Nigerose is also produced by acetolysis of B-421 dextran and, in trace amounts, from B-1229 dextran[143] and from B-142 dextran,[144] whereas kojibiose is obtained from B-1229 dextran,[143] B-1299 dextran,[145] and B-1424 dextran.[146]

Dextran B-1415 has 14% (1→4) linkages, whereas dextran B-1416 has 10% (1→3) and 7% (1→4) linkages.[147] Dextran B-1415, upon enzyme-catalyzed hydrolysis by dextranases from *Penicillium lilacinum, P. funiculosum,* and *Lactobacillus bifidus,* produces a series of oligosaccharides containing one D-glucopyranosyl unit joined by an α-D-(1→4) linkage to a D-glucopyranosyl unit of a homologue of isomaltose.[148]

Immunological reactions can give some indication of the degree of the type of branching. A dextran containing more than 90% of α-D-(1→6) linkages does not react with concanavalin A.[149] A dextran from *L. mesenteroides* strain C (NRRL B-1298), which is highly branched at the C-2 and C-3 positions, gives a strong precipitin reaction with type-12 *Pneumococcus* antiserum and with concanavalin A.[133] The proportion of (1→2) branch points can be correlated with the strength of the reaction with type-12 antiserum.[141]

The NMR spectra of branched dextrans contain a peak not found in the spectra of unbranched dextrans.[150] This peak is assigned to the C-1 protons of the non-(1→6) linkages and thus provides a quantitative measure of the ratio of (1→6) to non-(1→6) linkages. Introduction of sulfate groups[151] into unbranched dextran causes the appearance of a new signal, but the signal coincides with one caused by branching and is therefore not diagnostic of sulfate esters in branched molecules.

Infrared spectroscopy can be used to distinguish between α-D- and β-D-anomers of D-glucans, and to identify (1→3) linkages in dextrans.[152–158] A weak absorption at 788–794 cm^{-1} is attributable to α-D-(1→3) linkages,[153–157] whereas bands in the regions near 765, 905 and 915 cm^{-1} are assigned to α-D-(1→6) linkages.[157, 158]

From the shift in optical rotation of 16 degraded dextrans in water and cuprammonium solutions, (1→4) and (1→2) linkages could be distinguished.[159]

Physiocochemical studies can give useful information about dextrans. Mathematical relationships between viscosity and molecular weight have been formulated.[160–162] Through studies of the intrinsic viscosity of hydrolyzed dextrans, a branching parameter can be calculated that is an index to the type and degree of branching.[163, 164] Dextran fractions derived from methanol precipitation[165] and gel filtration[166] can be isolated and their molecular weights determined. From these values, the molecular dispersity is calculated. The value of M_w/M_n is calculated to be 1.3 by translational diffusion studies but 1.77 from a combination of light-scattering, osmotic pressure, and end-group analysis, the difference being attributed to polydispersity and shape of molecules.[167] Molecular weights with values as high as 6×10^8 for native dextrans are obtained by light-scattering and ultracentrifugation,[168] but the maximum number-average molecular weight was found to be only about 200,000. The high values are believed to be caused by the formation of very stable aggregates. The reducing end-groups of native dextrans can be reduced with tritium-labeled lithium borohydride and the number-average molecular weight can be calculated from the activity.[169] From calculations of radii of gyration, the effect of branching on dextran molecules is to decrease the effective solvated volume compared with linear dextran.[170] The radiuses of equivalent spheres and the molecular volumes of macromolecules, including dextrans, can be calculated from the point of precipitation with tobacco mosaic virus.[171] Dextran 2000 (m.w. 2,000,000) has a molecular volume of 264×10^6 A and a radius of equivalent sphere of 398 A, whereas in Dextran C (m.w. 60,000) the corresponding values are 6.84×10^6 and 118. Many of the properties of dextran solutions can be explained by the presence of molecular associations of helical structures.[172] Their solubility is improved by ultrasonic or acid treatment, which irreversibly destroys the helix.

IV. Uses

Dextran has been used mainly as a blood plasma extender and this application has received most attention from researchers. A review of the dependence of biological effects on the molecular weight has been written by Kjellman.[173] Two dextrans commonly used have molecular weights of 40,000 and 70,000. Kjellman[174] lists the U.S., Swedish, and British pharmacopeial specifications with regard to molecular weight dispersity of dextrans with average molecular weight 75,000 and states that such a dextran with predominantly (1→6) linkages causes few side effects, is serologically indifferent, and is completely metabolized. Dextran with molecular weight 70,000, on infusion, reduces hematocrit and depresses fibrinogen but slightly increases plasma viscosity and is used primarily for volume expansion.[175, 176] It has been reported to both increase erythrocyte aggregation[176] and to decrease it and platlet adhesiveness.[175] Dextran with molecular weight of 40,000 readily crosses the extravascular space, is rapidly excreted by the kidneys, and therefore has limited expansion powers. It is used to reduce blood viscosity and to improve microcirculation at low flow states. It reduces hematocrit and erythrocyte aggregation.[175, 176] Dextran has been of use in over 300 clinical applications. In human patients, 75,000 and 40,000 molecular weight dextrans cause no change in Duke or Ivy bleeding time, but in patients with hemorrhagic diathesis, only 40,000 molecular weight can be used safely.[177] Dextran of molecular weight 15,000 is reported to slightly reduce the sedimentation rate and viscosity of blood, but dextrans with molecular weights of 45,000 to 450,000 cause increases in both factors roughly proportional to the molecular weight.[178] On intraparenteral injection into rats, dextrans with molecular weights 40,000, 75,000, and 200,000 cause edema, whereas on slow intravenous infusion they cause arterial hypertension.[179] On an equimolar basis, on intraparenteral injection, the 200,000 molecular weight dextran was least edematogenic, but on intravenous infusion, it was most hypotensive. A review of the effect of low-molecular-weight dextrans on blood-sludging has been written by Lee and Walsh.[180] After examining the effects on blood coagulation of dextran (m.w. 40,000), they have concluded that the dextran should not be administered to people with hemorrhagic disorders.[181] Low-molecular-weight dextran does not lower the viscosity of blood but erythrocyte aggregation and rouleaux formation are rapidly decreased by its addition.[182] Reviews have been written on various aspects of the use of dextran as a blood plasma substitute, including preparation, testing, and applications;[183] chemical, physical, biological, and immunological properties of dextran; its metabolism in humans;[184] and the treatment of shock with dextran.[185] Radioactively labeled dextrans are used to follow its metabolism. [^{14}C]Dextran distributes itself in dogs to produce high concentrations in liver and lymph nodes.[186] After plasma levels fall to negligible amounts, $^{14}CO_2$ is still formed, indicating that it is metabolized after deposition in tissues. Single injections of [^{3}H]dextran administered to guinea pigs show that low-molecular-weight fractions are degraded more rapidly and that the declining order of retention by the various organs is spleen, skin, lungs, liver, kidneys, adipose tissue, skeletal muscle, brain.[187]

Dextrans are antigenic, and this antigenicity can be used to indicate their structures.[188, 189] Injection of dextran with molecular weight 800,000 causes necrosis of the liver and kidney cells of dogs, whereas humans injected with dextran of molecular weight 70,000 exhibit itching, urticaria, headache, and bronchospasm.[190] These effects are assigned to excessive branching because, in B-512 dextran, both branching and side effects are minimal.

There are many reports of the use of dextran as an antilipemic agent. Intravenous injection of dextran is reported to lower the cholesterol and total lipid levels in atherosclerotic patients.[191] It is suggested that this might cause the release of endogenous heparin, which then activates the clearing factor.[192] Other reports also indicate decrease in plasma lipid levels after infusion of dextrans.[193, 194] However, some recent reports refute the claims of antilipemic activity by dextrans. Repeated intravenous infusions are found to inhibit rather than activate lipoprotein lipase in human plasma,[195] and no lipid-lowering effect can be noted in experimental hypercholesterolemic rabbits.[196] Dextran is also reported to effectively lower lipid and cholesterol levels in rabbits fed a high-cholesterol diet.[197–199] General improvements and cholesterol lowering are noted in patients with an atherosclerotic condition[200–201] and in patients with myocardial infarction[202] after dextran treatment. However, using a large number of tests and statistical methods, other researchers show that dextran with a molecular weight of 40,000 has no effect on cholesterol levels in rats or humans.[203]

Dextran has a beneficial effect in aggregating clay soil particles. Low-molecular-weight dextran (5000–50,000) selectively aggregates sand particles in soil, whereas dextran with a molecular weight of 100,000 or higher aggregates silt or clay particles.[204] Dextran soil-conditioner, the molecular weight depending on the type of soil, can be combined with a fertilizer.[205] Certain dextran products when applied to the soil also enchance seedling emergence, rate of plant growth, and crop yields.[206] The dextran can be produced in the soil by the addition of appropriate organisms.[207] Dextran with molecular weight in excess of 1 million improves porosity in soil, water retentivity, and resistance to erosion.[208]

A list of examples of other uses to which dextrans have been applied is given in Table I.

V. Derivatives

Only the more important dextran derivatives will be discussed in detail. A list of some other derivatives and other uses is given in Table II.

TABLE I

Uses of Dextrans

Product	*Dextran Function*	*Ref.*
Pharmaceutical		
Cryoprotective	inhibits cell damage on freezing	209–211
X-ray opaque compositions	suspending agent	212
Water-insoluble vitamin preparations	stabilizing agent	213
Tablets	binding agent	214
Sustained-action tablets	protracts dissolution	215
Chloral–dextran complex	suppresses taste and stomach irritant action	216
Microcapsules of kerosine, menthol, aspirin, etc.	methylcellulose–dextran encapsulating substance	217
Cosmetic preparations	wrinkle smoothing	218
Foods		
Syrups and candies	improves moisture retentivity and body and inhibits crystallization	219
Gum and jelly confections	gelling agent	220
Ice cream	prevents shrinkage and ice formation	221
Icing compositions	stabilizing agent	222
Pudding compositions	bodying agent	223
Oil drilling		
Drilling fluids	dextran–aldehyde complex, inhibits water loss and coats well wall	224
Solutions for flooding underground reservoirs	increases viscosity of water	225, 226
Drilling muds	protective colloid	227, 228
Miscellaneous		
Olefinically polymerizable resins	filler and modifier	229, 230
Alumina manufacture	sedimentation agent	231, 232
Purification of caustic soda	iron–dextran complex precipitates	233
Metal powder production	gel precipitation suppresses crystal growth	234
Nuclear fuel production	complexing agent	235

1. Inorganic Esters

The anticoagulant properties of dextran sulfates are well documented. Toxic side effects are minimized by selecting suitable molecular weights.[236, 237] A dextran sulfate with reduced viscosity of 0.037 has both anticoagulant and anti-

TABLE II

Dextran Derivatives

Derivative	*Uses*	*Ref.*
Esters		
Triacetate	Film former	284, 285
Stearates and palmitates	chemical glove insoluble in water and soap and water, but soluble in non-polar solvents	286
	waterproof regenerated cellulose	287
	waterproofing fabrics	288
	improves cohesion of polyethylene coatings to metal surfaces	289
Dextran sulfate	treatment for simple goiter	290
	treatment for peptic ulcers	291
	photographic emulsions with increased covering power	292, 293
Dextran phosphate	high-viscosity gum	294
	inhibits peptic activity	295
Dextran nitrate or nitrate sulfate	gelled propellants and explosives	296, 297
Ethers		
Sulfoalkylated dextran	photographic emulsions with increased covering power	298
Benzyldextran	delayed release medicaments	299, 300
Hydroxyalkyldextrans	blood plasma extenders with improved storage qualities	301
Carboxymethyldextran (salt and acid forms)	auxiliary binding substance in tablets	302, 303
	fruit syrup thickener	304
	ice cream stabilizer	305
	carrier in antibiotics, pesticides, and germicides	306
	soil conditioner	307
	latex stabilizer	308
	water-emulsion paints	309
	disperses and stabilizes water-insoluble dyes	310
	water-soluble films for packaging of detergents, soaps, foods, and soil conditioners	311
	prevents precipitation of hydrous oxides of multivalent ions in alkaline solution	312
	aqueous gels	313
	organogels	314
Carboxymethylbenzyldextran	delayed-release medicaments	315

TABLE II—Continued

Derivative	*Uses*	*Ref.*
Miscellaneous		
Dialdehyde dextran	reacted with isoniazid gives low-toxicity, prolonged-reaction, tuberculostatic compound	316
	prolonged action drugs by reaction with novocaine, tubazid, etc.	317
Dicarboxy dextran (dialdehyde dextran oxidized with $NaClO_2$)	prolonged action drugs	317
Iododextran	water-soluble, nontoxic, nonirritant compounds of iodine	378
p-Iodobenzenesulfonic acid ester	increases electron-optical contrast without increasing toxicity of original dextran	319
Acylated amino acid esters	water-soluble gums and resins, impart affinity for acid wool dyes	320

lipemic properties.[238] Salts of pectin, xylan, and dextran sulfate and chondroitin sulfate have a synergistic effect in compositions with heparin and prolong blood-clotting time by up to eight times that obtained with heparin alone.[239] Low-molecular-weight polysaccharide sulfates, including those of dextran, exhibit very low anticoagulant activity while being useful in the treatment of lipemia and arteriosclerosis.[240] These low-molecular-weight sulfates can be administered orally and may be enteric coated for release in intestinal fluid.[241] Dextran sulfates containing 2.0–13.0% sulfur with intrinsic viscosities from 0.020 to 0.050 have no significant anticoagulant activity, but have lipolytic activity.[242] Another low-molecular-weght dextran sulfate lowers plasma lipid levels on oral or intravenous administration, has low anticoagulant activity, and has no adverse side effects.[243] Hyperlipemia that develops after burning of rabbits can be prevented by perfusion with dextran sulfate.[244]

The toxicity of dextran sulfate depends largely on its molecular weight. *In vivo,* fibrinogen is precipitated by dextran sulfate and results in embolism of the vessels of the lungs and brain.[245] The toxicity and precipitating action increases with the molecular weight of the ester. In the normal rat, intravenous injection of dextran sulfate with a molecular weight of 500,000 causes general arterial hypotension that sometimes results in death.[246] Intravenous injection of dextran sulfate and carboxymethyldextran sulfate with molecular weight 71,000 causes necrosis of the liver and kidneys of experimental animals, making them unsuitable as heparinoids.[247] Low-molecular-weight dextran sulfate injected intravenously daily for extended periods depresses weight gain and produces cachexia and general osteoporesis, effects that are not observed on heparin treatment.[248]

Degraded dextran phosphates are reported to have antilipemic activity,[249] but other work shows that the phosphoric esters of depolymerized dextran with molecular weight 8000 have decreased clarifying activity toward dog plasma compared with the parent dextran of molecular weight 80,000.[250]

2. *Cross-linked Dextran Gels for Molecular Sieves*

One of the most successful applications for dextrans is its use in the manufacture of molecular sieves. The reaction of dextran with an α, ω-diepoxy-, an α, ω-dihalo-, or an α-halo-ω-epoxy-compound and sodium hydroxide yields an insoluble hydrophilic gel that can be partially depolymerized to the required molecular weight by acid-catalyzed hydrolysis[251] or ground to the required particle size.[252] The gel was first produced in bead shape by dissolving the dextran in sodium hydroxide solution and then dispersing it in toluene by stirring, using poly(vinyl acetate) as a stabilizer.[253] Epichlorhydrin is then added and the temperature is maintained at 50° until the beads gel. Another process uses epichlorhydrin itself as the continuous phase in which the aqueous dextran solution is dispersed, using palmitic acid as an emulsifying agent.[254, 255] Molecular sieves are also produced using ethylene glycol glycidyl ether as the cross-linking agent.[256] In all these molecular sieves, the degree of cross-linking determines the water-regain value and pore size and hence the molecular exclusion limit.

Several modifications of this basic cross-linked dextran gel make it an even more useful product. By reacting with 1,2-epoxypropane, a 2-hydroxypropyl-substituted network is obtained that swells in water, ethanol, and chloroform, thereby extending molecular sieve operations to substances soluble in nonpolar solvents.[257] Non-cross-linked organophilic dextran gels suitable for molecular sieves are prepared simply by acetylation of dextran.[258] High-porosity ion exchangers can be produced by reacting the cross-linked dextran gel with suitable ionic groups,[256, 259, 260] or by cross-linking an already substituted dextran molecule, such as DEAE-dextran, with epichlorohydrin (1-chloro-2,3-epoxypropane).[261] Other groups can be substituted on cross-linked dextran, allowing it to act as a chelate[262] or an electron exchanger.[263] Trypsin and chymotrypsin coupled to cross-linked dextran by way of isocyanate derivatives retain their enzymic activity.[264] An organomercurial cross-linked dextran separates SH proteins,[265] whereas organophilic cross-linked dextran is used as the support in solid-state peptide synthesis.[266]

3. *Iron–Dextran Complexes*

The iron–dextran complexes have been very intensively studied in recent years. There have consequently been scores of patents issued for their preparations. Only a representative sample of these is given here.

The iron–dextran complex has great potential use in the treatment of iron-deficiency anemia, because it is soluble, nonionic, and nondialyzable and, therefore, suitable for injection. Dextran or dextran the molecular size of which has been reduced by partial hydrogenolysis and which has an intrinsic viscosity (25°) of 0.025–0.50, is treated with a water-soluble, weakly electrolytic iron(III) salt at 40°–115° and the pH is increased to 11.5 to give the complex containing from 5–30% by weight of iron.[267] The complex is stable on storage and in the pH range 4–11. A complex prepared from dextran, ferric chloride, and sodium carbonate is described as being useful orally, intramuscularly, or subcutaneously, in humans or animals.[268] Clinical studies of oral use on humans show an increase in protein-bound iron, but no toxicity.[269] More recently, mixed complexes of iron with dextran and citric acid or its alkali metal salts have been prepared.[270–272] The complex can be formed by reacting ferric hydroxide with hydrogenolyzed dextran in an alkaline solution to give a completely water-soluble product, after which citric acid or sodium citrate is added.[271, 272] Another treatment reacts alkali-modified dextran with ferric hydroxide to obtain products suitable for the treatment of iron-deficiency anemia.[273] Iron-dextran complexes are used as injectable iron preparations for the treatment of anemia in baby pigs.

Iron–dextran complexes were known to induce injection-site sarcomas in rats and mice.[274–276] However, in Wistar strain rats, weekly injection of the complex for up to 15 months did not induce sarcomas.[277] In 1961, one author advocated its withdrawal from routine use because of its side effects.[278] Problems associated with the use of iron–dextran were reviewed by Schen in 1962.[279] Iron–dextran injected intravenously in rats caused anaphylactoid reaction of lips and paws.[280] In a comparison of iron–sorbitol–citric acid complex and iron–dextran complex treatments of rats, the former did not induce injection-site sarcomas whereas the latter did in a number of cases.[281] Iron–dextran also induced hemolysis caused by peroxidation of erythrocyte lipids.[282, 283]

VI. References

(1) T. H. Evans and H. Hibbert, *Advan. Carbohyd. Chem.*, **2**, 204 (1946); *Sugar Res. Fdn., Sci., Rept. Ser.* No. **11** (1947).

(2) W. B. Neely, *Advan. Carbohyd. Chem.*, **15**, 341 (1960).

(3) C. R. Ricketts, *Prog. Org. Chem.*, **5**, 73 (1961).

(4) B. Ingelman, *Acta Acad. Regiae Sci. Upsal.*, **12**, 9 (1969); *Chem. Abstr.*, **72**, 6208g (1970).

(5) A. R. Jeanes, *in* "Encyclopedia of Polymer Science and Technology," Interscience, New York, Vol. 4, 1966, p. 805.

(6) A. R. Jeanes, "Dextran Bibliography, Pt. A: Patents on Production and Use of Native Dextrans, Partially Degraded Dextrans, and Their Derivatives," U.S. Dept. Agr., Agr. Res. Ser., Peoria, Ill. (1967); *Chem. Abstr.*, **68**, 22902 (1968).

(7) A. S. Aiyar, *Bombay Technol.*, **7,** 82 (1957); *Chem. Abstr.*, **54,** 5136 (1960).

(8) N. Fujiwara and A. Shinoda, *Kobunshi,* **16,** 1192 (1967); *Chem. Abstr.*, **68,** 103838n (1968).

(9) D. N. Sutherland, *Int. Sugar J.*, **70,** 355 (1968); *Chem. Abstr.*, **70,** 21154p (1969).

(10) R. J. Gibbons and S. B. Banghart, *Arch. Oral Biol.*, **12,** 11 (1967); *Chem. Abstr.*, **66,** 73439x (1967).

(11) B. F. Hammond, *Arch. Oral. Biol.*, **14,** 879 (1969); *Chem. Abstr.*, **71,** 88655s (1969).

(12) H. B. Woodruff and T. H. Stoudt, Brit. Patent 1,202,629 (1970); *Chem. Abstr.*, **73,** 123530n (1970).

(13) W. J. King and G. R. Miller, Germ. Offen. 1,940,223 (1970); *Chem. Abstr.*, **72,** 125077y (1970).

(14) G. R. Miller, Germ. Patent 1,963,013 (1970); *Chem. Abstr.*, **73,** 69863h (1970).

(15) P. J. Baker, Jr., *in* "Industrial Gums," R. L. Whistler, ed., Academic Press, New York, 1959, p. 531.

(16) K. H. Ebert and F. Patat, *Z. Naturforsch., B,* **17,** 738 (1962); *Chem. Abstr.*, **58,** 9367 (1963).

(17) K. H. Ebert, G. Schenk, and H. Stricker, *Ber. Bunsenges. Phys. Chem.*, **68,** 765 (1964); *Chem. Abstr.*, **62,** 8072 (1965).

(18) K. H. Ebert, G. Schenk, G. Rupprecht, M. Brosche, W.-W. Hsu, and D. Heinicke, *Makromol. Chem.*, **96,** 206 (1966); *Chem. Abstr.*, **65,** 15524 (1966).

(19) K. H. Ebert and M. Brosche, *Biopolymers,* **5,** 423 (1967).

(20) K. H. Ebert and G. Schenk, *Z. Naturforsch., B,* **23,** 788 (1968); *Chem. Abstr.*, **69,** 53045z (1968).

(21) E. Braswell, A. Goodman, and K. G. Stern, *J. Polymer Sci.*, **61,** 143 (1962).

(22) F. A. Bovey, *J. Polymer Sci.*, **35,** 167 (1959).

(23) E. J. Hehre and H. Suzuki, *Arch. Biochem. Biophys.*, **113,** 675 (1966).

(24) E. R. Ruckel and C. Schuerch, *J. Amer. Chem. Soc.*, **88,** 2605 (1966).

(25) E. R. Ruckel and C. Schuerch, *Biopolymers,* **5,** 515 (1967).

(26) E. T. Reese and F. W. Parrish, *Biopolymers,* **4,** 1043 (1966).

(27) U. Behrens and M. Ringpfeil, *Fortschr. Bot.*, **26,** 420 (1963/1964); *Chem. Abstr.*, **65,** 12398 (1966).

(28) F. H. Foster, *Process Biochem.*, **3,** 15 (1968); *Chem. Abstr.*, **69,** 26053u (1968).

(29) A. R. Jeanes, *Methods Carbohyd. Chem.*, **5,** 118 (1965).

(30) A. Vavra and I. Vavra, *Kem. Ind.*, **13,** 457 (1964); *Chem. Abstr.*, **61,** 16738 (1964).

(31) U. Behrens, M. Ringpfeil, A. Gabert, and K. Krueger, East Germ. Patent 35,367 (1965); *Chem. Abstr.*, **63,** 7624 (1965).

(32) Farbwerke Hoechst Akt.-Ges. vorm. Meister Lucius & Brüning, Germ. Patent 1,003,397 (1957); *Chem. Abstr.*, **54,** 6026 (1960).

(33) J. Malek and L. Lacko, Czech. Patent 96,077 (1960); *Chem. Abstr.*, **58,** 7338 (1963).

(34) U. Behrens and L. Wuensche, *Sobre Deriv. Cana Azucar,* **3,** 39 (1969); *Chem. Abstr.*, **73,** 57413y (1970).

(35) V. Jiricek, Czech. Patent 124,853 (1967); *Chem. Abstr.*, **69,** 34684r (1968).

(36) A. R. Jeanes, *Methods Carbohyd. Chem.*, **5,** 127 (1965).

(37) K. H. Ebert and G. Schenk, *Z. Naturforsch., B,* **17,** 732 (1962); *Chem. Abstr.,* **58,** 9367 (1963).

(38) Farbwerke Hoechst Akt.-Ges., Germ. Patent 1,089,711 (1960); *Chem. Abstr.,* **55,** 14758 (1961).

(39) H. M. Tsuchiya, *Bull. Soc. Chim. Biol.,* **42,** 1777 (1960); *Chem. Abstr.,* **55,** 23622 (1961).

(40) R. W. Bailey and A. E. Oxford, *Nature,* **182,** 185 (1958).

(41) E. J. Hehre, H. M. Tsuchiya, N. N. Hellman, and F. R. Senti, U.S. Patent 2,906,669 (1959); *Chem. Abstr.,* **54,** 2676 (1960).

(42) S. P. Rogovin, F. R. Senti, R. G. Benedict, H. M. Tsuchiya, P. R. Watson, R. Tobin, V. E. Sohns, and M. E. Slodki, *J. Biochem. Microbiol. Technol. Eng.,* **2,** 381 (1960); *Chem. Abstr.,* **55,** 10791 (1961).

(43) J. Cybulska and R. Pakula, *Med. Dosw. Mikrobiol.,* **15,** 273 (1963); *Chem. Abstr.,* **60,** 10982 (1964).

(44) U. Behrens and M. Ringpfeil, East Germ. Patent 21,324 (Appl. 1957); *Chem. Abstr.,* **56,** 3927 (1962).

(45) VEB-Serum-Werk Bernburg, Germ. Patent 1,083,020 (1960); *Chem. Abstr.,* **55,** 15827 (1961).

(46) K. Krueger and D. Brueckner, East Germ. Patent 31,300 (1965); *Chem. Abstr.,* **63,** 6288 (1965).

(47) U. Behrens, M. Ringpfeil, D. Brückner, M. Fröhlich, K. Krüger, H. Theil, and W. Torporski, *J. Biochem. Microbiol. Technol. Eng.,* **3,** 199 (1961); *Chem. Abstr.* **55,** 25150 (1961).

(48) VEB Serum-Werk Bernburg, East Germ. Patent 19,254 (1960); *Chem. Abstr.,* **55,** 21475 (1961).

(49) VEB Serum-Werk Bernburg, East Germ. Patent 35,269 (1965); *Chem. Abstr.,* **63,** 10635 (1965).

(50) VEB Serum-Werk Bernburg, Brit. Patent 888,298 (1962); *Chem. Abstr.,* **57,** 10360 (1962).

(51) E. Leibnitz, U. Behrens, and M. Ringpfeil, *Sitzungsber. Deut. Akad. Wiss. Berlin, Kl. Chem., Geol. Biol.,* 3 (1960); *Chem. Abstr.,* **55,** 13758 (1961).

(52) L. J. Novak and E. E. Witt, U.S. Patent 2,972,567 (1961); *Chem. Abstr.,* **55,** 13780 (1961).

(53) Aktiebolaget Pharmacia, Brit. Patent 583,378 (1946); *Chem. Abstr.,* **41,** 2540 (1947).

(54) A. J. T. Grönwall and B. G. A. Ingleman, U.S. Patent 2,437,518 (1948); *Chem. Abstr.,* **42,** 4311 (1948).

(55) G. H. Bixler, G. E. Hines, R. M. McGhee, and R. A. Shurter, *Ind. Eng. Chem.,* **45,** 692, 1377 (1953).

(56) *Chem. Eng. (New York),* **59,** No. 9, 215 (1952).

(57) *Chem. Eng. (New York),* **59,** No. 12, 240 (1952).

(58) V. E. Sohns, S. P. Rogovin, H. F. Conway, and C. T. Langford, *U.S. Dept. Agr., Agr. Res. Serv. Circ.* AIC-372 (1954); *Chem. Abstr.,* **49,** 16342 (1955).

(59) I. A. Wolff, C. L. Mehltretter, R. L. Mellies, P. R. Watson, B. T. Hofreiter, P. L. Patrick, and C. E. Rist, *Ind. Eng. Chem.,* **46,** 370 (1954).

(60) A. R. Lockwood and G. Swift, U.S. Patent 2,565,507 (1951); *Chem. Abstr.,* **45,** 9813 (1951).

(61) G. S. Stoycos, U.S. Patent 2,687,368 (1954); *Chem. Abstr.,* **49,** 564 (1955).

(62) R. S. Dalter, U.S. Patent 2,727,838 (1955); *Chem. Abstr.,* **50,** 5998 (1956).

(63) L. J. Novak, U.S. Patent 2,789,066 (1957); *Chem. Abstr.,* **51,** 12443 (1957).

(64) Commonwealth Eng. Co. of Ohio, Brit. Patent 811,770 (1959); *Chem. Abstr.,* **53,** 13520 (1959).

(65) R. M. Alsop, Brit. Patent 1,143,784 (1969); *Chem. Abstr.,* **70,** 98132n (1969).

(66) T. Kobayashi and Y. Tsukano, *Nippon Nogei Kagaku Kaishi,* **25,** 421, 424 (1951–1952); *Chem. Abstr.,* **46,** 10226 (1952).

(67) R. W. Jones, R. J. Dimler, and C. E. Rist, *J. Amer. Chem. Soc.,* **77,** 1659 (1955).

(68) E. Antonini, L. Bellelli, M. L. Bonacci, M. R. Bruzzesi, A. Caputo, E. Chiancone, and A. Rossi-Fanelli, *Biopolymers,* **2,** 35 (1964).

(69) I. Levi and E. Lozinski, *Can. J. Biochem. Physiol.,* **33,** 448 (1955); *Chem. Abstr.,* **49,** 10586 (1955).

(70) I. Levi and E. Lozinski, U.S. Patent 2,762,727 (1956); *Chem. Abstr.,* **50,** 16048 (1956).

(71) M. R. Bruzzesi and R. Zito, *G. Biochim.,* **16,** 136 (1967); *Chem. Abstr.,* **68,** 59819f (1968).

(72) F. B. Braun, Germ. Patent 1,000,964 (1957); *Chem. Abstr.,* **53,** 18399 (1959).

(73) Aktiebolaget Pharmacia, Brit. Patent 675,025 (1952); *Chem. Abstr.,* **46,** 11593 (1952).

(74) A. J. T. Grönwall and B. G. A. Ingelman, Can. Patent 520,216 (1956).

(75) P. Papini, M. Feroci, and G. Auzzi, *Ann. Chim. (Rome),* **59,** 890 (1969); *Chem. Abstr.,* **72,** 56959y (1970).

(76) E. Hultin and L. Nordström, *Acta Chem. Scand.,* **8,** 1296 (1954).

(77) Farbenfabriken Bayer Akt.-Ges., Patent 1,007,951 (1957); *Chem. Abstr.,* **54,** 16757 (1960).

(78) M. V. Lillienskiold, Germ. Patent 938,209 (1956); *Chem. Abstr.,* **52,** 17633 (1958).

(79) E. Hultin and L. Nordström, Swed. Patent 133,157 (1951); *Chem. Abstr.,* **46,** 3103 (1952).

(80) H. M. Tsuchiya, A. R. Jeanes, H. Bricker, and C. A. Wilham, U.S. Patent 2,742,399 (1956); *Chem. Abstr.,* **50,** 13378 (1956).

(81) E. J. Bourne, D. H. Hutson, and H. Weigel, *Biochem. J.,* **86,** 555 (1963).

(82) D. H. Hutson and H. Weigel, *Biochem. J.,* **88,** 588 (1963).

(83) T. Kobayashi, *Nippon Nogei Kagaku Kaishi,* **31,** 494 (1957); *Chem. Abstr.,* **52,** 1316 (1958).

(84) T. Kobayashi, *Nippon Nogei Kagaku Kaishi,* **31,** 497 (1957); *Chem. Abstr.,* **52,** 1316 (1958).

(85) R. G. Brown, *Can. J. Microbiol.,* **16,** 841 (1970); *Chem. Abstr.,* **73,** 117414m (1970).

(86) R. W. Bailey, D. H. Hutson, and H. Weigel, *Biochem. J.,* **80,** 514 (1961).

(87) K. Ishibashi, A. Hattori, and M. Arai, Germ. Patent 1,949,719 (1970); *Chem. Abstr.,* **73,** 23906w (1970).

(88) V. Whiteside-Carlson and W. W. Carlson, U.S. Patent 2,716,084 (1955); *Chem. Abstr.,* **49,** 16361 (1955).

(89) V. Whiteside-Carlson and W. W. Carlson, U.S. Patent 2,709,150 (1955); *Chem. Abstr.,* **49,** 12786 (1955).

(90) L. J. Novak and G. S. Stoycos, U.S. Patent 2,841,578 (1958); *Chem. Abstr.,* **52,** 17627 (1958).

(91) H. M. Tsuchiya, A. R. Jeanes, H. Bricker, and C. A. Wilham, *J. Bacteriol.,* **64,** 513 (1952).

(92) A. S. Saenko, A. B. Livshits, T. V. Polushina, and E. L. Rozenfel'd, *Dokl. Akad. Nauk SSSR,* **157,** 723 (1964); *Chem. Abstr.,* **61,** 10928 (1964).

(93) M. E. Preobrazhenskaya and E. L. Rozenfel'd, *Biokhimiya,* **35,** 753 (1970); *Chem. Abstr.,* **73,** 116651z (1970).

(94) A. R. Lockwood, A. E. James, and F. G. E. Pautard, *Research (London),* **4,** 46 (1951).

(95) M. Stacey, *Research (London),* **4,** 48 (1951).

(96) O. M. Zorina and I. E. El'piner, *Biokhimiya,* **28,** 781 (1963); *Chem. Abstr.,* **60,** 4379 (1964).

(97) H. Luecke, *Abhandl. Deut. Akad. Wiss. Berlin, Kl. Med.,* 109 (1964); *Chem. Abstr.,* **63,** 5937 (1965).

(98) S. A. Hawley and F. Dunn, *J. Chem. Phys.,* **50,** 3523 (1969).

(99) S. A. Hawley and F. Dunn, *J. Chem. Phys.,* **52,** 5497 (1970).

(100) P. R. Watson and I. A. Wolff, *J. Amer. Chem. Soc.,* **77,** 196 (1955).

(101) F. G. E. Pautard, *Chem. Ind. (London),* 1316 (1953); *Chem. Abstr.,* **48,** 5235 (1954).

(102) F. G. E. Pautard, Brit. Patent 732,923 (1955); *Chem. Abstr.,* **50,** 1270 (1956).

(103) G. O. Phillips and G. J. Moody, *J. Chem. Soc.,* 3534 (1958).

(104) K. A. Granath and P. O. Kinell, *Acta Chem. Scand.,* **15,** 141 (1961).

(105) S. V. Markevich, S. S. Kharamonenko, P. T. Gorbunov, E. V. Stakhovskii, A. I. Valakhanovich, and M. T. Bondarenko, *Vestsi Akad. Nauk Belarusk. SSR, Ser. Biyal. Nauk,* 107 (1964); *Chem. Abstr.,* **62,** 13481 (1965).

(106) R. L. Whistler and T. R. Ingle, *in* "Starch: Chemistry and Technology," R. L. Whistler and E. F. Paschall, eds., Academic Press, New York, Vol. 1, 1965, p. 409.

(107) F. P. Price, W. D. Bellamy, and E. J. Lawton, *J. Phys. Chem.,* **58,** 821 (1954).

(108) H. J. Koepsell, A. Kazenko, A. R. Jeanes, E. S. Sharpe, and C. A. Wilham, U.S. Patent 2,673,828 (1954); *Chem. Abstr.,* **48,** 8495 (1954).

(109) U. Behrens, M. Ringpfeil, A. Gabert, K. Krueger, D. M. Bella, and V. S. Guerra, East Germ. Patent 62,647 (1968); *Chem. Abstr.,* **71,** 20949k (1969).

(110) L. J. Novak and G. S. Stoycos, U.S. Patent 2,798,065 (1957); *Chem. Abstr.,* **51,** 15901 (1957).

(111) U. Behrens, A. Gabert, and M. Ringpfeil, East Germ. Patent 34,912 (1965); *Chem. Abstr.,* **63,** 4502 (1965).

(112) R. N. Sargent, U.S. Patent 3,184,334 (1965); *Chem. Abstr.,* **63,** 4502 (1965).

(113) E. L. Wimmer, U.S. Patent 2,686,777 (1954); *Chem. Abstr.,* **49,** 1353 (1955).

(114) American Sugar Refining Co., Brit. Patent 716,660 (1954); *Chem. Abstr.,* **49,** 10650 (1955).

(115) E. L. Wimmer, U.S. Patent 2,685,579 (1954); *Chem. Abstr.,* **49,** 7878 (1955).

(116) J. Malek, Czech. Patent 110,446 (1964); *Chem. Abstr.,* **61,** 12182 (1964).

(117) L. Lacko and J. Malek, Brit. Patent 877,688 (1961); *Chem. Abstr.,* **56,** 11717 (1962).

(118) Meito Sangyo Co., Ltd., Germ. Patent 1,246,166 (1967); *Chem. Abstr.,* **67,** 83186x (1967).

(119) A. R. Jeanes, W. C. Haynes, C. A. Wilham, J. C. Rankin, E. H. Melvin, M. Austin, J. E. Cluskey, B. E. Fisher, H. M. Tsuchiya, and C. E. Rist, *J. Amer. Chem. Soc.,* **76,** 5041 (1954).

(120) M. Abdel-Akher, J. K. Hamilton, R. Montgomery, and F. Smith, *J. Amer. Chem. Soc.*, **74,** 4970 (1952).

(121) J. W. Sloan, B. H. Alexander, R. L. Lohmar, I. A. Wolff, and C. E. Rist, *J. Amer. Chem. Soc.*, **76,** 4429 (1954).

(122) J. D. Moyer and H. S. Isbell, *Anal. Chem.*, **29,** 1862 (1957).

(123) B. Lindberg and S. Svensson, *Acta Chem. Scand.*, **22,** 1907 (1968).

(124) R. W. Bailey, *Biochem. J.*, **71,** 23 (1959).

(125) I. J. Goldstein, R. D. Poretz, L. L. So, and Y. Yang, *Arch. Biochem. Biophys.*, **127,** 787 (1968).

(126) J. K. N. Jones and K. C. B. Wilkie, *Can. J. Biochem. Physiol.*, **37,** 377 (1959); *Chem. Abstr.*, **53,** 10666 (1959).

(127) L. C. Brossard, *Sobre Deriv. Cana Azucar,* **2,** 41 (1968); *Chem. Abstr.*, **69,** 68444c (1968).

(128) A. Misaki, S. Yukawa, T. Asano, and M. Isono, *Takeda Kenkyusho Nempo,* **25,** 42 (1966); *Chem. Abstr.*, **66,** 54255t (1967).

(129) S. A. Barker, E. J. Bourne, G. T. Bruce, and M. Stacey, *J. Chem. Soc.*, 4414 (1958).

(130) J. W. Van Cleve, W. C. Schaefer, and C. E. Rist, *J. Amer. Chem. Soc.*, **78,** 4435 (1956).

(131) R. W. Jones, R. J. Dimler, A. R. Jeanes, C. A. Wilham, and C. E. Rist, *Abstr. Papers, Div. Carbohyd. Chem., Meeting Amer. Chem. Soc., New York,* **126th,** 13D (1954).

(132) R. J. Dimler, R. W. Jones, W. C. Schaefer, and J. W. Van Cleve, *Abstr. Papers, Div. Carbohyd. Chem., Meeting Amer. Chem. Soc., Dallas, Texas,* **129th,** 2D (1956)

(133) B. A. Lewis, M. J. St. Cyr, and F. Smith, *J. Org. Chem.*, **33,** 3136 (1968).

(134) B. Norman, *Acta Chem. Scand.*, **22,** 1381 (1968).

(135) M. Horisberger, *Carbohyd. Res.*, **10,** 379 (1969).

(136) W. J. Lewicki, L. W. Long, and J. R. Edwards, *Carbohyd. Res.*, **17,** 175 (1971).

(137) B. A. Lewis, M. J. St. Cyr, and F. Smith, *J. Org. Chem.*, **33,** 3139 (1968).

(138) W. J. Whelan, *Methods Carbohyd. Chem.*, **1,** 321 (1962).

(139) D. Abbott, E. J. Bourne, and H. Weigel, *J. Chem. Soc., C,* 827 (1966).

(140) F. Yamauchi and K. Matsuda, *Agr. Biol. Chem. (Tokyo)*, **33,** 103 (1969); *Chem. Abstr.*, **70,** 88153f (1969).

(141) H. Suzuki and E. J. Hehre, *Arch. Biochem. Biophys.*, **104,** 305 (1964).

(142) I. J. Goldstein and W. J. Whelan, *J. Chem. Soc.*, 170 (1962).

(143) K. Matsuda, H. Watanabe, K. Fujimoto, and K. Aso, *Nature,* **191,** 278 (1961).

(144) K. Matsuda, H. Watanabe, and K. Aso, *Tohoku J. Agr. Res.*, **12,** 351 (1961); *Chem. Abstr.*, **57,** 7590 (1962).

(145) K. Matsuda, K. Fujimoto, and K. Aso, *Tohoku J. Agr. Res.*, **12,** 359 (1961); *Chem. Abstr.*, **57,** 7590 (1962).

(146) M. Torii, E. A. Kabat, and S. Beychok, *Arch. Biochem. Biophys.*, **103,** 283 (1963).

(147) D. Abbott and H. Weigel, *J. Chem. Soc., C,* 816 (1966).

(148) D. Abbott and H. Weigel, *J. Chem. Soc., C,* 821 (1966).

(149) M. E. Preobrazhenskaya and E. L. Rozenfel'd, *Biokhimiya,* **33,** 784 (1968); *Chem. Abstr.*, **70,** 11921x (1969).

(150) W. M. Pasika and L. H. Cragg, *Can. J. Chem.*, **41,** 293 (1963).

(151) W. M. Pasika and L. H. Cragg, *Can. J. Chem.*, **41,** 777 (1963).

(152) S. C. Burket and E. H. Melvin, *Science,* **115,** 516 (1952).

(153) S. A. Barker, E. J. Bourne, M. Stacey, and D. H. Whiffen, *Chem. Ind.* (*London*), 196 (1953); *Chem. Abstr.,* **47,** 7322 (1953).

(154) S. A. Barker, E. J. Bourne, M. Stacey, and D. H. Whiffen, *J. Chem. Soc.,* 171 (1954).

(155) A. R. Jeanes, W. C. Haynes, C. A. Wilham, J. C. Rankin, E. H. Melvin, M. Austin, J. E. Cluskey, B. E. Fisher, H. M. Tsuchiya, and C. E. Rist, *J. Amer. Chem. Soc.,* **76,** 5041 (1954).

(156) B. Casu and M. Reggiani, *Staerke,* **18,** 218 (1966); *Chem. Abstr.,* **65,** 12395 (1966).

(157) E. Nemes-Nanasi, *Acta Biol. Debrecina,* **5,** 67 (1967); *Chem. Abstr.,* **68,** 111524x (1968).

(158) S. A. Barker, F. Pautard, I. R. Siddique, and M. Stacey, *Chem. Ind.* (*London*), 1450 (1955).

(159) T. A. Scott, N. N. Hellman, and F. R. Senti, *J. Amer. Chem. Soc.,* **79,** 1178 (1957).

(160) M. K. Hamdy, E. Gardner, Q. Van Winkle, and G. L. Stahly, *Ohio J. Sci.,* **58,** 177 (1958); *Chem. Abstr.,* **52,** 12438 (1958).

(161) V. Ya. Chernyak and T. V. Polushina, *Med. Prom. SSSR,* **15,** 39 (1961); *Chem. Abstr.,* **56,** 6097 (1962).

(162) M. Zebec, Gj. Dezelic, N. Dezelic, J. P. Kratohvil, and K. F. Schulz, *Croat. Chem. Acta,* **36,** 13 (1964); *Chem. Abstr.,* **61,** 12179 (1964).

(163) M. Wales, P. A. Marshall, and S. G. Weissberg, *J. Polymer Sci.,* **10,** 229 (1953); *Chem. Abstr.,* **47,** 6176 (1953).

(164) M. Wales, P. A. Marshall, S. Rothman, and S. G. Weissberg, *Ann. N.Y. Acad. Sci.,* **57,** 353 (1953); *Chem. Abstr.,* **48,** 4286 (1954).

(165) L. H. Arond and H. P. Frank, *J. Phys. Chem.,* **58,** 953 (1954).

(166) E. Farkas and E. Pandula, *Acta Pharm. Hung.,* **40,** 280 (1970); *Chem. Abstr.,* **74,** 24956v (1971).

(167) A. Ciferri and M. Duane, *J. Polymer Sci.,* **27,** 581 (1958); *Chem. Abstr.,* **52,** 13367 (1958).

(168) K. H. Ebert, G. Schenk, G. Rupprecht, M. Brosche, W.-W. Hsu, and D. Heinicke, *Makromol. Chem.,* **96,** 206 (1966); *Chem. Abstr.,* **65,** 15524 (1966).

(169) K. H. Ebert and G. Rupprecht, *Makromol. Chem.,* **94,** 153 (1966); *Chem. Abstr.,* **65,** 7435 (1966).

(170) J. H. S. Green, *Chem. Ind.* (*London*), 1036 (1961); *Chem. Abstr.,* **56,** 6680 (1962).

(171) W. F. Dudman, *Nature,* **211,** 1049, 1067 (1966).

(172) K. H. Ebert, *Monatsh. Chem.,* **98,** 1128 (1967); *Chem. Abstr.,* **67,** 79023z (1967).

(173) H. Kjellman, *Rev. Intern. Serv. Sante Armees Terre Mer Air,* **38,** 7 (1965); *Chem. Abstr.,* **62,** 14945 (1965).

(174) H. Kjellman, *Farm. Revy,* **64,** 85 (1965); *Chem. Abstr.,* **62,** 15989 (1965).

(175) M. Atik, *Arch. Surg.* (*Chicago*), **94,** 664, 671 (1967); *Chem. Abstr.,* **67,** 31427g (1967).

(176) C. G. Groth and G. Thorsen, *Acta Chim. Scand.,* **130,** 507 (1965); *Chem. Abstr.,* **64,** 14831 (1966).

(177) I. M. Nilsson and O. Eiken, *Thromb. Diath. Haemorrh.,* **11,** 38 (1964); *Chem. Abstr.,* **61,** 12407 (1964).

(178) M. I. Gregersen, B. Peric, S. Usami, and S. Chien, *Bibl. Anat.*, **4,** 58 (1962); *Chem. Abstr.*, **61,** 2111 (1964).

(179) J. Lecomte and J. Damas, *Bull. Soc. Roy. Sci. Liege,* **39,** 188 (1970); *Chem. Abstr.*, **73,** 129212g (1970).

(180) W. H. Lee, Jr., and N. S. Walsh, *Shock Hypotension, Pathogenesis Treat.*, 655 (1965); *Chem. Abstr.*, **65,** 7861 (1966).

(181) H. Wehinger, *Klin. Wochenschr.*, **45,** 1031 (1967); *Chem. Abstr.*, **68,** 11434z (1968).

(182) S. Eisenberg, *Amer. J. Med. Sci.*, **257,** 336 (1969); *Chem. Abstr.*, **71,** 37267j (1969).

(183) P. Frank, *Deut. Apoth.-Ztg.*, **108,** 1328 (1968); *Chem. Abstr.*, **70,** 27314n (1969).

(184) V. B. Koziner, *Usp. Sovrem. Biol.*, **62,** 197 (1966); *Chem. Abstr.*, **66,** 46w (1967).

(185) P. Lundsgaard-Hansen, *Vox Sang.*, **17,** 161 (1969); *Chem. Abstr.*, **71,** 111060e (1969).

(186) R. Terry, C. L. Yuile, A. Golodetz, C. E. Phillips, and R. R. White, III, *J. Lab. Clin. Med.*, **42,** 6 (1953); *Chem. Abstr.*, **47,** 10745 (1953).

(187) K. H. Ebert, G. Schenk, and R. Scholz, *Z. Klin. Chem. Klin. Biochem.*, **6,** 435 (1968); *Chem. Abstr.*, **70,** 35990v (1969).

(188) P. Z. Allen and E. A. Kabat, *J. Amer. Chem. Soc.*, **78,** 1890 (1956).

(189) R. G. Mage and E. A. Kabat, *J. Immunol.*, **91,** 633 (1963); *Chem. Abstr.*, **61,** 1109 (1964).

(190) A. Gronwall, *Acta Acad. Regiae Sci. Upsal.*, **12,** 24 (1969); *Chem. Abstr.*, **72,** 30188w (1970).

(191) G. Lusztig, L. Sajtos, J. Pataky, L. Jozsa, and M. Perneczky, *Z. Gesamte Inn. Med. Ihre Grenzgeb.*, **16,** 807 (1961); *Chem. Abstr.*, **56,** 4004 (1962).

(192) G. Lusztig. L. Josza, M. Perneczky, L. Sajtos, J. Pataky, and L. Szikulai, *Z. Gesamte Inn. Med. Ihre Grenzgeb.*, **17,** 114 (1962); *Chem. Abstr.*, **57,** 3964 (1962).

(193) E. Housset, J. P. Etienne, and J. Cottet, *Ann. Biol. Clin. (Paris)*, **24,** 971 (1966); *Chem. Abstr.*, **65,** 15958 (1966).

(194) M. Iida, *Acta Med. Nagasaki,* **12,** 112 (1968); *Chem. Abstr.*, **70,** 75807x (1969).

(195) E. Housset, J. Etienne, J. Loeper, and J. Cottet, *Pathol. Biol.*, **16,** 653 (1968); *Chem. Abstr.*, **69,** 85368y (1968).

(196) Y.-C. Lee, *Curr. Ther. Res., Clin. Exp.*, **12,** 242 (1970); *Chem. Abstr.*, **73,** 12918a (1970).

(197) G. Lusztig, L. Jozsa, M. Perneczky, and L. Sajtos, *Z. Gesamte Inn. Med. Ihre Grenzgeb.*, **17,** 152 (1962); *Chem. Abstr.*, **57,** 3987 (1962).

(198) J. Leblanc and L. Lafrance, *Union Med. Can.*, **94,** 90 (1965); *Chem. Abstr.*, **62,** 13741 (1965).

(199) D. M. Brahmankar and W. E. Connor, *Circ. Res.*, **21,** 817 (1967); *Chem. Abstr.*, **68,** 28319p (1968).

(200) J. Pataky, L. Szikulay, L. Sajtos, L. Jozsa, M. Perneczky, and G. Lusztig, *Z. Gesamte Inn. Med. Ihre Grenzgeb.*, **16,** 854 (1961); *Chem. Abstr.*, **56,** 9320 (1962).

(201) Y.-C. Lee and E. C. Layne, *Biochem. Med.*, **2,** 102 (1968); *Chem. Abstr.*, **70,** 95339z (1969).

(202) J. Ditzel and J. Dyerberg, *J. Atheroscler. Res.*, **10,** 5 (1969); *Chem. Abstr.*, **71,** 111364g (1969).

(203) H. Loesel and W. Appel, *Fortschr. Med.*, **85,** 517, 522 (1967); *Chem. Abstr.*, **68,** 1871a (1968).

(204) E. E. Witt and L. J. Novak, U.S. Patent 2,822,643 (1958); *Chem. Abstr.*, **52,** 7601 (1958).

(205) L. J. Novak and E. E. Witt, U.S. Patent 2,826,002 (1958); *Chem. Abstr.*, **52,** 8437 (1958).

(206) L. J. Novak, E. E. Witt, and M. J. Hiler, *J. Agr. Food Chem.*, **3,** 1028 (1955).

(207) L. J. Novak, U.S. Patent 2,756,134 (1956); *Chem. Abstr.*, **50,** 16013 (1956).

(208) M. J. Hiler, U.S. Patent 2,901,864 (1959); *Chem. Abstr.*, **53,** 22682 (1959).

(209) S. Seidl and W. Spielmann, *Bibl. Haematol.*, **10,** 205 (1969); *Chem. Abstr.*, **72,** 52731v (1970).

(210) A. M. Karow, Jr., and O. Carrier, Jr., *Surg., Gynecol. Obstet.*, **128,** 571 (1969); *Chem. Abstr.*, **70,** 85973n (1969).

(211) G. D. Moore and A. M. Karow, Jr., *Proc. Soc. Exp. Biol. Med.*, **133,** 106 (1970); *Chem. Abstr.*, **72,** 107656g (1970).

(212) S. V. Backman, P. G. Embring, and B. G.-A. Ingelman, Swed. Patent 174,789 (Appl. 1952); *Chem. Abstr.*, **56,** 4879 (1962).

(213) E. Janiszewski, Polish Patent 40,968 (1958); *Chem. Abstr.*, **54,** 13563 (1960).

(214) E. Darida, *Gyogyszereszet,* **11,** 171 (1967); *Chem. Abstr.*, **67,** 84817r (1967).

(215) Instytut Farmaceutyczny, Polish Patent 49,843 (1965); *Chem. Abstr.*, **65,** 578 (1966).

(216) Benger Laboratories Ltd., Brit. Patent 1,046,612 (1966); *Chem. Abstr.*, **66,** 5757s (1967).

(217) J. A. Bakan, Fr. Patent 1,453,745 (1966); *Chem. Abstr.*, **69,** 3890a (1968).

(218) W. H. Feinstone and A. S. Black, U.S. Patent 3,523,998 (1970); *Chem. Abstr.*, **73,** 91184j (1970).

(219) J. C. Mahoney, U.S. Patent 2,089,217 (1937); *Chem. Abstr.*, **31,** 6916 (1937).

(220) H. A. Toulmin, Jr., U.S. Patent 2,938,797 (1960); *Chem. Abstr.*, **54,** 20009 (1960).

(221) H. A. Toulmin, Jr., U.S. Patent 2,823,128 (1958); *Chem. Abstr.*, **52,** 7569 (1958).

(222) H. A. Toulmin, Jr., U.S. Patent 2,938,798 (1960); *Chem. Abstr.*, **54,** 20009 (1960).

(223) H. A. Toulmin, Jr., U.S. Patent 2,938,799 (1960); *Chem. Abstr.*, **54,** 20009 (1958).

(224) W. I. Owen, U.S. Patent 2,868,725 (1959); *Chem. Abstr.*, **53,** 7571 (1959).

(225) W. J. Sparks, U.S. Patent 3,053,765 (1962); *Chem. Abstr.*, **58,** 8838 (1963).

(226) J. D. Cypert and J. T. Patton, U.S. Patent 3,084,122 (1963); *Chem. Abstr.*, **59,** 1422 (1963).

(227) G. K. Dumbauld and P. H. Monaghan, U.S. Patent 3,065,170 (1962); *Chem. Abstr.*, **59,** 9707 (1963).

(228) P. Mueller, W. Klipp, B. Heyne, U. Behrens, M. Ringpfeil, A. Gabert, W. Toporski, K. Krueger, H. Werther, and H.-D. Werther, East Germ. Patent 47,403 (1966); *Chem. Abstr.*, **66,** 48031a (1967).

(229) L. J. Novak, U.S. Patent 2,885,373 (1959); *Chem. Abstr.*, **53,** 17573 (1959).

(230) L. J. Novak, U.S. Patent 2,933,463 (1960); *Chem. Abstr.*, **54,** 14782 (1960).

(231) C. A. Lesinski and E. J. Sullivan, U.S. Patent 3,085,853 (1963); *Chem. Abstr.*, **58**, 13600 (1963).

(232) A. Lux, E. László, J. Huszár, G. S. Gantner, and J. Holló, *Staerke*, **19**, 129 (1967) *Chem. Abstr.*, **67**, 84087c (1967).

(233) F. R. Minger and W. R. Bennett, U. S. Patent 2,958,585 (1960); *Chem. Abstr.*, **55**, 5884 (1961).

(234) J. H. Grimes and K. T. B. Scott, *Powder Met.*, **11**, 213 (1968); *Chem. Abstr.*, **70**, 39905u (1969).

(235) United Kingdom Atomic Energy Authority, Fr. Patent 1,519,066 (1968); *Chem. Abstr.*, **70**, 120434m (1969).

(236) C. R. Ricketts, Brit. Patent 695,787 (1953); *Chem. Abstr.*, **48**, 1636 (1954).

(237) A. E. James and C. R. Ricketts, Brit. Patent 715,821 (1954); *Chem Abstr.*, **49**, 3481 (1955).

(238) Chugai Pharmaceutical Co., Ltd., Belg. Patent 612,153 (1962); *Chem. Abstr.*, **58**, 11176 (1963).

(239) J. R. Geigy A.-G., Fr. Patent M2227 (1964); *Chem. Abstr.*, **61**, 543 (1964).

(240) Chugai Pharmaceutical Co., Ltd., Fr. Patent M1714 (1963); *Chem. Abstr.*, **59**, 3724 (1963).

(241) E. Morii, K. Iwata, and M. Shamoto, U.S. Patent 3,126,320 (1964); *Chem. Abstr.*, **60**, 15692 (1964).

(242) E. Morii, K. Iwata, and H. Kokkoku, U.S. Patent 3,141,014 (1964); *Chem. Abstr.*, **61**, 14478 (1964).

(243) J. Pares, L. Drobnic, M. Margarit, F. Taxonera, and J. Sabater-Tobella, *Therapie*, **24**, 1071 (1969); *Chem. Abstr.*, **72**, 41659b (1970).

(244) C. Harvengt and M. Jeanjean, *Rev. Fr. Etudes Clin. Biol.*, **12**, 47 (1967); *Chem. Abstr.*, **67**, 19871q (1967).

(245) S. Sasaki, T. Takemoto, and S. Oka, *C. R. Soc. Biol.*, **151**, 1789 (1957); *Chem. Abstr.*, **52**, 16589 (1958).

(246) J. Lecomte, *Bull. Soc. Roy. Sci. Liege*, **38**, 135 (1969); *Chem. Abstr.*, **71**, 69153k (1969).

(247) C. Ruechardt and E. Perlick, *Acta Biol. Med. Ger.*, **10**, 126 (1963); *Chem. Abstr.*, **59**, 6861 (1963).

(248) H. C. Hint and A. W. Richter, *Brit. J. Pharmacol.*, **13**, 109 (1958); *Chem. Abstr.*, **52**, 18893 (1958).

(249) Farmaceutici Italia Soc., Brit. Patent 800,484 (1958); *Chem. Abstr.*, **53**, 16989 (1959).

(250) A. Gallo, *G. Biochim.*, **10**, 366 (1961); *Chem. Abstr.*, **57**, 7593 (1962).

(251) Aktiebolaget Pharmacia, Brit. Patent 854,715 (1960); *Chem. Abstr.*, **56**, 14410 (1962).

(252) G. M. Flodin and B. G. A. Ingelman, U.S. Patent 3,042,667 (1962); *Chem. Abstr.*, **57**, 14042 (1962).

(253) Aktiebolag Pharmacia, Brit. Patent 974,054 (1964); *Chem. Abstr.*, **62**, 4175 (1965).

(254) A. Gabert and H. Seide, East Germ. Patent 55,649 (1967); *Chem. Abstr.*, **68**, 4195u (1968).

(255) VEB Serum-Werk Bernburg, Brit. Patent 1,087,964 (1967); *Chem. Abstr.*, **68**, 4157h (1968).

(256) E. B. Trostyanskaya and I. N. Murashko, *Tr., Vses. Nauch.-Issled. Inst. Khim. Reaktivov Osobo Chist. Khim. Veshchestv.*, No. **30**, 489 (1967); *Chem. Abstr.*, **68,** 115341h (1968).

(257) Aktiebolag Pharmacia, Brit. Patent 1,094,497 (1967); *Chem. Abstr.*, **68,** 41251e (1968).

(258) H. Wand, H. Seide, and A. Gabert, *Plaste Kaut.*, **14,** 572 (1967); *Chem. Abstr.*, **67,** 82466b (1967).

(259) Aktiebolag Pharmacia, Brit. Patent 936,039 (1963); *Chem. Abstr.*, **60,** 733 (1964).

(260) A. D. Virnik, O. P. Laletina, M. A. Penenzhik, K. P. Khomyakov, Z. A. Rogovin, and G. Ya. Rozenberg, *Vysokomol. Soedin., Ser. A,* **10,** 362 (1968); *Chem. Abstr.*, **69,** 3879d (1968).

(261) Aktiebolag Pharmacia, Brit. Patent 1,013,585 (1965); *Chem. Abstr.*, **64,** 8418 (1966).

(262) A. Gabert and B. Voigt, East Germ. Patent 44,124 (1965); *Chem. Abstr.*, **64,** 19901 (1966).

(263) P. Souchay and B. Charreton, *Chromatogr. Methods Immed. Separ., Proc. Meet., Athens, 1965,* **2,** 225 (1966); *Chem. Abstr.*, **68,** 99016y (1968).

(264) R. Axen and J. Porath, *Nature,* **210,** 367 (1966); *Chem. Abstr.*, **65,** 4188 (1966).

(265) L. Eldjarn and E. Jellum, *Acta Chem. Scand.*, **17,** 2610 (1963).

(266) G. P. Vlasov and A. Yu. Bilibin, *Izv. Akad. Nauk SSSR, Ser. Khim.*, **6,** 1400 (1969); *Chem. Abstr.*, **71,** 81710x (1969).

(267) Chemicals Inc., Brit. Patent 1,019,513 (1966); *Chem. Abstr.*, **64,** 10808 (1966).

(268) Armour Pharmaceutical Co., Brit. Patent 1,017,422 (1966); *Chem. Abstr.*, **64,** 15680 (1966).

(269) Chemicals Inc., Brit. Patent 1,025,735 (1966); *Chem. Abstr.*, **64,** 19334 (1966).

(270) H. M. Christensen, Germ. Offen. 1,954,960 (1970); *Chem. Abstr.*, **73,** 28933n (1970).

(271) Kutnowskie Zaklady Farmaceutyczne "Polfa," Brit. Patent 1,196,855 (1970); *Chem. Abstr.*, **73,** 80489m (1970).

(272) J. Mioduszewski, M. Mioduszewska, and A. Kornacki, Polish Patent 58,639 (1969); *Chem. Abstr.*, **73,** 7237y (1970).

(273) R. M. Alsop and I. Bremner, Brit. Patent 1,199,951 (1970); *Chem. Abstr.*, **73,** 89424u (1970).

(274) P. M. Landin, *Brit. J. Cancer,* **15,** 838 (1961); *Chem. Abstr.*, **57,** 3920 (1962).

(275) R. L. Carter, B. C. V. Mitchley, and F. J. C. Roe, *Brit. J. Cancer,* **22,** 521 (1968); *Chem. Abstr.*, **69,** 94488q (1968).

(276) R. L. Carter, *Brit. J. Cancer,* **23,** 559 (1969); *Chem. Abstr.*, **72,** 29698f (1970).

(277) J. Viallier and M. Rebouillat, *C. R. Soc. Biol.*, **156,** 691 (1962); *Chem. Abstr.*, **57,** 13087 (1962).

(278) D. Ben-Ishai, *Lancet,* **1,** 476 (1961); *Chem. Abstr.*, **55,** 11668 (1961).

(279) R. J. Schen, *Harokeach Haivri,* **9,** 137 (1962); *Chem. Abstr.*, **57,** 9143 (1962).

(280) G. Gabbiani and K. Nielsen, *Arch. Pathol.*, **78,** 626 (1964); *Chem. Abstr.*, **62,** 8249 (1965).

(281) F. J. C. Roe and A. Haddow, *Brit. J. Cancer,* **19,** 855 (1965); *Chem. Abstr.,* **64,** 11661 (1966).

(282) C. E. Mengel, H. E. Kann, Jr., and B. W. O'Malley, *Blood,* **26,** 74 (1965); *Chem. Abstr.,* **63,** 8935 (1965).

(283) K. A. Smith and C. E. Mengel, *J. Lab. Clin. Med.,* **72,** 505 (1968); *Chem. Abstr.,* **69,** 84927t (1968).

(284) A. R. Jeanes and C. A. Wilham, *J. Amer. Chem. Soc.,* **74,** 5339 (1952).

(285) A. R. Jeanes and C. A. Wilham, U.S Patent 2,587,623 (1952); *Chem Abstr.,* **46,** 9123 (1952).

(286) L. J. Novak and J. T. Tyree, U.S. Patent 2,954,372 (1960); *Chem. Abstr.,* **55,** 9799 (1961).

(287) L. J. Novak and J. T. Tyree, U.S. Patent 2,766,143 (1956); *Chem. Abstr.,* **51,** 3142 (1957).

(288) L. J. Novak, U.S. Patent 2,933,411 (1960); *Chem. Abstr.,* **54,** 14716 (1960).

(289) H. A. Toulmin, Jr., U.S. Patent 2,910,384 (1959); *Chem. Abstr.,* **54,** 3993 (1960).

(290) S. Umehara, S. Otsuka, A. Setoyama, T. Hiramine, and T. Tezuka, *Proc. Asia Oceania Congr. Endocrinol., 3rd, Manila, 1967,* **3,** 605 (1968); *Chem. Abstr.,* **70,** 86173p (1969).

(291) W. A. Barnes and S. F. Redo, U.S. Patent 3,487,150 (1969); *Chem. Abstr.,* **72,** 136405e (1970).

(292) E. Leberzammer, Germ. Patent 1,928,226 (1969); *Chem. Abstr.,* **72,** 95315e (1970).

(293) Meito Sangyo Co., Ltd., and Konishiroku Photo Industry Co., Ltd., Fr. Patent 1,553,411 (1969); *Chem. Abstr.,* **72,** 17269d (1970).

(294) Colonial Sugar Refining Co., Ltd., Brit. Patent 1,154,993 (1969); *Chem. Abstr.,* **71,** 72201e (1969).

(295) W. Anderson, A. J. Baillie, and J. E. Harthill, *J. Pharm. Pharmacol.,* **20,** 715 (1968); *Chem. Abstr.,* **69,** 83632f (1968).

(296) L. J. Novak, U.S. Patent 2,921,846 (1960); *Chem. Abstr.,* **54,** 7150 (1960).

(297) Midland Chemical Corp., Germ. Patent 1,106,659 (1959); *Chem. Abstr.,* **56,** 622 (1962).

(298) M. Juergens, H. Welzel, and G. Hartung, Germ. Patent 1,293,579 (1969); *Chem. Abstr.,* **71,** 44438u (1969).

(299) H. Heckel and R. T. Jefferson, U.S. Patent 3,130,125 (1964); *Chem. Abstr.,* **61,** 2915 (1964).

(300) E. E. Witt, U.S. Patent 3,196,146 (1965); *Chem. Abstr.,* **63,** 14967 (1965).

(301) Aktiebolag Pharmacia, Neth. Appl. 6,410,005 (1966); *Chem. Abstr.,* **65,** 578 (1966).

(302) L. Chwialkowska, J. Surowiecki, J. Z. Mioduszewski, and M. Mioduszewska, Polish Patent 54,650 (1968); *Chem. Abstr.,* **71,** 6546v (1969).

(303) L. Chwialkowska, J. Surowiecki, J. Z. Mioduszewski, and M. Mioduszewska, Polish Patent 54,383 (1968); *Chem. Abstr.,* **71,** 24744n (1969).

(304) H. A. Toulmin, Jr., U.S. Patent 2,813,797 (1957); *Chem. Abstr.,* **52,** 4885 (1958).

(305) H. A. Toulmin, Jr., U.S. Patent 2,834,684 (1958); *Chem. Abstr.,* **52,** 20769 (1958).

(306) L. J. Novak, U.S. Patent 2,997,423 (1961); *Chem. Abstr.,* **56,** 1538 (1962).

(307) L. J. Novak and E. E. Witt, U.S Patent 2,780,888 (1957); *Chem. Abstr.*, **52,** 635 (1958).

(308) L. J. Novak and E. P. Wenzelberger, U.S. Patent 2,856,372 (1958); *Chem. Abstr.*, **53,** 8680 (1959).

(309) E. P. Wenzelberger, U.S. Patent 2,854,421 (1958); *Chem. Abstr.*, **53,** 1775 (1959).

(310) Commonwealth Engineering Co. of Ohio, Brit. Patent 818,750 (1959); *Chem. Abstr.*, **54,** 12604 (1960).

(311) L. J. Novak, U.S. Patent 2,982,394 (1961); *Chem. Abstr.*, **55,** 16859 (1961).

(312) L. J. Novak and J. J. Bulloff, U.S. Patent 2,856,366 (1958); *Chem. Abstr.*, **53,** 2551 (1959).

(313) A. H. Clarkson, Brit. Patent 876,927 (1961); *Chem. Abstr.*, **56,** 9456 (1962).

(314) U. Behrens, M. Ringpfeil, and A. Gabert, U.S.S.R. Patent 154,369 (1963); *Chem. Abstr.*, **60,** 1546 (1964).

(315) Central Pharmacal Co., Brit. Patent 904,874 (1962); *Chem. Abstr.*, **58,** 3280 (1963).

(316) Deutsche Akademie der Wissenschaften zu Berlin, Brit. Patent 1,069,820 (1967); *Chem. Abstr.*, **67,** 76292a (1967).

(317) K. P. Khomyakov, A. D. Virnik, S. N. Ushakov, and Z. A. Rogovin, *Vysokomolekul Soedin.*, **7,** 1035 (1965); *Chem. Abstr.*, **63,** 16134 (1965).

(318) L. J. Novak, U.S. Patent 2,811,517 (1957); *Chem. Abstr.*, **52,** 3274 (1958).

(319) B. Ernst, W. Bostelmann, and C. Engelmann, *Acta Biol. Med. Ger.*, **23,** 345 (1969); *Chem. Abstr.*, **72,** 51631a (1970).

(320) L. J. Novak and A. E. Bishop, U.S. Patent 2,808,405 (1957); *Chem. Abstr.*, **52,** 3400 (1958).

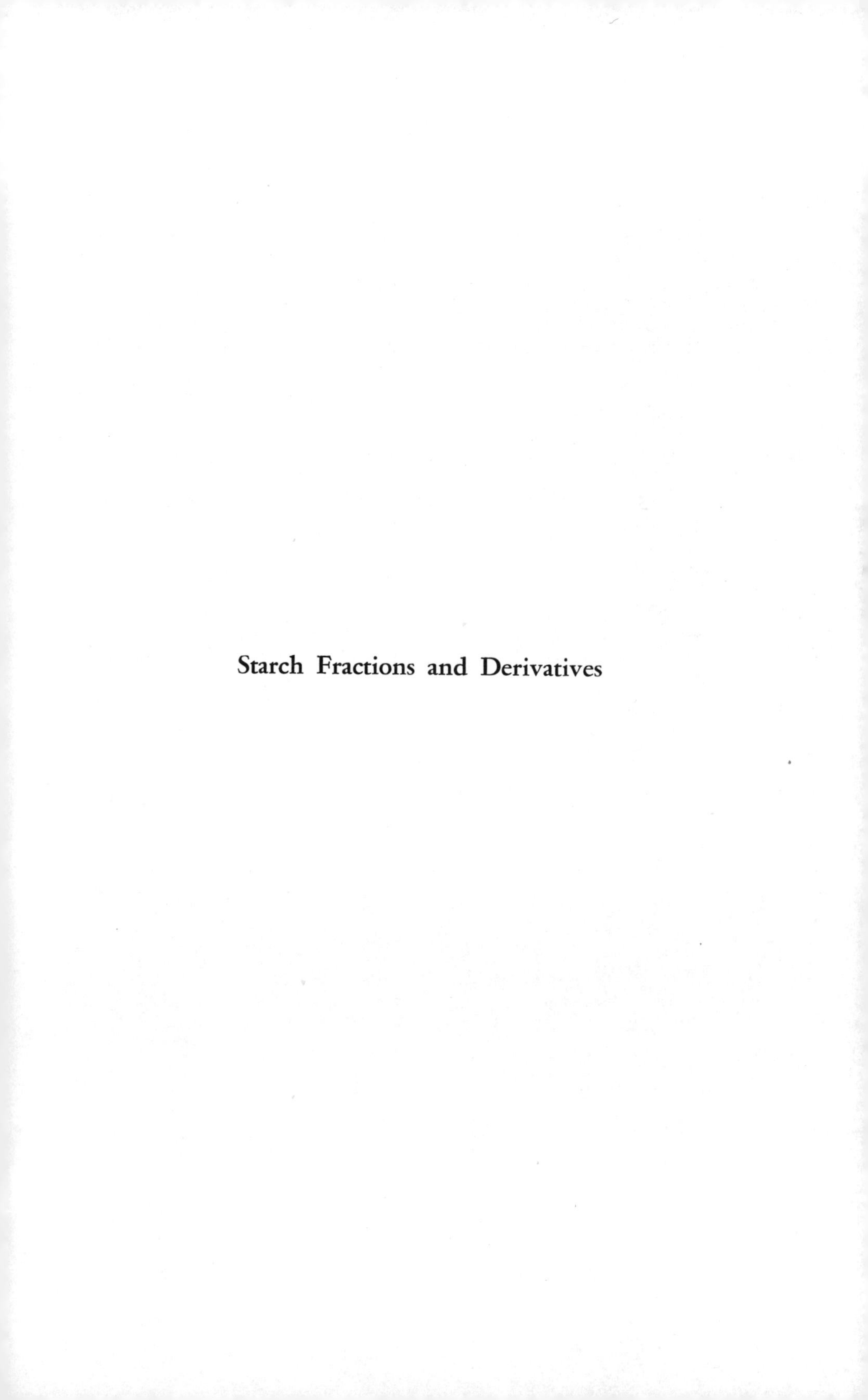

Starch Fractions and Derivatives

CHAPTER XXIV

STARCH AMYLOSE

J. N. BEMILLER

Department of Chemistry and Biochemistry, Southern Illinois University at Carbondale, Carbondale, Illinois

I. STRUCTURE

Amylose is a linear polymer of glucopyranosyl units joined by α-D-(1→4)-glycosidic linkages. Some molecules in an amylose preparation may be slightly branched, depending on the method of preparation, with linear chains of α-D-(1→4)-linked glucopyranosyl units linked to the main chain through α-D-(1→6) linkages. In all preparations, however, the branch chains are sufficiently long and the branches are sufficiently infrequent that the physical properties of the polymer are not affected. The average degree of polymerization depends upon the starch from which the amylose is isolated and upon the method of fractionation.

II. Production

1. *Development of High-Amylose Corn*

Most native starches contain only 20–30% of amylose.[1] No starches are known that contain amylose as the sole constituent. Even though industrial methods for the fractionation of starch into amylose and amylopectin are available, it would be more economical to be able to produce amylose starches in the same way that amylopectin (waxy) starches are produced and, perhaps, advantageous in some applications to have amylose in granular form.

In the late 1940's, a breeding program was established at the Agricultural Experiment Station at Purdue University with the goal of increasing the amylose content of corn (*Zea mays* L.) starch. This pioneering work of Kramer and Whistler[2–4] raised the amylose content, but starch of high amylose content was only produced in kernels of a low starch content. Since then, genetic control of the amylose content of corn starch has been thoroughly studied.[5] Using various combinations of specific amylose-enhancing and modifying genes, high-amylose single crosses and double-cross hybrids with a continuous range of amylose contents from 50% to >80% have been developed.[6] These hybrids are termed *amylomaize.*

The protein content of the amylomaize hybrids is substantially higher than that of ordinary dent corn.[6] This increased protein content has resulted in some difficulty in the separation of gluten and starch during wet milling of corn containing starch of >50% amylose.[7] Thus, it is an additional goal of the breeding programs to develop high-amylose populations with a relatively low endosperm protein content.[7]

At the same time, the recovery of starch from amylomaize hybrids by the wet milling procedure has been investigated.[6, 8–10]

High-amylose corn (amylomaize) starches are now commercially available. Increases in the variety of products of specific amylose content can be anticipated.

Amylomaize starches differ from starches of normal amylose content in that, in them, the loss of birefringence is not associated with marked granule swelling and development of viscosity at relatively low starch concentrations. In fact, they do not swell sufficiently in water to give a measureable viscosity in the conventional heating cycle of either a CIRF viscometer or a Brabender amylograph. However, amylomaize starches gelatinize below 95° upon addition of swelling agents, such as calcium chloride.[6]

Attrition grinding of dry amylomaize starch allows it to be dispersed (gelatinized) in water at 60°–100°. Such treatment results in the formation of firmer gels and clearer and more continuous films than can be made from untreated amylomaize starch.[11]

Amylomaize starch can be dissolved in potassium[12] or sodium[13] hydroxide solutions, formamide,[14] ethanolamine,[14] methyl sulfoxide (DMSO),[13, 15, 16] aqueous methyl sulfoxide,[15] and a concentrated aqueous solution of 5*M* lithium thiocyanate and 2*M* guanidinium thiocyanate.[17]

Smooth, white, stable dispersions can be made by partial acid-catalyzed hydrolysis of amylomaize starch.[18, 19]

The amylose of these high-amylose corn starches is somewhat smaller (lower molecular weight) than normal amylose.[6, 12, 20] The amylopectin is also of lower molecular size than normal corn amylopectin, but it has a much increased length of outer branches.[17, 20, 21]

Amylomaize starch granules can be oxidized, etherified, esterfied and otherwise derivitized in the normal manner,[6] and most of the reactions reported in the remainder of this chapter apply not only to isolated amylose but also to high-amylose starch granules.

2. *Separation from Starch*

Many methods for the preparation of amylose from starch have been developed.[1, 22] The three general methods most used are crystallization of amylose as complexes with polar organic compounds; crystallization of amylose in solutions of salts, particularly those of the alkaline earths; and crystallization of amylose from aqueous dispersions without added agents. The latter two methods have been used commercially.

The method first used commercially was the fractional crystallization of amylose from a solution of potato starch in 10–13% magnesium sulfate.[1, 23–25] The other method that has been used commercially is crystallization of amylose without added agents.[26, 27] This method depends on a carefully controlled temperature of crystallization after the starch has been dispersed to a high degree. Methods are available for purifying amylose produced by either of these methods.[28, 29]

Quantitative separation of amylose from starch by crystallization of complexes with organic compounds was first reported by Schoch[30] in 1941. Since then, the method has been widely used in laboratory preparations and there has been considerable interest in its use as a commercial process. Extensive reviews of its application are available.[1, 22, 31] Recent investigations have dealt with the use of less polar organic compounds,[32] fatty acids and alcohols,[33] carbon tetrachloride,[34] and *l*-menthone[35] as complexing agents. Amylose has been isolated from amylomaize starches by crystallization of amylose as 1-butanol complexes after freeze–thaw pretreatments[36] and as complexes with alcohols or acids containing 8–10 carbon atoms.[37]

Crystallization of amylose in the presence of both salts and alcohols can also be used.[38, 39]

Other patented processes include one based on a hydrodynamic separation of amylose and amylopectin. In this process, a starch gel is prepared. The starch gel is then separated into two distinct phases by a dispersing device. On cooling, amylose separates and can be collected by filtration or centrifugation.[1, 40, 41]

In another, amylose and amylopectin are precipitated together. Then amylopectin is preferentially dissolved by dilution and heating.[42]

Amylose has been prepared from normal or high-amylose corn starch after debranching the branched molecules with an amylo-1,6-glucosidase, such as pullulanase or isoamylase.[43–45] The amylose prepared in this way is water-soluble and has good absorbing power for gases and liquids.

III. Properties

Amylose has a number of unique properties that have been taken advantage of in developing uses and that may indicate additional uses.

1. *Solutions, Retrogradation, and Gels*

Amylose is normally not easily soluble (dispersible) in water. However, water dispersions can be made by heating an amylose slurry to about 150° in a continuous heat exchanger. About 5 min at maximum temperature is necessary for complete dispersion. Excessive heating can cause depolymerization. Under most conditions, amylose dispersions rapidly set to gels, even at elevated temperatures. The rheology of these amylose dispersions and gels is discussed by Langlois and Wagoner.[1]

The work of Foster and others, which has been reviewed,[46] has led them to the conclusion that amylose exists in solution as a relatively stiff, wormlike coil consisting largely of an imperfect or deformed helical backbone. However, Brant and Dimpfl[47] concluded that, in aqueous solution, amylose is a statistical coil without identifiable helical character in the absence of complexing agents.

The difficulty in dissolving amylose is related to retrogradation, a term used to describe the process by which amylose in a dissolved or dispersed hydrated state reverts to an insoluble form. The process of retrogradation probably involves interaction between molecules with mutual alignment, expulsion of water, and formation of new intermolecular forces to give insoluble microcrystalline aggregates that grow either by continued deposition of molecules or by coalition of particles until they attain sufficient size to precipitate in a semicrystalline state.

Factors that affect retrogradation have been a continuing source of investigation[46, 48] because of its importance. These investigations have shown that retrogradation is favored by low temperatures, neutral pH, relatively low degrees of polymerization, absence of branching, low degrees of polydispersity, high concentrations, presence of certain inorganic ions (according to the Hofmeister lyotropic series), presence of other dehydrating substances, and absence of surface-active (wetting) agents.[49–53]

Partial acid-catalyzed hydrolysis of amylose results in a maximum rate of retrogradation,[54] and the rheology of microcrystalline amylose so prepared has been investigated.[55]

Some amylose preparations will absorb approximately four times their weight of water. From them, very high viscosity slurries can be prepared. If these amylose preparations are allowed to stand with less than four times their weight of water, all the water will be absorbed and a thixotropic gel will result.[1]

Amylose gels are opaque, irreversible at normal temperatures, and without appreciable syneresis. Their gel strength is a function of both concentration and age.[1]

Amylose solutions (dispersions) can be stabilized for long periods of time by the addition of various aldehydes, particularly formaldehyde and glyoxal.[56–60] The quantity of aldehyde required and the method of preparation is determined by the amylose concentration and the desired time of solution stability.[56, 58] Adding aldehydes allows the preparation of amylose dispersions of high concentrations (up to 50%) for film and fiber production.

Amylose is also more soluble in alkaline solutions; there is an optimum amount of alkali required for solution that can be determined from phase diagrams.[1] Rigid, transparent, colorless gels can be made from alkaline solutions.[61]

Amylose is soluble in aqueous chloral hydrate, dichloroacetic acid, methyl sulfoxide, ethylenediamine, formic acid, formamide, pyrrolidine, acetamide, piperazine, and urea.[1]

Cold-water-dispersible amylose can be made by freeze-drying[82] and by drum or spray drying[63, 64] after dissolution by heating or in alkaline solutions, and by coprecipitation with an alkali metal salt.[65] A cold-water-dispersible starch composition for puddings can also be made by mixing solubilized amylose with pregelatinized starch.[66]

Dispersions of limited stability that are useful as adhesives and sizes and for esterification can be prepared by dissolving amylose in alkaline solutions, followed by neutralization.[67]

Amylose can be depolymerized by grinding.[68]

2. *Films and Fibers*

Transparent, unsupported, strong, pliable, lustrous films can be prepared from amylose by casting and by extrusion.[69–90] Highly oriented filaments can be spun from amylose and alkali amylose.[73–79, 83, 90–92] Films and fibers can also be prepared from amylose solutions stabilized with aldehydes.[1, 58, 61] Other cast and extruded articles, such as tubes,[73, 76, 83] can be made in a similar manner.

Films and fibers can be derivitized after preparation[93] or amylose derivatives can be used in their preparation. Waterproof and turpentine-impermeable laminated films can be prepared from amylose and polyolefins.[94]

Films and fibers can be plasticized with polyhydric compounds, such as glycerol[79] and acetylated poly(vinyl alcohol).[80] Their properties are greatly influenced by the molecular weight of the amylose and the plasticizer used.[1]

Because amylose is nontoxic and somewhat water-dispersible, amylose fibers can be used for making medical sutures and bandages[92] and inaccessible filters;[92] amylose sponges and gauze[95] can be made; and amylose tubes and films can be used for making edible sausage casings[73] and packaging such food products[72, 84] as instant coffee, soup, and tea.

Amylose films only swell in cold water[81] and have a low permeability to moisture vapor, organic vapors, and common gases.[1, 96] They are also grease resistant,[97] transparent to ultraviolet radiation,[97] and unoriented and have an elastic modulus and a tensile strength in all directions about equal to the tensile strength of Cellophane in the cross-machine direction.[1, 56] Irradiated films become less soluble in boiling water.[98] The tensile strength of films prepared from high-amylose cornstarches is correlated with the amylose content.[99]

3. *Complexes*

As pointed out in the section on fractionation (Section II.2), amylose combines with a large number of organic and inorganic substances. Some general aspects of these organic complexes and coordination compounds have been reviewed.[100] However, as pointed out by Langlois and Wagoner,[1] little industrial use has been made of this complexing ability.

The complex of amylose with the triiodide ion improves the fertility of a plant-growing medium and the quality of plants grown therein.[101] The amylose–iodine complex has strong antibacterial activity against both gram-positive and gram-negative organisms,[102] and the hydroxypropylamylose–iodine complex, when used to disinfect swimming pools, imparts a deep blue color to the pool.[103]

Amylose forms a water-insoluble complex with vitamin A palmitate,[104, 105] and this is a basis for its encapsulation to prevent oxidation.

Aluminum salt complexes of amylose sulfates are useful in oral compositions for the treatment of gastric ulcers.[106]

IV. Derivatives

1. *Esters*

Sulfate.—The preparation and properties of amylose sulfate has been a subject of investigation for some time.[107, 108] Newer methods involve sulfation with a sulfur trioxide–tertiary amine complex,[109, 110] with sulfur trioxide in methyl sulfoxide,[111] and with chlorosulfonic acid in the presence of a tertiary amine.[112]

When the former method is used, half of the ester groups are on C-6 of the D-glucopyranosyl units, and the products do not gel with potassium or ammonium ions.[110] Amylose sulfate has considerable antipeptic activity and can be used to treat stomach ulcers.[113–117] It prevents formation of gastric lesions, both by inhibiting the action of pepsin[113] and by inhibiting its formation,[116] but not as well as does amylopectin sulfate. Aluminum complexes of sulfates have also been used in oral treatment of gastric ulcers.[106]

Amylose sulfate will precipitate some proteins but not human plasma[110] and has been investigated as an anticoagulant. It does have anticoagulant activity,[118, 119] but again it is not as good an anticoagulant as is xylan sulfate[118] or amylopectin sulfate.[119]

Amylose sulfate increases the ability of certain antibiotics and drugs of basic character to enter the lymphatic system, making it possible to administer them by parenteral application.[120]

Organic esters.—A variety of amylose esters and mixed esters has been prepared.[56, 67, 121–133] Various processes, including continuous ones,[121, 122] for their production have been developed, and their properties have been investigated.[123–125]

A number of uses for these esters has been proposed. Some can be used to make thermoplastic molded articles or cast into transparent, lustrous, colorless films.[125]

Films, fibers, and a clear, transparent plastic can be made from amylose triacetate. Amylose triacetate films are high quality and transparent, with good pliability and low plasticizer requirements.[134, 135] Amylose ester films have about the same percentage of elongation as cellulose ester films but have less tensile strength.[125] Amylose acetate films are water resistant.[136]

Amylose triacetate can be spun into fibers that have a greater extensibility, a slightly greater flexibility, and less tensile strength than do cellulose triacetate fibers.[137, 138]

Various fatty acid esters (C_6–C_{26}) have been proposed as thickening agents[139] and as pour-point depressants and elevators for food oils,[139] lubricating oils,[140] and wax-containing crude or heavy oils;[141] as stabilizers and thickeners for paints and cosmetics;[140] as waterproofing agents for textiles and paper;[140] as stabilizers against changes in temperature in margarine and other fat-containing foods;[142, 143] as dough and bakery product improvers;[142,143] as a coating material for dehydrated foods;[144] and as a component of lotions, creams, deodorants, and sun screening creams that produce skin-protecting films.[145]

Aminoarenesulfonates and *m*-sulfobenzoates are useful in the photographic industry.[146, 147]

Phosphate.—Amylose phosphate, reported to be somewhat unstable,[148] can be prepared with 2-cyanoethyl phosphate[149] or with tetrapolyphosphoric acid and a triakylamine.[150] The former method gives phosphorylation primarily at C-6 of the D-glucopyranosyl units.

Amylose treated with an alkali metal orthophosphate can be used as a foundry core binder.[151]

Xanthate.—Amylose xanthate[152–156] has found uses in the paper industry. When used as a paper coating, it increases wet strength.[153, 154] Cross-linked amylose xanthate can be used as an adhesive for clay coats to give superior wet-rub resistance.[155] Amylose xanthate is also useful as an adhesive for paper, a carrier for inks, a binder in ion-exchange membranes, and a thickener.[156] High-wet-strength webs can be made by regenerating amylose from amylose xanthate after impregnating the paper web.[157] A xanthate derivative obtained by a nucleophilic addition reaction is useful in improving the wet and dry strength of paper, finishing textiles, and improving the dyeability and stability of textiles to ultraviolet radiation.[158]

Amylose xanthate can be converted into an olefin.[159]

Carbamate.—Amylose carbethoxymethyl carbamate can be prepared from ethyl iscocyanatoacetate.[160]

2. *Ethers*

Amylose hydroxyalkyl,[161–168] carboxymethyl,[161, 169–171] methyl,[161, 163, 172] diethylamino(hydroxy)propyl,[173] trimethylsilyl,[174, 175] 2-cyanoethyl,[168] α-methylcarboxymethyl,[168] and vinyl[161, 176] ethers have been prepared.

Hydroxyalkyl ethers can be prepared from ethylene oxide,[161, 163–168] ethylene chlorohydrin,[161, 163] propylene oxide,[161, 163, 166, 168] propylene chlorohydrin,[161, 163] and styrene oxide.[163] By altering the degree of substitution, products with a wide range of solubilities in water can be made.[163] Hot-water-soluble products can be used for paper coating and cold-water-soluble products for coating fruit.[163] The products have little or no tendency to gel or retrograde.[166]

Hydroxythylamylose is an effective binder for fragmented tobacco in reconstituted tobacco compositions[161] and, when used as a size and a lubricant coating for glass fibers, avoids migrations of solids.[167] It is resistant to moderate heating.[167]

Gels can be made from carboxymethylamylose,[171] and carboxymethylamylose increases the ability of some pharmaceuticals to enter the lymphatic system.[170]

Ethers plasticized with glycols form thermoplastic compositions that can be hot-worked to form water-soluble molded products.[177]

3. *Grafted Amylose*

Various vinyl monomers can be polymerized in the presence of amylose to form graft copolymers. These include acrylamide,[178–182] acrylonitrile,[182–184] ethyl and butyl acrylate,[185, 186] styrene,[184, 187] methyl methacrylate,[184] and vinyl acetate.[184] Polymerization can be initiated by gamma irradiation[178–180, 182, 187] or with ceric ions.[182, 186] It can be accomplished in solution, in the dry state, or on films.[184] Amylose or amylose derivatives, such as hydroxyethylamylose[185] and amylose esters,[186] can be used.

Amylose–polyacrylamide graft copolymers are good as an eletrolyte support material for dry batteries.[179] The properties of amylose–polyacrylamide graft copolymers that make them useful as pastes and dyeing assistants have been described.[188] Amylose–polyacrylate graft copolymers are useful in the preparation of a high-impact-strength graft copolymer blend.[185]

Amylose molecules have also been substituted (grafted) with poly[1-deoxy-1-(β-hydroxypropionamido)-D-glucitol][189] and with mono-, di-, and oligosaccharide units.[190, 191]

4. *Other Derivatives*

Amylose has been converted into 2-amino-2-deoxy,[192] and 3-amino-3-deoxy-6-*O*-triphenylmethyl,[193] and 6-amino-6-deoxy derivatives.[194] A 6-amino-6-deoxy derivative that has a cyclic sulfate ester at C-2 and C-3 of about one half of the D-glucopyranosyl units[195] and an animated amylose with some of the secondary hydroxyl groups replaced with amino groups[196] have also been prepared.

Starch containing >50% amylose has been converted into cationic nitrogenated products with calcium cyanamide.[197] Amylose-containing quaternary amine groups[198, 199] are useful as peptizers for photographic emulsions.[199]

3,6-Anhydroamylose,[200] 6-mercaptoamylose,[201] 6-deoxy-6-sulfonylamylose,[201] amylose 6-aldehyde (α-D-*gluco*-hexodialdo-1,5-pyranosylglycan),[202] and 6-*O*-triphenylmethyl-2,3-dideoxy-α-D-*erythro*-hex-2-enopyranosylglycan[202] have also been prepared from amylose. About 50% of the D-glucopyranosyl units of amylose have been oxidized to D-glucopyranosyluronic acid units.[203]

V. Uses

1. *Paper*

Amylose has found several applications in the paper industry. Already mentioned is the use of amylose xanthate for coating paper to increase its wet strength,[153, 154] as an adhesive for clay coats to make paper with superior wet-rub resistance,[155] as an adhesive for paper,[156] as a carrier for inks,[156] in the production of high-wet-strength webs,[157] and in the preparation of a derivative useful for improving the wet and dry strength of paper.[158]

Water-dispersible amylose prepared by freeze-drying[62, 65] or drum drying[63] is useful in paper sizes and coatings, as are amylose ethers.[163] A process for the continuous application of amylose coatings on paper has been described.[204]

Amylose esters of fatty acids with at least eight carbon atoms can be used as waterproofing agents for paper.[140]

Paper coated with amylose is grease resistant,[205] but a superior coating for greaseproof paper is made with a mixture of amylose and poly(vinyl alcohol).[206, 207]

Cationic 70% amylose corn starch plus amylose, when used to size paper, gives good strength, appearance, water resistance, printability, and optical properties and reduced water and printing ink absorption.[208]

When amylomaize starch is used in place of ordinary starch for the preparation of corrugating and laminating adhesives,[209–212] a stronger bond and higher water-resistance is obtained and faster production is allowed.

2. *Textiles*

As mentioned before, amylose triacetate and deacetylated amylose triacetate can be spun into highly oriented filaments. These fibers, however, at their highest tensile strength, are inferior to comparable fibers prepared from cellulose triacetate.

Amylose has found several uses in the textile industry, mainly in finish and size compositions, and its use therein has been reviewed.[213] Amylose is preferentially adsorbed by cellulose fibers from a solution of starch.[214] Thus, amylose itself is a more efficient size for cotton than is starch. To overcome the rapid gelling characteristics of unmodified amylose in aqueous solution, it can be applied from solutions stabilized with alkali or formaldehyde.[1] Water-dispersible amylose prepared by freeze-drying[62, 65] or drum drying[63] amylose or by derivitization[215] can be used as a size for both hydrophobic and hydrophilic textile yarns.

A xanthate derivative obtained by a nucleophilic addition reaction is useful in finishing textiles and improving the dyeability and stability of textiles to ultraviolet radiation. Cotton cloth treated with methoxyethylvinylsulfone-modified amylose has improved warp traction and warp tear resistance.[216] Amylose esters of fatty acids containing at least eight carbon atoms are reported to be effective waterproofing agents for textiles.[140]

Amylose[217, 218] and hydroxyethylamylose[167, 219] are good sizing agents and useful in the formulation of lubricant coatings for glass fibers. They are absorbed so strongly by the glass fibers that there is almost no migration during drying, and they are stable to moderate heating.

Polyacrylonitrile filaments prepared by the polymerization of acrylonitrile in a methyl sulfoxide solution of amylose are antistatic and have improved dyeing properties,[220] or amylose can be used in the precipitation bath to give polyacrylonitrile filaments 80% greater water absorption, making them antistatic.[221]

3. Food

Amylose has several potential uses in the food industry. Already mentioned (Section III.2) is the production of transparent amylose films for packaging food products, such as instant coffee, soup, and tea,[72, 84] and the production of edible sausage casings.[73, 222] Amylose succinate can also be used in the production of skinless frankfurters.[223]

Because amylose and its derivatives can be readily deposited as films and because the films have many desirable properties, such as low oxygen permeability, grease resistance, bland flavor, and edibility, several processes for using it to coat foods have been developed. Amylose[224] and amylose ethers[163] can be used to coat fruits, such as dates, figs, dried fruits, and candied fruits, to prevent them from clumping and sticking together.[224] An amylose coating on french-fried potatoes gives an improved product.[225] Amylose esters of 8- to 26-carbon fatty acids and ethyl or methyl cellulose are components of a water-resistant, edible coating for foods.[226] Esters of 12- to 26-carbon fatty acids alone can also be used to coat dehydrated foods.[144]

Because of its tendency to rapidly form stable gels, water-dispersible amylose is useful in making instant puddings and gravies.[62, 63, 65] A mixture of solubilized amylose and pregelatinized starch can also be used for this purpose.[66] High-amylose corn starch is useful as a thickener for cooked foods.[227]

Fatty acid esters of amylose can be used as thickening agents for food oils and increase their melting and pour points.[142, 143] They make fat-containing foods, such as margarine, peanut butter, chocolate, and baking fat, more stable against changes in temperature. They also improve the volume of dough and improve the crumb structure of bakery products.[142, 143]

High-amylose corn starch can be used in the manufacture of confectionery products by extrusion.[228]

4. Pharmaceuticals

Amylose has several unique properties that make it useful in the pharmaceutical industry. Its complexing ability allows it to be used for the encapsulation of such water-insoluble materials as vitamin A palmitate.[103, 104] The amylose–iodine complex is reported to have strong antibacterial activity against both gram-positive and gram-negative organisms.[102, 227]

Amylose has the properties of an ideal dry binder in direct compression tablet manufacture. It is compatible with many water-soluble dyes, flavoring agents, and sweeteners and is free flowing, self-lubricating, and self-disintegrating so that it, by itself, functions as the binder, filler, lubricant, and disintegrant and can be tableted directly.[1, 230, 231]

Already mentioned (Section IV.1) is the use of amylose sulfate in the treatment of gastric ulcers.[105, 113–116] Amylose sulfate and carboxymethylamylose increases the ability of certain antibiotics and drugs of basic character to enter the lymphatic system, making them suitable for parenteral application.[120, 170]

Nontoxic filaments and fibers for medical sutures and bandages can be made from amylose and hydroxypropylamylose.[92] Amylose sponges have high strength and water absorption and are absorbable by the body; hence, they can be used as absorptive stanching sponges in surgical operations.[95] By processing foamed amylose in thin layers, a highly absorbent gauzelike material, into which drugs may be incorporated, can be made.[95]

Esters of 6- to 22-carbon fatty acids give skin protecting films and, therefore, can be used in the formulation of lotions, creams, deodorants, and sunscreening creams.[145]

Highly chlorite-oxidized amylose is reported to have antiviral activity.[232, 233]

5. *Other*

Amylose is more effective than starch in the sedimentation of red mud to separate it from soluble aluminates in the treatment of bauxite during aluminum extraction by the Bayer process.[234, 235]

Amylose treated with an alkali metal orthophosphate can be used as a foundry core binder.[151]

Amylose film is permeable to electrolytes and, therefore, can be used to coat electrodes of dry batteries, that is, as a dividing agent.[236] A gamma-irradiated amylose–polyacrylamide graft copolymer is good as an electrolyte support material for dry batteries.[179]

Amylose xanthate is useful as a binder in ion-exchange membranes.[156]

Amylose *m*-sulfobenzoate can be used as a binder for light-sensitive silver salts in several photographic emulsions,[146] and quaternary ammonium derivatives can be used as polymeric peptizers in photographic emulsions.[199]

Amylose and carboxymethylamylose can be used to prime the surface of bowling pins and other wooden objects to improve their impact resistance.[237] A high-impact-strength graft copolymer blend can be made by blending polystyrene with a polymer made by graft copolymerization of ethyl or butyl acrylate, with or without acrylonitrile, on hydroxyethylamylose.[185]

Amylose esters with fatty acids of more than 15 carbon atoms improves the viscosity, pour point, and flow properties of wax-containing crude or heavy oils.[142] Amylose can be used to remove waxy components from hydrocarbons by adsorption to reduce pour points in the production of lubricating oils.[238] Amylose can also be used to prevent the deposition of solid hydrocarbons from oil onto surfaces that are normally susceptible to the accumulation of these deposits, for example, oil well tubing.[239]

Hydroxyethylamylose and cyanoethylamylose can be used in the preparation of cold-water-soluble films for packaging dry detergents and similar products.[166, 240]

Amylose is an effective binder for acoustical tile.[241]

Hydroxyethylamylose is an effective replacement for galactomannan gums as a binder for fragmented tobacco in reconstituted tobacco compositions.[165] It can also be used as a tobacco flavorant encapsulating agent used in coating cigarette paper.[242]

Aldehyde–amylose compounds can be used as binding agents in the manufacture of molded articles from artificial resin molding compounds.[243]

Nondeliquescent choline salt products can be made by their entrapment in gels of amylose or high-amylose corn starch for use in poultry feeds.[244]

Cold-water-dispersible amylose forms a stable gel and can be used as a base for bacteriological culture media.[62, 63, 65]

Amylose added to dry cement compositions gives added strength to products made from the composition.[245]

VI. References*

(1) D. P. Langlois and J. A. Wagoner, *in* "Starch: Chemistry and Technology," R. L. Whistler and E. F. Paschall, eds., Academic Press, New York, Vol. 2, 1967, p. 451.

(2) H. H. Kramer and R. L. Whistler, *Agron. J.*, **41,** 409 (1949).

(3) W. Dvonch, H. H. Kramer, and R. L. Whistler, *Cereal Chem.*, **28,** 270 (1951).

(4) G. M. Dunn, H. H. Kramer, and R. L. Whistler, *Agron. J.*, **45,** 101 (1953).

(5) R. G. Creech, *Advan. Agron.*, **20,** 275 (1968).

(6) F. R. Senti, *in* "Starch: Chemistry and Technology," R. L. Whistler and E. F. Paschall, eds., Academic Press, New York, Vol. 2, 1967 p. 499.

(7) M. S. Zuber, C. O. Grogan, V. L. Fergason, W. L. Deatherage, and M. M. MacMasters, *Cereal Chem.*, **37,** 212 (1960).

(8) R. A. Anderson, C. Vojnovich, and E. L. Griffin, Jr., *Cereal Chem.*, **37,** 334 (1960).

(9) R. A. Anderson, C. Vojnovich, and E. L. Griffin, Jr., *Cereal Chem.*, **38,** 84 (1961).

(10) R. A. Anderson, *Cereal Chem.*, **39,** 406 (1962).

(11) O. B. Wurzburg and W. Herbst, U.S. Patent 3,222,220 (1965); *Chem. Abstr.*, **64,** 6877 (1966).

(12) I. A. Wolff, B. T. Hofreiter, P. R. Watson, W. L. Deatherage, and M. M. MacMasters, *J. Amer. Chem. Soc.*, **77,** 1654 (1955)

(13) E. M. Montgomery and K. R. Sexson, *in* Ref. 6.

(14) J. W. Sloan, M. M. MacMasters, and F. R. Senti, *Cereal Chem.*, **36,** 196 (1959).

*The patent references given herein are not meant to be all inclusive but are only representative; that is, only one patent is given in most cases although the same or similar patents may have been issued in other countries.

(15) H. W. Leach and T. J. Schoch, *Cereal Chem.,* **39,** 318 (1962).

(16) M. W. Rutenberg, W. Jarowenko, and L. J. Ross, U.S. Patent 3,038,895 (1962); *Chem. Abstr.,* **57,** 10086 (1962).

(17) S. R. Erlander, R. Tobin, and R. J. Dimler, *Abstr. Papers, Amer. Chem. Soc.,* **114,** 15C (1963).

(18) O. A. Battista, U.S. Patent 3,351,489 (1967); *Chem. Abstr.,* **67,** 4190p (1967).

(19) FMC Corp., Fr. Patent 1,482,638 (1967); *Chem. Abstr.,* **68,** 14276k (1968).

(20) E. M. Montgomery, K. R. Sexson, and F. R. Senti, *Staerke,* **13,** 215 (1961).

(21) E. M. Montgomery, K. R. Sexson, R. J. Dimler, and F. R. Senti, *Staerke,* **16,** 345 (1964).

(22) R. L. Whistler, *in* "Starch: Chemistry and Technology," R. L. Whistler and E. F. Paschall, eds., Academic Press, New York, Vol. 1, 1965, p. 331.

(23) Cooperatieve Verkoop-en Productievereniging van Ardappelmeel en Derivaten "Avebe" G.A., Neth. Patent Appl. 98,152 (1953); 98,154 (1953); *Chem. Abstr.,* **56,** 1658 (1962).

(24) Cooperatieve Verkoop-en Productievereniging van Aardappelmeel en Derivaten "Avebe" G.A., (West) Ger. Patents 946,338 (1956); 946,339 (1956); 946,340 (1956); *Chem. Abstr.,* **53,** 7638 (1959).

(25) W. C. Bus, J. Muetgeert, and P. Hiemstra, U.S. Patent 2,822,305 (1958); *Chem. Abstr.,* **52,** 9635 (1958); U.S. Patent 2,829,987 (1958); *Chem. Abstr.,* **52,** 13295 (1958); U.S. Patent 2,829,988 (1958); *Chem. Abstr.,* **52,** 14204 (1958); U.S. Patent 2,829,989 (1958); *Chem. Abstr.,* **52,** 13296 (1958); U.S. Patent 2,829,990 (1958); *Chem. Abstr.,* **52,** 17768 (1958).

(26) O. R. Etheridge, J. A. Wagoner, J. W. McDonald, and D. A. Lippincott, U.S. Patent 3,067,067 (1962); *Chem. Abstr.,* **58,** 4723 (1963).

(27) Hoffmann's Staerkefabriken A.-G., Belg. Patent 628,265 (1963); *Chem. Abstr.,* **60,** 14724 (1964).

(28) A. E. Staley Mfg. Co., Neth. Patent Appl. 6,407,279 (1964); *Chem. Abstr.,* **62,** 16494 (1965).

(29) A. E. Staley Mfg. Co., Neth. Patent Appl. 6,411,642 (1965); *Chem. Abstr.,* **63,** 14968 (1965).

(30) T. J. Schoch, *Cereal Chem.,* **18,** 121 (1941).

(31) J. Muetgeert, *Advan. Carbohyd. Chem.,* **16,** 299 (1961).

(32) D. French, A. O. Pulley, and W. J. Whelan, *Staerke,* **15,** 349 (1963).

(33) R. A. Anderson, C. Vojnovich, and G. Soedomo, *Staerke,* **15,** 355 (1963).

(34) R. S. Leiser, D. P. Marcus, and J. A. Wagoner, U.S. Patent 3,323,949 (1967); *Chem. Abstr.,* **67,** 55325y (1967).

(35) T. Kuge and K. Tenichi, *Agr. Biol. Chem.* (*Tokyo*), **32,** 1232 (1968).

(36) E. M. Montgomery and K. R. Sexson, U.S. Patent 3,046,161 (1962); *Chem. Abstr.,* **57,** 11440 (1962).

(37) R. A. Anderson and C. Vojnovich, U.S. Patent 3,252,836 (1966); *Chem. Abstr.,* **65,** 2457 (1966).

(38) Cooperatieve Verkoop-en Productievereniging van Aardappelmeel en Derivaten "Avebe" G. A., Neth. Patent Appl. 98,155 (1953); *Chem. Abstr.,* **56,** 1658 (1962).

(39) Cooperatieve Verkoop-en Productievereniging van Aardappelmeel en Derivaten "Avebe" G.A., Neth. Patent Appl. 96,139 (1960); *Chem. Abstr.,* **56,** 14519 (1962).

(40) Hoffmann's Staerkefabriken A.G., Brit. Patent 1,014,105 (1965).

(41) M. Schmidt, U.S. Patent 3,307,975 (1967); *Chem. Abstr.,* **66,** 106129b (1967).

(42) Cooperatieve Verkoop-en Productievereniging van Aardappelmeel en Derivaten "Avebe" G.A., Neth. Patent Appl. 98,156 (1953); *Chem. Abstr.,* **56,** 1658 (1962).

(43) M. Seidman, U.S. Patent 3,532,602 (1970); *Chem. Abstr.,* **73,** 13225u (1970).

(44) M. Kurimoto and K. Sugimoto, Ger. Offen. 2,003,335 (1970); *Chem. Abstr.,* **73,** 89423t (1970).

(45) M. Yoshida and M. Hirao, S. African Patent 69 06,182 (1970); *Chem. Abstr.,* **73,** 111200s (1970).

(46) J. F. Foster, *in* "Starch: Chemistry and Technology," R. L. Whistler and E. F. Paschall, eds., Academic Press, New York, Vol. 1, 1965, p. 349.

(47) D. A. Brant and W. L. Dimpfl, *Macromolecules,* **3,** 655 (1970).

(48) R. L. Whistler, *in* "Starch and Its Derivatives," J. A. Radley, ed., John Wiley, New York, 3rd Ed., 1954, p. 213.

(49) J. F. Foster and M. D. Sterman, *J. Polym. Sci.,* **21,** 91 (1956).

(50) J. Holló, J. Szejtli, and G. Gantner, *Magyar Tudományos Akad. Kem Tudományok Osztályának Közleményei,* **11,** 465 (1959); *Chem. Abstr.,* **54,** 4009 (1960).

(51) J. Holló, J. Szejtli, and G. Gantner, *Periodica Polytech.,* **3,** 95 (1959); *Chem. Abstr.,* **54,** 5135 (1960).

(52) J. Holló, J. Szejtli, and G. S. Gantner, *Staerke,* **12,** 73 (1960).

(53) J. Holló, J. Szetli, and G. Gantner, *Staerke,* **12,** 106 (1960).

(54) R. L. Whistler and C. Johnson, *Cereal Chem.,* **25,** 418 (1948).

(55) N. Z. Erdi, M. M. Cruz, and O. A. Battista, *J. Colloid Interface Sci.,* **28,** 36 (1968).

(56) J. N. BeMiller and R. L. Whistler, *in* "Industrial Gums," R. L. Whistler, ed., Academic Press, New York, 1st Ed., 1959, p. 675.

(57) Cooperatieve Verkoop-en Productievereniging van Aardappelmeel en Derivaten "Avebe" G.A., Brit. Patent 822,587 (1959); *Chem. Abstr.,* **54,** 6148 (1960).

(58) A. Dekker, U.S. Patent 2,999,032 (Appl. 1959); Neth. Patent 98,574; *Chem. Abstr.,* **56,** 2626 (1962).

(59) J. Muetgeert, W. C. Bus, and P. Hiemstra, *J. Chem. Eng. Data,* **7,** 272 (1962).

(60) L. J. Hickey, U.S. Patent 3,222,199 (1965); *Chem. Abstr.,* **64,** 8468 (1966).

(61) H. W. Leach, U.S. Patent 3,265,632 (1966); *Chem. Abstr.,* **66,** 4066d (1967).

(62) B. R. Zeitlin, A. Sarko, and F. J. Germino, U.S. Patent 3,128,208 (1964); *Chem. Abstr.,* **60,** 14724 (1964).

(63) A. Sarko, B. R. Zeitlin, and F. J. Germino, U.S. Patent 3,086,890 (1963); *Chem. Abstr.,* **59,** 1829 (1963).

(64) A. Sarko, F. J. Germino, B. R. Zeitlin, and P. F. Dapas, *J. Appl. Polym. Sci.,* **8,** 1343 (1964).

(65) F. J. Germino, B. R. Zeitlin, and A. Sarko, U.S. Patent 3,128,209 (1964); *Chem. Abstr.,* **60,** 14724 (1964).

(66) J. R. Feldman, R. E. Klose, and R. V. MacAllister, U.S. Patent 3,515,591 (1970); *Chem. Abstr.,* **73,** 65210v (1970).

(67) R. B. Evans, U.S. Patent 3,130,081 (1964); *Chem. Abstr.,* **61,** 2040 (1964).

(68) S. Augustat, *Ernaehrungsforschung,* **13,** 475 (1968); *Chem. Abstr.,* **70,** 21149r (1969). See also reference 11.

(69) I. A. Wolff, H. A. Davis, J. E. Cluskey, and L. J. Gundrum, U.S. Patent 2,608,723 (1952); *Chem. Abstr.,* **47,** 2523 (1953).

(70) H. A. Davis, I. A. Wolff, and J. E. Cluskey, U.S. Patent 2,656,571 (1953); *Chem. Abstr.,* **48,** 1040 (1954).

(71) J. Muetgeert and P. Hiemstra, U.S. Patent 2,822,581 (1958); *Chem. Abstr.,* **52,** 8602 (1958).

(72) Cooperatieve Verkoop-en Productievereniging van Aardappelmeel en Derivaten "Avebe" G.A., Neth. Patent Appl. 88,996 (1958); *Chem. Abstr.,* **53,** 23030 (1959).

(73) Kalle & Co. Akt.-Ges., Brit. Patent 847,431 (1960); *Chem. Abstr.,* **55,** 6896 (1961).

(74) Tefag Etzkorn & Co. Technische Faser G.m.b.H., (West) Ger. Patent 1,063,325 (1959); *Chem. Abstr.,* **55,** 15951 (1961).

(75) D. Kudera, U.S. Patent 2,973,243 (1961); *Chem. Abstr.,* **55,** 17030 (1961).

(76) Wolff & Co. K.-G. auf Aktien, (West) Ger. Patent 1,127,074 (1962); *Chem. Abstr.,* **57,** 2460 (1962).

(77) W. B. Kunz, U.S. Patent 3,030,667 (1962); *Chem. Abstr.,* **57,** 3684 (1962).

(78) F. C. Wohlrabe, A.M. Mark, W.B. Roth, and C. L. Mehltretter, U.S. Patent 3,116,351 (1963); *Chem. Abstr.,* **60,** 9466 (1964).

(79) Department of Agriculture and Inspection, Nebraska, Brit. Patent 965,349 (1964); *Chem. Abstr.,* **61,** 12156 (1964).

(80) A. E. Staley Mfg. Co., Neth. Patent Appl. 6,406,933 (1964); *Chem. Abstr.,* **62,** 16478 (1965).

(81) American Machine and Foundry Co., Neth. Patent Appl. 6,501,338 (1965); *Chem. Abstr.,* **64,** 3805 (1966).

(82) A. E. Staley Mfg. Co., Neth. Patent Appl. 6,500,064 (1965); *Chem. Abstr.,* **64,** 5284 (1966).

(83) A. E. Staley Mfg. Co., Belg. Patent 641,123 (1964); *Chem. Abstr.,* **64,** 11429 (1966).

(84) American Machine and Foundry Co., Neth. Patent Appl. 6,414,395 (1965); *Chem. Abstr.,* **64,** 14404 (1966).

(85) A. H. Young, U.S. Patent 3,312,559 (1967); *Chem. Abstr.,* **67,** 3994w (1967).

(86) H. M. Walton, U.S. Patent 3,312,560 (1967); *Chem. Abstr.,* **67,** 12119s (1967).

(87) A. H. Young, M. T. Tetenbaum, and R. J. Pratt, U.S. Patent 3,320,081 (1967); *Chem. Abstr.,* **67,** 23095g (1967).

(88) M. T. Tetenbaum, U.S. Patent 3,318,715 (1967); *Chem. Abstr.,* **67,** 23096h (1967).

(89) F. E. Carevic, U.S. Patent 3,336,429 (1967); *Chem. Abstr.,* **67,** 101244w (1967).

(90) D. J. Bridgeford and D. M. Gallagher, U.S. Patent 3,497,584 (1970); *Chem. Abstr.,* **72,** 113005k (1970).

(91) P. Hiemstra and J. Muetgeert, U.S. Patent 2,902,336 (1959); *Chem. Abstr.,* **54,** 915 (1960).

(92) J. W. Barger and C. E. Mumma, U.S. Patent 3,499,074 (1970); *Chem. Abstr.,* **72,** 91407p (1970).

(93) S. A. Buckler and F. J. Germino, U.S. Patent 3,398,015 (1968); *Chem. Abstr.,* **69,** 97855t (1968).

(94) A. E. Staley Mfg. Co., Neth. Patent Appl. 6,506,669 (1965); *Chem. Abstr.,* **64,** 17826 (1966).

(95) M. W. Rutenberg and W. Jarowenko, Brit. Patent 900,868 (1962); *Chem. Abstr.,* **57,** 11440 (1962); U.S. Patent 3,081,181 (1963).

(96) J. C. Rankin, I. A. Wolff, H. A. Davis, and C. E. Rist, *Chem. Eng. Data Ser.,* **3,** 120 (1958).

(97) I. A. Wolff, H. A. Davis, J. E. Cluskey, L. J. Gundrum, and C. E. Rist, *Ind. Eng. Chem.*, **43,** 915 (1951).

(98) W. R. Grace and Co., Neth. Patent Appl. 6,600,774 (1966); *Chem. Abstr.*, **66,** 96462p (1967).

(99) N. E. Lloyd and L. C. Kirst, *Cereal Chem.*, **40,** 154 (1963).

(100) J. N. BeMiller, *in* "Starch: Chemistry and Technology," R. L. Whistler and E. F. Paschall, eds., Academic Press, New York, Vol. 1, 1965, p. 309.

(101) W. L. Minto, U.S. Patent 2,964,395 (1960); *Chem. Abstr.*, **55,** 7742 (1961).

(102) V. O. Makhnach, L. B. Borisov, M. A. Litvinov, and N. A. Matyko, *Tr. Leningr. Sanit.-Gigien. Med. Inst.*, **66,** 162 (1962); *Chem Abstr.*, **59,** 10498 (1963).

(103) C. L. Mehltretter and W. B. Roth, U.S. Patent 3,414,515 (1968); *Chem. Abstr.*, **70,** 31611z (1969).

(104) L. A. Anderson, Brit. Patent 1,072,795 (1967); *Chem. Abstr.*, **67,** 84877k (1967).

(105) O. B. Wurzburg, P. C. Trubiano, and W. Herbst, U.S. Patent 3,499,962 (1970); *Chem. Abstr.*, **72,** 113051x (1970).

(106) E. Morii, T. Numasawa, K. Iwata, H. Hanai, S. Yamagata, and A. Ishimori, Ger. Offen. 1,916,535 (1969); *Chem. Abstr.*, **72,** 35780t (1970).

(107) A. Wander A.-G., Swiss Patent 305,572 (1955); *Chem. Abstr.*, **50,** 15110 (1956).

(108) O. B. Wurzburg, M. W. Rutenberg, and L. J. Ross, U.S. Patent 2,786,833 (1957); *Chem. Abstr.*, **51,** 10936 (1957); *Brit. Patent* 755,461 (1957); *Chem. Abstr.*, **51,** 8460 (1957).

(109) National Starch Products Inc., Brit. Patent 755,461 (1956); *Chem. Abstr.*, **51,** 8460 (1957).

(110) R. L. Whistler and W. W. Spencer, *Arch. Biochem. Biophys.*, **95,** 36 (1961).

(111) R. L. Whistler, D. G. Unrau, and G. Ruffini, *Arch. Biochem. Biophys.*, **126,** 647 (1968).

(112) Spofa spojené farmaceutické závody, národní podnik, Brit. Patent 859,278 (1961); *Chem. Abstr.*, **55,** 11771 (1961).

(113) Z. Roubal, Z. Placer, and V. Vokáč, Czech. Patent 88,330 (1960); *Chem. Abstr.*, **54,** 2201 (1960).

(114) Z. Placer, Z. Roubal, and V. Vokáč, *Rev. Czech. Med.*, **4,** 11 (1958); *Chem. Abstr.*, **55,** 25036 (1961).

(115) L. J. Ravin, J. G. Baldinus, and M. L. Mazur, *J. Pharm. Sci.*, **51,** 857 (1962).

(116) SPOFA United Pharmaceuticals Works, Ger. Patent 1,200,277 (1965); *Chem. Abstr.*, **63,** 17807 (1965).

(117) V. Vokáč, Z. Placer, and Z. Roubal, *Československ. Gastroenterol. Výživa*, **11,** 266 (1957); *Chem. Abstr.*, **52,** 4033 (1958).

(118) E. Husemann, P. Pfannemüller, H. Schill, and W. Hertlein, *Z. Naturforsch.*, *B*, **12,** 427 (1957).

(119) W. M. Doane and R. L. Whistler, *Arch. Biochem. Biophys.* **101,** 436 (1963).

(120) M. Herold, J. Hofman, P. Málek, J. Čapkova, M. Heřmanský, M. Vondráček, and J. Kolc, Czech. Patent. 90,981 (1959); *Chem. Abstr.*, **54,** 14594 (1960).

(121) National Starch and Chemical Corp., Brit. Patent 969,711 (1964); *Chem. Abstr.*, **62,** 4203 (1965).

(122) R. B. Evans and W. G. Kunze, U.S. Patent 3,318,868 (1967); *Chem. Abstr.*, **67,** 55327a (1967).

(123) I. A. Wolff, D. W. Olds, and G. E. Hilbert, *Ind. Eng. Chem.*, **49,** 1247 (1957).

(124) A. T. Gros and R. O. Feuge, *J. Amer. Oil Chem. Soc.*, **39,** 19 (1962).

(125) I. A. Wolff, D. W. Olds, and G. E. Hilbert, *Ind. Eng. Chem.*, **43,** 911 (1951); *J. Amer. Chem. Soc.*, **73,** 346 (1951).

(126) J. Muetgeert, P. Hiemstra, and W. C. Bus, *Staerke*, **10,** 303 (1958).

(127) Upjohn Co., Brit. Patent 810,306 (1959); *Chem. Abstr.*, **53,** 18401 (1959).

(128) Gevaert Photo-Producten N.V., Belg. Patent 613,575 (1962); *Chem. Abstr.*, **57,** 16729 (1962).

(129) F. J. Germino and R. J. Moshy, U.S. Patent 3,184,335 (1965); *Chem. Abstr.*, **63,** 4502 (1965).

(130) R. J. Moshy, F. J. Germino, and H. W. P. Panzer, U.S. Patent 3,188,237 (1965); *Chem. Abstr.*, **63,** 5884 (1965).

(131) R. J. Moshy, F. J. Germino, and H. W. P. Panzer, U.S. Patent 3,326,893 (1967); *Chem. Abstr.*, **67,** 65703j (1967).

(132) F. J. Germino and R. J. Moshy, U.S. Patent 3,376,286 (1968); *Chem. Abstr.*, **68,** 96977b (1968).

(133) J. M. G. Cowie, P. M. Toporowski, and F. Costaschuk, *Makromol. Chem.*, **121,** 51 (1969).

(134) R. L. Whistler and G. E. Hilbert, *Ind. Eng. Chem.*, **36,** 796 (1944).

(135) E. Husemann and H. Bartl, *Makromol. Chem.*, **18–19,** 342 (1950).

(136) National Starch and Chemical Corp., Brit. Patent 871,634 (1961); *Chem. Abstr.*, **57,** 1139 (1962); M. W. Rutenberg, W. Jarowenko, and L. J. Ross, U.S. Patent 3,038,895 (1962); *Chem. Abstr.*, **57,** 10086 (1962).

(137) R. L. Whistler and G. N. Richards, *Ind. Eng. Chem.*, **50,** 1551 (1958).

(138) I. A. Wolff, *Ind. Eng. Chem.*, **50,** 1552 (1958).

(139) Unilever N.V., (West) Ger. Patent 1,205,806 (1965); *Chem. Abstr.*, **64,** 7289 (1966).

(140) Rootry Expolitatie Maatschappij N. V., Neth. Patent Appl. 93,356 (1960); *Chem. Abstr.*, **55,** 6736 (1961).

(141) A. H. Wagenaar and P. H. Van der Meij, Ger. Offen. 1,963,567 (1970); *Chem. Abstr.*, **73,** 89848k (1970).

(142) Rootry Exploitatiemij N.V., Brit. Patent 826,226 (1959); *Chem. Abstr.*, **54,** 11328 (1960); 826,940 (1960).

(143) Rootry Expolitatie Maatschappij N.V., Neth. Patent Appl. 93,354 (1960); 93,355 (1960); *Chem. Abstr.*, **55,** 9718 (1961); 93,357 (1960); *Chem. Abstr.*, **55,** 9719 (1961).

(144) M. S. Cole, U.S. Patent 3,479,191 (1969); *Chem. Abstr.*, **72,** 30444b (1970).

(145) K. Laden, P. E. Sokol, H.-C. Tsai, and B. A. Rogers, Ger. Offen. 1,926,068 (1970); *Chem. Abstr.*, **72,** 82911w (1970).

(146) Gavaert Photo-Producten N.V., Belg. Patent 623,091 (1963); *Chem. Abstr.*, **59,** 13011 (1963).

(147) Gevaert Photo-Producten N.V., Belg. Patent 606,123 (1961); *Chem. Abstr.*, **58,** 6943 (1963).

(148) F. Schierbaum, *Ernaehrungsforschung*, **13,** 485 (1968); *Chem. Abstr.*, **70,** 27705r (1969).

(149) J. L. Sanella and R. L. Whistler, *Arch. Biochem. Biophys.*, **102,** 226 (1963).

(150) R. L. Whistler and G. A. Towle, *Arch Biochem. Biophys.*, **135,** 396 (1969).

(151) J. W. Sietsema, U.S. Patent 2,974,048 (1961); *Chem. Abstr.*, **55,** 14267 (1961).

(152) D. J. Bridgeford, U.S. Patent 3,330,820 (1967); *Chem. Abstr.*, **68,** 70348x (1968).

(153) D. J. Bridgeford, U.S. Patent 3,335,023 (1967); *Chem. Abstr.*, **67,** 83151g (1967).

(154) D. J. Bridgeford and D. M. Gallagher, U.S. Patent 3,336,144 (1967); *Chem. Abstr.*, **67,** 83150f (1967).

(155) C. R. Russell, R. A. Buchanan, and C. E. Rist, U.S. Patent 3,160,552 (1964); *Chem. Abstr.*, **62,** 5442 (1965).

(156) Tee-Pak, Inc., Neth. Patent Appl. 6,512,208 (1966); *Chem. Abstr.*, **65,** 17157 (1966).

(157) D. J. Bridgeford, U.S. Patent 3,429,735 (1969); *Chem. Abstr.*, **70,** 107683x (1969).

(158) D. J. Bridgeford, Ger. Offen. 1,942,676 (1970); *Chem. Abstr.*, **72,** 123215t (1970).

(159) N. I. Burke, D. J. Bridgeford, and A. F. Turbak, *Ind. Eng. Chem., Prod. Res. Develop.*, **9,** 230 (1970).

(160) E. Husemann, R. Resz, and R. Werner, *Makromol. Chem.*, **47,** 48 (1961).

(161) A. E. Staley Mfg. Co., Belg. Patent 667,137 (1966); *Chem. Abstr.*, **65,** 15642 (1966).

(162) E. Husemann and R. Resz, *J. Polym. Sci.*, **19,** 389 (1956).

(163) Cooperatieve Verkoop-en Productievereniging van Aardappelmeel en Derivaten "Avebe" G.A., Brit. Patent 869,192 (1961); *Chem. Abstr.*, **55,** 22878 (1961).

(164) A. J. Staverman, J. Muetgeert, and A. Schors, *Mezhdunarod. Simpozium Makromol. Khim., Dokl., Moscow, 1960,* **3,** 326 (1960); *Chem. Abstr.*, **55,** 10331 (1961).

(165) M. M. Samfield and M. G. Christy, U.S. Patent 3,009,835 (Appl. 1960); 3,009,836 (Appl. 1959); *Chem. Abstr.*, **56,** 7566 (1962).

(166) A. E. Staley Mfg. Co., Belg. Patent 609,702 (1962); *Chem. Abstr.*, **57,** 10086 (1962).

(167) Owens-Corning Fiberglas Corp., Belg. Patent 616,790 (1962); *Chem. Abstr.*, **57,** 16166 (1962).

(168) A. E. Staley Mfg. Co., Belg. Patent 635,222 (1963); *Chem. Abstr.*, **61,** 13404 (1964).

(169) T. Inoue, *Aichi Gakugei Daigaku Kenkyu Hokoku,* No. **9,** 69 (1960); *Chem. Abstr.*, **54,** 2420 (1960).

(170) J. Hoffmann, P. Málek, M. Herold, J. Čapková, M. Heřmanský, M. Vondráček, and J. Kolc, Czech. Patent 90,980 (1959); *Chem. Abstr.*, **55,** 3014 (1961).

(171) P. A. Inklaar, U.S. Patent 3,345,358 (1967); *Chem. Abstr.*, **68,** 99049m (1968).

(172) J. E. Hodge, S. A. Karjala, and G. E. Hilbert, *J. Amer. Chem. Soc.*, **73,** 3312 (1951).

(173) J. W. Wood and P. T. Mora, *J. Polym. Sci., Pt. A,* **1,** 3511 (1963).

(174) G. Keilich, K. Tihlarik, and E. Husemann, *Makromol. Chem.*, **120,** 87 (1968).

(175) K. Bredereck, K. Strunk, and H. Menrad, *Makromol. Chem.*, **126,** 139 (1969).

(176) M. O. Weaver and C. R. Russell, U.S. Patent 3,294,781 (1966); *Chem. Abstr.*, **66,** 47435y (1967).

(177) E. D. Klug, U.S. Patent 3,117,014 (1964); *Chem. Abstr.*, **60,** 14695 (1964).

(178) A. Mishina, *Mem. Inst. Sci. Ind. Res., Osaka Univ.*, **18,** 93 (1961); *Chem. Abstr.*, **56,** 4947 (1962).

(179) A. Mishina, Z. Nikuni, and T. Shimizu, *Nippon Nogei Kagaku Kaishi,* **37,** 389 (1963); *Chem. Abstr.*, **61,** 14179 (1964).

(180) A. Mishina, *Nippon Nogei Kagaku Kaishi,* **35,** 40 (1961); *Chem. Abstr.*, **62,** 10653 (1965); *Nippon Nogei Kagaku Kaishi,* **36,** 617 (1962); *Chem. Abstr.*, **62,** 4200 (1965).

(181) A. Mishina and Z. Nikuni, *Nippon Nogei Kagaku Kaishi,* **37,** 46 (1963); *Chem. Abstr.*, **63,** 4503 (1965).

(182) Z. Reyes and C. R. Russell, U.S. Patent 3,518,176 (1970); *Chem. Abstr.*, **73,** 57423b (1970).

(183) H. J. Hagemeyer, Jr., U.S. Patent 2,763,627 (1956); *Chem. Abstr.*, **51,** 1657 (1957).

(184) N. Geacintov, V. T. Stannett, E. W. Abrahamson, and J. J. Hermans, *Proc. Cellulose Conf., 2nd, Syrcause, 1959,* 142 (1959); *Chem. Abstr.*, **54,** 882 (1960).

(185) N. R. Segro and W. Hodes, U.S. Patent 3,044,972 (1962); *Chem. Abstr.*, **57,** 11401 (1962).

(186) D. K. Ray-Chaudhuri, *Staerke,* **21,** 47 (1969).

(187) A. Mishina and Z. Nikuni, *Nippon Nogei Kagaku Kaishi,* 38, 372 (1964); *Chem. Abstr.*, **63,** 1893 (1965).

(188) A. Mishina and Y. Yoshimoto, *Nippon Nogei Kagaku Kaishi,* **37,** 43 (1963); *Chem. Abstr.*, **62,** 10583 (1965).

(189) R. L. Whistler and H. J. Roberts, *J. Org. Chem.*, **24,** 2458 (1961).

(190) E. Husemann and M. Reinhardt, *Makromol. Chem.*, **57,** 109 (1962); **57,** 129 (1962).

(191) E. Husemann and A. P. O. Schmidt, *Makromol. Chem.*, **65,** 114 (1963).

(192) M. L. Wolfrom and P. Y. Wang. *Chem. Commun.*, 113 (1968); *Carbohyd. Res.*, **12,** 109 (1970).

(193) R. G. Krylova, S. N. Ryadovskaya, and O. P. Golova, *Vysokomol. Soedin., Ser. B,* **11,** 562 (1969); *Chem. Abstr.*, **72,** 21871m (1970).

(194) R. L. Whistler and D. G. Medcalf, *Arch. Biochem. Biophys.*, **104,** 150 (1964).

(195) M. L. Wolfrom, K. C. Gupta, K. K. De, A. K. Chatterjee, T. Kinoshita, and P. Y. Wang, *Staerke,* **21,** 39 (1969).

(196) M. L. Wolfrom, M. I. Taha, and D. Horton, *J. Org. Chem.*, **28,** 3553 (1963).

(197) L. H. Elizer, G. C. Glasscock, and J. M. Seitz, U.S. Patent 3,051,700 (1962); *Chem. Abstr.*, **58,** 11564 (1963).

(198) Eastman Kodak Co., Fr. Patent 1,472,741 (1967); *Chem. Abstr.*, **68,** 25426k (1968).

(199) E. J. Perry and D. D. Reynolds, U.S. Patent 3,425,836; *Chem. Abstr.*, **70,** 120008a (1969).

(200) R. L. Whistler and S. Hirase, *J. Org. Chem.*, **26,** 4600 (1961).

(201) R. L. Whistler and D. G. Medcalf, *Arch. Biochem. Biophys.*, **105,** 1 (1964).

(202) D. M. Clode, D. Horton, M. H. Meshreki, and H. Shoji, *J. Chem. Soc., D,* 694 (1969).

(203) Y. Hirasaka, *Yakugaku Zasshi,* **83,** 976 (1963); *Chem. Abstr.*, **60,** 4233 (1964).

(204) A. E. Staley Mfg. Co., Neth. Patent Appl. 6,408,175 (1965); *Chem. Abstr.*, **63,** 11876 (1965).

Popular Mechanics
BUILD 'EM, FLY 'EM!
SKY SCOOTERS
SPECIAL SECTION
BOATING '93

(205) R. W. Best and R. M. Powers, U.S Patent 3,329,523 (1967); *Chem. Abstr.*, **67,** 74685g (1967).

(206) A. E. Staley Mfg. Co., Neth. Patent Appl. 6,410,751 (1965); *Chem. Abstr.*, **63,** 10168 (1965).

(207) R. M. Powers, U.S. Patent 3,329,525 (1967); *Chem. Abstr.*, **67,** 65645s (1967).

(208) G. Brown and E. D. Mazzarella, Ger. Offen. 1,918,415 (1969); *Chem. Abstr.*, **72,** 45240f (1970).

(209) L. J. Hickey, R. H. Williams, and E. D. Mazzarella, U.S. Patent 3,284,381 (1966); *Chem. Abstr.*, **66,** 12114r (1967).

(210) National Starch and Chemical Corp., Neth. Patent Appl. 6,608,467 (1966); *Chem. Abstr.*, **67,** 34043q (1967).

(211) D. L. Wilhelm, U.S. Patent 3,532,648 (1970); *Chem. Abstr.*, **73,** 13229y (1970).

(212) National Starch and Chemical Corp., Neth. Patent Appl. 6,515,704 (1966); *Chem. Abstr.*, **65,** 12400 (1966).

(213) P. Hiemstra and B. Nijland, *Tex De,* **18,** 1534 (1959); *Chem. Abstr.*, **54,** 11486 (1960).

(214) E. Pacsu and J. W. Mullen, *J. Amer. Chem. Soc.*, **63,** 1168 (1941).

(215) W. A. Scholten's Chemische Fabrieken N.V., (West) Ger. Patent 1,118,151 (1961); *Chem. Abstr.*, **56,** 14501 (1962).

(216) J. P. Stevens and Co., Inc., Belg. Patent 626,797 (1963); *Chem. Abstr.*, **61,** 8493 (1964).

(217) Owens-Corning Fiberglas Corp., Brit. Patent 923,058 (1963); *Chem. Abstr.*, **58,** 13581 (1963).

(218) A. Marzocchi and G. E. Rammel, U.S. Patent 3,462,254 (1969); *Chem. Abstr.*, **71,** 103268p (1969).

(219) C. W. Charon and L. C. Renaud, U. S. Patent 3,108,891 (1963).

(220) Union Rhenische Braunkholen Kraftstoff A.-G., Brit. Patent 1,005,685 (1965); *Chem. Abstr.*, **63,** 16578 (1965).

(221) Union Rheinische Braunkohlen Kraftstoff A.-G., (West) Ger. Patent 1,207,040 (1965); *Chem. Abstr.*, **64,** 9883 (1966).

(222) R. E. O'Brian and E. D. O'Brian, U.S. Patent 2,729,565 (1956); *Chem. Abstr.*, **50,** 5945 (1956).

(223) J. L. Louis, U.S. Patent 2,627,466 (1953); *Chem. Abstr.*, **47,** 4520 (1953).

(224) C. O. Moore and J. W. Robinson, U.S. Patent 3,368,909 (1968); *Chem. Abstr.*, **68,** 77068q (1968).

(225) National Starch and Chemical Corp., Neth. Patent Appl. 6,514,900 (1966); *Chem. Abstr.*, **65,** 11251 (1966).

(226) M. M. Mokhtar and H. S. White, U.S. Patent 3,471,303 (1969); *Chem. Abstr.*, **72,** 11507h (1970).

(227) E. P. L. de Muynck and A. E. Van Branteghem, S. African Patent 68 02,001 (1968); *Chem. Abstr.*, **70,** 95600c (1969).

(228) O. B. Wurzburg and W. G. Kunze, U.S. Patent 3,265,509 (1966); *Chem. Abstr.*, **65,** 14343 (1966).

(229) W. M. Malisoff, U.S. Patent 2,022,729, (1935); *Chem. Abstr.*, **30,** 819 (1936).

(230) K. C. Kwan and G. Milosovich, *J. Pharm. Sci.*, **55,** 340 (1966).

(231) G. K. Nichols and R. W. P. Short, U.S. Patent 3,490,742 (1970); *Chem. Abstr.*, **72,** 103756s (1970).
(232) P. Claes, A. Billiau, E. De Clercq, J. Desmyter, E. Schonne, H. Vanderhaege, and P. De Somer, *J. Virol.*, **5,** 313 (1970).
(233) A Billiau, J. Desmyter, and P. De Somer, *J. Virol.*, **5,** 321 (1970).
(234) J. Holló, J. Szejtli, E. László, G. S. Gantner, Z. Toth, J. Huszar, and A. Lux, *Staerke,* **16,** 118 (1964).
(235) J. Holló, E. László, J. Szejtli, and A. Lux, *Staerke,* **16,** 167 (1964).
(236) Matsushita Denki Sangyo K.K., Japan. Patent 2966 (1960); *Chem. Abstr.*, **55,** 9119 (1961).
(237) F. J. Germino, J. R. Infantino, R. J. Moshy, and R. A. Smith, U.S. Patent 3,220,731 (1965); *Chem. Abstr.*, **64,** 6904 (1966).
(238) I. Cheema and H. N. Weinberg, U.S. Patent 3,355,378 (1967); *Chem. Abstr.*, **68,** 61485n (1968).
(239) O. F. Parks, J. E. Strassner, and F. W. Burtch, U.S. Patent 3,096,777 (1963).
(240) A. E. Staley Mfg. Co., Belg. Patent 611,052 (1961); *Chem. Abstr.*, **57,** 14042 (1962).
(241) National Starch and Chemical Co., Brit. Patent 916,710 (1962); Can. Patent 648,053 (1962).
(242) R. J. Reynolds Tobacco Co., Can. Patent 658,442 (1961).
(243) K. Seidel, (West) Ger. Patent 825,746 (1951); *Chem. Abstr.*, **49,** 6653 (1955).
(244) A. T. Derivan, U.S. Patent 3,163,579 (1964); *Chem. Abstr.*, **62,** 10304 (1965).
(245) M. P. Ptasienski and J. W. Gill, U.S. Patent 3,003,979 (1961); *Chem. Abstr.*, **56,** 1160 (1962).

CHAPTER XXV

STARCH AMYLOPECTIN (WAXY CORN AND WAXY SORGHUM)

EUGENE L. POWELL

American Maize-Products Company, Hammond, Indiana

I. INTRODUCTION

Amylopectin is not a name applied to a single well-defined compound. It is, rather, the designation for a broad group of branched starch molecules. Starches found in the root, stems, tubers, leaves, and seeds of plants contain polymeric molecules of α-D-glucopyranosyl units, some of which have a multiple branching structure. Amylopectin is the name applied to those glucans presumed by many to have a branch on branch or bushlike structure. Even amylopectin molecules from a single plant species vary in molecular size, differ in the number of monomer units in a branch, and differ in the frequency of branch points along the chains. Furthermore, amylopectin molecules in the natural state are always aggregated and organized into granules of microscopic size, most often with varying amounts of the nonbranched polymer, *amylose,* present. This chapter is confined to a discussion of amylopectins that are free of amylose, either by virtue of their natural occurrence, or because separation has been effected.

Even when purity and molecular structure of two amylopectin samples is the same, the presence or absence of a granule structure will profoundly affect the properities displayed. To keep this distinction clear, the word *amylopectin* will be used to designate isolated branched molecules of starch that have been freed of amylose. The term *amylopectin starch* will indicate granular starch composed of branched molecules and obtained from a waxy mutant of one of the cereal grains. The nature and condition of the granules has an important effect on the properties of the starch. Although exact production figures are not published, it is believed that amylopectin starches greatly exceed the amount of isolated amylopectins used commercially.

II. Isolated Amylopectins

1. Chemistry and Properties

Most of the D-glucopyranosyl units of an amylopectin molecule are joined by α-D-(1→4) linkages. In addition, there is an α-D-(1→6) linkage at each branch point. Each D-glucopyranosyl unit is most probably in the $^4C_1$(D) chair conformation.[1] There is only one potentially free aldehyde group per molecule, with numerous nonreducing end-groups terminating the outer chains. Most amylopectins have a branch for each 15–30 sugar units. Distribution of the branch points is irregular within a given molecule; branches may occur close together or separated by over 40 sugar units.[2, 3] The interior chains tend to be shorter and more closely spaced, giving the molecule a rather dense core with longer, more flexible outer branches.[4]

The great number of primary and secondary hydroxyl groups and the ring oxygen provide ample opportunities for both inter- and intramolecular hydrogen bonding. Hydrogen bonding also promotes solvation and complex formation with suitable reagents. However, the changes observed when complexing agents, such as surfactants, are added to an amylopectin are much less profound than when they are added to a starch containing amylose. The α-D-(1→4) linkage coupled with the $^4C_1$(D) conformation tends to give the linear segments of the polymer chains a natural twist. Under certain conditions hydrogen bonding between the C-2 hydroxyl on one sugar ring and the C-3 hydroxyl on an adjacent ring may tend to hold portions of the chain in a tightly coiled helix.[5]

Once all aggregation has been destroyed by disrupting intermolecular hydrogen bonding, the amylopectins become water soluble; that is, they dissolve readily in cold water to give solutions that are stable indefinitely and not subject to retrogradation, gelling, or precipitation. When an amylopectin solution is dried rapidly, the product is water soluble; but if a solution is dried slowly, intermolecular bonds will form and the amylopectin becomes cold water insoluble.[6] Methyl sulfoxide, ethylenediamine (hydrate or anhydrous), chloral hydrate, and hydrazine hydrate are amylopectin solvents.[4] When granular or aggregated amylopectins are to be dispersed in water, hydrogen bond breaking agents, such as alkalis, calcium chloride, lithium bromide, urea, or guanidinium salts, are sometimes employed.[5]

Foster[4] pictures the amylopectin molecule in solution as a rather compact body with a dense core and as roughly spherical in shape when not distorted by mechanical shear or added chemical agents. However, under a shearing force, the flexible, highly solvated outer branches deflect and give the molecule an ellipsoidal shape. Measured molecular weights range from 4.5×10^4 to 4.2×10^8 (DP 280–1,450,000), depending on the method used and on the particular amylopectin preparation examined.[7] Foster[4] interprets the wide difference between weight-average and number-average molecular weights (ratios of 300 to 500) as proof of polydispersity, that is, proof that individual molecules within a given amylopectin vary in size.

2. *Commercial Isolation Processes*

Marquenne and Roux[8] coined the word *amylopectin* in 1905. During the 1930's, K. H. Meyer and associates[9] produced amylopectin-enriched starch granules by leaching the amylose from them with warm water. Schoch[10] separated the two starch polymers by precipitating the amylose from an autoclaved starch paste as a complex with a polar, sparingly water soluble, organic compound. This separation led to a patented process[11] for the preparation of amylopectin and, more importantly, provided starch fractions for fundamental research. Whistler and Hilbert[12] showed that a number of hydrogen bonding agents could act as fractionating agents. Pacsu and Bauer[13] further modified the procedure by adding acid or alkali before autoclaving to aid dispersion, a procedure that partially degrades the amylopectin. Cantor and Wimmer[14] ultilized a strong calcium chloride solution to disperse the starch and then used addition of alkali to precipitate calcium hydroxide and at the same time fractionate the two polymers, because of their different solubilities. However, separation on a commercial scale by these methods is costly.

Muetgeert and others,[6] working in The Netherlands, have shown that a variety of complexing agents and precipitants for amylose function best at certain critical concentration levels; excessive amounts also precipitate the amylopectin. Thus, control of the concentration–temperature relationship gives reasonable purity of the two isolated polymers. Commercially feasible procedures using sodium sulfate to selectively salt out the starch polymers have evolved.[15] The salt solutions can be recycled to lower costs. A low concentration of sodium sulfite is added to prevent oxidative discoloration of the polymers. It is claimed that crude flours as well as purified starch can be used to yield a commercially acceptable amylopectin. The Netherlands company Cooperatieve Verkoop-en Productievereniging van Aardappelmeel en Derivaten, "Avebe" G. A., has been producing amylopectin from potato starch. Since 1957, this product has been imported into the United States under the trade name Ramalin by Stein Hall and Co., Inc.

More recently, a research group at the A. E. Staley Mfg. Co. demonstrated that salts and complexing agents are not needed to achieve a simplified fractionation process.[16] They dispersed a 2.5% corn starch slurry by means of a continuous, steam injection cooker operating at 150° or higher. By holding the resulting solution at 50°–100° for about 8 hr, a nearly quantitative precipitation of amylose is achieved. Amylopectin is recovered from the supernatant by drum or spray drying. On a semicommercial scale, this process yields amylopectin of about 90% purity. Some degradation occurs, as the intrinsic viscosity averages about half that of the best laboratory preparations. Interestingly, most of this degradation is caused by shear in the spray dryer atomizer rather than during the steam injection cooking.[17] According to this group, addition of neutral salts or changing the pH has little effect on solutions of amylopectin but addition of borax greatly increases the viscosity. Addition of certain specific agents, such as common soap, polyethylene glycol 1000 monostearate, or polyethylene glycol 400 monolaurate, to solutions of their product increases the viscosity by a factor of 20 to 30, whereas other surfactants, both anionic and cationic, have little or no effect.[17]

Operation of this semicommercial process has been discontinued. No doubt, the ready availability of practically unlimited quantities of pure amylopectin from waxy starches and the more recent availability of high-amylose starches make the fractionation approach commercially unattractive in the United States. In addition, the amylopectin obtained by fractionation cannot compete with granular starch in the many uses where properties of granules are essential. Furthermore, granules of an amylopectin starch can be distentegrated to simulate isolated amylopectin products.

III. Amylopectin Starches

1. *Waxy Maize*

In 1908, the Rev. J. M. W. Farnham sent a peculiar kind of corn to the U.S. Department of Agriculture. He had observed this corn growing near Shanghai, China, and described it as being "much more glutenous" than ordinary corn. Collins[18] suggested that it be named *cereous* or *waxy* because the cut surface of the endosperm looked waxy. Thus, the waxy misnomer has been used to describe this particular recessive genotype. In 1922, Weatherwax[19] noted that waxy maize stained red with iodine rather than blue. By the late 1930's, it became obvious that the starch of waxy corn (maize) was pure amylopectin, a fact that Caldwell and Hixon[20] confirmed by fractionation. In rapid succession, the waxy variant was found in barley, millet, grain sorghum, and rice, as well as in corn.[21] Waxy or glutenous rice has been grown in China for centuries.

Disruption of tapioca imports during World War II prompted the American Maize-Products Company and National Starch and Chemical Corporation to jointly commercialize waxy corn starch (maize) as a substitute. Iowa State College and The University of Nebraska cooperated in developing corn hybrids of waxy types. The first commercial crop was processed at American Maizes' factory by the conventional wet milling process in 1942.[22] Waxy corn has enjoyed a sustained growth since, with at least four different companies now competing vigorously for a share of the amylopectin starch market. Many millions of pounds are produced each year.

Current breeding practice is to select the best single-cross, three-way cross, or double-cross hybrids available for a given area and then to convert these to a waxy type. Thus, such things as agronomic properties, yield per acre, and oil and protein content of this grain equal those of the best corn belt hybrids. The only difficulty in production involves segregation to avoid contamination with ordinary corn pollen in the field or mixing with other grain during shipment, storage, and processing.

2. *Waxy Sorghum*

The success of waxy maize starches prompted the Corn Products Company to add waxy starches to their line of sorghum starches of the milo type produced at Corpus Christy, Texas. The waxy mutant of milo sorghum already at hand was further developed agronomically under the name White Milo.[23] Yield per acre, composition of the grain, and wet milling characteristics were reported to be comparable to those of the ordinary grain sorghum. Waxy sorghum starch granules resembled those from waxy maize microscopically and their properties were essentially the same except for about a 3° higher gelatinization temperature and 30% less granule swelling.[24]

3. *Useful Properties of Amylopectin Starches*

Commercial waxy starches normally contain 10–12% moisture, 0.4% protein, 0.3% ash, and 0.2% or less bound fatty acids. The latter figure contrasts with the 0.6% fatty acid bound in regular corn and sorghum starches. The balance is amylopectin. The granules resemble those of common corn or sorghum, except that they stain red-brown instead of blue with iodine. Unmodified waxy starches cook in water to give pastes that are clear and high in viscosity but extremely stringy and cohesive. These pastes do not retrograde, set back, or gel on cooling and aging; hence they are of great value in foods, sizes, and adhesives where shelf-life or pot-life is a problem. The unmodified amylopectin starch pastes are, however, readily thinned by acidic foods, autoclaving procedures, or high mechanical shear during processing.

Very early in this commercial development it became apparent that chemical and physical modifications of the amylopectin granules, and blending with other starches, was the key to a large scale marketing. By such combinations, its weaknesses were overcome.

Evans[25] has described the modified waxy maize starches and Watson[23] has done the same for the waxy sorghum. Modification procedures applied to other starches are used to control the granule properties of the amylopectin starches. Cross-linked, acid thin-boiling, hydroxyalkyl ether, ester, pyrodextrin, oxidized, and pregelatinized modifications are commercially available. In some cases, two or three of these modifications may be applied in concert to give the properties desired. The methodology and procedures applied have been reviewed.[26] In addition to Leaches' discussion of granule swelling,[24] the chapter by Myers and Knauss[27] should prove helpful to those engaged in applied research with amylopectin starches. Problems of retrogradation and gel formation are minimal with waxy starches, but there is a problem associated with stringiness, cohesiviness, or anomalous viscosity.

IV. Applications of Amylopectins and Amylopectin Starches

The advent of the amylopectins has added new dimensions to starch technology. The unusual optical and rheological properties of these pastes and the greatly reduced tendency to gel in dispersions or crystallize in dried films make possible products and formulations previously unattainable or achieved only with more costly gums. Table I gives an overall view of the use of these thickeners, stabilizers, and adhesives. The majority of the uses listed represent well-established, continuing markets for the amylopectin starches. In a few cases, particularly the most recent patent references and those from Europe, the writer has been unable to confirm that the proposed use has actually been reduced to commercial practice. However, these proposed uses are included as indicators of current trends and are witness to the technological possibilities for future developments.

As a general rule, the modified waxy starches used in foods are digested more rapidly than are ordinary cereal and root starches. Rogols and Meites[67] confirmed this by exposing waxy maize starch to various amylases of animal origin. In all studies of starch digestion and metabolism, *in vitro* or *in vivo,* starches under examination should be properly cooked or converted to a pregelatinized form before testing so as to conform to the condition of the starches as consumed by humans. In all too many cases, raw starches are mixed into the synthetic diets of test animals or the starches are subjected to rather artificial laboratory procedures, such as lintnerization.

The various producers of amylopectin and amylopectin starches do not publish figures on their annual sales volume. However, the annual production of these amylopectin gums must total many millions of pounds. Current (1972)

TABLE I

Uses of Amylopectin and Amylopectin Starches

Modification	Uses	References
Unmodified waxy starches (often blended with ordinary starches or flours)	Salad dressings	25
	Canned and frozen foods: infant, oriental, soups, white sauces, gravies, puffed cereals, and snack foods	23, 25, 28
	Paper manufacture: surface sizing, coating adhesive, corrugating, boxboard	23, 29
	Textiles: warp sizing, printing pastes, laundry starches	23, 25, 30
Pregelatinized waxy starches or isolated amylopectin	Bakery goods: cake and icing mixes,	31
	pie fillings (bakery and frozen),	25, 32
	sterilized bread	33
	Salad dressings	34
	Confections	35
	Pudding mixes	36, 37
	Textiles: sizing, creaseproofing	17, 38
	Paper: calendar size, wet-end additive	17, 39
	Adhesive: bags, wallpaper, corrugating,	40, 41
	foundry core binders	42, 43
	Waterproof dynamite	44
	Therapeutic dusting powders	45
	Ore refining, oil well drilling and fracturing	17, 46, 47
	Photographic films and emulsions	48, 49
	Fermentation media	50
	Paint remover	51
Pyrodextrinized waxy starches	Adhesives: remoistening types,	23, 25, 52, 53
	liquid, pastes	23, 25
Acid thin-boiling waxy starches	Textile sizing, paper sizing	23, 39
	Synthetic latex stabilizer and extender, confections, glazes and coatings for foods	25
Oxidized waxy starches	Adhesives, textile sizing, paper sizing	25
	Photographic film	54
Cross-linked waxy starches	Pie fillings (bakery, canned, and frozen), white sauces, and gravies in canned and frozen foods, puddings, salad dressings, soups, confections, oriental foods, harvard beets, sandwich spreads, infant foods	23, 25
	Textile printing pastes	23, 25

TABLE I—Continued

Modification	Uses	References
Ethers of waxy starches:		
hydroxyethyl	Textile sizing, thermoplastics	55
	Blood plasma extender	56
hydroxypropyl	Frozen and refrigerated foods, canned foods	
carboxymethyl	Emulsion stabilizer, dispersions of SiO_2	57
Esters of waxy starches:		
acetates	Frozen and refrigerated foods, infant foods, canned foods, oriental foods	23, 25
	Photographic film	58
succinates and adipates	Frozen and refrigerated foods, canned foods, encapsulated flavors	59
sulfates	Thickener and emulsion stabilizer	60
	Anticoagulant	61
	Ulcer treatment (pepsin inhibitor)	62–68

prices for amylopectin products range from 10 cents per pound to 25 cents per pound, depending on the nature and degree of the chemical and/or physical modifications involved. Isolated amylopectin and pregelatinized amylopectin starches, that is, the cold-water-soluble forms, command the higher end of this price range.

V. REFERENCES

(1) W. Brock Neely, *J. Med. Chem.*, **12,** 16 (1969).

(2) E. Y. C. Lee, C. Mercier, and W. J. Whelan, *Arch. Biochem. Biophys.*, **125,** 1028 (1968).

(3) G. K. Adkins, W. Banks, and C. T. Greenwood, *Carbohyd. Res.*, **2,** 502 (1966).

(4) J. F. Foster, *in* "Starch: Chemistry and Technology," R. L. Whistler and E. F. Paschall, eds., Academic Press, New York, Vol. 1, 1965, p. 349.

(5) S. R. Erlander and R. Tobin, *Makromol. Chem.*, **111,** 194, 212 (1968).

(6) J. Muetgeert, *Advan. Carbohyd. Chem.*, **16,** 299 (1961).

(7) C. T. Greenwood, *Advan. Carbohyd. Chem.*, **11,** 335 (1956).

(8) L. Marquenne and E. Roux, *C. R. Acad. Sci.*, **140,** 1303 (1905).

(9) K. H. Meyer, M. Weitheim, and P. Bernfeld, *Helv. Chim. Acta.*, **23,** 865 (1940).

(10) T. J. Schoch, *Cereal Chem.*, **18,** 121 (1941).

(11) T. J. Schoch, U.S. Patents 2,515,095 and 2,515,096 (1950); *Chem. Abstr.*, **44,** 11141 (1950).

(12) R. L. Whistler and G. E. Hilbert, *J. Amer. Chem. Soc.*, **67,** 1161 (1945).

(13) E. Pacsu and A. W. Bauer, U.S. Patents 2,779,693 (1957); *Chem. Abstr.*, **51,** 8459 (1957); 2,779,694 (1957); *Chem. Abstr.*, **51,** 8460 (1957).

(14) S. M. Cantor and E. L. Wimmer, U.S. Patent 2,779,692 (1957); *Chem. Abstr.*, **51,** 8460 (1957).

(15) W. C. Bus, J. Muetgeert, and P. Hiemstra, U.S. Patents 2,822,305; 2,829,987; 2,829,988; 2,829,989; and 2,829,990 (1958); *Chem. Abstr.*, **52,** 9635, 13295, 13296, 14204, 17768 (1958).

(16) O. R. Etheridge, J. A. Wagoner, J. V. McDonald, and D. A. Lippincott, U.S. Patent 3,067,067 (1962); *Chem. Abstr.*, **58,** 4723 (1962).

(17) T. F. Protzman, A. E. Staley Mfg. Co., Decatur, Illinois, personal communication.

(18) G. N. Collins, *U.S. Dep. Agr., Bur. Plant Ind. Bull.*, **161** (1909).

(19) P. Weatherwax, *Genetics*, **7,** 568 (1922).

(20) C. G. Caldwell and R. M. Hixon, *J. Amer. Chem. Soc.*, **63,** 2876 (1941).

(21) R. M. Hixon and G. F. Sprague, *Ind. Eng. Chem.*, **34,** 959 (1942).

(22) H. H. Schopmeyer, G. E. Felton, and C. L. Ford, *Ind. Eng. Chem.*, **35,** 1168 (1943).

(23) S. A. Watson, *in* "Sorghum Production and Utilization," J. S. Wall and W. M. Ross, eds., AVI Publishing Co., Westport, Conn., 1969, p. 193.

(24) H. W. Leach, *in* "Starch: Chemistry and Technology," R. L. Whistler and E. F. Paschall, eds., Academic Press, New York, Vol. 1, 1965, p. 292.

(25) J. W. Evans, *Cereal Sci. Today*, **3,** 81 (1958); *Can. Food Ind.*, 1 (July 1956).

(26) "Starch: Chemistry and Technology," R. L. Whistler and E. F. Paschall, eds., Academic Press, New York, Vol. 2, 1967.

(27) R. R. Myers and C. J. Knauss, *in* "Starch: Chemistry and Technology," R. L. Whistler and E. F. Paschall, eds., Academic Press New York, Vol. 1, 1965, p. 393.

(67) S. Rogols and S. Meites, *Staerke*, **20,** 256 (1968).

(28) J. G. Davis, J. H. Anderson, and H. L. Hanson, *Food Technol.* (*Chicago*), **9,** 13 (1955).

(29) J. Weaver, *Paper Trade J.*, **144,** 30 (1960).

(30) A. Roscelli and R. L. High, U.S. Patent 3,081,183 (1963); *Chem. Abstr.*, **58,** 11520 (1963).

(31) D. H. Hughes, W. T. Bedenk, and N. B. Howard, U.S. Patent 3,366,487 (1968); *Chem. Abstr.*, **68,** 86282g (1968).

(32) K. A. Scheick, *Baker's Dig.*, **40,** 50 (1966).

(33) R. Dehne, U.S. Patent 3,193,389 (1965); *Chem. Abstr.*, **63,** 12238 (1965).

(34) A. S. Szezesniak and E. Engel, U.S. Patent 3,300,318 (1967); *Chem. Abstr.*, **66,** 64528y (1967).

(35) L. A. Wollermann and E. W. Makstell, *Cereal Sci. Today*, **3,** 244 (1958).

(36) M. A. Reinders, K. Hiddema, and K. F. Gotlieb, Neth. Patent Appl. 67 07,893 (1968); *Chem. Abstr.*, **71,** 11932x (1969).

(37) E. A. Beck, U.S. Patent 3,350,162 (1967); *Chem. Abstr.*, **67,** 118052e (1967).

(38) P. Hiemstra and B. Nijland, *Tex De*, **18,** 1534 (1959); *Chem. Abstr.*, **54,** 11486 (1960).

(39) D. V. B. Miller, *Bull. Brit. Pap. Bd. Ind. Res. Ass.*, **20,** 10, (1960).

(40) H. Otto, Germ. Patent 1,003,382 (1957); *Chem. Abstr.*, **53,** 23105 (1959).

(41) J. W. Horner, Jr., U.S. Patent 2,999,028 (1961); *Chem. Abstr.*, **56,** 2625 (1962).

(42) J. W. Sietsema, U.S. Patent 2,974,048 (1961); *Chem. Abstr.*, **55,** 14267 (1961).

(43) T. E. Barlow, U.S. Patent 2,974,050 (1961); *Chem. Abstr.*, **55,** 14268 (1961).

(44) L. E. Sentz and M. A. Curtis, U.S. Patent 3,003,862 (1960); *Chem. Abstr.,* **56,** 623 (1962).

(45) H. B. Bernstein and G. N. Cyr, U.S. Patent 3,025,217 (1962).

(46) Societe d'Etudes Chimiques pour l'Industrie et l'Agriculture, Fr. Patent 1,489,195 (1967); *Chem. Abstr.,* **68,** 88379t (1968).

(47) I. Iwasaki and R. W. Lai, *Trans. AIME,* **232,** 364 (1965).

(48) Gevaert Photo-Producten N.V., Brit. Patent 860,632 (1961); *Chem. Abstr.,* **55,** 14978 (1961).

(49) Gevaert Photo-Producten N.V., Belg. Patent 568,154 (1958); *Chem. Abstr.,* **55,** 24340 (1961).

(50) W. Moses, U.S. Patent 3,352,761 (1967); *Chem. Abstr.,* **68,** 11718v (1968).

(51) G. Mitschke, Germ. Patent 1,052,609 (1959); *Chem. Abstr.,* **55,** 6736 (1961).

(52) W. B. Roberson and H. W. Craig, U.S. Patent 2,997,404 (1961); *Chem. Abstr.,* **55,** 27993 (1961).

(53) R. S. Hatch and W. B. Roberson, U.S. Patent 2,791,512 (1957); *Chem. Abstr.,* **51,** 11748 (1957).

(54) Gevaert Photo-Producten N.V., Belg. Patent 566,352 (1958); *Chem. Abstr.,* **55,** 24341 (1961).

(55) E. D. Klug, U.S. Patent 3,117,014 (1964); *Chem. Abstr.,* **60,** 14695 (1964).

(56) C. T. Greenwood and D. J. Hourston, *Staerke,* **19,** 243 (1967).

(57) A. Poenisch, East Germ. Patent 32,810 (1965); *Chem. Abstr.,* **64,** 6912 (1966).

(58) Gevaert Photo-Producten N.V., Belg. Patent 623,091 (1963); *Chem. Abstr.,* **59,** 13011 (1963).

(59) N. G. Maratta, R. M. Boettger, B. H. Nappen, and C. P. Szymanski, U.S. Patent 3,455,838 (1969); *Chem. Abstr.,* **71,** 126285f (1969).

(60) R. L. Whistler, D. G. Unrau, and G. Ruffini, *Arch. Biochem. Biophys.,* **126,** 647 (1968).

(61) R. L. Whistler and W. W. Spencer, *Arch. Biochem. Biophys.,* **95,** 36 (1961).

(62) W. M. Doane and R. L. Whistler, *Arch. Biochem. Biophys.,* **101,** 436 (1963).

(63) P. S. Cammarata and S. Eich, U.S. Patent 3,271,388 (1966); *Chem. Abstr.,* **66,** 12112p (1967).

(64) D. L. Cook and V. A. Drill, *Ann. N.Y. Acad. Sci.,* **140,** 724 (1967).

(65) D. C. H. Sun, *Ann. N.Y. Acad. Sci.,* **140,** 747 (1967).

(66) L. H. Kruger and O. B. Wurzburg, U.S. Patent 3,441,558 (1969); *Chem. Abstr.,* **71,** 14394j (1969).

(67) R. E. Ranney and S. J. Eich, *Toxicol. Appl. Pharmacol.,* **15,** 593 (1969).

(68) K. Nagasama, Japan. Patent 71 19,613 (1971); *Chem. Abstr.,* **75,** 153173g (1971).

CHAPTER XXVI

STARCH DEXTRINS

ROBERT W. SATTERTHWAITE AND DONALD J. IWINSKI

Technical Service Dept., Industrial Division of CPC International Inc., Argo Illinois

I. INTRODUCTION

1. Definition

The term *dextrin* generally refers to products produced by the dry heating or roasting of unmodified starches. However, the term *dextrin* is sometimes interpreted to include products that result from controlled hydrolysis of starch by certain enzymes or by controlled acid-catalyzed hydrolysis of wet starch. From the industrial viewpoint, degradation products resulting from enzyme or wet-acid conversions are least important. These include the crystalline Schardinger dextrins, residual or limit dextrins produced by liquefying or saccharifying enzymes, and certain oligosaccharides from wet-acid hydrolysis, such as those that are present in some corn syrups. In the following presentation, primary emphasis is placed on pyrodextrins (torrefaction dextrins) that are prepared by heating dry starch at 79°–190° for 3–24 hr, usually in the presence of small amounts of an acid catalyst. Occasionally, an alkaline catalyst or, rarely, an oxidizing agent is used. A large number of products can be prepared that, depending on conditions of manufacture, range in color from white to tan, have low to high solubilities in water, and afford pastes or solutions of low to medium viscosities. In general, these products are more soluble than is the parent starch and have lower viscosities, higher reducing powers, and different adhesive characteristics.

A large number of dextrin grades and types can be manufactured by controlling the starch source, moisture, catalyst, temperature, and duration of roasting, which are the variables in the process. Because precise chemical characterization of the dextrins has not been achieved, these products are generally classified in terms of parent starch source, manufacturing method, dry color, solubility in cold water, and relatively viscosity of the aqueous solution. It must be noted, however, that the distinction between groups is quite arbitrary. There are instances where overlapping between grades is evident.

It has become common practice to classify pyrodextrins into three primary categories (see Table I) although many minor subdivisions have been introduced.

White dextrins.—White dextrins are prepared by mildly heating starch with an acidic catalyst, such as hydrochloric or acetic acid. Because this can be done

TABLE I

Classifications of Dextrins

	White Dextrins	Yellow or Canary Dextrins	British Gums
Manufacturing conditions			
Usual catalyst	hydrochloric acid	hydrochloric acid	none or an alkali
Temperature, °C	79–121	149–190	135–190
Time, hr	3–7	6–20	10–24
Product characteristics			
Color	white to light cream	buff to dark yellow or brown	buff to dark brown
Solubility, %[a]	1–98	95–100	1–100
Useful dilution, parts of water	2–5	1 or less	3–8

[a]See section III.2., page 585

in the presence of relatively large amounts of acid, conversion is rapid even at comparatively low temperatures. Thus, colors remain light, and the water solubility can be varied from low (short conversion time and/or low temperature) to high (longer cooking times at higher temperatures). The low-solubility types might retain much starchlike behavior in regard to setback and other paste properties.

Yellow or canary dextrins.—Yellow or canary dextrins are prepared by heating starch with acid at higher temperatures than are used for white conversions. As a rule, these dextrins are over 95% soluble and can be used at concentrations of about 60%.

British gums.—British gums are prepared by heating starch without added acid. The conversion is catalyzed by the traces of acids naturally present in the starch (or that formed during pyrolysis) or by alkaline materials, such as sodium carbonate, sodium bicarbonate, and ammonia. In general, the products have dark colors, a wide range of cold-water solubilities, and high viscosities.

2. *History*

Dextrin development has been reviewed in detail by Bloede,[1] who reports that this type of product was discovered about 1804. Bouillon-Lagrange, seeking substitutes for gum arabic and other plant gums, discovered that gumlike products are formed during the roasting of starch. These products were found to be particularly suitable for the manufacture of black dyes and inks. A number of reports describing the characteristics of roasted starches were soon published, but the word *dextrin* appears to have first been used by Biot and Persoz in 1833.

The extensive industrial use of torrefaction products from starch (British gums) can be traced to a fire in a textile plant near Dublin in 1821. An observant worker noted that potato starch that had been scorched was soluble in cold water and that the resultant syrup was a good adhesive.

Today, dextrins are used primarily as adhesives. However, they are so versatile and adaptable that they find a wide variety of other applications. Because of the diversity of dextrins, it is possible to indicate only the class of each found most suitable for each application and to show how properties can be modified in manufacturing or by the addition of various adjuncts.

3. *Supply and Price*

Corn starch is the principal source of dextrins in the United States (Table II). The basic price of corn white dextrins was 10.6¢/lb in 1971, which is an increase of over 2¢/lb since 1955. This change has been caused by economy rather than by supply. Although no similar price tabulations are available for other kinds of dextrins, the price of imported potato dextrins fluctuates widely and, as a rule, is slightly above that of comparable corn dextrins. Prices of tapioca

TABLE II

United States' Supply of Starches and Dextrins[a]

Product	Millions of Pounds						
	1951	1952	1953	1954	1955	1968	1971
Corn Starch							
Domestic sales							
Total	1865	1867	2015	1972	2122	3197	3388
Starch	1560	1606	1740	1729	1841	2969	3150
Dextrin	188	169	182	169	187	188	171
Miscellaneous	117	92	93	74	94	40	67
Potato Starch							
Production	31	54	108	72	92	76	75[a]
Imports							
Starch	2	33	38	20	32	1	5
Dextrins	3	5	8	11	11	14	12
Tapioca starch							
Imports	110	61	67	62	130	194	182
Other starches and dextrins[b]							
Imports	10	16	26	21	19	21	18
Total	2021	2036	2253	2158	2384	2503	3680
Corn Dextrin price,[c] ¢/lb	7.43	7.83	7.78	7.78	7.88	9.08	10.6

[a]Estimate.
[b]Includes sago, arrowroot, corn, rice, wheat, and other starches and dextrins.
[c]Average base price of basic white dextrin (CPC International).

dextrins are also higher and more variable than those of corn dextrins because there are wide fluctuations in the price and quality of imported tapioca starch.

Corn dextrins have the largest sales volume of all dextrins because of their availability and low price.

II. PRODUCTION

1. *Raw Materials*

Dextrins can be produced from all commercial grain and tuber starches. Although the conversion methods are mainly the same for all starches, the ease of conversion and the usefulness of the dextrins vary appreciably, depending on the nature and quality of the starch.

Potato starch is generally regarded as the easiest to convert. The potato dextrins have excellent adhesiveness, clarity, and stability of their solutions. However, their application has been limited by a disagreeable, cucumber-like odor that is characteristic of the parent starch. Recently, ways have been found to minimize odor (for example, by drying the starch under reduced pressure) and, within the last few years, some potato dextrins imported from The Netherlands have been substantially free from objectionable odor.

Tapioca and sago starches convert nearly as easily as does potato starch. Dispersions of these tapioca dextrins also have excellent clarity and stability and have the added advantage of being virtually tasteless. Commercial sago starch is usually of inferior quality and requires washing or other processing before dextrinization. The best quality sago dextrins compare favorably with potato and tapioca products. Most sago dextrins are manufactured outside the United States.

Corn and wheat starches require longer processing and higher temperature for conversion to dextrins than does potato starch. Corn dextrins have good adhesive properties, but their pastes do not have the clarity of those from potato or tapioca dextrins and the pastes tend to retrograde or thicken rapidly during storage.

Recently, dextrins produced from waxy maize and waxy milo (sorghum) starches have been marketed. These closely approach the characteristics of high quality tapioca and potato products. The stability of their high solids adhesives and their lack of retrogradation or setback are also comparable. Adhesive strength is comparable to that of potato and tapioca dextrins.

The application of new refining techniques in the wheat-starch industry has been reflected in an improvement in the quality of wheat dextrins. Their characteristics are very similar to those of corn dextrins.

2. *Manufacture*

The nature of a dextrin is determined by the source of the starch and its moisture content, the presence of contaminants, the rates of heating and agitation, the kind and concentration of catalyst, the duration of heating, the maximum temperature, and the efficiency of removal of volatile products. During dextrinization, solubility and fluidity are increased; there is less tendency for the dextrin to gel, setback, or retrograde; and, in general, the utility of the product for adhesive formulations is improved. These physical changes are produced by chemical changes manifested by an alteration in molecular structure, an increase in the proportion of lower-molecular-weight, more-water-soluble materials, an increase in the amount of reducing substances, and a darkening of color. The extent of dextrinization can be followed by determinations of paste viscosity, solubility, alkali lability, reducing sugar value, and/or color.

Because of process variables, dextrins from different manufacturers might be alike in some properties but will almost surely differ in others. For these reasons, dextrins will be considered in terms of generally recognized classes rather than as individual products.

Dry conversion processes.—For most pyrodextrin productions, the starch is ground, sieved through fine silk, and redried. A moisture content of 5% is usually required. Sometimes the starch may have a moisture content as high as 12%, particularly if the dextrin cooker is designed so that water vapor can be removed easily during the early stages of heating.

If a catalyst is used, it can be added to the starch either in a separate vessel or in the roasting vessel. Various types of acids, including trichloroacetic and hypochlorous acids, are used. However, hydrochloric acid is generally preferred because it is a strong acid, disperses easily and uniformly through the starch, and tends to volatilize during the last stages of dextrinization. For alkaline conversions, solutions of sodium carbonate, ammonium hydroxide, or urea are blended with the starch in quantities sufficient to maintain the pH of the mixture at the desired level throughout the heating cycle.

For batch processes, vessels can vary in capacity from 100 to 10,000 lbs and can be horizontal or vertical. All vessels have means for controlling heat, agitation, and removal of volatile products. Recently, fluid-bed techniques have been proposed for dextrinization.

Because dry, powdered starch is highly combustible, indirect heating of the cookers is preferred. The vessels, although commonly heated with steam (either through jackets or coils in the walls of the kettle), can also be heated with hot oil, Dowtherm, or other heat-transfer media. Uniform heating is necessary. Although batch operations are commonly used for dextrin manufacture, several continuous processes are used. In these, dextrinization occurs either on heated, endless, stainless steel belts or in a series of rotary-drier-type units.

Conversion is stopped by cooling usually by transferring the dextrin to a cold-water-jacketed drum in which agitation is provided. Acidic dextrins may be neutralized after conversion by the addition of ammonia or other material; however, the dextrin is frequently packed and shipped without neutralization. As a result, some dextrins contain residual free acid and, thus, their solutions have a low pH. Although acidity might decrease the stability of the product during storage, it is advantageous to the user if he wishes to continue to modification.

When the dextrin comes from the cooker, it is dry and contains large amounts of adsorbed gases (primarily carbon dioxide formed during the heating). Some manufacturers humidify the product with moist air, whereas others allow the product to absorb atmospheric moisture in storage. The latter process is preferable because the adsorbed gases are released as water is adsorbed. In a "green" or unaged dextrin, these gases tend to create undesirable foam when the product is dissolved in water.

Wet conversion with acid.—For certain purposes, dextrins are produced by a wet process using an acidic catalyst. These conditions favor hydrolysis with the production of large amounts of oligosaccharides and low-molecular-weight polysaccharides. Two common procedures are employed. In one, a starch slurry in acid is heated gently until the desired degree of conversion is reached; then, the product is immediately neutralized and quickly dried on heated rolls or in a spray drier. In the other procedure, the partially converted starch is separated from the converting solution by filtration without neutralization and the wet, acidic cake is held in warm kilns for several days, during which time dehydration and further conversion occurs.

Wet dextrinization with enzymes.—Although enzymic dextrinization is frequently performed by the user, some liquid or semidehydrated pastes and glues are made either directly from enzymically converted starch or by mixing such a product with other types of dextrins. The liquifying or dextrinogenic enzymes are the α-amylases, which catalyze the hydrolysis of starch to low-molecular-weight fragments. Most commercial enzymes of this type also contain saccharifying enzymes (for example, β-amylase and glucoamylase) that continue the hydrolysis to maltose and/or D-glucose. To avoid this secondary reaction, bacterial α-amylases are used, because they can be employed at temperatures at which β-amylase is inactivated. Most adhesive manufacturers use high starch concentrations, raise the temperature slowly, and add the enzyme periodically in small portions.

III. Product Description

1. *Chemical Structure*

Although superficially simple, dextrinization produces fundamental changes in molecular structure that lead to complex mixtures of products. The primarily reactions appear to be (*1*) hydrolytic cleavage[2] of the starch molecule, resulting in the formation of lower-molecular-weight, more-water-soluble compounds; (*2*) transglucosylation,[3] leading to the formation of more-highly-branched molecules with little net change in molecular weight; and (*3*) some repolymerization or condensation of the smaller molecules, [2, 3] particularly during the manufacture of highly converted products. In addition, some carbon dioxide and other gaseous degradation products may be formed.

Transglucosylation is an important reaction in pyrodextrinization. It probably is an intermolecular exchange in which some of the predominant α-D-(1→4) glucosidic linkages of native starch are changed to α-D-(1→6) linkages (Fig. 1). Geerdes and coworkers[4] suggest that the extent of transglucosylation can be

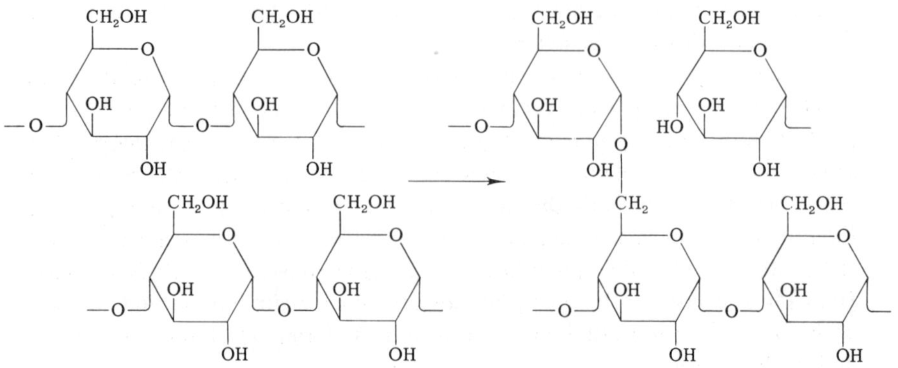

FIG. 1.—Transglucosylation reaction in which an α-D-(1→4)-glucosidic bond becomes an α-D-(1→6) bond with the formation of a branch point.

estimated from the percentage of D-glucose units resistant to periodate oxidation, for the reaction makes both the linear and the branched polysaccharides of starch progressively more branched. As this occurs, the dextrins disperse more readily and form clearer, less viscous, and more stable solutions than did the original starch.

White dextrins.—Formation of a typical white dextrin is, essentially, a a hydrolytic reaction that leads to a steady increase in water solubility and reducing values and a corresponding decrease in viscosity (Fig. 2). Sparingly soluble white dextrins retain the retrogradative characteristics of the original starch.

Canary dextrins.—Conditions for the preparation of the yellow or canary dextrins favor more extensive structural changes. At first, hydrolysis predominates. However, as the temperature increases, the mixture becomes drier and conditions become more favorable for transglucosylation and polymerization reactions. Thus, there is initially an increase in the alkali lability followed by a decrease (Fig. 3), indicating a decrease and then an increase in average molecular size. Examination[3, 4, 6] of the structures of several highly converted canary dextrins has shown that they contain a greater proportion of branches than does the original starch. Their formation can be ascribed to transglucosylation.

British gums.—British gums are produced during prolonged heating of dry starch, either without added catalyst or in the presence of small amounts of alkaline buffers. British gums have higher viscosities than do either of the two preceding classes of dextrins. During dextrinization, the viscosity and alkali lability do not change as rapidly with time of conversion as in the manufacture of white dextrins. Only moderate increases in reducing values are observed. Significantly, British gums have relatively high molecular weights, are disper-

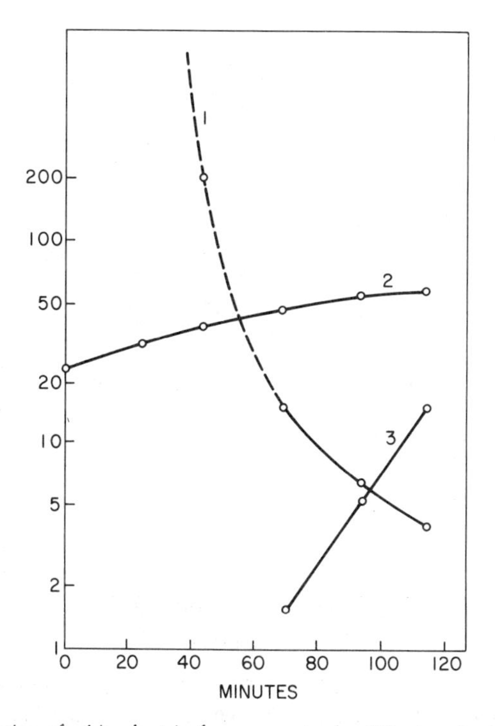

FIG. 2.—Formation of white dextrin from corn starch: (1) viscosity in centistokes; (2) alkali lability; (3) solubility, %. The abscissa is the conversion time in minutes. (Redrawn from Caesar.[5])

sible, and still retain the properties of imparting high viscosity and good water-carrying power.

Some hydrolysis probably occurs in the absence of added catalyst, particularly during the early stages of the process. Hydrolysis is caused by residual water in the starch and the traces of acidic materials present or formed during heating, the action of which is prevented by alkaline buffers which can also promote atmospheric oxidation at the high temperatures used. The major reaction appears to be transglucosylation with resultant formation of highly branched molecules.[3]

2. *Physical Properties*

Solutions.—Dextrins have the abilities to absorb large amounts of water and to form colloidal solutions. Each dextrin is a mixture of many molecular species differing both in size and in degree of branching. Molecular size and aggregation in dextrin systems are such that the term *dispersion, hydrosol,* and *hydrogel* should be used. However, it is common to refer to starch and dextrin

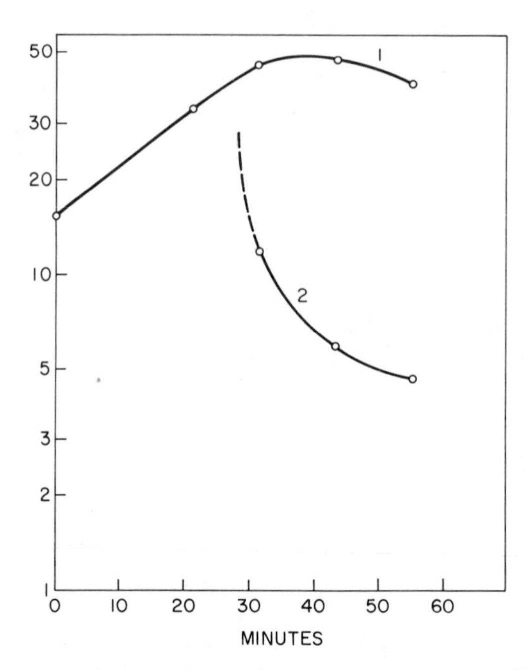

FIG. 3.—Formation of canary dextrin from tapioca starch: (1) alkali lability; (2) viscosity in centistokes. The abscissa is the conversion time in minutes. (Redrawn from Caesar.[5])

dispersions as solutions or pastes and to discuss them in terms of solubility, viscosity, and fluidity.

Slightly converted white dextrins contain only small amounts of low-molecular-weight material. The starch molecules are so little modified that they behave much as in normal starch. The highly-branched, water-soluble molecules present in considerable quantities in highly converted canary dextrins form fluid solutions that do not gel. British gums also contain a high proportion of easily dispersible molecules but do not contain much low-molecular-weight material.

Viscosity.—Solution viscosity is one of the most important factors in all dextrin applications. It governs both the working characteristics of the paste and the adhesiveness of the film. Unfortunately, because these are colloidal systems instead of true solutions, viscosity is difficult to define and control. There is a tendency for molecules to aggregate as density, temperature, pH, or other characteristics change; however, with the more soluble dextrins and British gums, aggregation is not a problem.

Actually, measurement of the absolute viscosity of a dextrin solution would have little practical significance. Industrialists are interested in the concentration and temperature necessary to achieve a certain viscosity and effects specific to an application. Detailed descriptions of apparatus and procedures are not presented in this discussion because these are readily available in standard references.[7, 8] However, a brief review of the types of information that can be obtained from commonly used instruments is given.

Some instruments for the determination of viscosity depend on a measurement of the rate of delivery of a given volume of solution from a pipet (Dudley, Alexander, Zahn cup) or from an orifice funnel (Buel, Hot Scott viscometer, Clinton fluidity funnel, Saybolt viscometer), or on data from a rotational instrument (Brookfield viscometer or Hercules Hi-Shear viscometer). These instruments permit rapid comparisons of solutions at given temperatures and are convenient for quality control purposes. However, they provide little information about the viscosity behavior of solutions under conditions of actual use. More information on the effects of paste concentration, temperature, and time on viscosity can be obtained by the use of recording viscometers, such as the Corn Industries and Brabender viscometers. No single instrument or procedure is universally accepted; each has its advantages and disadvantages. Unfortunately, data from one type of instrument is not comparable to that from another.

The source of the starch used affects the viscosity and retrogradation characteristics of the dextrin. Dextrinization decreases the viscosity of a starch solution and its tendency to retrograde. Dextrins, however, retain some of the setback characteristics of the parent starch. This is particularly true of the less-converted grades (the slightly soluble white dextrins). For equivalent degrees of conversion, as measured by viscosity and solubility, the decreasing order of setback is the same as that of the parent starches; namely, corn > potato > tapioca > waxy maize > waxy (white) milo (sorghum).

The temperatures to which a solution has been heated and at which the viscosity is measured also affect the recorded viscosity. Brabender recording viscometer curves for three different dextrins are given in Figure 4. These patterns were obtained by suspending the dextrin in water (concentration is expressed as grams of dextrin per 500 ml of a slurry), heating the stirred suspension to 95° at a uniform rate, holding the temperature at 95° for 1 hr, cooling the dispersion to 50°, and finally holding it at this temperature for another hour.

Dextrin A is a light-tan corn-starch gum (2% soluble in cold water). This slightly converted dextrin was cooked at a concentration of 52.5 gm per 500 ml; that is, 10.2% by weight. The curve shows the high peak viscosity reached during pasting and the marked setback on cooling characteristic of this type of dextrin.

Dextrin B is a corn white dextrin (45% soluble in cold water) that was cooked at a concentration of 283 gm per 500 ml; that is, 46.9% by weight. The

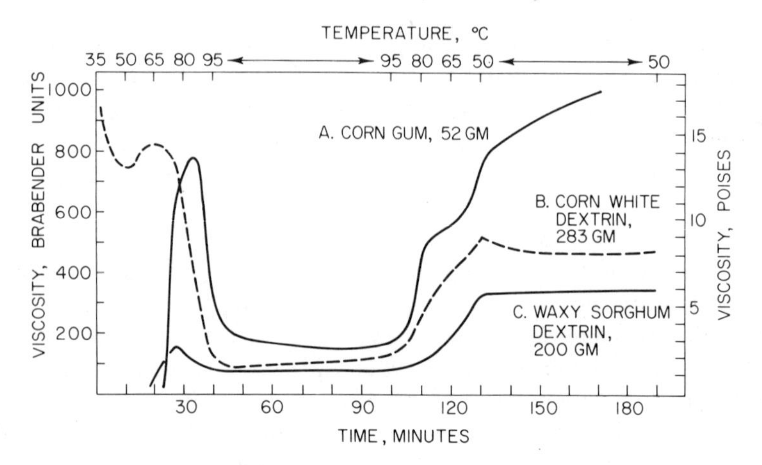

FIG. 4.—Brabender recording viscometer curves. (Data supplied by Mr. E. G. Mazurs.)

curve shows properties characteristic of the more highly converted types of dextrins; that is, high initial viscosity (as a result of cold-water solubility), marked thinning during cooking as the dextrin dissolves completely, and very low setback on cooling.

Dextrin C is a white dextrin prepared from waxy sorghum (white milo) starch and is 4% soluble in cold water. It was cooked at a concentration of 200 gm per 500 ml; that is, 35.0% by weight. The low peak viscosity and minimal setback result largely from the absence of a linear fraction in the parent starch.

Differences between dextrin types are also evident in the relationship between solution concentration (solids content) and viscosity. Thus, to prepare solutions of equivalent viscosities, a canary dextrin might be cooked with 1 part of water, whereas a British gum of comparable solubility would require 3 parts. By modifications in dextrin manufacture, the viscosity of a solution of a given concentration of product can be varied over a wide range from low solids–high viscosity to high solids–low viscosity or high solids–high viscosity.

Acidity is also an important determinant of product properties. Acids present in dextrins can cause further hydrolysis and result in a gradual thinning of solutions. Although some dextrins are neutralized with sodium carbonate or ammonia before packaging, others are shipped containing residual acid from the manufacturing process. Residual acid can be both an advantage, if the user wishes to continue conversion as the dextrin is dissolved, and a disadvantage, in that the colloidal properties of the dextrin may change during storage. In general, neutralization is made before shipment. The more highly converted canary dextrins are the least sensitive to variations in pH.

Increase in viscosity caused by gelation or retrogradation as solutions age is particularly noticeable in the less soluble corn-starch dextrins. The viscosity increase caused by retrogradation, generally a disadvantage, is considerably less evident in dextrins and British gums than in starches. Instead, dextrin systems are often thixotropic; that is, they are fluidified under pressure or motion but change to a soft paste or gel when allowed to stand.

Films (coatings).—Dextrins produce brittle films that must be supported on paper or fibers. The nature of the parent starch has a great influence on the characteristics of dextrin films. Tapioca dextrin films are the clearest and have the highest gloss. Corn dextrin films, particularly those of white dextrins, are usually more brittle and cloudy. Tapioca dextrin films are usually tougher and easier to remoisten than are films made with corn dextrins. The elasticity of the tapioca, potato, and waxy maize starch dextrin films is generally greater than that of corn-starch dextrin films. The properties of dextrin films are altered by modifiers.

Adhesiveness.—Adhesives must form a temporary bond to hold two surfaces together until a permanent bond is produced by the gradual loss of solvent and the transition of the colloidal solution to a gel and finally to a continuous noncrystalline film. Adhesiveness varies directly with viscosity and concentration. Because the more highly branched polysaccharide molecules impart less solution viscosity than do linear molecules (Chapter 1, Section III), dextrin pastes are less viscous than those of the parent starches at the same concentration and, consequently, are more suitable for adhesive applications.

Color.—White dextrins range in color from white to light cream, canary dextrins from cream to a dark yellow or brown, and British gums from light tan to dark brown.

Storage stability.—Pyrodextrins, particularly the white ones, may contain a considerable amount of residual acid. Residual acid is most often found in the less-soluble dextrins in which the conversion is of short duration and a relatively large amount of acid is used. The unspent acid causes a reduction in viscosity during dry storage. To eliminate this problem, dextrin manufacturers neutralize dextrins of low solubility with ammonia or sodium carbonate in the cooling vessel.

3. Dextrin Modification

Dextrins can be used either by themselves or in conjunction with other materials. When a dextrin is to be the sole binding agent in a particular application, its behavior can be rather easily predicted from its viscosity and solubility characteristics.

As the minor component in a system, dextrin finds use as an extender for poly (vinyl acetate) resin emulsions, natural rubber emulsions, poly (vinyl alcohol), natural and synthetic gums, animal glue, and, occasionally, starch. When combined with poly (vinyl acetate), each resin–dextrin system is individually examined because the properties of both dextrin and emulsion may vary considerably. With poly (vinyl alcohol), dextrin must be used in low concentrations because of the high viscosities developed and their poor compatibility. Water-soluble, yellow dextrins appear to be compatible in all proportions with poly (vinyl alcohol), provided that the total solids is not above 15%. Animal glue, natural and synthetic gums, or starch are extended with dextrin to either reduce cost or impart control on viscosity and solids in specific systems.

When dextrin is the major or primary component in a system, there are many modifiers with which it can be combined. The following discussion describes types of adjuncts and explains their function as well as their limitations.

A plasticizer (flexibilizer) is a material that affects the finished film characteristics of a dextrin-based paste. There are three effective types: solvents that form solid solutions of the dextrin, humectants that control the amount of moisture in the film, and fatty compounds that lubricate the film.

Such substances as urea, sodium nitrate, dicyandiamide, salicylic acid, thiocyanates, formaldehyde, iodides, and guanidine salts at 1–10% concentration belong to the first group because they each decrease the viscosity of dextrin solutions. In addition, these compounds tend to hold the dextrin in solution and, thus, give plasticity to the film. Of these additives, urea seems to be most generally applicable. Fluidifiers have the effect of both decreasing viscosity and improving stability.

Humectants that are used include glycerol, ethylene glycol, invert sugar, D-glucose, and sorbitol. Because they retain moisture in the film, they cause it to dry slowly and thus prevent it from drying completely and becoming brittle. The quantity of these humectant plasticizers incorporated in the film must be limited to that level which will give the desired film flexibility without excessive stickiness or blocking. Little information about the relative effectiveness of these materials is available. Traditionally, sugars are assumed to retard the drying rate of dextrin films more than do urea and sodium nitrate. However, glycerol and ethylene glycol are better humectants than sugars. Films plasticized with sugars may darken with age.

Lubricants include sulfonated castor oil, sulfated alcohols, and soluble soaps. Although these have the advantage of imparting permanent flexibility regardless of atmospheric conditions, they must be used at low levels because they weaken adhesive bonds.

Examples of tackifiers are sodium tetraborate, sodium hydroxide (0.5–1.5% based on the weight of dextrin), sodium silicate, and other alkali chemicals. Increasing the concentration of sodium tetraborate increases the viscosity as well as the wet tack (adhesiveness) of a dextrin-based paste until a maximum value is reached. Above an optimum concentration, borated systems become difficult to spread in a continuous film, become rubbery, and lose tackiness. When added to borated solutions, sodium hydroxide and sodium silicate also increase tack and penetration in certain types of paper stock. However, the use of these materials at greater than optimum concentrations causes stringiness or crumbling of the film and a gradual increase in color.

Fillers are normally used to increase solids, lower cost, and control penetration into a substrate. The most common are inert materials, such as clay and bentonite.

Some resins have been already mentioned as major components in systems in which dextrin is used as an extender. An exception is urea–formaldehyde resin, in which the synthetic material is present as the minor component. One of the inadequacies of dextrin adhesives is the inability to develop water-resistant films. This problem can be partially overcome by adding a number of resinous materials, but the most commonly used is urea–formaldehyde. Amounts of 10–20% are required along with an acid that will maintain the paste pH between 4.5 and 6.0.

Organic solvents are employed in conjunction with dextrin-based pastes when adhesion to certain types of stock, such as waxed-treated or waxed-impregnated paper, is required. When a solvent is introduced to a dextrin-based paste, it is normally added to the cooked solution, which has been previously cooled to room temperature. Examples of such chemicals are toluene, carbon tetrachloride, and trichloroethylene.

Defoamers, preservatives, whiteners, deodorizers, and penetrants are also added to modify properties. Both defoamers and preservatives are very important adjuncts in dextrin-based systems. It is critical to choose a defoamer compatible with all the components of the cooked paste. Whitening agents, such as titanium dioxide, are used to impart a white appearance to dextrin pastes and films. In addition, other chemicals, such as sodium bisulfite and sodium peroxide, are used as bleaches. They are used primarily with white dextrins of high solubility, canary dextrins, and British gums to impart light film color that is critical in some applications. Since some dextrins exhibit a strong cereal-type odor (particularly the British gums), deodorizers are often needed. Penetrants are incorporated into dextrin-based solutions to allow sufficient or controlled penetration into a specific substrate. When dextrins are used in adhesives, it is important that the paste penetrates into the paper stock to help bond formation.

IV. Applications

The nature of the parent starch used, as well as the conditions under which it is dextrinized, greatly affect the characteristics of dextrin with respect to gelation (setback) tendencies, paste clarity, coating ability, viscosity, and adhesiveness. For these reasons, dextrin finds utility in many applications. The main purpose of starch dextrinization is to increase fluidity and solubility and to decrease the tendency of the starch to retrograde so that high solids levels can be employed.

Dispersions of dextrins, with or without modifiers, are prepared for use in either of two ways: (*1*) by cooking a slurry at a designated solids content, or (*2*) by slurrying dextrin in water until complete colloidal dispersion is achieved. The latter is normally accomplished with room-temperature water; however, if preparation time is limited, the water is warmed to effect faster dispersion.

The nature of the surface being coated or bonded and the operating characteristics of the equipment impose definite restrictions on the choice of a dextrin or gum for a specific application. The greatest amount of dextrin is used as a base component for adhesives in the paper industry. The higher the concentration of solution, the better the paste will cover a paper-base sheet without unduly wetting the stock or being absorbed by it. In laminating, high solids solutions give a stronger bond; in remoistening gums, high solids result in a quicker drying, glossier coat that is easy to remoisten. As the gum dries, the soluble portion of the dextrin is drawn into the surface of the porous material being bonded and contributes to the strength of the bond.

A list of dextrin uses is given in Table III.

Because the major areas of dextrin application are in adhesives, paper, and textiles, and expansion, including equipment, systems, and critical properties, is presented separately for each of these areas.

1. Adhesives

Laminating.—General laminating means the bonding of either a combination of one ply of paperboard with one or more sheets of paper or the bonding of two sheets of paper. Mounting board, outdoor poster board, paper towels, and special container boards are examples. Solid fiber lamination involves the joining of several plies of paperboard fillers and liners to form multiple laminations. In this area, in which speed and water resistance as well as adhesion to specific stock are needed, resin-based adhesives are commonly used. Dextrin systems do not normally exhibit sufficient tack, bond strength, and water resistance to be successful in this phase of laminating.

In general laminating, dextrin is usually extended with modifiers to give good lay-flat, controlled penetration, fast set, and non slip properties. The dextrin used can be white, canary, or British gum; but, because of its color, a white dextrin is preferred, especially when lightweight paper is used. If water resistance is necessary, the formulas are modified with urea–formaldehyde resin. The solids content range from 20% to 60% in viscosity levels of 2000 to 10,000 cps. The adhesive is applied either at room temperature or slightly above.

Tube winding.—Paper tubes, cans, cones, and even cores are formed by

TABLE III

Dextrin Applications

Application	Suitable Dextrin		
	White	Canary	British gum
Adhesives			
Bag bottom	X		X
Bag seam	X		
Bill posting	X		
Carton sealing	X	X	X
Cigarette Seam	X		
Container sealing	X	X	
Envelope gum	X	X	
Envelope seal		X	
Gummed tape		X	
Laminating paste	X	X	X
Library paste	X		
Lining paste	X		
Linoleum cement		X	
Paper tube and cone winding	X	X	
Postage stamp		X	
Ceramics			
Pottery and tile glazes	X		
Cord polishing	X		
Crayons	X	X	
Detergents	X		
Dyes	X	X	
Feed blocks		X	
Flares and fireworks		X	
Food uses	X	X	
Foundries			
Sand binder		X	
Core paste		X	
Core wash		X	
Fuel briquettes		X	

TABLE III—Continued

Application	Suitable Dextrin		
	White	Canary	British gum
Insecticides	X		
Leather			
Adhesives	X		
Shoe counter pastes	X		
Shoe polishes	X		
Linoleum		X	
Paints			
Cold-water	X		
Poster	X		
Paper			
Calender sizing	X		
Coatings	X		
Tub sizing	X		
Wallpaper	X		
Patching plaster	X		
Plaster board	X		
Printing			
Ink vehicle			X
Textile	X		X
Protective colloids	X	X	
Textiles			
Cloth printing	X		X
Dyeing	X		X
Glass fiber sizing	X		
Felt sizing	X		
Finishing	X		X
Rug sizing			X
Stiffening	X		
Warp sizing (spun rayon)	X		X
Welding rods		X	
Window shades	X		

winding several plies of board on a revolving mandrel. Products can be formed by spiral or convolute winding to make soda straws, rug cores, and fiber drums.

Adhesion must be rapid so that the bond sets before the tube is removed from the mandrel. Setting speed is critical because of the short compression time involved in the formation of the tube. Borated dextrins, either white or canary, are normally used in combination with a small amount of alkali to provide more rapid development of tack and strong bond. The paste must not be too cohesive and the film must be rigid when dry. The adhesive is usually applied between 50° and 70°. The solids level may vary from 25% to 50% and the Brookfield viscosity can range from 500 to 3000 cps. Some users also add animal glue or sodium silicate to the formulation to promote more rapid and firmer adhesion with fast machines or with special grades of paper. Silicate will impart additional rigidity to the tube.

Bag seam and bottom.—White dextrins are used in bag seam formulas. Depending on the properties desired, the systems may vary in solids content from 15% to 45% and the Brookfield viscosity from 1000 to 4000 cps. The dextrin is usually borated to provide adequate tack, and the entire system usually contains modifiers that insure non slipping, controlled penetration, and no discoloration to the paper stock. Acid-modified starch can be used in combination with dextrin to give proper solids–viscosity relationships in a particular formula. If water resistance is required, 5–15% of urea–formaldehyde resin is included.

Cross pastes for multiwall bags require much the same type of adhesive as that used for seam applications. Controlled penetration is the most critical property needed for satisfactory performance.

Use of dextrin in bag bottom adhesives is limited because many manufacturers have found starch acceptable. However, there is some usage of dextrin in combination with starch for this application. Adjuncts are incorporated into these formulas to control penetration (clay) and to insure good lay-flat (plasticizer) and soft texture (soap). Urea–formaldehyde resin is also used for bag bottom adhesives when water resistance is required. Satisfactory bag bottom adhesives are short breaking, tacky, and resistant to viscosity breakdown under shear. The solids content of these types of adhesives range from 10% to 50% with a Brookfield viscosity as low as 5000 cps to approximately 150,000 cps.

Case and carton sealing.—Adhesives used in this type of sealing are applied by automatic machines and, as such, the bond is under pressure on the conveyer belt for only a brief time. Therefore, a fast setting speed as well as good machinability at high speed is critical. Borated white and canary dextrins, usually containing a small amount of alkali, are used. Fast tack and fast setting (speed of bond development) are critical properties required of a dextrin-based adhesive to be successfully used in case and carton sealing. Solids content can range from 30% to 40% and Brookfield viscosity from 1000 to 3000 cps.

Envelope seams and flaps.—A seam adhesive should dry slowly so that no strains develop that might deform the envelope. The bond should lie flat and not soften, pucker, or discolor the paper. A highly soluble, light-colored dextrin is suitable in most instances if it is plasticized with either urea, sucrose, D-glucose or sodium nitrate. This type of paste is used at a high solids level (60–70%) at a Brookfield viscosity of 2500 to 5000 cps. It must also exhibit a minimum amount of setback.

The remoistening adhesive that is used for the flaps of envelopes must have excellent viscosity stability and not cause curling, puckering, or discoloration of the paper. In addition, it must dry rapidly and be nonstringing because machine rates can be as high as 100 envelopes per minute. The finished film must have the desired color, gloss, smoothness, and clarity. Furthermore, it must not block (become sticky) during storage at normal humidities but must remoisten readily to form a strong seal with the body of the envelope. The highly soluble, canary dextrins derived from tapioca starch, along with those from waxy maize and waxy sorghum (milo), have been found to be the most suitable for the preparation of envelope front seal adhesives. Films of these dextrins yield coatings of light color, high gloss, and good remoistenability. A typical formula for this application is prepared at 60% to 70% solids in a viscosity range of 5000 to 15,000 cps. To achieve the high solids required along with the desired film properties, it is necessary to incorporate various plasticizers; for example, 2% of glycerol or diethylene glycol are usually added to prevent curling, but sorbitol or its esters can also be used. Care must be taken so that the quantity used of these hygroscopic plasticizers is not so great that blocking occurs. Poly (vinyl acetate) emulsions are sometimes used to impart machinability and film clarity to the base adhesive.

Library paste.—White dextrins of low to medium solubility are combined with certain starches and modifiers to produce library paste. The smooth, stiff texture of library pastes is obtained by the natural setback properties of the starch. Good wet tack is a critical property, and the dried film should not wrinkle after being applied to paper. The adhesive should exhibit a white, opaque color and be edible, because it often finds its way into children's mouths. Library pastes are prepared at solids levels from 20% to 40% in a viscosity range of 80,000 to 180,000 cps as measured by a Brookfield viscometer. Humectants, such as corn syrup, D-glucose, or glycerol, are often added to keep the paste moist. Oil of wintergreen is incorporated to mask the bland, cereal odor of the starch products used.

Gummed paper (stamps, labels, tape).—The requirements for the adhesive on the backs of labels, stamps, stickers, and paper tape are similar to those for an envelope flap gum. However, the brittleness of the dried dextrin adhesive is less of a disadvantage with gummed paper than with envelope flap adhesives and might actually be an advantage. In manufacturing gummed paper, the finished sheet is usually run over a bar that fragments the adhesive coating so that the strain between the gummed and ungummed surfaces is reduced and the sheet lies flat.

There are two types of gumming processes involved in the manufacture of labels and stamps: continuous (in which the solvent is water) and discontinuous (in which the solvent is an organic material, such as toluene). For continuous gumming, highly soluble, light-colored white or canary dextrins from corn, tapioca or potato starches are used at solids levels of about 50%. The dextrin is either completely or partially cooked prior to application. In the discontinuous process, canary dextrin is only dispersed, not cooked. In either case, the finished gummed sheet may be subjected to further processing, such as printing and perforating. Therefore, a major requirement for the sheet is lack of curl. Because the dextrin is just dispersed in the discontinuous gumming process and not actually a colloidal suspension, the dried film poses no problem in this respect. However, sheets prepared by continuous gumming require further mechanical processing to achieve *lay-flat*. In addition to having a light, glossy film, in most instances gummed sheets must not stick to each other (blocking) under conditions of high humidity. Application temperature for continuous gumming is approximately 60°–72°; however, with discontinuous gumming, there is no heat applied to the slurry prior to or during application. If a high degree of bond strength is necessary, animal glue is the primary binder for either process and dextrin is employed only as an extender.

A completely different remoistening application is that of gummed tape. Specifications for gummed paper tape are usually so stringent that most adhesives used in the past have been prepared from animal glue or gelatin and contained canary dextrins only as extenders. Recently, however, formulas based on waxy maize, oxidized potato, and waxy sorghum starches have satisfactorily replaced animal glue in many areas of gummed tape manufacture. In these instances, a dextrin may be used as an extender for the starch product. In this capacity, highly soluble white or canary dextrins allow higher solids, impart better gloss, improve curl, and in some instances reduce cost.

2. *Paper*

Sizing.—Only limited quantities of specially prepared torrefaction dextrins are used for the tube sizing of paper, but white dextrins of low solubility have been used extensively for the surface sizing of fine papers at the size press or calender.

Coatings.—Several low-solubility, medium-viscosity dextrins are particularly suitable as suspending agents in the application of clay coatings and other heavily pigmented paper coatings. They form thixotropic mixtures that flow well under the action of the applicator roll and form a smooth surface on the paper. Because these dextrins have good suspending power, they can be used at higher pigment solids levels than can most of the other coating adhesives. They are very effective pigment-binding agents.

3. Textiles

Sizing.—Dextrins are well adapted to the sizing of certain filament and spun yarns as well as glass fibers on which conventional starch or gelatin sizes are not satisfactory. One of the obvious advantages of dextrins and gums, as compared with native starch, is that they are more convenient to handle because they are more soluble and do not gel readily. More important, however, is the fact that they dissolve at lower temperatures and, consequently, the fabric can be more easily desized. This is particularly important in the case of synthetic fibers.

White dextrins and British gums form films that are strong enough to last under the conditions of weaving. For the sizing of viscose rayon, slightly soluble white dextrins from either corn or tapioca starch can be used alone or can be added to a gelatin–glue size mixture. However, the canary dextrins are not used because they tend to remain tacky and cause the yarn to stick to the drying cylinder of the slasher. Yellow dextrins may discolor yarn. British gums are particularly satisfactory for the sizing of spun yarn because they form viscous pastes that smooth the filaments without excessive stiffening of the yarn. A small amount of a softener, such as a sulfonated oil, is usually added to the size bath.

Printing.—Printing pastes for woven fabrics essentially consist of a dye and other chemicals in a paste of starch, a gum or another suitable thickener. These pastes generally have a fairly short, heavy consistency. Viscosity is adjusted to give the desired degree of penetration into the fabric. British gums are preferred for alkaline printing pastes because they are relatively alkali stable. The natural gums, such as gum tragacanth, form excellent printing pastes with high viscosity at low solids content but are expensive and are generally extended with starch products. The dextrins that form thicker pastes and British gums are frequently used in textile printing. As compared to starch, they have the advantages of being easier to remove by washing and of not adversely affecting the feel of the finished fabric if not removed after printing.

Finishing.—Finished fabrics are sized to improve smoothness and sheen or to increase weight, water resistance, or opacity. Depending on the amount of size to be deposited, white dextrins of various degrees of solubility and viscosity are used. Dextrins or gums, the solutions of which have little tendency to retrograde and become cloudy, are preferred for finishing dyed goods or glossy fabrics. They are used to give a flexible finish to preshrunk fabrics and, in combination with various synthetic resins, to form water-resistant finishes. The weight of goods from which rough working clothes are made can be increased by as much as 25% by loading the cloth with highly converted dextrins.

The filling of a rug backing or base is a special case in which the purpose of the loading is not to increase weight but, instead, to fill the interstices between threads. Traditionally, tapioca dextrins have been used in back-filling the bases of Wilton rugs.

Opacity of window shades is achieved by coating and filling fabrics with dispersons of clay, inorganic salts, or other pigments in a plasticized dextrin adhesive.

V. References

(1) V. J. Bloede, *in* "A Comprehensive Survey of Starch Chemistry," R. P. Walton, ed., Chemical Catalog Co., New York, 1928, Vol. 1, p. 158.

(2) B. Brimhall, *Ind. Eng. Chem.,* **36,** 72 (1944).

(3) R. W. Kerr and F. C. Cleveland, *Staerke,* **5,** 261 (1953).

(4) J. D. Geerdes, A. Lewis, and F. Smith, *J. Amer. Chem. Soc.,* **79,** 4209 (1957).

(5) G. V. Caesar, *in* "Chemistry and Industry of Starch," R. W. Kerr, ed., Academic Press, New York, 2nd Ed., 1950, p. 346.

(6) G. M. Christensen and F. Smith, *J. Amer. Chem. Soc.,* **79,** 4492 (1957).

(7) "Starch Chemistry and Technology," R. L. Whistler and E. F. Paschall, eds., Academic Press, New York, Vols. I and II (1965–1966).

(8) M. J. Mason and W. R. Fetzer, *in* "Starch and Starch Products in Paper Coatings," TAPPI Monograph Series No. **17,** Technical Association of the Pulp and Paper Industry, New York, 1957, p. 65.

CHAPTER XXVII

STARCH HYDROXYETHYL ETHERS AND OTHER STARCH ETHERS

ERLING T. HJERMSTAD

Penick and Ford Ltd., Cedar Rapids, Iowa

I. INTRODUCTION

Use of starch as a raw material for the synthesis of derivatives possessing characteristics similar to those of the water-soluble natural gums has been seriously considered for several decades. Starch is abundant, is low in cost, and has a relatively high degree of purity. One approach, which has already achieved remarkable commercial success, is the production of partially etherified starch. This success is because of the ease with which starch can be alkylated without degradation and the stability of the ether linkage in the products.

II. HISTORY

Alkyl and substituted alkyl ethers of gelatinized or dispersed starch have been known since about 1920. Considerable information exists in the technical and patent literature published during the period 1920–1950.[1–4]

During recent years, considerable attention has been given to chemical derivatization of starch in its native, ungelatinized granule form. It was found that starch in this form was surprisingly reactive to a wide variety of reagents; and, when starch was reacted in this form, purification and other production operations were greatly simplified and the development of substantially undegraded products, having the inherent viscosity and colloidal characteristics of natural starch plus the added effect of the attached groups was made possible.

A large variety of commercial grades of starches, modified starches, and dextrins have for many years been available for use in industrial applications requiring low-cost, pure hydrocolloids exhibiting adhesive, thickening, and filming properties. All varieties of starch occur in plants in the form of cold-water-insoluble granules. Granule insolubility is a major factor responsible for the economical, large-volume production of commercial starches. Starch occurs in nature in intimate association with other materials, such as protein, fiber, mineral salts, lipids, and other carbohydrates. Because of the insolubility of starch granules, these materials are fairly easily removed by various wet-processing operations, and water suspensions of unswollen starch will undergo various substitution and addition reactions without significant alteration in the physical appearance or structure of the granules. This has resulted in large-scale production of highly purified, low-cost starch hydroxyalkyl ether derivatives. The production of hydroxyethyl ethers of starch by treatment with ethylene oxide has become a bulk-carload business, with production levels of over 100 million pounds per year.[5]

Such starch derivatives in which the hydrogen atom of a hydroxyl group is replaced by a hydroxyalkyl group with the formation of a hydroxyalkyl ether are referred to as *substituted starches,* for they have a hydroxyalkoxyl group in place of the hydroxyl group. In such starches, the average number of molecules of ethylene or propylene oxide that have reacted per D-glucopyranosyl unit is referred to as the *amount of substitution.*

III. Low-Substituted Starch Hydroxyalkyl Ethers

1. Preparation

Cold-water-insoluble hydroxyalkyl ethers of starch with an average of up to 0.1 mole of alkyl group per D-glucopyranosyl unit are prepared by reacting starch in water suspension with ethylene or propylene oxide without appreciable alteration of the granule structure. The product is easily dewatered and washed to a high degree of purity. A process now in general use involves reacting ethylene oxide with starch in a 35–40% solids suspension in water containing an alkali, such as an alkali metal hydroxide or alkaline earth metal hydroxide, at temperatures below the swelling temperature of the starch.[6]

The surprising efficiency of the reaction is caused by the concentration of alkali on the starch caused by adsorption. This specifically promotes starch etherification rather than formation of ethylene glycol.

The addition of hydroxyethyl groups to starch results in a lowering of the starch gelatinization temperature range. To avoid swelling of the starch in the alkaline water, especially when substitutions of up to 0.1 hydroxyethyl group per D-glucopyranosyl unit are desired, alkali metal salts are added in amounts of 4–10% based on the water present. Soluble alkali metal salts have been found to be very effective for the prevention of swelling of starch in aqueous alkali; and, in certain proportions, sodium hydroxide with sodium chloride or sodium sulfate exhibits a synergistic effect in preventing starch gelatinization. By the use of the above proportions of alkali and salt, substitutions of up to 0.1 hydroxyethyl group per D-glucopyranosyl unit are regularly obtained in commercial production, and the products can be dewatered, washed free of salt, and dried in the usual starch processing equipment. Higher substitutions, for example, an average of 0.6 hydroxyethyl group per D-glucopyranosyl unit, can be achieved in these alkaline water suspensions, provided the proportion of salt is increased. The resulting suspensions can be readily dewatered but are difficult to wash because hydroxyethylstarch gelatinizes in the filter cake as the salt is removed, rendering further purification or drying extremely difficult. The higher-substituted products can be dried at low temperatures without gelatinization if the salt remaining in the filter cake is not removed.

2. *Grades*

Low-substituted, high purity, ungelatinized hydroxyethylstarch is at present produced chiefly with two levels of hydroxyethyl group substitution — ~0.05 and ~0.10 of a hydroxyethyl group per D-glucopyranosyl unit. By the use of acidic or oxidative depolymerization of the starch, a variety of viscosity grades are available with a fluidity range similar to those of the thin-boiling starches. The efficiency and product yield is improved by efficient addition and distribution of the alkali before etherification, especially when the etherified starch is to be subsequently depolymerized to high fluidities by acids or oxidizing agents.[7]

Dextrinized hydroxyethylstarch retains hydroxyethyl groups and exhibits the improvement in properties imparted by such groups.

Hydroxyethylstarches are produced chiefly from corn, milo, wheat, and potato starches. At present, they are priced only a few cents per pound above the commercial base starch, depending on the viscosity grade and degree of hydroxyethyl group substitution. Because ethylene oxide is available in large quantities and at relatively low cost, these low-substituted hydroxethylstarches could probably be produced in quantities equal to that of the common starch modifications if the demand should develop. Other starches, such as waxy maize starch, can be etherified with ethylene or propylene oxide to obtain improvement in physical properties, for example, in freeze–thaw resistance. The properties of high-amylose corn starch are also improved by hydroxyalkylation.

3. *Properties*

Substitution in corn starch with an average of 0.05–0.10 of a hydroxyalkyl group per D-glucopyranosyl unit results in a drastic alteration of the paste and film properties of the starch. The gelatinization temperature is lowered 5°–18°, and the starch granules readily disintegrate toward a true colloidal state when heated in water. On cooling, the hydrosols of low-substiuted hydroxyethylstarch form translucent, very cohesive pastes with high stabilities and indefinite storage lives if properly preserved. These hydrosols dry to relatively clear, flexible, water-soluble films. The films are nontacky at high humidities, even though they are readily remoistened by water.

The ether linkages by which the hydroxyethyl groups are attached to the starch molecules are strong and resist cleavage by acids, alkalies, and mild oxidizing agents. This is a useful property in alkaline adhesives, and acid- or alkali-catalyzed, water-resistant resin–starch adhesives and coatings. Hydroxyethyl groups remain intact even after the starch has undergone conversion to dextrins, acid hydrolysis to monosaccharides, hypochlorite oxidation, and enzymic liquefaction and saccharification.

Introduction of the hydroxyethyl ether group into the starch molecule greatly increases the hydrophilic character of starch, even though proportions as low as an average of one hydroxyethyl group per 20 D-glucopyranosyl units are present. The great increase in hydrophilic properties is believed caused in part by the hydrophilic nature of the hydroxyethyl groups themselves and in part to an inhibition of reassociation of the starch molecular chains to form crystalline regions that resist hydration. The inhibiting effect of such low proportions is almost great enough to eliminate the well-known retrogradation and gelling properties of the cereal starches.

4. *Uses*

These derivatives are used chiefly in the paper and textile industries. They are used as beater sizes because their low gelatinization temperature gives them an increased tendency to swell or gelatinize as the paper furnish passes over the dryers. The smooth, relatively flexible, and continuous nature of the films of low-substituted hydroxyethylstarch is effective in increasing the resistance of paper surfaces to penetration of hydrophobic materials such as greases, waxes, varnishes, and inks and has led to its extensive use in paper sizing and coating. It also improves the performance of high-gloss inks. Paper coating colors have improved flow, leveling properties, and viscosity stability when low-substituted, depolymerized hydroxyethylated starches are used as binders. Better control of adhesive penetration from coating colors into paper stock is obtained because of the improved water-holding of such starch derivatives, and coating failures caused by adhesive loss in the coating layer are minimized. The high adhesive strength of these starch derivatives gives high coating pick values and ink receptivity, and printing properties are improved. Hydroxyethylstarch has been found to be a highly effective starch derivative for use with glyoxal to produce coatings with increased water resistance. These derivatives are compatible with various wax emulsions, polyvinyl alcohols, proteins, latices, galactomannans, water-soluble resins, water-soluble cellulose derivatives, and other natural or synthetic water-soluble film formers used to obtain a variety of finishes. The increased hydration characteristics of hydroxyethylstarch are valuable in various adhesives and coatings because they tend to prevent excessive penetration into surfaces. Thus, they are used in solid fiber adhesives, bag pastes, case-sealing glues, waterproof and nonwaterproof veneer glues, label and envelope adhesives, and, to some extent, in corrugating adhesives.

The textile industry uses low-substituted hydroxyethylstarch chiefly in warp sizing. The characteristics that make it suitable for this use are increased rate of hydration, reduced gelling tendency, compatibility with sizing adjuncts, ease of desizing, film smoothness, film flexibility, and film clarity. Formulations with polyvinyl alcohol are used for sizing blended yarns, such as polyester–viscose blends. It is also used in textile printing and finishing because of its relatively clear films and high reactivity with resins.

Low-substituted hydroxethylstarch is also used in liquid laundry starches, commercial laundry starch, and in the cooked-dried form as instant laundry starches. Detailed information on characteristics of various grades of hydroxyethylstarch and their formulation in industrial applications can be found in the advertising brochures of several corn wet-milling companies.

IV. Higher-Substituted Hydroxyalkyl Ethers

1. Preparation

Ethylene and propylene oxide can be readily combined with starch in the native, ungelatinized granule in proportions of 0.8–1.0 mole per D-glucopyranosyl unit without altering the appearance of the granules. Because starch becomes cold-water-swellable when an average of 0.3–0.4 or more of the hydroxyl groups of each D-glucopyranosyl unit are derivatized, it is necessary to conduct the hydroxyalkylation reaction in the presence of a limited amount of moisture or in a liquid, such as an alcohol, that has little or no tendency to swell the product. The process can also be used to prepare lower substituted derivatives of water-soluble dextrins.

Ethylene and propylene oxides react readily with commercially dried starch or dextrins (5–12% moisture) containing alkaline catalysts.[6] Suitable catalysts are alkali metal hydroxides, alkali metal carbonates[8] and phosphates, tertiary and quarternary ammonium bases,[9] and alkali metal salts, such as sodium chloride, sodium sulfate, potassium chloride, and potassium sulfate. The neutral alkali metal salts are particularly convenient to use because of their low cost and the ease with which they can be incorporated into ungelatinized starch. When starch containing sodium chloride and moisture is exposed to ethylene oxide, sodium hydroxide and ethylene chlorohydrin are formed. Sodium chloride, therefore, acts as a latent alkaline catalyst and, as long as an excess of ethylene oxide is present, the starch remains at an alkalinity sufficient to catalyze the hydroxyalkylation reaction. Sodium hydroxide alone will catalyze the reaction equally well, but it is difficult to prepare a dry starch containing sodium hydroxide because of its conversion to sodium carbonate during drying of the alkali starch. In practice, starch containing about 1% of sodium chloride or sodium sulfate and 7–10% of moisture is heated and agitated in a closed reactor in contact with ethylene oxide gas at temperatures of 8° to 22° and at pressures of 5 to 15 psig. Under these conditions, reaction rates equivalent to a 1% per hour increase in product weight are obtained. Negligible color change occurs and a hydroxyethylstarch in the ungelatinized granule form is obtained.

Other processes involve hydroxyalkylation of starch suspended in lower aliphatic alcohols or ketones.[10, 11] The hydroxyalkylating agent, for example, ethylene oxide, propylene oxide or ethylene chlorohydrin, reacts readily and efficiently at high solids concentrations with highly alkaline starch suspended in an alcohol or ketone, such as isopropanol or acetone. A small amount of water, insufficient to cause swelling of the product, should be present. Alkylene oxides and chlorohydrins react preferentially with the moist alkaline starch rather than with the alcohol or ketone, which has little or no alkali in solution. Substitution by progressively higher proportions of ethylene oxide results in a progressive increase in thermoplasticity and solubility of the product in alcohol.

Other processes involve hydroxyalkylation of starch in higher alcohols[12] and in mixtures of organic liquids.[13] Highly substituted hydroxyalkylstarch in granular form can be produced in water suspension if the starch is first cross-linked to inhibit swelling.[14] Such derivatives, when prepared with the usual cross-linking agents, are difficult to gelatinize and form pastes of partially swollen granules with low adhesive power and poor film-forming character. However, high-substituted hydroxyalkylstarch derivatives cross-linked with divinyl sulfone can be gelatinized to a high degree of granule dispersion by cooking under slightly alkaline conditions.[15]

Several processes have been disclosed for producing hydroxyalkylstarch in the drum-dried, cold-water-soluble form.[16–18] Other processess describe drum-drying hydroxyalkylstarches with additives, such as boric acid, borax, gum arabic, and various sulfates.[19–22]

2. *Grades*

There has so far been very little commercial development of higher-substituted hydroxyalkyl starch. As with the low-substituted derivatives, a wide range of viscosities can be produced. Each viscosity grade can be produced with a wide range of hydroxyalkyl contents.

3. *Properties*

The gelatinization temperature of hydroxyalkylstarch is dependent on the degree of substitution. As the average number of hydroxyalkyl groups is increased from 0.1 to 0.4 per D-glucopyranosyl unit, progressively lower gelatinization temperatures are obtained. As the substitution level is increased from 0.4 to 1.0 hydroxyalkyl group per D-glucopyranosyl unit, the products show increasingly rapid cold-water gelatinization. With further substitution, the product becomes more easily solvated by lower alcohols, such as methanol or ethanol.

It is possible for starch to react with an average of more than 3 moles of alkylene oxide per D-glucopyranosyl unit,[23] for polyoxyalkyl chains can be formed by repeated hydroxyalkylation of the hydroxyl group of the alkyl substituent. In fact, polyoxyalkylene graft copolymers of starches and dextrins that are solvent-soluble and thermoplastic have been prepared.[24]

Higher-substituted hydroxyethylstarch produces solutions with higher adhesive values, high paste stability on aging, and resistance to biological spoilage. Their sols have no tendency to retrograde or gel, have a high degree of freeze–thaw stability, and are nonionic and therefore relatively unaffected by electrolytes. Highly substituted hydroxyethylstarch resists enzymic hydrolysis when injected into the blood stream.

4. *Uses*

Because these products have been developed recently, their commercial applications have not been extensively evaluated. Many uses will depend on the demand for relatively low-cost, low-ash, cold-water-soluble thickening agents and film formers. One use that has shown merit is in low biochemical oxygen demand (BOD) textile sizing. It has been demonstrated that, as the hydroxyethyl content of starch is increased, the BOD of water containing the starch removed during desizing is decreased. Other uses for higher-substituted hydroxyalkylstarch ethers include textile back filling and printing pastes, special adhesives,[25] suspending agents in dermatological lotions,[26] and as intermediate water-soluble layers in photographic multilayer material.[27–29] Cold-water soluble hydroxyalkylstarches have been suggested for use as soil conditioners.[30]

Encouraging results have been obtained with depolymerized hydroxyethylstarch as a plasma volume expander.[31–44] It has been found to be nontoxic to animals when injected into the bloodstream and is effective in restoring and maintaining arterial pressures. Ordinary starch solutions are physically unstable and undergo rapid enzymic destruction or degradation in the bloodstream. Etherified starch, particularly amylopectin types having an average of ~0.7 of a hydroxyethyl group per D-glucopyranosyl unit, has a degree of enzyme resistance, high physical stability, and good storage stability, and can be manufactured in suitable viscosity grades.

Low- and medium-substituted hydroxypropyl ethers of starch are produced by processes similar to those in use for the preparation of hydroxyethyl ethers. Hydroxypropylstarch has been cleared for use as Food Starch Modified under Food Additives Regulation No. 121.1031 of the Federal Food, Drug and Cosmetics Act.[45] Commercial hydroxypropyl ethers of corn, waxy corn, and sorghum starch are marketed in considerable quantities for use in foods, that require a clear or translucent thickener with adequate freeze–thaw resistance.[46, 47] The starches are generally produced with a slight degree of cross-linking to control texture and to stabilize against thinning caused by high temperatures, high shear, and acidity during the formulation and processing of foods.[48–50] Relatively highly hydroxypropylated starches that are purifiable by cold-water washing are prepared by the hydroxypropylation of heat–moisture stabilized potato starch. These derivatives exhibit a very high degree of freeze–thaw resistance.[51]

Hydroxyalkylation of high-amylose starch modifies its properties for certain uses. Edible films from hydroxyalkylated starch have been developed, and excellent film properties are reported.[52–54] Amylose has been hydroxyalkylated to enable dispersion under normal cooking conditions. Hydroxyethylamylose ethers are said to be effective in paper coating[55] and for sizing textile fibers.[56, 57]

V. Viscosity, Gel, and Film Properties of Hydroxyalkylstarch

Addition of low proportions of hydroxyethyl groups onto starch does not appreciably change the hot-paste viscosity as compared with untreated starch, provided the starches are thoroughly dispersed during gelatinization. However, the cold-paste viscosity is considerably changed, because of increased hydration and removal of retrogradation and gelation tendencies. Because hydroxyethylstarch is produced in numerous viscosity grades from several varieties of starch, it is not feasible to enumerate the effects of all combinations of factors on viscosity and other rheological properties. The very wide range of viscosities that are available with commercial hydroxyethylstarches is illustrated in Table I, in which commercially produced hydroxyethylstarches at the extreme limits of the viscosity range are compared.

TABLE I

Viscosity Range of Commercial Low-Substituted Hydroxyethylstarch

Trade Name	Substitution[a]	Paste Concentration,[b] % Solids	Brookfield Viscosity at 25°, cps
Penford Gum 300 [c]	0.10	2	250
Pen Cote [c]	0.07	36	6000

[a] Determined by a Modified Zeisel alkoxyl analysis [59] and calculated as the average number of hydroxyethyl groups per D-glucopyranosyl unit.
[b] Dispersed by heating to 99° in water of moderate hardness.
[c] Trademark of products manufactured by Penick and Ford Ltd., Cedar Rapids, Iowa.

There are several methods available for determining the hydroxyalkyl group content of hydroxyalkylated starch. These include modifications of the Zeisel alkoxyl procedure[58, 59] and spectroscopic procedures.[60, 61] A method based on pyrolysis–gas chromatography has also been developed.[62]

By hydroxyethylation of dextrins or dextrinization of hydroxyethylstarch, derivatives that are fluid in concentrations considerably higher than the range shown in Table I can be produced. Hydroxyethyldextrins with an average of more than 0.1 hydroxyethyl group per D-glucopyranosyl unit have high viscosity stability on aging. The effect of increasing concentration on the viscosity of a low-substituted hydroxyethyl derivative of undeploymerized corn starch is shown in Table II.

Increasing the hydroxyethyl content of starch does not result in an appreciably changed degree of polymerization. The change in rheological properties may be

TABLE II

Approximate Viscosity Level of Corn Starch Containing an Average of 0.1 of a Hydroxyethyl Group per D-Glucopyranosyl Unit

Paste concentration,[a] % solids	Brookfield Viscosity, cps	
	90°	25°
3	500	1,000
6	4,000	30,000
9	7,500	100,000

[a] Dispersed by heating to 99° in water of moderate hardness at pH 7.0.

reflected in the viscosity values obtained using different types of viscometers. Generally, increasing substitution results in increased rate of granule dispersion, increased paste cohesiveness or glutinous character, and decreased reassociation tendencies.

The paste viscosity of high- and low-substituted hydroxyethylstarches is greatly decreased when the starches are dispersed by heating in water of low pH because of the normal acid-catalyzed depolymerization of starch. They are relatively stable when heated under alkaline conditions. Variations in pH of the cooled dispersions have little effect on paste viscosity, which is also relatively unaffected by dissolved electrolytes, such as sodium chloride or sodium sulfate, that have no specific complexing effect with starch.

The pastes of high- and low-substituted hydroxyethylstarches are remarkably stable on aging. Those with hydroxyethyl substitutions of >0.1 per D-glucopyranosyl unit can be dispersed to pastes that show no viscosity change on standing.

With increasing substitution up to an average of 0.1 per D-glucopyranosyl unit, the gelling starches, such as corn- and wheat starch, show progressively lower gel strengths. Higher substitution does not significantly change the nature of the cold paste. Substitution with ~ 0.1 of a hydroxyethyl group per D-glucopyranosyl unit in acid-converted starches, dextrins, or oxidized starches results in pastes that have stable, fluid characteristics on cooling, even in fairly high concentrations, for example, 10–60%.

Reaction of starch with ethylene oxide produces a beneficial effect on the filming characteristics,[63] especially in the case of the common cereal starches, such as corn and wheat. As the substitution is increased up to an average of 0.1 of a hydroxyethyl group per D-glucopyranosyl unit, the films become increasingly flexible, clear, and smooth. Shrinkage during drying is also substantially reduced. As the substitution with hydroxyethyl groups is increased above 0.1 per D-glucopyranosyl unit, the films attain increasing flexibility, although their clarity is not significantly increased. The increased flexibility is probably caused by a plasticizing effect of the substituent groups. Higher substitution does not appreciably change the affinity for absorption of water vapor. Films made from higher-substituted starches are not hygroscopic. Starches that have been hydroxyethylated and subjected to mild oxidation, such as oxidation by sodium hypochlorite, produce films having exceptional clarity. Hydroxyalkyl potato starch films are also exceptionally clear and glossy. Although acid conversion of potato starch results in cloudy and high-gelling paste character, hydroxyalkylation of acid-converted potato starch results in a product that gelatinizes to clear, nongelling pastes and that forms exceptionally clear and glossy films.

VI. Structure of Alkylated Starch

Substitution of ungelatinized starch by alkyl and substituted alkyl groups by the processes described above does not alter the basic starch molecular structure. Substitution can take place at the hydroxyl groups on the 2, 3, or 6 positions of the D-glucopyranosyl units without depolymerization, giving a product that has hydroxyalkoxyl or substituted hydroxyalkoxyl groups[64] distributed at intervals along the polymer chains. Hydroxyalkylation to a very high degree results in the formation of polyoxyalkyl side chains.

In view of the marked effect of low substitution, for example, an average of 0.05 of a hydroxyethyl group per D-glucopyranosyl unit, it is presumed that there is little or no formation of polyoxyalkyl chains in low-substituted hydroxyalkylstarch ethers. This is in agreement with observations made in the author's laboratory that equimolar proportions of methyl or hydroxyethyl groups result in similar changes in starch properties.

Several workers have studied the distribution of hydroxyethyl groups in hydroxyethylstarches having hydroxyethyl group substitutions ranging from 0.1 to 0.6 per D-glucopyranosyl unit by different analytical techniques. Their results indicate that substitution is predominantly in the C-2 position.[65–67] Studies on hydroxyethylamylose having hydroxyethyl substitutions up to 1.1 per D-glucopyranosyl unit also showed substitution to be chiefly at C-2.[68, 69] However, there was some disagreement as to the relative distribution of hydroxyethyl groups on the C-3 and C-6 positions. Very little polyoxyethylene substitution was found, especially with starches having low hydroxyethyl contents.

VII. Other Commercial Alkyl and Substituted Alkyl Ethers of Starch

Following the introduction of commercial hydroxyethyl ethers of starch in the native granule form, several other alkyl ethers of granular starch have been produced commercially and marketed. Low-substituted, purified sodium carboxymethylstarch[70, 71] has been produced in commercial quantities for several years and is used as a paper furnish additive to increase strength and other properties of paper. Low-substituted sodium carboxymethyl ethers of ungelatinized granular starch can be produced with a wide range of viscosities similar to those of commercial hydroxyethylstarches. Highly anionic, low-cost derivatives can be produced by oxidation of sodium carboxymethylstarch with oxidants such as sodium hypochlorite. Such derivatives have increased pigment or soil dispersing action, and they exhibit increased compatibility with certain other natural and synthetic polymers.

Low-substituted cationic ethers of ungelatinized granular starch have been produced commercially for several years. These contain tertiary amino or quaternary ammonium alkyl groups, such as diethylaminoethyl,[72] or quaternary ammonium alkyl groups, such as (hydroxypropyl)trimethylammonium chloride[73] or (2,3-butenyl)trimethylammonium chloride.[74–76] They are used chiefly in the paper industry as furnish additives to increase strength of paper and as pigment retention aids because their flocculation action results in greater retention of added pigments when the furnish is dewatered. Cationic starches are also used at present, to some extent, for paper sizing and coating, for flocculation, and for textile warp sizing.[77]

Low-substituted cyanoethyl ethers of ungelatinized granular starch have been produced in commercial quantities and are used for paper sizing and coating.[78, 79]

Another type of starch ether derivative that has recently been produced commercially is a medium-substituted hydrophobic starch ether, containing ether groups such as benzyl. Benzylstarch having a degree of substitution sufficient to prevent gelatinization when cooked under atmospheric pressure can be dispersed to a high degree by heating in water at superatmospheric pressure in a jet cooker. A very fluid sol containing submicron-sized particles is formed.[80] This sol dries to form a continuous, clear film. At present it is used for sizing glass fiber warp yarns.

Although there have been a great number of other different starch alkyl and substituted alkyl ether derivatives described in the technical and patent literature,[81] there has been very little commercialization of any save those described above, probably because of the similarity in properties imparted to starch by low degrees of substitution regardless of the nature of the group that is introduced. Thus, the manufacturers of starch derivatives tend to use etherifying reagents that are lowest in cost and that are convenient to handle in their processing equipment. At present the commercially produced ether derivatives of starch fall into three general types. These are the nonionic derivatives, such as hydroxyalkyl and cyanoethyl; the anionic derivatives, such as sodium carboxyalkyl; and the cationic derivatives, such as tertiary amino and quaternary ammonium alkyl ethers of starch.

VIII. References

(1) E. F. Degering, *in* "Chemistry and Industry of Starch," R. W. Kerr. ed., Academic Press, New York, 2nd Ed., 1950, p. 259.

(2) M. L. Cushing and N. F. Kennedy, *in* "Starch and Starch Products in Paper Coatings," TAPPI Monograph Series No. **17**, Technical Association of the Pulp and Paper Industry, New York, 1957, p. 165.

(3) E. F. Degering, *in* "Starch and Its Derivatives," J. A. Radley, ed., Chapman and Hall, London, 3rd Ed., 1953, Vol. 1, p. 326.

(4) C. G. Caldwell and C. C. Kesler, *in* "Starch and Starch Products in Paper Coatings," TAPPI Monograph Series No. **17**, Technical Association of the Pulp and Paper Industry, New York, 1957, p. 31.

(5) S. M. Parmenter, *in* "Kirk-Othmer Encyclopedia of Chemical Technology," A. Standen, ed., John Wiley and Sons, Inc., New York, 2nd Ed., 1969, Vol. 18, pp. 687–691.

(6) E. T. Hjermstad, *in* 'Starch: Chemistry and Technology," R. L. Whistler and E. F. Paschall, eds., Academic Press, New York, 1967, Vol. 2, pp. 423–432.

(7) E. T. Hjermstad and O. J. Rajtora, U.S. Patent 3,632,803 (1972).

(8) C. G. Caldwell and I. Martin, U.S. Patent 2,802,000 (1957); *Chem. Abstr.,* **51,** 172215 (1957).

(9) R. W. Kerr and W. A. Faucette, U.S. Patent 2,773,238 (1956); *Chem. Abstr.,* **50,** 6824 (1956).

(10) C. C. Kesler and E. T. Hjermstad, U.S. Patent 2,845,417 (1958); *Chem. Abstr.,* **53,** 2657 (1959).

(11) E. T. Hjermstad and L. C. Martin, U.S. Patent 3,135,739 (1964); *Chem. Abstr.,* **61,** 4580 (1964).

(12) A. E. Broderick, U.S. Patent 2,682,535 (1954); *Chem. Abstr.,* **48,** 11100 (1954).

(13) D. B. Benedict and A. E. Broderick, U.S. Patent 2,744,894 (1956); *Chem. Abstr.,* **50,** 16865 (1956).

(14) L. O. Gill and J. A. Wagoner, U.S. Patent 3,014,901 (1961); *Chem. Abstr.,* **56,** 7564 (1962).

(15) E. T. Hjermstad, U.S. Patent 3,438,913 (1969); *Chem. Abstr.,* **71,** 4729 (1969).

(16) R. A. Brobst, U.S. Patent 3,049,538 (1962); *Chem. Abstr.,* **58,** 651 (1963).

(17) J. Lolkema, U.S. Patent Reissue 23,443 (1951); *Chem. Abstr.,* **46,** 2830 (1952).

(18) E. D. Klug, U.S. Patent 3,033,853 (1962); *Chem. Abstr.,* **57,** 2484 (1962).

(19) R. W. Kerr, U.S. Patent 2,732,309 (1956); *Chem. Abstr.,* **50,** 6824 (1956).

(20) R. W. Kerr, U.S. Patent 2,903,391 (1959); *Chem. Abstr.,* **54,** 1850 (1960).

(21) W. A. Scholten's Chemische Fabrieken, Brit. Patent 816,049 (1959); *Chem. Abstr.,* **53,** 20868 (1959).

(22) G. Moes, Germ. Patent 1,117,510 (1960); *Chem. Abstr.,* **56,** 8984 (1962).

(23) P. W. Morgan, *Ind. Eng. Chem., Anal. Ed.,* **18,** 500 (1946).

(24) A. Zilkha, M. Tahan, and G. Ezra, U.S. Patent 3,414,530 (1968).

(25) J. Lolkema and W. A. Van der Meer, Neth. Patent 81,917 (1956); *Chem. Abstr.,* **51,** 9192 (1957).

(26) D. H. Cronk and L. C. Zopf, *J. Amer. Pharm. Ass., Pract. Pharm. Ed.,* **14,** 302 (1953).

(27) Gevaert Photo-Producten N.V., Belg. Patent 617,036 (1962); *Chem. Abstr.,* **58,** 9789 (1963).

(28) T. A. Scott, U.S. Patent 3,047,392 (1962); *Chem. Abstr.,* **57,** 16044 (1962).

(29) J. D. Overman, Germ. Patent 1,161,478 (1964); *Chem. Abstr.,* **60,** 10100 (1964).

(30) F. A. Moller and J. Lolkema, Germ. Patent 963,690 (1957); *Chem. Abstr.,* **54,** 10210 (1960).

(31) M. Wiedersheim, *Arch. Int. Pharmacodyn.,* **111,** No. 3, 353 (1957).

(32) W L. Thompson, J. J. Britton, and R. P. Walton, *J. Pharmacol. Exp. Ther.,* **136,** 125 (1962).

(33) W. L. Thompson and R. P. Walton, *J. Pharmacol. Exp. Ther.,* **146,** 359 (1964).

(34) G. P. Murphy, D. E. Demaree, and J. A. Gagnon, *J. Urol.,* **93,** 534 (1965).

(35) W. L. Thompson, D. H. Wayt, and R. P. Walton, *Proc. Soc. Exp. Biol. Med.,* **115,** 474 (1964).

(36) W. L. Thompson and R. H. Gadsden, *Transfusion (Philadelphia),* **5,** 440 (1965).

(37) R. P. Walton, A. L. Hauck, and E. H. Herman, *Proc. Soc. Exp. Biol. Med.*, **121,** 272 (1966).

(38) C. T. Greenwood and D. J. Hourston, *Staerke,* **19,** 243 (1967).

(39) T. F. Solanke, *Brit. Med. J.*, **3,** 783 (1968).

(40) J. E. Arrants, N. Cooper, and W. H. Lee, Jr., *Amer. Surg.*, **35,** 465 (1969).

(41) K. A. Granath, R. Stromberg, and A. N. de Belder, *Staerke,* **21,** 251 (1969).

(42) H. Hershenson, J. M. Brake, and M. Roberts, U.S. Patent 3,523,938 (1970).

(43) M. Takaori, P. Safar, and S. J. Galla, *Arch. Surg.*, **100,** 263 (1970).

(44) B. S. Evangelista, T. J. Green, D. J. Gwilt, and T. J. Brown, Jr., *Arch. Int. Pharmacodyn. Ther.*, **180,** 57 (1969).

(45) *Code of Federal Regulations,* Title 21, *Food and Drugs,* U.S. Government Printing Office, Washington, D.C., 1966, Chapt. I, Part 121, Subpart D, Sec. 121.1031.

(46) O. B. Wurzburg and C. D. Szymanski, *J. Agr. Food Chem.*, **18,** 997 (1970).

(47) D. C. Leegwater and J. B. Luten, *Staerke,* **23,** 430 (1971).

(48) J. V. Tuschhoff and C. E. Smith, U.S. Patent 3,238,193 (1966).

(49) A. M. Goldstein, A. F. Kohl, and P. Kaplan, U.S. Patent 3,278,522 (1966); *Chem. Abstr.*, **66,** 4065 (1967).

(50) J. V. Tuschhoff, G. L. Kessinger, and C. E. Hansen, U.S. Patent 3,422,088 (1969).

(51) E. T. Hjermstad, U.S. Patent 3,577,407 (1971).

(52) W. B. Roth and C. L. Mehltretter, *Food Technol.* (*Chicago*), **21,** 72 (1967).

(53) F. J. Mitan and L. Jokay, U.S Patent 3,427,951 (1969); *Chem. Abstr.*, **70,** 105300 (1969).

(54) F. R. Senti, *in* "Starch: Chemistry and Technology," R. L. Whistler and E. F. Paschall, eds., Academic Press, New York, 1967, Vol. 2, pp. 517–519.

(55) Cooperatieve Verkoop-en Productiever-eniging van Aardappelmeel en Derivaten "Avebe" G.A., Brit. Patent 869,192 (1961); *Chem. Abstr.*, **55,** 22878 (1961).

(56) J. Lolkema, G. Moes, and W. F. Vogel, Germ. Patent 1,118,151 (1961); *Chem. Abstr.*, **56,** 14501 (1962).

(57) D. P. Langlois and J. A. Wagoner, *in* "Starch: Chemistry and Technology," R. L. Whistler and E. F. Paschall, eds., Academic Press, New York, 1967, Vol. 2, p. 490.

(58) P. W. Morgan, *Ind. Eng. Chem., Anal. Ed.*, **18,** 500 (1956).

(59) H. J. Lortz, *Anal. Chem,* **28,** 892 (1956).

(60) D. P. Johnson, *Anal. Chem.*, **41,** 859 (1969).

(61) H. Stahl and R. P. McNaught, *Cereal Chem.*, **47,** 345 (1970).

(62) H. Tai, R. M. Powers, and T. F. Protzman, *Anal. Chem.*, **36,** 108 (1964).

(63) T. J. Schoch, *TAPPI,* **35,** No. 7, 22A (1952).

(64) H. J. Roberts, *in* "Starch Chemistry and Technology," R. L. Whistler and E. F. Paschall, eds., Academic Press, New York, 1967, Vol. 1, pp. 461–463.

(65) H. C. Srivastava and K. V. Ramalingam, *Staerke,* **19,** 295 (1967).

(66) G. N. Bollenback, R. S. Golik, and F. W. Parrish, *Cereal Chem.*, **46,** 304 (1969).

(67) H. C. Srivastava, K. V. Ramalingam, and N. M. Doshi, *Staerke,* **21,** 181 (1969).

(68) E. Husemann and M. Kafka, *Makromol. Chem.*, **41,** 208 (1960).

(69) C. E. Lott and K. M. Brobst, *Anal. Chem.*, **38,** 1767 (1966).

(70) E. T. Hjermstad and C. C. Kesler, U.S. Patent 2,773,057 (1956); *Chem. Abstr.*, **51,** 4746 (1957).

(71) E. T. Hjermstad and C. C. Kesler, U.S. Patent 3,062,810 (1962); *Chem Abstr.*, **58,** 3593 (1963).

(72) C. G. Caldwell and O. B. Wurzburg, U.S. Patent 2,813,093 (1957); *Chem. Abstr.,* **52,** 2438 (1958).

(73) E. F. Paschall, U.S. Patent 2,867,217 (1959); *Chem. Abstr.,* **53,** 12720 (1959).

(74) K. W. Kirby, U.S. Patent 3,336,292 (1967); *Chem. Abstr.,* **67,** 91920 (1967).

(75) C. P. Patel, M. A. Jaeger, and R. E. Pyle, U.S. Patent 3,378,547 (1968); *Chem. Abstr.,* **70,** 59126 (1969).

(76) W. G. Hunt, U.S. Patent 3,624,070 (1971).

(77) E. F. Paschall, *in* "Starch: Chemistry and Technology," R. L. Whistler and E. F. Paschall, eds., Academic Press, New York, Vol. 2, 1967, pp. 403–422.

(78) R. M. Powers, U.S. Patent 3,387,998 (1968).

(79) T. E. Yeates, M. E. Carr, and C. E. Mehltretter, U.S. Patent 3,224,891 (1965); *Chem. Abstr.,* **64,** 6893 (1966).

(80) E. T. Hjermstad, L. C. Martin, and K. W. Kirby, U.S. Patent 3,462,283 (1969).

(81) H. J. Roberts, *in* "Starch Chemistry and Technology," R. L. Whistler and E. F. Paschall, eds., Academic Press, New York, Vol. 2, 1967, pp. 293–350.

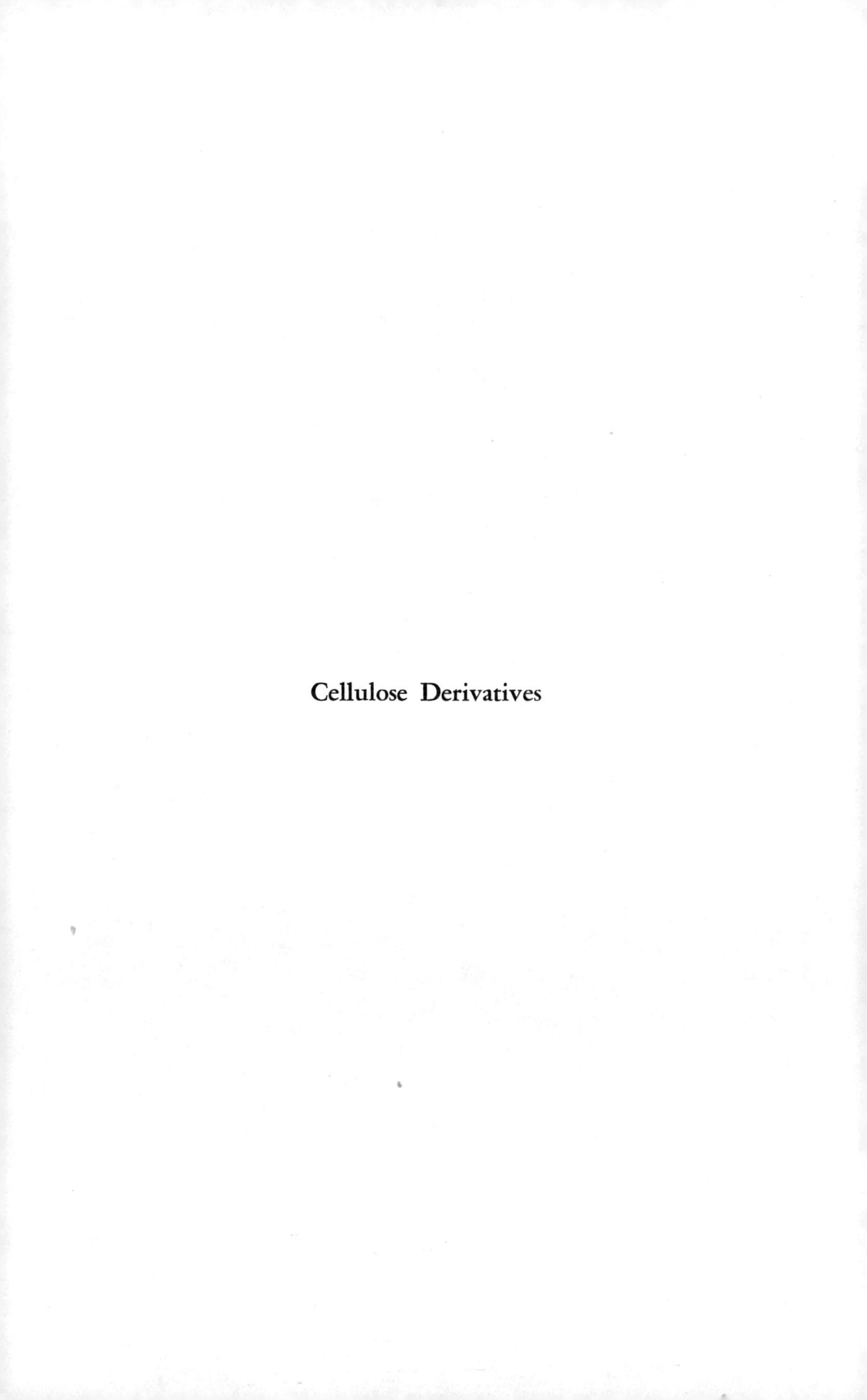

Cellulose Derivatives

Chapter XXVIII

METHYLCELLULOSE AND ITS DERIVATIVES

G. K. Greminger, Jr. and A. B. Savage

The Dow Chemical Company, Midland, Michigan

I. Introduction

Chemical modification of cellulose to produce water-soluble products has created many new markets, products, and techniques that affect almost every phase of our daily life. Methylcelluloses and hydroxypropylmethylcelluloses are a group of synthetic water-soluble, cellulose-based products that can be tailored to a wide variety of commercial uses. Methylcelluloses and hydroxypropylmethylcelluloses possess characteristic thermal gelling, surfactant, nonionic, thickening, film-forming, and adhesive properties coupled with extremely low toxicity. It is possible to adapt the methylcelluloses to specific end uses by controlling the type and ratio of their modifying substituents. Thus, the common grades of methylcelluloses are soluble in cold water but form gels when their aqueous solutions are heated. Products that contain fewer methoxyl groups are soluble only in aqueous alkali solutions, whereas highly substituted products are soluble in polar organic solvents. Modification of the method of manufacture yields products that give water solutions that can be heated to high temperatures, in some instances approaching their boiling points, without gelation. Chemical modification of the methylcelluloses leads to products that are soluble in certain mixtures of polar organic solvents, and such materials become thermoplastic when properly plasticized.

It is estimated that the production of methylcellulose and its derivatives will exceed 50 million pounds (22.5 million kilograms) per year in 1972. The price ranges from $0.69 to $0.84/lb ($1.52 to $1.85/kg). Methylcellulose and its derivatives are manufactured in the United States, the German Federal Republic, Great Britain, and Japan. Uses continue to expand.

II. Source

1. *Types*

Methylcellulose and its derivatives are modified D-glucans. Substituents used to alter the properties of methylcellulose include ethyl, hydroxyethyl, and hydroxypropyl ether groups. The specific products that are commercially available will vary with the particular manufacturer. Methylcellulose contains methoxyl substituent groups. Modified methylcelluloses of United States origin also contain hydroxypropoxyl or hydroxybutoxyl groups; those of German origin, hydroxyethoxyl groups; those of British origin, ethoxyl, hydroxyethoxyl or hydroxypropoxyl groups.

2. *Process*

Chemically, the methylcelluloses are cellulose ethers wherein methoxyl groups have replaced the hydroxyl groups of cellulose. The effect of such groups is to disorder and to spread apart the cellulose chains so that water or other solvents can enter to solvate them. These ethers are obtained by the reaction of chemical cellulose with suitable etherifying agents.[1, 2] The properties of the resultant cellulose ether vary with the degree to which the cellulose is substituted, the uniformity of substitution, and the nature of the substituent group. The degree of substitution (DS) is the number of hydroxyl groups of a β-D-glucopyranosyl unit of cellulose reacted. The maximum is three.

At any particular degree of substitution, the solubility of the product will depend upon its history. Thus, a decrease in the molecular weight, as indicated by intrinsic viscosity,* will result in improved solubility. The homogeneity of distribution of the substituent groups along the cellulose chain will also affect solubility of the material. These factors will determine whether aqueous solutions of the cellulose ether will gel when heated.[3] Introduction of groups other than methoxyl, such as ethoxyl, methoxyl, hydroxyethoxyl, hydroxypropoxyl, and hydroxybutoxyl, will also affect its solubility.

Although cellulose contains three active hydroxyl groups per D-glucopyranosyl unit, these hydroxyl groups are not all normally available for reaction[4] because of the crystallinity of native cellulose. The problem is to etherify these hydroxyl groups and yet avoid excessive degradation that would injure the physical properties of the product. In practice, the cellulose is first treated with a base, sodium hydroxide, and then with a solvating agent, water, to form alkali cellulose that reacts with the etherifying agent. Although dilute sodium hydroxide is used to prepare viscose and alkali-soluble cellulose ethers, methylcelluloses are commonly prepared with 35–60% sodium hydroxide solutions. Three moles of sodium hydroxide per D-glucopyranosyl unit are commonly used.

It is very important that the sodium hydroxide penetrate uniformly through the cellulose so that both the swelling and the sodium hydroxide concentration will be uniform. The water distributes the etherifying agents. There is little etherification in the absence of water.[5]

The degree of substitution that is reached depends upon the sodium hydroxide concentration late in the reaction.[6] The rate of methylation depends upon the ratio of sodium hydroxide to cellulose remaining at a given time, whereas by-product formation depends upon the sodium hydroxide concentration above the stoichiometric minimum and upon the reaction temperature.

Alkali cellulose can be prepared by dipping a cellulose sheet in sodium hydroxide solution, by spraying or mixing cellulose with sodium hydroxide solution, by liquid removal from a cellulose–sodium hydroxide solution slurry, or by adding sodium hydroxide solution to a slurry of cellulose fibers in an inert diluent. Alkali cellulose for methylcelluose production can be prepared from wood cellulose or from cotton linters. The latter are more amenable to uniform caustic distribution, whereas the former retains less of the morphology of the fibers after reaction. Very high-viscosity grades are made from pulp having a TAPPI Standard T230 viscosity at 0.5% concentration in 0.5% cupriethylenediamine hydroxide solution of 40–60 centipoises.

In American practice, methylcellulose is prepared from alkali cellulose that contains a limited amount of sodium hydroxide solution.[7–9] The cellulose sheet

*Intrinsic viscosity is the limit of the ratio of the specific viscosity to the concentration as the concentration approaches zero. For practical purposes, it has a bilogarithimic relation to the 2%-solution viscosity (Fig. 1). Intrinsic viscosity is represented by [η].

passes continuously through sodium hydroxide solution, is aged with warm air so that a mild oxidation with resultant intrinsic viscosity decrease takes place, and is then reacted with the etherifying agents.

In Europe, it is common to prepare alkali cellulose in a Werner–Pfleiderer shredder, in which dentated, sigma-shaped arms revolving at differential speeds knead and mix the alkali cellulose. Such shredding, or other mixing, can be aided by the addition of an inert diluent. Methylation is accomplished in agitated, jacketed, nickel-clad autoclaves at a pressure of about 200 psig. The heat of reaction is removed by the condensation of the solvents.

The relation of the composition of the alkali cellulose to the properties of the methylcellulose made from it was studied by Swinehart and Maasberg.[10, 11] An alkali cellulose, containing a weight ratio of sodium hydroxide to cellulose of 0.9 : 1.2 and a weight ratio of water to cellulose of 0.9 : 1.5, that was prepared from 37.8–57.2% sodium hydroxide solution, gave, on methylation with a slight excess of methyl chloride at 50°–100°, a cold-water soluble methylcellulose of DS 1.6–2.0 that could be washed with hot water at 85°–100° and dried. The physical properties of the product were regulated by pressure–temperature control during processing.

When an alkylating agent is added, only part reacts with alkali cellulose; the remainder is converted to by-products; for example, in the case of methyl chloride the products are methanol and dimethyl ether. Sodium chloride is quantitatively formed. Reactions in which etherifying agents other than methyl chloride are used are conducted in a similar manner, but due allowance must be made for their different reactivity and stability. Thus, methyl chloride reacts at 50°–100°, but ethyl chloride reacts at 120°. Dimethyl sulfate[12] and diethyl sulfate have been used in Europe under mild conditions, but much by-product formation occurs. Ethylmethylcellulose has been prepared in Great Britain by typical European stagewise methods.

Hydroxyalkoxyl substitution is obtained when methylcellulose is reacted with alkene oxides. Alkene oxides readily polymerize,[13] and ethylene oxide will react with the hydroxyl groups of its substituent chain at least as readily as with the hydroxyl groups of the cellulose. Hydroxyethylmethylcellulose was prepared in Germany[14, 15] by successive reaction of alkali cellulose with ethylene oxide and methyl chloride. Hydroxypropylmethylcellulose production was disclosed in patents issued to Savage.[16–18]

III. Uses

1. *History*

Commercial production of methylcellulose under the registered trademark Methocel was established by The Dow Chemical Company in 1938. Other current producers include British Celanese (Celacol) and Imperial Chemical Industries, Ltd. (Methofas and Edifas) of Great Britain; Kalle & Company A.–G. (Tylose S and Tylose SL), Henkel & Cie G.m.b.H. (Culminal), and Wolff A.–G. of Germany; and Shin-Etsu Chemical Products, Ltd. (Metalose) of Japan.

2. *Pharmaceuticals*

The pharmaceutical industry is consistently a large user of methylcellulose. Practically all the properties of methylcellulose products are utilized in therapeutic items or procedures. Several reviews of these applications are available.[19–23]

Bulking agent.—Because methylcellulose in aqueous media can swell to 40 times its dry volume, it is used in the treatment of various intestinal ailments, such as diarrhea, where bulking action is desirable.[24–28] The nonnutritive bulking action of methylcellulose is accompanied by demulcent and surfactant properties, and it contains no protein contaminants that could cause allergies. Constipation, often a chronic complaint of bedridden patients, can be relieved by the use of a hydrophilic colloid of methylcellulose.[29] Methylcellulose can be administered in the form of granules, tablets, solutions, or slurries.[19–23, 30, 31] Methylcellulose is not metabolized and is useful in the formulation of products for use in low-calorie or diabetic diets.[30, 31]

Ointments and lotions.—Many types of jellies and lotions are prepared with methylcellulose products. Examples are cystoscope jellies, water-based burn ointments, barrier creams, neocalamine lotion, ophthalmic preparations, and general ointment bases for drugs. Methylcellulose is particularly suitable in such products because its surface activity allows excellent spreading on body tissues. Because it has a nonionic structure, it is compatible with a wide range of drugs, including those that contain quaternary ammonium compounds or heavy-metal salts, that would precipitate carboxyl-containing gums. Film-forming properties permit a combination of protection and visibility in dressings for wounds and burns. Methylcellulose is also very effective in minimizing loss of blood and plasma, particularly from burned tissues. Its dressings can be removed by a warm water soak with little discomfort to the patient. Methylcellulose gels also have superior slip, which is valuable in the formulation of jellies required for surgical, diagnostic, or therapeutic procedures.[19–23, 32–48]

Emulsions.—Methylcellulose is used both in emulsions and in suspensions. A combination of surfactant, thickening, and protective colloid action is possible. Hydroxypropylmethylcellulose is preferred for oil emulsions. The choice of intrinsic viscosity type will depend upon the composition of the finished emulsion; low-viscosity types give better results with flavor oil emulsions. Oils readily emulsified include mineral oil, vitamin A, olive oil, soya oil, and flavor oils.

Suspensions.—The surfactant and thickening properties of methylcelluloses are used to stabilize suspensions against settling and to inhibit caking of suspended materials. As little as 0.1% based on total weight has proved effective in suspending magnesium trisilicate and aluminum hydroxide. Acceptance and palatability of ion-exchange resins is also improved.[49] Combinations of methylcellulose and barium sulfate produce sharper x-ray photographs than does barium sulfate alone, because the filming and suspending action of methylcellulose results in more uniform distribution of the barium sulfate over tissues.[50]

Tablets.—Methylcellulose is currently used as a granulating agent, as a coating agent, and as a disintegrating agent in tablet manufacture. Methylcellulose can be added to a tablet mix as a 10–15% slurry at 80°–90° or as a water solution if the quantity of water is not objectionable. The water solution can also be sprayed on the tablet mix. Solutions for spraying can be water systems using 15 cps methylcellulose at a concentration of 5%. Organic solvent systems, such as chloroform–ethanol or dichloromethane–ethanol, can be used with hydroxypropylmethylcellulose.[51, 52] Further modification of hydroxypropylmethylcellulose coatings to obtain controlled solubility can be achieved by the addition of 50 cps ethylcellulose. Because both of these materials are organic solvent soluble, it is possible to combine them in order to obtain retarded disintegration times in water.

Drug carrier.—The bland taste and demulcent, nonionic, and surfactant properties of methylcelluloses result in their widespread use as carriers for drugs of many types. Some of the products that are reported to be formulated with methylcelluloses include anesthetics, digitalis, sodium pantothenate, penicillin, sodium gentisate, sulfonamides, antibiotics, and antihistamines.[53–59] The combined water-soluble and thermoplastic properties of hydroxpyropylmethylcellulose formulations have been utilized in molding and extruding radioactive needles and sutures that can be sewn into tumors to achieve localized radiation of diseased tissues when surgical procedures are restricted or undersirable.[60]

Diagnostic techniques.—Solutions of methylcellulose gums are used as mounting media for the study of protozoa. By adjusting the viscosity of the carrying solution, it is possible to provide any desired retardation of movement of these organisms without interfering with their metabolic processes.[61–65] Methylcellulose solutions are used to provide a water-soluble base for the water-soluble contrast media that are used in bronchography, because they clear from the lungs much more rapidly than do oil solutions.[66, 67]

3. Foods

Methylcellulose gums have many properties that make them functional additives to food products in order to obtain better customer acceptance. They are useful as edible surfactants, thickeners, stabilizers, and film formers.[68]

Bakery products.—Methylcellulose gums have functional properties that are of advantage in baked goods, particularly in foods baked of flours that are low or lacking in gluten, such as rice, corn, and rye flours. Methylcellulose contributes to cell structure and reduces crumbling. The wide range of gel points permit the choice of a methylcellulose best suited for a particular product. Use of methylcellulose also allows more water to be carried in a dough and, thus, permits more complete gelatinization of the starch. Syneresis of additional water during gelation of the methylcellulose also helps. Commercial products include doughnut mix, corn bread, cake mixes, and speciality breads.

Canned-fruit pie fillings.—The clarity and sheen of pie fillings is improved when hydroxypropylmethylcellulose, 1500 or 4000 cps, at a level of 0.5% based upon the weight of the filling, including drained fruit juices and additional water, is added to cherry, blueberry, and raspberry products. The stabilizing properties of the nonionic methylcelluloses are not affected by the fruit acids.

Dipping batters for deep-fried breaded foods.—The heat-gelation, adhesive, and grease-barrier properties of the methylcellulose gums provide a combination of effects especially desirable in a dipping batter for deep-fried breaded foods. Hydroxpyropylmethylcellulose at a concentration of 1.0%, based upon the total weight of the wet batter, is suggested. The thermal gelation of the hydroxypropylmethylcellulose, which occurs when it comes in contact with the hot fat, creates a barrier around the fried article that greatly reduces fat absorption. Hydroxypropylmethylcellulose can also be used in a conventional batter to control viscosity and to permit dipping at lower batter solids.

Dehydrated fruits and vegetables.—Methylcellulose of 15 or 25 cps, at a concentration of 0.5–1.0% based upon the weight of solids in a dehydrated fruit or vegetable product, aids in redispersion and provides better consistency in the reconstituted product. The reversible thermal gel point, plus the viscosity and suspending properties, results in improved product performance.[69]

Dietetic foods.—Methylcellulose gums are effective as edible surfactants and thickeners in the formulation of dietetic food products. Methylcellulose gums are not metabolized and do not contain sodium as part of their chemical structure. Methylcellulose is an essential ingredient of special breads and pastries developed by the Childrens' Hospital of Los Angeles for use as part of the therapeutic diet of children with phenylketonuria, an inherited metabolic defect that causes mental retardation in children.[70]

Frozen foods.—Methylcellulose gums possess the very desirable property of stability at temperatures near the freezing point and lack of syneresis of water at subzero temperatures. These properties are used in a group of food products that were recently disclosed in a series of patents. The foods are patties containing a mixture of meat, fish, or fowl with vegetables. Combinations of frozen fruits and vegetables are also possible. Methylcelluloses are used to hold these patties together by gelation during frying and to provide the moisture required to prevent burning of the vegetable components.[71]

4. *Adhesives*

The increasing number of leather pasting operations in the tanning industry has necessitated the development of efficient adhesives. The hides are dried after being pasted onto large frames made of glass, porcelain, resin-coated plywood, or metals. Methylcellulose and its derivatives offer a combination of properties that make them ideal adhesives for this operation. As the hides pass through the heating zone, the methylcellulose pastes gel instead of thinning as do other types of adhesives. Pastes containing methylcellulose form excellent wet adhesive bonds between the leather and the pasted plate, resulting in fewer drop offs.

A typical starting formula contains 2.0% methylcellulose of 4000 cps, 0.2% plasticizer (*N*-acetylethanolamine), 0.3% casein, and 97.5% water. The casein content can be reduced or eliminated for light leathers, such as goat or sheepskin. Heavy hides will require a stronger adhesive, which can be obtained by increasing the concentrations of methylcellulose and casein. This might necessitate the use of 1500 cps material to maintain the viscosity of the paste. When casein is used, the addition of a preservative is suggested. On high-temperature pasting units in which hot water is circulated through the plate, hydroxypropylmethylcellulose (90° gel point) has proved to be the best material.

Methylcellulose solutions are also used as temporary bonding agents for glass fibers and as thickeners in phenolic resin adhesives. Additional adhesive applications include wallpaper paste, prepasted wallpaper, library paste, paper laminates, cigars, bookbinding adhesives, starch adhesives, pencil-lead binder, and latex adhesives.

5. *Agricultural Uses*

The effectiveness of pesticides can be increased by the use of methylcellulose as a spreader–sticker or as a dispersing agent. The addition of 25–50% of methylcellulose of 15 cps, based on the weight of dry protectant, to a slurry treating formula for seeds will increase the coverage of seeds and reduce exposure hazards caused by dusting of the protectant. The surfactant and adhesive properties of methylcellulose provide improved spreader–sticker action in insecticidal and fungicidal sprays. Hydroxypropylmethylcellulose of 4000 cps is generally used at a level of 2–4 oz per 100 gal (0.15–0.30 g/1) of water in foliage sprays. This same material is also added to agricultural dusts as a level of 1–2 oz/lb (63–125 g/kg) of dust to obtain better adhesion when the dust is wetted by rain or dew. Methylcellulose is useful as a dispersing agent with wettable powders. A level of 0.5–2% of methylcellulose of 15 cps, based on the weight of dry powder, is commonly used.

6. *Ceramics*

Refractory materials used in specialty mortars often lack cohesiveness or green strength. Nonionic, hydrophilic methylcellulose imparts the desired troweling properties without loss of melting point. The concentration required will depend upon the raw materials used and the properties desired. Generally, 0.1–0.3% of methycellulose of 4000 cps will suffice.

Mark-off or chipping of glazed ware prior to firing can be a serious problem. The binding properties of methylcellulose have proved useful in reducing loss from these causes. The quantity required varies from 0.3% to 0.5% of methylcellulose of 15 cps based upon the total solids. Higher viscosity grades should be used when running or sagging of the glaze occurs. When radiant heat is used in the spray method of application, clogging of the spray nozzles can result unless hydroxypropylmethylcellulose gums of higher gel points are used.

7. *Cement Formulations*

The use of methylcellulose and its derivatives in inorganic cementitious compositions is a known art. Compositions have included both Portland cement and gypsum-based products, such as masonry mortars, stuccos, grouts, cement coatings, plasters, and jointing compounds. All these end uses require water retention in order to permit proper hydration of the cementitious product and consequent development of high bond strength. Performance and utility are drastically reduced if water is removed prematurely from the mix by a porous, absorbtive building component, such as a brick or a block. The artisan needs good workability if he is to use the mix properly and efficiently.

Methylcellulose and its derivatives can also act as air entraining agents, reducing weight and eliminating the need of specific additives.

When methylcellulose is added to a masonry mortar, it is possible, depending upon the quantity used, to extend the ratio of sand to cement to as high as 7 : 1, thus offsetting the cost of the methylcellulose and achieving a three- to fourfold increase in the bonding strength of the mortar.[72–78]

8. *Cosmetics*

Cosmetic formulation requires a wide range of gum properties. Methylcellulose gums provide an especially desirable combination of these properties, for they can act as thickeners, emulsifiers, stabilizers, binders, film-formers, suspending agents, and surfactants. They can be used in hypoallergenic preparations because they contain no protein fraction that could be the source of allergies. They are bland, neutral, and stable over the pH range 2–12. In general formulating practice, the use of hydroxypropylmethylcellulose of 65° gel point is recommended. When special compatibility with organic compounds and solvent systems is required, 60° gel point is preferred. When ethanol is used in a formulation, rapid dissolution of the methylcellulose gum can be obtained by slurrying it in the alcohol prior to addition of the other ingredients. Use of a similar technique with oils will result in more stable emulsions. Cosmetic preparations formulated with methylcellulose include hand creams, vanishing creams, shampoos, hair dressings, depilatories, shaving creams, hand lotions, bubble bath liquids, and sun creams.

Because methylcellulose provides a barrier against oily materials, it has been used in protective creams to provide protection against irritants ranging from tear gas to paint. Methylcellulose-based barrier creams have served effectively and satisfactorily in industrial practice.

9. *Protective Coatings*

Paint.—Hydroxypropylmethylcellulose is widely used in latex paint formulations in which it functions as a protective colloid, as a thickener, and as a pigment dispersant, and contributes both to flow and leveling and to wet-edge retention. Its degree of substitution confers superior resistance to enzymic degradation, thus minimizing viscosity loss during the manufacture and storage of paint. A recent patent disclosed a product, with a methoxyl DS in the range of 0.4 to 1.3, with a hydroxypropoxyl MS* of 0.2 to 1.4, higher than the usual value, and having a total substitution of at least 1.0 MS, which is especially designed for latex paint.[79]

The recent development of surface-treated, glyoxal cross-linked powdered products permits the use of a shift from an acid to an alkaline pH to determine the moment of dissolution. The paint manufacturer may add the hydroxypropylmethylcellulose at any convenient point in the paintmaking process. It can be added at a concentration of up to 15% at a pH below 7.5 without viscosity development, but a shift to an alkaline pH achieved by the addition of alkaline pigments, an alkaline latex, or a base such as ammonium hydroxide destroys the crosslinking through a Cannizzaro reaction; rapid viscosity development then occurs.

Multicolor finishes.—Methylcellulose is the critical ingredient in decorative coatings that provide, in a single application, a finish of two, three, or more discrete colors. It acts as a protective colloid and stabilizing agent to maintain the particular nature of the individual drops of colored lacquer so that they can be intermixed without coalescence. The recommended concentration of methylcellulose or hydroxypropylmethylcellulose of 4000 cps is 0.3–2% by weight of the dispersing medium.[80]

Miscellaneous latex applications.—Methylcellulose gums are used to thicken

*MS is the molar substitution of reagent per β-D-glucopyranosyl unit of cellulose.

latices, adhesives, paper coatings, dipped rubber goods, and fabric coatings. Because methylcellulose is nonionic, it is compatible with the metallic salts used to vulcanize latices.

10. Textiles

Methylcellulose gums are used as thickeners in a wide variety of fabric-coating compositions. In addition, hydroxypropylmethylcellulose provides an unusual combination of properties for thickening textile printing pastes. Although it is a hydrophilic gum, it is also soluble in a number of organic solvent systems.

The adhesive coatings now used in carpet backsizing contain two primary ingredients, latex and filler. The latex may be a modified rubber used with a clay filler or a synthetic, such as a styrene–butadiene latex, used with a calcium carbonate (whiting) filler. Additional ingredients include thickeners, surfactants, dispersants, and antifoaming agents. Thickeners are used primarily as viscosity control agents to control the positioning of the adhesive on the fabric, and the penetration of the adhesive into the tufts and jute. The thermally gelling, thickening, and surfactant properties of hydroxypropylmethylcellulose are used to achieve quick-set and improved production rates.

11. Mold-Release Coatings

Because methylcellulose gums melt only at high temperatures, they have found many applications as mold-release agents. The thermal gelation of the methylcelluloses also contributes to their effectiveness. It is possible to spray such solutions into hot surfaces without having them run.

12. Paper

The high strength, flexibility, clarity, oil and wax resistance, and binding power of methycellulose films have proved particularly useful to the paper industry in a wide range of sizing and coating applications. Methylcellulose sizing solutions are easily applied with conventional equipment, such as size presses, calender stacks, and off-machine coaters. These coatings are tasteless and odorless and normally do not require preservatives. Equally suitable for food packaging or industrial packaging, methylcellulose is also used as an adhesive for a number of paper products.

Light surface sizings of methylcellulose are especially effective in resisting the penetration of gloss-ink vehicles, varnishes, lacquers, and waxes. Paper board treated at the calendar stack in this way yields a surface that is characterized by high gloss, low mottling, and low ink consumption when printed with high-gloss inks. Reduction of wax penetration in waxed boards serves both to maintain brightness of the liner and to reduce wax consumption. Mills have reported that savings in wax costs alone more than pay for the treatment in which a level of whiteness not otherwise obtainable results.

Paper, as well as board, is frequently treated in this way for improved lacquering and waxing performance. Carbonizing tissue requiring particularly high resistance to strike-through presents a typical example in which the sizing properties imparted by methylcellulose can be utilized. Continuous films and coatings of methylcellulose are impervious to practically all types of greases, waxes, oils, and solvents. This resistance is unimpaired by aging or temperature.

Methylcellulose is an excellent parting agent. Methylcellulose coatings as parting agents are also valuable as linings for fiber drums and containers for hot asphalt or other tacky materials that tend to adhere to the container wall.

Because calender stacks generally operate at 62° or higher, hydroxypropylmethylcellulose (90° gel point) is recommended for use on this equipment. The coatings are generally applied at a rate of 0.03–0.3 lb/1000 ft^2 (0.15–1.5 g/m^2) and are suitable for gloss ink wax holdout. When appreciable grease resistance is required, coatings of 1.5–3.0 lb/1000 ft^2 (73–150 g/m^2) are generally used.

IV. Properties

1. *Powder*

The properties of methylcellulose and of hydroxypropylmethylcellulose are a function of the type of and the degree of substitution of methoxyl and other modifying substituent groups. By controlling these, it is possible to modify such properties as surface activity, solubility in organic solvents, and compatibility with additives. By controlling the intrinsic viscosity, it is possible to obtain a wide range of thickening power at a given concentration. In describing the viscosities of the methylcelluloses, the absolute viscosities in centipoises of 2% aqueous solutions, measured at 20°, are used. Chemical compositions are given in Table I and physical properties, in Table II.

2. *Solutions*

A cellulose ether is most soluble in solvents that best match the ether in cohesive energy density.[81] As a cellulose ether is prepared, the gradual increase in degree of substitution is accompanied by a progression from insoluble cellulose through solubility in aqueous alkali, to solubility in water, and finally to solubility in water–alcohol or hydrocarbon–alcohol mixtures. Methylcellulose is soluble in cold water at 1.3–2.6 DS and increasingly soluble in ethanol about 2.1 DS. Ethoxyl modification results in solubility at a lower total DS (methoxyl DS, 0.4; ethoxyl DS, 0.9).[82] Hydroxyalkoxyl substitution leads to cold-water

TABLE I

Chemical Compositions of Methylcelluloses and Hydroxypropylmethylcelluloses[a]

Methoxyl, %	Hydroxypropoxyl, %	Soluble in:	Nominal Gel Temperature, °C
27.5–32.0	0	water	54–56
26–30	7–12	water and organic solvents[b]	60
27–29	4.0–7.5	water	65
19–24	4–12	water	70–90

[a] Courtesy of The Dow Chemical Company, Midland, Michigan.
[b] Thermoplastic.

solubility above 1.5 total DS and increasing solubility in ethanol at a slightly higher substitution.

Swelling is considered to be solution of the solvent in the ether. There is little difference between a swollen gel and a solution. Ethers of a given substitution are more soluble, the lower their intrinsic viscosity; degraded cellulose ethers are more soluble than are undegraded ethers of the same kind. Solubility

TABLE II

Physical Properties of Methylcelluloses

Color	White
Apparent density, g/cm^3	
4000 cps	0.35 ± 0.06
400 cps	0.41 ± 0.05
15 cps	0.46 ± 0.04
Charring temperature, °C	
Methylcellulose	300–305
Hydroxypropylmethylcellulose	295–300
Particle size	95% through 40 mesh
Moisture, as packaged	3% maximum
Moisture, 2 weeks open exposure at 20° and 75% relative humidity	11%

of the more nearly crystalline, or less substituted, portions of the cellulose chain is greater near 0° than at higher temperatures. Chilling appears to aid hydration of the hydroxyl groups of cellulose, and water solubility appears to stem from the wedging apart of the cellulose chains by the substituent groups so that the remaining hydroxyl groups are available for hydration.[83] Bulky substituent groups permit water solubility at lower substitutions than do smaller groups, although water solubility occurs at about the same weight percent of added substituent in all cases.

Methylcellulose gums, although substituted and, in the case of powdered materials, gelled, retain the morphology of the original cellulose. The cellulose was presumably laid down in layers, and the methylcellulose gums dissolve by successive hydration of these layers. Optimum solution rates and solution quality can only be obtained by adhering to recommended methods of solution preparation. Failure to follow such procedures will lead to lumping and to slow solubility rates. Methylcellulose gums are insoluble in hot water, so if hot water is used first to wet all portions of the methylcellulose, the latter will dissolve rapidly in cold water. If, however, cold water is used first, a gelatinous layer will form on the outside of the particles and lumping and slow diffusion of water into the interior of the particles will result. A methylcellulose is dissolved most easily by first mixing it with 20–50% of the required water heated to 80°–90° and allowing it to wet out for a short time. The rest of the water can then be added as cold water or as ice. The mixture is stirred until smooth. Cooling will yield solutions of maximum clarity. Alternatively, the methylcellulose can be slurried in high-purity methanol, ethanol, or low-molecular-weight glycols. The slurry will dissolve readily in cold water. The methylcellulose can be dry-blended with other powders or salts so that the resultant dispersion will have an improved solubility rate in cold water. Slurries in hot water or in salt or sugar solutions can be used.

Improvement in the handling and dissolution of powdered methylcelluloses may be obtained by treating the material with glyoxal in order to obtain temporary crosslinking and resultant insolubilization. The product may be dispersed at high concentrations at an acid pH but develops viscosity rapidly when the pH is made alkaline and the temporary crosslinking is destroyed by a Cannizzaro reaction.

The properties that best describe a cellulose ether are its degree of substitution and its viscosity. The viscosities of solutions of finite concentration determine commercial utility. From a theoretical point of view the limiting or intrinsic viscosity better describes the average length of the molecule. The intrinsic viscosity of methylcellulose in water is related to its 2% solution viscosity in water at 20°, the conditions used for grading methylcellulose gums. Figure 1 shows the relation of solution viscosity to concentration in water at 20° for

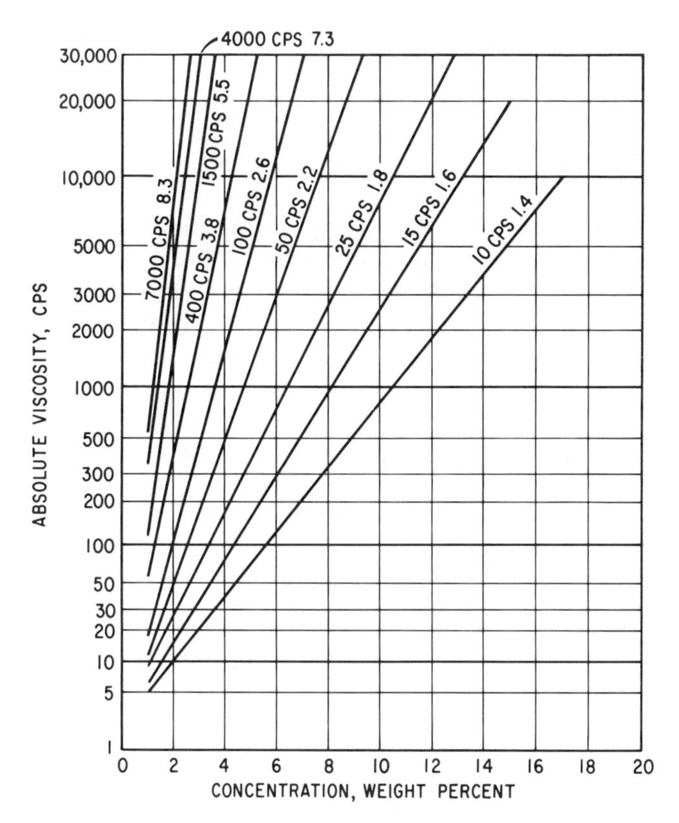

FIG. 1.—Viscosity–concentration chart for methylcelluloses. Notations on lines are the viscosity in centipoises of a 2% solution in water at 20° and the intrinsic viscosity at 20°. (Courtesy of The Dow Chemical Co.)

methylcellulose gums having intrinsic viscosities of 1.4–8.3. The difference in intercepts between types is caused by the processing of the granular form.

Constants suitable for the conversion of intrinsic viscosity to degree of polymerization (DP) or to molecular weight are a matter of personal opinion because suitable data on samples closely fractionated to uniform DS and uniform DP have not yet been prepared, although methylcelluloses have been fractionated.[84, 85]

Staudinger[86] proposed an expression for the relation of intrinsic viscosity to weight-average molecular weight, which for convenience may be written as $[\eta] = K_m M$, when M is the molecular weight and K_m is a constant of proportionality. For methylcellulose,[87] approximately, $[\eta] = 2.8 \times 10^{-4} M^{0.63}$. Thus, if intrinsic viscosities range from 1.4 to 7.3, DP's of unfractionated commercial methylcelluloses can, in the authors' opinion, range in order of magnitude from 170 to 900 and molecular weights in order of magnitude from 30,000 to 170,000. It remains to be shown by careful experimental work whether these estimates of magnitude hold in the case of closely fractionated methylcelluloses. Any estimate of DP or molecular weight should report the intrinsic viscosity as well (see Fig. 2).

The degree of substitution of the methylcellulose and the regularity of distribution of the substituent groups affect the solubility and utility of a cellulose ether. Spurlin[88] proposed the hypothesis that all hydroxyl groups have an equal chance to react and that the distribution of substituents in a cellulose ether is random and can be calculated by statistical methods. Mahoney and Purves[89] and Timell[90] have investigated substituent group distribution in methylcellulose gums. By comparing the actual frequency of 2,3-glycol groups with the calculated frequency of such groups, they found that methylation yields

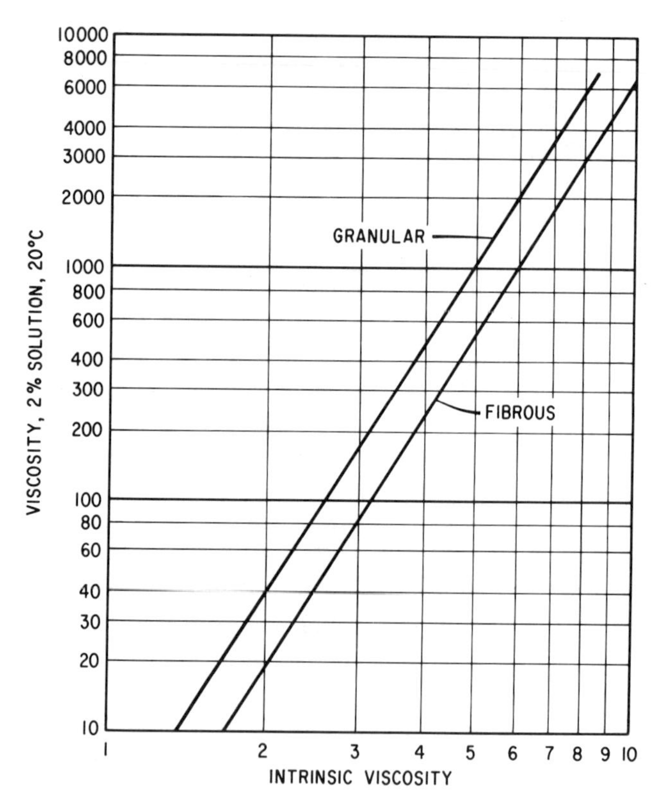

FIG. 2.—Relation of viscosity, in centipoises, of 2% solutions of methylcellulose in water at 20° to intrinsic viscosity. (Courtesy of The Dow Chemical Co.)

heterogeneous products. However, the products from methylation with methyl chloride are more nearly homogeneous than those from dimethyl sulfate, and homogeneity is approached with increasing degree of substitution in the water-soluble range.

Homogeneously-prepared, water-soluble methylcellulose gums dissolve to give aqueous solutions that can be heated almost to their boiling points without gelation. These ethers have been prepared from alkali cellulose under special research conditions. Such preparations are water soluble at lower degrees of substitution than are needed to make the corresponding heterogeneously prepared ethers soluble in water. It is believed that the failure of their solutions to gel is caused by lack of trisubstitution in the case of methylcellulose[91] and modified methylcellulose gums (Fig. 3) or to the polarity of the special substituents. However, gelation may be caused by a hydrophobic substituent (Fig. 4).

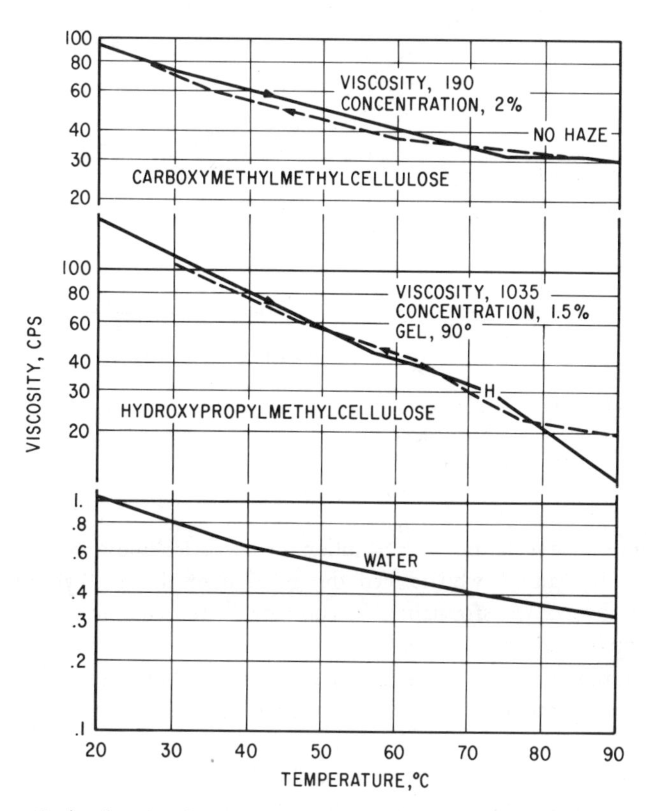

FIG. 3.—Brabender Amylograph viscosity *vs.* °C for carboxymethylmethylcellulose (Methocel CAM) and hydroxypropylmethylcellulose (Methocel 90 HG). Cooling line is broken. Water is shown for comparison. (Courtesy of The Dow Chemical Co.)

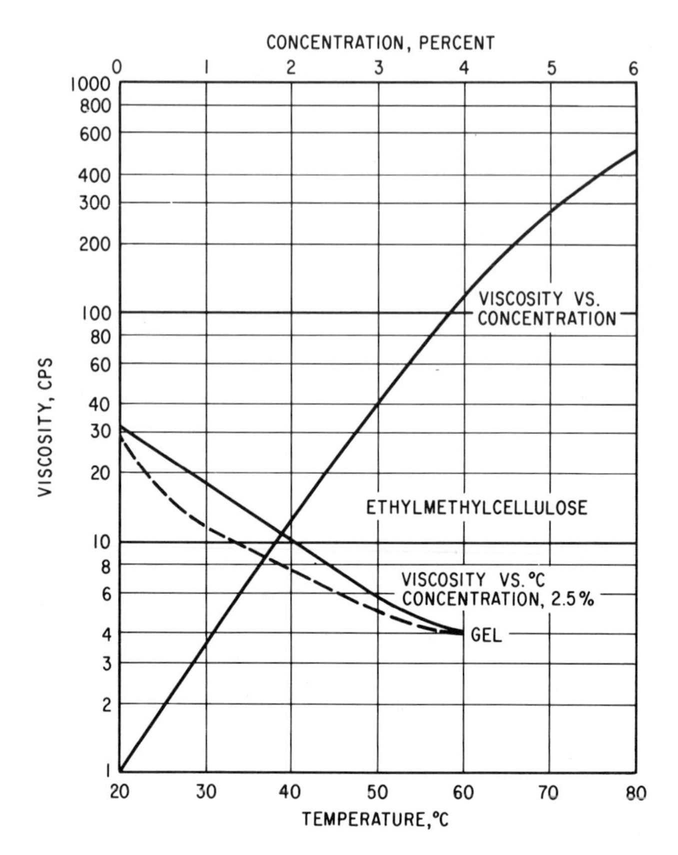

FIG. 4.—Ethylmethylcellulose (Edifas A): upper abscissa, viscosity *vs.* concentration at 22°; lower abscissa, viscosity *vs.* °C. Cooling line is broken. (Courtesy of Nobel Division, Imperial Chemical Industries, Ltd.)

The reversible gelation of a solution of methylcellulose of DS 2.22 was studied by Heymann,[92] who termed the gelation of this methylcellulose upon heating an inverse transformation in contradistinction to the gelation of agar or gelatin upon cooling.

Temperature–viscosity relationships for methylcellulose gums (Figs. 3 and 5) over a wide range of intrinsic viscosity and substitution were studied by Savage,[3] who used the Brabender Amylograph, a constant-speed torsional viscometer with elaborate temperature control. This work indicated that the water solubility of water-soluble cellulose ethers depended upon uniform distribution of low-substituted or unsubstituted units throughout the ether. These units hold water, which serves as a lubricant, and the viscosity parallels that of water but

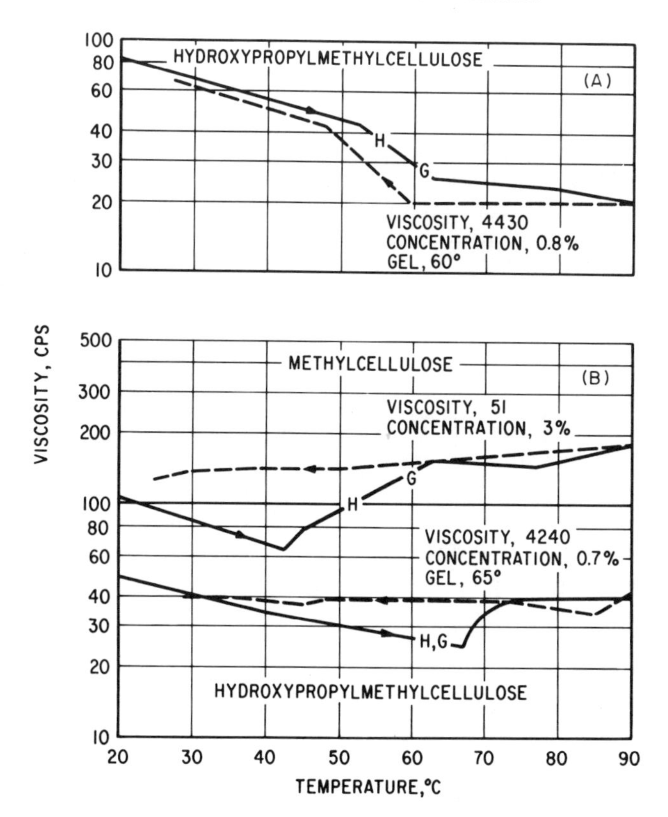

FIG. 5.—Brabender Amylograph viscosity *vs.* °C for methylcellulose (Methocel MC) and hydroxypropylmethylcelluloses (Methocel 65 HG and 60 HG). Cooling line is broken. (Courtesy of The Dow Chemical Co.)

is higher by the degree of chain interference. When the temperature is raised, the more highly substituted or less polar units tend to lose water, slippage is less, and association and aggregation can occur.

The viscosity of a solution can be increased at lower temperatures by residual crystalline components of the initial cellulose and at higher temperatures by association, aggregation, and, perhaps, crystallization of the more highly substituted units.

The gelation temperature decreases with increasing concentration of methylcellulose of a given intrinsic viscosity. At a given concentration, the gel temperature decreases with increasing intrinsic viscosity of the cellulose ether.

3. *Surface Activity*

Methylcellulose reduces the surface tension and the interfacial tension values of aqueous systems. This property results in effective emulsification in a two-phase system and moderate foaming in a single-phase water system. Should foaming be undesirable, several defoamers are available. Among these are Nopco KFS (Nopco Chemical Co.), Anti-foam A, AF, or B (Dow Corning Corp.), Polyglycol P–1200 (The Dow Chemical Co.), lauryl alcohol and tri-*n*-butyl phosphate. The concentration should be kept to the minimum required.

The properties of aqueous solutions of methylcellulose are summarized in Table III.

TABLE III

Properties of Aqueous Solutions of Methylcelluloses[a]

Property	Value
Surface tension[b]	
Methylcellulose	47–53 dynes/cm (25°)
Hydroxypropylmethylcellulose	44–56 dynes/cm (25°)
Interfacial tension (paraffin oil)	
Methylcellulose, 65° gel temp.	19–23 dynes/cm (25°)
Hydroxypropylmethylcellulose, 60° gel temp.	18–19 dynes/cm (25°)
Hydroxypropylmethylcellulose, 90° gel temp.	26–28 dynes/cm (25°)
Bulking density	
Methylcellulose	0.725 cc/g (0.087 gal/lb)
Hydroxypropylmethylcellulose, 65° gel temp.	0.734 cc/g (0.088 gal/lb)
pH	Neutral
Solution stability	Stable from pH 2 to pH 12 at 20°
Freezing point, all types 2%	0.0°
Specific gravity (20°/4°) (all types)	
1%	1.0012
5%	1.0117
10%	1.0245
Refractive index (2%), n_D^{20}	
Methylcellulose	1.3355
Hydroxypropylmethylcellulose, 65° gel temp.	1.3368

[a] Courtesy of The Dow Chemical Company, Midland, Michigan.

[b] For solutions below 500 cps. In many cases, concentrations of 0.001% will give the same value as 1% solutions.

4. *Compatibility*

Methylcellulose is nonionic, so it is not precipitated from its solutions as an insoluble salt if multivalent metal ions are added. In common with other colloidal materials, however, methylcellulose products can be gelled or salted from solution when the concentration of electrolytes or other dissolved materials exceeds certain limits. Thus, Heymann[92] found that soluble inorganic salts usually lower the gelation temperature of the solution to which they are added because they decrease the hydration of the cellulose ether. The effect of salt addition on gelation is similar to the effect of heating, in that tolerance for added materials decreases with increasing concentration or with increased intrinsic viscosity of the methylcellulose. The salt tolerance is dependent upon the ions present; thus, tolerance is higher for thiocyanates and iodides than for other ions in the order of the lyotropic series: phosphate > sulfate > tartrate > acetate > chloride > nitrite > nitrate. Salt tolerance is increased by modifying methylcellulose with sodium carboxymethoxyl substitution or alkoxyl substitution; thus, the tolerance is higher for hydroxypropylmethylcellulose products than for unmodified methylcellulose. Table IV gives the concentrations of additives that will cause 2% solutions of methylcellulose to gel.

TABLE IV

Grams of Additive Tolerated by 100 ml of a 2% Solution without Salting Out[a]

Additive	Methylcellulose		Hydroxypropyl-methylcellulose	
	15 cps	4000 cps	50 cps	4000 cps
NaCl	11	7	17	11
$MgCl_2$	11	8	35	25
$FeCl_3 \cdot 6H_2O$	190	100	200	135
Na_2SO_4	6	4	6	4
$Al_2(SO_4)_3 \cdot 18H_2O$	6	5	8	7
Na_2CO_3	4	3	5	4
$Na_3PO_4 \cdot 12H_2O$	9	6	9	8
Sucrose	67[b]	38	67	67

[a] Courtesy of The Dow Chemical Company, Midland, Michigan.
[b] Maximum solubility of sucrose in water at 25°.

5. *Sheeting*

High-strength, water-soluble sheeting is prepared from formulations of methylcellulose gums. This sheeting is clear, smooth, and impervious to oils, greases, and many organic solvents. Its properties are shown in Table V. In order

TABLE V

Properties of Sheeting[a]

Properties	Methylcellulose (Unmodified)	Hydroxypropylmethylcellulose (Plasticized with 30 Voranol CP 260[b]
Specific gravity	1.29	1.2330
Area factor, in^2/lb/mil	21,400	22,400
Melting point, °C	290–305	240–260
Moisture vapor transmission rate at 100° F, 90–100% R.H.	67.5 g/100 in^2/24 hr/mil	65 g/100 in^2/24 hr/mil
Oxygen transmission rate at 75° F	25 cm^3/100 in^2/24 hr/mil	70 cm^3/100 in^2/24 hr/mil
Tensile strength, 75° F, 50% RH	600–800 kg/cm^2 (8500–11,400 lb/in^2)	700 kg/cm^2 (10,000 lb/in^2)
Elongation, 75° F, 50% RH	10–15%	45–50%
Stability to ultraviolet	excellent	excellent
Resistance to oils and most solvents	excellent	excellent
Mullen Burst	23.9 psi/mil	17.9 psi/mil
MIT double-fold, 2 mil film		
Load	12,000	11,000
No Load	32,000	25,000
Ultraviolet transmission, 2-mil film		
400/nm	54.6%	65.7%
290/nm	49.0%	55.3%
210/nm	25.7%	24.3%
Equilibrium moisture content	6.5% water	5–6% water

[a] Properties of only two film formulations are given. Others can be made. Courtesy of The Dow Chemical Company, Midland, Michigan.

[b] Voranol CP 260 is the registered trademark of a hydroxypropylglycerol product manufactured by The Dow Chemical Company.

to increase production rates, sheeting of organic solvent-soluble hydroxypropylmethylcellulose can be cast from such solvent systems as 1 : 1 v/v methanol–benzene or 1 : 1 v/v methanol–water. Sheeting of this material will dissolve in water more rapidly than will sheeting of methylcellulose. Sheeting of hydroxypropylmethylcellulose can be heat sealed at 168° or electronically sealed at 27–40 megacycles.

Plasticization.—The mechanical properties of sheeting or plastic depend upon the plasticization, either internal by the substituent group or external by an added plasticizer, and upon the intrinsic viscosity of the material. For greater flexibility, various plasticizers can be used with methylcellulose to obtain up to 80% elongation.

Compatibility.—Methylcellulose can be combined with a wide range of materials, such as starches, glues, soaps, and other water-soluble gums, in coatings and sheeting. Methylcellulose is also compatible with many water-soluble resins. A maximum of 40% of oxidized and chlorinated starches and dextrin can be incorporated in a methylcellulose film. Methylcellulose sheeting and coatings are unaffected by animal and vegetable oils or by greases and petroleum hydrocarbons. The resistance to most organic solvents is excellent. Some of the oils and organic solvents that have no effect upon methylcellulose sheeting are listed below:

Peanut oil	Hydrogenated vegetable oil
Machine oil	Toluene
Petroleum ether	Castor oil
Mineral oil	Animal fat
Cyclohexane	Benzene

Thermoplasticity.—Because the water-soluble natural and synthetic gums lack a melting point in the thermoplastic range and because it has been difficult to find a satisfactory plasticizer–solvent for these materials, the use of either of these classes of gums in the techniques of solvent-casting and thermoplastic fabrication has been negligible. The thermoplastic properties of a cellulose ether, however, are an extension of its solution properties. Thermoplasticity depends upon the nature of the substituent group, upon the degree of substitution, and upon the chain length of the cellulose as measured by the intrinsic viscosity. Modification of methylcellulose with alkene oxides yields products that tend toward organic solvent solubility and thermoplasticity.

New utility for synthetic water-soluble gums has resulted from the synthesis of hydroxypropylmethylcellulose with a 60° gel point. This modified methylcellulose combines solubility in certain hot and cold organic solvents with the hydrophilic properties of methylcellulose. By using the proper techniques, it is possible to extrude, injection mold, compression mold, and hot cast this water-soluble cellulose derivative using standard plastic fabricating equipment. In addition, this material can be used to prepare water-soluble sheeting in organic solvent-casting systems. All the thermoplastic techniques are based upon the fact that the derivative is soluble in certain hot organic solvents. When such solutions are cooled, a gel forms. The types of compounds that are solvents are definitely restricted to certain classes and chemical structures.[93] Typical solvents include propylene glycol, polyethylene glycols up to a molecular weight of 600, glycerol acetates, ethylene glycol monoethyl ether (Methyl Cellosolve, Dowanol EM), the ethanolamines, and ethylene carbonate.

Extrusion and molding.—The thermoplasticity and organic solvent solubility of hydroxypropylmethylcellulose (60° gel point) permit forming techniques that are not normally a part of water-soluble gum technology. By the proper choice of a hot organic solvent, it is possible to extrude or compression mold the derivative in the form of sheeting. The dry mixes can be compounded in a ribbon-type blender at room temperatures and handled satisfactorily by feeders designed for powders. Optimum feeder performance is obtained, however, by predensification either on press rolls or by pelleting. The extrusion or molding temperatures required range from 120° to 190°, depending upon the composition of the dry mix. The proper choice of these solvents also imparts heat-sealing properties to sheeting so that water-soluble bags and envelopes can be fabricated on standard packaging machinery.

Hydroxypropylmethylcellulose is useful when quick drying coatings, applied from organic systems but removable by water, such as parting compounds and printing pastes, are desired. Silk screen printing techniques have been improved through the use of multilayer sheeting prepared from the derivative.[94]

Increasing water resistance.—Methylcellulose can be rendered virtually insoluble by condensation reactions that involve the free hydroxyl groups of the D-glucopyranosyl units. The formation of a few cross-links between chains is sufficient to render methylcellulose coatings and sheeting insoluble. Substances capable of performing this condensation reaction include urea–formaldehyde resins, melamine–formaldehyde resins, polybasic acids, and dialdehydes.

6. *General*

Physiological.—The list of food additives that are usually considered safe[95] includes methylcellulose U.S.P. (F.D.A. Regulations, Section 121.101) and hydroxypropylmethylcellulose N.F. (Section 121.1021). These materials also meet the specifications of the United States Food Chemicals Codex.[96] Additional regulations permit the use of these materials in standardized foods, such as salad dressing, in the types of coatings used in food packaging, and (Meat Inspection Division) as a carrier and to stabilize meat and vegetable patties under regulation.

Methylcellulose and hydroxypropylmethylcellulose are listed as permitted emulsifying agents by the food Standards Committee of the Ministry of Agriculture, Fisheries and Food of Great Britain.

Monographs on methylcellulose and hydroxypropylmethylcellulose are in cluded in the FAC/WHO Codex Alimentarius.

Stability.—Degradation of the methylcelluloses leads primarily to cellulose chain breakage and resultant intrinsic viscosity reduction. Hydrolysis of methylcellulose and identification of the products was a major part of the classical work that elucidated the structure of cellulose.[1] The positions of the hydroxyl groups in cellulose were determined by methylation, methanolysis and/or hydrolysis, and identification of the product. The corresponding methylated D-glucoses were prepared by synthesis.[97] The results showed that the maximum number of hydroxyl groups on a D-glucopyranosyl unit of cellulose is three and that, during hydrolysis, the methoxyl groups retain their positions. Hydroxyl groups and corresponding methoxyl groups are found in the C-2, C-3, and C-6 positions.

Homogeneous acid-catalyzed hydrolysis of methylcellulose was studied by Gibbons,[98] who found that the kinetics in the range of measurement were essentially pseudo first-order but that plots of the rate constants against acidity were not linear. Factors effecting the rates of acid-catalyzed hydrolysis of polysaccharides have been reviewed.[99, 100] Decrease in DP is accompanied by an increase in reducing power.

Methylcellulose is stable under alkaline conditions in the absence of oxygen but will undergo intrinsic viscosity reduction when oxygen is present. This process, which is similar to the aging of alkali cellulose before methylation, does not decrease the intrinsic viscosity as much as does acid-catalyzed hydrolysis. It is accelerated by such oxygen carriers as manganese and cobalt compounds. Formation of carboxyl and carbonyl groups occurs. The type of attack depends upon the oxidizing agent and upon the conditions.

When methylcellulose is heated to 135° in a stream of oxygen, it does not appear to take up oxygen, but intrinsic viscosity decreases. Under such conditions, the ethoxyl groups of an ethylmethycellulose will undergo oxidative attack by a free-radical mechanism.[101] The effect of oxidation upon methylcellulose modified with alkene oxides is not known.

Methylcelluloses of DS 2 do not support the growth of bacteria that elaborate cellulolytic enzymes. However, if such enzymes are introduced into a solution or produced therein, attack can occur. Such enzymolysis causes viscosity loss and can lead ultimately to association at room temperature as the result of the removal of low-substituted components of the methylcellulose.[3] Enzymolysis of methylcellulose was studied by Husemann[102] and by Reese[103–105] and coworkers. Uniform substitution confers resistance to attack, but residual components of the initial cellulose can serve as focuses for cellulolytic attack. Furthermore, if the nature of the substituent group, through size or polarity, is such that it permits solubility at such a low DS that a considerable number of unsubstituted D-glucopyranosyl units are present, the latter can be attacked.

Analysis.—The instrument most used for the measurement of methylcellulose solution viscosity is the Ubbelohde viscometer.[106] The Brookfield rotational viscometer is used to measure the viscosity of certain thick pastes. The Brabender Amylograph has been used to study gelation of dilute solutions and the properties of heavy pastes in food uses.

Methoxyl and ethoxyl contents are determined by Samsel and McHard's modification of the Zeisel method;[107] hydroxyethoxyl, by the method of Morgan[108] and hydroxypropoxyl, by a modification of the chromic acid oxidation method of Lemieux and Purves.[109]

Cellulose ethers can be recognized in mixtures by Black's method,[110] which applies the anthrone reagent of Dreywood,[111] or by the use of diphenylamine.

V. References

(1) The history of cellulose and classical experiments that led to the present understanding of its structure are reviewed by C. B. Purves, *in* "High Polymers," Vol. 5, "Cellulose and Cellulose Derivatives," E. Ott, H. M. Spurlin, and M. W. Grafflin,eds., Interscience, New York, 2nd Ed., 1954, pp. 54–59.

(2) The general principles of cellulose etherification are reviewed by A. B. Savage, A. E. Young, and A. T. Maasberg, *in* "High Polymers," Vol. 5, "Cellulose and Cellulose Derivatives," E. Ott, H. M. Spurlin, and M. W. Grafflin, eds., Interscience, New York, 2nd Ed., 1954, pp. 882–958.

(3) A. B. Savage, *Ind. Eng. Chem.*, **49,** 99 (1957).

(4) Reactivity of the hydroxyl groups in cellulose is reviewed by H. M. Spurlin, *in* "High Polymers," Vol. 5, "Cellulose and Cellulose Derivatives," E. Ott, H. M. Spurlin, and M. W. Grafflin, eds., Interscience, New York, 2nd Ed., 1954, pp. 673–712.

(5) J. Chédin and A. Tribot, *Mem. Serv. Chim. État (Paris)*, **33,** 168 (1947).

(6) E. J. Lorand, *Ind. Eng. Chem.*, **31,** 891 (1939).

(7) S. L. Bass, U.S. Patent 2,145,855 (1939); *Chem. Abstr.*, **33,** 3150 (1939).

(8) W. R. Collings, L. DePree, and M. H. Weymouth, U.S. Patent 2,143,863 (1939); *Chem. Abstr.*, **33,** 3150 (1939).

(9) W. R. Collings, L. DePree, and M. H. Weymouth, U.S. Patent 2,142,862 (1939); *Chem. Abstr.*, **33,** 3586 (1939).

(10) A. T. Maasberg, U.S. Patent 2,160,782 (1939); *Chem. Abstr.*, **33,** 7563 (1939).

(11) R. W. Swinehart and A. T. Maasberg, U.S. Patents 2,331,864 and 2,331,865 (1943); *Chem. Abstr.*, **38,** 1693 (1944).

(12) The laboratory use of dimethyl sulfate is described by E. Heuser, *in* "The Chemistry of Cellulose," John Wiley, New York, 1944, pp. 379–391.

(13) H. Staudinger and O. Sweitzer, *Ber.*, **62,** 2345 (1929).

(14) D. Traill and S. Brown, *FIAT, Final Rep.* No. **486** (January 10, 1946), through Library of the U.S. Dept. Agr., Washington, D.C.

(15) M. Hagedorn and E. Rossback, U.S. Patent 1,877,856 (1932); *Chem. Abstr.*, **27,** 413 (1933).

(16) A. B. Savage, U.S. Patent 2,831,852 (1958); *Chem. Abstr.,* **52,** 13261 (1958).
(17) A. B. Savage, U.S. Patent 2,835,666 (1958); *Chem. Abstr.,* **52,** 21101 (1958).
(18) A. B. Savage, U.S. Patent 2,949,452 (1960); *Chem. Abstr.,* **53,** 22948 (1959).
(19) G. A. Bergy, *Amer. Prof. Pharm.,* **5,** 691 (1939).
(20) M. A. Lesser, *Drug Cosmet. Ind.,* **62,** 612, 670, 692, 750, 830 (1948).
(21) E. Bactowsky and J. Presto, *Bull. Amer. Soc. Hosp. Pharm.,* **7,** 65 (1950).
(22) J. Stawitz, *Pharm. Ind.,* **12,** 39, 71, 90 (1950).
(23) "A Bibliography of Pharmaceutical and Medical Uses of Methocel," The Dow Chemical Co., Midland, Michigan, March 1958.
(24) M. L. Tainter, *Proc. Soc. Exp. Biol. Med.,* **54,** 77 (1943).
(25) K. Schweig, *N.Y. State J. Med.,* **48,** 1822 (1948).
(26) J. A. Bargen, *Gastroenterology,* **13,** 275 (1949).
(27) V. H. Musick, *J. Okla. State Med. Ass.,* **43,** 360 (1950).
(28) A. L. Monaco and E. J. Dehner, *J. Amer. Pharm. Ass., Sci. Ed.,* **44,** 237 (1955).
(29) R. C. Batterman, *Ann. N.Y. Acad. Sci.,* **58,** 520 (1954).
(30) C. W. Bauer and L. A. Wasson, *J. Amer. Pharm. Ass., Pract. Ed.,* **10,** 296 (1949).
(31) A. D. Jonas, *Amer. Pract. Dig. Treat.,* **1,** 933 (1950).
(32) H. Clemens and H. W. Read, *Australas. J. Pharm.,* **20,** 191 (1939).
(33) J. G. Downing, L. M. Ohmart, and M. J. Stoklosa, *Arch. Dermatol. Syphilol.,* **50,** 8 (1944).
(34) R. H. Draeger and M. Pijoan, *J. Amer. Med. Ass.,* **128,** 1096 (1945).
(35) K. E. Avis, *Amer. J. Pharm.,* **119,** 271 (1947).
(36) R. E. M. Davies, *Pharm. J.,* **160,** 82 (1948).
(37) J. Cohen, G. Klein, K. Ross, and E. P. Guth, *J. Amer. Pharm. Ass., Pract. Ed.,* **14,** 292 (1953).
(38) H. K. Stephens, *Chem. Prod.,* **16,** 299 (1953).
(39) W. J. Campbell, D. H. M. Burlage, and W. R. Lloyd, *J. Amer. Pharm. Ass., Pract. Ed.,* **16,** 38 (1955).
(40) R. E. Singiser and H. H. Hutchins, *J. Amer. Pharm. Ass., Pract. Ed.,* **16,** 477 (1955).
(41) K. L. Pickrell, *Bull. Johns Hopkins Hosp..* **71,** 304 (1942).
(42) K. L. Pickrell, *J. Amer. Pharm. Ass., Pract. Ed.,* **3,** 386 (1942).
(43) J. K. Dixon and R. L. Morgan, U.S. Patent 2,429,404 (1947); *Chem. Abstr.,* **42,** 1030 (1948).
(44) K. C. Swan, *Arch. Ophthalmol.,* **33,** 378 (1945).
(45) T. E. Obrig, *Arch. Ophthalmol.,* **38,** 668 (1947).
(46) G. L. Phillips, *Bull. Amer. Soc. Hosp. Pharm.,* **6,** 118 (1949).
(47) G. A. Bergy, *Amer. Prof. Pharm.,* **18,** 340, 367 (1952).
(48) R. R. Feinstein and A. W. Bechtold, U.S. Patent 2,703,777 (1955); *Chem. Abstr.,* **49,** 8569 (1955).
(49) A. E. Heming, U.S. Patent 2,611,730 (1952); *Chem. Abstr.,* **47,** 3528 (1953).
(50) M. M. Marks, *Amer. J. Surg.,* **81,** 6 (1951).
(51) R. E. Singiser, U.S. Patent 3,256,111 (1966); *Chem. Abstr.,* **65,** 5312, 8681 (1966).
(52) N. E. Brendamour, U.S. Patent 3,383,236 (1968).
(53) M. Bockmühl and W. Ludwig, U.S. Patent 2,061,544 (1936); *Chem. Abstr.,* **31,** 816 (1937).

(54) M. Bockmühl and W. Ludwig, Can. Patent 358,980 (1936); *Chem. Abstr.,* **30,** 6135 (1936).
(55) E. A. Ferguson, U.S. Patent 2,486,937 (1949); *Chem. Abstr.,* **44,** 5546 (1950).
(56) H. H. Bogin and R. D. Feick, U.S. Patent 2,555,463 (1951); *Chem. Abstr.,* **45,** 7753 (1951).
(57) Farbewerke Hoechst, A.-G., Brit. Patent 701,621 (1953); *J. Appl. Chem., Abstr.,* 585 (1954).
(58) N. E. Clarke, R. E. Mosher, and C. N. Clarke, *Circulation,* **7,** 247 (1953).
(59) W. F. Hamilton, M. F. George, Jr., and E. Simon, U.S. Patent 2,361,624 (1944); *Chem. Abstr.,* **39,** 2179 (1945).
(60) H. C. Dudley and T. G. Mitchell, *J. Amer. Pharm. Ass., Pract. Ed.,* **6,** 467 (1955).
(61) J. B. Gerberich, *J. Econ. Entomol.,* **38,** 393 (1945).
(62) A. Pijper, *J. Pathol. Bacteriol.,* **58,** 325 (1946).
(63) G. M. Lehrer, *Biol. Rev., Coll. City New York,* **8,** 23 (1946).
(64) A. Pijper, *J. Bacteriol.,* **53,** 257 (1947).
(65) K. A. Stiles and D. A. Hawkins, *Science,* **105,** 101 (1947).
(66) E. Solzman, M. E. Peck, and A. J. Neerken, *Radiology,* **58,** 209 (1952).
(67) G. Jacobson and K. A. Heitmann, *Radiology,* **62,** 241 (1954).
(68) G. K. Greminger, *Food Technol.,* **11,** No. 5, 20 (1957).
(69) C. W. Eddy, U.S. Patent 2,496,278 (1950); *Chem. Abstr.,* **44,** 3632 (1950).
(70) *Dow Western,* 5 (Jan. 1969).
(71) E. J. Rivoche, U.S. Patent 2,798,814 (1957).
(72) C. R. Nelson and R. E. Madison, U.S. Patent 2,600,018 (1952); *Chem. Abstr.,* **41,** 10640 (1952).
(73) G. B. Heijmer, U.S. Patent 2,700,615 (1955); *Chem. Abstr.,* **49,** 4258 (1955).
(74) H. B. Wagner, *Ind. Eng. Chem.,* **52,** 233 (1960).
(75) H. B. Wagner, U.S. Patent 2,820,713 (1958); *Chem. Abstr.,* **52,** 6576 (1958).
(76) H. B. Wagner and J. V. Fitzgerald, U.S. Patent 2,990,382 (1961); *Chem. Abstr.,* **56,** 4378 (1962).
(77) K. G. Cunningham and A. J. Baille, U.S. Patent 3,057,742 (1962).
(78) E. Anerein and H. R. Gloor, *Schweiz. Ton. Ind.,* **11,** 3 (1967).
(79) S. M. Rodgers, Jr., B. F. Wakeman, and A. B. Savage, U.S. Patent 3,388,082 (1968); *Chem. Abstr.,* **69,** 28673v (1968).
(80) J. C. Zola, U.S. Patent 2,591,904 (1952); *Chem. Abstr.,* **46,** 6404 (1952).
(81) H. M. Spurlin, *J. Polym. Sci.,* **3,** 714 (1948).
(82) L. Forshan, Nobel Division, Imperial Chemical Industries, Ltd., Stevenston, Ayrshire, Scotland, personal communication, August 27, 1957.
(83) F. D. Farrow and S. M. Neale, *J. Text. Inst.,* **15,** T157 (1924).
(84) R. Signer and P. von Tavel, *Angew. Chem.,* **50,** 902 (1937); **51,** 635 (1938).
(85) R. Steele and E. Pacsu, *Text. Res. J.,* **19,** 771, 784 (1949).
(86) H. Staudinger, "Die hochmolekularen organischen Verbindungen," Springer, Berlin, 1936, p. 32.
(87) K. Uda and G. Meyerhoff, *Makromol. Chem.,* **47,** 168 (1961).
(88) H. M. Spurlin, *J. Amer. Chem. Soc.,* **61,** 2222 (1939).
(89) J. F. Mahoney and C. B. Purves, *J. Amer. Chem. Soc.,* **64,** 15 (1942).
(90) T. Timell, "Studies on Cellulose Reactions," Esselte, Stockholm, Sweden, 1950.
(91) J. Vacher, *Chim. Ind. (Paris),* **43,** 347 (1940).

(92) E. Heymann, *Trans. Faraday Soc.,* **31,** 846 (1935); **32,** 462 (1936).

(93) G. K. Greminger, R. W. Swinehart, and A. T. Maasberg, *Ind. Eng. Chem.,* **47,** 156 (1955).

(94) E. G. Conley, U.S. Patent 2,731,356 (1956); *Chem. Abstr.,* **50,** 6070 (1956).

(95) *Fed. Regist.,* **24,** 21CFR, 6711 (August 19, 1959); **25,** 21CFR, 8498 (Sept. 17, 1960); **25,** 21CFR, 8949 (Sept. 17, 1960); **26,** 21CFR, 938–940 (Jan. 31, 1961).

(96) "Food Chemicals Codex," Publ. 1406, Natl. Acad. Sci.–Natl. Res. Council, Washington, D.C., 1st Ed., 1966, pp. 335–336, 432–433.

(97) J. C. Irvine, *J. Chem. Soc.,* **123,** 898 (1923).

(98) G. C. Gibbons, *J. Text. Inst.,* **43,** T25 (1952).

(99) J. N. BeMiller, *in* "Starch: Chemistry and Industry," R. L. Whistler and E. F. Paschall, eds., Academic Press, New York, Vol. 1, 1965, p. 495.

(100) J. N. BeMiller, *Advan. Carbohyd. Chem.,* **22,** 25 (1967).

(101) L. F. McBurney, *Ind. Eng. Chem.,* **41,** 1251 (1949).

(102) E. Husemann, *Papier,* **8,** 157 (1954).

(103) H. S. Levinson and E. T. Reese, *J. Gen. Physiol.,* **33,** 601 (1950).

(104) E. T. Reese, R. G. U. Siu, and H. G. Levinson, *J. Bacteriol.,* **59,** 485 (1950).

(105) E. T. Reese, *Ind. Eng. Chem.,* **49,** 104 (1957).

(106) L. Ubbelohde, *Ind. Eng. Chem., Anal. Ed.,* **9,** 85 (1937).

(107) E. P. Samsel and J. A. McHard, *Ind. Eng. Chem., Anal. Ed.,* **14,** 750 (1942).

(108) P. W. Morgan, *Ind. Eng. Chem., Anal. Ed.,* **18,** 500 (1946).

(109) R. U. Lemieux and C. B. Purves, *Can. J. Res., Sect. B,* **25,** 485 (1947). The unpublished modification by E. P. Samsel is known as Dow Method 200.

(110) H. C. Black, Jr., *Anal. Chem.,* **23,** 1792 (1951).

(111) R. Dreywood, *Ind. Eng. Chem., Anal. Ed.,* **18,** 499 (1946).

CHAPTER XXIX

HYDROXYALKYL DERIVATIVES OF CELLULOSE[a]

A. J. DESMARAIS

Hercules Incorporated, Wilmington, Delaware

I. INTRODUCTION

The principal commercial hydroxyalkyl derivatives of cellulose are water-soluble hydroxyethylcellulose and hydroxypropylcellulose. These polymers are nonionic cellulose ethers and their properties differ in many respects from those

[a] The technical information contained in this chapter, other than that acknowledged by specific literature references, has for the most part been developed in the laboratories of Hercules Incorporated, Wilmington, Delaware.

of anionic cellulose ethers, such as cellulose gum (sodium carboxymethylcellulose). These hydroxyalkyl derivatives of cellulose are readily soluble in water and their primarly use is to produce solutions having a wide range of viscosity. Such solutions are pseudoplastic; that is, they vary in viscosity depending upon the amount of stress applied.

Hydroxyethylcellulose is useful as a water thickener, protective colloid, binder, stabilizer, and suspending agent in a number of industrial applications, including pharmaceuticals, textiles, paper adhesives, decorative and protective collids, emulsion polymerization, and ceramics.

Commercially available hydroxypropylcellulose differs in that it is insoluble in hot water, above 40°–50°, and it has excellent solubility in a broad range of polar organic liquids. It is compatible with several waxes and oils, so it can also be used to modify the properties of these materals. Moreover, commercal hydroxypropylcellulose is thermoplastic and can be thermoformed into rigid articles or flexible film. Fillers, plasticizers, and other internal additives in a wide range of types and amounts can be used to modify the properties, reduce material costs, or improve solubility rates of thermoformed products.

Hydroxyalkylcelluloses are manufactured by reacting alkali cellulose with alkylene oxides (ethylene oxide or propylene oxide) at elevated temperatures and pressures. Each D-glucopyranosyl unit in the cellulose molecule has three reactive hydroxyl groups. The average number of hydroxyl groups per D-glucopyranosyl unit substituted in any reaction is known as the degree of substitution or DS. Theoretically, all three hydroxyl groups can be substituted, and the product from such a reaction would have a DS of 3.0. The reaction of ethylene oxide or propylene oxide with cellulose also leads to formation of new hydroxyl groups with which the alkylene oxide will react further to form side chains. The average number of moles of ethylene or propylene oxide that become attached to each D-glucopyranosyl unit in the two ways described is called the moles of substituent combined, or MS. Generally speaking, the MS controls the solubility and compatibility properties of these products; however, the DS is also a factor. Commercial, water-soluble grades of hydroxyethylcellulose have an MS value between about 1.6 and 3.0. Hydroxyethylcelluloses having an MS of 0.3–1.0 are soluble in 5–10% aqueous sodium hydroxide. These low MS products have been used as sizing agents in the textile industry and as the raw material for packaging film. Commercial hydroxpropylcellulose has an MS of greater than 2.0. Typical values range from 3.5 to 4.2. Techniques for determining the MS and DS are outlined in Section VII.

Because the solution viscosity, processability, and physical properties of these derivatives vary considerably depending upon their molecular weights, several grades are produced commercially. The degree of polymerization (DP) or number of D-glucopyranosyl units in the molecule varies from about 150 in the lowest molecular weight types to about 4000 in the highest. The products are normally characterized as to molecular weight type by the Brookfield viscosity[1, 2] of aqueous solutions (Table I). The values will vary from about 100 cps for 10% solutions of the lowest molecular weight types to about 4000 cps for 1% solutions of the highest molecular weight types. Many intermediate viscosity grades are available.

The viscosity ranges of solutions of hydroxypropylcellulose in anhydrous ethanol (Table II) demonstrate the solubility and thickening power of this derivative in polar organic solvents.

TABLE I

Viscosity Range of Aqueous Solutions of Hydroxyalkylcelluloses of Various Concentrations at 25°

	Centipoises			
Concentration, % by weight:	1	2	5	10
Viscosity type				
Extra high	2500–4500			
High	1500–2500			
Medium		4500–6500		
Intermediate types		1000–2000		
		150–400		
			150–400	
Low			60–150	
Extra low				200–300

TABLE II

Viscosity Range of Solutions of Hydroxypropylcelluloses of Various Concentrations in Anhydrous Ethanol at 25°

	Centipoises			
Concentration, % by weigtht:	1	2	5	10
Viscosity type				
High	1000–2500			
Medium		3000–6500		
Intermediate types		75–400		
			75–400	
Low			25–150	
Extra Low				100–300

Production of hydroxyethyl cellulose was started in the United States of America in about 1937. The volume has grown markedly and is a substantial portion of the 60 million pounds of noncarboxymethylated, water-soluble cellulose derivatives produced in the U.S. and reported in the latest U.S. Tariff Commission Report.

II. Preparation of Hydroxyalkylcellulose

In the preparation of water-soluble hydroxyethylcellulose, alkali cellulose is first made by treating cotton linters or wood cellulose high in alpha content with aqueous sodium hydroxide. This alkali cellulose is then reacted with ethylene oxide in the presence of a water miscible diluent, such as isopyropyl alcohol or *tert*-butyl alcohol.[3] In a typical preparation,[4] 52 parts of finely divided chemical wood pulp are dispersed into 450 parts of isopropyl alcohol. Then 126 parts of 22% aqueous caustic (sodium hydroxide) are added. After 1 hr, 51 parts of ethylene oxide are added and the dispersion is heated to 30° and held at that temperature for 1 hr. The temperature is then increased to 35° and held there for 3 hr. The reaction mixture is filtered to recover the product. The product is washed with a methanol–acetone solution, neutralized with acetic acid, and dried.

Modifications of the original processes have been made. For example, a two-stage process for the hydroxalkylation of cellulose has been used to obtain a more easily purified, salt-free product.[5] In the first stage, hydroxyethylcellulose of low MS is prepared and purified. This step serves to decrease the crystallinity of the cellulose so that the subsequent further reaction with ethylene oxide can be effected with much less caustic.

Products of controlled lower molecular weights can be prepared by proper selection of the cellulose raw material or by oxidative degradation using such reagents as hydrogen peroxide.[6]

Recent studies have shown that resistance[7,8] to enzymic degradation is related to the uniformity of substitution (DS) on the cellulose molecule.

Hydroxypropylcellulose can be prepared in a similar manner and purified by hot water washing.[9] The reaction can be accomplished in the absence of inert organic diluents.[10]

III. Physical Properties of Hydroxyethylcellulose

Hydroxyethylcelluloses having an MS greater than 1.6 are readily soluble in either hot or cold water.[1, 2] As with all water-soluble polymers, the particles have a tendency to agglomerate or lump when first wetted with water. Therefore, the time required to achieve complete solution of the thickener is usually governed by the degree of lumping that is allowed to develop during the solution process. The low-viscosity types of hydroxyethylcellulose are more rapidly soluble than the high-viscosity types. Solutions of high clarity can be prepared easily by one of several techniques, such as by (*a*) sifting the thickener into the vortex of vigorously agitated water slowly enough to prevent lumping, (*b*) prewetting the polymer with a water-miscible, nonsolvent, organic liquid before the powder is added to water, thereby separating the particles, or (*c*) dry blending with other dry materials that are to be added to the solution. The last technique effectively separates particles of hydroxyethylcellulose so that no lumping is experienced when the dry mix is added to water.

1. Fast Dissolving Grades

Fast dissolving, nonlumping grades of hydroxyethylcellulose are available. A chemical modification on the surface of the particles temporarily reduces the hydration rate of the polymer. This allows the particles to separate and form lump-free dispersions. Agitation should be used to make the dispersion and to prevent settling of the particles. Clear, smooth solutions are readily obtained within a short time. The rate of solution is affected by the temperature and the pH of the water (Fig. 1). Typical grades have a hydration time of 6 to 20 min at a pH of 7.0 and 25°.

2. Solubility in Organic Compounds

Water-soluble hydroxyethylcellulose is insoluble in most organic solvents. It swells or partially dissolves in a few of the polar solvents that are miscible with water (Table III).

3. Effect of Temperature on Solution Viscosity

The viscosity of solutions of hydroxyethylcellulose changes with tempera-

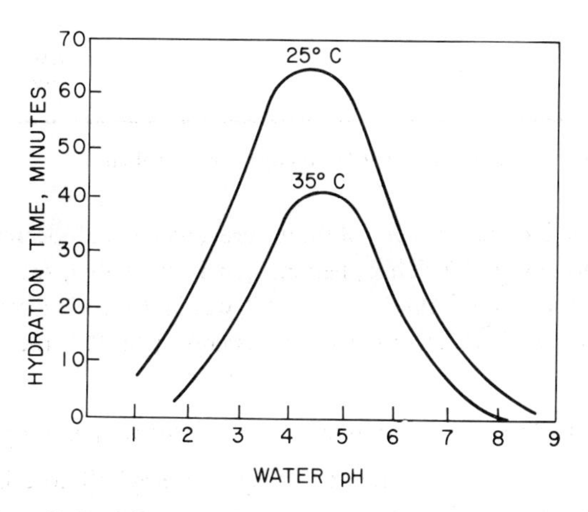

Fig. 1.—Effect of pH and temperature on the hydration time of fast dissolving grades of hydroxyethylcellulose.

TABLE III

Solubility of Hydroxyethylcellulose [a]

Alcohols	
Methanol	I
Ethanol	I
Ethanol–Water (70 : 30 by weight)	PS
Ethanol–Water (30 : 70 by weight)	S
1-Butanol	I
Aldehydes	
Butyraldehyde	PS
Formalin	S
Esters	
Ethyl acetate	I
Methyl salicylate	I
Ethers	
Cellosolve (2-Ethoxyethanol)	I
Ethyl ether	I
Acids	
Formic acid (88%)	S
Acetic acid (glacial)	I
Miscellaneous	
Methyl sulfoxide	S
Aniline	I
Mineral spirits	I
Hexane	I
Ethylene diamine	S
Ethylene glycol	SW
Glycerol	SW

[a] I, insoluble; S, soluble; PS, partly soluble; SW, swollen.

ture. Viscosities increase when solutions are cooled and decrease when solutions are warmed (Fig. 2.). This phenomenon is reversible; for example, a solution having a viscosity of 1200 cps at 25° would have a viscosity of approximately 200 cps at 80°. When the solution is cooled to 25°, the viscosity returns to 1200 cps.

4. *Effect of Concentration on Solution Viscosity*

The viscosity of aqueous solutions of hydroxyethylcellulose increases rapidly with concentration (Fig. 3). The rate of increase is not a linear relationship and differs for the varying molecular weight types.

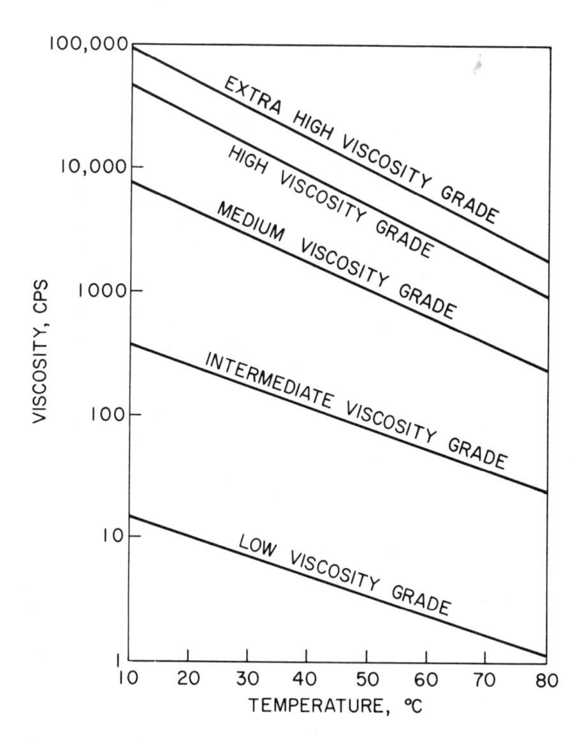

Fig. 2.—Effect of temperature on the viscosity of 2% aqueous solutions of hydroxyethylcellulose.

5. *Effect of pH on Solution Viscosity*

Hydroxyethylcellulose is a nonionic polymer and, therefore, undergoes little viscosity change over a pH range of from 2 to 12. Solutions show the greatest viscosity stability at pH 6.5–8.0. Solutions may show a drop in viscosity resulting from acid-catalyzed hydrolysis at pH values below 3.0. At very high pH, alkaline oxidation may occur; it is accelerated by heat and light.

6. *Effect of Salt on Solution Viscosity*

The salt tolerance of hydroxyethylcellulose is much greater than that of most other water-soluble cellulose derivatives; however, it can be salted out of solution. Hydroxyethylcellulose is not precipitated from solution by reaction with di- and trivalent ions as are such anionic derivatives as carboxymethylcellulose. The compatibility of hydroxyethylcellulose with several inorganic salts is shown in Table IV.

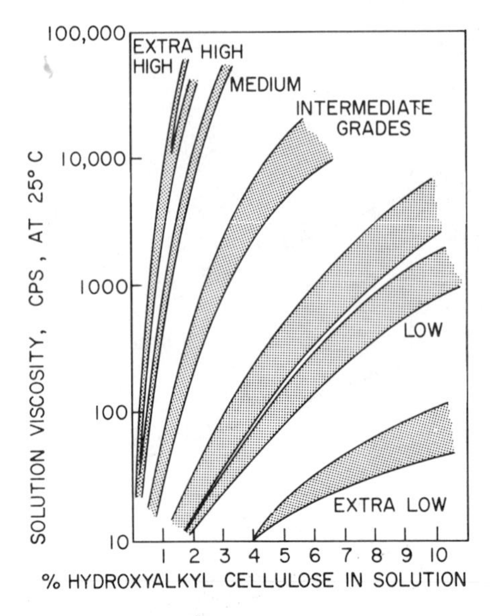

Fig. 3.—Effect of concentration on the viscosity of aqueous hydroxyalkylcellulose solutions.

7. *Rheology*

All solutions of hydroxyethylcellulose are pseudoplastic or non-Newtonian. This means that a single solution appears to have a wide range of viscosities when different conditions of physical force are imposed upon the solution (Fig. 4). The medium- and high-viscosity grades show a high degree of pseudo-

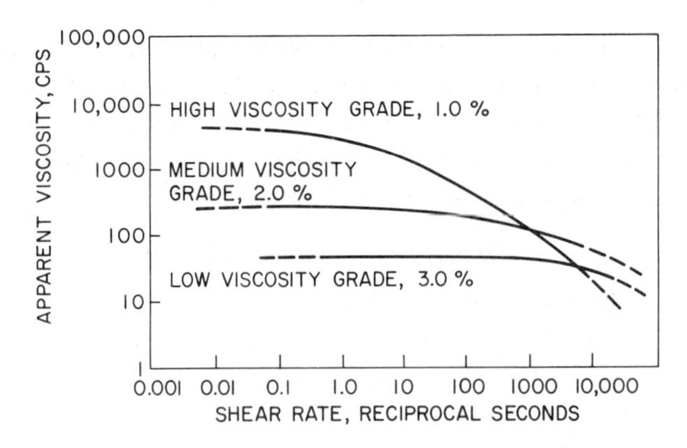

Fig. 4.—Effect of shear rate on the viscosity of aqueous hydroxyalkylcellulose solutions.

TABLE IV
Solubility of Hydroxyethylcellulose in Aqueous Salt Solutions [a]

	Salt Concentration	
	10%	50%
Aluminum nitrate	C	C
Aluminum sulfate	C	P
Ammonium nitrate	C	C
Ammonium sulfate	C	P
Calcium chloride	C	C
Chromic sulfate	C	C
Disodium hydrogen phosphate	P	P
Iron(III) sulfate	C	C
Magnesium chloride	C	C
Magnesium nitrate	C	C
Silver nitrate	C	P
Sodium chloride	C	C
Sodium nitrate	C	P
Sodium sulfate	P	P
Tin(II) (stannous) chloride	C	C
Zinc chloride	C	C
Zinc nitrate	C	C
Zinc sulfate	C	P

[a] C, compatible; P, precipitate.

plasticity, whereas solutions of the low-molecular-weight types show a low degree of pseudoplasticity and are nearly Newtonian.

At low rates of shear, the molecular arrangement of hydroxyethylcellulose is random, resulting in a high degree of chain entanglement and consequent high viscosity. At high rates of shear, the molecules become oriented with respect to the direction of flow, thereby decreasing the resistance to flow resulting in lower viscosities. Shear rate is a very important criterion in selecting a grade of thickener for a specific application. For example, some degree of pseudoplasticity is required in a thickened latex paint. In the can, the paint has to have sufficient viscosity to suspend the paint pigments; however, for ease of application it is desirable that the viscosity be considerably lower during brushing or rolling of the paint. Once the paint is applied and the shearing stresses removed, the paint should return to its original viscosity to prevent sagging.

8. *FDA Status*

Hydroxyethylcellulose is approved by the United States Food and Drug Administration (FDA) for use in indirect food applications such as in packaging adhesives and in resinous and polymeric coatings employed on metal, paper or paperboard articles intended for use in food packaging. It has also been cleared for use in film intended for food packaging. Hydroxyethylcellulose has not been cleared for use as a direct food additive. Specific uses under the Food Additives amendment to the Federal Food, Drug and Cosmetic act are as follows:

(*a*) Regulation 121.2514—As the basic polymer in resinous and polymeric coatings applied as a continuous film over a metal substrate.

(*b*) Regulation 121.2520—Component of food packaging adhesives.

(*c*) Regulation 121.2526—For use as components of resinous coatings for paper and paperboard in contact with fatty and aqueous foods.

(*d*) Regulation 121.2571—Component of the food-contact surface of paper and paperboard in contact with dry food.

IV. Applications of Hydroxyethylcellulose

The largest single volume use for water-soluble hydroxyethylcellulose is latex paint thickening.[11–15] Its ease of solution, low foaming characteristics, thickening efficiency, and contribution to color development[16] are reasons why a considerable quantity of the latex paint produced in the United States is thickened with this ether. The viscosity obtained provides excellent application properties and suspension of pigment particles. Rheological properties, such as leveling and degree of brush drag, can be controlled by selection from the various molecular weight types available. The flow characteristics of the paint can be varied from pseudoplastic behavior, as in the case of dripless paints, to nearly Newtonian flow in the semigloss paints.

Hydroxyethylcellulose is a thickener and a protective colloid in latex emulsion polymerization.[1, 17–19] Such factors as latex viscosity, particle size, grit formation, and freeze–thaw, chemical, and mechanical stability can be controlled by selecting the proper molecular weight and MS of the ether. The latex particle size partially controls the viscosity of the latex and also the degree of its ability to coalesce. If coalescence is not sufficient, the resultant film will have poor physical properties.

Most of the housing industry is based on wallboard construction. Hydroxyethylcellulose is widely used in joint cements that seal the spaces between the sheets of wallboard[1, 20] It imparts viscosity and improves the workability of these materials, lessening the need for sanding after application. The thickener binds substantial quantities of water, allowing a longer working or open time. Other benefits derived from its use are reduced cracking and shrinkage because of the binding of excess water and control of the rate of water release.

In mortar cements,[21–25] hydroxyethylcellulose extends the open time or workability of the cement, improves water retention, increases adhesion to masonry surfaces, and reduces capillary absorption into the brick or block to retain more water in the mortar for higher strength. Depending on the MS and the molecular weight type, it can have a great or no effect on the set retardation of the cement.

In 1970, over 15% of the ready-mix concrete used by the construction industry was pumped. The use of this technique is growing at the rate of 25% per year. The addition of about 0.25 lb (100 gm) of hydroxyethylcellulose per cubic yard (764 liters) of concrete results in faster pumping rates at lower pressures, with less blockage and at lower slumps.[26] It also permits the pumping of slag and lightweight concrete that previously had been considered unpumpable.

A variety of adhesives such as billboard,[27] corrugating, plywood, and wallpaper adhesives can be prepared from hydroxyethylcellulose. The ether has been approved for use in adhesives for packages containing foods.

Clear films[28–33] of moderate strength can be cast from aqueous solutions. The films are odorless, tasteless, flexible, and resistant to oils and greases. These films can be insolubilized[1, 2] by reaction with a variety of resins, such as dimethylolurea, using such catalysts as ammonium chloride. The films can also be rendered photosensitive[1] by chrome and by azo dyes and may then be rendered insoluble by exposure to ultraviolet light.

A variety of pharmaceutical and cosmetic products[34–36] contain hydroxyethylcellulose. A few of these are pharmaceutical emulsions, jellies and ointments; dental impression compounds;[37] suppositories; and tablet binders.[38–40] Cosmetic products include bubble bath formulations, hair dressing agents, roll-on deodorants, shampoos, shaving creams, toothpastes, and wave set lotions.

Miscellaneous uses include a thickener and suspension agent for solid toxicants in agricultural chemicals applied as water-based sprays.[1, 2] The polymer helps to bind the toxicant to the foliage and to reduce drift[41] in spraying operations. Sprays can, therefore, be placed directly on foliage with increased efficiency. The polymer has been used in caulking compounds, and it acts as a non-fermenting protective colloid and friction reducing agent in drilling fluids and muds.[42–47] Hydroxyethylcellulose has been included as a thickening agent in synthetic detergents to assist in keeping dirt particles in suspension.[48, 49] Other current and proposed uses are as a beer foam stabilizer,[50] a dip coating[51] in which is incorporated antibacterial or mold retarding materials for hams and other meat products, an elastic and fracture resistant subcoating for medicinal tablets,[52] a coating for grain pellets or wafers of hardened fat or edible substances to protect their vitamin content,[53] a coagulation aid in water treatment,[54] an additive in making an improved stereotype mat,[55] and a linoleum binding composition.[56] In addition, hydroxyethylcellulose is used in conjunction with methylcellulose in the emulsion polymerization of poly(vinyl chloride).[57] Numerous other uses include a battery separator,[58–60] a binder for reconstituted tobacco,[61, 62] a film former in planographic printing plates,[63–66] a thickener in cement grouts,[67] a binder for charcoal briquettes,[68] and a sausage casing.[69–76] It is also being used as a film former in the photographic industry.[77–85].

Although references to the use of hydroxyethylcellulose in food products have been published and cited above, it should be repeated that this polymer is not approved as an intentional food additive by the U.S. Food and Drug Administration.

The use of alkaline solutions of low-substituted hydroxyethyl cellulose in the art of textile finishing is well documented.[86–100] A method of making a clear filterable solution of the water-insoluble ether having a range of ethylene oxide substitution varying from 3.8% to 4.2% in dilute sodium hydroxide has been been published.[101] Water clarification using alkali-soluble, water-insoluble cellulose derivatives as a coagulation aid has also been attempted.[102]

V. Physical Properties of Hydroxypropylcellulose

Commercially available hydroxypropylcelluloses have an MS greater than 2.0 and are completely soluble in water at temperatures below 40°. The polymer, however, is insoluble in hot water and is precipitated as a highly swollen floc at a temperature between 40° and 45°. This phenomenon is completely reversible; for example, if the polymer is once dissolved in water and then precipitated from solution by heating, the polymer will redissolve upon cooling the water. Particles of hydroxypropylcellulose have a tendency to lump when first wet with water, although to a lesser extent than do those of hydroxyethylcellulose, and similar methods are recommended to prepare lump-free solutions of the polymer, that is (*a*) add the powder into the vortex of vigorously agitated water at room temperature, (*b*) dry blend with other dry materials that are to be added to the solution, or (*c*) preslurry the powder in a water-miscible nonsolvent such as hot water or glycerol.

1. Solubility in Organic Compounds

Hydroxypropylcellulose has excellent solubility in a wide range of polar organic compounds (Table V), and there is no tendency for precipitation of the polymer when its solutions in these organic compounds are heated.

2. Effect of Temperature on Solution Viscosity

The viscosity of aqueous hydroxypropylcellulose solutions decreases as the temperature is increased. This behavior is typical for most polymers in solution. However, the polymer will precipitate from water at a temperature between 40 and 45°, causing a rapid loss in viscosity. This phenomenon is completely reversible (Fig. 5).

TABLE V

Solution Quality of Hydroxypropylcellulose in Organic Compounds

Clear and Smooth Solutions	
Water	Dioxane
Methyl alcohol	Methyl sulfoxide
Ethyl alcohol	Dimethylformamide
Isopropyl alcohol (95%)	Ethylene chlorohydrin
Propylene glycol	Tetrahydrofuran
Methyl Cellosolve (2-Methoxyethanol)	Cyclohexanone
Cellosolve (2-Ethoxyethanol)	*tert*-butyl alcohol–water (9:1 v/v)
Chloroform	Acetone–water (9:1 v/v)
Formic acid (88%)	Glycerol–water (3:7 v/v)
Acetic acid (glacial)	Benzene–methyl alcohol (1:1 v/v)
Pyridine	Toluene–ethyl alcohol (3:2 v/v)
Morpholine	Methylene chloride–methyl alcohol (9:1 v/v)
Moderately Granular and/or Hazy Solutions	
tert-Butyl alcohol	Methylene chloride
Cyclohexyl alcohol	Butyl acetate
Acetone	Butyl Cellosolve
Methylethyl ketone (butanone)	Lactic acid
Methyl acetate	Naphtha–ethyl alcohol (1:1 v/v)
Isopropyl alcohol (99%)	Xylene–isopropyl alcohol (1:3 v/v)
Insoluble	
Aliphatic hydrocarbons	Methyl chloroform
Glycerol	Carbon tetrachloride
Benzene	Gasoline
Toluene	Kerosene
Xylene	Mineral oils
Dichlorobenzene	Soybean oil
Trichloroethylene	Linseed oil

The presence of other dissolved solids and miscible liquids has a pronounced effect on the precipitation temperature, which is lower in the presence of relatively high concentrations of some salts and sugars. For example, the addition of 5% NaCl reduces the precipitation temperature to 30°, and 40% sucrose reduces it to 20°. Organic liquids that are solvents for the polymer cause an increase in the precipitation temperature (Fig. 6).

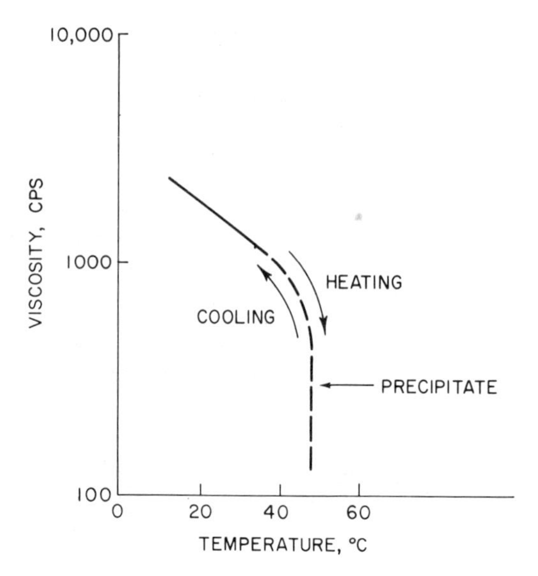

Fig. 5.—Effect of temperature on the viscosity of aqueous hydroxypropylcellulose solutions.

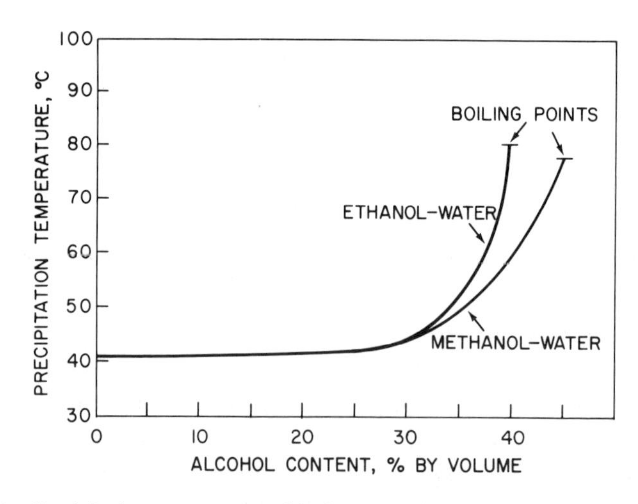

Fig. 6.—Precipitation temperature of hydroxypropylcellulose in aqueous alcohols.

3. *Effect of Concentration on Solution Viscosity*

The viscosity of aqueous solutions of hydroxypropylcellulose increases rapidly with concentration (Fig. 4).

4. *Effect of pH on Solution Viscosity*

As is hydroxyethylcellulose, this polymer is nonionic and, therefore, undergoes little viscosity change over the pH range of from 2 to 12.

5. *Compatibility with Inorganic Salts*

Compatibility of hydroxypropylcellulose with dissolved inorganic salts varies with the type and concentration of the salt (Table VI).

6. *Thermoplasticity*

Hydroxypropylcellulose is a true thermoplastic in that it can be thermoformed into rigid articles or flexible film without the aid of plasticizers or other

TABLE VI

Compatibility of Hydroxypropylcellulose with Some Inorganic Salts and Sucrose at 25°

	Compatibility [a]			
Concentration, % by weight:	2	5	10	50
Disodium hydrogen phosphate	I			
Sodium carbonate	C	I		
Aluminum sulfate	C	I		
Ammonium sulfate	C	I		
Sodium sulfate	C	I		
Sodium sulfite	C	I		
Sodium thiosulfate	C	I		
Sodium acetate	C	C	I	
Sodium chloride	C	C	I	
Potassium ferrocyanide	C	C	I	
Calcium chloride	C	C	C	I
Sodium nitrate	C	C	C	I
Iron(II) chloride	C	C	C	I
Ammonium nitrate	C	C	C	I
Silver nitrate	C	C	C	I
Sucrose	C	C	C	I

[a] C, compatible: I, incompatible. Tests were conducted by adding a 2% solution of hydroxypropylcellulose to solutions of various concentrations to produce a final polymer concentration of approximately 0.1% by weight.

internal additives. For plastic applications, the material is available in either a powder or pellet form. All molecular weight grades ranging from 50,000 to 1,250,000 are thermoplastic (Fig. 7).

Normally, small amounts of plasticizer, internal lubricant, and antioxidants are added to improve the melt flow, antiblock, or mold-release properties and the heat stability of the polymer. Melt flow (melt index)[103] decreases with increasing molecular weight. Practical experience indicates that the higher molecular weight grades have excellent melt flow but are somewhat pseudoplastic. The standard I_2 test method does not impart sufficient work to the melt to overcome the pseudoplastic yield point. Thus, it does not indicate the true melt viscosities found in high-shear extrusion or injection machinery.

Molecular weight also affects the physical properties of molded hydroxypropylcellulose articles. Table VII illustruates these effects on molded tensile bars.

Fabrication methods that have been studied include compression molding, injection molding, foam molding, blow molding, vacuum forming, and extrusion of blown and slit die film. The choice of the fabrication method usually dictates the type of hydroxypropylcellulose most suitable. Low- to medium-molecular-weight types are preferred in injection and blow molding to take advantage of their inherent hardness, rigidity, dimensional stability, and rapid solubility. Extruded film or sheet requires the improved flexibility and increased physical strength of the medium- to high-molecular-weight polymers.

The use of fillers that can withstand the heat of extrusion has been studied in hydroxypropylcellulose melts. Simple extenders, such as talc, clay, or starch, have been added to reduce material costs and to improve water solubility rates.

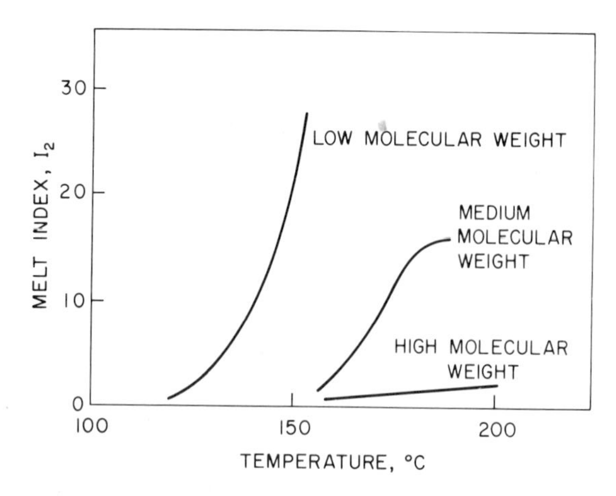

Fig. 7.—Effect of molecular weight on the melt index of hydroxypropylcellulose.

TABLE VII

Properties of Injection-Molded Tensile Bars of Hydroxypropylcellulose

Properties	ASTM Test Method	Viscosity type High	Medium	Low
Average molecular weight		1,250,000	300,000	75,000
Melt flow at 200°	D-1238	0.8		
I_2 test at 180°			13.6	
I_2 test at 140°				15.0
Softening point, °C		150	130	110
Tensile modulus, psi	D-638	92,000	267,000	340,000
Tensile yield strength, psi	D-638	3,200	6,400	6,600
Tensile elongation, %	D-638	4	4	4
Flexural strength, psi	D-790	800	2,030	3,600
Density		1.15		

Functional fillers, such as dry food stuffs, detergents, fragrances, and water-treatment chemicals, have been added as an integral part of the extruded or molded article.

Although films and articles thermoformed from hydroxypropylcellulose are soluble in water, they are nontacky and do not readily fingerprint during handling, even at high humidities. This property is related to the relatively low equilibrium moisture content of the polymer. This relationship is shown in Figure 8 for the low-molecular-weight polymer, which is the most water sensitive. Values for the high-molecular-weight polymer would be somewhat lower.

7. *FDA Status*

Purified hydroxypropylcellulose having an MS of 4.6 or lower is approved as a direct food additive as set forth in Section 121.1160 of the FDA Regula-

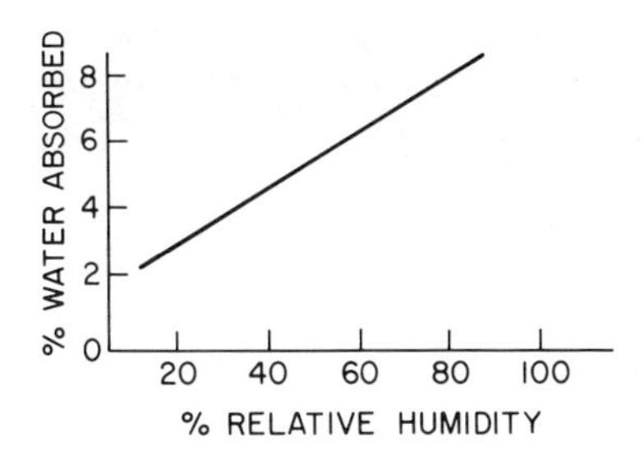

Fig. 8.—Equilibrium moisture content for low-molecular-weight hydroxypropylcellulose.

tions. It is permitted in foods for human and animal consumption except for the restrictions listed for standardized foods. Toxicity tests indicate that the polymer is physiologically inert. Insult patch tests on humans disclose no evidence that it is either a primary skin irritant or skin sensitizing agent.

VI. Applications of Hydroxypropylcellulose

Hydroxypropylcellulose is a surface active polymer, and water solutions display greatly reduced surface and interfacial tension. Because of this, the polymer can function both as an emulsifying agent and a whipping aid. The polymer also provides protective colloid action. It is, therefore, used as a stabilizer and whipping aid in whipped toppings (whipped cream substitutes)[104] and cosmetic emulsions and as a suspending agent in poly(vinyl chloride) polymerization.[103, 105]

Hydroxypropylcellulose forms strong films that act as barriers to air and moisture;[103, 105, 106] therefore, it can be used to coat pharmaceutical tablets,[105] nuts[105] (to prevent oxidative rancidity), and candies[105, 107] and other confections. These coatings can impart gloss and they do not become sticky even at high humidities because of their low equilibrium moisture content.

Because of its organic solvent solubility,[103, 105] the polymer is used as a thickener in solvent-based adhesives, alcohol-based hair dressings,[108] grooming aids, perfumes, colognes, inks, and methylene chloride–methanol-based paint removers.[105, 109]

The low ash content of hydroxypropylcellulose makes it useful as a burnout binder in ceramics, glazes, and electrical insulators.

As a plastic,[103, 105, 110, 111] hydroxypropylcellulose has been used as the base material for blow-molded containers that are water-soluble, disposable, and biodegradable. Vacuum formed tubs that may someday be used in packaging margarine or shortening have been made on a developmental basis. Packaging film produced on either a vertical or horizontal film train in a blown film tower has been produced. Potential application of the film include soluble portion-packaging of food products and industrial products, such as chlorine chemicals for swimming pool sanitizing or wettable herbicide powders for spray application. A number of rigid and semi-rigid injection-molded articles have been made.

VII. Analytical Methods

Generally speaking, analytical methods for determining the MS of hydroxyalkylcelluloses are straightforward. However, the formation of side chains during the hydroxyalkylation of cellulose makes it very difficult to determine accurately the DS of these polymers.

The MS of hydroxyethyl ethers of cellulose is generally determined by the

Morgan modification[112] of the Zeisel method. A 60 min modification of the Zeisel method for the quantitative microdetermination of hydroxyethyl groups has also been reported.[113]

Phthalation of the unreacted primary alcohol groups of low-substituted ethers has been used for the determination of the DS[114–116]. The basis for this method is that phthalation occurs at hydroxyethyl hydroxyl groups and not at the cellulose hydroxyl groups. However, investigations of the method indicate that this technique is questionable when applied to water-soluble hydroxyethylcellulose having an MS greater than 1.5.[117]

A degradation profile technique for the qualitative analysis of water-soluble cellulose ethers based on mass spectra obtained by thermal degradation of the ethers in a time-of-flight mass spectrometer has been developed.[118]

The MS of hydroxypropylcellulose can be determined by a procedure in which the terminal methyl groups are measured based on the recovery of acetic acid formed when such units are oxidized with hot chromic acid solution.[119] Nuclear magnetic resonance procedures appear promising for measuring the MS and DS of hydroxypropylcellulose.[120]

Fractionation data indicate that hydroxypropylcellulose substitution is fairly uniform along the cellulose backbone, although there is a tendency for shorter chains to become more highly substituted than longer ones.[121] The molecular weight and the molecular weight distribution reflect the properties of the cellulose furnish used in the sample preparation.

A solid-state characterization of the structure and deformation behavior of water-soluble hydroxypropylcellulose has been accomplished.[122, 123]

VIII. References

(1) "Natrosol, Hydroxyethyl Cellulose," Hercules Inc., Wilmington, Del. (1970).

(2) "Cellosize Hydroxyethyl Cellulose," Union Carbide Corp., New York (1968).

(3) E. D. Klug and H. G. Tennent, U.S. Patent 2,572,039 (1951); *Chem. Abstr.,* **46,** 1256 (1952).

(4) A. E. Broderick, U.S. Patent 2,682,535 (1954); *Chem. Abstr.,* **48,** 11100 (1954).

(5) E. D. Klug and W. D. Roberson, U.S. Patent 3,049,537 (1962); *Chem. Abstr.,* **58,** 2565 (1963).

(6) K. Ward, *J. Poly. Sci., Part C,* **124,** 137 (1968).

(7) M. G. Wirick, *J. Poly. Sci., Part A-1,* **6,** 1705 (1968).

(8) M. G. Wirick, *J. Poly, Sci., Part A-1,* **6,** 1965 (1968).

(9) E. D. Klug, U.S. Patent 3,278,521 (1966).

(10) E. D. Klug, U.S. Patent 3,278,520 (1966).

(11) R. Sterling, U.S. Patent 2,778,740 (1957); *Chem. Abstr.,* **51,** 7037 (1957).

(12) E. C. Scholl, U.S. Patent 2,880,104 (1959); *Chem. Abstr.,* **53,** 19405 (1959).

(13) J. C. Becker and J. H. Bress, U.S. Patent 3,150,110 (1964); *Chem. Abstr.,* **61,** 13531 (1964).

(14) J. G. Brodynan, *J. Paint Technol.,* **40,** 138 (1968).

(15) C. Schneider, *Paint Varn. Prod.*, **60,** 83 (1970).

(16) B. B. Kine, U.S. Patent 2,886,474 (1959); *Chem. Abstr.*, **53,** 15586 (1959).

(17) W. R. Peterson and D. A. Walker, U.S. Patent 3,303,155 (1967); *Chem. Abstr.*, **66,** 66452t (1967).

(18) J. G. Brodynan, *Amer. Chem. Soc., Div. Polymer Chem. Preprints,* **7,** 827 (1966).

(19) H. Bauer and J. Heckmaier, U.S. Patent 3,375,238 (1968).

(20) H. Elden, U.S. Patent 3,303,147 (1967); *Chem. Abstr.*, **66,** 68658p (1967).

(21) F. J. Shell, U.S. Patent 2,852,402 (1958); *Chem. Abstr.*, **53,** 2572 (1959).

(22) H. G. Cutforth, U.S. Patent 2,995,189 (1961); *Chem. Abstr.*, **55,** 27844 (1961).

(23) M. T. Ptasienski and J. W. Gill, U.S. Patent 3,003,979 (1961); *Chem. Abstr.*, **56,** 1160 (1962).

(24) C. F. Weisend, U.S. Patent 3,132,693 (1964); *Chem. Abstr.*, **61,** 1596 (1964).

(25) G. Selden, U.S. Patent 3,243,307 (1966); *Chem. Abstr.*, **64,** 15560 (1966).

(26) J. B. Batdorf and A. J. Desmarais, *Amer. Chem. Soc., Div. Chem. Market. Econ.*, CMEC 17 (1971).

(27) E. M. Barrentine, T. S. Hodgins, and E. J. Shelton, U.S. Patent 2,862,897 (1958); *Chem. Abstr.*, **53,** 6694 (1959).

(28) D. R. Erickson, U.S. Patent 2,817,114 (1957).

(29) D. T. Milne, U.S. Patent 2,994,924 (1961); *Chem. Abstr.*, **56,** 3706 (1962).

(30) M. F. Fillius and J. S. Bruce, U.S. Patent 3,086,900 (1963).

(31) Union Carbide Corp., Can. Patent 811,098 (1970).

(32) R. L. Mitchell, Deutsche ITT Ind., Neth. Appl. 70 01017 (1970).

(33) Hooker Chem. Co., Can. Patent 790,106 (1968).

(34) J. E. Haberle and W. B. Swafford, *Amer. J. Pham.*, **133,** 58 (1961).

(35) U. Bogs and H. Naumann, *Pharmazie,* **17,** 231 (1962).

(36) Soc. Detudes De. R., Belg. Patent 743,526 (1970).

(37) G. B. Eberhard and F. T. Friedl, U.S. Patent 2,997,399 (1961); *Chem. Abstr.*, **56,** 2520 (1962).

(38) D. W. Doerr, E. R. Serles, and D. R. Deardorff, U.S. Patent 2,816,062 (1957); *Chem. Abstr.*, **52,** 4938 (1958).

(39) R. P. Tansey, U.S. Patent 3,133,863 (1964); *Chem. Abstr.*, **61,** 2915 (1964).

(40) E. J. deJong, U.S. Patent 3,266,992 (1966).

(41) G. McWhorter and W. L. Barrentine, *Weed Sci.*, **18,** 500 (1970).

(42) C. M. Himel and E. G. Lee, U.S. Patent 2,570,947 (1951); *Chem. Abstr.*, **46,** 245 (1952).

(43) P. W. Fisher and R. A. Rogers, U.S. Patent 2,606,151 (1952); *Chem. Abstr.*, **46,** 11661 (1952).

(44) P. W. Fisher and R. A. Rogers, U.S. Patent 2,687,375 (1954); *Chem. Abstr.*, **48,** 14175 (1954).

(45) A. Park, U.S. Patent 3,081,260 (1963).

(46) C. E. McCoy, Jr., U.S. Patent 3,280,026 (1966); *Chem. Abstr.*, **66,** 12030k (1967).

(47) W. N. Epler and W. T. Malone, U.S. Patent 3,417,820 (1968); *Chem. Abstr.*, **70,** 39507r (1969).

(48) L. R. Bacon and C. E. Smith, U.S. Patent 2,602,781 (1952); *Chem. Abstr.*, **46,** 10650 (1952).

(49) Algemeene Kunstzijde, Neth. Patent 73,088 (1953); *Chem. Abstr.*, **48,** 3713 (1954).

(50) M. A. Weaver, U.S. Patent 2,712,500 (1955); *Chem. Abstr.*, **49,** 14266 (1955).

(51) R. T. K. Cornwell, U.S. Patent 2,558,042 (1951); *Chem. Abstr.*, **45,** 8166 (1951).

(52) A. B. Spradling, U.S. Patent 2,693,436 (1954); *Chem. Abstr.*, **49,** 1284 (1955).

(53) R. P. Dunmirr, U.S. Patent 2,685,537 (1954); *Chem. Abstr.*, **48,** 14038 (1954).

(54) F. N. Kemmer, C. B. Houser, R. E. Steeper, and B. W. Dickerson, *Water Sewage Works,* **99,** 16 (1952).

(55) A. W. Schorger, U.S. Patent 2,533,145 (1950); *Chem. Abstr.*, **45,** 2210 (1951).

(56) F. G. Oswald, U.S. Patent 2,603,572 (1952); *Chem. Abstr.*, **46,** 9868 (1952).

(57) C. Schneider, *Paint Varn. Prod.*, **60,** 83 (1970).

(58) P. Robinson, U.S. Patent 2,788,383 (1957).

(59) J. Salauze, U.S. Patent 2,819,962 (1958); *Chem. Abstr.*, **52,** 8800 (1958).

(60) T. L. Pfluger, NASA Accession No. N65-33136, Reprint NASA-CR-64607 (1964).

(61) D. Bandel, U.S. Patent 2,769,734 (1956); *Chem. Abstr.*, **51,** 3941 (1957).

(62) J. Townend, U.S. Patent 3,416,537 (1968).

(63) F. E. Brimnick and A. Reilly, U.S. Patent 2,778,301 (1957); *Chem. Abstr.*, **51,** 5608 (1957).

(64) T. U. Marron, U.S. Patent 2,800,077 (1957).

(65) L. E. Richard, U.S. Patent 3,020,839 (1962); *Chem. Abstr.*, **56,** 12462 (1962).

(66) D. A. Newman, U.S. Patent 3,274,928 (1966); *Chem. Abstr.*, **65,** 20371 (1966).

(67) S. J. Rehmar, U.S. Patent 3,197,323 (1965).

(68) R. W. Swinehart and H. W. Bull, U.S. Patent 2,822,251 (1958); *Chem. Abstr.*, **52,** 6768 (1958).

(69) American Viscose Corp., Brit. Patent 724,465 (1955).

(70) F. H. Reichel and A. O. Russell, U.S. Patent 2,575,467 (1951).

(71) W. Wadr, U.S. Patent 2,638,624 (1953).

(72) J. Conti, U.S. Patent 2,686,128 (1954).

(73) J. Conti, U.S. Patent 2,685,769 (1954).

(74) F. H. Reichel, U.S. Patent 2,757,495 (1956).

(75) C. W. Tasker, U.S. Patent 2,766,568 (1956).

(76) J. Wilhelm VonWilzleben, U.S. Patent 3,369,911 (1968).

(77) A. J. Bachelder and W. J. McCune, U.S. Patent 2,698,243 (1954).

(78) E. H. Land, U.S. Patent 2,759,825 (1956); *Chem. Abstr.*, **51,** 116 (1957).

(79) E. H. Land, U.S. Patent 2,866,705 (1958); *Chem. Abstr.*, **53,** 7846 (1959).

(80) M. T. Ptasienski and J. W. Gill, U.S. Patent 3,003,979 (1961); *Chem. Abstr.*, **56,** 1160 (1962).

(81) H. G. Rogers, U.S. Patent 3,249,435 (1966); *Chem. Abstr.*, **65,** 9996 (1966).

(82) W. J. Weyerts and W. M. Salminen, U.S. Patent 3,266,894 (1966); *Chem. Abstr.*, **65,** 14700 (1966).

(83) E. H. Land, U.S. Patent 3,415,644 (1968); *Chem. Abstr.*, **70,** 33239h (1969).

(84) E. H. Land, U.S. Patent 3,415,645 (1968); *Chem. Abstr.*, **70,** 33240b (1969).

(85) Andrews Paper Co., Brit. Patent 1,113,398 (1968); *Chem. Abstr.,* **67,** 77891 (1967).

(86) I. G. Farbenind, Brit. Patent 341,516 (1929); *Chem. Abstr.,* **25,** 4720 (1931).

(87) H. Dreyfus, Brit. Patent 344,528 (1931).

(88) A. E. Broderick, Brit. Patent 592,210 (1947); *Chem. Abstr.,* **42,** 770 (1948).

(89) A. E. Broderick, U.S. Patent 2,469,431 (1949); *Chem. Abstr.,* **43,** 5640 (1949).

(90) I. H. Welch and J. A. Woodruff, *Amer. Dyestuff Reptr.,* **41,** 880 (1952).

(91) Y. Kajitani, U.S. Patent 3,418,405 (1968); *Chem. Abstr.,* **70,** 48566m (1969).

(92) L. Lilienfeld, U.S. Patent 1,722,928 (1929); *Chem. Abstr.,* **23,** 4572 (1929).

(93) J. A. Clark, *Textile Mfr.,* **64,** 252 (1938).

(94) J. A. Clark, *Laundry Age,* **18,** 73 (1938).

(95) J. A. Clark, *Cotton (Atlanta),* **102,** 53 (1938).

(96) J. A. Clark, *Amer. Dyest. Rep.,* **29,** 549 (1940).

(97) R. T. K. Cornwell, *Amer. Dyest. Rep.,* **35,** 304 (1946).

(98) G. A. Richter, U.S. Patent 2,103,639 (1938); *Chem. Abstr.,* **32,** 1929 (1938).

(99) L. Lilienfeld, Brit. Patent 507,137 (1939); *Chem. Abstr.,* **34,** 622 (1940).

(100) N. P. Solovev, U.S.S.R. Patent 65,199 (1945); *Chem. Abstr.,* **40,** 6830 (1946).

(101) R. L. Mitchell and A. A. E. Couninis, U.S. Patent 3,312,685 (1967).

(102) W. E. Gloor, U.S. Patent 2,728,724 (1955); *Chem. Abstr.,* **50,** 5956 (1956).

(103) J. Rossman and A. J. Desmarais, *Hercules Chem.,* No. **61,** 9 (1970).

(104) A. J. Ganz, U.S. Patent 3,479,190 (1969); *Chem. Abstr.,* **72,** 53916w (1970).

(105) E. D. Klug, *Food Technol.,* **24,** 51 (1970).

(106) "Klucel, Hydroxypropyl Cellulose," Hercules Inc., Wilmington, Del. (1970).

(107) C. A. Signorino, *Food Sci. Technol. Abstr.,* **2,** 667 (1970).

(108) E. D. Klug, U.S. Patent 3,210,251 (1965); *Chem. Abstr.,* **63,** 17794 (1965).

(109) "Klucel, Hydroxypropyl Cellulose in Paint and Varnish Removers," VC-477, Hercules Inc., Wilmington, Del. (1970).

(110) A. J. Desmarais, "Klucel—A New Water-Soluble Edible Thermoplastic," to be published by SPI, Canada.

(111) E. D. Klug, U.S. Patent 3,314,809 (1967).

(112) P. W. Morgan, *Ind. Eng. Chem., Anal. Ed.,* **18,** 500 (1946).

(113) M. A. Shlionskaya, *Zh. Anal. Khim.,* **23,** 640 (1968).

(114) R. Senju, *Nippon Nogei Kagaku Kaishi,* **22,** 58 (1948).

(115) G. Froment, *Ind. Chim. Belge,* **23,** 115 (1958).

(116) J. Quinchon, *Compt. Rend.,* **248,** 225 (1959).

(117) Unpublished data, Hercules Inc., Wilmington, Del.

(118) H. R. Harless and R. L. Anderson, *Textile Res. J.,* **40,** 448 (1970).

(119) R. U. Lemieux and C. B. Purves, *Can. J. Res.,* **25B,** 485 (1947).

(120) F. F. L. Ho, R. R. Kohler, and G. A. Ward, *Anal. Chem.,* **44,** 178 (1972).

(121) M. G. Wirick and M. H. Waldman, *J. Appl. Poly. Sci.,* **14,** 579 (1970).

(122) J. H. Elliott, *J. Appl. Poly. Sci.,* **13,** 755 (1969).

(123) R. J. Samuels, *J. Poly. Sci., Part A-2,* **7,** 1197 (1969).

IX. Additional Reference Books and Reviews

E. C. Worden, "Technology of Cellulose Ethers," Worden Laboratory and Library, Millburn, N.J., 1933.

O. Faust, "Cellulose Verbindungen," Julius Springer, Berlin, 1935.

"Cellulose and Cellulose Derivatives," E. Ott, H. M. Spurlin, and M. W. Grafflin, eds., Interscience Publishers, Inc., New York, 2nd Ed., 1954, Part 2, p. 945.

R. E. Kirk and D. F. Othmer, "Encyclopedia of Chemical Technology," The Interscience Encyclopedia, Inc., New York, 1945, Vol. 3.

P. P. Shorygin and I. Rymashevskaya, "The Hydroxyethyl Ether of Cellulose and Its Acetate," *Ber.*, **66,** 1014 (1933); *Chem. Abst.*, **27,** 4529 (1933).

D. Traill, "The Ethers of Cellulose," *J. Soc. Chem. Ind.*, **53,** 337 (1934).

A. W. Schorger and M. J. Shoemaker, "Hydroxyalkyl Ethers of Cellulose," *Ind. Eng. Chem.*, **29,** 114 (1937).

P. P. Shorygin and Y. A. Rymashevskaya, "Glycerol Ethers of Cellulose," *Zh. Obshch. Khim.*, **7,** 2428 (1937); *Chem. Abstr.*, **32,** 1925 (1938).

T. P. Clemens, "Hydroxyalkyl Ethers of Cellulose," *Rusta-Rayonne,* **12,** 299 (1937); *Chem. Abstr.*, **32,** 2341 (1938).

C. L. Mantell, "Some Notes on Special Textile Finishes," *Textile Res. J.*, **16,** 481 (1946).

W. A. Caldwell and A. J. Watters, "Water-soluble Cellulose Ethers," *Res.*, **1,** 248 (1948); *Chem. Abstr.*, **42,** 3953 (1948).

R. Hebermehl, "Cellulose Ethers and Chlorinated Rubber," *Farben, Lacke, Anstrichst.*, **3,** 105 (1949); *Chem. Abstr.*, **43,** 5588 (1949).

W. E. Gloor, B. H. Mathlman, and R. D. Ulbrich, 'Hydroxyethyl Cellulose and Its Uses," *Ind. Eng. Chem.*, **42,** 2150 (1950).

Courtaulds Ltd. and J. H. Macgregor, "Water-soluble Mixed Cellulose Ethers," Brit. Patent 636,295 (1950); *Chem. Abstr.*, **44,** 6624 (1950).

P. L. E. Fournier, "Preparation and Properties of Hydroxyethyl Ethers of Cellulose and Their Derivatives," *Ann. Chim. (Paris)*, **7,** 75 (1952); *Chem. Abstr.*, **47,** 1610 (1953).

S. Kuriyama, E. Shiratsuchi, Y. Kondo, and S. Mori, "Preparation of Hydroxyethyl Cellulose," *Kogyo Kagaku Zasshi,* **56,** 210 (1953); *Chem. Abstr.*, **48,** 9682 (1954).

S. G. Cohen, H. C. Haas, L. Farney, and C. Valle, Jr., "Preparation and Properties of Some Ether and Ester Derivatives and Hydroxyethyl Cellulose," *Ind. Eng. Chem.*, **45,** 200 (1953).

I. Jullander, "Cellulose Ethers Soluble in Water-Manufacture, Properties and Fields of Utilization," *Chim. Ind. (Paris)*, **71,** 288 (1954).

R. W. Butler, "Natrosol 250 (Hercules Powder Company)," *Chem. Prod. Chem. News,* **23,** 303 (1960).

E. D. Klug, "Purification of Cotton Cellulose-Hydroxyethyl Ethers of Cellulose and their Analytical Determination," *Methods Carbohyd. Chem.*, **3,** 315 (1963).

E. D. Klug, "Some Properties of Water-Soluble Hydroxyalkyl Celluloses and their Derivatives," *J. Poly. Sci., Part C,* **36,** 491 (1971).

E. D. Klug, "Cellulose Derivatives," *in* "Encyclopedia of Chemical Technology," John Wiley & Sons, New York, Vol. 4, 2nd Ed., 1964, p. 616.

E. D. Klug, "Cellulose Derivatives," *Hercules Chem.*, **54,** 8 (1966).

T. Z. Ball, "Sea Water Destructive Booster Assembly," U.S. Patent 3,415,188 (1968).

R. L. Davidson, M. Sittig, and F. E. Windover, "Water Soluble Resins," 2nd Ed., Reinhold Publishing Corp., New York, 1968, p. 52.

A. J. Desmarais, "Protective Coating Composition of Binder Latex in Water-Glycol Vehicle," U.S. Patent 3,483,148 (1969); *Chem. Abstr.*, **72,** 45098r (1970).

"Recent Advances in the Chemistry of Cellulose and Starch," J. Honeyman, ed., Interscience Publishers, New York; Heywood, London, 1959.

R. G. Zhbankov, "Infrared Spectra of Cellulose and its Derivatives," New York, Plenum Press, 1966.

W. D. Paist, "Cellulosics," New York, Reinhold Publishing Co., 1958.

"Treatise on Coatings," R. R. Myers and J. S. Long, eds., New York, Marcel Dekker, Vol. 1, Part 2, 1968.

CHAPTER XXX

ETHYLHYDROXYETHYLCELLULOSE

S. LINDENFORS AND I. JULLANDER

Research and Development Department, MoDoKemi AB, Örnsköldsvik, Sweden and Swedish Forest Products Laboratory, Stockholm, Sweden

I. INTRODUCTION

Ethylhydroxyethylcellulose* is a nonionic cellulose ether first manufactured in Sweden in 1945. Products with widely different solubility properties are obtained by varying the amounts of and the ratio between ethyl and hydroxyethyl groups. The commercially available types are soluble in cold water, but insoluble in hot water. They are also soluble in mixtures of water and some polar organic liquids. Few pure organic compounds dissolve ethylhydroxyethylcellulose,

*Ethylhydoxyethylcellulose manufactured by MoDoKemi AB is known by the registered trademark MODOCOLL, designated MDC in this chapter.

but a large number of mixtures of organic liquids are solvents for the more highly ethylated grades. The physical structure and surface properties facilitate rapid dispersion and dissolution in cold water.

Experience has shown that the use of the product varies greatly, especially from country to country. For this reason, this chapter is mainly devoted to a description of properties.

II. History

Three patent applications of 1912 by Leuchs,[1] Lilienfeld,[2] and Dreyfus[3] laid the foundation of cellulose ether chemistry. The use of ethylene oxide or ethylene chlorohydrin as an etherifying agent for the production of hydroxyethylcellulose was first mentioned in two patents applied for in 1920.[4,5] Ethylhydroxyethylcellulose was mentioned by Dreyfus,[5] who also gave an example of how to make an ether of this type which is water insoluble.[6] Later the synthesis of water-soluble and alkali-soluble ethylhydroxyethylcellulose was described.[7,8] No evidence has been found that ethylhydroxyethylcellulose of the water-soluble type was actually produced and studied prior to World War II.

Water-soluble ethylhydroxyethylcellulose was developed by Mo och Domsjo AB in Sweden.[9–12] Properties of ethylhydroxyethylcellulose have been described.[13–18]

Patents have appeared in the United States for the production of ethylhydroxyethylcellulose.[19,20] A highly substituted water-insoluble grade and a water-soluble grade (available in two substitution types) are manufactured in the U.S.

III. Manufacture

Manufacture of ethylhydroxyethylcellulose starts with a dissolving grade, bleached, sulfite pulp. Sheets of pulp are steeped in sodium hydroxide, pressed, shredded, and placed in an autoclave with a mixture of ethylene oxide and ethyl chloride. Ethylene oxide is more reactive than ethyl chloride and, with increasing temperature, ethoxylation of the cellulose chain takes places prior to ethylation. The raw product is washed with hot water, dewatered by centrifugation, ground, and dried.

IV. Structure

The average degree of substitution (DS) of ethylhydroxyethylcellulose (Fig. 1) by ethyl and hydroxyethyl groups cannot be estimated from the amounts of ethyl chloride and ethylene oxide consumed in the etherification, as it is well known that undersirable side reactions always occur.[12,21] The procedure used is to determine the ethoxyl content by oxidation to acetic acid with chromic acid by the method of Lemieux and Purves[22] and the total ethoxyl plus hydroxy-

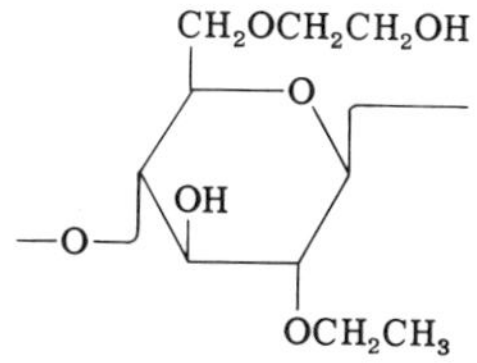

FIG. 1.—Schematic representation of ethylhydroxyethylcellulose.

ethoxyl content by alkoxyl analysis by the method of Morgan.[23] The hydroxyethoxyl content is the difference between the two analyses. However, when an ethylene oxide molecule reacts with a hydroxyl group of cellulose, a new hydroxyl group is created that can react further with ethylene oxide to form a polyoxyethylene chain. A nomogram for calculating DS and MS from analytically determined ethoxyl and ethylene oxide contents can be prepared (Fig. 2). The molar substitution (MS) is defined as the average number of molecules of ethylene oxide combined per D-glucopyranosyl unit.

A simple qualitative reagent is a 5% solution of tannin in water added dropwise at room temperature to an aqueous solution of ethylhydroxyethylcellulose. A white precipitate is formed immediately. This identification reaction is not specific. Other nonionic cellulose ethers react in the same way but it is a convenient method for distinguishing ethylhydroxyethylcellulose from sodium carboxymethylcellulose.

It is valuable to know the position of the substituents in ethylhydroxyethylcellulose both from theoretical and practical points of view. This is, however, a complex problem. By chromatographic separation[24] of the hydrolysis products of hydroxyethylcellulose (MS 0.6), 15 different monosaccharide derivatives have been obtained, ten of which are identified. Of the total amount of ethylene oxide reacted with cellulose, 25% reacts with an hydroxyethyl group. By statistical analysis of the results, the relative reaction rate constants k_2, k_3, and k_6 for ethoxylation at C-2, C-3, and C-6 and the rate constant k_E for ethoxylation at a hydroxyethyl group are calculated. Ethylhydroxyethylcellulose has also been investigated with the same chromatographic technique.[25, 26] The composition of the solution obtained upon hydrolysis is even more complex in this case, and a complete separation has not been obtained. Both investigations have been made on cellulose ethers from cotton linters, that is, the purest possible cellulose source.

Lindenfors[27] has calculated the theoretical composition of an ethylhydroxyethylcellulose (MS 0.81 and DS ethyl 0.91) using the values of the relative reaction rate constants available at that time.[24–26] Even though combinations with more than three ethylene oxide units in the polyoxyethylene chains have

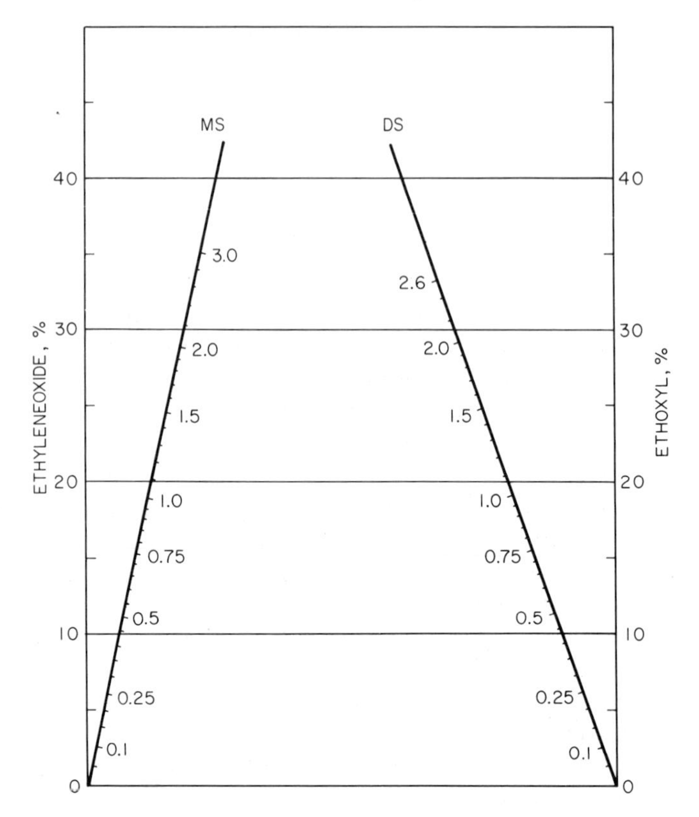

FIG. 2.—Nomogram for calculation of degree of substitution, DS, by ethyl groups and molecular substitution, MS, by hydroxyethyl groups in ethylhydroxyethylcellulose from the ethoxyl and ethylene oxide percentages, respectively. (From Jullander.[13])

been excluded in his calculations, a total of 242 different monosaccharide derivatives have been found to be possible.

A thorough investigation of the hydrolyzates of hydroxyethylcellulose was done by Ramnas and Samuelson,[28] who used a method involving group separation on an anion-exchange resin followed by gas-liquid chromatography of the trimethylsilylated fractions. They found that, with the hydrolysis conditions used, 2-*O*-hydroxyethyl-D-glucose (S_2) underwent an intramolecular glucosidation that gave rise to 1,2-*O*-ethylene-D-glucose (S_{2A}). This compound escaped detection by previous methods for the study of hydroxyethylcellulose. It was important that the total amount of S_2 and S_{2A} was determined. If the amount of S_{2A} were not taken into account, the value for the substitution at C-2 would be too low.

Manley[29] made a thorough investigation of the molecular properties of ethylhydroxyethylcellulose. A sample with MS 0.7 and DS ethyl 0.9 was fractionated. The chemical composition of the fractions was constant, within reasonable limits, except for the first fractions (low alkoxyl value). Four fractions were chosen for further study. The intrinsic viscosity was measured in an Ubbelohde-type capillary viscometer provided with four bulbs. The limiting viscosity number decreased with increasing temperature (Fig. 3.), probably because of a decrease in hydration at the higher temperatures.

Table I gives some of the average molecular weights obtained. Values from sedimentation equilibrium measurements, $(M_w)_{sed}$, and sedimentation + diffusion, M_{ww}, agree well. The lower values of M_n from osmotic pressure data are an indication of the heterogeneity of the fractions. The modified Staudinger relation between limiting viscosity number at 25° and molecular weight assumes the form $[\eta] = 3.7 \times 10^{-4} \times M_{ww}^{0.8}$ dl/g.

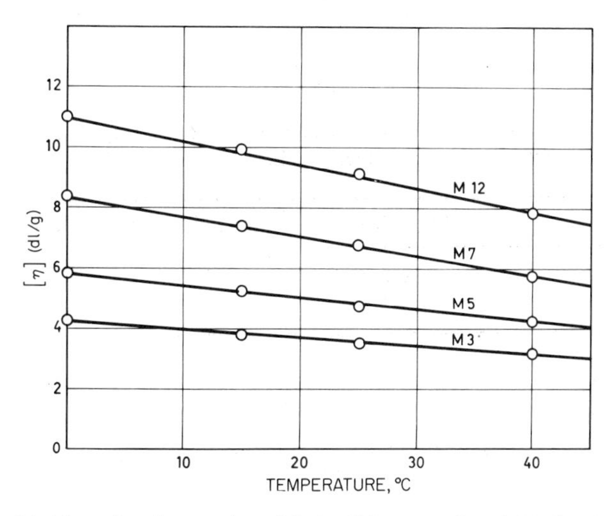

FIG. 3.—Limiting viscosity number, $[\eta]$ in dl/g, as a function of temperature for four fractions of an ethylhydroxyethylcellulose (see Table I). (Redrawn from Manley.[29])

TABLE I

Molecular Weights for Fractions of Ethylhydroxyethylcellulose[29]

$M_n \times 10^{-3}$	$M_{ww} \times 10^{-3}$	$(M_w)_{sed} \times 10^{-3}$
176	310	270
128	200	200
77.6	140	140
52	92	90

V. Properties of Commerically Available Products

Ethylhydroxyethylcellulose is manufactured in two basic qualities and several viscosity grades (Table II). The grade with the higher ethoxyl content is also soluble in a number of solvent mixtures of chlorinated hydrocarbons, hydrocarbons, esters, and alcohols. The solubility parameter concept, discussed later in this chapter, is useful when predicting solubility in a specific solvent mixture.

1. Solid State (Table III)

Etherification is a topochemical reaction that does not change the gross fibrous structure of the cellulose. However, in the manufacture of granular types, the washing and milling operations are conducted in such a way as to destroy almost completely the fibrous structure. A free-flowing granular powder is obtained that dissolves rapidly in water.

TABLE II

Average Chemical Composition of two Commercially Available Ethylhydroxyethylcelluloses of the MoDoKemi Company

	Ethyl, DS	Hydroxyethyl, MS	Sodium Chloride, %
MDC E	0.9	0.8	< 2
MDC M	1.4	0.5	< 0.5

TABLE III

Physical Data for Two Ethylhydroxyethylcelluloses of the MoDoKemi Company

	MDC E	MDC M
Physical appearance	Granules/Powder	Fibrous/Granules
Bulk density	400–600/300–500 g/liter	100–200/400–600 g/liter
Specific gravity of film	1.33	1.24
Refractive index of film	1.49	1.49
Tensile strength of film, 65% relative humidity, 20°	450–550 kg/cm²	250–350 kg/cm²
Elongation of film, 65% relative humidity, 20°	25–35%	5–15%

Ethylhydroxyethylcellulose shows a moisture hysteresis behavior typical of cellulose.[30] The moisture content is approximately the same as that for the pulp used as starting material and far below that of ionic cellulose derivatives, such as sodium carboxymethylcellulose.

Transparent films of ethylhydroxyethylcellulose are easily made in the laboratory by pouring a 1–3% aqueous solution onto a glass plate and drying. The transmission in the visible and ultraviolet regions is good down to about 300 nm. Adhesion toward metal surfaces, unless they are very rough, is poor, but it is good on paper, concrete, mortar, and plaster.

Suitable plasticizers are glycerol, ethylene glycol, and polyglycols; 10–30% of a plasticizer can be included if necessary, but with high plasticizer content, there may be syneresis at high humidities.

Prolonged heating of films to high temperatures (for example, 170° for 6 hr) develops partial insolubility in water. A considerable delay in the dissolution of a film can be obtained by adding a few percent of glyoxal to an ethylhydroxyethylcellulose solution and drying the film at room temperature (Fig. 4).

2. Preparation of Aqueous Solutions

Even though ethylhydroxyethylcellulose is water soluble, the solution rate depends on particle size. Menkart and Allan[31] and Jullander[32] developed proce-

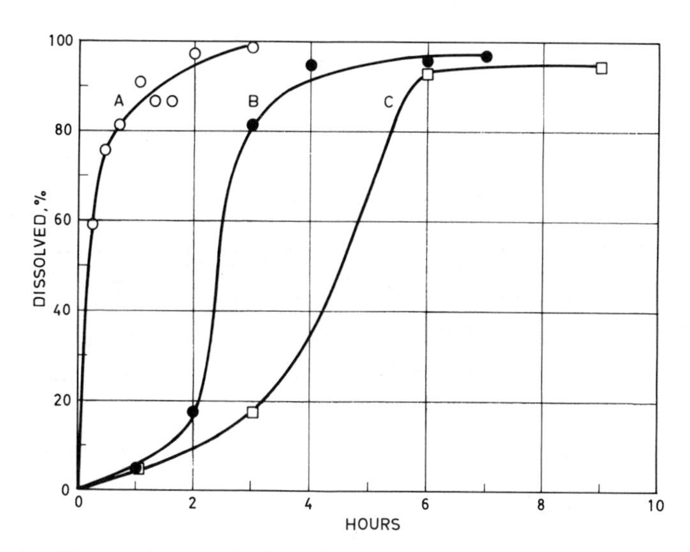

FIG. 4.—Water resistance of films of ethylhydroxyethycellulose. Films made from water solution, pH 6.0, and dried at 20°. Percentages of glyoxal calculated on cellulose ether: A, 0%; B, 2%; C, 5%. (From Jullander.[15])

dures for facilitating the dissolution. The effect of variations in pH on the viscosity and light transmission is shown in Figure 5.

Preparaton of concentrated (5–10%) stock solutions of high viscosities should be avoided, for careful and lengthy mixing is necessary to prepare properly diluted solutions from the jellylike stock solution.

As long as reasonable care and cleanliness are exercised in the preparation of solutions, addition of preservatives is usually unnecessary. Substituted phenols and their more readily soluble salts, organomercurials and quaternary ammonium salts may be used as preservatives.[33] Except for the possibility of microbiological degradation, neutral, aqueous solutions of ethylhydroxyethylcellulose seem to be indefinitely stable at room temperature. Solutions have been stored for 18 months without any viscosity diminution.

The water solubility of pure ethylcellulose is confined to a small range of DS around 1.2 and is not satisfactory even at the optimum value.[34] Coetherifica-

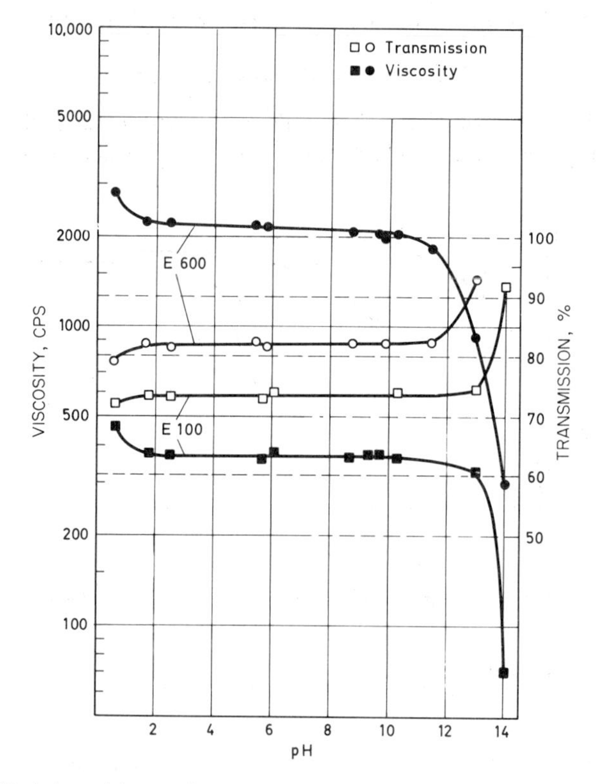

FIG. 5.—Variation with pH of the viscosity and turbidity of 2% aqueous solutions of ethylhydroxyethylcellulose at 20°.

tion with ethylene oxide improves the water solubility and at the same time displaces the gel point toward higher temperatures.[18] Ethylhydroxyethylcellulose of DS ethyl 1.4 and MS 0.5 is soluble in water but the gel point is rather low (about 40°). Because of the rather high substitution by ethyl groups, the product is also soluble in a number of organic solvents and solvent mixtures. The solubility may be predicted by the method of Hansen and Skaarup.[35, 36]

The total solubility parameter, δ_{tot}, can be divided into contributions from dispersion forces, δ_d, polar forces, δ_p, and hydrogen-bonding forces, δ_h. The relation is $\delta^2_{tot} = \delta_d^2 + \delta_p^2 + \delta_h^2$. The total solubility parameter will be a point in a three-dimensional system (Fig. 6). If the total solubility parameter of a liquid (or a mixture of liquids) is located within a certain maximum distance from this point, it will dissolve ethylhydroxyethylcellulose. This distance is called the radius of interaction, *R*. Choosing the unit distance along the δ_d axis as twice that along the δ_p and δ_h axes gives an essentially spherical volume with radius *R*. Liquids or mixtures of liquids with a solubility parameter within this sphere will dissolve ethylhydroxyethylcellulose. The parameters for ethylhydroxyethylcellulose are $\delta_h = 6.5$, $\delta_p = 5.5$, $\delta_d = 8.6$, and $R = 3.4$ (1.7 units along the δ_d axis).[37]

3. *Viscosity*

The logarithm of viscosity plotted against concentration gives a slightly curved line (Fig. 7). An even better approach to linearity is given by the Martin equation.[13] Viscosities of mixtures of low- and high-viscosity types of ethylhydroxyethylcellulose can be predicted from a lineal plot of log viscosity against composition of mixture (Fig. 8).

Solutions of the concentrations usually encountered in practice (1–4%)

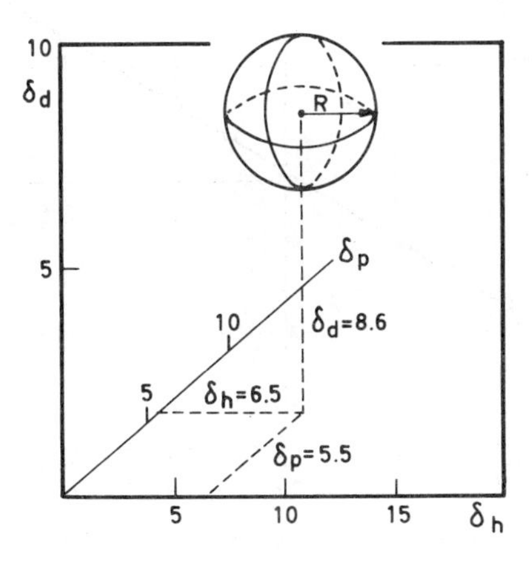

FIG. 6.—The solubility parameter system of ethylhydroxyethylcellulose.

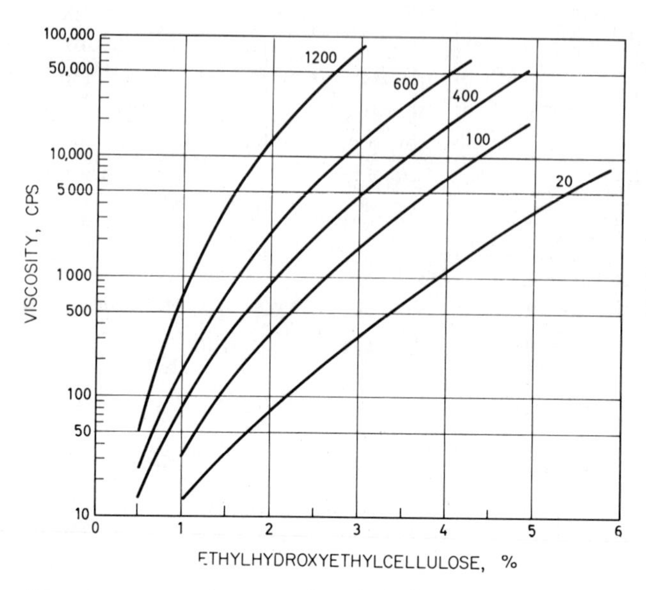

FIG. 7.—Viscosity–concentration curves for ethylhydroxyethylcellulose at 20°.

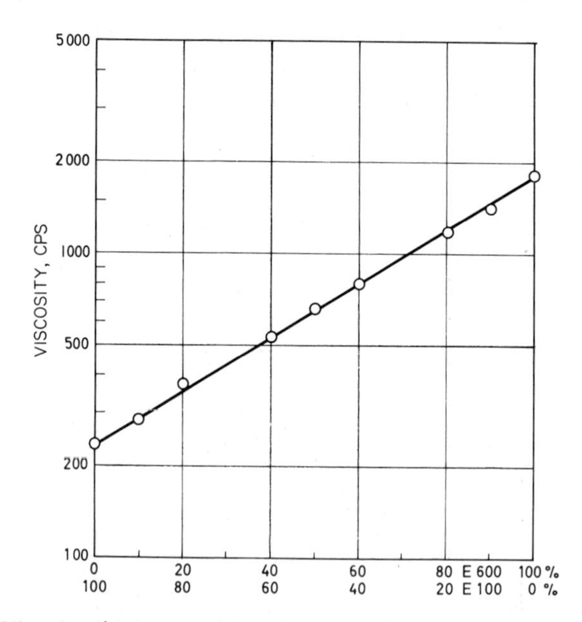

FIG. 8.—Viscosity of mixtures of low-viscosity and high-viscosity ethylhdroxyethylcellulose at 20°; Brookfield viscometer, type LVF.

are highly pseudoplastic (Fig. 9); consequently, the determined viscosities vary considerably with the type of viscometer used (Table IV).

Upon heating a solution, the viscosity first decreases by an amount more than that corresponding to the temperature dependence of the viscosity of water[38] (see also Fig. 3). In derivatives of high hydroxyethyl content, the viscosity changes in the vicinity of the gel point are moderate. The derivative with higher ethoxyl content has a much stronger tendency to gel when heated.[18, 38] A solution that has been flocculated by heating regains its original properties after cooling to room temperature.

The effects of pH on viscosity and turbidity are negligible until very high values are reached (Fig. 5). Unlike the ionic cellulose ethers, ethylhydroxyethylcellulose solutions are stable in dilute acids except for a slow but unavoidable hydrolysis of the glucosidic linkages (Fig. 10). As long as the solutions are not to be kept for several weeks, ethylhydroxyethylcellulose is a convenient thickener for dilute acid solutions. The viscosity drop in highly alkaline solutions is, at least partly, irreversible and is probably due to scission of alkali-sensitive linkages developed during etherification.

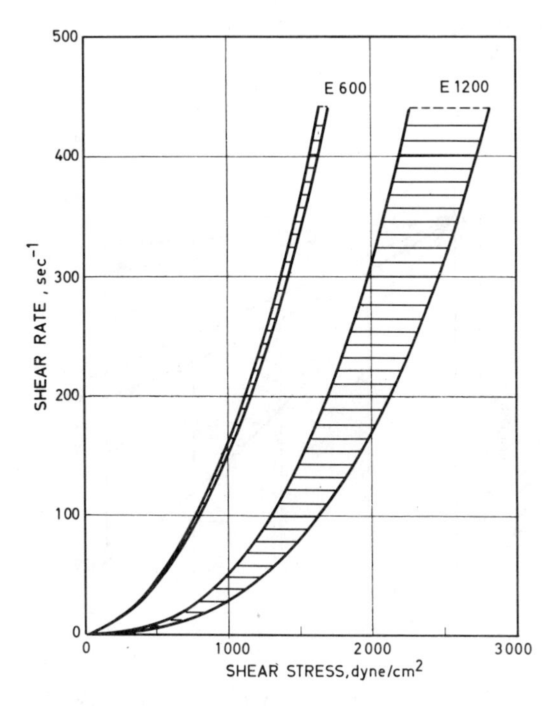

FIG. 9.—Consistency curves for 2% aqueous solutions of ethylhydroxyethylcellulose, 20°, Rotovisko viscometer.

TABLE IV

Measurements on 2% Ethylhydroxyethylcellulose Solutions in Water with Different Viscometers[27] at 20°

Viscometer		Relative Viscosity	
		MDC E 100	MDC E 600
Hoeppler		1.00	1.00
Ubbelohde		1.67	1.73
Falling-ball	1/16 in[a]	2.81	3.39
	3/32 in	2.67	2.93
	1/8 in	2.68	2.35
Brookfield (Type LVF)	1/12[b]	2.61	—
	2/12		3.44
	3/12		2.46
	4/12		2.42

[a] Diameter of steel ball.
[b] Spindle No./rpm.

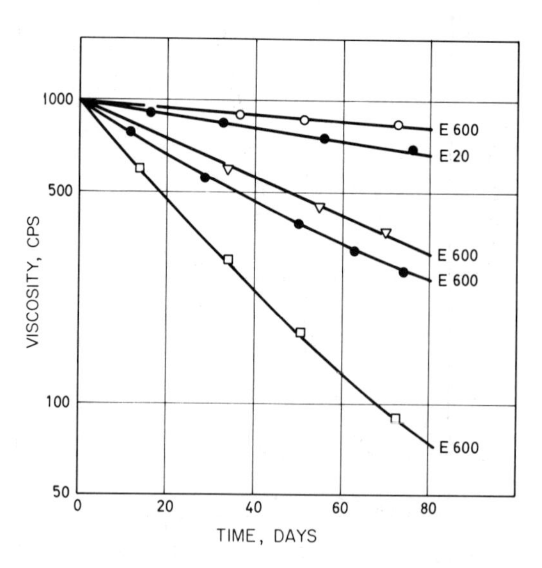

FIG. 10.—Viscosity drop of ethylhydroxyethylcellulose (MDC E 20 and E 600) in 0.1*N* sulfuric acid (•), 1*N* acetic acid (O), 1*N* oxalic acid (□), and 1*N* phosphoric acid (∇) during storage at 20°, Brookfield viscometer.

4. *Surface Tension and Foaming*

In ethylhydroxyethylcellulose, the ethyl groups have hydrophobic properties, and the hydroxyethyl groups have hydrophilic properties. Hence, the product acts as a surfactant and reduces the surface tension of aqueous solutions to less than 50 dynes/cm at concentrations below $\sim$0.3%.[15] Ethylhydroxyethylcellulose thus has good wetting and dispersing properties. Foaming is the result of the combination of surface activity and the thickening effect. Effective antifoaming agents are tributylphosphate, Bevaloid defoamer 581 B (R. Hodgson & Sons, U.K.), Nopco NDW and NXZ (Nopco Chemical Co., U.S.A.), and Hercules defoamer 340 and 340 M (Hercules Inc., U.S.A.). A water-soluble alcohol, such as ethanol or 2-ethylhexanol, may also be used.

5. *Salt Effects*

The behavior of aqueous ethylhydroxyethylcellulose solutions upon addition of salt has been examined in detail.[14] Because of the nonionic character of the product, insoluble salts are not formed with heavy metals. With a few salts, for example with potassium thiocyanate and potassium iodide, moderate salting-in at low salt concentrations has been observed; in other cases, salting-out occurs. The flocculation point has been defined as that concentration of a salt at which ethylhydroxyethylcellulose, added as a 2% aqueous solution, begins to precipitate. Salts at concentrations below the flocculation point, sometimes increase the viscosity of the solution considerably. Flocculation points for a large number of salts and other compounds have been determined.[14, 38] As soon as the salt concentration is reduced below the critical point, ethylhydroxyethylcellulose passes into solution again.

6. *Addition of Substantive Dyes*

The viscosity of aqueous solutions of ethylhydroxyethylcellulose can be increased by adding a few percent (calculated on the cellulose ether) of some types of substantive dyes (Fig. 11). Spectrophotometric measurements[18] demonstrate an interaction between the dyestuff and the cellulose ether. A probable explanation of the viscosity increase is a loose cross-linking of the cellulose chains by the dyestuff molecules.

7. *Mixtures of Ethylhydroxyethylcellulose and Carboxymethylcellulose*

The logarithm of the viscosity of mixtures of two viscosity grades of ethylhydroxyethylcellulose at constant total concentration varies linearly with the percentage composition, However, mixtures of ethylhydroxyethylcellulose and carboxymethylcellulose behave differently,[39] there often being a pronounced maximum in viscosity (Fig. 12).[27]

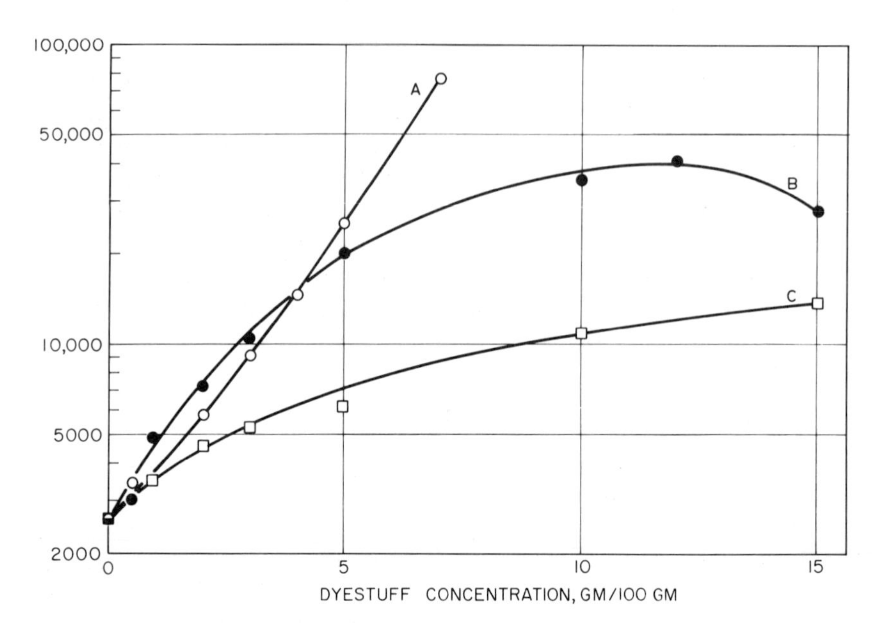

FIG. 11.—Viscosity increase obtained by adding substantive dyes to aqueous solutions of ethylhydroxyethylcellulose (2% MDC E 600) at 20°. Brookfield viscometer, type LVF. A, Du Pont Brilliant Paper Yellow, Extra Concentrated 125%; B, Congo Red (for microscopy); C. Chlorazol Blue G 150, ICI, England.

8. Solutions in Solvents Other than Water

As mentioned earlier, the solubility parameter is useful for predicting solubility. However, solutions in solvent mixture sometimes display unexpected behavior. For example, water and methyl sulfoxide, are completely miscible and are both solvents for ethylhydroxyethylcellulose,[40] yet ethylhydroxyethylcellulose is insoluble in an intermediate concentration range (Fig. 13). Also notable is the sudden change from solubility to insolubility as a result of a minor change in concentration. Friedberg and coworkers[41] have observed a similar effect of the same solvent solution on hydroxyethylcellulose. Ethylhydroxyethylcellulose with a high ethoxyl substitution (DS 1.3–1.4) and low ethylene oxide content (MS 0.4–0.5) behaves similarly in mixtures of water and water-miscible alcohols, especially propanol.[17] It is well known that the physical properties of methyl sulfoxide–water mixtures and alcohol–water mixtures are anomalous in many respects. These anomalies are caused by the formation of loose aggregates,[27, 42] composed, for example, of methyl sulfoxide and water in the molar proportion 1 : 2, and ethanol and water in the molar proportion of 1 : 3.

As already mentioned, the composition of ethylhydroxyethylcellulose can be

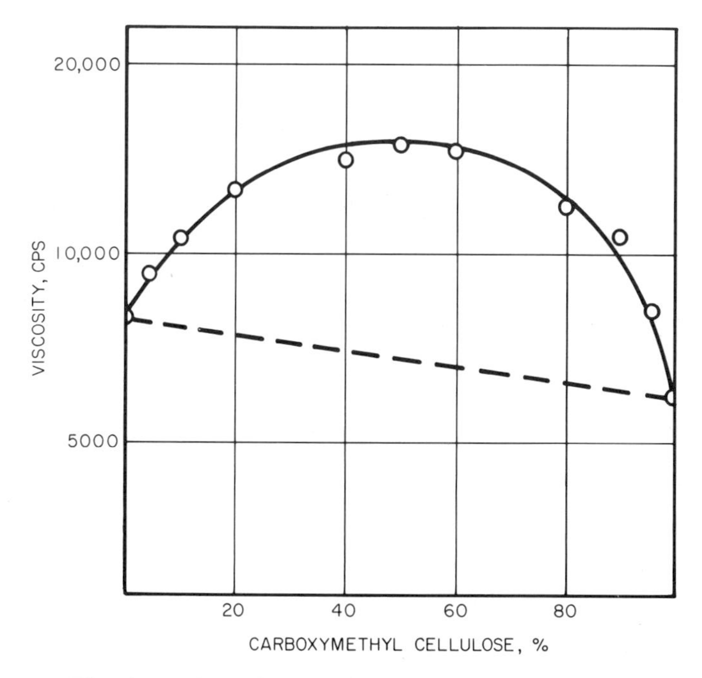

FIG. 12.—Viscosity maximum in aqueous solutions (2% total concentration) of mixtures of ethylhydroxyethylcellulose and carboxymethylcellulose at 20°. Brookfield viscometer. (From Jullander.[27])

adjusted so that the ether swells strongly in chlorinated hydrocarbons, such as dichloromethane or 1,2-dichloroethane, and then passes into solution on adding methanol, ethanol, or propanol. The viscosity passes through a maximum as a function of the alcohol concentration. At the maximum, the cellulose ether solution is in an intermediate stage between highly swollen suspension and true solution. The maximum is displaced toward higher alcohol concentration in the series methanol–ethanol–propanol.[37]

VI. USES

Ethylhydroxyethylcellulose is usable wherever a thickening or binding agent is needed or where a combined emulsifying or dispersing and thickening effect is desired.[38] Its use as a thickening agent is sometimes limited by its insolubility in boiling water and by the risk of flocculation in the presence of salts, especially sulfates. On the other hand, it is advantageous that a thickening effect is obtained over a very large range in pH.

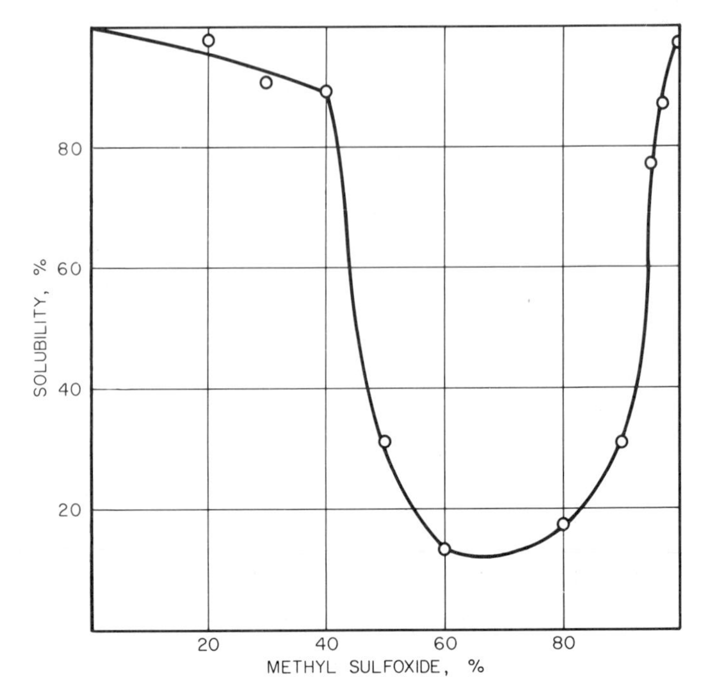

FIG. 13.—Solubility of ethylhydroxyethylcellulose (MDC E 600) in mixtures of water and methyl sulfoxide at 20°. (From Jullander.[27])

1. Paints

A large part of the ethylhydroxyethylcellulose production goes to the paint industry. It is used in paints in a range from latex paints containing small amounts as a stabilizer and thickener to paints consisting entirely of pigments, extenders, ethylhydroxyethylcellulose, and water. The use of ethyhydroxyethylcellulose in paints containing an alkyd-modified acrylic and in the manufacturing of alkyd emulsions are examples of the use of this ether in paints for exterior use. Water solutions of ethylhydroxyethylcellulose exhibit pseudoplastic flow that produces superior flow and leveling in latex paints. The ether also gives very good brushability (low viscosity at high rates of shear) to a paint. Because only small amounts of ethylhydroxyethylcellulose are required to give the desired viscosity, it has no effect on the washability of ordinary latex paints. As painting habits differ widely between countries, so does the use of ethylhydroxyethylcellulose. The properties of ethylhydroxyethylcellulose as a thickener and stabilizer in latex paints have been investigated.[43–48]

To illustrate its use in this industry, three typical formulas are given below in parts by weight.

A. Latex paint, pigment volume concentration 40%: 246 parts of titanium dioxide (Rutile type); 38 parts of blanc fixe; 80 parts of talc; 9 parts of sodium hexametaphosphate, 10% solution; 150 parts of water; 154 parts of MDC E 600 ethylhydroxyethylcellulose, 2.0% solution; 2 parts of defoamer; 1 part of preservative; 10 parts of film former; and 310 parts of PVA acrylic emulsion (56% solids content).

B. Distemper for interior use: 50 parts of whiting; 25 parts of water; and 25 parts of MDC E 100 ethylhydroxyethylcellulose, 4% solution.

C. Cement plaster paint: 25 parts of chalk; 40 parts of white cement; 25 parts of water; and 10 parts of MDC E 100 ethylhydroxyethylcellulose, 10% solution.

In paint removers of the chlorinated hydrocarbon type, the gum is used as a thickener. The following formula is proposed for the scrape-off type: 857 parts of methylene chloride; 110 parts of methanol; 13 parts of MDC M ethylhydroxyethylcellulose; 10 parts of paraffin wax, mp about 40°; and 10 parts of wetting agent.

2. *Wallpaper Pastes*

An important part of the ethylhydroxyethyl production is used as wallpaper paste. In contrast to starch wallpaper paste, ethylhydroxyethylcellulose leaves no stains when inadvertently spilled. The paste is easily prepared and easy to apply. It is suitable for all ordinary types of wallpaper. For hanging vinyl wall coverings, canvas back or paper back, and most other fabric and heavy-duty wallcovering, special pastes based on ethylhydroxyethylcellulose and synthetic polymers have been developed.

3. *Building Industry*

Ethylhydroxyethylcellulose is used in some special mortars, such as mortars for fixing and grouting ceramic tiles, to improve their plasticity and water-retaining properties.[49–51]

These special mortars can be used in very thin layers without risk of too rapid drying on highly absorbent materials. They also have a longer pot life, that is, the time during which work can be continued without noticeable changes in their consistency. Ethylhydroxyethylcellulose, like air-entraining agents, causes, in amounts $> 0.2\%$ a considerable increase in the air content of the mortars but investigations have shown that this does not have a detrimetal effect on bond strength. Bond strength to highly absorbent tiles and backgrounds is improved considerably by the addition of ethylhydroxyethylcellulose.

Mixtures of ethylhydroxyethylcellulose and sodium pentachlorophenate impart many desirable and valuable properties to mortars.[52] These admixtures give mortars special water-retaining properties and a pot life of a whole day.

Combinations of ethylhydroxyethylcellulose and D-gluconates are valuable set-retarding additives for mortars and concrete.[53]

4. *Foundry Cores*

Ethylhydroxyethylcellulose is used as core binder, either alone in the form of a fine powder or admixed with heat-setting resins, such as melamine, urea–formalydehyde, phenol–formaldehyde, or poly(vinyl acetate).[54–60] It is also used in conjunction with other binding agents, such as core oil and cement. Ethylhydroxyethylcellulose is also recommended for preparation of foundry cores according to the carbon dioxide process.[60–62] Among its advantages when used in foundry cores are low moisture uptake by cores in humid atmosphere, low gas content, and high collapsibility after pouring.[60]

5. *Tobacco Industry*

The manufacture of reconstituted tobacco sheet has become a very important part in the tobacco industry during the past 15 years.[63] Ethylhydroxyethylcellulose has been found to give many desirable properties to the final tobacco sheet.[64] Grades with low and with high ethyl content are both of interest. Ethylhydroxyethylcellulose has also found an extensive use as binding agent for wrappers in cigar manufacture.

6. *Miscellaneous*

Ethylhydroxyethylcellulose is used in the food industry as a thickening and consistency improving agent,[65] as a bulk laxative,[66, 67] as a substrate in the brewing industry for determining the cellulose activity in barley and malt,[68–70] and as an ingredient in a number of pharmaceutical preparations.[71, 72]

Ethylhydroxyethylcellulose has also been proposed as a thickening agent in preparations containing pentachlorophenol and similar agents used to prevent decay in wood,[73, 74] as a means of preventing drying of plant roots during reforestation,[75] and as a dispersing agent in fiber suspension.[76]

The ether is also used in spinning emulsions in glass fiber manufacture,[77] as a binder in decoration of ceramics by decalcomania,[78] and as an agent improving the quality of wallboard.[79]

Ethylhydroxyethylcellulose solutions have also been used as one component in two-phase systems for the fractionation of cellular particles.[80]

Ethylhydroxyethylcellulose improves release of the board from the gloss plate in the hot pressing step.[81] It is also a good dust preventing agent on powdered materials.[82]

VII. References

(1) O. Leuchs, Ger. Patent 322,586 (1920); *Chem. Zentralbl.*, **91,** IV, 416 (1920).
(2) L. Lilienfeld, Brit. Patent 12,854 (1913); *Chem. Abstr.*, **7,** 3839 (1913).

(3) H. Dreyfus, Fr. Patent 462,274 (1914); *Chem. Abstr.*, **8,** 3859 (1914).

(4) E. Hubert, Ger. Patent 363,192 (1922); *Chem. Zentralbl.*, **94,** II, 276 (1923).

(5) H. Dreyfus, Brit. Patent 166,767 (1921); *Chem. Abstr.*, **16,** 830 (1922).

(6) H. Dreyfus, Brit. Patent 327,157 (1930); *Chem. Abstr.*, **24,** 5157 (1930).

(7) L. Lilienfeld, U.S. Patent 2,265,913 (1941); *Chem. Abstr.*, **36,** 2141 (1942).

(8) L. Lilienfeld, U.S. Patent 2,306,451 (1942); *Chem. Abstr.*, **37,** 3270 (1943).

(9) S. Sönnerskog, *Sv. Papperstidn.*, **48,** 413 (1945).

(10) S. Sönnerskog, *Sv. Papperstidn.*, **49,** 409 (1946).

(11) S. Sönnerskog, *Tek. Tidskr.*, **77,** 133 (1957); *Chem. Abstr.*, **41,** 5714 (1947).

(12) S. Sönnerskog, "Some Ethers of Cellulose and Starch," Dissertation, Stockholm, 1952; *Acta Polytech.*, No. 157, *Chem. Met. Ser.*, **4,** No. 4 (1954); *U.S. Govt. Research Rept.* PB 117,119; *Chem. Abstr.*, **48,** 5487 (1954); **49,** 5830 (1955).

(13) I. Jullander, *Sv. Papperstidn.*, **55,** 197 (1952).

(14) I. Jullander, *Sv. Papperstidn.*, **56,** 443 (1953).

(15) I. Jullander, *Sv. Kem. Tidskr.*, **65,** 223 (1953); *Chem. Abstr.*, **48,** 6114 (1954).

(16) I. Jullander, *Chim. Ind.* (*Paris*), **71,** 288 (1954).

(17) I. Jullander, *Acta Chem. Scand.*, **9,** 1620 (1955).

(18) I. Jullander, *Ind. Eng. Chem.*, **49,** 364 (1957).

(19) E. D. King, U.S. Patent 2,610,180 (1952); *Chem. Abstr.*, **47,** 3564 (1953).

(20) S. G. Cohen, H. C. Haas, L. Farney, and C. Valle, *Ind. Eng. Chem.*, **45,** 200 (1953).

(21) A. B. Savage, A. E. Young, and A. T. Maasberg, in "Cellulose and Cellulose Derivatives," E. Ott, H. M. Spurlin, and M. W. Grafflin, eds., Interscience Publishers, Inc., New York, N.Y., 2nd Ed., 1954, p. 917.

(22) R. U. Lemieux and C. B. Purves, *Can. J. Res.*, **25B,** 485 (1947).

(23) P. W. Morgan, *Ind. Eng. Chem., Anal. Ed.*, **18,** 500 (1946).

(24) I. Croon and B. Lindberg, *Sv. Papperstidn.*, **59,** 794 (1956).

(25) I. Croon and B. Lindberg, *Sv. Papperstidn.*, **60,** 82 (1957).

(26) I. Croon and E. Flamm, *Sv. Papperstidn.*, **61,** 963 (1958).

(27) I. Jullander, *Papier* (*Darmstadt*), **19,** 166, 224 (1965).

(28) O. Ramnäs and O. Samuelson, *Sv. Papperstidn.*, **71,** 674 (1968).

(29) J. Manley, *Ark. Kemi*, **9,** 519 (1956).

(30) J. A. Howsmon, in "Cellulose and Cellulose Derivatives," E. Ott, H. M. Spurlin, and M. W. Grafflin, eds., Interscience Publishers, New York, 2nd Ed., 1954, p. 393.

(31) J. H. Menkart and R. S. Allan, U.S. Patent 3,072,635 (1963).

(32) I. E. Jullander, U.S. Patent 2,879,268 (1959); *Chem. Abstr.*, **53,** 9669 (1959).

(33) Technical Information No. 81, MoDoKemi AB, Stenungsund, Sweden, 1971.

(34) I. Jullander, *Acta Chem. Scand.*, **9,** 1291 (1955).

(35) C. M. Hansen, "The Three Dimensional Solubility Parameter and Solvent Diffusion Coefficient," Dissertation, Copenhagen, Danish Technical Press, 1967.

(36) C. M. Hansen and K. Skaarup, *J. Paint Technol.*, **39,** 511 (1967).

(37) Technical Information No. 110 E, MoDoKemi AB, Stungsund, Sweden, 1971.

(38) "The Modocoll E Manual," Mo och Domsjö AB Örnsköldsvik, Sweden, 1960.

(39) F. P. Nyberg, Brit. Patent 846,692 (1960); *Chem. Abstr.*, **55,** 5952 (1961). Corresponding patents have also been issued in Denmark, Finland, France, Germany, Italy, Norway, Sweden, and the United States.

(40) J. Manley, *Sv. Papperstidn.*, **61,** 96 (1958).

(41) F. Friedberg, W. Brown, D. Henley, and J. Öhman, *J. Makromol. Chem.*, **66,** 168 (1963).

(42) S. Lindenfors, *Papier* (*Darmstadt*), **21,** 65 (1967).

(43) J. I. Evans, *J. Oil Colour Chem. Ass.*, **39,** 553 (1956).

(44) G. A. Baseden, *J. Oil Colour Chem. Ass.*, **40,** 37 (1957).

(45) F. P. Grimshaw and R. A. W. Pateman, *J. Oil Colour Chem. Ass.*, **43,** 34 (1960).

(46) E. F. Parker, *Paint, Oil Colour J.*, **143,** 455 (1963).

(47) A. Reveley and H. Bates, *J. Oil Colour Chem. Ass.*, **50,** 16 (1967).

(48) Technical Information No. 111 E, MoDoKemi AB, Stenungsund, Sweden, 1971.

(49) S. Nycander, *Byggmästaren,* **33B,** 97 (1954).

(50) S. Nycander and A. Backman, *Byggmästaren,* **33B,** 129, (1954).

(51) S. Nycander, in "Tekno's byggarbeten," Teknografiska Institutet, Stockholm, Sweden, 1962, p. 375.

(52) J. Ericson, U.S. Patent 3,215,549 (1965); *Chem. Abstr.*, **64,** 430 (1966).

(53) J. Ericson, and U. Palm, U.S. Patent 3,528,832 (1970).

(54) R. E. Morén and L. H. V. Villner, Brit. Patent 740,323 (1955).

(55) R. E. Morén, Swed. Patent 153,803 (1956); *Chem. Abstr.*, **50,** 13708 (1956); Brit. Patent 754,823 (1956).

(56) R. E. Morén, Swed. Patent 154,468 (1956); *Chem. Abstr.*, **50,** 13708 (1956); U.S. Patent 2,838,406 (1958); *Chem. Abstr.*, **52,** 14511 (1958).

(57) R. E. Morén, Swed. Patent 155,354 (1956); *Chem. Abstr.*, **51,** 4253 (1957).

(58) R. E. Morén, Swed. Patent 159,340 (1957); *Chem. Abstr.*, **51,** 17712 (1957).

(59) R. E. Morén and S. Bergman, *Gjuteriet,* **45,** 51 (1955); *Chem. Zentralbl.*, **127,** 242 (1956).

(60) R. E. Morén, *Gjuteriet,* **47,** 157 (1957); *Foundry Trade J.*, **103,** 759 (1957).

(61) R. E. Morén, Swed. Patent 160,468 (1957).

(62) R. E. Morén, *Gjuteriet,* **49,** No. 1 (1959).

(63) R. J. Moshy, in "Tobacco and Tobacco Smoke," E. L. Wynder and D. Hoffmann, eds., Academic Press, New York, 1967, p. 47.

(64) R. E. Moren and S. E. H. Hammar, U.S. Patent 3,427,236 (1969).

(65) R. E. Moren and H. Anderson, *Chokolade, Konfekture Sukkervareindustrien,* **15,** 1 (1952).

(66) O. Alm. *Amer. J. Dig. Dis.*, **2,** 493 (1957).

(67) J. Tomenius, *Amer. J. Dig. Dis.*, **2,** 508 (1957).

(68) E. Sandegren, *Eur. Brew. Conv., Proc. Congr. Brighton,* 228 (1951); *Chem. Abstr.*, **46,** 6793 (1952).

(69) E. Sandegren and L. Enebo, *J. Inst. Brew., London,* **58,** 198 (1952).

(70) R. Thomas, *Aust. J. Biol. Sci.*, **9,** 159 (1956); *Chem. Abstr.*, **50,** 8797 (1956).

(71) O. Kylin, *Nord. Med.*, **49,** 358 (1953).

(72) E. Madsen, *Arch. Pharm. Chemi,* **61,** 817 (1954); *Chem. Abstr.*, **49,** 1282 (1955).

(73) E. Björkman, *Kgl. Skogshogskolans Skr., Stockholm, Bull. Roy. School Forestry,* No. 16 (1953).

(74) I. Jullander, Swed. Patent 152,437 (1955); *Chem. Abstr.*, **50,** 7460 (1956).

(75) K.-R. Samuelson, *Skogen,* **41,** 166 (1954).

(76) A. J. de Roos, *Tappi,* **41,** 354 (1958).

(77) H. W. Anderson, Swed. Patent 154,122 (1956); *Chem. Abstr.,* **50,** 14200 (1956); Brit. Patent 730,228 (1955).

(78) R. E. Morén, Swed. Patent 142,357 (1953).

(79) B. Unger and R. E. Morén, Brit. Patent 796,033 (1958).

(79) B. Unger and R. E. Morén, Brit. Patent 796,033 (1958).

(80) P.-Å. Albertsson, "Partition of Cell Particles and Macromolecules," Almqvist & Wiksell, Sweden, and John Wiley & Sons, New York, 1969, p. 32.

(81) R. E. Morén, Can. Patent 661,529 (1963).

(82) K. G. Görling, R. E. Morén, and A. F. Scholander, Swed. Patent 183,374 (1963).

CHAPTER XXXI

SODIUM CARBOXYMETHYLCELLULOSE[a]

JACK B. BATDORF AND JAMES M. ROSSMAN

Hercules Incorporated, Wilmington, Delaware

I. INTRODUCTION

Sodium carboxymethylcellulose, a water-soluble cellulose ether, is manufactured by reacting sodium monochloroacetate with alkali cellulose. Its ability to suspend solids in and control the viscosity of aqueous solutions and to form strong, tough films has accounted for the rapid growth of its use in industry since 1947. Sodium carboxymethylcellulose was first used as a substitute for natural gums but new markets based on its own properties have since been developed. Large volumes of this cellulose derivative are employed in paper,

[a] The technical information in this chapter, other than that acknowledged by specific literature references, has for the most part been developed in the laboratories of Hercules Incorporated, Wilmington, Delaware.

textile processing, detergents, drilling fluids, and protective coatings. The purified grade, known as cellulose gum, is used extensively in the pharmaceutical, cosmetic, and food industries.

II. Manufacture

It is believed that each of the several companies that manufacture sodium carboxymethylcellulose employs different techniques to accomplish the reaction.[1-12] Only one of these companies has disclosed its process.[6] Essentially, this consists of concurrently spraying aqueous sodium hydroxide and monochloroacetic acid onto powdered cellulose as it passes slowly through a revolving drum. The resulting mixture is aged to complete the reaction, oven-dried, and packed for shipment.

The traditional process is accomplished in a sigma-blade mixer. Either the cellulose is steeped in sodium hydroxide, pressed, and shredded, or the alkali cellulose is prepared entirely in the shredder.[2-4] The sodium monochloroacetate or chloroacetic acid can be mixed with the cellulose either before or after the alkali is added. If a technical product is desired, the reaction mass is neutralized, dried, and packaged. The crude product can be purified by the use of alcohol–water mixtures to extract the salts without dissolving the gum.[13]

1. Grades

Hercules Incorporated began commercial production of sodium carboxymethylcellulose in the United States in the fall of 1943.[14] Today, E. I. Du Pont de Nemours and Co., Inc.;[11] Wyandotte Chemicals Corp.;[7] Hercules Inc.; H. Kohnstamm & Co., Inc.; Essex International, Inc.; and Buckeye Cellulose Corp.[8] produce and market a technical-grade product in the United States.[6, 15] In addition, Hercules and Du Pont manufacture a purified product that has been cleared by the Food and Drug Administration for use in foods and pharmaceuticals. This material should be distinguished from the technical grades, often referred to as CMC, that can be used only for industrial applications. Technical grades are best known by their trade names: Du Pont's Sodium CMC, Wyandotte's Carbose, Hercules' CMC-T. The food-grade sodium carboxymethylcellulose manufactured by Hercules and Du Pont is commonly known as cellulose gum. Chemical Developments of Canada, Ltd., manufactures both a purified- and a technical-grade product. There are about 40 manufacturers throughout the world.

2. Production

United States Tariff Commission Reports[16] indicate that the U.S. production of sodium carboxymethylcellulose has increased from approximately 2 million lb in 1947 to nearly 60 million lb in 1968 and an estimated 62 million lb in

1969 (Fig. 1). Table I summarizes published information[17] on the end-use distribution for sodium carboxymethylcellulose, comparing 1965 and 1969 consumption patterns.

III. Applications

1. History

Sodium carboxymethylcellulose was developed in Germany shortly after World War I and was first introduced as a substitute for gelatin.[18, 19] How-

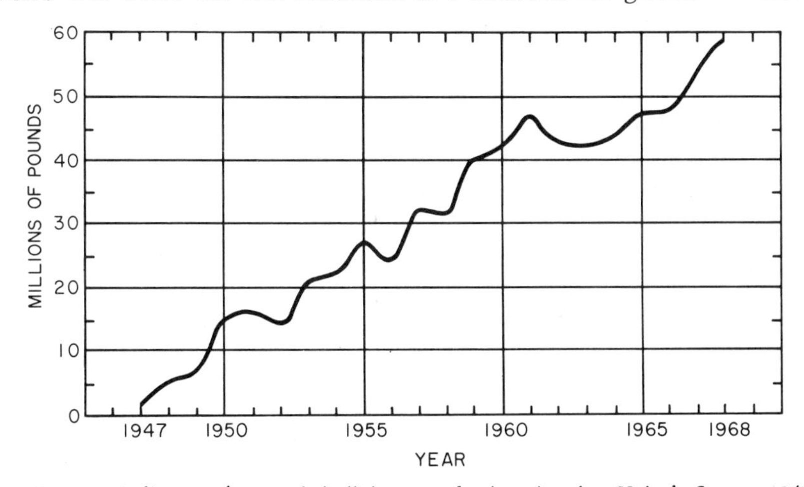

Fig. 1.—Sodium carboxymethylcellulose production in the United States, 1947–1968.[16]

Table I[17]

Estimated Consumption of Sodium Carboxymethylcellulose by End Use

End Use	1965	1969
Textiles	15	18
Detergents	13	15
Food and dentifrices	8	13
Drilling muds	6	7
Paper	3	4
Other	3	5
	48	62

ever, commercial-scale production was delayed by high production costs and technical difficulties. The discovery that the polymer improved the performance of synthetic detergents, coupled with the shortage of water-soluble gums in Germany during World War II, promoted large-scale manufacture.

2. *Commercial Value*

Sodium carboxymethylcellulose is useful in applications in which hydrophilic colloids are indicated. The basic properties that enhance its commercial value are its abilities to thicken water, suspend solids in aqueous media, stabilize emulsions, absorb moisture from the atmosphere, and form films. These properties can be, and have been, utilized in widely divergent applications.[15, 17, 20–22]

Textiles.—Sodium carboxymethylcellulose has received considerable attention in the textile industry because of its ready solubility and excellent film-forming characteristics.[23, 38] It has been particularly useful as a warp size for yarns and filaments.[29–32, 37, 38] Sodium carboxymethylcellulose has replaced starch in many warp sizing operations because of its marked advantage in weaving efficiency and in reducing stream pollution at the finishing mill.[20, 31] Other advantages, such as reduced shedding on the loom, elimination of the need for desizing enzymes, and lowered water and steam consumption, have also been realized.

This cellulose ether has also been used successfully for many years as a thickener for textile printing pastes.[24, 38] It is not sensitive to acids or alkalies within the pH range of these pastes and can be used, therefore, with a variety of dyes. It dissolves readily in both hot and cold water, which facilitates formula preparation and removal of the gum after printing. Because viscosity can be closely controlled, sharper prints can be obtained at higher machine speeds.

Fabrics sized with sodium carboxymethylcellulose are difficult to soil and easy to clean.[35, 36] This unique property has sparked considerable interest in its utilization as a domestic laundry size.[23, 25, 33] Other textile applications include its use as a scaffolding fiber,[26] as a binder for nonwoven fabrics,[39–43] and, when insolubilized, as a permanent finish.[29, 43]

Detergents.—The ability of sodium carboxymethylcellulose to improve detergency was discovered in Germany in 1935.[44] Laboratory investigations have demonstrated that this cellulose derivative is selectively absorbed by cotton fibers through hydrogen bonding. The soil-suspending properties are caused by electrostatic repulsion between the negatively charged dirt particles and the negativity induced by the carboxymethyl group.[45] This property has been used to upgrade synthetic detergents and builders.[46–60]

Investigations have shown that sodium carboxymethylcellulose can replace much of the fatty acid soaps in washing compounds on both an economic and a performance basis.[61, 62] In addition, it reduces the skin irritation caused by various synthetic detergents.[56, 57]

Food.—Chronic and acute toxicity investigations show that sodium carboxymethylcellulose is physiologically inert. As a binder and thickener, cellulose gum functions to control consistency in jellies,[63] pie fillings,[64] cheese spreads,[56, 66] custards, puddings, and both dry and semimoist pet foods.[67] Its water binding action retards undesirable crystal growth in ice cream,[68–70] sherbets,[71, 72] and sugar products, such as glazes, icings, syrup, and dessert toppings.[73–76] It increases volume and prolongs freshness of cakes[77–80] and other baked goods[81] and is effective as a texturizing agent[82] and as a protective colloid in flavor emulsions[83] and salad dressings.[84]

Aqueous systems of fats, flavors, and fruit juices can be converted to dry, stable products by dehydration in the presence of cellulose gum.[85, 86] It improves palatability and helps retain carbon dioxide (carbonation) in low-calorie beverages.[87, 88] As a carrier for vitamin C, it is useful in the preservation of fruit.[89] Foodstuffs, such as meats, vegetables, fruits, and eggs, have been protected from bacteria and molds by coating them with a film of sodium carboxymethylcellulose containing a preservative. It has also been shown that a fine powder coating of sodium carboxymethylcellulose applied to bacon slices or cheese slices will prevent sticking in subsequent packaging and handling.[90] In addition, the gum has found application as a processing aid in literally hundreds of specialty food products.[76, 91–96]

Pharmaceuticals and cosmetics.—Because of its uniform quality, sodium carboxymethylcellulose, which is insoluble in stomach acid but soluble in alkaline intestinal fluids, has replaced other water-soluble gums as an enteric coating for powders and tablets.[97, 98] It is a mild but effective bulk laxative because of its hydrophilic nature. Its suspending action makes it valuable for aqueous systems containing calamine, antibiotics, and barium salts (for x-ray contrast media).[99–106]

It is used as a stabilizer for hand lotions[107] and vitamin–oil emulsions and is widely used as a binder in cosmetic products.[109–112] It has been found to have the necessary demulcent properties for cough syrups. When used with gelling agents and fillers, a dental impression material can be made.[143] Its grease resistance is utilized in formulating protective hand creams.[110, 114] The adhesive qualities of sodium carboxymethylcellulose enable it to be readily tableted.[115–118] In toothpaste, it is employed as a suspending agent for mild abrasives and as a thickener.

Drilling fluids.—In the oil industry, drilling muds are used to lubricate and cool the bit and to carry cuttings away from the bore hole. These fluids are primarily aqueous dispersions of clay, bentonite, and weighting agents, such as barytes. Water-loss inhibitors are used in these systems to facilitate viscosity

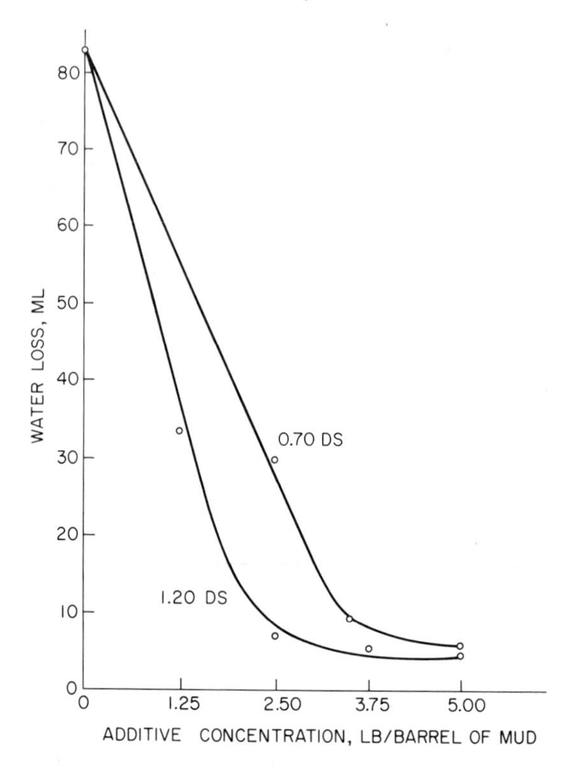

FIG. 2.—Effect of sodium carboxymethylcellulose concentration on the water loss of a saturated saltwater mud.

control and to minimize water loss into porous strata (Fig. 2). The excellent deflocculating characteristics and thickening action of sodium carboxymethylcellulose make it an efficient water-loss agent.[119–123] Its use has enabled drilling companies to decrease weighting, reduce mud volumes, and lower drilling costs.[124]

Paper.—The grease-resistant, film-forming characteristics of sodium carboxymethylcellulose have made it a useful additive for paper and paperboard products. It can be used alone or in conjunction with starch on the size press to give increased dry strength properties and improved surface characteristics.[125–128] Size press application of sodium carboxymethylcellulose on lightweight papers can decrease porosity (higher densitometer reading) and increase grease- and oil-resistance, wax pick, and smoothness.

In coating operations, the presence of sodium carboxymethylcellulose aids in controlling the viscosity and flow properties of the coating color and suspends and deflocculates the pigment.[129, 130] It also binds the coating to the paper more efficiently, thus providing a more suitable printing surface with less dusting.[131, 132]

Sodium carboxymethylcellulose can be economically and successfully applied at the calender stack to increase dry strength and wax pick, decrease porosity, impart greater oil and grease resistance, and improve surface and gloss ink printing characteristics (Fig. 3). Paperboard finish is improved because abrasion resistance, stiffness, and smoothness are increased.[133]

Special types of sodium carboxymethylcellulose have been developed for addition to the wet end of the papermaking process to promote chemical hydration of the fibers. Greater hydration is reflected in increased dry strength (Fig. 4). In Fourdrinier paper or paperboard, increased internal bonding and, in cylinder machine products, improved ply bonding are realized without loss of sheet speed.[134–138]

Adhesives.—The film-forming ability, easy solubility in cold water, and specific adhesion to a variety of surfaces make sodium carboxymethylcellulose an excellent adhesive. Because it is readily water soluble, it can be removed easily with a wet sponge and possesses no wet tack; it has found extensive use as a nonstaining wallpaper and billboard adhesive.[139, 140] For those applications

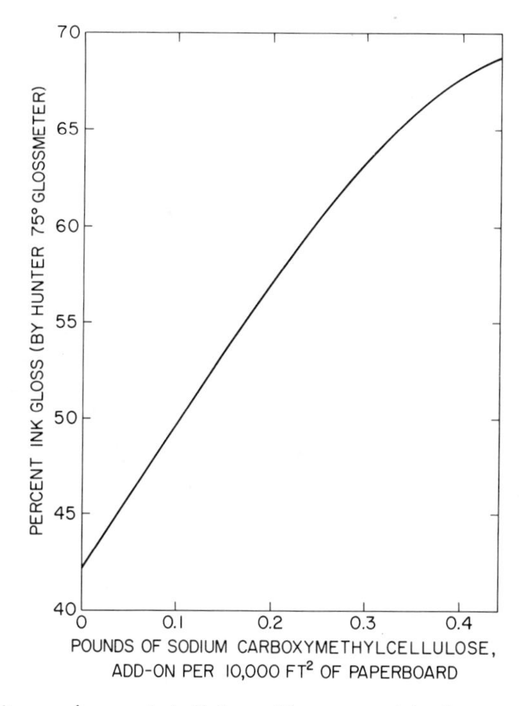

FIG. 3.—Sodium carboxymethylcellulose add-on versus ink gloss.

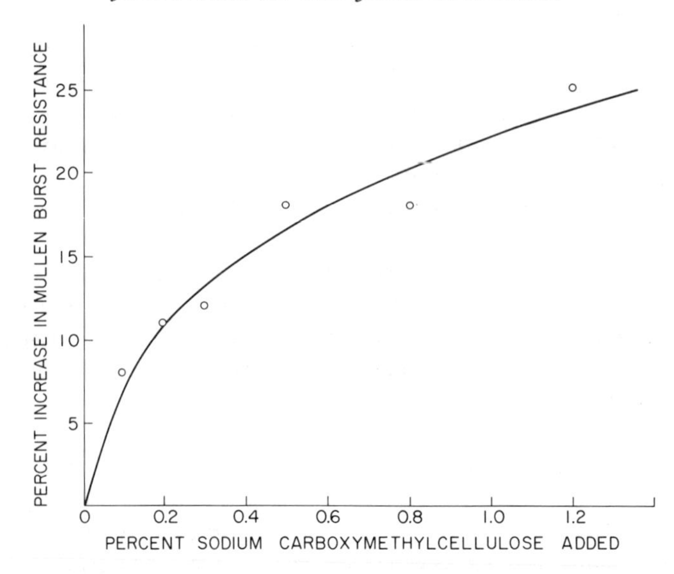

FIG. 4.—Effect of addition of sodium carboxymethylcellulose on the mullen burst resistance of 40 lb, unbleached, kraft handsheets.

requiring wet tack, there can be added gum arabic, hide glue, dextrin, or glucose.[141, 142] Sodium carboxymethylcellulose is an effective viscosity-controlling agent for water-soluble and emulsion-type resin adhesives. It is also being successfully used as an adhesive for leather, wood, and abrasive particles.

Protective coatings.—The thickening action and suspending properties of sodium carboxymethylcellulose are mainly responsible for its widespread use in emulsion paints. Brushing properties and pigment dispersion are also improved through its use.[143–145] Intumescent paints, that is, those that make self-extinguishing or flame-retardant coatings, can be formulated by combining it with urea, formalydehyde, phosphates, and pigments.[146, 147] Water-resistant, greaseproof films can be prepared by reacting it with polyvalent cations or some other insolubilizing agent.

Agricultural.—Sodium carboxymethylcellulose acts as a protective colloid to stabilize and control the viscosity of pesticide emulsions.[148] After application, its adhesive properties function to hold the active ingredients without impairing their effectiveness. It also finds use as a thickener for water and water–chemical systems to produce "sticky water" for use in fire-fighting.[149, 150]

Soil aggregant.—Field studies have shown that sodium carboxymethylcellulose improves the structure and air–water relationship of soils.[154] The initial problem of microbial deterioration has been overcome by using ethers with degrees of substitution greater than 0.7.

Ceramics.—Sodium carboxymethylecllulose is used by the ceramic industry because of its ability to suspend solids and act as a green strength binder prior to firing.[151, 152] Subsequently, the gum is completely burned out in the kiln. Its most important applications are in glazes for sanitary ware and structural tile, in refractories, as a jigger-body additive, and as a plasticizer for electrical porcelain.[153]

Foundry.—The German foundry industry pioneered in the use of sodium carboxymethylcellulose as a core binder.[155] Its advantages over other binders are based on a combination of the following properties: low drying temperatures, good hot strength (overcomes tearing), low decomposition temperature (excellent knockout characteristics), low gas evolution, economy of use.[156]

Lithography.—Sodium carboxymethylcellulose has been widely used as a desensitizing agent for lithographic plates. Its hydrophilic nature imparts desirable ink-repellent properties to the treated areas of the plate. It has better adhesion, is more durable, and necessitates less fountain water to desentitize the plate than does gum arabic, which has been the standard gum in the industry. As a result, lithographic plates last longer, production can be increased, and ink colors are brighter and sharper.[157]

Cement.—Sodium carboxymethylcellulose is used in grouting cements to suspend mineral fillers and metallic powders.[158] It acts as a set retardant for neat cements employed under conditions of high temperature and pressure by reducing the water loss.[159] When used in conjunction with a wetting agent, it is valuable in forming cellular cement for structural work.[160]

Hydrosols.—Sodium carboxymethylcellulose is useful in hydrosols in that it can eliminate dilatancy, prevent settling, adjust viscosity, and regulate penetration into porous surfaces.

Miscellaneous.—Sodium carboxymethylcellulose has a variety of other uses. It can be used in auto polish,[161] in suspension polymerization,[162] to make water-soluble films,[163] in leak detectors for gaseous systems,[164] in molding extrusion of lead pencils and coal briquettes,[165] in lubricants for friction reduction,[166] for water-proofing of explosives, for suspending solids in aqueous gelled explosives,[167] in emulsions for photographic film,[168] as clarifying agents for liquids, in leather dressings, and in embalming fluids.

3. *Chemical Modification*

Carboxymethylcellulose.—When a mineral acid is added to a sodium carboxymethylcellulose solution, the free acid, carboxymethylcellulose, is formed.[26] The free acid can also be formed by removing cations from the solution with suitable ion-exchange resins. By this latter method, a salt-free, stable, water dispersion of the acid form of carboxymethylcellulose can be prepared. This dispersion can be used to form water-insoluble films (Fig. 5),[169] a property valuable in

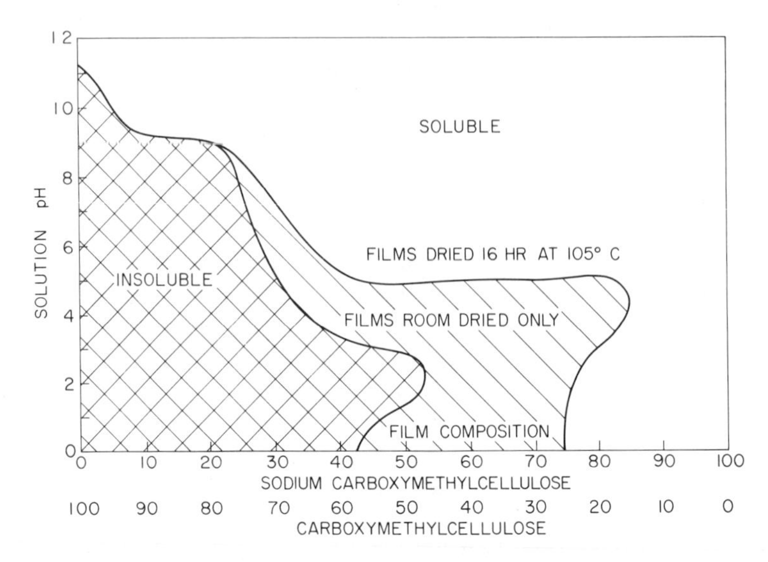

FIG. 5.—Solubility of various sodium carboxymethylcellulose–carboxymethylcellulose films at various pH's.

applications in which pigment dispersion is needed prior to application and in which water insolubility is required after the water base has evaporated, for example, in inks and shoe polishes. Films of carboxymethylcellulose are unaffected by greases and organic solvents.

Typical reactions.—The unsubstituted hydroxyl groups in sodium carboxymethylcellulose provide centers for the typical reactions of cellulose.[170] For example, esters and ethers can be formed, and the primary carbon atoms having unsubstituted hydroxyl groups can be converted to aldehyde or acid groups.[171] In aqueous solutions, sodium carboxymethylcellulose forms insoluble compounds with some cations.[172] Strong oxidizing agents cleave the cellulose molecule and lower the molecular weight.[173]

IV. CHEMISTRY

Despite the different techniques used by the various manufacturers of sodium carboxymethylcellulose, the basic chemistry of the etherification is essentially unchanged from that described by the inventor.[19] Cellulose is steeped in sodium hydroxide solution, and the alkali cellulose is then etherified with sodium monochloroacetate to form sodium carboxymethylcellulose and sodium chloride.[176]

$$R_{cell}(OH)_3 + ClCH_2COONa + NaOH \rightarrow R_{cell}(OH)_2(OCH_2COONa) + NaCl + H_2O$$

Because three reactive hydroxyl groups are present on each D-glucopyranosyl unit, it is theoretically possible to introduce three sodium carboxymethyl groups per unit. Such a product would be described as having a degree of substitution of 3.0.[175-177] Commercial sodium carboxymethylcellulose generally has an average degree of substitution of less than 1.5. The most common substitution range is between 0.4 and 0.8 (Fig. 6).

The properties of sodium carboxymethylcellulose can be controlled by varying the uniformity of substitution, the degree of substitution (DS), and the degree of polymerization (DP).[163] In general, low-substituted types (DS 0.3 or less) are insoluble in water but soluble in alkali. Water solubility is achieved with a degree of substitution greater than $\sim$ 0.4. It should be noted, however, that water solubility can be affected at a degree of substitution lower than 0.4 in cases in which special preparative conditions are employed to produce uniform substitution. The degree of polymerization of the cellulosic molecule is largely responsible for its thickening action; that is, the longer the average chain length of the polymer, the more viscous is the solution. The DP for most commercial types ranges from 500 to 2000.[178]

Sodium carboxymethylcellulose itself forms strong, clear films that are unaffected by oils, greases, and organic solvents. The addition of a plasticizer increases elongation and flexibility.

Feeding tests on rats, dogs, and guinea pigs have established that purified sodium carboxymethylcellulose is physiologically inert.[179, 180]

V. Properties

1. *ASTM Methods*

The ASTM methods, entitled "Testing Sodium Carboxymethylcellulose," designation D 1439–65, cover the analysis of sodium carboxymethylcellulose under the jurisdiction of ASTM Committee D-23 on Cellulose and Cellulose Derivatives. (Copies can be obtained from the American Society for Testing Materials, 1916 Race St., Philadelphia, Pennsylvania 19103.)

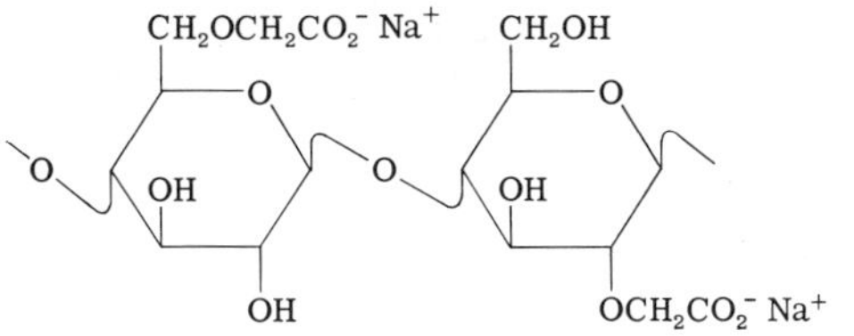

$CH_2OCH_2CO_2^- Na^+$ CH_2OH O OH O OH O OH $OCH_2CO_2^- Na^+$

FIG. 6.—Structure of a typical disaccharide fragment from DS 1 sodium carboxymethylcellulose.

2. Safety in Food, Pharmaceuticals and Cosmetics

Cellulose gum is classified under "substances that are generally recognized as safe" by Section 121.101 of the Food and Drug Regulations published in the January 31, 1961 issue of the Federal Register. The Food and Drug Administration (FDA) definies cellulose gum as the sodium salt of carboxymethylcellulose, not less than 99.5% on a dry-weight basis, with a maximum substitution of 0.95 carboxymethyl groups per D-glucopyranosyl unit, and with a minimum viscosity of 25 cps in a 2% by weight aqueous solution at 25°. Detailed reports on investigations conducted to establish the safety of sodium carboxymethylcellulose are on file at the FDA.

The Definitions and Standards for Food, Title 21 of the Federal Food, Drug and Cosmetic Act, permit the use of cellulose gum in numerous standardized foods, such as cheeses and cheese products;[181] ice cream and other frozen desserts;[182] salad dressings;[183] fruit butters, jellies, and preserves;[184] and nonalcoholic beverages.[185]

Cellulose gum is listed by the Food Chemicals Codex and is listed as a pharmaceutical aid, in the U.S. Pharmacopoeia under U.S.P. XVIII. The Cosmetic, Toiletry, Fragrance Association's Standard No. 34 also lists specifications for sodium carboxymethylcellulose.

3. Dissolution

A good solution is formed from a uniform colloidal dispersion of the gum particles in either hot or cold water. Sodium carboxymethylcellulose solutions can be most readily prepared when the gum concentration is low, the water is heated, and vigorous agitation is applied. The dry gum particles should be separated, not clumped, as they come in contact with the water in order to expedite swelling of the individual particles, which is the first step in dissolution. If the dry particles are bunched as they come in contact with the water and if the clump is not immediately broken up, a swollen, tough skin forms over the surface of the clump and prevents rapid hydration of the interior. This slows the solution rate. To easily produce solutions of the gum, one of the following methods should be used.

Method 1.—Add the gum to the vortex of vigorously agitated water. The rate of addition must be slow enough to permit the particles to separate and their surfaces to become individually wetted. However, it should be fast enough to minimize viscosity buildup of the aqueous phase while the gum is being added.

Method 2.—Wet the powder with a water-miscible liquid that will not cause the gum to swell (such as ethanol, ethylene glycol, or glycerol), prior to addition to water. Two to three parts per part of polymer should be sufficient.

Method 3.—Dry blend the gum with a dry, nonpolymeric material, such as sugar, that will be used in the formulation. Preferably, the gum should comprise less than 20% of the total blend.

Method 4.—The most rapid method uses a special mixing device in which the sodium carboxymethylcellulose is fed through a funnel into a water-jet eductor, where it is dispersed by water flowing at high velocity. Each granule is instantaneously wet by water, producing a uniform solution.

When a polymer is dispersed in a solvent, the degree of disaggregation,[186] that is, the separation of the polymer molecules, is affected by the chemical composition of the polymer, the solvating power of the solvent, and the shear history of the resulting solution.

When a polymer, such as sodium carboxymethylcellulose, is added to a liquid and its degree of disaggregation reaches equilibrium, the polymer may remain as a suspended powder, neither swelling nor dissolving (I,. Fig. 7A), swell to a point of maximum viscosity without dissolving (II), reach maximum dissolution (III), or exist in an intermediate state (Ia, Ib, IIa).

The effect of various factors on the disaggregation of sodium carboxymethylcellulose as measured by solution viscosity is shown in Table II. Increasing the DS makes sodium carboxymethylcellulose more hydrophilic; hence, types having high DS are more disaggregated in water. Plotting solution viscosity at constant shear against DS, from DS 0.4 to DS 1.2, produces a curve similar in slope to that of the curve of Figure 7A. Increasing electrolyte concentration reduces disaggregation, as evidenced by the lower viscosity in salt water of DS 0.4 and 0.7 types. For DS 0.9 and 1.2, the slight viscosity increase noted in saturated salt solutions is caused by the viscosity bonus effect, which will be discuused in Sec-

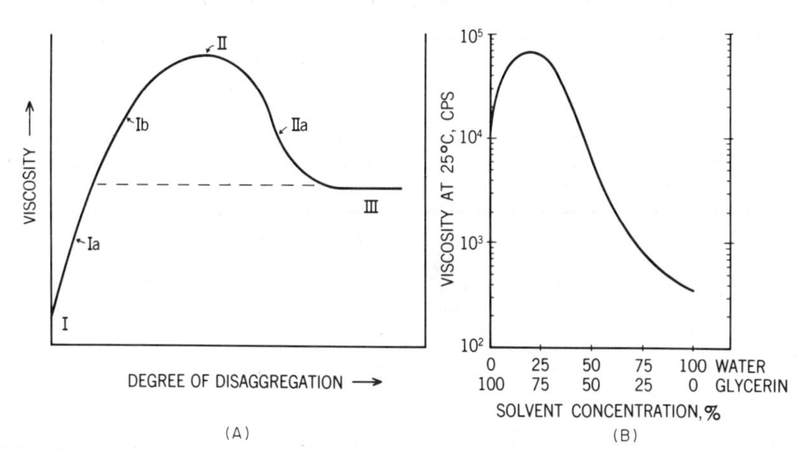

FIG. 7.—(A) Idealized curve showing effect of disaggregation on viscosity of solution of polymer in a solvent. (B) Effect of solvent strength on disaggregation of DS 0.7 sodium carboxymethylcellulose (1.3% solution, medium-viscosity type).

tion V, 6. The high shear imparted by the Waring Blendor enhances viscosity development or disaggregation. The effect of solvent strength on the disaggregation of sodium carboxymethyl cellulose is shown in Figure 7B. Note the similarity of the curve to that in Figure 7A.

An increase in the solvating power (Fig. 7B) and an increase in mechanical shear (Table II) break internal associations of the gel centers and promote disaggregation.

A closely standardized method for the preparation of sodium carboxymethylcellulose solutions and a closely standardized test method are necessary to obtain reproducible viscosity measurements; both are available.[144]

4. *Rheology*

A single carboxymethylcellulose solution will appear to have a different viscosity when different conditions of physical force are imposed on the solution.[187–190] These conditions of physical force may be conveniently referred to as high, intermediate, or low shear stress. For example, rolling or spreading a liquid as if it were an ointment or lotion would be high shear stress. On the other hand, after the liquid has been applied, gravity and surface tension control flow. These forces are conditions of low stress. Intermediate stress can be typified by pouring a liquid out of a bottle. Thus, if a solution of high-viscosity sodium carboxylmethylcellulose appears to be a viscous syrup as it is poured

TABLE II

Factors Affecting Disaggregation of Sodium Carboxymethylcellulose

		Viscosity, cps [a]					
		Anchor stirrer			Waring Blendor		
DS	Viscosity Type	Distilled water	4% NaCl	Saturated NaCl	Distilled water	4% NaCl	Saturated NaCl
0.4	High	900	11	6	4000	65	16
0.7	High	1680	140	45	760	1040	2440
0.7	High[b]	1680	570	165	760	750	1720
0.9	Medium	215	160	225	125	95	235
1.2	Medium	175	80	180	100	55	140

[a] This table shows the effect of polymer composition, solvent strength, and mechanical shear on disaggregation as measured by solution viscosity. All data are in centipoises at 25°.

[b] Special, nonthixotropic type.

from a bottle, it will behave as a runny liquid when applied as a lotion and yet, when high shear stress is removed, it will instantly revert to a nonflowing character. This type of flow behavior is termed *pseudoplasticity,* a form of non-Newtonian flow; it differs from the time-dependent consistency change that is called *thixotropy.*

Solutions of this gum exhibit pseudoplastic behavior because long-chain molecules tend to orient themselves in the direction of flow; as the applied force (shear stress) is increased, the resistance to flow (viscosity) is thereby decreased. Using a smaller force on the same solution, the apparent viscosity is higher because a random orientation of molecules presents increased resistance to flow. If shear rate is plotted versus shear stress, a Newtonian fluid will produce a straight line passing through the origin; a pseudoplastic liquid will yield a curved line.

Solutions of low-molecular-weight sodium carboxymethylcellulose, are low-viscosity types and less pseudoplastic than are solutions of high-molecular-weight sodium carboxymethylcellulose. However, at very low shear rates, all sodium carboxymethylcellulose solutions approach Newtonian flow (Fig. 8).

If dissolved long-chain polymers interact, they will develop a three-dimen-

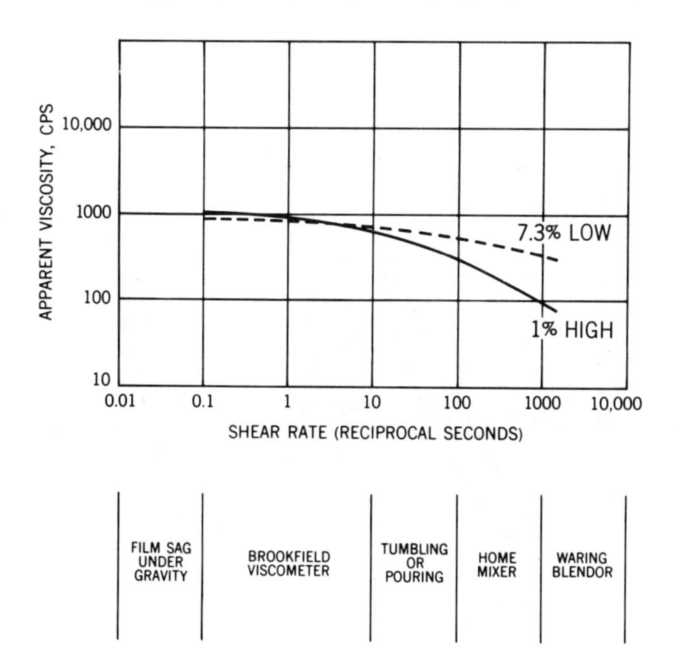

FIG. 8.—Effect of shear rate on apparent viscosity of sodium carboxymethylcellulose (DS 0.7) solutions.

sional structure and exhibit a phenomenon known as thixotropy. Thixotropy is a time-dependent consistency change. It is characterized by an increase in apparent viscosity when a solution remains at rest for a period of time. In certain cases, the solution may develop some gel strength or even set to an almost solid gel. If sufficient force is exerted on a thixotropic solution, the structure can be broken and the apparent viscosity reduced.

Thixotropic solutions form a hysteresis loop when shear rate is plotted against shear stress (Fig. 9A). The increased shear stress required to break the thixotropic structure reduces the viscosity. If a solution has gel strength, a spur forms in the hysteresis loop (Fig. 9B), an indication of the stress necessary to break the gel structure and cause the solution to revert to its normal apparent viscosity.

Thixotropic solutions are desirable, or even essential, for certain uses, such as the suspension of solids. The high- and medium-viscosity types of sodium carboxymethylcellulose generally exhibit thixotropic behavior. Other uses require a polymeric solution having no thixotropic properties. Special smooth types have been developed for those applications that require clear solutions free of thixotropy. Figure 10 illustrates the difference in appearance between solutions of regular and nonthixotropic sodium carboxymethylcellulose.

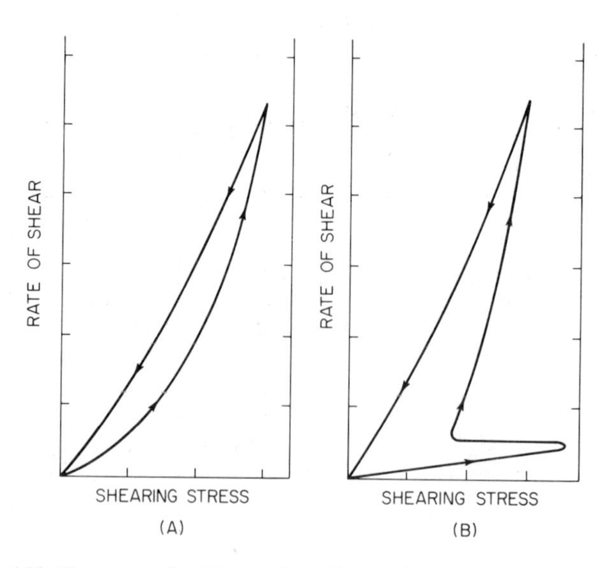

FIG. 9.—(A) Rheogram of a thixotropic sodium carboxymethylcellulose solution. (B) Rheogram of a highly thixotropic sodium carboxymethylcellulose solution with gel strength.

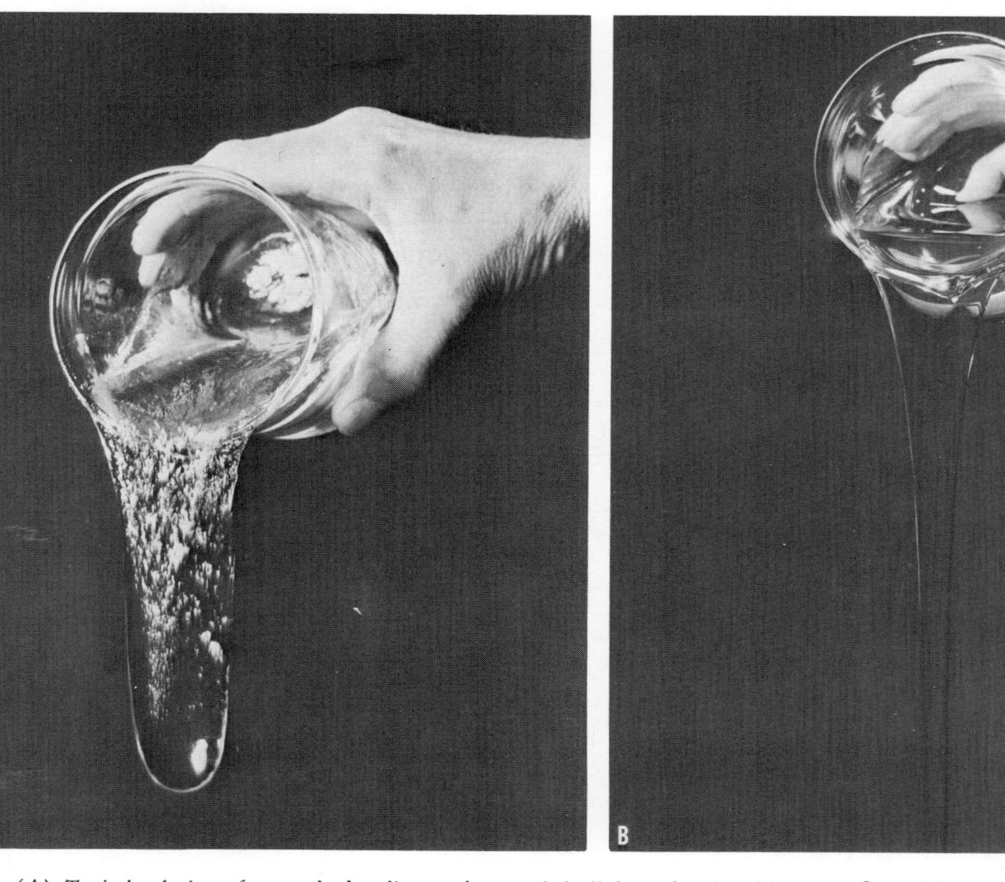

FIG. 10.—**(A)** Typical solution of a standard sodium carboxymethylcellulose, showing thixotropic flow. (B) Typical solution of a nonthixotropic sodium carboxymethylcellulose.

5. *Stability*[163, 170]

Effect of temperature.[192]—Viscosities of sodium carboxymethylcellulose solutions become lower with increasing temperature (Fig. 11). Under normal conditions, the effect of temperature on viscosity is reversible, and raising or lowering the solution temperature has no permanent effect on the viscosity characteristics of the solution. However, long periods of heating at high temperatures will tend to depolymerize sodium carboxymethylcellulose and reduce its viscosity.

Effect of pH.[191]—Solutions of sodium carboxymethylcellulose maintain their normal viscosity over a wide pH range. In general, solutions exhibit their maximum viscosity and best stability at pH 7–9. Below pH $\sim$ 3, precipitation of the free-acid form of carboxymethylcellulose may occur. Above pH 10, a slight decrease in viscosity is observed.

In acidic systems, the order of mixing sodium carboxymethylcellulose and solvent effects solution properties. If a solution of the gum is prepared prior

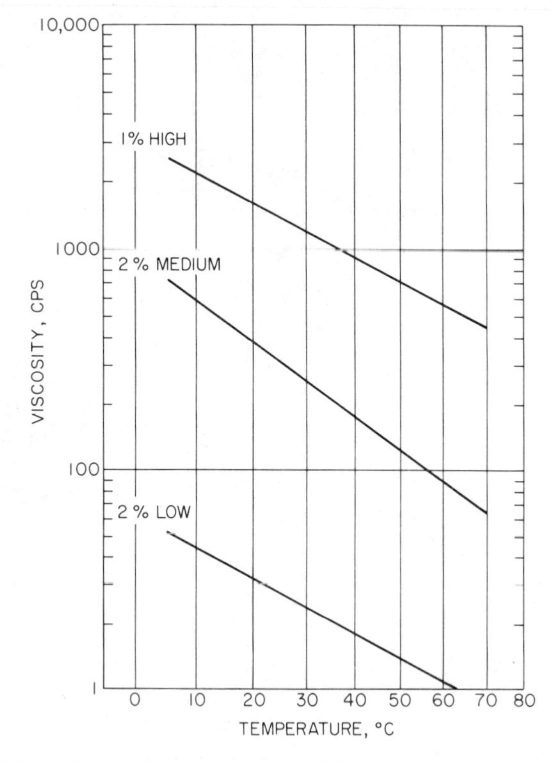

FIG. 11.—Effect of temperature on viscosity of three viscosity types of sodium carboxymethylcellulose solutions.

to the addition of acid, a higher viscosity is obtained than when dry sodium carboxymethylcellulose is dissolved in an acidic solution. Carboxymethylcellulose may undergo acid-catalyzed hydrolysis in acidic solutions, resulting in permanent loss of viscositiy. This reaction is accelerated by increasing the temperature and/or lowering the pH. A special type of acid-stable gum that provides some resistance to viscosity degradation and precipitation in low pH systems is manufactured.

Oxidative degradation occurs under alkaline conditions in the presence of oxygen. The rate of viscosity loss is also increased by heat and/or ultraviolet radiation. Inclusion of an antioxidant, exclusion of oxygen, and avoidance of highly alkaline conditions are obvious preventive measures.

Effect of concentration.—The increase in viscosity exhibited by sodium carboxymethylcellulose in water can be explained by the following mechanism. At concentrations of 1% or more, molecules are in close contact so that there is little tendency for the counterions (in this case, sodium) to move out of the sphere of influence of the fixed charges on the polymer molecules. Upon dilution, the cations tend to move away from the fixed charges into the aqueous interpolymer regions, leaving a net charge on the molecules. These unsatisfied charges along the polymer chains set up repulsions that cause the molecules to become less coiled. As dilution continues, the charge density on the chains increases and the chains may uncoil to their maximum lengths (Fig. 12). Because viscosity is determined, in part, by the size and length of the polymer molecules,

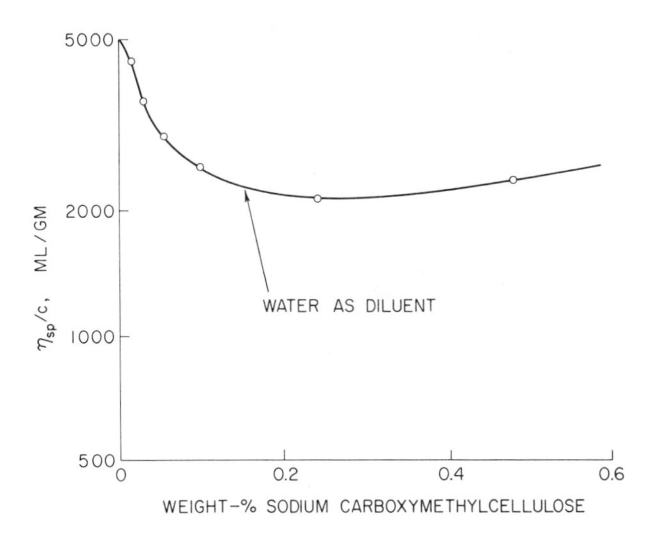

FIG. 12.—Dilute solution viscosity data for sodium carboxymethylcellulose. (Redrawn from Tait and coworkers.[206])

there will be a corresponding rise in viscosity as the effective chain length is increased.

Effect of biological agents.[191–194]—Sodium carboxymethylcellulose solutions can be stored at room temperatures for indefinite periods with no loss in viscosity, provided they remain free of mold and bacterial growths. Heat treatment may be used to destroy many microorganisms, having little effect on the properties of sodium carboxymethylcellulose. Heating for 30 min at 80° or for 1 min at 100° is generally sufficient. For storage of solutions, a preservative should be added to prevent viscosity degradation (Table III). Preservatives that are cleared by the Food and Drug Administration for food, cosmetic, and pharmaceutical products are listed separately (Table IV).

If cellulases have been introduced by microbial action, preservatives do not effectively prevent degradation. Therefore, it is important to preserve solutions

TABLE III

Fungus and Mold Inhibitors for Sodium Carboxymethylcellulose Solutions

Preservative	Manufacturer	Concentration, % of total solution
Chlorinated phenols	several	
Dowicide A	Dow Chemical Co.	0.10
Santobrite	Monsanto Chemical Co.	0.10
Formaldehyde	several	0.10
Hydroxyquinoline	several	0.10
Iodine	several	0.002
K-Ortho 50	Buckman Labs, Inc.	
Methyl parasept	Heyden Chemical Corp.	0.20
Phenol	several	0.30
Phenylmercuric acetate	several	0.001
Phenylmercuric nitrate	several	0.002
Pine Oil	Hercules Inc.	0.20
Preventol	General Dyestuf Corp.	0.10
Thimerosal	Eli Lilly & Co.	0.01
Thymol	several	0.10

TABLE IV

Preservatives Cleared by the FDA for Food, Cosmetics, and Pharmaceuticals

Sorbic acid, and its potassium, calcium, and sodium salts
Methyl *p*-hydroxybenzoate
Propyl *p*-hydroxybenzoate
Sodium benzoate
Sodium propionate

as soon as possible after preparation. Cellulase in water solutions can usually be destroyed by heating the solution for 15 min at 50° at pH 9–10.

Biological oxygen demand for sodium carboxymethylcellulose as determined by a standard procedure is lower than that of corn starch (Fig. 13).

6. *Compatibility*[163, 170]

Water-soluble polymers.—Sodium carboxymethylcellulose is compatible with most water-soluble gums over a wide range of concentrations. In many instances, the low-viscosity types are compatible over a broader range than the high-viscosity types. When a solution of the anionic polymer sodium carboxy-

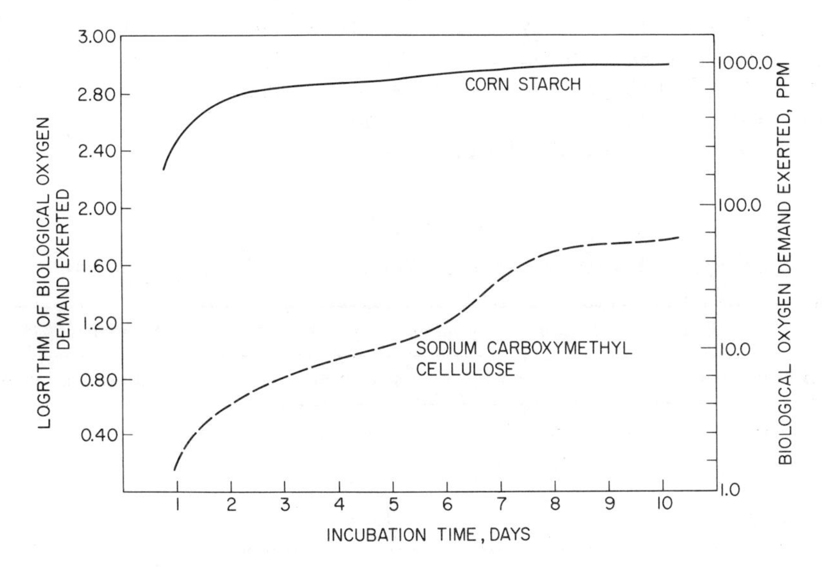

FIG. 13.—Comparison of biological oxygen demand exerted by corn starch and sodium carboxymethylcellulose. (Redrawn from Dickerson.[24])

methylcellulose is blended with a solution of a nonionic polymer, such as hydroxyethylcellulose or hydroxypropylcellulose, a synergistic viscosity effect occurs and solution viscosities are considerably higher than normally expected.

Solvents.—Although sodium carboxymethylcellulose is insoluble in organic solvents, it is soluble in mixtures of water and water-miscible solvents, such as ethanol or acetone. Aqueous solutions of low concentration containing up to 50% ethanol or 40% acetone can be made. Aqueous solutions of sodium carboxymethylcellulose tolerate subsequent addition of even higher quantities of acetone or ethanol. The low viscosity types are considerably more tolerant than the high viscosity types, as shown in Table V.

Salts.—As a general rule, monovalent cations form soluble salts of carboxymethylcellulose, divalent cations form hazy solutions of reduced viscosity, and trivalent cations form insoluble salts.

The effect of salts varies with the particular salt, the concentration of the salt, the pH of the solution, the degree of substitution of the sodium carboxymethylcellulose, and the manner in which the salt and the gum come into contact. Highly substituted materials (DS 0.9–1.2) have a greater tolerance for most salts. Increased salt tolerance may be obtained by mixing solutions of sodium carboxymethylcellulose and solutions of salt or by adding the salt to a gum solution, rather than adding dry gum to a salt solution, or by dissolving the salt and gum simultaneously.

Generally, divalent cations will not form crosslinked gels with sodium carboxymethylcellulose. Viscosity reduction occurs when divalent cations are added to a sodium carboxymethycellulose solution and may be accompanied by the formation of haze. Calcium, barium, cobalt(II), magnesium, iron(II), and manganese(II) cations perform in such a manner.

Gelation.—Gelation of sodium carboxymethylcellulose solutions can be con-

TABLE V

Tolerance of Sodium Carboxymethylcellulose (DS 0.7) Solutions for Ethanol

Viscosity Type	Volume Ratio of Ethanol to Gum Solution (1%) [a]	
	First evident haze	First distinct precipitate
Low	2.4 to 1	3.6 to 1
Medium	2.1 to 1	2.7 to 1
High	1.6 to 1	1.6 to 1

[a] In these tests, 95% ethanol was added slowly at ~25° to the vigorously stirred 1% gum solution.

trolled to produce stable food- or industrial-grade gels, ranging in consistency from soft and pourable to very firm. Trivalent metal ions from such compounds as aluminum sulfate (hydrated), basic aluminum acetate, and ferric phosphate can be used to produce gelation. Chelating the trivalent metal ions with certain anions, such as citrate or malate, and adjustment of solution pH or temperature provide complete control over the rate of gel formation and the consistency of the resultant gel.

Examples of gels from a 1% solution of high-viscosity sodium carboxymethylcellulose (DS 0.7) are illustrated in Table VI.

Protein.—A 0.5% solution of high-viscosity, DS 0.7 sodium carboxymethylcellulose will provide solubilization of an equal amount of soy protein in the isoelectric range, as evidenced by a large increase in apparent viscosity (Table VII). This interaction between sodium carboxymethylcellulose and soy protein is dependent upon the degree of substitution and the degree of polymerization of the gum. A 0.5% solution of high-viscosity, DS 0.4 sodium carboxymethylcellulose will solubilize 0.5% soy protein, whereas it requires 1.5% of a medium viscosity, DS 1.2 gum to solubilize 0.5% soy protein. A low viscosity, DS 0.7 type does not solubilize the protein. Heating the gum–protein complex does not insolubilize the protein or irreversibly dissociate the complex.

The reaction between casein and sodium carboxymethylcellulose is similar to that between soy protein and sodium carboxymethylcellulose. Sodium carboxymethylcellulose inhibits precipitation of casein in its isoelectric region, and abnormally high viscosities are obtained. Casein is apparently more readily solubilized by sodium carboxymethylcellulose than is soy protein. At higher soy concentrations, a suspending action is evident for the gum and/or the gum–

TABLE VI

Gelation of Sodium Carboxymethylcellulose Solutions [a]

% Gum	Solution pH	Reagent Solution [b] Citrate : Al^{3+}	pH	Results
1.0	7.0	0.33	2	immediate, firm gel
1.0	7.0	0.50	2	immediate, pourable gel
1.0	7.0	0.30	6.5	soft gel in 100 min
1.0	5.5	0.50	6.5	medium gel in 60 min
1.0	5.0	0.50	7.0	soft gel in 180 min

[a] Gels are prepared by adding the reagent solution directly to the gum solution with hand stirring for several seconds to ensure uniform dispersion.

[b] Aluminum sulfate (hydrated) is the source of trivalent cation.

TABLE VII

Interaction of Sodium Carboxymethylcellulose (DS 0.7) and Soy Protein

	Viscosity of Aqueous Solution,[a] cps		
pH	0.5% Soy Protein	0.5% Gum	0.5% Soy Protein + 0.5% Gum
6.9	3	180	160
6.5	3	160	160
5.9	3	160	170
5.6	precipitate	125	330
4.7	precipitate	96	990
4.2	precipitate	82	920
3.5	precipitate	62	600

[a] A high-viscosity, nonthixotropic type of sodium carboxymethylcellulose was used.

protein complex. The increase in viscosity is more pronounced with sodium caseinate than with calcium caseinate as the pH decreases.

Carbohydrates.—In solutions of sugars, sodium carboxymethycellulose increases the viscosity.

The viscosity behavior of regular types of sodium carboxymethylcellulose in sugar solutions is dependent upon the order of addition and the method of solution preparation. For example, if the gum is first thoroughly dissolved in water and sugar is then added, the viscosity of the gum solution is only slightly affected. However, if sugar is first dissolved in water and dry gum is then added to the sugar solution, the sugar tends to inhibit disaggregation of the sodium carboxymethylcellulose and a lower viscosity is obtained. The special smooth types and highly substituted types of sodium carboxymethylcellulose do not show this effect. In a mixed-solvent system of sugar and water, these types of gum provide a viscosity bonus effect, such as is illustrated by the glycerol–water system (Fig. 14).

In addition to thickening sugar solutions, sodium carboxymethylcellulose retards the rate of crystallization of sugar from supersaturated solutions. It also aids in retarding the growth of large crystals (Fig. 15). Of the types with DS 0.7, low-viscosity grades are most effective as crystallization inhibitors whereas the high viscosity grades are most effective in preventing water loss by evapora-

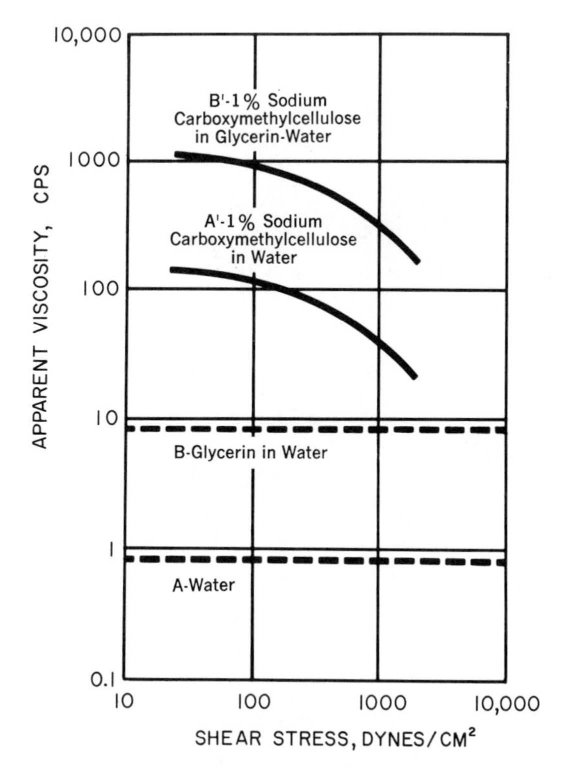

FIG. 14.—Effect of mixed solvents on the viscosity of sodium carboxymethylcellulose (DS 1.2) solutions.

tion. The relative water-retention ability of various viscosity types of sodium carboxymethylcellulose in a fondant is compared in Figure 16.

In starch systems, sodium carboxymethylcellulose is used to increase viscosity economically, as well as to control syneresis or water separation. It imparts viscosity to starch systems prior to gelatinization of the starch.

7. *Films*[163]

Sodium carboxymethylcellulose is seldom used to prepare free or unsupported films. However, its ability to form strong, oil-resistant films is important to many applications. Clear films can be obtained by evaporation of a sodium carboxymethylcellulose solution. These strong, fairly flexible films are unaffected by oils, greases, or organic solvents. Typical properties of such films are given in Table VIII.

Plasticizers.—Where improved flexibility and elongation are desired, plasticizer is added to the casting solution. By including 10–30% glycerol in a formula-

FIG. 15.—For legend see facing page.

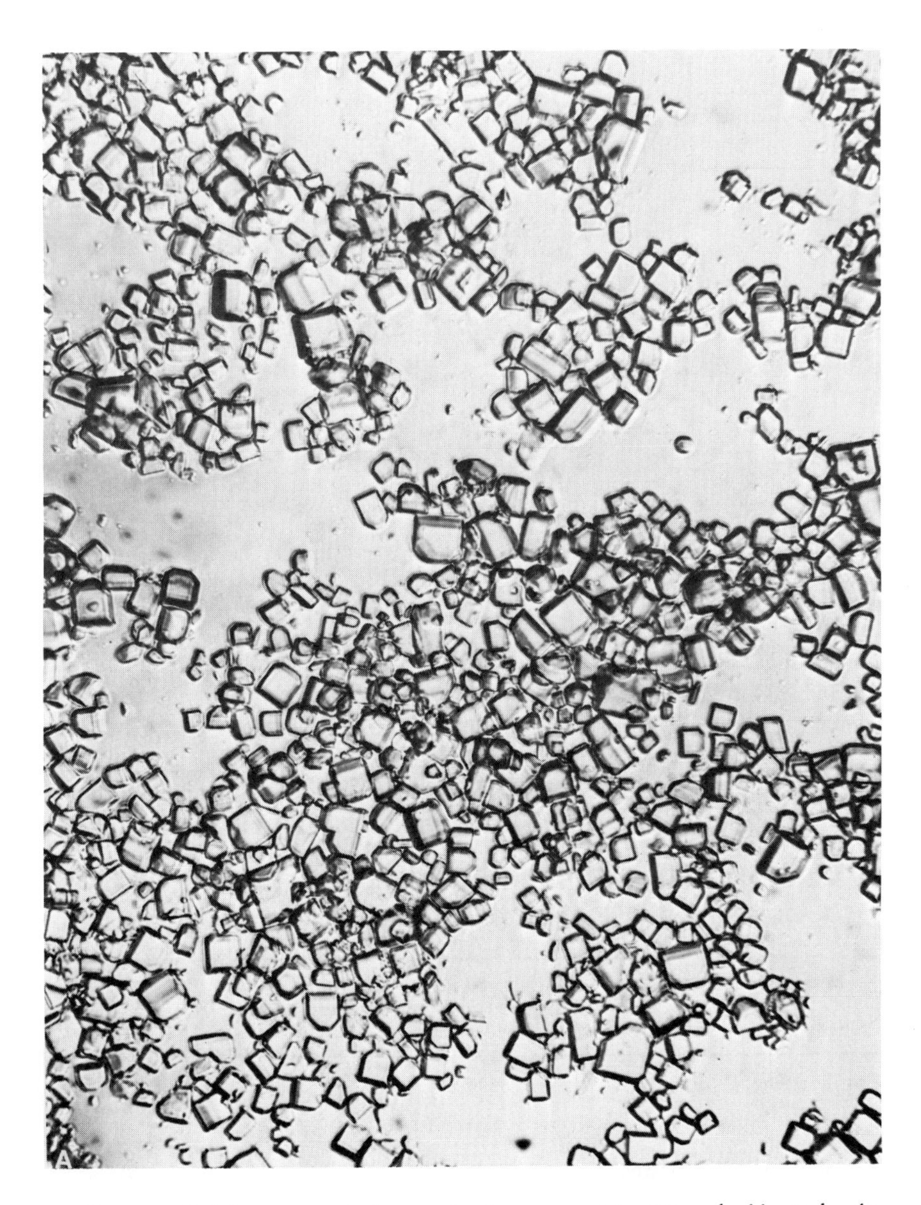

FIG. 15.—Cellulose gum retards growth of sugar crystals. When freshly made, the crystal sizes in these all-sucrose fondants were the same. After 2 months, fondant at left without cellulose gum contained crystals measuring up to 80 μ. Fondant at right, containing 0.25% of DS 0.7 cellulose gum, had crystals measuring up to 18 μ and was not substantially different from a freshly made fondant.

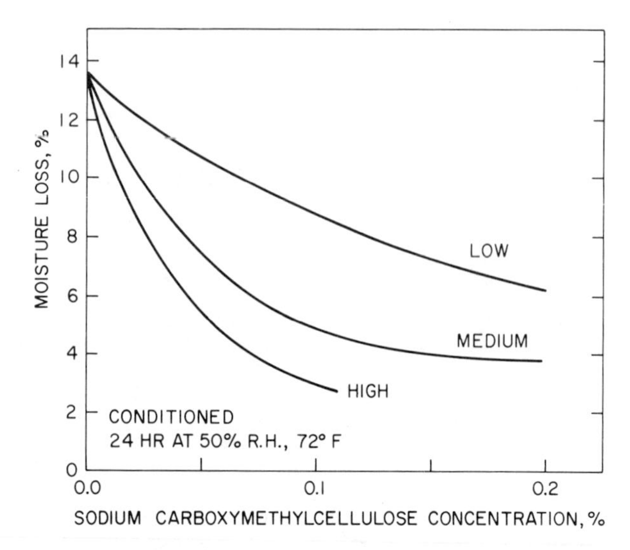

FIG. 16.—Effect of sodium carboxymethylcellulose (DS 0.7) on moisture retention in 85% sucrose system.

tion, elongation can be increased by 40–50% and the folding endurance to 10,000 MIT doublefolds. Other plasticizers that have proved effective with sodium carboxymethylcellulose are listed in Table IX.

Insolubilization.—Sodium carboxymethylellulose films can be insolubilized by crosslinking the hydroxyl or carboxyl groups of the gum with a polyfunctional

TABLE VIII

Typical Properties of Sodium Carboxymethylcellulose (DS 0.7) Films

Property Tested	Typical Values		
	Low-viscosity	Medium-viscosity	High-viscosity
Tensile strength, psi	8000	13,000	15,000
Elongation, % at break	8.3	14.3	14.3
Flexibility, MIT double folds	93	131	513
Electrostatic charge	negative	negative	negative
Refractive index	1.515	1.515	1.515
Specific gravity	1.59	1.59	1.59

TABLE IX

Plasticizers for Sodium Carboxymethylcellulose Films

Diglycerol	2,3-Butylene glycol
Ethanolamines	1,2,6-Hexanetriol
Ethylene glycol	2-Methyl-2-nitro-1,3-propanediol
Propylene glycol	Glycerol α-monomethyl ether
Glycerol monochlorohydrin	

resin, such as a polyamide, melamine–formaldehyde, or dimethylol urea. An aqueous solution of sodium carboxymethylcellulose and the resin can be air dried at room temperature, or at elevated temperature, to obtain insoluble films of good strength and flexibility. Clear films with varying degrees of water insolubility can be prepared by controlling the amount of added resin and the time and temperature at which the film is cured.

Water-insoluble films may also be produced from a dry film of sodium carboxymethylcellulose by treating it with a solution of an appropriate salt, usually of a trivalent cation, that can replace the sodium ions.

8. *Moisture*[190–196]

Sodium carboxymethylcellulose absorbs moisture from the atmosphere. The effect of relative humidity on the equilibrium moisture content of three different DS levels of sodium carboxymethylcellulose is shown in Fig. 17.

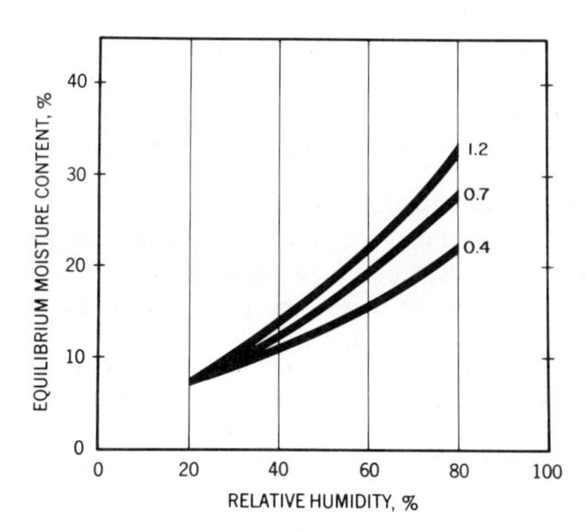

FIG. 17.—Effect of relative humidity at 25° on equilibrium moisture content of sodium carboxymethylcellulose at different DS levels.

VI. Analysis

Numerous[197–205] analytical procedures are available for determination of carboxymethylcellulose and its degree of substitution.

VII. References

(1) J. M. DeBell, H. M. Richardson, and W. E. Gloor, *in* "German Plastics Practice," DeBell and Richardson, Springfield, Mass., 1946, p. 203.

(2) E. D. Klug, U.S. Patent 2,523,377 (1950); *Chem. Abstr.,* **45,** 1344 (1951).

(3) J. F. Haskins and R. W. Maxwell, U.S. Patent 2,131,733 (1938); *Chem. Abstr.,* **32,** 9496 (1938).

(4) R. W. Swinehart and S. T. Allen, U.S. Patent 2,524,024 (1950); *Chem. Abstr.,* **45,** 1344 (1951).

(5) W. R. Collings, R. M. Freeman, and R. C. Anthonisen, U.S. Patent 2,278,612 (1942); *Chem. Abstr.,* **36,** 5013 (1942).

(6) R. N. Hader, W. F. Waldeck, and F. W. Smith, *Ind. Eng. Chem.,* **44,** 2803 (1952).

(7) W. F. Waldeck, U.S. Patent 2,510,355 (1950); *Chem. Abstr.,* **44,** 7538 (1950).

(8) L. N. Rogers, W. A. Mueller, and E. E. Hembree, U.S. Patent 2,553,725 (1951); *Chem. Abstr.,* **45,** 8247 (1951).

(9) E. D. Klug and J. S. Tinsley, U.S. Patent 2,517,577 (1950); *Chem. Abstr.,* **44,** 10318 (1950).

(10) V. R. Grassie and C. R. Wallis, U.S. Patent 2,680,737 (1954); *Chem. Abstr.,* **48,** 14205 (1954).

(11) O. H. Paddison and R. W. Somers, U.S. Patent 2,976,278 (1961); *Chem. Abstr.,* **55,** 15929 (1961).

(12) G. Goethals, Belg. Patent 572,851 (1962).

(13) B. T. Lamborn, U.S. Patent 2,513,807 (1950); *Chem. Abstr.,* **44,** 8656 (1950).

(14) L. H. Burt, *Drug Stand.,* **19,** 106 (1951).

(15) "Chemical Profile: CMC," *Oil, Paint, Drug Rep.,* **194,** 9 (1968).

(16) United States Tariff Commission Reports 1968, U.S. Government Printing Office, Washington, D.C., 1968.

(17) "Chemical Economics Handbook," Stanford Research Institute, Menlo Park, California, 1970.

(18) B. Sundberg, *Farm. Revy,* **45,** 8247 (1946).

(19) Deutsche Celluloid Fabrik Eilenberg, Germ. Patent 332,203 (1918).

(20) G. S. Baird and J. K. Speicher, *in* "Water-Soluble Resins," R. L. Davidson and M. Sittig, eds., Reinhold Publishing Corp., New York, 1962, p. 69.

(21) Y. L. Meltzer, *Soap Chem. Spec.,* **44,** No. 11, 72 (1968).

(22) Y. L. Meltzer, *Soap Chem. Spec.,* **44,** No. 12, 122 (1968).

(23) Unilever Research Laboratories, *J. Appl. Chem.,* **17,** 276 (1967).

(24) B. W. Dickerson, *Ind. Wastes,* **1,** 10 (1955).

(25) J. G. Wiegerink, U.S. Patent 2,645,584 (1953); *Chem. Abstr.,* **47,** 10866 (1953).

(26) S. F. Dieckman, J. G. Jarrell, and R. S. Voris, *Ind. Eng. Chem.,* **45,** 2287 (1953).

(27) E. K. Bolton, U.S. Patent 2,087,237 (1937); *Chem. Abstr.,* **31,** 6482 (1937).

(28) C. Zehender, *Papier (Darmstadt),* **14,** 631 (1960).

(29) G. S. Baird and A. L. Griffiths, *Mod. Text. Mag.*, **46,** 34 (1965).
(30) G. S. Baird, *in* "The American Cotton Handbook," Interscience Publishers, New York, Vol. 2, 3rd Ed., 1966, p. 550.
(31) *Text. Ind.*, **124,** 161 (1960).
(32) J. H. Elliott, U.S. Patent 2,650,887 (1953); *Chem. Abstr.*, **48,** 384 (1954).
(33) R. T. Gardner, Jr., *Soap Chem. Spec.*, **38,** No. 6, 98 (1962).
(34) *Text. World,* **113,** 60 (1963).
(35) *Soap Chem. Spec.*, **35,** No. 4, 135 (1959).
(36) T. A. Seequist, U.S. Patent 2,979,373 (1961); *Chem. Abstr.*, **55,** 17053 (1961).
(37) E. Wurz and O. Wurz, *Text. Rundsch.*, **8,** 557 (1953).
(38) H. B. Bush and H. B. Trost, *Hercules Chem.*, **60,** 14 (1970).
(39) R. B. Seymour and G. M. Schroder, U.S. Patent 2,486,803 (1949); *Chem. Abstr.*, **44,** 3263 (1949).
(40) *Pap. Trade J.*, **154,** 28 (1970).
(41) A. Portolani, E. Di Giulio, and M. Milano, U.S. Patent 3,427,183 (1969).
(42) Johnson & Johnson Co., Neth. Patent 69 15317 (1970).
(43) *Chem. Week,* **100,** 81 (1967).
(44) R. Dicker, *Rayonne,* **6,** No. 2, 73; No. 3, 77 (1950).
(45) E. Bartholomé and K. F. Buschmann, *Melliand Textilber.*, **30,** 249 (1949).
(46) H. L. Sanders and J. M. Lambert, *Text. Res. J.*, **21,** 680 (1951).
(47) E. F. Hill, C. E. Smith, and T. H. Vaughn, U.S. Patent 2,568,334 (1951); *Chem. Abstr.*, **46,** 764 (1952).
(48) A. S. Weatherburn, *Text. Res. J.*, **20,** 510 (1950).
(49) W. P. Untermohlen, M. E. Ryan, and D. O. Young, *Text. Res. J.*, **21,** 510 (1951).
(50) F. W. Smith, U.S. Patent 2,542,215 (1951); *Chem. Abstr.*, **45,** 4392 (1951).
(51) K. J. Niewenhuis, *J. Polym. Sci.*, **12,** 237 (1954).
(52) J. B. Batdorf, *Soap Chem. Spec.*, **38,** No. 1, 58 (1962).
(53) F. C. Hall, *Chem. Ind. (London)*, 314 (1961).
(54) J. G. Jarrell and H. B. Trost, *Soap Sanit. Chem.*, **28,** No. 7, 40; No. 8, 50 (1952).
(55) N. Pilpel, *Res. Appl. Ind.*, **14,** 319 (1961).
(56) H. Stupel, *SVF Fachorgan Textilveredl.*, **7,** 154 (1952).
(57) H. Stupel, *Seifen-Öle-Fette-Wachse,* **82,** 99 (1956).
(58) J. Berch and G. L. Drake, *Text. Res. J.*, **34,** 29 (1964).
(59) B. J. Rutkowski, *J. Amer. Oil Chem. Soc.*, **45,** 266 (1968).
(60) G. A. Johnson and F. G. Foster, *J. Appl. Chem.*, **18,** 235 (1968).
(61) C. H. Bayley, A. S. Weatherburn, and G. R. F. Rose, *Laundry Dry Cleaning J. Can.* (Dec. 1948); *Nat. Res. Coun. Can. Rep.*, No. 1847.
(62) L. J. Armstrong, *Amer. Dyest. Rep.*, **37,** No. 18, *Proc. Amer. Ass. Text. Chem. Color.*, 596 (1948).
(63) D. T. Berglund, *Sv. Papperstidn.*, **51,** 555 (1948).
(64) E. E. Anderson, W. B. Esselen, and C. R. Fellers, *J. Amer. Diet. Ass.*, **29,** 770 (1953).
(65) Z. D. Roundy and N. R. Osmond, U.S. Patent 2,956,885 (1960); *Chem. Abstr.*, **55,** 3871 (1961).
(66) National Dairy Products, Brit. Patent 916,104 (1963).
(67) V. D. Ludington, R. E. Schara, and R. E. Mohlie, U.S. Patent 3,119,691 (1964).

(68) L. H. Burt, U.S. Patent 2,548,865 (1951); *Chem. Abstr.*, **45,** 6320 (1951).

(69) J P. Julien, *Ice Cream Trade J.*, **49,** No. 9, 44 (1953).

(70) F. E. Potter and D. H. Williams, *Milk Plant Mon.*, **39,** No. 4, 76 (1950).

(71) M. Landers, U.S. Patent 2,423,600 (1947); *Chem. Abstr.*, **41,** 5998 (1947).

(72) Onderzoekinginstituut "Research" N.V., Neth. Patent 73,644 (1953); *Chem. Abstr.*, **48,** 6617 (1954).

(73) A. J. Ganz, U.S. Patent 3,009,812 (1960); *Chem. Abstr.*, **56,** 3870 (1962).

(74) A. J. Ganz, *Candy Ind. Confect. J.*, **128,** No. 9, 5 (1967).

(75) A. J. Desmarais and A. J. Ganz, *Mfg. Confect.*, **42,** No. 10, 33 (1962).

(76) J. B. Batdorf, *Food Eng.*, **36,** 66 (1964).

(77) J. B. Batdorf and J. B. Klis, *Food Process.*, **24,** No. 5, 32 (1963).

(78) E. G. Bayfield, *Baker's Dig.*, **36,** No. 2, 50 (1962).

(79) W. E. Young and E. G. Bayfield, *Cereal Chem.*, **40,** 195 (1963).

(80) B. S. Miller and R. M. Sandstedt, *Food Technol.*, **21,** 59A (1967).

(81) R. H. McGreer, *J. Amer. Diet. Ass.*, **51,** 534 (1967).

(82) A. J. Ganz, *Mfg. Confect.*, **46,** No. 10, 23 (1966).

(83) *Food Technol.* (*Chicago*), **19,** 155 (1965).

(84) A. Partyka, U.S. Patent 3,093,485 (1963).

(85) H. H. Bogin and R. D. Feick, U.S. Patent 2,555,465 and 2,555,467 (1951); *Chem. Abstr.*, **45,** 7726, 7722 (1951).

(86) R. Perech, U.S. Patent 2,393,561 and 2,393,562 (1964); *Chem. Abstr.*, **40,** 2558 (1946).

(87) L. D. Dunn, U.S. Patent 3,073,703 (1963).

(88) *Chem. Week,* **106,** No. 8, 93 (1970).

(89) D. F. Mason, U.S. Patent 3,472,662 (1969); *Chem. Abstr.*, **72,** 11506g (1970).

(90) F. Long, U.S. Patent 2,927,029 (1960).

(91) Pillsbury Company, Brit. Patent 937,566 (1963).

(92) R. G. Hyldon, U.S. Patent 3,526,513 (1970).

(93) J. T. Collins and R. G. Hyldon, U.S. Patent 3,526,512 (1970); *Chem. Abstr.*, **73,** 119427s (1970).

(94) J. T. McKnight, U.S. Patent 3,123,482 (1964).

(95) *Food Eng.*, **33,** No. 5, 92 (1961).

(96) A. S. Kiratsous, F. J. Francis, and J. W. Zahradnik, *Food Technol.* (*Chicago*), **16,** 111 (1962).

(97) E. Bactowsky and J. Presto, *Bull. Amer. Soc. Hosp. Pharm.*, **7,** 65 (1950).

(98) B. F. Allen, *Md. Pharm.*, **37,** 612 (1961).

(99) R. H. Blythe, J. J. Gulesich, and H. L. Tuthill, *J. Amer. Pharm. Ass.*, **38,** 59 (1949).

(100) S. W. Goldstein, *J. Amer. Pharm. Ass., Prac. Pharm. Ed.*, **13,** 550 (1952).

(101) D. Etchells, *Pharm. J.*, **172,** 372 (1954).

(102) H. G. DeKay, *Bull. Amer. Soc. Hosp. Pharm.*, **9,** 520 (1952).

(103) K. Ishii, H. Nakatani, and F. Imai, Jap. Patent 3297 ('64) (1964); *Chem. Abstr.*, **61,** 2518 (1964).

(104) T. W. Schwarz, *Amer. Perfum. Cosmet.*, **77,** No. 10, 85 (1962).

(105) P. Hedner, *Acta Endocrinol.* (*Copenhagen*), **43,** 499 (1963).

(106) Cilag, Ltd., West Gr. Patent 806,701 (1951); *Chem. Abstr.*, **45,** 9229 (1951).

(107) R. J. Mehoffey, U.S. Patent 2,678,902 (1954); *Chem. Abstr.*, **48,** 9634 (1954).

(108) A. J. Watters and D. M. Shepherd, U.S. Patent 2,520,805 (1950); *Chem. Abstr.*, **45,** 3654 (1951).

(109) A. H. Knechtel, *Amer. Perfum.*, **78,** 95 (1963).

(110) G. Scapparino, *Riv. Ital. Essenze, Profumi, Piante Offic., Olii Veg., Saponi,* **34,** 343 (1952).

(111) M. Wotzilka and R. Samel, U.S. Patent 3,122,481 (1964).

(112) A. J. Mueller, U.S. Patent 3,072,535 (1963); *Chem. Abstr.*, **58,** 6644 (1963).

(113) J. A. Cornell and H. A. Kemmler, U.S. Patent 2,733,156 and 2,733,157 (1956); *Chem. Abstr.*, **50,** 6840 (1956).

(114) M. K. Gluzman, I. B. Levitskaya, and G. S. Bashura, *Vestn. Dermatol. Venerol.*, **35,** 40 (1961).

(115) D. W. Doerr, E. R. Serles, and D. L. Deardorff, *J. Amer. Pharm. Ass.*, **43,** 98 (1954).

(116) R. P. Tansey, U.S. Patent 3,133,863 (1964); *Chem. Abstr.*, **61,** 2915 (1964).

(117) G. L. Christenson and L. B. Dale, U.S. Patent 3,065,143 (1962); *Chem. Abstr.*, **58,** 5462 (1963).

(118) S. Long, U.S. Patent 3,043,747 (1962); *Chem. Abstr.*, **57,** 8670 (1962).

(119) A. C. Nestle, U.S. Patent 2,601,050 (1952); *Chem. Abstr.*, **46,** 9294 (1952).

(120) C. S. Scanley, *Amer. Chem. Soc., Div. Petrol. Chem. Preprints,* **7,** 65 (1962).

(121) H. Diener, *Erdoel Kohle,* **16,** 841 (1963).

(122) S. A. Williams, *J. Appl. Polym. Sci.*, **8,** 1015 (1964).

(123) T. E. Watkins, West Gr. Patent 1,041,892 (1958); *Chem. Abstr.*, **55,** 966 (1961).

(124) R. Bobo, *Oil Gas J.*, **69,** No. 5, 78 (1971).

(125) J. P. Casey, "Pulp and Paper," Vols. 1–3, 2nd Ed., Interscience Publishers, New York, 1960–1961.

(126) R. A. Raff and M. F. Adams, *Tappi,* **50,** 45A (1967).

(127) E. J. Barber and H. O. Ware, *Tappi,* **40,** 365 (1957).

(128) O. Wurz, *Papier (Darmstadt)*, 377 (1953).

(129) A. R. Hurst, unpublished paper given at 1957 annual convention of the Technical Association of the Pulp and Paper Industry (1957).

(130) J. D. Lohnas and A. B. Carlson, U.S. Patent 3,060,044 (1962); *Chem. Abstr.*, **58,** 3603 (1963).

(131) H. Kohne, Jr., *Tappi,* **42,** 294 (1959).

(132) P. W. Sherwood, *Adhes. Age,* **5,** No. 6, 32 (1962).

(133) A. R. Hurst, *Hercules Chem.*, **29,** 22 (1957).

(134) P. B. Davidson, *Tappi,* **37,** 18 (1954).

(135) N. A. Afonchikov and E. A. Terent'ev, *Bumazhn. Prom.*, **38,** 8 (1963).

(136) E. J. Barber, *Tappi,* **44,** No. 2, 179A (1961).

(137) N. Millman and M. E. Fretz, U.S. Patent 3,130,063 (1964); *Chem. Abstr.*, **61,** 1587 (1964).

(138) *Chem. Week,* **101,** No. 17, 92 (1967).

(139) R. H. Jenkins, *Chem. Eng. News,* **32,** 3310 (1954).

(140) J. P. Casey and E. R. Lehman, U.S. Patent 3,015,572 (1962); *Chem. Abstr.*, **56,** 7565 (1962).

(141) Borden Co., U.S. Patent 3,011,985 (1961); *Chem. Abstr.,* **56,** 11818 (1962).
(142) National Lead Co., Brit. Patent 929,004 (1963); see U.S. Patent 3,015,572 (1962); *Chem. Abstr.,* **56,** 7565 (1962).
(143) H. L. Jaffee, *J. Paint Technol.,* **32,** 706 (1960).
(144) Dow Chemical Co., Brit. Patent 912,068 (1962); *Chem. Abstr.,* **58,** 7042 (1963).
(145) G. A. Westwood, Brit. Patent 922,456 (1963); *Chem. Abstr.,* **59,** 1842 (1963).
(146) G. L. Cunningham, U.S. Patent 2,679,489 (1954); *Chem. Abstr.,* **48,** 12424 (1954).
(147) F. P. Liberti, *Paint Varn. Prod.,* **51,** No. 12, 57 (1961).
(148) Dow Chemical Co., Brit. Patent 948,185 (1964); *Chem. Abstr.,* **61,** 2420 (1964).
(149) C. B. Phillips, *Fire Control Exp. Calif., Div. Forestry,* 2 (1961).
(150) J. B. Batdorf and K. J. Fletcher, *Hercules Chem.,* **47,** 11 (1963).
(151) J. E. Charas, *Ceramica (Sao Paulo),* **7,** 54 (1961).
(152) R. R. Danielson, *Brick Clay Rec.,* **121,** No. 3, 53 (1952).
(153) J. Stawitz, *Tonind.-Ztg.,* **77,** 14 (1953).
(154) W. P. Martin and G. S. Taylor, *Agr. Eng.,* **34,** 550 (1953).
(155) B.I.O.S. Final Rept. No. 1219 (PB Report 79449); *Foundry Trade J.,* **82,** 135 (1947).
(156) R. E. Moren, *Int. Foundry Congr., Congr. Pap., 29th,* Detroit, 410 (1962).
(157) P. J. Hartsuch, M. H. Bruno, and J. W. White, Instr., Bull. No. 801, Lithographic Technical Foundation, Inc. (1950).
(158) L. S. Wertz, U.S. Patent 2,655,004 (1953); *Chem. Abstr.,* **48,** 974 (1954).
(159) G. Sucetti, U.S. Patent 3,063,853 (1962).
(160) S. Wikne, U.S. Patent 3,027,266 (1962); *Chem. Abstr.,* **57,** 1866 (1962).
(161) L. Chalmers, *Paint Mfr.,* **36,** 58 (1966).
(162) I. Taubel, J. Rahm, and F. Lesek, *Chem. Prum.,* **12,** 389 (1962).
(163) "Chemical and Physical Properties of Hercules Cellulose Gum (CMC)," Hercules Inc., Wilmington, Del., 1968.
(164) T. W. Potter, U.S. Patent 2,668,120 (1954); *Chem. Abstr.,* **48,** 6736 (1954).
(165) C. D. Greaves and B. R. Atkins, U.S. Patent 2,673,812 (1954); *Chem. Abstr.,* **48,** 9655 (1954).
(166) H. C. Hershey and J. L. Zakin, *Ind. Eng. Chem., Fundam.,* **6,** 381 (1967).
(167) S. H. Davidson, U.S. Patent 2,680,067 (1954); *Chem. Abstr.,* **48,** 11062 (1954).
(168) H. G. Rogers and H. W. Lutes, U.S. Patent 3,007,400 (1963).
(169) R. W. Butler, U.S. Patent 3,064,313 (1962); *Chem. Abstr.,* **58,** 5860 (1963).
(170) "Cellulose and Cellulose Derivatives," E. Ott, H. M. Spurlin, and M. W. Grafflin, eds., Interscience Publishers, New York, 2nd Ed., 1954, Chap. 9.
(171) C. J. Brown and A. A. Houghton, *J. Soc. Chem. Ind., London,* **60,** 254 (1941).
(172) F. Prakke, *Verfkroniek,* **23,** 181 (1950).
(173) H. Vink, *Makromol. Chem.,* **67,** 105 (1963).
(174) E. D. Klug, *Encycl. Polym. Sci. Technol.,* **3,** 520 (1965).
(175) R. R. McLaughlin and J. H. E. Herbst, *Can. J. Res.,* **28B,** 737 (1950).
(176) T. E. Timell and H. M. Spurlin, *Sv. Paperstidn.,* **55,** 700 (1952).
(177) K. G. Ghosh and K. J. Balakrishna, *J. Sci. Ind. Res., Sect. B,* **21,** 194 (1962).
(178) T. E. Timell, *Sv. Papperstidn.,* **56,** 311 (1953).

(179) H. A. Shelanski and A. M. Clark, *Food Res.,* **13,** 29 (1948).
(180) W. Kern, *Pharm. Ind.,* **21,** 45 (1959).
(181) Code of Federal Regulations, Title 21, No. 19.530, 19.775–19.776, 19.780–19.783, U. S. Govt. Printing Office, Washington, D.C.
(182) Code of Federal Regulations, Title 21, No. 20.1–20.5, U.S. Govt. Printing Office, Washington, D.C.
(183) Code of Federal Regulations, Title 21, No. 25.2–25.3, U.S. Govt. Printing Office, Washington, D.C.
(184) Code of Federal Regulations, Title 21, No. 29.4–29.5, U.S. Govt. Printing Office, Washington, D.C.
(185) Code of Federal Regulations, Title 21, No. 31.1, U.S. Govt. Printing Office, Washington, D.C.
(186) P. S. Francis, *J. Appl. Polym. Sci.,* **5,** 261 (1961).
(187) E. H. deButts, J. A. Hudy, and J. H. Elliott, *Ind. Eng. Chem.,* **49,** 94 (1957).
(188) E. Ott and J. H. Elliott, *Makromol. Chem.,* **18–19,** 352 (1956).
(189) J. B. Batdorf and P. S. Francis, *J. Soc. Cosmet. Chem.,* **14,** 117 (1963).
(190) R. A. Hermes and A. G. Fredrickson, *AIChE J.,* **13,** 253 (1967).
(191) P. Ingram and H. G. Jerrard, *Nature,* **196,** 57 (1962).
(192) M. G. Wirick, *J. Polym. Sci., Part A-1,* **6,** 1965 (1968).
(193) S. Linderfors, *Acta Chem. Scand.,* **16,** 1111 (1962).
(194) T. Cayle, *Wallerstein Lab. Commun.,* **25,** 349 (1962).
(195) N. Shikazono and T. Shiina, *Tokyo Kogyo Shikensho Hokoku,* **48,** 169 (1953).
(196) M. Masuzawa and C. Sterling, *J. Appl. Polym. Sci.,* **12,** 2023 (1968).
(197) "Analytical Procedures for Assay of CMC and its Determination in Formulations," Hercules Inc., Wilmington, Del., 1966.
(198) I. Croon and C. B. Purves, *Sv. Papperstidn.,* **62,** 876 (1959).
(199) R. W. Eyler and R. T. Hall, *Pap. Trade J.,* **125,** No. 15, 59 (1947).
(200) R. W. Eyler, E. D. Klug, and F. Diephuis, *Anal. Chem.,* **19,** 24 (1947).
(201) H. C. Black, *Anal. Chem.,* **23,** 1792 (1951).
(202) E. Wurz, *Oesterr. Chem.-Ztg.,* **56,** 285 (1955).
(203) A. Z. Conner and R. W. Eyler, *Anal. Chem.,* **22,** 1129 (1950).
(204) C. V. Francis, *Anal. Chem.,* **25,** 941 (1953).
(205) S. H. Newberger, J. H. Jones, and G. R. Clark, *Proc. Sci. Sect. Toilet Goods Ass.,* **19,** 25 (1953).
(206) C. W. Tait, R. J. Vetter, J. M. Swanson, and P. Debye, *J. Polym. Sci.,* **7,** 261 (1951).

AUTHOR INDEX

—B—

—D—

—E—

—F—

—G—

—H—

—K—

—L—

—M—

—N—

—O—

—P—

—Q—

—R—

—S—

—T—

—U—

—Y—

—Z—

SUBJECT INDEX

D

E

G

H

M

N

O

Q

T

U

V

W